Operator Theory: Advances and Applications
Vol. 146

Editor:
I. Gohberg

Operator Approach to Linear Problems of Hydrodynamics

Volume 2: Nonself-adjoint Problems for Viscous Fluids

Nikolay D. Kopachevsky
Selim G. Krein (†)

Springer Basel AG

Nikolay D. Kopachevsky
Mathematical Department
Taurida National V. Vernadsky University
ul. Yaltinskaya 4
95007 Simferopol
Ukraine

e-mail: kopachevsky@tnu.crimea.ua

2000 Mathematics Subject Classification 76-02, 76M22, 76M99, 47N20, 47N99

A CIP catalogue record for this book is available from the
Library of Congress, Washington D.C., USA

Bibliographic information published by Die Deutsche Bibliothek
Die Deutsche Bibliothek lists this publication in the Deutsche Nationalbibliografie; detailed bibliographic data is available in the Internet at <http://dnb.ddb.de>.

ISBN 978-3-0348-9425-8 ISBN 978-3-0348-8063-3 (eBook)
DOI 10.1007/978-3-0348-8063-3

Originally published by Birkhäuser Verlag, Basel-Boston-Berlin in 2003
Softcover reprint of the hardcover 1st edition 2003
Printed on acid-free paper produced from chlorine-free pulp. TCF ∞
Cover design: Heinz Hiltbrunner, Basel

ISBN 978-3-0348-9425-8

9 8 7 6 5 4 3 2 1 www.birkhauser-science.com

Preface to Volume II

This book is the second volume of the monograph entitled "Operator Approach to Linear Problems of Hydrodynamics." The first volume, "Self-adjoint Problems for an Ideal Fluid," presented evolution and spectral problems arising from the study of small motions and eigenoscillations of an ideal fluid in a bounded region. All these problems are self-adjoint in a specially chosen Hilbert space or in a Pontryagin space.

The second volume is basically a collection of nonself-adjoint problems on small motions and normal oscillations of a viscous fluid filling a bounded region (container). There are no modifications to the table of contents of the second volume as it was presented in the first volume. All the chapters in this volume can be read independently from Chapters III–VI in Volume I. Chapters I and II may be used as a reference book.

* * *

We would like to express our gratitude to I. T. Gohberg and A. S. Markus, who encouraged us to publish the book and who offered many helpful suggestions. Our gratidude goes also to our colleagues T. Ya. Azizov, O. A. Ladyzhenskaya, N. N. Moiseev, A. D. Myshkis, F. L. Chernousko, V. I. Yudovich, and Ngo Zuy Kan with whom we had fruitful discussions during the process of obtaining some mathematical results and of writing the final version of the book.

We would also like to acknowledge the work of E. D. Volodkovich and E. A. Danilova who drafted a preliminary version of the English translation of the book.

The staff in the Department of Mathematics and Theoretical Mechanics at the Voronezh Forestry Engineering University, Russia, and in the Department of

Mathematical Analysis at the Simferopol State University, Ukraine, offered us their professional assistance during the time of writing the book. Their help is greatfully acknowledged.

Our particular gratitude goes to Mircea Martin and his wife, Larisa, for their professional and qualified work in editing the English translation and typesetting the monograph in TeX.

The Authors
Simferopol, Voronezh
October 30, 1995

While the book was prepared for press, its co-author, Selim Grigoryevich Krein, a dedicated teacher, an elder friend of mine, an outstanding mathematician and person, and a great specialist in functional analysis and its applications, passed away. In the process of investigating initial boundary value and spectral problems of mathematical physics and hydrodynamics with his students and colleagues, he used to inquire how the problem under investigation would look in operator form. This book answers the question in the case of linear problems of hydrodynamics. S. G. Krein dreamed of this book being published, not only in Russia and in the former USSR, but also for the English-speaking readers as well. I hope they will not be disappointed by this two volume monograph.

Nikolay D. Kopachevsky
May 18, 2001

Table of Contents

Volume II

Preface v
Table of Contents
Volume II vii
Volume I xvi
Introduction 1

Part III: Motion of Bodies with Cavities Containing Viscous Incompressible Fluids 5

Chapter 7: Motion of Bodies with Cavities Completely Filled with Viscous Incompressible Fluids

7.1 Motion of Fluids Completely Filling a Cavity in a Stationary Body 8
7.1.1 Statement of the Problem and the Basic Equations 8
7.1.2 Reducing the Problem to a Differential Equation in a Hilbert Space. Existence of Solutions 8
7.1.3 Structure of the Spectrum of the Problem 9
7.1.4 Perturbation of the Stationary Motion of a Fluid 10
7.1.5 Small Movements of a Fluid in a Rotating Container 12

7.2 Small Movements of a Gyrostate Around a Fixed Mass Center . . 15
7.2.1 Statement of the Problem and the Basic Equations 15

7.2.2 Transition to a Differential Equation in a Hilbert Space . . . 16
7.2.3 Properties of the Translation Operator 17
7.2.4 Existence of Solutions of the Evolution Problem 19
7.2.5 Normal Oscillations 20
7.3 Rotating Motion of a Gyrostate 21
7.3.1 Statement of the Problem and the Basic Equations 21
7.3.2 Transition to a Differential Equation in a Hilbert Space . . . 22
7.3.3 Properties of the Operators of the Problem 23
7.3.4 Normal Oscillations 23
7.3.5 Solvability of the Nonstationary Problem 27
7.4 Asymptotic Solutions for High Viscosity 28
7.4.1 Solving the Hydrodynamics Problem 29
7.4.2 Asymptotic Equations of the Motion of a Rigid Body 30
7.4.3 An Example 33
7.5 Oscillations of a Pendulum With a Cavity Completely Filled With a Viscous Fluid 36
7.5.1 Towards the Statement of the Problem 36
7.5.2 Transition to a System of Operator Equations 37
7.5.3 The Indefinite Metric Approach 39
7.5.4 Other Properties of Solutions of the Spectral Problem . . . 41
7.5.5 On Riesz and p-Basicity of Modes of Dissipative Waves . . . 47
7.6 Problems on Fluids Flowing Through a Given Cavity 50
7.6.1 The Basic Equations 50
7.6.2 Application of the Abstract Scheme 50
7.6.3 Transition to Operator Equations in Orthogonal Subspaces 52
7.6.4 Theorem on Existence of a Generalized Solution 53
7.7 Convective Movements of Fluids in a Closed Cavity 54
7.7.1 Equations of Thermal Convection 54
7.7.2 Conditions of Mechanical Equilibrium 55
7.7.3 Final Statement of the Problem 55
7.7.4 Transition to an Operator Equation 56
7.7.5 Solvability of the Initial Boundary Value Problem 58
7.7.6 Normal Movements of a System Heated from Below 58
7.7.7 Normal Oscillations for Heating from Above 60
7.7.8 On Transition of Eigenvalues to the Left Half-Plane in the Case of Heating from Below 62

Chapter 8: Motion of Viscous Fluids in Open Containers

8.1 Small Movements of Viscous Fluids in an Open Immovable Container . . . 67
8.1.1 Classical Statement of the Problem . . . 68
8.1.2 Auxiliary Boundary Value Problems . . . 69
8.1.3 Generalized Solutions of the Homogeneous Nonstationary Problem . . . 72
8.1.4 Motions with Small Mass Forces . . . 76
8.1.5 Equation of Energy Balance . . . 77
8.1.6 Equation of Normal Oscillations . . . 78

8.2 The Main Operator Pencil . . . 79
8.2.1 Structure of the Spectrum of the Problem . . . 79
8.2.2 Linearization of the Pencil . . . 81
8.2.3 Mutual Relationships between Eigen- and Associated Elements of the Two Pencils . . . 83
8.2.4 Transformation to a Nondegenerate Pencil . . . 85
8.2.5 The Property of Two-Multiple Basicity . . . 86
8.2.6 Spectral Factorization of the Overdamped Pencil. Separate Basicity . . . 88
8.2.7 Double-Sided Inequalities for the Two Branches of Eigenvalues . . . 92
8.2.8 The General Case . . . 95

8.3 Normal Oscillations and the Spectrum of the Hydrodynamics Problem . . . 97
8.3.1 General Properties of the Spectrum . . . 97
8.3.2 Influence of Fluid Viscosity on the Structure of the Spectrum of the Problem . . . 100
8.3.3 Properties of Surface and Internal Waves . . . 101
8.3.4 On the Basicity of Modes of Normal Oscillations . . . 104

8.4 Oscillations of a Heavy Rotating Fluid . . . 107
8.4.1 Statement of the Problem . . . 107
8.4.2 Transition to a System of Operator Equations . . . 109
8.4.3 Solvability of the Nonstationary Problem . . . 109
8.4.4 Equations of Normal Oscillations . . . 112
8.4.5 Investigation of the Spectral Problem . . . 113
8.4.6 On the Completeness of Systems of Modes of Normal Oscillations of the Initial Problem . . . 120

8.5 Asymptotic Solutions for High Viscosity 121
8.5.1 The Cauchy Problem 121
8.5.2 Normal Oscillations 124
8.5.3 Motions under Mass Forces 127
8.5.4 The Case of Rotating Viscous Fluids 130
8.6 Oscillations of a System of Nonmixing Fluids 133
8.6.1 Statement of the Problem on Small Oscillations 134
8.6.2 Transition to a System of Operator Equations 135
8.6.3 The Theorem on Existence of a Generalized Solution 137
8.6.4 Normal Oscillations of a System of Fluids 138
8.6.5 On the Stability of the Relative Equilibrium State 139
8.7 Small Motions Around a Fixed Point of a Body with a Cavity Partially Filled with Fluid 140
8.7.1 Basic Equations 141
8.7.2 Transition to a System of Operator Equations 141
8.7.3 Auxiliary Results 143
8.7.4 Existence of Solution of the Boundary Value Problem 145
8.8 Normal Oscillations of a Pendulum Partially Filled with a Fluid (The Plane Problem) 148
8.8.1 Statement of the Problem 148
8.8.2 Transition to a Differential Equation in a Hilbert Space 149
8.8.3 Properties of Operator Coefficients of the Evolution Equation 151
8.8.4 Normal Oscillations. Properties of the Spectral Problem 154
8.8.5 Theorem on Instability 159
8.8.6 On the Solvability of the Initial Boundary Value Problem . . 161
8.9 Convection in a Partially Filled Container 162
8.9.1 Statement of the Problem 162
8.9.2 Transition to a System of Operator Equations 164
8.9.3 Solvability of the Evolution Problem 165
8.9.4 Normal Convective Movements. Reduction to an Operator Pencil 167
8.9.5 Dissipatively Thermal and Surface Waves under the General Law of Heat Transfer 169
8.9.6 Surface and Internal Waves for Heating From Above 173

8.9.7 Normal Oscillations for Heating from Below and for a Given Heat Flow on the Free Surface 174

8.10 Sufficient Conditions of Instability for Convective Movements of a Fluid 176

8.10.1 Transition to a Two-Parameter Pencil 176

8.10.2 On the Structure of the Kernels of Operator Coefficients 180

8.10.3 On the Existence of Eigenvalues in the Left Complex Half-Plane 181

8.10.4 Double-Sided Estimates for Eigenvalues 182

8.10.5 Derivation of a Sufficient Condition for Instability 183

8.10.6 Remarks 185

Chapter 9: Oscillations of Capillary Viscous Fluids

9.1 Statement of the Problem 187

9.1.1 Basic Equations and Boundary Conditions 187

9.1.2 Some Properties of Solutions to the Normal Oscillation Problem 189

9.1.3 On the Spectrum Structure of Normal Oscillations 191

9.2 Oscilations of Capillary Fluids in Arbitrary Containers 192

9.2.1 Transition to a System of Operator Equations 192

9.2.2 Normal Oscillations. Properties of the Operators of the Oscillations Problem 194

9.2.3 Normal Oscillations of a Rotating Fluid 200

9.2.4 Normal Oscillations of a Nonrotating Fluid 202

9.2.5 The Matrix Structure of the Main Operator 205

9.2.6 On the Finiteness of the Number of Nonreal Eigenvalues . . 206

9.2.7 Heuristic Considerations. The Abstract Spectral Problem 210

9.2.8 Heuristic Considerations. Physical Conclusions 212

9.2.9 The Solvability of the Evolution Problem 215

9.3 The Inverse of the Lagrange Theorem on Stability 217

9.3.1 Formulating the Theorem 218

9.3.2 Auxiliary Propositions 219

9.3.3 The Principle of Changing Stability 222

9.3.4 Transition to an Equation with a Compact Operator 223

9.3.5 Application of Perturbation Theory 225

9.3.6 The Existence of Eigenvalues in the Left Complex Half-Plane for an Arbitrary Viscosity Value 228
9.4 Motions of a Rigid Body Containing a Cavity Filled with a Capillary Fluid under Conditions of Complete Low Gravity . . . 229
9.4.1 Statement of the Problem 229
9.4.2 Transition to a System of Operator Equations 230
9.4.3 Normal Oscillations. Transition to an Equation with a Dissipative Operator 231
9.4.4 Properties of Normal Movements 233
9.4.5 The Existence of a Generalized Solution to the Nonstationary Problem 233

Appendix C: Remarks and Reference Comments to Part III
C.1 Chapter 7 237
C.2 Chapter 8 240
C.3 Chapter 9 243

Part IV: Small Oscillations of Complex Hydrodynamic Systems . . 249

Chapter 10: Oscillations of Partially Dissipative Hydrosystems
10.1 Statement of the Problem 252
10.1.1 The Classical Statement of the Problem 252
10.1.2 The Law of Full Energy Balance. Definition of a Generalized Solution 254
10.1.3 Normal Oscillations. Statement of the Problem 257
10.2 Studying an Initial Boundary Value Problem 260
10.2.1 Projections of Euler and Navier-Stokes Equations on Orthogonal Subspaces 260
10.2.2 Auxiliary Boundary Value Problems 263
10.2.3 Properties of Matrix Blocks and Their Physical Meanings 266
10.2.4 Theorem on Correct Solvability of the Initial Boundary Value Problem 269
10.3 Model Problem on Normal Oscilations of Partially Dissipative Hydrosystems 271
10.3.1 Statement of the Model Problem 271
10.3.2 Obtaining the Characteristic Equation 273
10.3.3 Studying the Characteristic Equation 275

10.3.4 General Conclusions and Hypotheses on the Structure of the Spectrum of the Hydrodynamics Problem 281

10.4 Normal Oscillations of a Partially Dissipative Hydrosystem in an Arbitrary Domain 283

10.4.1 Transition to an Operator Pencil with Bounded Operator Coefficients 283

10.4.2 General Properties of the Spectrum 286

10.4.3 The Theorem on Spectrum Location 293

10.5 On the Completeness of the System of Modes of Normal Oscillations 300

10.5.1 Auxiliary Results 300

10.5.2 Theorem on Completeness. Keldysh Scheme Realization 302

10.5.3 On the Existence of Branches of Eigenvalues with a Limit Point at Infinity. Heuristic Arguments 307

10.5.4 Concluding Remarks 314

Chapter 11: Oscillations of Visco-Elastic and Relaxing Media

11.1 Visco-Elastic Fluids in Completely Filled Containers 316

11.1.1 A Model of a Visco-Elastic Fluid 316

11.1.2 Statement of the Initial Boundary Value Problem 318

11.1.3 On the Solvability of the Initial Boundary Value Problem 319

11.1.4 Normal Oscillations 321

11.2 Abstract Evolution and Spectral Problems Generated by Small Motions of a Visco-Elastic Fluid 323

11.2.1 Statement of the Problem. Transition to an Equation with a Dissipative Operator 324

11.2.2 On the Solvability of the Cauchy Problem for Integro-Differential Equations 325

11.2.3 Spectral Problem. Transition to an Equation with a Bounded Operator 326

11.2.4 Properties of Operator Coefficients of the Spectral Problem 330

11.2.5 Properties of Solutions of the Spectral Problem 333

11.3 Small Motions and Normal Oscillations of a Visco-Elastic Fluid in an Open Container 342
11.3.1 Mathematical Statement of the Problem 342
11.3.2 Transition to a System of Operator Equations 343
11.3.3 On the Solvability of the Initial Boundary Value Problem 345
11.3.4 Normal Oscillations. Main Operator Pencil 346
11.4 Multiple Basicity of the System of Eigen- and Associated Elements for the Problem on Normal Oscillations of a Visco-Elastic Fluid in an Open Container 349
11.4.1 The Linearization of the Pencil 349
11.4.2 The Theorem on Basicity of the System of Root Elements of the Linear Pencil 354
11.4.3 Connection between the System of Root Elements of the Two Problems 355
11.4.4 The Theorem on Basicity of Special Form Elements . . . 363
11.5 Additional Properties of Solutions of the Spectral Problem . . . 374
11.5.1 On the Existence of Different Branches of Eigenvalues . . 374
11.5.2 On the Location of Nonreal Eigenvalues in the Complex Plane 377
11.5.3 Multiple Basicity and p-Basicity. An Indefinite Metric Approach Using Krein Space Theory 384
11.6 Oscillations of Relaxing Fluids 390
11.6.1 Classical Statement of the Problem on Small Motions of a Relaxing Fluid 390
11.6.2 Transition to an Initial Boundary Value Problem for One Scalar Function 392
11.6.3 The Simplest Problem on Oscilations of a Relaxing Fluid 394
11.6.4 On the Solvability of the Cauchy Problem for an Abstract Integro-Differential Equation Connected with Small Motions of a Relaxing Fluid 397
11.6.5 Normal Oscilations of a Relaxing Fluid with Variable Medium Characteristics 400
11.6.6 Physical Conclusions 410

Appendix D: Remarks and Reference Comments to Part IV

D.1 Chapter 10 411

D.2 Chapter 11 412

Standard Reference Texts 415

Bibliography 417

List of Symbols 433

Subject Index 437

Table of Contents

Volume I

Preface v

Table of Contents

Volume I vii

Volume II xvi

Introduction 1

Part I: Mathematical Foundations of Linear Hydrodynamics 5

Chapter 1: Operators on Hilbert Spaces

1.1 General Facts 7

1.1.1 The Concept of a Hilbert Space 7

1.1.2 The Space $L_2(\Omega)$ 8

1.1.3 Orthogonality. Projection onto a Subspace 9

1.1.4 Equivalent Norms 10

1.1.5 Linear Functionals. Riesz Theorem 10

1.1.6 Embeddings of Spaces. Riesz Theorem for Equipments . . . 11

1.1.7 Orthonormal Systems and Bases 12

1.1.8 Bounded Linear Operators 12

1.1.9 Adjoint Operators 14

1.1.10 Self-Adjoint Operators 14

1.1.11 Self-Adjoint Compact Operators 15

1.1.12 Compact operators. s-numbers 16

1.1.13 Riesz Bases and p-Bases 17
1.1.14 Direct Sum of Subspaces. Invariant Subspaces 18
1.1.15 Eigen- and Associated (Root) Elements. Root Subspaces . . 19
1.1.16 Unbounded Linear Operators 20
1.1.17 Resolvent and Spectrum of a Linear Operator 21
1.1.18 Classification of Points in the Spectrum of a Linear Operator 22
1.1.19 Spectrum of a Self-Adjoint Operator. Weyl Theorem 23
1.1.20 Riesz Projections 24
1.1.21 Symmetric and Self-Adjoint Operators 25
1.1.22 Spectral Decomposition of Self-Adjoint Operators. Functions of Operators 26
1.1.23 Spaces with Degenerate Scalar Products. Seminorms 28
1.1.24 Equivalent Corrections of Seminorms 29

1.2 Sobolev Spaces 29
1.2.1 Finite Functions 30
1.2.2 Generalized Derivatives 30
1.2.3 The Definition of Sobolev Spaces 31
1.2.4 The Space $L_2^1(\Omega)$. Regions of the First Type 31
1.2.5 The Subspace $H_0^1(\Omega)$ 32
1.2.6 Embedding $H^1(\Omega)$ into $L_2(\Omega)$. Regions of the Second Type 34
1.2.7 The Trace Operator. Regions of the Third Type 34

1.3 Spaces With Indefinite Metrics 35
1.3.1 J-Spaces 35
1.3.2 Uniformly Definite Subspaces 37
1.3.3 J-Orthonormal Systems and Bases 37
1.3.4 Linear Operators on J-Spaces 38
1.3.5 Invariant Subspaces of J-Self-Adjoint Operators 39
1.3.6 Pontryagin Spaces 39
1.3.7 On Completeness and Basicity for the System of Root Elements of a $\boldsymbol{J}$-Self-Adjoint Operator 41

1.4 Eigenvalue Problems 43
1.4.1 Operator B is the Identity Operator 43
1.4.2 Operator B is Positive Definite 44
1.4.3 Positivity Condition for a Matrix Operator 47
1.4.4 Simplifying Equations with an Alternating Operator 48

1.4.5 Equations in Spaces with Indefinite Metrics 49

1.5 Evolution Equations in Hilbert Spaces 51

1.5.1 First Order Linear Differential Equations with Bounded Operator Coefficient 51

1.5.2 The Cauchy Problem for Equations with Unbounded Operators 52

1.5.3 Equations With a Negative Self-Adjoint Operator 53

1.5.4 Equations With a Dissipative Operator 54

1.5.5 Equations With Perturbed Operators 55

1.5.6 Stability 56

1.5.7 Nonhomogeneous Equations 57

1.5.8 Linear Differential Equations of the Second Order 58

1.5.9 Volterra Integral Equations 62

1.6 Spectral Theory of Operator Pencils 63

1.6.1 Eigen- and Associated Elements of an Operator Pencil . . . 63

1.6.2 Root Functions 65

1.6.3 Fredholm Holomorphic Operator-Valued Functions 66

1.6.4 Linear Pencils. Theorems on Completeness of the System of Eigen- and Associated Elements 66

1.6.5 Keldysh-Type n-Multiple Completeness 68

1.6.6 Spectral Factorization of an Operator Pencil 69

1.6.7 Completeness With Finite Defect of a System of Eigen- and Associated Elements of an Operator-Valued Function 70

1.6.8 Asymptotic Behavior of Branches of Eigenvalues 71

1.6.9 Self-Adjoint Operator Pencils 72

1.6.10 On Riesz Basicity of the System of Eigenelements of a Self-Adjoint Operator-Valued Function 73

1.6.11 Variational Methods for Investigating Continuous Operator-Valued Functions 75

1.7 Asymptotic Methods for Solving Evolution Equations With a Small Parameter Attached to the Derivative 76

1.7.1 Equations With a Small Parameter Attached to the Derivative 77

1.7.2 Splitting of a Homogeneous Equation 78

1.7.3 Solvability of Commutator Equations 79

1.7.4 Asymptotic Expansions of Solutions 82

1.7.5 The Special Case of a Splitable Operator Kernel 86
1.7.6 The Case of a Nonstationary Perturbation 87
1.7.7 Nonhomogeneous Equations 87
1.7.8 Eigenvalue Problems 90

1.8 A General Scheme for Solving Boundary Value Problems 93
1.8.1 Hilbert Pairs. Generating Operators 93
1.8.2 Hilbert Pairs Connected With the Spaces $H^1(\Omega)$ and $L_2(\Omega)$ 95
1.8.3 Hilbert Scale of Spaces. Space $E^{-1/2}$ 98
1.8.4 Self-Adjoint Extensions of Positive Definite Symmetric Operators. Generalized and Weak Solutions of Equations . . 99
1.8.5 Nonhomogeneous Boundary Value Problems 101
1.8.6 Spaces of Harmonic Functions 103
1.8.7 Embedding and Mapping. The Abstract Green Formula . . . 107

Chapter 2: Fundamental Spaces and Operators of Linear Hydrodynamics

2.1 Fundamental Spaces and Hydrodynamics Operators for an Ideal Fluid 109
2.1.1 Fields with Finite Kinetic Energy 110
2.1.2 Potential Fields 110
2.1.3 Divergence of Fields with Finite Kinetic Energy 111
2.1.4 The Space of Solenoidal Fields 112
2.1.5 Laplace Operator on the Space $H^1(\Omega)$ 112
2.1.6 Normal Component of a Field on the Boundary 113
2.1.7 Green Formula for the Laplace Operators. Harmonic Fields 114
2.1.8 Weyl Decomposition 115
2.1.9 Approximation by Smooth Fields 116
2.1.10 The Space of Velocity Fields for an Ideal Fluid in an Open Container 117
2.1.11 Systems of Nonmixing Fluids 118
2.1.12 Spaces of Velocity Fields for Systems of Nonmixing Ideal Fluids 124

2.2 Spaces and Hydrodynamics Operators for a Viscous Fluid 125
2.2.1 Forces of Internal Friction. Energy Dissipation 126
2.2.2 Divergence Operator. Solenoidal Fields 127
2.2.3 Vector Laplace Operator. Green Formula 128

2.2.4 Movement of a Viscous Fluid in a Closed Container. Korn Identity and Korn Inequality 129
2.2.5 Stokes Operator 131
2.2.6 Spaces of Velocity Fields for a Viscous Incompressible Fluid in an Open Container 132
2.2.7 Main Boundary Value Problems for the Fluid Movement in an Open Container 134
2.2.8 Spaces of Velocity Fields for a System of Viscous Fluids . . . 138

Appendix A: Remarks and Reference Comments to Part I
A.1 Chapter 1 143
A.2 Chapter 2 146

Part II: Motion of Bodies With Cavities Containing Ideal Fluids . . 149

Chapter 3: Oscillations of a Heavy Ideal Fluid in Stationary and Nonstationary Containers
3.1 Equations of the Motion of a Rigid Body with a Cavity Filled with an Incompressible Fluid 151
3.1.1 Basic Concepts of Kinematics 152
3.1.2 Equations of Motion for an Incompressible Fluid 153
3.1.3 Boundary Conditions 154
3.1.4 Motion Equations for a Gyrostate 155
3.1.5 Dynamics Equations of the System "Body + Fluid" With a Partially Filled Cavity 158
3.1.6 Transition to Undimensional Variables 163
3.2 Motion of an Ideal Fluid in a Closed Stationary Container 164
3.2.1 Basic Equations 164
3.2.2 Existence of Solutions 165
3.3 Small movements of an Ideal Fluid in an Open Immovable Container 166
3.3.1 Statement of the Problem and the Basic Equations 166
3.3.2 Projection of Euler Equations 167
3.3.3 Existence of Solutions 168
3.3.4 Proper Oscillations 170
3.4 Small Joint Movements of a Fluid and a Container 171
3.4.1 Statement of the Problem and the Basic Equations 171
3.4.2 Finding the Velocity Field and the Pressure 172

3.4.3 Defining the Motion Law of the Body 173
3.4.4 Zhukovsky Potentials 174

3.5 Small Joint Movements Around a Fixed Point of a Body and a Fluid Partially Filling the Cavity 174
3.5.1 Statement of the Problem and the Basic Equations 175
3.5.2 The Law of Full Energy Balance 176
3.5.3 Projecting Euler Equations 177
3.5.4 Kinetic Moment Equation 178
3.5.5 Investigating the Complete System of Motion Equations . . 179
3.5.6 Proper Oscillations 184
3.5.7 Solving the Evolution Problem 187

3.6 Oscillations of a System of Fluids in an Immovable Container . . . 189
3.6.1 Statement of the Problem 190
3.6.2 Orthogonal Projection Method 191
3.6.3 Transition to Operator Equation 193
3.6.4 Proper Osccilations 194
3.6.5 Small Movements of Stable Systems 196

Chapter 4: Problems on Oscillations of Capillary Fluids and Problems on Hydroelasticity in Immovable Containers

4.1 Oscillations of a Capillary Fluid in a Rigid Container 199
4.1.1 On the Equilibrium State 200
4.1.2 Statement of the Problem on Small Oscillations 201
4.1.3 Law of Energy Balance 203
4.1.4 Transition to Operator Equation 204
4.1.5 Properties of the Potential Energy Operator 205
4.1.6 Proper Oscillations 206
4.1.7 Instability Conditions of the System 208
4.1.8 Solvability of the Evolution Problem 208
4.1.9 Oscillations of a System of Capillary Fluids 209

4.2 Oscillations of Fluids in Containers With Elastic Ends 213
4.2.1 Solving the Static Problem 213
4.2.2 Formulation of the Problem on Small Oscillations 216
4.2.3 Energy-Preserving Law 217
4.2.4 Operator Equation of the Problem 219
4.2.5 Properties of the Operators of the Problem 220
4.2.6 Proper Oscillations 221

4.2.7 Evolution Problem 223
4.2.8 Oscillations of a Fluid in a Container With One Elastic End 224
4.2.9 Oscillations of a System of Fluids in a Container With Elastic Plane Partitions 226

4.3 Oscillations of a Fluid in a Partially Filled Container With an Elastic Bottom 232
4.3.1 Determining the Equilibrium State 232
4.3.2 Formulation of the Problem on Small Oscillations 234
4.3.3 On Solvability of the Evolution and Spectral Problems . . . 234
4.3.4 Oscillations of a Capillary Fluid in a Partially Filled Elastic Container 237
4.3.5 Systems of Heavy Fluids in a Container With Elastic Plates 241
4.3.6 Systems of Capillary Fluids in a Container With Elastic Ends 244
4.3.7 Compound Systems of Plate-Partitions and Nonmixing Fluids 246

Chapter 5: Other Operator Approaches to Hydrodynamics Problems of Ideal Fluids

5.1 Plane Problems on Proper Oscillations of a Heavy Fluid in a Channel. An Application of the Stream Function 249
5.1.1. Spectral Problem for the Stream Function 249
5.1.2. Properties of Nodal Lines of Stream Eigenfunctions 252
5.1.3. Estimates of Eigenvalues 256

5.2 Shallow Water Theory in Problems on Oscillations of Heavy Ideal Fluids in Bounded Regions 257
5.2.1. Formulation of the Problem with a Small Parameter 257
5.2.2. Asymptotic Solution in First Approximation 259
5.2.3. Formulas for Calculating Second Order Approximations . . 261
5.2.4. Plane Problems 263
5.2.5. Examples 264
5.2.6. Systems of Nonmixing Fluids 265

5.3 Oscillations of a System "Fluid-Gas" in a Bounded Region 270
5.3.1. Formulation of the Initial Boundary Value Problem 270
5.3.2. Formulation of the Spectral Problem 272

5.3.3. Transition to a System of Operator Equations 274
5.3.4 Theorem on Spectrum 276
5.3.5 Variational Principles for Eigenvalues 278

Chapter 6: Oscillations of an Ideal Rotating Fluid

6.1 Motion of Fluids in Rotating Containers 281
6.1.1 Statement of the Problem and the Main Equations 281
6.1.2. Existence of Solutions 282
6.1.3 Normal Oscillations 284

6.2 Motion of a Gyrostate Similar to Uniform Rotation About a Fixed Axis 287
6.2.1 Statement of the Problem 287
6.2.2 Transition to the Evolution Equation in a Hilbert Space . . . 288
6.2.3 Properties of the Operators in the Problem 289
6.2.4 Existence of Solution to the Boundary Value Problem 289

6.3 Rotation of a Fluid in a Partially Filled Container 290
6.3.1 On the Equilibrium State 290
6.3.2 Statement of the Problem on Small Oscillations 292
6.3.3 Method of Orthogonal Projection 294
6.3.4 Properties of the Operators of the Problem 297
6.3.5 Systems of Nonmixing Fluids 299
6.3.6 Transition to an Operator Equation and Properties of the Operators of the Problem 302

6.4 Solving the Initial Boundary Value Problem 310
6.4.1 Generalized Solution of the Operator Equation 310
6.4.2 Small Movements of Fluid in a Partially Filled Container . . 313
6.4.3 On the Structure of the Spectrum of a Vortical Operator . . 313
6.4.4 Classes of Free Movements 316
6.4.5 Free Movements of a System of Fluids 318

6.5 Self-adjoint Operator Pencils Generated by Problems on Oscillations of a Rotating Ideal Fluid 321
6.5.1 The Main Operator Pencil 321
6.5.2 On the Spectrum of the Operator Pencil 322
6.5.3 Operator Pencils with Analytic Perturbations 323
6.5.4 Factorization of the Operator Pencil 325
6.5.5 Systems of Eigenelements With Defect Basicity 327

6.5.6 Double-Sided Estimates of Positive and Negative Eigenvalues 328
6.5.7 On the Essential Spectrum of the Problem 331
6.6 Proper Oscillations of a Rotating Fluid 333
6.6.1 Surface and Internal Waves 333
6.6.2 Properties of the Surface Waves 335
6.6.3 On Existence and Properties of Internal Waves 338
6.6.4 Oscillations of a System of Nonmixing Fluids 341

Appendix B: Remarks and Reference Comments to Part II
B.1 Chapter 3 345
B.2 Chapter 4 347
B.3 Chapter 5 350
B.4 Chapter 6 350

Standard Reference Texts 355

Bibliography 357

List of Symbols 375

Subject Index 379

Introduction

As mentioned in the Introduction to Volume I, the present monograph is intended both for mathematicians interested in applications of the theory of linear operators and operator-functions to problems of hydrodynamics, and for researchers of applied hydrodynamic problems, who want to study these problems by means of the most recent achievements in operator theory.

The second volume considers nonself-adjoint problems describing motions and normal oscillations of a homogeneous viscous incompressible fluid. These initial boundary value problems of mathematical physics include, as a rule, derivatives in time of the unknown functions not only in the equation, but in the boundary conditions, too. Therefore, the spectral problems corresponding to such boundary value problems include the spectral parameter in the equation and in the boundary conditions, and are nonself-adjoint. In their study, we widely used the theory of nonself-adjoint operators acting in a Hilbert space and also the theory of operator pencils. In particular, the methods of operator pencil factorization and methods of operator theory in a space with indefinite metric find here a wide application.

We note also that this volume presents both the now classical problems on oscillations of a homogeneous viscous fluid in an open container (in an ordinary state and in weightlessness) and a new set of problems on oscillations of partially dissipative hydrodynamic systems, and problems on oscillations of a visco-elastic or relaxing fluid. Some of these problems need a more careful additional investigation and are rather complicated.

The second volume has two parts, Parts III and IV. Part III, consisting of Chapters 7, 8, and 9, studies the problems on small motions and normal oscillations of a homogeneous viscous incompressible fluid. Chapter 7 presents the problems on the motion of a solid body with a cavity completely filled by a fluid. In particular, problems on fluid motion in a fixed or uniformly rotating with respect to a fixed axis container, small motions of a gyrostate, problems on oscillations of a pendulum with fluid, problems of a fluid flowing through a given region, and problems on convection are considered.

In Chapter 8—the longest in Volume II—we describe in detail the problem on small motions and normal oscillations of a viscous fluid in an open container. Here, the properties of an associated meromorphic operator pencil are studied and, based on them, the physical conclusions on properties of the solutions of the spectral problem are formulated. In particular, the properties of surface and interior waves arising in a viscous fluid and also the asymptotic behavior of the corresponding branches of eigenvalues are investigated. The problems on the motion of a body with a cavity partially filled with a fluid, pendulum oscillations, and convection problems in the presence of a free surface of the fluid are also considered. Sufficient conditions of instability of convective fluid motions in an arbitrary container are obtained.

Chapter 9 considers the problems on oscillations of a viscous capillary fluid, partially filling a container and being under conditions close to weightlessness. Here it is stated that the consideration of capillary forces influences significantly the structure of the spectrum of frequencies of normal oscillations of a viscous fluid in an open container. The problem, which has not been studied yet, is rather complicated and interesting, and the theory of operators in spaces with indefinite metrics could be applied to study it. The general theory of linear operator perturbation is used here to prove the inverse of Lagrange theorem on stability in hydrodynamic systems, that is, systems with an infinite number of degrees of freedom.

Part IV of the second volume considers the problems on oscillation theory of complex hydrodynamic systems. Chapter 10 studies the problem of small motions of a system "viscous fluid–ideal fluid," that is, a partially dissipative hydrosystem. The classical statement of the problem, its model variant, and the operator evolution equation arising from the problem are considered. Normal oscillations of such system are studied on the basis of the theory of nonself-adjoint operator-functions and the approach proposed by M. V. Keldysh. Chapter 11 is devoted to a number of problems on oscillations of a visco-elastic, or relaxing, fluid. First, we study the oscillations of a visco-elastic fluid in a completely filled container and also an analogue of the evolution problem generated by this problem. Further, we consider the problem on normal oscillations of a visco-elastic fluid in an open container generalizing the problem of Chapter 8. The meromorphic operator pencil, which arises in this problem, is studied both by means of the theory of Pontryagin spaces and the general operator theory in a Krein space. Finally, the last section of Chapter 11 studies a new problem on oscillations of an ideal relaxing fluid in a bounded region.

As a last general remark, we should mention that in the second volume we use the same notations and method of labeling the chapters, sections and subsections as in the first volume.

We also note that the first volume consists of Parts I and II and comprises six chapters. The interested reader may find the entire table of contents of Volume I listed right before this introduction and after the table of contents of Volume II.

In both volumes each part ends with concluding remarks and bibliographical comments.

Part III

MOTION OF BODIES WITH CAVITIES CONTAINING VISCOUS INCOMPRESSIBLE FLUIDS

In this part of the book, we investigate a number of problems on the motion of a rigid body with a cavity partially or completely filled with a viscous incompressible fluid. In Chapter 7, we consider the case of a cavity completely filled with fluid, that is, the system "body + fluid" is a gyrostate. In this case, we investigate the motion in an immovable container, the small oscillations of the gyrostate around its mass centre, and the small deviations from uniform rotation of the gyrostate. The new problem on the flowing of a fluid through a given region is also considered. Finally, we pay some attention to convective movements.

Chapter 8 is central to Part III. In the process of investigating the normal oscillations of a viscous fluid in an immovable container, we come upon an operator pencil that was studied by many mathematicians. Here we deal with surface and internal waves interacting with each other. The problem on convection in a partially filled container is also considered.

In Chapter 9, we investigate the movements of a capillary fluid, that is, those movements for which the surface tension plays a significant role. The capillary forces affect essentially the structure of spectrum of those problems.

It should be pointed out that, unlike in other published papers, in Chapter 9 we manage to consider the case when the free surface of the fluid intersects the surface of the body. Several heuristic assumptions are advanced and a few related problems are stated in this chapter as well.

Chapter 7

Motion of Bodies with Cavities Completely Filled with Viscous Incompressible Fluids

First of all, in this chapter we address problems that originate in investigations of the motion of a viscous incompressible fluid filling completely the cavity of an immovable rigid body. The main operator connected with such problems is the Stokes operator described in Section 2.2.

Later on, we will study the equations of joint motion of a rigid body and a fluid that are linearized near various movements of the system "body + fluid." At first, the linearization is performed near the immovable state, that is, small oscillations of this system around its immovable mass center are considered. Here, a new bounded self-adjoint operator appears, which is connected with translational movements, and, therefore, is called the *translation operator*. Further, we consider the linearization near a uniform rotation of the system "body + fluid." Here, once again we find the gyroscopic operator that was already used in problems on the motion of an ideal fluid.

For all these problems we prove theorems of existence of solutions of the nonstationary equations and investigate the properties of normal oscillations.

For the problem on the motion of a rotating fluid, we build two terms of the asymptotic expansion of a solution for high viscosity (small Reynolds number). While building approximate solutions, it is necessary to solve a number of stationary boundary value problems that depend only on the shape of the cavity filled with fluid.

Finally, in the last section, we consider the motion of a body with a cavity completely filled with a nonuniformly heated fluid. Here, it is not difficult to prove the theorems of existence, and, therefore, our attention is focused mainly on the spectrum of the problem.

7.1 Motion of Fluids Completely Filling a Cavity in a Stationary Body

In this section, we study the theorem of existence of solutions of the Cauchy problem for the linearized nonstationary system of Navier–Stokes equations and describe the spectral properties of the Stokes operator. Movements close to a given stationary motion are considered and the completeness of the sytem of fading normal oscillations is proved.

7.1.1 Statement of the Problem and the Basic Equations

Suppose that a heavy viscous homogeneous fluid fills completely a cavity Ω of an immovable body with the boundary $S = \partial\Omega$. After linearization, the Navier–Stokes equations have the following form,

$$\frac{\partial \boldsymbol{u}}{\partial t} = \nu\Delta\boldsymbol{u} - \frac{1}{\rho}\nabla p + \boldsymbol{f}, \quad \operatorname{div}\boldsymbol{u} = 0 \qquad \text{in } \Omega. \tag{1.1}$$

As usual, here $\boldsymbol{u}(t,x)$ is the velocity field in a stationary coordinate system $Ox_1x_2x_3$ with the axis Ox_3 directed vertically upright, that is, against the acceleration of the gravitation force, $p(t,x)$ is the dynamic pressure, $\boldsymbol{f}(t,x)$ is a small mass field of external forces that differ from the gravitation force, and ν and ρ are the kinematic coefficient of viscosity and the density of the fluid, respectively.

On the boundary S of the region Ω, the viscous fluid should satisfy the stickiness condition

$$\boldsymbol{u} = \boldsymbol{0} \qquad \text{on } S, \tag{1.2}$$

as well the initial condition of the form

$$\boldsymbol{u}(0,x) = \boldsymbol{u}^0(x) \qquad \text{in } \Omega. \tag{1.3}$$

7.1.2 Reducing the Problem to a Differential Equation in a Hilbert Space. Existence of Solutions

If we assume that, for every t, all the terms in the Navier–Stokes equation belong to $\boldsymbol{L}_2(\Omega)$, then we can apply the operator P_0 of orthogonal projection onto the subspace $\boldsymbol{J}_0(\Omega)$ to both sides of this equation. In virtue of the incompressibility condition and the boundary condition (1.2), it is natural to assume that the field $\boldsymbol{u}(t,x)$ belongs to the space $\boldsymbol{J}_0^1(\Omega)$, and, therefore, to the space $\boldsymbol{J}_0(\Omega)$. After performing the orthogonal projection, we obtain

$$\frac{\partial \boldsymbol{u}}{\partial t} = \nu P_0\Delta\boldsymbol{u} + P_0\boldsymbol{f}. \tag{1.4}$$

Extending the operator $-P_0\Delta$ to the Stokes operator A_0 (see Section 2.2), we obtain in the space $\boldsymbol{J}_0(\Omega)$ the abstract equation

$$\frac{d\boldsymbol{u}}{\mathrm{d}t} = -\nu A_0\boldsymbol{u} + P_0\boldsymbol{f}. \tag{1.5}$$

If the field $P_0\boldsymbol{f}$ satisfies a Hölder condition with respect to t in the norm of the space $\boldsymbol{L}_2(\Omega)$, then, according to the results in Section 1.5, equation (1.5) has a unique weak solution for any initial velocity field $\boldsymbol{u}^0(x)$ in $\boldsymbol{J}_0(\Omega)$. This solution is expressed through the analytic semigroup $\exp(-\nu t A_0)$ by the formula

$$\boldsymbol{u}(t) = \exp(-\nu t A_0)\boldsymbol{u}^0 + \int_0^{\mathrm{t}} \exp(-\nu(t-s)A_0)(P_0\boldsymbol{f})(s)\mathrm{d}s. \tag{1.6}$$

For every $t > 0$, the field $\boldsymbol{u}$ belongs to the space $\boldsymbol{J}_0^1(\Omega)$.

Equation (1.4) is not equivalent to equations (1.1) and (1.2). On the other hand, the system consisting of (1.4) and the following equation is equivalent to (1.1) and (1.2):

$$\boldsymbol{0} = \nu(I - P_0)\Delta\boldsymbol{u} - \frac{1}{\rho}\nabla p + (I - P_0)\boldsymbol{f}. \tag{1.7}$$

If the solution $\boldsymbol{u}$ in equation (1.4) is found, then the field ∇p and the function p can be found from equation (1.7).

7.1.3 Structure of the Spectrum of the Problem

Let us consider the issue on normal oscillations of a fluid in a cavity, that is, those solutions of the homogeneous problem (1.1)–(1.2) that depend on time according to the exponential law

$$\boldsymbol{u}(t,x) = \exp(-\nu t\lambda)\boldsymbol{v}(x). \tag{1.8}$$

From equation (1.5), for $\boldsymbol{f}(t) \equiv \boldsymbol{0}$, we obtain

$$A_0\boldsymbol{v} = \lambda\boldsymbol{v}. \tag{1.9}$$

Hence, the fields $\boldsymbol{v}$ are eigenelements of the Stokes operator A_0 and the numbers λ are its eigenvalues. Since, according to the results in Section 2.2.5, the Stokes operator is positive definite and has a compact inverse operator, we conclude that problem (1.9) has a discrete positive spectrum of normal eigenvalues $\lambda = \lambda_n$, with $\lambda_n \to +\infty$ as $n \to \infty$, and a system of eigenfields $\{\boldsymbol{v}_n(x)\}_{n=1}^{\infty}$ that forms an orthonormal basis in the space $\boldsymbol{J}_0(\Omega)$.

To find out properties of the eigenvalues λ_n, let us recall that they can be determined as consecutive minima of the variational ratio

$$\frac{(A_0 \boldsymbol{v}, \boldsymbol{v})_{\boldsymbol{L}_2(\Omega)}}{(\boldsymbol{v}, \boldsymbol{v})_{\boldsymbol{L}_2(\Omega)}} = \frac{\displaystyle\int_\Omega |\mathrm{rot}\boldsymbol{v}|^2 d\Omega}{\displaystyle\int_\Omega |\boldsymbol{v}|^2 d\Omega}, \tag{1.10}$$

which is considered in the set of all nonzero elements of the space $\boldsymbol{J}_0^1(\Omega)$. Hence, for a region Ω of the first type, we obtain the following asymptotic formula,

$$\lambda_n(A_0) = \left(\frac{\mathrm{mes}\,\Omega}{3\pi^2}\right)^{-2/3} n^{2/3}[1 + \mathrm{o}(1)], \qquad n \to \infty. \tag{1.11}$$

Using more complicated methods, for regions Ω with a smooth boundary $\partial\Omega$ we obtain the following estimate,

$$n(\lambda) := \sum_{\lambda_n \leq \lambda} 1 = \frac{1}{3\pi^2}(\mathrm{mes}\,\Omega)\lambda^{3/2} + \mathrm{O}(\lambda), \qquad \lambda \to +\infty. \tag{1.12}$$

7.1.4 Perturbation of the Stationary Motion of a Fluid

Let us assume that, under the influence of some field $\boldsymbol{F}_0(x)$, the fluid performs a smooth stationary motion with the velocity field $\boldsymbol{v}_0(x)$ in the immovable cavity. Then the field $\boldsymbol{v}_0$ satisfies the stationary nonlinear Stokes equations

$$(\boldsymbol{v}_0 \cdot \nabla)\boldsymbol{v}_0 = \nu \Delta \boldsymbol{v}_0 - \frac{1}{\rho}\nabla P_0 + \boldsymbol{F}_0,$$

where $P_0 = P_0(x)$ is the stationary field of pressures.

Now let us consider movements that are close to the stationary motion. The linearization of the nonstationary system of Navier–Stokes equations leads to the following problem for the deviations of perturbed fields,

$$\frac{\partial \boldsymbol{u}}{\partial t} + (v_0 \cdot \nabla)\boldsymbol{u} + (\boldsymbol{u} \cdot \nabla)\boldsymbol{v}_0 = \nu \Delta \boldsymbol{u} - \frac{1}{\rho}\nabla p + \boldsymbol{f}, \quad \mathrm{div}\boldsymbol{u} = 0 \qquad \text{in } \Omega, \tag{1.13}$$

$$\boldsymbol{u} = 0 \qquad \text{on } S = \partial\Omega, \quad \boldsymbol{u}(0, x) = \boldsymbol{u}^0(x). \tag{1.14}$$

If we apply the orthoprojector P_0 again and extend the operator $-P_0\Delta$ to the Stokes operator, then (1.13) and (1.14) give the Cauchy problem

$$\frac{d\boldsymbol{u}}{dt} = -\nu A_0 \boldsymbol{u} - R\boldsymbol{u} + P_0 \boldsymbol{f}, \qquad \boldsymbol{u}(0) = \boldsymbol{u}^0, \tag{1.15}$$

where $R\boldsymbol{u} := P_0((\boldsymbol{v}_0 \cdot \nabla)\boldsymbol{u} + (\boldsymbol{u} \cdot \nabla)\boldsymbol{v}_0)$. The operator R maps continuously the space $\boldsymbol{J}_0^1(\Omega)$ to the space $\boldsymbol{J}_0(\Omega)$. Since for the operator A_0 the domain of definition $\mathcal{D}(A_0^{1/2})$ coincides with $\boldsymbol{J}_0^1(\Omega)$, then $RA_0^{-1/2}$ is a bounded operator in $\boldsymbol{J}_0(\Omega)$, that is,

$$\|RA_0^{-1/2}\boldsymbol{u}\|_{\boldsymbol{L}_2(\Omega)} \leq c\|\boldsymbol{u}\|_{\boldsymbol{L}_2(\Omega)}, \qquad \boldsymbol{u} \in \boldsymbol{J}_0(\Omega),$$

or

$$\|R\boldsymbol{v}\|_{\boldsymbol{L}_2(\Omega)} \leq c\|A_0^{1/2}\boldsymbol{v}\|_{\boldsymbol{L}_2(\Omega)},$$

for all $\boldsymbol{v} \in \mathcal{D}(A_0^{1/2}) = \boldsymbol{J}_0^1(\Omega)$.

Hence, operator R is subordinated to operator $A_0^{1/2}$. According to Section 1.5.5, operator $-\nu A_0 - R$ is a generating operator of the analytic semigroup $\exp(-t(\nu A_0 + R))$, and, therefore,

$$\boldsymbol{u}(t) = \exp(-t(\nu A_0 + R))\boldsymbol{u}^0 + \int_0^t \exp(-(t-s)(\nu A_0 + R))(P_0)\boldsymbol{f})(s)\mathrm{d}s \qquad (1.16)$$

is a solution of problem (1.15).

Considering free normal oscillations, we obtain from (1.15) for $\boldsymbol{f} \equiv \boldsymbol{0}$ the following equation for modes of oscillations,

$$(\nu A_0 + R)\boldsymbol{v} = \lambda \boldsymbol{v}. \qquad (1.17)$$

Let us perform the substitution $\boldsymbol{v} = A_0^{-1}\boldsymbol{w}$. Then

$$L(\lambda)\boldsymbol{w} := \left(I + \nu^{-1}RA_0^{-1} - \lambda\nu^{-1}A_0^{-1}\right)\boldsymbol{w} = \boldsymbol{0} \qquad (1.18)$$

Hence, we just obtained a spectral problem for the linear operator pencil $L(\lambda)$; this problem satisfies the conditions of the second Keldysh theorem (see Section 1.6.4). Indeed, operator $\nu^{-1}A_0^{-1}$ in (1.18) is a completely positive compact operator that belongs to class $\mathfrak{S}_p$ for $p > 3/2$, in virtue of the asymptotic formula (1.11). Further, operator $RA_0^{-1} = (RA_0^{-1/2})A_0^{-1/2}$ is compact, because operator $RA_0^{-1/2}$ is bounded and $A_0^{-1/2}$ is compact.

From the second Keldysh theorem, for equation (1.18), the existence of a sequence of normal eigenvalues λ_k with $\operatorname{Re}\lambda_k \to +\infty$ as $k \to \infty$ follows; here, only a finite number of eigenvalues λ_k can be situated outside any sector $|\arg\lambda| < \varepsilon$ of the complex plane. Further, to each eigenvalue λ_k there correspond d_k eigenelements $\boldsymbol{w}_k^{(j,0)}$, $1 \leq j \leq d_k$, of problem (1.18), and to each eigenelement $\boldsymbol{w}_k^{(j,0)}$ there corresponds possibly a finite collection of associated elements $\boldsymbol{w}_k^{(j,p)}$, $1 \leq p \leq m_{jk} - 1$, where $m_{jk} \geq 1$ is the length of the Jordan chain consisting of the eigenelement $\boldsymbol{w}_k^{(j,0)}$ and its associated elements.

The union of all such canonical systems of elements (see Section 1.6.2), for all k, forms according to Keldysh theorem a complete system in $\boldsymbol{J}_0(\Omega)$. This means that, for any element $\boldsymbol{w} \in \boldsymbol{J}_0(\Omega)$, there exists a linear combination of elements $\boldsymbol{w}_k^{(j,p)}$ that approximates the element $\boldsymbol{w}$ in the norm of the space $\boldsymbol{J}_0(\Omega)$ with any accuracy,

$$\left\| \boldsymbol{w} - \sum_{k=1}^{N(\varepsilon)} \sum_{j=1}^{d_k} \sum_{p=0}^{m_{jk}-1} c_{kjp}(\varepsilon) \boldsymbol{w}_k^{j,p} \right\|_{\boldsymbol{L}_2(\Omega)} < \varepsilon. \tag{1.19}$$

The fields $\boldsymbol{v}_k^{(j,p)} = A_0^{-1} \boldsymbol{w}_k^{(j,p)}$ are solutions of the equation

$$\begin{aligned} \nu A_0 \boldsymbol{v}_k^{(j,0)} + R \boldsymbol{v}_k^{(j,o)} &= \lambda_k \boldsymbol{v}_k^{(j,0)}, \qquad 1 \le j \le d_k, \ k = 1, 2, \ldots, \\ \nu A_0 \boldsymbol{v}_k^{(j,p)} + R \boldsymbol{v}_k^{(j,p)} &= \lambda_k \boldsymbol{v}_k^{(j,p)} + \boldsymbol{v}_k^{(j,p-1)}, \qquad 1 \le p \le m_{jk} - 1. \end{aligned} \tag{1.20}$$

If $\boldsymbol{v} \in \mathcal{D}(A_0)$, then denoting $\boldsymbol{w} = A_0 \boldsymbol{v}$, we obtain from (1.19)

$$\left\| A_0 \left(\boldsymbol{v} - \sum_{k=1}^{N(\varepsilon)} \sum_{j=1}^{d_k} \sum_{p=0}^{m_{jk}-1} c_{kjp}(\varepsilon) \boldsymbol{v}_k^{j,p)} \right) \right\|_{\boldsymbol{L}_2(\Omega)} < \varepsilon. \tag{1.21}$$

Accordingly, the system $\{\boldsymbol{v}_k^{(j,p)}\}$ is complete in the domain of definition $\mathcal{D}(A_0)$ of operator A_0 in the graph norm. Therefore, it is complete in $\mathcal{D}(A_0)$ in the norm of the space $\boldsymbol{J}_0^1(\Omega) = \mathcal{D}(A_0^{1/2})$ and also in the norm of $\boldsymbol{J}_0(\Omega)$. Since $\mathcal{D}(A_0)$ is dense in $\boldsymbol{J}_0^1(\Omega)$ and $\boldsymbol{J}_0^1(\Omega)$ is dense in $\boldsymbol{J}_0(\Omega)$, then this system is dense both in $\boldsymbol{J}_0^1(\Omega)$ and $\boldsymbol{J}_0^1(\Omega)$.

Let us note that to each collection $\boldsymbol{v}_k^{(j,p)}$, $0 \le p \le m_{jk} - 1$, there corresponds a collection of particular solutions of the homogeneous equation

$$\frac{\mathrm{d}\boldsymbol{u}}{\mathrm{d}t} = -\nu A_0 \boldsymbol{u} - R\boldsymbol{u} \tag{1.22}$$

of the form

$$e^{-\lambda_k t} \left(\boldsymbol{v}_k^{(j,p)} - \frac{t}{1!} \boldsymbol{v}_k^{(j,p-1)} + \cdots + (-1)^p \frac{t^p}{p!} \boldsymbol{v}_k^{(j,0)} \right). \tag{1.23}$$

Hence, near any stationary motion of a viscous fluid in a closed cavity Ω, there exists the complete system of oscillations (1.23).

7.1.5 Small Movements of a Fluid in a Rotating Container

The problem on movements and normal oscillations of a viscous fluid filling a container and performing movements close to a uniform rotation, is related to the previously considered problems.

Suppose that the angular velocity $\boldsymbol{\omega}_0 = \omega_0 \boldsymbol{e}_3$ of the container is constant in the nonperturbed state and the fluid performs a rigid rotation. Then, in the nonstationary coordinate system $Ox_1x_2x_3$, rigidly connected to the container, the fluid does not move, that is, the velocity field equals zero, and the nonperturbed pressure is

$$P_0(x) = -\rho g x_3 + \frac{1}{2}\rho\omega_0^2\left(x_1^2 + x_2^2\right) + \text{const}. \tag{1.24}$$

Let us next consider those small movements of a fluid in a cavity that are close to a uniform rotation. For the velocity field $\boldsymbol{u}(t,x)$ and the dynamic pressure $p(t,x)$, we obtain instead of (1.1)–(1.4) the following problem,

$$\frac{\partial \boldsymbol{u}}{\partial t} - 2\omega_0 \boldsymbol{u} \times \boldsymbol{e}_3 = \nu\Delta\boldsymbol{u} - \frac{1}{\rho}\nabla p + \boldsymbol{f}, \quad \operatorname{div}\boldsymbol{u} = 0 \qquad \text{in } \Omega, \tag{1.25}$$

$$\boldsymbol{u} = \boldsymbol{0} \qquad \text{on } S = \partial\Omega, \quad \boldsymbol{u}(0,x) = \boldsymbol{u}^0(x). \tag{1.26}$$

Projecting the first equation in (1.25) onto $\boldsymbol{J}_0(\Omega)$ and introducing the Stokes operator A_0, we obtain the Cauchy problem

$$\frac{\mathrm{d}\boldsymbol{u}}{\mathrm{d}t} = -\nu A_0 \boldsymbol{u} + K\boldsymbol{u} + P_0\boldsymbol{f}, \qquad \boldsymbol{u}(0) = \boldsymbol{u}^0 \in \boldsymbol{J}_0(\Omega), \tag{1.27}$$

where $K\boldsymbol{u} := 2\omega_0 P_0(\boldsymbol{u}\times\boldsymbol{e}_3)$ is the gyroscopic or Coriolis operator that was mentioned in Section 6.1 and in other chapters. Let us recall that this operator is skew-summetric, that is, $K^* = -K$, and its spectrum $\sigma(K)$ coincides with the interval on the imaginary axis $[-2\mathrm{i}\omega_0, 2\mathrm{i}\omega_0]$.

Since K is a bounded operator, then $-\nu A_0 + K$ is the generating operator of the analytic semigroup $\exp(t(-\nu A_0 + K))$, and the weak solution of problem (1.27) is given by the formula

$$\boldsymbol{u}(t) = \exp(t(-\nu A_0 + K))\boldsymbol{u}^0 + \int_0^t \exp\left((t-s)(-\nu A_0 + K)\right)(P_0\boldsymbol{f}(s))\mathrm{d}s. \tag{1.28}$$

As in the previous section, we obtain that, if $\boldsymbol{u}^0 \in \boldsymbol{J}_0(\Omega)$ and $(P_0\boldsymbol{f})(t)$ satisfies a Hölder condition with respect to t in the norm of $\boldsymbol{L}_2(\Omega)$, then problem (1.25)–(1.26) is univalently solvable, its solution $\boldsymbol{u}(t,x)$ can be found by formula (1.28), and, for $t > 0$, the function $\boldsymbol{u}(t,x)$ belongs to the space $J_0^1(\Omega)$. Besides, the law of kinetic energy balance is valid for any $t > 0$,

$$\begin{aligned} &\rho\|\boldsymbol{u}(t)\|^2_{\boldsymbol{L}_2(\Omega)} \\ &\quad = \rho\|\boldsymbol{u}^0\|^2_{\boldsymbol{L}_2(\Omega)} + 2\rho\int_0^t \left[-\nu\|\operatorname{rot}\boldsymbol{u}(s)\|^2_{\boldsymbol{L}_2(\Omega)} + (\boldsymbol{f}(s),\boldsymbol{u}(s))_{\boldsymbol{L}_2(\Omega)}\right]\mathrm{d}s. \end{aligned} \tag{1.29}$$

For the normal oscillations $\boldsymbol{u}(t,x) = \exp(-\lambda t)\boldsymbol{v}(x)$, if $\boldsymbol{f}(t) \equiv \mathbf{0}$, we obtain from (1.27) the following spectral problem

$$(\nu A_0 + 2\mathrm{i}\omega_0 T)\boldsymbol{v} = \lambda \boldsymbol{v}, \qquad \boldsymbol{v} \in \mathcal{D}(A_0) \subset \boldsymbol{J}_0(\Omega), \tag{1.30}$$

where $T := [\mathrm{i}/(2\mathrm{i}\omega_0)]K = T^*$, and the spectrum of operator T coincides with the interval $[-1, 1]$.

Equation (1.30) belongs to the same type of equations as (1.17), but here $R := 2\mathrm{i}\omega_0 T$ is a bounded operator and $\|R\| = 2\omega_0$. That is why, in case of problem (1.30), all general conclusions in the previous section concerning the structure of its spectrum and the completeness of the system of eigen- and associated elements are still valid.

The explicit form of operator R allows us to clarify some of the properties of the eigenvalues λ_k. Namely, for problem (1.30), these eigenvalues are located in the halfband

$$\nu\lambda_1(A_0) \le \operatorname{Re}\lambda_k < +\infty, \qquad |\operatorname{Im}\lambda_k| \le 2\omega_0, \tag{1.31}$$

where $\lambda_1(A_0)$ is the smallest eigenvalue of the Stokes operator. Indeed, if λ_k and $\boldsymbol{v}_k$ are an eigenvalue and a normalized eigenelement in $\boldsymbol{J}_0(\Omega)$ of the problem (1.30), then in virtue of the properties $T = T^*$, $\|T\| = 1$, the following relations hold true,

$$\begin{aligned} \nu\left(A_0\boldsymbol{v}_k, \boldsymbol{v}_k\right)_{\boldsymbol{L}_2(\Omega)} &= \operatorname{Re}\lambda_k, \\ 2\omega_0(T\boldsymbol{v}_k, \boldsymbol{v}_k)_{\boldsymbol{L}_2(\Omega)} &= \operatorname{Im}\lambda_k, \end{aligned}$$

from which the estimates (1.31) follow.

Hence, for the oscillating fading modes that appear in the given spectral problem, the fading decrement is bounded from below by the positive constant $\nu\lambda_1(A_0)$ that depends only on the shape of region Ω and the kinematic viscosity ν, and the frequencies of oscillations $\operatorname{Im}\lambda_k$ are bounded in module by twice the angular velocity of the system's rotation. Moreover, let us note that, for $\omega_0 \to 0$, the eigenvalues $\lambda_k = \lambda_k(\omega_0)$ of problem (1.30) converge to the eigenvalues $\nu\lambda_k(A_0)$ of the Stokes problem (see Section 7.1.3), and the frequencies of normal oscillations $\operatorname{Im}\lambda_k$ converge uniformly to zero.

In Section 1.6.10 we obtained results on the asymptotics of the spectrum of an operator pencil of the form (1.6.49). From those results and in virtue of the asymptotics (1.11) for the eigenvalues of the Stokes operator, we obtain the following result for the eigenvalues of problem (1.30):

$$\lambda_n = \nu\lambda_n(A_0)\left[1 + \mathrm{o}(1)\right], \qquad n \to \infty. \tag{1.32}$$

7.2 Small Movements of a Gyrostate Around a Fixed Mass Center

In this section, we investigate the linearized equations of the motion of a body and the fluid near an immovable state. Here, we focus on an operator that is related to the translational motions of the fluid. Its properties wil be studied in detail. We prove a theorem of existence of a solution of the Cauchy problem by introducing a new equivalent norm in the space $\boldsymbol{L}_2(\Omega)$. Some of the properties of normal oscillations will be considered as well.

7.2.1 Statement of the Problem and the Basic Equations

Let us assume that the hydromechanical system "body + fluid" is a gyrostate, that is, the fluid fills completely the cavity Ω. We choose the pole O of the nonstationary coordinate system $Ox_1x_2x_3$ rigidly connected with the body in the mass center C. Moreover, let us assume that in a nonperturbed state the whole system is immovable and then, under the influence of small external forces and initial conditions, it starts to perform a small movememnt around the fixed mass center.

The moment of the amount of motion is described by the linearized equation (3.1.25) in which $\omega_0 = 0$, that is, in which there is no rotation in the nonperturbed state,

$$\boldsymbol{J}\boldsymbol{\varepsilon} + \rho\frac{\mathrm{d}}{\mathrm{d}t}\int_{\Omega}(\boldsymbol{r}\times\boldsymbol{u})\mathrm{d}\Omega = \boldsymbol{M}, \tag{2.1}$$

where $\boldsymbol{M} = \boldsymbol{M}(t)$ is the moment relatively to the mass center C of small external forces influencing the system (this moment is written down in the stationary coordinate system $Oy_1y_2y_3$, see Section 3.1), $\boldsymbol{\varepsilon} = \mathrm{d}\boldsymbol{\omega}/\mathrm{d}t$ is the angular acceleration of the body, $\boldsymbol{u}(t,x)$ is the field of relative velocity, and $\boldsymbol{J} = \boldsymbol{J}_b + \boldsymbol{J}_f$ is the tensor of inertia of the whole system.

The linearized Navier–Stokes equations that describe the motion of the viscous fluid in the cavity Ω can be obtained from (3.1.26) for $\omega_0 = 0$:

$$\frac{\partial \boldsymbol{u}}{\partial t} + \boldsymbol{\varepsilon}\times\boldsymbol{r} = -\frac{1}{\rho}\nabla p + \nu\Delta\boldsymbol{u} + \boldsymbol{f}, \qquad \operatorname{div}\boldsymbol{u} = 0 \quad \text{in } \Omega, \tag{2.2}$$

$$\boldsymbol{u} = \boldsymbol{0} \quad \text{on } S = \partial\Omega. \tag{2.3}$$

The initial conditions are given by the following equations

$$\boldsymbol{u}(0,x) = \boldsymbol{u}^0(x), \quad \boldsymbol{\omega}(0) = \boldsymbol{\omega}^0. \tag{2.4}$$

We next derive the law of full energy balance for a classical solution of problem (2.1)–(2.4). Let us scalarly multiply equation (2.1) by $\boldsymbol{\omega}(t)$ and equation (2.2) by $\rho\boldsymbol{u}$ and then integrate over the region Ω. Taking into consideration the orthogonality of the fields ∇p and $\boldsymbol{u}$ in $\boldsymbol{L}_2(\Omega)$, we obtain

$$\frac{1}{2}\frac{\mathrm{d}}{\mathrm{d}t}(\boldsymbol{J}\boldsymbol{\omega}\cdot\boldsymbol{\omega})+\rho\int_{\Omega}\left(\boldsymbol{r}\times\frac{\partial\boldsymbol{u}}{\partial t}\right)\cdot\boldsymbol{\omega}\mathrm{d}\Omega=\boldsymbol{M}\cdot\boldsymbol{\omega},$$
$$\frac{1}{2}\frac{\mathrm{d}}{\mathrm{d}t}\int_{\Omega}|\boldsymbol{u}|^2\mathrm{d}\Omega+\rho\int_{\Omega}\left(\frac{\mathrm{d}\boldsymbol{\omega}}{\mathrm{d}t}\times\boldsymbol{r}\right)\cdot\boldsymbol{u}\mathrm{d}\Omega=\nu\rho\int_{\Omega}\Delta\boldsymbol{u}\cdot\boldsymbol{u}\mathrm{d}\Omega+\rho\int_{\Omega}\boldsymbol{f}\cdot\boldsymbol{u}\mathrm{d}\Omega.$$

Adding up these identities and using formulas (2.2.13), (2.2.14) and the identity $\sum_{i,j=1}^{3}|r_{ij}|^2=2|\mathrm{rot}\boldsymbol{u}|^2$, we obtain the relation

$$\frac{1}{2}\frac{\mathrm{d}}{\mathrm{d}t}\left(\rho\int_{\Omega}|\boldsymbol{u}|^2\mathrm{d}\Omega+\boldsymbol{J}\boldsymbol{\omega}\cdot\boldsymbol{\omega}+2\rho\int_{\Omega}(\boldsymbol{\omega}\times\boldsymbol{r})\cdot\boldsymbol{u}\mathrm{d}\Omega\right)$$
$$=-\rho\nu\int_{\Omega}|\mathrm{rot}\boldsymbol{u}|^2\mathrm{d}\Omega+\rho\int_{\Omega}\boldsymbol{f}\cdot\boldsymbol{u}\mathrm{d}\Omega+\boldsymbol{M}\cdot\boldsymbol{\omega},\tag{2.5}$$

which shows that the modifications of the kinetic energy of the system (the potential energy of a gyrostate fixed at the mass centre C is constant) is caused by its dissipation in the fluid and by the work of external forces in the system.

7.2.2 Transition to a Differential Equation in a Hilbert Space

If we find the angular velocity $\boldsymbol{\varepsilon}=\mathrm{d}\boldsymbol{\omega}/\mathrm{d}t$ from (2.1) and substitute it into the equation (2.2), we obtain

$$\frac{\partial\boldsymbol{u}}{\partial t}-\rho\left(\boldsymbol{J}^{-1}\int_{\Omega}\left(\boldsymbol{r}\times\frac{\partial\boldsymbol{u}}{\partial t}\right)\mathrm{d}\Omega\right)\times\boldsymbol{r}=-\frac{1}{\rho}\nabla p+\nu\Delta\boldsymbol{u}+\boldsymbol{f}_1,$$
$$\mathrm{div}\,\boldsymbol{u}=0,\quad \boldsymbol{f}_1:=\boldsymbol{f}-(\boldsymbol{J}^{-1}\boldsymbol{M})\times\boldsymbol{r}.\tag{2.6}$$

Let us analyse in details the second term in the left side. Fields of the form $\boldsymbol{c}\times\boldsymbol{r}$, where $\boldsymbol{c}$ is a constant vector, form a three-dimensional linear space E_3 with the basis fields $\boldsymbol{w}_i:=\boldsymbol{e}_i\times\boldsymbol{r}$, $i=1,2,3$. The term we are considering has that form, and therefore, the expression

$$\rho\left(\boldsymbol{J}^{-1}\int_{\Omega}(\boldsymbol{r}\times\boldsymbol{v})\mathrm{d}\Omega\right)\times\boldsymbol{r}\tag{2.7}$$

can be viewed as a linear operator acting, for example, from $\boldsymbol{J}_0(\Omega)$ into the three-dimensional subspace $E_3\subset\boldsymbol{L}_2(\Omega)$. If the projector P_0 onto the subspace $\boldsymbol{J}_0(\Omega)$ is applied to expression (2.7), then the formula

$$B\boldsymbol{v}:=P_0\left[\rho\left(\boldsymbol{J}^{-1}\int_{\Omega}(\boldsymbol{r}\times\boldsymbol{v})\mathrm{d}\Omega\right)\times\boldsymbol{r}\right]\tag{2.8}$$

defines a finite-dimensional operator B that maps $\boldsymbol{J}_0(\Omega)$ into another three-dimensional space. We will refer to B as the *translation operator.*

Applying P_0 to both sides of the first equation in (2.6) and extending the operator $-P_0\Delta$ to the Stokes operator A_0, we obtain the problem

$$(I-B)\frac{\mathrm{d}\boldsymbol{u}}{\mathrm{d}t} = -\nu A_0\boldsymbol{u} + P_0\boldsymbol{f}_1, \qquad \boldsymbol{u}(0) = \boldsymbol{u}^0. \tag{2.9}$$

7.2.3 Properties of the Translation Operator

Let us now investigate the properties of operator B. For any $\boldsymbol{u}$ and $\boldsymbol{v}$ from $\boldsymbol{J}_0(\Omega)$, using properties of the triple scalar product of vectors, we have

$$\begin{aligned}
(B\boldsymbol{u},\boldsymbol{v})_{\boldsymbol{L}_2(\Omega)} &= \int_\Omega P_0\left[\rho\left(\boldsymbol{J}^{-1}\int_\Omega(\boldsymbol{r}\times\boldsymbol{u})\mathrm{d}\Omega\right)\times\boldsymbol{r}\right]\cdot\boldsymbol{v}\mathrm{d}\Omega \\
&= \rho\int_\Omega \boldsymbol{J}^{-1}\left(\int_\Omega(\boldsymbol{r}\times\boldsymbol{u})\mathrm{d}\Omega\right)\times\boldsymbol{r}\cdot\boldsymbol{v}\mathrm{d}\Omega \\
&= \rho\int_\Omega \boldsymbol{J}^{-1}\int_\Omega(\boldsymbol{r}\times\boldsymbol{u})\mathrm{d}\Omega\cdot(\boldsymbol{r}\times\boldsymbol{v})\mathrm{d}\Omega \\
&= \rho\boldsymbol{J}^{-1}\left(\int_\Omega(\boldsymbol{r}\times\boldsymbol{u})\mathrm{d}\Omega\right)\cdot\left(\int_\Omega(\boldsymbol{r}\times\boldsymbol{v})\mathrm{d}\Omega\right).
\end{aligned} \tag{2.10}$$

Since the tensor of inertia $\boldsymbol{J}$ is symmetric, then the right side of (2.10) depends symmetrically on $\boldsymbol{u}$ and $\boldsymbol{v}$, and, therefore, B is a self-adjoint operator. Further, for $\boldsymbol{v}=\boldsymbol{u}$ we obtain

$$(B\boldsymbol{u},\boldsymbol{u})_{\boldsymbol{L}_2(\Omega)} = \rho\boldsymbol{J}^{-1}\left(\int_\Omega(\boldsymbol{r}\times\boldsymbol{u})\mathrm{d}\Omega\right)\cdot\left(\int_\Omega(\boldsymbol{r}\times\boldsymbol{u})\mathrm{d}\Omega\right) \geq 0, \tag{2.11}$$

that is, B is a nonnegative operator.

The proof of positive definiteness and invertibility of operator $I-B$ requires a more detailed reasoning. Let us recall that the tensor of intertia $\boldsymbol{J}$ equals the sum of the tensor of inertia of the body $\boldsymbol{J}_b$ and the tensor of inertia of the hardened fluid $\boldsymbol{J}_f$, which for any vector $\boldsymbol{a}$ can be defined by formula (3.1.25):

$$\boldsymbol{J}_f\boldsymbol{a} := \rho\int_\Omega(\boldsymbol{r}\times(\boldsymbol{a}\times\boldsymbol{r}))\,\mathrm{d}\Omega. \tag{2.12}$$

We use the following identity from vector analysis:

$$(\boldsymbol{a}\times\boldsymbol{r})\cdot(\boldsymbol{a}\times\boldsymbol{r}) = \boldsymbol{a}\cdot(\boldsymbol{r}\times(\boldsymbol{a}\times\boldsymbol{r}))\,.$$

Assuming that

$$\boldsymbol{a} := \rho \boldsymbol{J}^{-1} \int_{\Omega} (\boldsymbol{r} \times \boldsymbol{u}) \mathrm{d}\Omega,$$

we obtain

$$\rho\left((\boldsymbol{a} \times \boldsymbol{r}), (\boldsymbol{a} \times \boldsymbol{r})\right)_{\boldsymbol{L}_2(\Omega)} = \rho\left(\boldsymbol{a}, \boldsymbol{r} \times (\boldsymbol{a} \times \boldsymbol{r})\right)_{\boldsymbol{L}_2(\Omega)} = \boldsymbol{a} \cdot \boldsymbol{J}_f \boldsymbol{a}.$$

Let us calculate

$$\begin{aligned}
&\rho\left((I - B)\boldsymbol{u}, \boldsymbol{u}\right)_{\boldsymbol{L}_2(\Omega)} \\
&\quad = \rho(\boldsymbol{u}, \boldsymbol{u})_{\boldsymbol{L}_2(\Omega)} - \rho(B\boldsymbol{u}, \boldsymbol{u})_{\boldsymbol{L}_2(\Omega)} \\
&\quad = \rho(\boldsymbol{u}, \boldsymbol{u})_{\boldsymbol{L}_2(\Omega)} - \boldsymbol{a} \cdot \boldsymbol{J}\boldsymbol{a} \\
&\quad = \rho\|\boldsymbol{u} - (\boldsymbol{a} \times \boldsymbol{r})\|^2_{\boldsymbol{L}_2(\Omega)} - \rho\left((\boldsymbol{a} \times \boldsymbol{r}), (\boldsymbol{a} \times \boldsymbol{r})\right)_{\boldsymbol{L}_2(\Omega)} + 2\rho\left(\boldsymbol{u}, (\boldsymbol{a} \times \boldsymbol{r})\right)_{\boldsymbol{L}_2(\Omega)} - \boldsymbol{a} \cdot \boldsymbol{J}\boldsymbol{a} \\
&\quad = \rho\|\boldsymbol{u} - (\boldsymbol{a} \times \boldsymbol{r})\|^2_{\boldsymbol{L}_2(\Omega)} - \boldsymbol{a} \cdot \boldsymbol{J}_f \boldsymbol{a} + 2\boldsymbol{J}\boldsymbol{a} \cdot \boldsymbol{a} - \boldsymbol{a} \cdot \boldsymbol{J}\boldsymbol{a} \\
&\quad = \rho\|\boldsymbol{u} - (\boldsymbol{a} \times \boldsymbol{r})\|^2_{\boldsymbol{L}_2(\Omega)} + \boldsymbol{a} \cdot \boldsymbol{J}_b \boldsymbol{a} \geq 0.
\end{aligned}$$

We obtain that $I - B$ is a nonnegative operator. Further, we show that the left side equals zero if and only if $\boldsymbol{u} \equiv \boldsymbol{0}$. Indeed, from the fact that the term $\boldsymbol{a} \cdot \boldsymbol{J}_a \boldsymbol{a}$ equals zero, it follows, in virtue of positive definiteness of the tensor of inertia $\boldsymbol{J}_b$, that

$$\boldsymbol{a} = \rho \boldsymbol{J}^{-1} \int_{\Omega} (\boldsymbol{r} \times \boldsymbol{u}) \mathrm{d}\Omega = \boldsymbol{0}.$$

Then, according to (2.8), $B\boldsymbol{u} = \boldsymbol{0}$; therefore $\boldsymbol{u} = \boldsymbol{0}$, because the term $\rho\|(I - B)\boldsymbol{u}\|^2$ equals zero.

Hence, operator $I - B$ equals zero at zero only. Since B is a finite-dimensional operator, according to Fredholm theorem, the operator $I - B$ has a bounded inverse operator, $(I - B)^{-1}$.

To conclude, the translation operator B is finite-dimensional, self-adjoint, and satisfies $0 \leq B < I$. These properties of operator B make it possible to perform a transition from equation (2.9) to an equivalent equation,

$$\frac{\mathrm{d}\boldsymbol{u}}{\mathrm{d}t} = -\nu(I - B)^{-1} A_0 \boldsymbol{u} + (I - B)^{-1} P_0 f_1. \tag{2.13}$$

7.2.4 Existence of Solutions of the Evolution Problem

We introduce a new scalar product in the space $\boldsymbol{J}_0(\Omega)$ defined by the formula

$$\langle \boldsymbol{u}, \boldsymbol{v} \rangle := ((I - B)\boldsymbol{u}, \boldsymbol{v})_{\boldsymbol{L}_2(\Omega)}\,. \tag{2.14}$$

In virtue of the properties of the translation operator, there exists a constant $c > 0$ such that for all $\boldsymbol{u}$ from $\boldsymbol{J}_0(\Omega)$

$$c^2(\boldsymbol{u}, \boldsymbol{u})_{\boldsymbol{L}_2(\Omega)} \le ((I - B)\boldsymbol{u}, \boldsymbol{u})_{\boldsymbol{L}_2(\Omega)} \le (\boldsymbol{u}, \boldsymbol{u})_{\boldsymbol{L}_2(\Omega)}.$$

Hence the norm

$$\|\boldsymbol{u}\|_B := \langle \boldsymbol{u}, \boldsymbol{u} \rangle^{1/2}$$

that corresponds to the scalar product (2.14), is equivalent to the ordinary norm in $\boldsymbol{J}_0(\Omega)$, that is,

$$c\|\boldsymbol{u}\|_{\boldsymbol{L}_2(\Omega)} \le \|\boldsymbol{u}\|_B \le \|\boldsymbol{u}\|_{\boldsymbol{L}_2(\Omega)}. \tag{2.15}$$

The operator $(I - B)^{-1}A_0$, whose domain of definition coincides with $\mathcal{D}(A_0)$, is self-adjoint in the scalar product (2.14). Indeed, if for some $\boldsymbol{v} \in \boldsymbol{J}_0(\Omega)$ and any $\boldsymbol{u} \in \mathcal{D}(A_0)$ the following inequality holds true

$$|\langle (I - B)^{-1}A_0\boldsymbol{u}, \boldsymbol{v} \rangle| \le k\|\boldsymbol{u}\|_B,$$

then

$$|(A_0\boldsymbol{u}, \boldsymbol{v})_{\boldsymbol{L}_2(\Omega)}| \le k\|\boldsymbol{u}\|_B \le k\|\boldsymbol{u}\|_{\boldsymbol{L}_2(\Omega)},$$

and, therefore, $\boldsymbol{v} \in \mathcal{D}(A_0)$ and

$$\left\langle (I - B)^{-1}A_0\boldsymbol{u}, \boldsymbol{v} \right\rangle = (A_0\boldsymbol{u}, \boldsymbol{v})_{\boldsymbol{L}_2(\Omega)} = (\boldsymbol{u}, A_0\boldsymbol{v})_{\boldsymbol{L}_2(\Omega)} = \left\langle \boldsymbol{u}, (I - B)^{-1}A_0\boldsymbol{v} \right\rangle.$$

Further,

$$\left\langle (I - B)^{-1}A_0\boldsymbol{u}, \boldsymbol{u} \right\rangle = (A_0\boldsymbol{u}, \boldsymbol{u})_{\boldsymbol{L}_2(\Omega)} \ge \lambda_1(A_0)\|\boldsymbol{u}\|^2_{\boldsymbol{L}_2(\Omega)} \ge \lambda_1(A_0)\|\boldsymbol{u}\|^2_B > 0$$

for $\boldsymbol{u} \ne \boldsymbol{0}$ and hence the operator $(I - B)^{-1}A_0$ is positive definite.

Considering now the Cauchy problem for equation (2.13) we conclude that to the self-adjoint operator $\nu(I - B)^{-1}A_0$ there corresponds the semigroup $\exp(-\nu t(I - B)^{-1}A_0)$ that is analytic in the right half-plane. If the function $(P_0 f_1)(t)$ satisfies a Hölder condition with respect to t in the norm of the space $\boldsymbol{L}_2(\Omega)$, then the function

$$\begin{aligned}\boldsymbol{u}(t) &= \exp\left(-\nu t(I - B)^{-1}A_0\right)\boldsymbol{u}^0 \\ &\quad + \int_0^t \exp\left(-\nu(t - s)(I - B)^{-1}A_0\right)(I - B)^{-1}(P_0\boldsymbol{f}_1)(s)ds\end{aligned} \tag{2.16}$$

for any $\boldsymbol{u}^0 \in \boldsymbol{J}_0(\Omega)$ is a weak solution of the Cauchy problem for equation (2.13) with the initial condition $\boldsymbol{u}(0) = \boldsymbol{u}^0$.

Hence we obtain the solvability of the initial problem (2.1)–(2.4) on small movements of a gyrostate fixed at the mass center C. Besides, the law of kinetic energy balance, which follows from (2.5), holds true under the above mentioned conditions:

$$\begin{aligned}
&\frac{1}{2}\left(\rho\|\boldsymbol{u}(t)\|^2_{\boldsymbol{L}_2(\Omega)} + \boldsymbol{J}\boldsymbol{\omega}(t)\cdot\boldsymbol{\omega}(t) + 2\rho(\boldsymbol{\omega}(t)\times\boldsymbol{r},\boldsymbol{u})_{\boldsymbol{L}_2(\Omega)}\right)\\
&\quad= \frac{1}{2}\left(\rho\|\boldsymbol{u}^0\|^2_{\boldsymbol{L}_2(\Omega)} + \boldsymbol{J}\boldsymbol{\omega}^0\cdot\boldsymbol{\omega}^0 + 2\rho(\boldsymbol{\omega}^0\times\boldsymbol{r},\boldsymbol{u}^0)_{\boldsymbol{L}_2(\Omega)}\right)\\
&\quad+\int_0^t\left(-\rho\nu\|\mathrm{rot}\boldsymbol{u}(s)\|^2_{\boldsymbol{L}_2(\Omega)} + \rho(\boldsymbol{f}(s),\boldsymbol{u}(s))_{\boldsymbol{L}_2(\Omega)}\right)\mathrm{d}s\\
&\quad+\int_0^t \boldsymbol{M}(s)\cdot\boldsymbol{\omega}(s)\mathrm{d}s
\end{aligned}\tag{2.17}$$

7.2.5 Normal Oscillations

Let us consider here solutions of the homogeneous equation (2.13) that depend on time according to the law $\boldsymbol{u}(t,x) = \exp(-\lambda t)\boldsymbol{v}(x)$, where $\boldsymbol{v}(x)$ are modes of normal oscillations. We obtain the equation

$$\nu(I-B)^{-1}A_0\boldsymbol{v} = \lambda\boldsymbol{v}. \tag{2.18}$$

Since, according to Section 7.2.4, operator $\nu(I-B)^{-1}A_0$ is self-adjoint and positive definite with respect to the scalar product (2.14) and the inverse operator is compact, then problem (2.18) has a discrete spectrum $\{\lambda_n\}_{n=1}^{\infty}$ that consists of normal eigenvalues with the limit point $\lambda = +\infty$ and a system of eigenfields $\{\boldsymbol{v}_n(x)\}_{n=1}^{\infty}$ such that

$$\nu(I-B)^{-1}A_0\boldsymbol{v}_n = \lambda_n\boldsymbol{v}_n. \tag{2.19}$$

These fields form a basis in the space $\boldsymbol{J}_0(\Omega)$ that is orthonormal in the sense of (2.14):

$$\begin{aligned}
&\int_\Omega (I-B)\boldsymbol{v}_n\cdot\boldsymbol{v}_m \mathrm{d}\Omega = \delta_{nm},\\
&\nu\int_\Omega \mathrm{rot}\boldsymbol{v}_n\cdot\mathrm{rot}\boldsymbol{v}_m\mathrm{d}\Omega = \lambda_n\delta_{nm}.
\end{aligned}\tag{2.20}$$

Several formulas in Section 2.2.4 were used to obtain (2.20).

The eigenvalues λ_n can be found as consecutive minima of the variational ratio

$$\frac{\nu\displaystyle\int_\Omega |\mathrm{rot}\boldsymbol{v}|^2\mathrm{d}\Omega}{\displaystyle\int_\Omega (I-B)\boldsymbol{v}\cdot\boldsymbol{v}\mathrm{d}\Omega}, \tag{2.21}$$

considered in the set of functions $\boldsymbol{J}_0^1(\Omega)$ dense in $\boldsymbol{J}_0(\Omega)$. In virtue of the obvious inequalities

$$(1-\lambda_1(B))\,\|\boldsymbol{v}\|^2_{\boldsymbol{L}_2(\Omega)} \le ((I-B)\boldsymbol{v},\boldsymbol{v})_{\boldsymbol{L}_2(\Omega)} \le \|\boldsymbol{v}\|^2_{\boldsymbol{L}_2(\Omega)}, \tag{2.22}$$

where $\lambda_1(B)=\lambda_{\max}(B)<1$ is the maximal eigenvalue of operator B, we obtain from (2.21) that the following inequalities take place,

$$\nu\lambda_n(A_0) \le \lambda_n \le \frac{\nu\lambda_n(A_0)}{(1-\lambda_1(B))}, \qquad n=1,2,\dots, \tag{2.23}$$

where $\lambda_n(A_0)$ are the eigenvalues of the Stokes operator A_0 (see Section 7.1). From (2.21) and the finite dimension of operator B, it follows that the asymptotic formula

$$\lambda_n = \nu\lambda_n(A_0)[1+\mathrm{o}(1)], \qquad n\to\infty \tag{2.24}$$

is satisfied, where the asymptotic behavior of the numbers $\lambda_n(A_0)$ is defined by formula (1.11).

7.3 Rotating Motion of a Gyrostate

Here we consider a problem that, on one hand, is a generalization of the problem in Section 7.2 on a rotating gyrostate and, on the other hand, generalizes the problem in Section 7.1.5, where the rotation axis of the gyrostate was fixed.

7.3.1 Statement of the Problem and the Basic Equations

Let us assume that under the influence of the long time acting viscous forces and some external forces, the system "body + fluid" (gyrostate) rotates around an immovable pole as a single rigid body with angular velocity $\boldsymbol{\omega}_0=\omega_0\boldsymbol{e}_3$. In addition, let us assume that the moment of the gravitational forces influencing the system equals zero. This is possible only in one of the two following cases:

(1) The angular velocity of the system is so large that the centrifugal forces are much bigger than the gravitational ones, and the latter can be omitted. In particular, this situation will take place under conditions of low gravity, or for small gravitation.

(2) The mass center C of the system coincides with the immovable point O as in Section 7.2.

We will consider small movements of the system that are close to a uniform rotation. Let us denote the deviation of the angular velocity of the system from $\boldsymbol{\omega}_0$ by $\boldsymbol{\omega}(t)$, the small moment of external forces relatively to the point O by $\boldsymbol{M}(t)$, and the given field of external forces influencing the system by $\boldsymbol{f}(t,x)$. Assuming that $\boldsymbol{u}(t,x)$, $\boldsymbol{\omega}(t)$, $\boldsymbol{M}(t)$, $\boldsymbol{f}(t,x)$, and the dynamic pressure $p(t,x)$ have small magnitudes of the first order, let us write down the linearized Navier–Stokes equations (3.1.26) together with the stickiness condition, the equation of moments (3.1.25), and the initial conditions for the velocity field and for the angular velocity of system's rotation:

$$\frac{\partial \boldsymbol{u}}{\partial t} + 2\boldsymbol{\omega}_0 \times \boldsymbol{u} + \frac{\mathrm{d}\boldsymbol{\omega}}{\mathrm{d}t} \times \boldsymbol{r} = -\frac{1}{\rho}\nabla p + \nu\Delta\boldsymbol{u} + \boldsymbol{f},$$

$$\mathrm{div}\boldsymbol{u} = 0 \quad \text{in } \Omega, \qquad \boldsymbol{u} = \boldsymbol{0} \quad \text{on } S, \tag{3.1}$$

$$\boldsymbol{J}\frac{\mathrm{d}\boldsymbol{\omega}}{\mathrm{d}t} + \boldsymbol{\omega} \times \boldsymbol{J}\boldsymbol{\omega}_0 + \boldsymbol{\omega}_0 \times \boldsymbol{J}\boldsymbol{\omega} + \rho\int_\Omega \left(\boldsymbol{r} \times \frac{\partial \boldsymbol{u}}{\partial t}\right)\mathrm{d}\Omega + \boldsymbol{\omega}_0 \times \left(\rho\int_\Omega (\boldsymbol{r} \times \boldsymbol{u})\mathrm{d}\Omega\right) = \boldsymbol{M}, \tag{3.2}$$

$$\boldsymbol{u}(0,x) = \boldsymbol{u}^0(x), \qquad \boldsymbol{\omega}(0) = \boldsymbol{\omega}^0. \tag{3.3}$$

7.3.2 Transition to a Differential Equation in a Hilbert Space

Let us assume that the fields $\boldsymbol{u}(t,x)$ and $\nabla p(t,x)$ are functions with values in the space $\boldsymbol{L}_2(\Omega)$. Projecting (3.1) onto $\boldsymbol{J}_0(\Omega)$ by the orthoprojector P_0, introducing the Stokes operator A_0, which is an extension of operator $-P_0\Delta$, and the gyroscopic operator $K = -2\mathrm{i}\omega_0 T$ (see Section 7.1), instead of (3.1) we obtain the following

$$\frac{\mathrm{d}\boldsymbol{u}}{\mathrm{d}t} + P_0\left(\frac{\mathrm{d}\boldsymbol{\omega}}{\mathrm{d}t} \times \boldsymbol{r}\right) + \nu A_0\boldsymbol{u} + 2\mathrm{i}\omega_0 T\boldsymbol{u} = P_0\boldsymbol{f}, \tag{3.4}$$

$$T\boldsymbol{u} := \mathrm{i}P_0(\boldsymbol{u} \times \boldsymbol{e}_3), \qquad T = T^*,\ \sigma(T) = [-1,1]. \tag{3.5}$$

We consider equations (3.4), (3.2) as a system of differential equations relatively to the column vector $\boldsymbol{v}(t) := (\boldsymbol{u}(t,x);\boldsymbol{\omega}(t))^{\mathrm{t}}$, which is a function with values in the Hilbert space $\boldsymbol{H} := \boldsymbol{J}_0(\Omega)\oplus\mathbb{R}^3$. (As we did earlier in some similar problems, we assume that a transition to nondimensional variables has already been performed.) In short, this system can be written down as an equation

$$\frac{\mathrm{d}}{\mathrm{d}t}\tilde{I}\boldsymbol{v} + (A+B)\boldsymbol{v} = \varphi(t), \qquad \varphi(t) := ((P_0\boldsymbol{f})(t), \boldsymbol{M}(t))^{\mathrm{t}}, \tag{3.6}$$

$$\tilde{I}\boldsymbol{v} := \left(\boldsymbol{u} + P_0(\boldsymbol{\omega} \times \boldsymbol{r}); \rho\int_\Omega (\boldsymbol{r} \times \boldsymbol{u})\mathrm{d}\Omega + \boldsymbol{J}\boldsymbol{\omega}\right)^{\mathrm{t}},$$

$$A := \mathrm{diag}(\nu A_0; I), \qquad T_1\boldsymbol{\omega} := -\mathrm{i}(\boldsymbol{e}_3 \times \boldsymbol{J}\boldsymbol{\omega} + \boldsymbol{\omega} \times \boldsymbol{J}\boldsymbol{e}_3), \tag{3.7}$$

$$B\boldsymbol{v} := \begin{pmatrix} 2\mathrm{i}\omega_0 T\boldsymbol{u} \\ \omega_0\boldsymbol{e}_3 \times \left(\rho\int_\Omega (\boldsymbol{r} \times \boldsymbol{u})\mathrm{d}\Omega\right) + \mathrm{i}\omega_0 T_1\boldsymbol{\omega} - \boldsymbol{\omega} \end{pmatrix} \tag{3.8}$$

The initial conditions (3.3) generate the initial element

$$\boldsymbol{v}(0) = \left(P_0\boldsymbol{u}^0;\boldsymbol{\omega}^0\right)^{\mathrm{t}}, \tag{3.9}$$

and therefore, the initial boundary value problem (3.1)–(3.3) is equivalent to the Cauchy problem (3.6), (3.9).

7.3.3 Properties of the Operators of the Problem

First of all let us note that operator A is self-adjoint and positive definite, because $A_0 \gg 0$ in $\boldsymbol{J}_0(\Omega)$ and $I \gg 0$ in $\mathbb{R}^3$. Since the Stokes operator A_0 has a compact inverse operator A_0^{-1}, and I is the three-dimensional identity operator, then $A^{-1} = \operatorname{diag}(\nu^{-1}A_0^{-1}; I)$ is a compact positive operator in $\boldsymbol{H}$. We also note that operator A has a discrete spectrum that consists of a three-multiple eigenvalue $\lambda_0 = 1$ and the eigenvalues $\lambda_k(A) = \nu\lambda_k(A_0)$, $k = 1, 2, \ldots$. For the numbers $\lambda_k(A_0)$ the asymptotic formula (1.11) holds true, therefore, $A^{-1} \in \mathfrak{S}_p$ for $p > 3/2$.

Let us next point out other properties of the operators from (3.7) and (3.8). It is easy to prove that the operator T_1 is self-adjoint in $\mathbb{R}^3$. Further, operator B is bounded in H, because T is bounded in $\boldsymbol{J}_0(\Omega)$ and all the other operators (elements of the matrix B) are finite-dimensional.

We note that operator $\tilde{I}$ in (3.7) was mentioned already in the problem described in Section 6.2. As it was shown in Section 6.2.3, this operator is bounded and positive definite.

$$m\|\boldsymbol{v}\|^2_{\boldsymbol{H}} \le \left(\tilde{I}\boldsymbol{v}, \boldsymbol{v}\right)_{\boldsymbol{H}} \le M\|\boldsymbol{v}\|^2_{\boldsymbol{H}}, \qquad 0 < m \le M < \infty, \tag{3.10}$$

in the Hilbert space $\boldsymbol{H} = \boldsymbol{J}_0(\Omega) \oplus \mathbb{R}^3$ with the squared norm

$$\|\boldsymbol{v}\|^2_{\boldsymbol{H}} := \rho \int_\Omega |\boldsymbol{u}|^2 \mathrm{d}\Omega + |\boldsymbol{\omega}|^2. \tag{3.11}$$

Hence it follows, that the inverse operator $\tilde{I}^{-1}$ is also bounded and positive definite

$$M^{-1}\|\boldsymbol{w}\|^2_{\boldsymbol{H}} \le \left(\tilde{I}^{-1}\boldsymbol{w}, \boldsymbol{w}\right)_{\boldsymbol{H}} \le m^{-1}\|\boldsymbol{w}\|^2_{\boldsymbol{H}}. \tag{3.12}$$

7.3.4 Normal Oscillations

Let us consider solutions of the homogeneous equation (3.6) that depend on t according to the law $\exp(-\lambda t)$. We have

$$(A + B)\boldsymbol{v} = \lambda\tilde{I}\boldsymbol{v}, \quad \boldsymbol{v} \in \boldsymbol{H}, \tag{3.13}$$

and in components this equation has the following form (with regard to (3.7) and (3.8)),

$$\begin{gathered} \nu A_0\boldsymbol{u} + 2\mathrm{i}\omega_0 T\boldsymbol{u} = \lambda\left(\boldsymbol{u} + P_0(\boldsymbol{\omega} \times \boldsymbol{r})\right), \\ \mathrm{i}\omega_0 T_1\boldsymbol{\omega} + \omega_0\boldsymbol{e}_3 \times \left(\rho\int_\Omega (\boldsymbol{r} \times \boldsymbol{u})\mathrm{d}\Omega\right) = \lambda\left(\rho\int_\Omega (\boldsymbol{r} \times \boldsymbol{u})\mathrm{d}\Omega + J\boldsymbol{\omega}\right). \end{gathered} \tag{3.14}$$

Let us prove that equation (3.1) has the number $\lambda = 0$ as an eigenvalue, and find out the physical sense of the corresponding solutions $\boldsymbol{v}_0$. In (3.14), we assume that $\lambda = 0$. Then from the first equation in (3.14) we have

$$\nu \| A_0^{1/2} \boldsymbol{u} \|^2_{\boldsymbol{J}_0(\Omega)} + 2\mathrm{i}\omega_0 (T\boldsymbol{u}, \boldsymbol{u})_{\boldsymbol{J}_0(\Omega)} = 0.$$

Therefore, by the self-adjointness of T and the positive definiteness of A_0, we obtain that $\boldsymbol{u} = \boldsymbol{0}$. Then the second equation in (3.14) leads to the relation

$$\omega_0 T_1 \boldsymbol{\omega} = \boldsymbol{0}. \tag{3.15}$$

If equation (3.15) has a nonzero solution $\boldsymbol{\omega} = \boldsymbol{\omega}_*$, then equation (3.13) has the zero eigenvalue and the corresponding eigenelement of the form $(\boldsymbol{0}; \boldsymbol{\omega}_*)^{\mathrm{t}}$, which describes a rotation of the whole system with the additional constant angular velocity $\boldsymbol{\omega}_*$. Let us show, that in the most important special case this fact takes place.

Suppose that the cavity and the considered rigid body are axially symmetrical relatively to the axis Ox_3, and the tensor of inertia $\boldsymbol{J} = \boldsymbol{J}_b + \boldsymbol{J}_f$ in the system $Ox_1x_2x_3$ has a diagonal form, with the elements $J_{11} = J_{22} \neq J_{33}$. Then, by the definition (3.8), equation (3.15) leads to the following relation for $\boldsymbol{\omega} = \sum_{k=1}^{3} \omega_k \boldsymbol{e}_k$:

$$\omega_2 (J_{33} - J_{22}) \boldsymbol{e}_1 - \omega_1 (J_{33} - J_{11}) \boldsymbol{e}_2 = \boldsymbol{0}.$$

From the latter, it follows that $\omega_1 = \omega_2 = 0$ and thus problem (3.15) has solutions of the form $\boldsymbol{\omega}_* = \alpha \boldsymbol{e}_3$ with an arbitrary α. To these solutions there corresponds a rotation of the whole system around the former rotation axis Ox_3, with the additional angular velocity $\alpha \boldsymbol{e}_3$.

Now let us assume that $\lambda \neq 0$ and consider nontrivial solutions of problem (3.14). Substituting $\boldsymbol{\omega}$ in the form $\boldsymbol{\omega} = \boldsymbol{\omega}_\perp + \alpha \boldsymbol{e}_3$, $\boldsymbol{\omega}_\perp := \sum_{k=1}^{2} \omega_k \boldsymbol{e}_k$, and using the relation $T_1 \boldsymbol{\omega} \cdot \boldsymbol{e}_3 = 0$, we obtain the following expression for the projection of the second equation (3.14) onto Ox_3:

$$\lambda \left(\rho \int_\Omega (\boldsymbol{r} \times \boldsymbol{u} \cdot \boldsymbol{e}_3) \mathrm{d}\Omega + \alpha J_{33} \right) = 0. \tag{3.16}$$

Since $\lambda \neq 0$ and $J_{33} > 0$, then the next formula for determinig α can be obtained by means of the known solution $\boldsymbol{u}(x)$,

$$\alpha = -\rho J_{33}^{-1} \int_\Omega (\boldsymbol{r} \times \boldsymbol{u} \cdot \boldsymbol{e}_3) \mathrm{d}\Omega. \tag{3.17}$$

Now, substituting α into the equations (3.14), we obtain the system

$$\nu A_0\boldsymbol{u} + 2\mathrm{i}\omega_0 T\boldsymbol{u} = \lambda\left(\boldsymbol{u} + P_0(\boldsymbol{\omega}_\perp \times \boldsymbol{r}) - \rho\left(J_{33}^{-1}\int_\Omega (\boldsymbol{r}\times\boldsymbol{u}\cdot\boldsymbol{e}_3)\mathrm{d}\Omega\right)P_0(\boldsymbol{e}_3\times\boldsymbol{r})\right),$$

$$\mathrm{i}\omega_0 T_1\boldsymbol{\omega}_\perp + \omega_0\boldsymbol{e}_3 \times \left(\rho\int_\Omega (\boldsymbol{r}\times\boldsymbol{u})\mathrm{d}\Omega\right) = \lambda\left(\rho P_\perp \int_\Omega (\boldsymbol{r}\times\boldsymbol{u})\mathrm{d}\Omega + \boldsymbol{J}\boldsymbol{\omega}_\perp\right), \tag{3.18}$$

where $P_\perp$ is the orthoprojector onto the two-dimensional subspace formed by the axial vectors $\boldsymbol{e}_1$ and $\boldsymbol{e}_2$; in (3.18) the column vector $\boldsymbol{v}_\perp := (\boldsymbol{u};\boldsymbol{\omega}_\perp)^{\mathrm{t}}$ from the space $\boldsymbol{H}_\perp = \boldsymbol{J}_0(\Omega)\oplus\mathbb{R}^2$, $\mathbb{R}^2 = P_\perp\mathbb{R}^3$, is unknown.

Similarly to (3.13), the system of equations (3.18) can be written down in the form

$$(A_\perp + B_\perp)\,\boldsymbol{v}_\perp = \lambda I_\perp \boldsymbol{v}_\perp, \tag{3.19}$$

where the operators with the subscript $\perp$ act in $\boldsymbol{H}_\perp$ and were obtained as restrictions of the operators (3.7) and (3.8) to the subspace $\boldsymbol{H}_\perp$ by applying the orthoprojector $\operatorname{diag}(I;P_\perp)$.

From properties of the operators of problem (3.13) that were proved in Section 7.3.3, it follows that, in problem (3.19), $I_\perp$ is a bounded positive definite operator with a bounded and positive definite inverse operator $(I_\perp)^{-1}$. Further, operator $A_\perp := \operatorname{diag}(\nu A_0; P_\perp)$ is positive definite, has a discrete spectrum with asymptotics (1.11), and a compact positive inverse operator $(A_\perp)^{-1}$ in the class $\mathfrak{S}_p$ for $p > 3/2$. Finally, operator $B_\perp$ (as operator B) is bounded in the space $\boldsymbol{H}_\perp$.

Let us transform equation (3.19) using the properties of the considered operators. In (3.19), we perform the substitution

$$I_\perp^{1/2}\boldsymbol{v}_\perp = \boldsymbol{\psi} \tag{3.20}$$

and apply operator $I_\perp^{-1/2}$ to the left. We obtain the following equation,

$$(F+G)\boldsymbol{\psi} = \lambda\boldsymbol{\psi}, \qquad F = I_\perp^{-1/2}A_\perp I_\perp^{-1/2},\ G = I_\perp^{-1/2}B_\perp I_\perp^{-1/2}. \tag{3.21}$$

Since $I_\perp^{-1/2} \gg 0$ and $A_\perp \gg 0$, then $F \gg 0$ is an unbounded invertible operator, with a compact inverse operator $F^{-1} = I_\perp^{-1/2}A_\perp^{-1}I_\perp^{1/2}$ from the class $\mathfrak{S}_p$ for $p > 3/2$. The operator $G \in \mathcal{L}(\boldsymbol{H}_\perp)$ because it equals a product of bounded operators. Let us note that the eigenvalues $\lambda_k(F)$ of operator F are consecutive minima of the variational ratio

$$\frac{\|F^{1/2}\boldsymbol{\psi}\|^2_{\boldsymbol{H}_\perp}}{\|\boldsymbol{\psi}\|^2_{\boldsymbol{H}_\perp}} = \frac{\|A^{1/2}\boldsymbol{v}_\perp\|^2_{\boldsymbol{H}_\perp}}{\|I_\perp^{1/2}\boldsymbol{v}_\perp\|^2_{\boldsymbol{H}_\perp}} = \frac{\rho\nu\|A_0^{1/2}\boldsymbol{u}\|^2_{\boldsymbol{J}_0(\Omega)} + |\boldsymbol{\omega}_\perp|^2}{\rho\|\boldsymbol{u}\|^2_{\boldsymbol{J}_0(\Omega} + (\cdots)}.$$

Because of the finite dimension of $\mathbb{R}^2$, the asymptotic behavior of the eigenvalues $\lambda_k(F)$ coincides with the spectrum asymptotics of the ratio $\nu\|A_0^{1/2}\boldsymbol{u}\|^2_{\boldsymbol{J}_0(\Omega)}/\|\boldsymbol{u}\|^2_{\boldsymbol{J}_0(\Omega)}$, that is, has the form (1.11).

Further consideration of problem (3.21) can be performed according to the scheme that was already used in Section 7.1. This scheme deals with the transition from equation (1.17) or (1.30) to a linear pencil that satisfies all the conditions of the second Keldysh theorem. Specifically, after performing the substitution $F\psi = \varphi$ in (3.21), we obtain the problem

$$\left(I + GF^{-1} - \lambda F^{-1}\right)\varphi = 0, \tag{3.22}$$

where $GF^{-1} \in \mathfrak{S}_\infty$, and F^{-1} is a complete positive operator from the class $\mathfrak{S}_p$ for $p > 3/2$. Similar to the considerations in Sections 7.1.4 and 7.1.5, the following conclusions follow from Keldysh theorem.

1° Problem (3.22) together with the problems (3.19) and (3.14), has a discrete spectrum $\{\lambda_k\}_{k=1}^{\infty}$ that consists of isolated eigenvalues with finite algebraic multiplicity and has the limit point $\lambda = \infty$. Moreover, in the axially symmetric case, problem (3.14) has the one-multiple eigenvalue $\lambda_0 = 0$, to which there corresponds a rotation of the system as a rigid body with the additional angular velocity $\boldsymbol{\omega}_* = \alpha \boldsymbol{e}_3$.

2° All eigenvalues λ_k are located in the halfband

$$\operatorname{Re}\lambda_k \geq \lambda_1(F) - \|\operatorname{Re}G\|, \quad |\operatorname{Im}\lambda_k| \leq \|\operatorname{Im}G\|, \qquad k = 1, 2, \dots, \tag{3.23}$$

where $\lambda_1(F)$ is the first eigenvalue of operator F, $\operatorname{Re}G := (G + G^*)/2$, and $\operatorname{Im}G := (G - G^*)/(2\mathrm{i})$. In particular, from (3.23) it follows that $\operatorname{Re}\lambda_k \to +\infty$ as $k \to \infty$.

The proof of properties (3.23) follows from equation (3.21), if we write G in the form $\operatorname{Re}G + \mathrm{i}\operatorname{Im}G$ and observe that G is a bounded operator.

3° The eigen- and associated elements $\{\varphi_{k,q}\}_{k=1}^{\infty}$ corresponding to the eigenvalues $\{\lambda_k\}_{k=1}^{\infty}$ of problem (3.22) form a complete system of elements in the space $\boldsymbol{H}_\perp = \boldsymbol{J}_0(\Omega) \oplus \mathbb{C}^2$. Hence it appears that the system of elements $(\boldsymbol{v}_\perp)_{k,q} = A_\perp^{-1}(I_\perp^{1/2}\varphi_{k,q})$, $k = 1, 2, \dots$, is complete in $\mathcal{D}(A_\perp)$ in the graph norm of operator $A_\perp$. Hence, the mentioned property of completeness of elements of the form $(\boldsymbol{v}_\perp)_{k,q} = (\boldsymbol{u}_{k,q}; (\boldsymbol{\omega}_\perp)_{k,q})^t$ takes place in the space $\mathcal{D}(A_0) \oplus \mathbb{C}^2$, with the squared norm $\|(A_0\boldsymbol{u}\|^2_{\boldsymbol{J}_0(\Omega)} + |\boldsymbol{\omega}_\perp|^2$.

4° For the eigenvalues of problem (3.14), the asymptotic formula

$$\lambda_k = \nu\lambda_k(A_0)[1 + \mathrm{o}(1)], \qquad k \to \infty, \tag{3.24}$$

holds true, where the asymptotic behavior of the numbers $\lambda_k(A_0)$ is defined by formula (1.11).

Assertion (3.24) (as the similar assertion (1.32) in Section 7.1.5) follows from the results of Section 1.6.8 and the power asymptotics of the eigenvalues of the Stokes operator A_0.

7.3.5 Solvability of the Nonstationary Problem

We consider here the Cauchy problem (3.6)–(3.9). As in Section 7.3.4, let us characterize the solution that corresponds to a forced rotation around the axis Ox_3, namely, let us look for $\boldsymbol{\omega}(t)$ of the form

$$\boldsymbol{\omega}(t) = \boldsymbol{\omega}_\perp(t) + \alpha(t)\boldsymbol{e}_3. \tag{3.25}$$

Also, let us represent $\boldsymbol{M}(t)$ in the form $\boldsymbol{M}(t) = \boldsymbol{M}_\perp(t) + \gamma(t)\boldsymbol{e}_3$.

Substituting these functions into (3.6), we obtain the following relations after projecting onto $\boldsymbol{e}_3$,

$$\begin{aligned} &\frac{\mathrm{d}}{\mathrm{d}t}\left(\alpha(t)J_{33} + \beta(t)\right) = \gamma(t), \\ &\beta(t) = \rho\int_\Omega (\boldsymbol{r}\times\boldsymbol{u}(t,x)\cdot\boldsymbol{e}_3)\,\mathrm{d}\Omega = \rho\int_\Omega P_0(\boldsymbol{e}_3\times\boldsymbol{r})\cdot\boldsymbol{u}\mathrm{d}\Omega. \end{aligned} \tag{3.26}$$

Integrating (3.26) between the limits 0 and t we have

$$\alpha(t) = \boldsymbol{\omega}^0\cdot\boldsymbol{e}_3 - J_{33}^{-1}\left(\rho\int_\Omega(\boldsymbol{e}_3\times\boldsymbol{r})\cdot\left(P_0\boldsymbol{u}(t,x) - P_0\boldsymbol{u}^0\right)\mathrm{d}\Omega + \int_0^t\gamma(s)\mathrm{d}s\right). \tag{3.27}$$

This formula allows us to find the unknown function $\alpha(t)$ from the known solution $\boldsymbol{u}(t,x)$ and initial data (3.9).

For the solution $\boldsymbol{v}_\perp(t) = (\boldsymbol{u}(t);\boldsymbol{\omega}_\perp(t))^{\mathrm{t}}$ we obtain the following Cauchy problem from (3.6)–(3.9) and (3.27),

$$\frac{\mathrm{d}}{\mathrm{d}t}I_\perp\boldsymbol{v}_\perp + (A_\perp + B_\perp)\boldsymbol{v}_\perp = \varphi_\perp(t), \qquad \boldsymbol{v}_\perp(0) = (P_0\boldsymbol{u}^0;\boldsymbol{\omega}_\perp^0)^{\mathrm{t}}, \tag{3.28}$$

where operators $I_\perp$, $A_\perp$ and $B_\perp$ are the same as in problem (3.19). Applying to (3.28) the bounded positive definite operator $(I_\perp)^{-1}$, we have

$$\frac{\mathrm{d}\boldsymbol{v}_\perp}{\mathrm{d}t} + \left((I_\perp)^{-1}A_\perp + (I_\perp)^{-1}B_\perp\right)\boldsymbol{v}_\perp = (I_\perp)^{-1}\varphi_\perp(t). \tag{3.29}$$

Since operator $A_\perp$ is positive definite in $\boldsymbol{H}_\perp$, then operator $(I_\perp)^{-1}A_\perp$ is self-adjoint and positive definite in the space $\boldsymbol{H}_{I_\perp}$ of functions with the squared norm $(I_\perp \boldsymbol{v}_\perp, \boldsymbol{v}_\perp)_{\boldsymbol{H}_\perp}$. This fact insures that $(I_\perp)^{-1}A_\perp$ is the generating operator of the analytic semigroup $\exp(-t(I_\perp)^{-1}A_\perp)$. Since $(I_\perp)^{-1}B_\perp$ is bounded, then operator $(I_\perp)^{-1}(A_\perp + B_\perp)$ is also a generating operator of the analytic semigroup $\exp(-t(I_\perp)^{-1}(A_\perp + B_\perp))$. Therefore, the solution of the Cauchy problem corresponding to equation (3.29) is given by the formula

$$\begin{aligned} \boldsymbol{v}_\perp(t) &= \exp\left(-t(I_\perp)^{-1}(A_\perp + B_\perp)\right) \boldsymbol{v}_\perp(0) \\ &\quad + \int_0^t \exp\left(-(t-s)(I_\perp)^{-1}(A_\perp + B_\perp)\right) (I_\perp)^{-1} \varphi_\perp(s) ds. \end{aligned} \tag{3.30}$$

From (3.30) it follows that if, in the initial problem (3.1)–(3.3), $\boldsymbol{u}^0(x) \in \boldsymbol{J}_0(\Omega)$, $P_0 \boldsymbol{f}(t,x)$ satisfies a Hölder condition in t in the norm of $\boldsymbol{J}_0(\Omega)$, and $\boldsymbol{M}_\perp(t)$ satisfies a Hölder condition in t, then problem (3.1)–(3.3) is univalently solvable and its generalized solution $\{\boldsymbol{u}(t); \boldsymbol{\omega}(t)\}$ can be found by the formulas (3.30), (3.25), and (3.27). Here, $\boldsymbol{u}(t,x)$ for $t > 0$ belongs to the space $\boldsymbol{J}_0^1(\Omega)$ and is a continuous function of t; $\boldsymbol{\omega}(t)$ is also a continuous function of t. For this solution, the law of full energy balance holds true. Finding its expression is left to the reader [in particular, see (2.5)].

In conclusion, let us notice that the method used in this section can also be applied to the problem considered in Section 7.2. For that purpose, one should assume that $\omega_0 = 0$ in the initial problem. Let us point out that in this case rotations of the system as a rigid body around an arbitrary axis, not just around the axis Ox_3, should be taken as trivial solutions.

7.4 Asymptotic Solutions for High Viscosity

In this section, we consider the same problem as in Section 7.3 in the case of large viscosity values. Since the coefficient of kinematic viscosity ν is a dimensional quantity, it is natural to consider the nondimensional Reynolds number Re$= ul/\nu$, where l is the characteristic size of the region Ω, and u is the characteristic velocity. However, one should assume that the units of length and time are chosen in such a way that the Reynolds number coincides numerically with ν^{-1}. Further, let us consider the case of a large parameter ν. For this assumption we can build the first two terms of the formal asymptotic expansion of the solution of the problem by powers of ν^{-1}. The basics of this method were stated in Section 1.7.6.

7.4.1 Solving the Hydrodynamics Problem

Let us consider equation (3.4) while assuming that the field $\boldsymbol{f}$ is potential and $\boldsymbol{\omega}$ is a known function. We have

$$\nu^{-1}\frac{\mathrm{d}\boldsymbol{u}}{\mathrm{d}t} = -A_0\boldsymbol{u} - \nu^{-1}P_0\left(\frac{\mathrm{d}\boldsymbol{\omega}}{\mathrm{d}t}\times\boldsymbol{r}\right) - 2\nu^{-1}P_0\left(\boldsymbol{\omega}^0\times\boldsymbol{u}\right). \tag{4.1}$$

If the viscosity ν is high, then (4.1) is an equation with a small parameter attached to the derivative. Here, operator $(-A_0)$ has a bounded inverse operator. Therefore, according to the theory stated in Section 1.7, the approximate solution of equation (4.1) can be looked for in the form

$$\boldsymbol{u}_N(t;\nu^{-1}) = \sum_{k=0}^{N}\nu^{-k}\boldsymbol{u}^{(k)}(t); \tag{4.2}$$

$\boldsymbol{u}^{(k)}(t)$ can be defined in such a way that the function $\boldsymbol{u}_N(t;\nu^{-1})$ differs from an accurate solution of equation (4.1) in some magnitude of order $\nu^{-(N+1)}$. Substituting (4.2) into (4.1) and identifying the coefficients of equal powers of ν^{-1}, we get to the following formulas for $\boldsymbol{u}^{(k)}(t)$:

$$\begin{aligned}
\boldsymbol{u}^{(0)} &= \boldsymbol{0},\\
\boldsymbol{u}^{(1)} &= -A_0^{-1}P_0\left(\frac{\mathrm{d}\boldsymbol{\omega}}{\mathrm{d}t}\times\boldsymbol{r}\right)\\
\boldsymbol{u}^{(2)} &= -A_0^{-1}\frac{\mathrm{d}\boldsymbol{u}^{(1)}}{\mathrm{d}t} - 2A_0^{-1}P_0\left(\boldsymbol{\omega}_0\times\boldsymbol{u}^{(1)}\right),\\
&\dots\\
\boldsymbol{u}^{(N)} &= -A_0^{-1}\frac{\mathrm{d}\boldsymbol{u}^{(N-1)}}{\mathrm{d}t} - 2A_0^{-1}P_0\left(\boldsymbol{\omega}_0\times\boldsymbol{u}^{(N-1)}\right).
\end{aligned} \tag{4.3}$$

In particular, these formulas show that in first approximation only translational terms influence the motion of the fluid; the influence of Coriolis forces becomes noticeable only in second approximation by ν^{-1}.

Further, we consider only the first two approximations. Let us denote by ε_i the projections of angular acceleration $\boldsymbol{\varepsilon} = \mathrm{d}\boldsymbol{\omega}/\mathrm{d}t$ onto the axes Ox_i and assume (as in Section 7.3), that $\boldsymbol{\omega}_0 = \omega_0\boldsymbol{e}_3$. Then formulas (4.3) can be transformed as follows:

$$\begin{aligned}
\boldsymbol{u}^{(1)} &= -\sum_{i=1}^{3}\varepsilon_i A_0^{-1}P_0(\boldsymbol{e}_i\times\boldsymbol{r}),\\
\boldsymbol{u}^{(2)} &= \sum_{i=1}^{3}\frac{\mathrm{d}\varepsilon_i}{\mathrm{d}t}A_0^{-2}P_0(\boldsymbol{e}_i\times\boldsymbol{r}) + 2\omega_0\sum_{i=1}^{3}\varepsilon_i A_0^{-1}P_0\left(\boldsymbol{e}_3\times A_0^{-1}P_0(\boldsymbol{e}_3\times\boldsymbol{r})\right).
\end{aligned}$$

Hence, it follows that the problem of determining $\boldsymbol{u}^{(2)}$ can be reduced to solving three sets of stationary problems for finding the functions

$$\begin{aligned}\boldsymbol{w}_i^{(1)} &= -A_0^{-1} P_0 \left(\boldsymbol{e}_i \times \boldsymbol{r}\right),\\ \boldsymbol{w}_i^{(2)} &= -A_0^{-1} \boldsymbol{w}_i^{(1)},\\ \boldsymbol{v}_i^{(2)} &= -2A_0^{-1} P_0 \left(\boldsymbol{e}_3 \times \boldsymbol{w}_i^{(1)}\right).\end{aligned} \tag{4.4}$$

Let us write down these problems in their classical form.

(1) The problem for determing $\boldsymbol{w}_i^{(1)}$:

$$\begin{aligned}\Delta \boldsymbol{w}_i^{(1)} &= -\nabla s_i^{(1)} + \boldsymbol{e}_i \times \boldsymbol{r}, \qquad \operatorname{div} \boldsymbol{w}_i^{(1)} = 0 \quad \text{in } \Omega,\\ \boldsymbol{w}_i^{(1)} &= 0 \quad \text{on } S = \partial\Omega, \qquad i = 1, 2, 3.\end{aligned} \tag{4.5}$$

(2) The problem for determining $\boldsymbol{w}_i^{(2)}$:

$$\begin{aligned}\Delta \boldsymbol{w}_i^{(2)} &= -\nabla s_i^{(2)} + \boldsymbol{w}_i^{(1)}, \qquad \operatorname{div} \boldsymbol{w}_i^{(2)} = 0 \quad \text{in } \Omega,\\ \boldsymbol{w}_i^{(2)} &= 0 \quad \text{on } S, \qquad i = 1, 2, 3.\end{aligned} \tag{4.6}$$

(3) The problems for determing $\boldsymbol{v}_i^{(2)}$:

$$\begin{aligned}\Delta \boldsymbol{v}_i^{(2)} &= \nabla s_i^{(3)} + 2\left(\boldsymbol{e}_3 \times \boldsymbol{w}_i^{(1)}\right), \qquad \operatorname{div} \boldsymbol{w}_i^{(2)} = 0 \quad \text{in } \Omega,\\ \boldsymbol{v}_i^{(2)} &= 0 \quad \text{on } S, \qquad i = 1, 2, 3.\end{aligned} \tag{4.7}$$

The following fact is of a significant, practical value: The solutions of these problems depend only on the region Ω and do not depend on time and characteristics of the motion.

The approximate solution of equation (4.1) that satisfies that equation with accuracy up to $\mathrm{O}(\nu^{-3})$ has now the following form,

$$\boldsymbol{u}_2 = \nu^{-1} \sum_{i=1}^{3} \varepsilon_i \boldsymbol{w}_i^{(1)} + \nu^{-2} \sum_{i=1}^{3} \frac{\mathrm{d}\varepsilon_i}{\mathrm{d}t} \boldsymbol{w}_i^{(2)} + \omega_0 \nu^{-2} \sum_{i=1}^{3} \varepsilon_i \boldsymbol{v}_i^{(2)}. \tag{4.8}$$

7.4.2. Asymptotic Equations of the Motion of a Rigid Body

To get an equation of motion of the body, one should replace $\boldsymbol{u}$ in equation (3.2) by $\boldsymbol{u}_2$ given in (4.8). Let us calculate the gyrostatic moment that is included in (3.2):

$$\boldsymbol{L} = \rho \int_{\Omega} (\boldsymbol{r} \times \boldsymbol{u})\, \mathrm{d}\Omega =: \nu^{-1} \boldsymbol{L}_1 + \nu^{-2} \boldsymbol{L}_{21} + \nu^{-2} \boldsymbol{L}_{22}. \tag{4.9}$$

Here,

$$\boldsymbol{L}_1 = -\rho \boldsymbol{P}^{(1)} \boldsymbol{\varepsilon} = -\rho \sum_{i,j=1}^{3} P_{ij}^{(1)} \varepsilon_i \boldsymbol{e_j},$$

where the components of tensor $\boldsymbol{P}^{(1)}$ have the form

$$P_{ij}^{(1)} = \int_\Omega \boldsymbol{e}_j \cdot \left(\boldsymbol{w}_i^{(1)} \times \boldsymbol{r} \right) \mathrm{d}\Omega. \tag{4.10}$$

Similarly,

$$\boldsymbol{L}_{21} = \rho \boldsymbol{P}^{(2)} \frac{\mathrm{d}\boldsymbol{\varepsilon}}{\mathrm{d}t} = \rho \sum_{i,j=1}^{3} P_{ij}^{(2)} \frac{\mathrm{d}\varepsilon_i}{\mathrm{d}t} \boldsymbol{e}_j,$$

where

$$P_{ij}^{(2)} = \int_\Omega \boldsymbol{e}_j \cdot \left(\boldsymbol{r} \times \boldsymbol{w}_i^{(2)} \right) \mathrm{d}\Omega, \tag{4.11}$$

and

$$\boldsymbol{L}_{22} = \rho \omega_0 \boldsymbol{Q}^{(2)} \boldsymbol{\varepsilon} = \rho \omega_0 \sum_{i,j=1}^{3} Q_{ij}^{(2)} \varepsilon_i \boldsymbol{e}_j,$$

where

$$Q_{ij}^{(2)} = \int_\Omega \boldsymbol{e}_j \cdot \left(\boldsymbol{r} \times \boldsymbol{v}_i^{(2)} \right) \mathrm{d}\Omega. \tag{4.12}$$

Let us point out some properties of the tensors $\boldsymbol{P}^{(1)}$, $\boldsymbol{P}^{(2)}$, and $\boldsymbol{Q}^{(2)}$. The tensors $\boldsymbol{P}^{(1)}$ and $\boldsymbol{P}^{(2)}$ are symmetric and there are some positive definite quadratic forms that correspond to them. Indeed,

$$\begin{aligned} P_{ij}^{(1)} &= \int_\Omega \boldsymbol{e}_j \cdot \left(\boldsymbol{w}_i^{(1)} \times \boldsymbol{r} \right) \mathrm{d}\Omega \\ &= -\int_\Omega (\boldsymbol{e}_j \times \boldsymbol{r}) \cdot \boldsymbol{w}^{(1)} \mathrm{d}\Omega \\ &= \int_\Omega (\boldsymbol{e}_j \times \boldsymbol{r}) \cdot A_0^{-1} P_0 (\boldsymbol{e}_i \times \boldsymbol{r}) \, \mathrm{d}\Omega \\ &= \int_\Omega P_0 (\boldsymbol{e}_j \times \boldsymbol{r}) \cdot A_0^{-1} P_0 (\boldsymbol{e}_i \times \boldsymbol{r}) \, \mathrm{d}\Omega. \end{aligned} \tag{4.13}$$

Similarly,

$$\begin{aligned} P_{ij}^{(2)} &= \int_\Omega \boldsymbol{e}_j \cdot \left(\boldsymbol{r} \times \boldsymbol{w}_i^{(2)} \right) \mathrm{d}\Omega = \int_\Omega (\boldsymbol{e}_j \times \boldsymbol{r}) \cdot \boldsymbol{w}_i^{(2)} \mathrm{d}\Omega \\ &= \int_\Omega (\boldsymbol{e}_j \times \boldsymbol{r}) \cdot A_0^{-2} P_0 (\boldsymbol{e}_i \times \boldsymbol{r}) \, \mathrm{d}\Omega \\ &= \int_\Omega A_0^{-1} P_0 (\boldsymbol{e}_j \times \boldsymbol{r}) \cdot A_0^{-1} P_0 (\boldsymbol{e}_i \times \boldsymbol{r}) \, \mathrm{d}\Omega, \end{aligned} \tag{4.14}$$

and all the assertions follow from the fact that the operatorA_0^{-1} is self-adjoint and positive. Further, from the equality

$$\begin{aligned} Q_{ij}^{(2)} &= \int_\Omega \boldsymbol{e}_j \cdot \left(\boldsymbol{r} \times \boldsymbol{v}_i^{(2)}\right) \mathrm{d}\Omega = \int_\Omega (\boldsymbol{e}_j \times \boldsymbol{r}) \cdot \boldsymbol{v}_i^{(2)} \mathrm{d}\Omega \\ &= 2 \int_\Omega (\boldsymbol{e}_j \times \boldsymbol{r}) \cdot A_0^{-1} P_0 \left(\boldsymbol{e}_3 \times A_0^{-1}(\boldsymbol{e}_i \times \boldsymbol{r})\right) \mathrm{d}\Omega \\ &= 2 \int_\Omega A_0^{-1} P_0(\boldsymbol{e}_j \times \boldsymbol{r}) \cdot \left(\boldsymbol{e}_3 \times A_0^{-1} P_0(\boldsymbol{e}_i \times \boldsymbol{r})\right) \mathrm{d}\Omega, \end{aligned} \tag{4.15}$$

it follows that $Q^{(2)}$ is an antisymmetric tensor.

Let us note that equalities (4.14) and (4.15) can be written down as follow

$$\begin{aligned} P_{ij}^{(2)} &= \int_\Omega \boldsymbol{w}_i^{(1)} \cdot \boldsymbol{w}_j^{(1)} \mathrm{d}\Omega, \\ Q_{ij}^{(2)} &= 2 \int_\Omega \boldsymbol{w}_j^{(1)} \cdot \left(\boldsymbol{e}_3 \times \boldsymbol{w}_i^{(1)}\right) \mathrm{d}\Omega. \end{aligned} \tag{4.16}$$

Therefore, in order to set up an equation of motion for the body, it is sufficient to solve only the first set of boundary value problems (1)–(3), that is, the problem of finding the functions $\boldsymbol{w}_i^{(1)}$. The tensors $\boldsymbol{P}^{(2)}$ and $\boldsymbol{Q}^{(2)}$ can be expressed in terms of these functions.

We call $\boldsymbol{P}^{(1)}$ and $\boldsymbol{P}^{(2)}$ the *tensors of translation of the first* and *second order*, respectively, and $\boldsymbol{Q}^{(2)}$ the *Coriolis tensor*. All these tensors are defined only by the shape of region Ω. The magnitudes $\boldsymbol{P}_{ij}^{(1)}$ have dimension l^7, and $P_{ij}^{(2)}$ and $Q_{ij}^{(2)}$ have dimension l^9, where l is the characteristic size of the cavity.

Having the representation (4.9) for the gyrostatic moment and the formulas (4.10)–(4.16) for its terms, let us substitute (4.9) into equation (3.2) of the kinetic moment. We obtain the following

$$\begin{aligned} &J\frac{\mathrm{d}\boldsymbol{\omega}}{\mathrm{d}t} + \rho\frac{\mathrm{d}}{\mathrm{d}t}\left(\nu^{-1}\boldsymbol{P}^{(1)}\frac{\mathrm{d}\boldsymbol{\omega}}{\mathrm{d}t} + \nu^{-2}\boldsymbol{P}^{(2)}\frac{\mathrm{d}^2\boldsymbol{\omega}}{\mathrm{d}t^2} + \omega_0\nu^{-2}\boldsymbol{Q}^{(2)}\frac{\mathrm{d}\boldsymbol{\omega}}{\mathrm{d}t}\right) \\ &\quad + \omega_0\boldsymbol{\omega} \times \boldsymbol{J}\boldsymbol{e}_3 + \omega_0\boldsymbol{e}_3 \times \boldsymbol{J}\boldsymbol{\omega} \\ &\quad + \rho\omega_0\boldsymbol{e}_3 \times \left(-\nu^{-1}\boldsymbol{P}^{(1)}\frac{\mathrm{d}\boldsymbol{\omega}}{\mathrm{d}t} + \nu^{-2}\boldsymbol{P}^{(2)}\frac{\mathrm{d}^2\boldsymbol{\omega}}{\mathrm{d}t^2} + \omega_0\nu^{-2}\boldsymbol{Q}^{(2)}\frac{\mathrm{d}\boldsymbol{\omega}}{\mathrm{d}t}\right) = \boldsymbol{M}, \end{aligned} \tag{4.17}$$

with accuracy up to $\mathrm{O}(\nu^{-3})$. This is an ordinary differential equation of the third order with respect to the vector-valued function $\boldsymbol{\omega}(t)$. By the positive definiteness of matrix $\boldsymbol{P}^{(2)}$, this equation is solvable relatively to the third order derivative $\mathrm{d}^3\boldsymbol{\omega}/\mathrm{d}t^3$. After performing this substitution, we obtain a differential equation that contains the small parameter ν^{-2} attached to the derivative of the highest order. The theory of integration of such equations is well developed.

Hence, the initial problem (3.1)–(3.3) on dynamics of a rotating rigid body with a cavity completely filled with a fluid with high viscosity is reduced to the problem on solving asymptotically the ordinary differential equation (4.17), that is, it is reduced to a much simpler problem.

7.4.3 An Example

Let us consider the case $i = 3$ as an example of solving the boundary value problems (4.5). Then,

$$\Delta \boldsymbol{w}_3^{(1)} = -\nabla s_3^{(1)} + x_1 \boldsymbol{e}_2 - x_2 \boldsymbol{e}_1, \quad \operatorname{div} \boldsymbol{w}_3^{(1)} = 0 \quad \text{in } \Omega.$$

The first equation can be reduced to two scalar equations if its solution is sought in the form $\boldsymbol{w}_3^{(1)} = \varphi_1(x)\boldsymbol{e}_1 + \varphi_2(x)\boldsymbol{e}_2$, $s_3^{(1)} = s(x_1, x_2)$. Then

$$\Delta\varphi_1 = -\frac{\partial s}{\partial x_1} - x_2, \quad \Delta\varphi_2 = -\frac{\partial s}{\partial x_2} + x_1.$$

Eliminating s from these equations we obtain

$$\Delta\left(\frac{\partial\varphi_1}{\partial x_2} - \frac{\partial\varphi_2}{\partial x_1}\right) = -2.$$

In order to satisfy the incompressibility condition $\partial\varphi_1/\partial x_1 + \partial\varphi_2/\partial x_2 = 0$, we assume that

$$\varphi_1 = \frac{\partial\psi_3}{\partial x_2}, \quad \varphi_2 = -\frac{\partial\psi_3}{\partial x_1}.$$

For the function $\psi_3(x)$, we obtain the following boundary value problem:

$$\Delta\left(\Delta_2\psi_3\right) = -2 \quad \text{in } \Omega, \qquad \Delta_2 := \frac{\partial^2}{\partial x_1^2} + \frac{\partial^2}{\partial x_2^2},$$
$$\frac{\partial\psi_3}{\partial x_1} = \frac{\partial\psi_3}{\partial x_2} = 0 \quad \text{on } S. \tag{4.18}$$

If a solution $\psi_3(x)$ of this problem is found, then

$$\boldsymbol{w}_3^{(1)} = \frac{\partial\psi_3}{\partial x_2}\boldsymbol{e}_1 - \frac{\partial\psi_3}{\partial x_1}\boldsymbol{e}_2 = \nabla\psi_3 \times \boldsymbol{e}_3 \tag{4.19}$$

The problems for the functions $\psi_1(x)$ and $\psi_2(x)$ can be formulated similarly. Then

$$\boldsymbol{w}_i^{(1)} = \nabla\psi_i \times \boldsymbol{e}_i, \qquad i = 1, 2.$$

For some regions Ω, the problem (4.18) admits an accurate solution. Let us consider as an example an ellipsoidal cavity Ω with the equation of the boundary S in the form

$$\frac{x_1^2}{a_1^2} + \frac{x_2^2}{a_2^2} + \frac{x_3^2}{a_3^2} - 1 = 0. \tag{4.20}$$

A solution of problem (4.18) can be found as a polynomial of degree 4,

$$\psi_3(x) = b_1 x_1^4 + 2b_2 x_1^2 x_2^2 + b_3 x_2^4 + b_4 x_1^2 x_3^2 + b_5 x_2^2 x_3^2 + b_6 x_1^2 + b_7 x_2^2. \tag{4.21}$$

Then, equation (4.18) leads to the relation

$$12b_1 + 8b_2 + 12b_3 + 2b_4 + 2b_5 = -1, \tag{4.22}$$

and the boundary conditions lead to the equalities

$$\begin{aligned} 2x_1 \left(2b_1 x_1^2 + 2b_2 x_2^2 + b_4 x_3^2 + b_6\right) &= 0, \\ 2x_2 \left(2b_2 x_1^2 + 2b_3 x_2^2 + b_5 x_3^2 + b_7\right) &= 0 \end{aligned}$$

that are satisfied on the ellipsoid's surface. For this purpose, it is sufficient to take the coefficients in the two expressions within parentheses proportional to the coefficients in the expression in the left side of (4.20). Such an approach gives raise to six equations for determining the coefficients b_i and together with equation (4.22) it allows us to determine the seven coefficients

$$\begin{aligned} b_1 &= -\frac{b}{4a_1^4}, \\ b_2 &= -\frac{b}{4a_1^2 a_2^2}, \\ b_3 &= -\frac{b}{4a_2^4}, \\ b_4 &= -\frac{b}{2a_1^2 a_3^2}, \\ b_5 &= -\frac{b}{2a_2^2 a_3^2}, \\ b_6 &= \frac{b}{2a_1^2}, \\ b_7 &= \frac{b}{2a_2^2}, \\ b :&= \frac{a_1^4 a_2^4 a_3^2}{3a_3^2 \left(a_1^4 + a_2^4\right) + a_1^2 a_2^2 \left(a_1^2 + a_2^2 + 2a_3^2\right)}. \end{aligned} \tag{4.23}$$

Let us notice that, for such a choice of coefficients, we have

$$\frac{\partial \psi_3}{\partial x_1} = -\frac{b}{a_1^2} x_1 \left(\frac{x_1^2}{a_1^2} + \frac{x_2^2}{a_2^2} + \frac{x_3^2}{a_3^2} - 1 \right),$$

$$\frac{\partial \psi_3}{\partial x_2} = -\frac{b}{a_2^2} x_2 \left(\frac{x_1^2}{a_1^2} + \frac{x_2^2}{a_2^2} + \frac{x_3^2}{a_3^2} - 1 \right). \tag{4.24}$$

If the function $\psi_3(x)$ is known, then one can find the components of tensors $\boldsymbol{P}^{(1)}$ and $\boldsymbol{P}^{(2)}$ with subscript 33. Indeed, from (4.10) and (4.19) we obtain

$$P_{33}^{(1)} = \int_\Omega \boldsymbol{e}_3 \cdot \left(\boldsymbol{w}_3^{(1)} \times \boldsymbol{r} \right) \mathrm{d}\Omega = -\int_\Omega (x_1 \boldsymbol{e}_2 - x_2 \boldsymbol{e}_1) \cdot \boldsymbol{w}_3^{(1)} \mathrm{d}\Omega$$

$$= -\int_\Omega (x_1 \boldsymbol{e}_2 - x_2 \boldsymbol{e}_1) \cdot \left(\frac{\partial \psi_3}{\partial x^2} \boldsymbol{e}_1 - \frac{\partial \psi_3}{\partial x_1} \boldsymbol{e}_2 \right) \mathrm{d}\Omega.$$

Using formulas (4.24), we obtain the following:

$$P_{33}^{(1)} = b \int_\Omega \left(\frac{x_1^2}{a_1^2} + \frac{x_2^2}{a_2^2} \right) \left(1 - \frac{x_1^2}{a_1^2} - \frac{x_2^2}{a_2^2} - \frac{x_3^2}{a_3^2} \right) \mathrm{d}x_1 \mathrm{d}x_2 \mathrm{d}x_3.$$

A transition to generalized spherical coordinates $x_1 = a_1 r \sin\theta \cos\varphi$, $x_2 = a_2 r \sin\theta \times \sin\psi$, $x_3 = a_3 r \cos\theta$ allows us to calculate the integral and obtain

$$P_{33}^{(1)} = \frac{16}{105} \pi a_1 a_2 a_3 b.$$

To calculate $P_{33}^{(2)}$, we use formulas (4.16) and (4.19). We have

$$P_{33}^{(2)} = \int_\Omega w_3^{(1)} \cdot w_3^{(1)} \mathrm{d}\Omega = \int_\Omega \left(\left| \frac{\partial \psi_3}{\partial x_1} \right|^2 + \left| \frac{\partial \psi_3}{\partial x_2} \right|^2 \right) \mathrm{d}\Omega$$

$$= b^2 \int_\Omega \left(\frac{x_1^2}{a_1^4} + \frac{x_2^2}{a_2^4} \right) \left(1 - \frac{x_1^2}{a_1^2} - \frac{x_2^2}{a_2^2} - \frac{x_3^2}{a_3^2} \right) \mathrm{d}\Omega.$$

The result of this calculation is

$$P_{33}^{(2)} = \frac{32}{105} \pi a_1 a_2 a_3 b^2 \left(a_1^{-2} + a_2^{-2} \right).$$

Since the ellipsoid's axes have similar roles, the functions $\psi_1(x)$ and $\psi_2(x)$ can be calculated from our previous formulas by using the circular permutation of indexes $(1,2,3) \mapsto (2,3,1)$ for $\psi_1(x)$, and $(1,2,3) \mapsto (3,1,2)$ for $\psi_2(x)$. Then, we can calculate the components of tensor $\boldsymbol{Q}^{(2)}$. For example,

$$Q_{12}^{(2)} = 2 \int_\Omega e_3 \cdot \left(w_1^{(1)} \times w_2^{(1)} \right) \mathrm{d}\Omega$$

$$= 2 \int_\Omega \frac{\partial \psi_1}{\partial x_3} \cdot \frac{\partial \psi_2}{\partial x_3} \mathrm{d}\Omega = \frac{128}{945} \pi a_1 a_2 b^2 a_3^{-1}.$$

7.5 Oscillations of a Pendulum With a Cavity Completely Filled With a Viscous Fluid

The problem considered here generalizes the problem in Section 7.2 to the case when the fixed point O of the gyrostate does not coincide with the mass center C of the system. First, we will study the plane (two-dimensional) case in which a Pontryagin space Π_1 occurs. In the space (three-dimensional) problem, that is, the problem on oscillations of the so-called spherical pendulum, we naturally get the Pontryagin space Π_2 after several transformations.

7.5.1 Towards the Statement of the Problem

Let us suppose that a hydromechanical system is a gyrostate with a fixed point O and it is situated on a plane Oy_2y_3. A fluid with density ρ and kinematic viscosity ν fills completely a region $\Omega \subset \mathbb{R}^2 := \{(y_2; y_3) : -\infty < y_i < \infty,\ i = 2, 3\}$, where the mass center C of this system does not coincide with the pole O. The system is affected by a uniform gravitational field of intensity $\boldsymbol{g} = -g\boldsymbol{e}_3'$ ($\boldsymbol{e}_i'$ are the axial vectors of the system Oy_2y_3, $i = 2, 3$). Thus, the given system is a plane pendulum with a fluid filling.

Being in an equilibrium state, the center of masses C and the pole O are on the vertical OC, which is parallel to vector $\boldsymbol{g}$. We consider small oscillations of the pendulum that are close to the equilibrium state. The pendulum position at any time is characterized by the vector of angular displacement $\boldsymbol{\delta} = \delta_1 \boldsymbol{e}_1$, where $\boldsymbol{e}_1 = \boldsymbol{e}_1' = \boldsymbol{e}_2' \times \boldsymbol{e}_3'$. In the nonstationary coordinate system Ox_2x_3, rigidly connected to the body, the radius-vector $\boldsymbol{r}_C$ of the center of masses C is clearly equal to $\boldsymbol{r}_C = -l\boldsymbol{e}_3$, where $l > 0$ is the length of segment OC (the length of the mathematical pendulum). Therefore, the moment of external gravitational forces $\boldsymbol{M}(t)$ with respect to the pole O is equal to

$$\boldsymbol{M}(t) = m\,(\boldsymbol{r}_C \times \boldsymbol{g}) = -mgl\,(\boldsymbol{e}_3 \times \boldsymbol{e}_3') = -mgl \sin\delta_1\, \boldsymbol{e}_1,$$

where $m = m_b + m_f$ is the total mass of body and fluid, that is, of the whole gyrostate.

For a small pendulum displacement from the vertical, $\sin\delta_1 \approx \delta_1$. Introducing the field $\boldsymbol{w} = \boldsymbol{w}(t, x)$ of related displacements of the fluid connected to the field of related velocities $\boldsymbol{u}(t, x)$ by the relationship $\partial\boldsymbol{w}/\partial t = \boldsymbol{u}(t, x)$, and also the field of dynamical pressures $p(t, x)$, $x = (x_2; x_3) \in \Omega$, from equations (2.1)–(2.4) [see also (3.1.25), (3.1.26) when $\omega_0 = 0$] we get the following initial boundary value problem,

$$\rho\frac{\partial^2 \boldsymbol{w}}{\partial t^2} + \rho\left(\frac{d^2\boldsymbol{\delta}}{dt^2} \times \boldsymbol{r}\right) = -\nabla p + \mu\Delta\frac{\partial \boldsymbol{w}}{\partial t}, \quad \operatorname{div} \boldsymbol{w} = 0 \qquad \text{in } \Omega,$$

$$J_1 \frac{d^2 \boldsymbol{\delta}}{\mathrm{d}t^2} + \rho \frac{\mathrm{d}^2}{\mathrm{d}t^2} \int_\Omega (\boldsymbol{r} \times \boldsymbol{w}) \mathrm{d}\Omega + mgl\boldsymbol{\delta} = \mathbf{0}, \quad \boldsymbol{w} = \mathbf{0} \quad \text{on } \partial\Omega,$$

$$\boldsymbol{w}(0,x) = \boldsymbol{w}^0(x), \quad \frac{\partial \boldsymbol{w}}{\partial t}(0,x) = \boldsymbol{u}^0(x), \quad \boldsymbol{\delta}(0) = \boldsymbol{\delta}^0, \quad \frac{d\boldsymbol{\delta}(0)}{\mathrm{d}t} = \boldsymbol{\omega}^0, \tag{5.1}$$

where $J_1 > 0$ is the tensor component of the inertia system with reference $Ox_1 = Oy_1$, and $\mu = \rho\nu$ is the dynamical fluid viscosity.

We note that, for solutions of the problem on small plane gyrostate oscillations, the law of a full energy balance is valid,

$$\frac{1}{2} \frac{\mathrm{d}}{\mathrm{d}t} \left\{ \rho \int_\Omega \left| \frac{\partial \boldsymbol{w}}{\partial t} \right|^2 \mathrm{d}\Omega + J_1 \left| \frac{d\boldsymbol{\delta}}{\mathrm{d}t} \right|^2 + 2\rho \int_\Omega \left(\frac{d\boldsymbol{\delta}}{\mathrm{d}t} \times \boldsymbol{r} \cdot \frac{\partial \boldsymbol{w}}{\partial t} \right) \mathrm{d}\Omega \right\}$$
$$+ \frac{1}{2} \frac{\mathrm{d}}{\mathrm{d}t} \left(mgl|\boldsymbol{\delta}|^2 \right) = -\mu \int_\Omega \left| \mathrm{rot} \left(\frac{\partial \boldsymbol{w}}{\partial t} \right) \right|^2 \mathrm{d}\Omega, \tag{5.2}$$

and it can be derived in a similar way to the derivation of law (2.5). Here, the second term on the left characterizes the change of the potential energy of the system, and the first, just like in (2.5), is the change of kinetic energy.

Formally, if we think that in the case of the pendulum movement the fluid in the cavity does not perform any motion ($\boldsymbol{w}(t,x) \equiv 0$, $p(t,x) \equiv 0$), then, for $\boldsymbol{\delta}$ in equation in (5.1), we get that $\boldsymbol{\delta}(t)$ depends on t according to the law $\exp(\mathrm{i}\omega_0 t)$, where $\omega_0^2 := mgl/J_1 > 0$ is the squared frequency of oscillations of the pendulum with a hardened fluid. Furthermore, we will assume that $\omega_0 > 0$.

7.5.2 Transition to a System of Operator Equations

Let us assume that the fields $\boldsymbol{w}(t,x)$ and $\nabla p(t,x)$ in (5.1) are functions of the variable t with values in $\boldsymbol{L}_2(\Omega)$. Then it is obvious that $\boldsymbol{w}(t,x) \in \boldsymbol{J}_0(\Omega)$ and $\nabla p(t,x) \in \boldsymbol{G}(\Omega)$. Introducing the ortoprojector P_0 onto $\boldsymbol{J}_0(\Omega)$ and applying it to the first equation in (5.1) we get the system of equations

$$\frac{\mathrm{d}^2}{\mathrm{d}t^2} (I_{11}\boldsymbol{w} + I_{12}\boldsymbol{\delta}) + \mu A_0 \frac{\mathrm{d}\boldsymbol{w}}{\mathrm{d}t} = \mathbf{0},$$
$$\frac{\mathrm{d}^2}{\mathrm{d}t^2} (I_{21}\boldsymbol{w} + I_{22}\boldsymbol{\delta}) + J_1\omega_0^2 \boldsymbol{\delta} = \mathbf{0}, \tag{5.3}$$

$$I_{11}\boldsymbol{w} : = \rho\boldsymbol{w}, \quad I_{12}\boldsymbol{\delta} := \rho P_0(\boldsymbol{\delta} \times \boldsymbol{r}),$$
$$I_{21}\boldsymbol{w} : = \rho \int_\Omega (\boldsymbol{r} \times \boldsymbol{w}) \mathrm{d}\Omega, \quad I_{22}\boldsymbol{\delta} := J_1\boldsymbol{\delta}, \tag{5.4}$$

where A_0 is the Stokes operator (see, for example, Section 7.1), and $\omega_0 > 0$ is the previously introduced frequency of oscillations of a pendulum with a hardened fluid.

In a vector-matrix form, problem (5.3)–(5.4) together with the initial conditions (5.1) can be interpreted as a Cauchy problem for a differential equation of the second order in the Hilbert space $\boldsymbol{H} := \boldsymbol{J}_0(\Omega) \oplus \mathbb{R}$:

$$\tilde{I}\frac{d^2\boldsymbol{v}}{dt^2} + \tilde{A}_0\frac{d\boldsymbol{v}}{dt} + J_1\omega_0^2\tilde{P}\boldsymbol{v} = \boldsymbol{0}, \qquad \boldsymbol{v}(0) = \boldsymbol{v}^0,\ \boldsymbol{v}'(0) = \boldsymbol{v}^1, \tag{5.5}$$

$$\tilde{I} := (I_{ik})_{i,k=1}^2, \quad \tilde{A}_0 := \operatorname{diag}(A_0; 0), \quad \tilde{P} := \operatorname{diag}(0; 1), \tag{5.6}$$

$$\boldsymbol{v}(t) := (\boldsymbol{w}(t); \boldsymbol{\delta}(t))^{\mathrm{t}}, \quad \boldsymbol{v}^0 = \left(\boldsymbol{w}^0; \boldsymbol{\delta}^0\right)^{\mathrm{t}}, \quad \boldsymbol{v}^1 = \left(\boldsymbol{u}^0; \boldsymbol{\omega}^0\right)^{\mathrm{t}}.$$

We should mention here that operator $\tilde{I}$ [in a slightly different form that corresponds to another determination of the scalar product in $\boldsymbol{H}$, see (6.2.8)] was studied in Sections 6.2.2 and 6.2.3 (see also Sections 7.3.2, 7.3.3). In those sections $\tilde{I}$ was assumed to be a bounded positive definite operator.

We consider normal oscillations of the gyrostate, that is, such solutions of problem (5.5) that depend on t according to the law $\exp(-\lambda t)$. For an amplitude function $\boldsymbol{v} = (\boldsymbol{w}; \boldsymbol{\delta})^{\mathrm{t}} \in \boldsymbol{H}$ we get the spectral problem

$$\lambda^2\tilde{I}\boldsymbol{v} - \lambda\mu\tilde{A}_0\boldsymbol{v} + J_1\omega_0^2\tilde{P}\boldsymbol{v} = \boldsymbol{0}, \tag{5.7}$$

with the operator coefficients (5.6).

If we formally assume in (5.7) that $I_{12} = I_{21}^* = 0$, then this problem splits into two independent problems

$$\begin{aligned} \rho\left(\lambda^2\boldsymbol{w} - \lambda\nu A_0\boldsymbol{w}\right) &= \boldsymbol{0}, \\ J_1\left(\lambda^2\boldsymbol{\delta} + \omega_0^2\boldsymbol{\delta}\right) &= \boldsymbol{0}, \end{aligned} \tag{5.8}$$

corresponding, respectively, to dissipative internal waves in a completely filled immovable container (see Section 7.1) and one-dimensional oscillations of a mathematical pendulum (a gyrostate with a hardened fluid) with the eigenoscillation frequency ω_0.

As it will be stated later, in the general case, when $I_{12} = I_{21}^* \neq 0$, many common properties of the solutions of problems (5.8) are the same for the problem (5.7).

7.5.3 The Indefinite Metric Approach

To study both the Cauchy problem (5.5)–(5.6) and the spectral problem (5.7) we can use the indefinite methods presented in Section 1.3. With this aim in mind, we introduce in (5.5) the field of fluid velocity $\boldsymbol{u}(t,x) = \partial \boldsymbol{w}/\partial t$ and the angular pendulum rotation velocity $\omega(t) = \mathrm{d}\boldsymbol{\delta}/\mathrm{d}t$. Then (5.5) may be interpreted as a Cauchy problem in the Hilbert space $\tilde{\boldsymbol{H}} := \boldsymbol{H} \oplus \mathbb{R} = \boldsymbol{J}_0(\Omega) \oplus \mathbb{R} \oplus \mathbb{R}$, as follows:

$$\mathcal{J}\frac{d\boldsymbol{z}}{\mathrm{d}t} + \mathcal{A}\boldsymbol{z} = \boldsymbol{0}, \qquad \boldsymbol{z}(0) = \boldsymbol{z}^0, \tag{5.9}$$

$$\mathcal{J} := \begin{pmatrix} I_{11} & I_{12} & O \\ I_{21} & I_{22} & O \\ O & O & -J_1\omega_0^2 \end{pmatrix}, \quad \mathcal{A} := \begin{pmatrix} \mu A_0 & O & O \\ O & O & J_1\omega_0^2 \\ O & J_1\omega_0^2 & O \end{pmatrix},$$

$$\boldsymbol{z}(t) := (\boldsymbol{u}(t); \boldsymbol{\omega}(t); \boldsymbol{\delta}(t))^{\mathrm{t}}, \qquad \boldsymbol{z}^0 = \left(\boldsymbol{u}^0; \boldsymbol{\omega}_0; \boldsymbol{\delta}^0\right)^{\mathrm{t}}. \tag{5.10}$$

Here, the field of displacement

$$\boldsymbol{w}(t,x) = \boldsymbol{w}^0(x) + \int_0^{\mathrm{t}} \boldsymbol{u}(\tau,x)\mathrm{d}\tau, \quad x = (x_2; x_3) \in \Omega,$$

can be obtained from the solution $\boldsymbol{u}(t,x)$ of the problem (5.9)–(5.10).

For the spectral problem (5.7) with $\lambda \neq 0$, we make the substitutions $\boldsymbol{u} = -\lambda \boldsymbol{w}$, $\boldsymbol{\omega} = -\lambda\boldsymbol{\delta}$. Applying the same substitutions to (5.9) we get

$$\mathcal{A}\boldsymbol{z} = \lambda\mathcal{J}\boldsymbol{z}, \qquad \boldsymbol{z} = (\boldsymbol{u}; \boldsymbol{\omega}; \boldsymbol{\delta})^{\mathrm{t}}. \tag{5.11}$$

Let us consider the properties of operators $\mathcal{J}$ and $\mathcal{A}$, defined by formulas (5.10). Since the operator $\tilde{I} = (I_{ik})_{i,k=1}^2$ is bounded and positive definite in $\boldsymbol{H} := \boldsymbol{J}_0(\Omega) \oplus \mathbb{C}$, and $J_1\omega_0^2 > 0$, then the operator $\mathcal{J}$:=diag$(\tilde{I}; -J_1\omega_0^2)$, acting in $\tilde{\boldsymbol{H}} := \boldsymbol{H} \oplus \mathbb{C}$, is self-adjoint and bounded, and its form $(\mathcal{J}\boldsymbol{z}, \boldsymbol{z})_{\tilde{\boldsymbol{H}}}$ has only one negative square. The inverse operator $\mathcal{J}^{-1}$ exists and equals $\mathcal{J}^{-1}$ =diag$(\tilde{I}^{-1}; -(J_1\omega_0^2)^{-1})$.

As for $\mathcal{A}$, it is an unbounded self-adjoint operator, with a discrete spectrum, having a compact inverse operator,

$$\mathcal{A}^{-1} = \begin{pmatrix} (\mu A_0)^{-1} & 0 & 0 \\ 0 & 0 & \left(J_1\omega_0^2\right)^{-1} \\ 0 & \left(J_1\omega_0^2\right)^{-1} & 0 \end{pmatrix}. \tag{5.12}$$

The form $(\mathcal{A}\boldsymbol{z}, \boldsymbol{z})_{\tilde{\boldsymbol{H}}}$, defined on $\mathcal{D}(\mathcal{A}) = \mathcal{D}(A_0) \oplus \mathbb{C} \oplus \mathbb{C}$, has only one negative square because $A_0 \gg 0$.

To obtain properties of the solutions to problem (5.11) we represent $\mathcal{J}$ as

$$\begin{aligned}
&\mathcal{J} = |\mathcal{J}|^{1/2} J |\mathcal{J}|^{1/2} = J|\mathcal{J}| = |\mathcal{J}|J, \\
&|\mathcal{J}| = \operatorname{diag}\left(\tilde{I}; J_2\omega_0^2\right) \gg 0, \\
&J := \operatorname{diag}\left(I_{\boldsymbol{H}}; -1\right), \ I_{\boldsymbol{H}} = \operatorname{diag}\left(I_\Omega; 1\right) \gg 0,
\end{aligned} \tag{5.13}$$

and we make in (5.11) the following substitutions

$$\boldsymbol{z} = \mathcal{A}^{-1}|\mathcal{J}|^{1/2}\psi, \quad \lambda = \varepsilon^{-1}. \tag{5.14}$$

Instead of (5.11) we get the next problem on eigenvalues,

$$\mathcal{K}\psi = \varepsilon\psi, \qquad \mathcal{K} := J|\mathcal{J}|^{1/2}\mathcal{A}^{-1}|\mathcal{J}|^{1/2}, \tag{5.15}$$

for a compact J-self-adjoint operator $\mathcal{K}$ acting in the Pontryagin space $\Pi_1 = \tilde{\boldsymbol{H}}$ with the scalar product $[\boldsymbol{\varphi}, \boldsymbol{\psi}]_{\tilde{\boldsymbol{H}}} := (J\boldsymbol{\varphi}, \boldsymbol{\psi})_{\tilde{\boldsymbol{H}}}$.

Since $\mathcal{K}$ is an invertible operator because its factors in (5.15) are invertible, then the root subspace $\boldsymbol{L}_0(\mathcal{K})$ corresponding to the value $\varepsilon = 0$, is the zero subspace. Therefore, according to a criterion formulated in Section 1.3.7, problem (5.15) has a discrete spectrum consisting of eigenvalues $\varepsilon_k > 0$ of finite multiplcity with a limit point at zero as well as a pair of eigenvalues ε_0^+ and ε_0^- that can be two complex conjugate nonreal numbers. The system of eigenelements $\boldsymbol{\psi}_k$ corresponding to the numbers ε_k together with the eigenelements $\boldsymbol{\psi}_0^+$ and $\boldsymbol{\psi}_0^-$ corresponding to the numbers ε_0^+ and ε_0^- forms a Riesz basis in $\tilde{\boldsymbol{H}} = \boldsymbol{J}_0(\Omega) \oplus \mathbb{C} \oplus \mathbb{C}$. Here, the elements $\{\boldsymbol{\psi}_k\}_{k=1}^{\infty}$ form a J-orthogonal basis for a uniformly positive subspace $\boldsymbol{L}_+$ invariant with respect to $\mathcal{K}$, and the expansion $\tilde{\boldsymbol{H}} = \boldsymbol{L}_+[+]\boldsymbol{M}$, where $\boldsymbol{M}$ is the two-dimensional subspace with basis $\boldsymbol{\psi}_0^+$ and $\boldsymbol{\psi}_0^-$, takes place. If $\overline{\varepsilon_0^+} = \varepsilon_0^-$, then the elements $\boldsymbol{\psi}_0^+$ and $\boldsymbol{\psi}_0^-$ are J-neutral: $[\boldsymbol{\psi}_0^\pm, \boldsymbol{\psi}_0^\pm] = (J\boldsymbol{\psi}_0^\pm, \boldsymbol{\psi}_0^\pm)_{\tilde{\boldsymbol{H}}} = 0$.

By (5.14) and going from (5.15) back to problem (5.11), from the properties mentioned previously we conclude that problem (5.11) has a discrete spectrum consisting of positive eigenvalues λ_k, with a limit point $\lambda = +\infty$, and two more eigenvalues λ_0^+ and λ_0^- that are possibly complex conjugate numbers. The eigenelements $\{\boldsymbol{z}_k\}_{k=1}^{\infty}$ corresponding to the numbers $\{\lambda_k\}_{k=1}^{\infty}$ form a $\mathcal{J}$-orthogonal basis for a uniformly positive subspace $\mathcal{L}_+$ invariant with respect to operator $\mathcal{J}^{-1}\mathcal{A}$. In this regard, $\tilde{\boldsymbol{H}} = \mathcal{L}_+[+]\mathcal{M}$, where $\mathcal{M}$ is a two-dimensional subspace with basis $\boldsymbol{z}_0^+$ and $\boldsymbol{z}_0^-$ and the symbol $[+]$ represents the $\mathcal{J}$-orthogonal sum. The elements $\{\boldsymbol{z}_k\}_{k=1}^{\infty}$ can be chosen to be $\mathcal{J}$-orthogonalized,

$$[\boldsymbol{z}_k, \boldsymbol{z}_m] = (\mathcal{J}\boldsymbol{z}_k, \boldsymbol{z}_m)_{\tilde{\boldsymbol{H}}} = \delta_{km}. \tag{5.16}$$

If λ_0^+ and λ_0^- are complex conjugate, then the elements $\boldsymbol{z}_0^+$ are $\mathcal{J}$-neutral,

$$\left[\boldsymbol{z}^\pm + o, \boldsymbol{z}_0^\pm\right] = \left(\mathcal{J}\boldsymbol{z}_0^\pm, \boldsymbol{z}_0^\pm\right)_{\tilde{\boldsymbol{H}}} = 0. \tag{5.17}$$

7.5.4 Other Properties of Solutions of the Spectral Problem

Let us return to problem (5.7) and get those properties of its solutions that are not based on the approach in Section 7.5.3, but can be deduced from properties of the coefficients $\tilde{I}$, $\tilde{A}_0$, and $\tilde{P}$.

1° Problem (5.7) has the obvious solution

$$\lambda = 0, \quad \boldsymbol{\delta} = \mathbf{0}, \qquad \boldsymbol{w} \in \boldsymbol{J}_0(\Omega). \tag{5.18}$$

This solution corresponds to an infinitely multiple eigenvalue and a state in problem (5.5), related to a new—different from the original—position of the fluid particles in region Ω.

Excluding such physically obvious solutions, we will further consider the case $\lambda \neq 0$.

2° All (nonzero) eigenvalues of problem (5.7) are located in the half-plane $\operatorname{Re}\lambda > 0$. (Due to the results proved in Section 7.5.3, this property means that condition $\operatorname{Re}\lambda_0^{\pm} > 0$ is satisfied for the pair of eigenvalues $\lambda_0^{\pm}$ mentioned at the end of Section 7.5.3.)

To prove this, we divide (5.7) by λ ($\neq 0$), scalarly multiply in $\boldsymbol{H}$ by $\boldsymbol{v}$, and then take the real part. Thus we get

$$\operatorname{Re}\lambda\left(\left(\tilde{I}\boldsymbol{v},\boldsymbol{v}\right)_{\boldsymbol{H}} + J_1\omega_0^2|\lambda|^{-2}\left(\tilde{P}\boldsymbol{v},\boldsymbol{v}\right)_{\boldsymbol{H}}\right) = \mu\left(A_0\boldsymbol{v},\boldsymbol{v}\right)_{\boldsymbol{H}}. \tag{5.19}$$

Since $\tilde{I} \gg 0$, $\tilde{P} \geq 0$, and $\tilde{A}_0 \geq 0$, from (5.19) it follows that $\operatorname{Re}\lambda \geq 0$. If $\lambda = \mathrm{i}\alpha \neq 0$, $\alpha \in \mathbb{R}$, and from (5.7) we immediately deduce that $\alpha\mu(A_0\boldsymbol{v},\boldsymbol{v}) = 0$, whence it follows that $\boldsymbol{v} = (\mathbf{0};\boldsymbol{\delta})^{\mathrm{t}}$ with arbitrary $\boldsymbol{\delta} = \delta_1\boldsymbol{e}_1$. In this case from (5.7) it also follows that

$$\alpha^2 I_{22}|\boldsymbol{\delta}|^2 + J_1\omega_0^2|\boldsymbol{\delta}|^2 = |\boldsymbol{\delta}|^2 J_1(\omega_0^2 - \alpha^2) = 0.$$

If $\alpha \neq \pm\omega_0$, from it we get $\boldsymbol{\delta} = \delta_1\boldsymbol{e}_1 = 0$, and, therefore, problem (5.7), with $\lambda = \mathrm{i}\alpha$, $\alpha \in \mathbb{R}$, has trivial solution.

We consider now the case when $\alpha = \pm\omega_0$. The second equation of the vector-matrix expression (5.7) is satisfied in this case, and the first equation gives the condition [see the definition of I_{12} in (5.4)]

$$\begin{aligned} -\omega_0^2 I_{12}\boldsymbol{\delta} &= -\omega_0^2\rho P_0\left(\delta_1\boldsymbol{e}_1 \times (x_2\boldsymbol{e}_2 + x_3\boldsymbol{e}_3)\right) \\ &= -\delta_1\omega_0^2\rho P_0\left(x_2\boldsymbol{e}_3 - x_3\boldsymbol{e}_2\right) = \mathbf{0}, \end{aligned}$$

where P_0 is the orthoprojector onto $\boldsymbol{J}_0(\Omega)$, and $\boldsymbol{e}_j$ are the axial vectors of the non-stationary coordinate system Ox_2x_3. Due to the orthogonal expansion $\boldsymbol{L}_2(\Omega) = \boldsymbol{J}_0(\Omega) \oplus \boldsymbol{G}(\Omega)$, here we get that

$$\delta_1 \left(x_2 \boldsymbol{e}_3 - x_3 \boldsymbol{e}_2 \right) = \nabla \Phi(x_2, x_3) \in \boldsymbol{G}(\Omega),$$

and, therefore, $\partial \Phi / \partial x_2 = -\delta_1 x_3$, $\partial \Phi / \partial x_3 = \delta_1 x_2$. These relations are possible only when $\delta_1 = 0$. Thus, the numbers $\lambda = \pm i\omega_0$ can not be eingenvalues of problem (5.7).

The proof of Property 2° also shows that a pendulum filled with fluid does not have the same oscillation frequency as the pendulum filled with a hardened fluid.

To obtain the next properties of the solutions of problem (5.7), we represent it as a system of equations:

$$\begin{aligned} &\lambda^2 \left(I_{11} \boldsymbol{w} + I_{12} \boldsymbol{\delta} \right) - \lambda \mu A_0 \boldsymbol{w} = \boldsymbol{0}, \\ &\lambda^2 \left(I_{21} \boldsymbol{w} + I_{22} \boldsymbol{\delta} \right) + J_1 \omega_0^2 \boldsymbol{\delta} = \boldsymbol{0}. \end{aligned} \tag{5.20}$$

The real positive spectrum $\{\lambda_k\}_{k=1}^{\infty}$ of problem (5.20) mentioned in Section 7.5.3, can be obtained by studying the graph of the scalar characteristic equation of the problem. Let $\lambda \neq \nu \lambda_k(A_0)$, where $\lambda_k(A_0)$ are the eigenvalues of the Stokes operator A_0, with $\lambda_k(A_0) \to +\infty$ as $k \to \infty$. Taking into account the definitions (5.4) of operators I_{ik} we can represent (5.20) as follows

$$\begin{aligned} &\left(\nu A_0 - \lambda I_0 \right) \boldsymbol{w} = \lambda P_0 \left(\boldsymbol{\delta} \times \boldsymbol{r} \right) \qquad \left(= \lambda \rho^{-1} I_{12} \boldsymbol{\delta} \right), \\ &\left(\lambda^2 + \omega_0^2 \right) \boldsymbol{\delta} = -\lambda^2 \rho J_1^{-1} \int_\Omega \left(\boldsymbol{r} \times \boldsymbol{w} \right) \mathrm{d}\Omega \qquad \left(= -\lambda^2 J_1^{-1} I_{21} \boldsymbol{w} \right). \end{aligned} \tag{5.21}$$

Expressing $\boldsymbol{w}$ as a solution of the first equation and substituting it in the second one we get

$$\left(\lambda^2 + \omega_0^2 \right) \boldsymbol{\delta} + \lambda^3 \rho J_1^{-1} \int_\Omega \boldsymbol{r} \times \left(\left(\nu A_0 - \lambda I_0 \right)^{-1} P_0 \left(\boldsymbol{\delta} \times \boldsymbol{r} \right) \right) \mathrm{d}\Omega = 0. \tag{5.22}$$

Here, the second term is proportional to $\boldsymbol{\delta} = \delta_1 \boldsymbol{e}_1$, since the expression $\boldsymbol{r} \times \boldsymbol{w}$ is collinear to $\boldsymbol{e}_1$. Denoting the integral term in (5.22) by $\boldsymbol{a}$, we have

$$\begin{aligned} \boldsymbol{a} :&= \delta_1 \int_\Omega \boldsymbol{r} \times \left(\left(\nu A_0 - \lambda I_0 \right)^{-1} P_0 \left(\boldsymbol{e}_1 \times \boldsymbol{r} \right) \right) \mathrm{d}\Omega = \left(\boldsymbol{a} \cdot \boldsymbol{e}_1 \right) \boldsymbol{e}_1 \\ &= \left(\delta_1 \boldsymbol{e}_1 \right) \left(\int_\Omega \left[\boldsymbol{r} \times \left(\nu A_0 - \lambda I_0 \right)^{-1} P_0 \left(\boldsymbol{e}_1 \times \boldsymbol{r} \right) \right] \cdot \boldsymbol{e}_1 \mathrm{d}\Omega \right) \\ &= \left(\int_\Omega \left(\nu A_0 - \lambda I_0 \right)^{-1} P_0 \left(\boldsymbol{e}_1 \times \boldsymbol{r} \right) \cdot \left(\boldsymbol{e}_1 \times \boldsymbol{r} \right) \mathrm{d}\Omega \right) \boldsymbol{\delta} \\ &= \left(\left(\nu A_0 - \lambda I_0 \right)^{-1} P_0 \left(\boldsymbol{e}_1 \times \boldsymbol{r} \right),\ P_0 \left(\boldsymbol{e}_1 \times \boldsymbol{r} \right) \right)_{\boldsymbol{L}_2(\Omega)} \boldsymbol{\delta}. \end{aligned} \tag{5.23}$$

Using (5.23), from (5.22) we get the characteristic equation for obtaining the eigenvalues,

$$\lambda^2 + \omega_0^2 + f_\nu(\lambda) = 0,$$
$$f_\nu(\lambda) := \lambda^3 \rho J_1^{-1} \left((\nu A_0 - \lambda I_0)^{-1} P_0 (\boldsymbol{e}_1 \times \boldsymbol{r}) \,,\; P_0 (\boldsymbol{e}_1 \times \boldsymbol{r}) \right)_{\boldsymbol{L}_2(\Omega)}. \tag{5.24}$$

Furthermore, for simplicity, we suppose that all the eigenvalues of operator A_0 are simple. Then,

$$(\nu A_0 - \lambda I_0)^{-1} \boldsymbol{w} = \sum_{k=1}^{\infty} \frac{(\boldsymbol{w}, \boldsymbol{u}_k (A_0))_{\boldsymbol{L}_2(\Omega)} \boldsymbol{u}_k (A_0)}{\nu \lambda_k (A_0) - \lambda}, \qquad \| \boldsymbol{u}_k (A_0) \|_{\boldsymbol{L}_2(\Omega)} = 1, \; (5.25).$$

and $0 < \lambda_1(A_0) < \cdots < \lambda_k(A_0) < \cdots$, $\lambda_k(A_0) \to +\infty$ as $k \to \infty$. In this case, the function $f_\nu(\lambda)$ from (5.24) takes the form

$$f_\nu(\lambda) = \lambda^3 \rho J_1^{-1} \sum_{k=1}^{\infty} \frac{|(\boldsymbol{e}_1 \times \boldsymbol{r}, \boldsymbol{u}_k(A_0))_{\boldsymbol{L}_2(\Omega)}|^2}{\nu \lambda_k(A_0) - \lambda}. \tag{5.26}$$

Again, for simplicity, we assume that

$$(\boldsymbol{e}_1 \times \boldsymbol{r}, \boldsymbol{u}_k (A_0))_{\boldsymbol{L}_2(\Omega)} \neq 0, \qquad k = 1, 2, \ldots. \tag{5.27}$$

Equation (5.24), where the function $f_\nu(\lambda)$ is given by (5.26) with the assumptions (5.27), can be studied graphically by finding its roots as points of intersection of the graph of the functions $z_1 = \lambda^2 + \omega_0^2$ and $z_2 = -f_\nu(\lambda)$. Schematically, without any regard to proportions, the graphs are shown in Figure 7.5.1. The result of this investigation is formulated in the following statements that refine the properties of the spectrum of problem (5.7).

3° The real eigenvalues λ_k of problem (5.7) are situated in the intervals $(\nu\lambda_k(A_0); \lambda_k^0)$. The numbers λ_k^0 satisfy the inequalities $\lambda_k^0 < \lambda_k^1$, where $\lambda_k^1 < \lambda_{k+1}(A_0)$ and λ_k^1 are zeros of the function $f_\nu(\lambda)$. The physical meaning of the values λ_k^0 lies in the fact that they are equal to the eigenvalues of the problem on normal oscillations of a gyrostate in the case when the pole O and the mass center C of a system coincide ($\omega_0^2 = mgl/J_1 = 0$, since $l = 0$). In the three-dimensional case, this problem was analyzed in Section 7.2.

For the number λ_k we have the asymptotic formula

$$\lambda_k = \nu \lambda_k (A_0) + \mathrm{o}(1), \qquad k \to \infty. \tag{5.28}$$

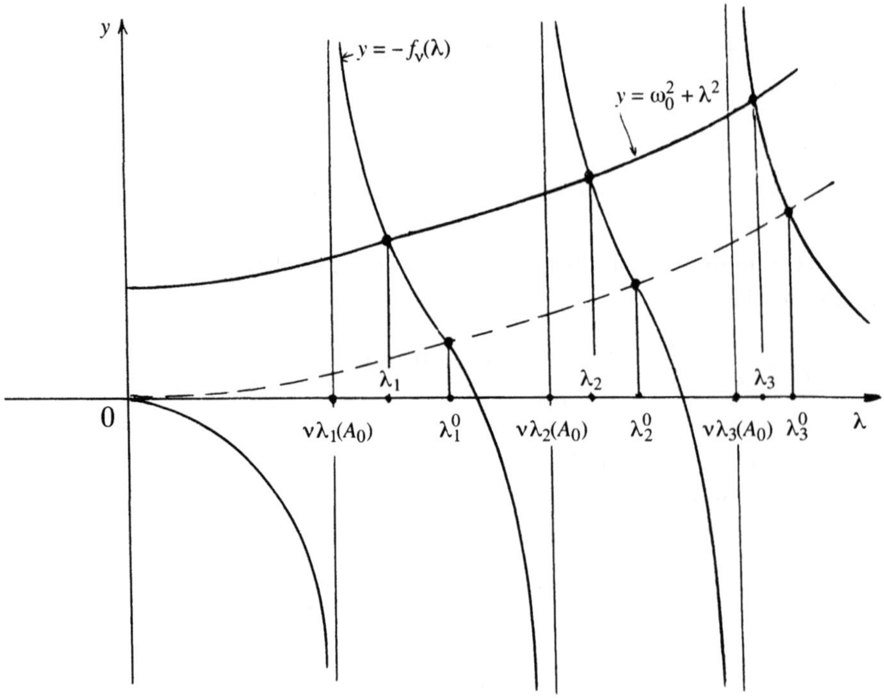

Figure 7.5.1

4° For a sufficiently large viscosity $\nu > 0$, equation (5.24) has only one pair of nonreal eigenvalues situated in the neighborhood of the points $\lambda = \pm i\omega_0$, where ω_0 is the oscillation frequency of the pendulum filled with a hardened fluid.

To prove this property we use the Rouché theorem. Let us consider the circle C_r of radius $r > 0$ and with center at the point $\lambda = i\omega_0$. It is obvious that the function $z_1 = \lambda^2 + \omega_0^2$ for any $r < \omega_0$ has only one zero inside the contour C_r. We evaluate the modulus of the functions $z_1(\lambda)$ and $z_2(\lambda) = -f_\nu(\lambda)$ on the circle C_r.

For $z_1(\lambda)$, with $\lambda = i\omega_0 + re^{i\psi}$, we have

$$|\lambda^2 + \omega_0^2| = |(\lambda - i\omega_0)(\lambda + i\omega_0)| = r|2i\omega_0 + re^{i\psi}| \geq r(2\omega_0 - r). \tag{5.29}$$

To estimate the modulus of the function $z_2(\lambda) = -f_\nu(\lambda)$ we use, for the self-adjoint operator A, the inequality

$$\| (A - \lambda I)^{-1} \| \leq \frac{1}{d(\lambda)}, \tag{5.30}$$

where $d(\lambda)$ =dist$(\lambda; \tau(A))$ is the distance from a point $\lambda \in \mathbb{C}$ to the spectrum $\sigma(A)$ of operator A. Thus, we can represent $f_\nu(\lambda)$ using operator I_{12} [see (5.24, (5.4)] in the following form

$$f_\nu(\lambda) = \lambda^3 \rho^{-1} J_1^{-1} \left((\nu A_0 - \lambda I)^{-1} I_{12} \boldsymbol{e}_3, I_{12} \boldsymbol{e}_3 \right)_{\boldsymbol{L}_2(\Omega)}, \tag{5.31}$$

and obtain on C_r the next estimate

$$\begin{aligned}|f_\nu(\lambda)| &\le |\lambda|^3 \rho^{-1} J_1^{-1} \|I_{12}\|^2 \left\|(\nu A_0 - \lambda I_0)^{-1}\right\| \\ &\le \frac{|\lambda|^3 \rho^{-1} J_1^{-1} \|I_{12}\|^2}{\min\limits_{\lambda \in C_r} \sqrt{|\mathrm{Im}\,\lambda|^2 + |\nu\lambda_1(A_0) - \mathrm{Re}\lambda|^2}} \\ &\le \frac{\rho^{-1} J_1^{-1} (\omega_0 + r)^3 \|I_{12}\|^2}{\sqrt{\omega_0^2 + \nu^2 \lambda_1^2(A_0)} - r} \\ &= \mathrm{O}\left(\nu^{-1}\right), \qquad \nu \to \infty. \end{aligned} \tag{5.32}$$

With fixed r and large enough ν, from (5.29) and (5.32) it follows that the inequality $|z_2(\lambda)| = |f_\nu(\lambda)| < |z_1(\lambda)|$ is valid on C_r. Therefore, by the Rouché theorem we get that equation (5.24) for the chosen ν has only one root inside the circle C_r. A similar statement takes place for a circle of radius r and with center at the point $\lambda = -\mathrm{i}\omega_0$.

5° When the viscosity ν is arbitrary, equation (5.24) has only one pair of nonreal roots.

Indeed, according to Property 4° and the conclusions of Section 7.5.3, this property takes place for large values of ν. Further, by changing continuously the parameter ν from a large enough value to any given positive value $\nu = \nu_0$, all the eigenvalues of problem (5.24) change continuously. The graphical study of the solutions of the equation shows that, because the graph of the function $-f_\nu(\lambda)$ is monotone, there are no solutions in the interval $(-\infty, \nu\lambda_1(A_0))$ for any $\nu > 0$, and in the intervals $(\nu\lambda_k(A_0),\ \nu\lambda_{k+1}(A_0))$ for any $k \in \mathbb{N}$ and $\nu > 0$ there is only one solution λ_k. Thus, by changing ν from large values to a given value ν_0, there will be no new real roots added to the previous ones.

6° The asymptotic formulas

$$\begin{aligned}\lambda_0^{\pm}(\nu) &= \pm\mathrm{i}\omega_0 + \mathrm{o}(1), \\ \lambda_k(\nu) &= \nu\left[\gamma_k - \varepsilon_k(\nu)\right], \qquad \nu \to \infty,\end{aligned} \tag{5.33}$$

take place, where γ_k are the positive roots of equation

$$1 + \gamma\rho J_1^{-1}\left((A_0 - \gamma I_0)^{-1} P_0\left(\boldsymbol{e}_1 \times \boldsymbol{r}\right), \left(\boldsymbol{e}_1 \times \boldsymbol{r}\right)\right)_{\boldsymbol{L}_2(\Omega)} = 0, \tag{5.34}$$

and $\varepsilon_k(\nu) > 0$ and tend to zero for $\nu \to \infty$.

The first formula in (5.33) follows from the proof of Property 4° if we suppose that $r > 0$ is any possible small number. The second formula in (5.33) is proved as follows. In (5.24), we make the substitution of $\lambda = \nu\gamma$. To find the roots γ we employ the equation

$$\gamma^2 + \omega_0^2\nu^{-2} + \gamma^3\rho J_1^{-1}\left((A_0 - \gamma I_0)^{-1} P_0\left(\boldsymbol{e}_1 \times \boldsymbol{r}\right), P_0\left(\boldsymbol{e}_1 \times \boldsymbol{r}\right)\right)_{\boldsymbol{L}_2(\Omega)} = 0, \qquad (5.35)$$

which is graphically studied in the same way as (5.24). For $\nu \to \infty$, the positive roots $\gamma_k(\nu)$ of equation (5.35) become the roots γ_k of equation (5.34). Here, $\gamma_k(\nu) = \gamma_k - \varepsilon_k(\nu)$, and $\varepsilon_k(\nu)$ have the previously described properties. Hence the second formula in (5.33) follows.

7° If conditions (5.27) are not satisfied for some values $k \in \mathbb{N}$, that is, if $(\boldsymbol{e}_1 \times \boldsymbol{r}, \boldsymbol{u}_k(A_0))_{\boldsymbol{L}_2(\Omega)} = 0$, then studying graphically the characteristic equation (5.24) it reduces to the fact that there are no corresponding vertical asymptotes of the function $f_\nu(\lambda)$, just as the similar terms in formula (5.26). Furthermore, if the eigenvalues $\lambda_k(A_0)$ of operator A_0 are not simple, then instead of using the expressions $(\boldsymbol{w}, \boldsymbol{u}_k(A_0))_{\boldsymbol{L}_2(\Omega)}\boldsymbol{u}_k(A_0)$ and $(\boldsymbol{e}_1 \times \boldsymbol{r}, \boldsymbol{u}_k(A_0))_{\boldsymbol{L}_2(\Omega)}$, in (5.25) and (5.26) we use the expressions $P_k\boldsymbol{w}$ and $\|P_k(\boldsymbol{e}_1 \times \boldsymbol{r})\|^2_{\boldsymbol{L}_2(\Omega)}$, respectively, where P_k is the orthoprojector corresponding the eigenvalue $\lambda_k(A_0)$.

Hence the next important conclusion follows: The multiplicities of the eigenvalues of the two problems, namely, of the positive eigenvalues $\lambda_k(\omega_0)$ of problem (5.24) on oscillations of a two-dimensional pendulum when the pole O and the mass center C do not coincide, and of the eigenvalues $\lambda_k(0)$ of the same problem when O and C coincide, that is, the normal gyrostate oscillations around the fixed mass center (see Section 7.2 in the three-dimensional case), are the same.

Indeed, in (5.35) the roots $\gamma_k = \gamma_k(\omega_0)$ are depending continuously on ω_0, so, in particular, by changing ω_0 from a given value to zero, there is neither a junction nor a splitting of the roots.

8° For nonreal roots $\lambda_0^\pm(\nu;\omega_0)$, a typical graph of their changes in the complex plane that results from changes in the viscosity ν is shown for $\lambda_0^+(\nu;\omega_0)$ in Figure 7.5.2.

For $\nu \to \infty$, the roots $\lambda_0^\pm(\nu;\omega_0)$ coincide, as it follows from (5.33), with the eigenvalues $\pm \mathrm{i}\omega_0$ of the problem on oscillations of a mechanical pendulum with a hardened fluid. For $\nu = 0$, that is, for an ideal fluid, we again obtain purely imaginary eigenvalues $\lambda = \pm \mathrm{i}\omega_1$. Here, $\omega_1 > 0$ is the frequency of eigenoscillations of a pendulum with a cavity completely filled with an ideal fluid. As it was shown in Section 3.4 for the three-dimensional problem, the motion of such a system is equivalent to the motion of a mechanical system with a changed inertia moment, that is, in the given case, a

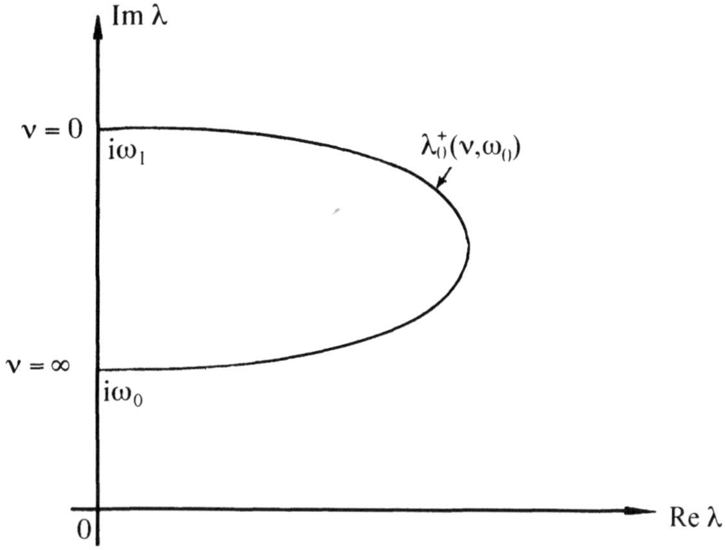

Figure 7.5.2

mathematical pendulum with a changed length to which the oscillation frequency ω_1 corresponds.

The Properties 1°–8° of solutions of the problem (5.7) allow to formulate the following main conclusions.

In the problem on normal oscillations of a two-dimensional pendulum with a cavity completely filled with a viscous fluid, all normal oscillation modes are divided into two classes:

(i) *dissipative waves with fading decrements* $\{\lambda_k\}_{k=1}^{\infty}$ *such that* $\lambda_k \to +\infty$ *as* $k \to \infty$, *and*

(ii) *two oscillation modes with eigenvalues* $\lambda_0^{\pm}$ *that correspond to oscillating fading oscillations of the pendulum with frequency* $Im\,\lambda_0^+ > 0$ *and fading decrement* $Re\lambda_0^+ = Re\lambda_0^- > 0$.

7.5.5 On Riesz and p-Basicity of Modes of Dissipative Waves

Here, we will state one more property of the solutions of problem (5.7) characterizing the basicity of eigenmodes of the velocity fields of the fluid in region Ω.

The following statement takes place. The first components of the eigen-elements $\boldsymbol{v}_k = (\boldsymbol{w}_k; \delta_k)^t$ of problem (5.7), that is, the functions $\{\boldsymbol{w}_k(x)\}_{k=1}^{\infty}$, form in the space $J_0^1(\Omega)$ a p-basis with a finite defect, and for a large enough viscosity ν, a p-basis with $p > \tilde{p} = 3/2$.

To prove this fact, we eliminate $\boldsymbol{\delta}$ in the system of equations (5.20); using the definitions of operators I_{11} and I_{22} from (5.9) we have

$$\left(\nu A_0 - \lambda I_0 + \lambda^3\left(\lambda^2+\omega_0^2\right)^{-1}(\rho J_1)^{-1} I_{12}I_{21}\right)\boldsymbol{w} = \mathbf{0}. \tag{5.36}$$

Performing next the substitutions

$$A_0^{1/2}\boldsymbol{w} = \boldsymbol{u}, \quad \lambda = \nu\gamma^{-1}, \tag{5.37}$$

we get the problem on eigenvalues

$$L(\gamma)u := \left(\gamma I_0 - A_0^{-1} + \left(1+\gamma^2\omega_0^2/\nu^2\right)^{-1}(\rho J_1)^{-1} A_0^{-1/2} I_{12}I_{21}A_0^{-1/2}\right)\boldsymbol{u} = \mathbf{0} \tag{5.38}$$

for a self-adjoint operator pencil $L(\gamma)$.

Since for the function $L(\gamma)$ the following conditions are fulfilled,

$$\begin{aligned} L(0) &= -A_0^{-1} + (\rho J_1)^{-1} A_0^{-1/2} I_{12}I_{21}A_0^{-1/2} \\ &= -A_0^{-1/2}\left[I_0 - (\rho J_1)^{-1} I_{12}I_{21}\right] A_0^{-1/2} \\ &=: A_0^{-1/2}\left(I_0 - \tilde{\Pi}_1\right)A_0^{-1/2} \in \mathcal{S}_\infty, \qquad L'(0) = I_0 \gg 0, \end{aligned} \tag{5.39}$$

then, according to statement 3° of Section 1.6.10, for any $\varepsilon > 0$ the set of eigenelements $\{\boldsymbol{u}_k\}$ of the problem (5.38) corresponding to eigenvalues γ_k, $0 < \gamma_k < \varepsilon$, forms a Riesz basis with the finite defect in the space $\boldsymbol{J}_0(\Omega)$. Since $\mathcal{D}(A_0^{-1/2}) = \boldsymbol{J}_0^1(\Omega)$, then after the inverse substitution (5.37) we get that the eigenfunctions $\{\boldsymbol{w}_k(x)\}_{k=1}^\infty$ corresponding to eigenvalues $\lambda_k > \nu\varepsilon^{-1}$, form a Riesz basis with a finite defect in $\boldsymbol{J}_0^1(\Omega) = \mathcal{D}(A_0^{1/2})$. Furthermore, since the analytic disturbance in (5.38) is an operator-function taking on values in the set of one-dimensional operators, then it can be represented as a Taylor series in powers of γ, with coefficients belonging to the class of compact operators $\mathfrak{S}_p$ with any $p > 0$. Then, by the formula for the number $\tilde{p}_1$ from Statement 4° of Section 1.6.10, we get that elements $\{\boldsymbol{w}_k(x)\}_{k=1}^\infty$ form a p-basis with a finite defect in $\boldsymbol{J}_0^1(\Omega)$ with $p > \tilde{p} = 3/2$, since operator A_0^{-1} belongs to such a class [see formula (7.1.11)].

To prove the property of basicity without defect for large viscosity, we consider the interval $\gamma \in [-\varepsilon_1, \varepsilon_2]$ with some $\varepsilon_1 > 0$, $\varepsilon_2 > 0$. Since for $\varepsilon_1 \to +0$,

$$L(-\varepsilon_1) = -\varepsilon_1 I_0 + L(0) + O\left(\varepsilon_1^2\right),$$

then the condition $L(-\varepsilon_1) \ll 0$ with small $\varepsilon_1 > 0$ will be fulfilled if $L(0) \leq 0$. We check whether this fact is valid. By virtue of (5.39) and the property $A_0^{-1/2} > 0$ it is sufficient to check that $I_0 - \tilde{\Pi}_1 \gg 0$. According to the definition of operator I_{12}, we have

$$\begin{aligned} \|I_{12}\boldsymbol{\delta}\|^2_{\boldsymbol{L}_2(\Omega)} &= \rho^2\int_\Omega |P_0(\boldsymbol{\delta}\times\boldsymbol{r})\|^2\,d\Omega \leq \rho^2|\delta_1|^2\int_\Omega |\boldsymbol{e}_1\times\boldsymbol{r}|^2 d\Omega \\ &= \rho|\delta_1|^2(J_1)_f, \end{aligned}$$

where $(J_1)_f$ is the component of the tensor of inertia of the hardened fluid (see Section 7.2.3). Hence we get that

$$\|I_{12}\| \leq \sqrt{\rho (J_1)_f}.$$

Therefore,

$$I_0 - \tilde{\Pi}_1 = I_0 - (\rho J_1)^{-1} I_{12} I_{21} \geq \left(1 - (\rho J_1)^{-1} \|I_{12}\|^2\right) I_0$$
$$\geq (I - q) I_0 \gg 0, \qquad q := \frac{(J_1)_f}{J_1} < 1. \quad (5.40)$$

Thus, for sufficiently small $\varepsilon_1 > 0$ we have $L(-\varepsilon_1) \ll 0$.

Furthermore, since the last left term in (5.38) is a non-negative operator for $\gamma \geq 0$, then

$$L(\varepsilon_2) \geq \varepsilon_2 I_0 - A_0^{-1} \geq \left(\varepsilon_2 - \|A_0^{-1}\|\right) I_0 \gg 0, \qquad \varepsilon_2 > \|A_0^{-1}\|. \quad (5.41)$$

Now we adjust the viscosity ν in such a way that on the segment $[-\varepsilon_1, \varepsilon_2]$ the condition $L'(\gamma) \gg 0$ is fulfilled. We have

$$L'(\gamma) = I_0 - \frac{2\dfrac{\gamma\omega_0^2}{\nu^2}}{\left(1 + \dfrac{\gamma^2\omega_0^2}{\nu^2}\right)^2} \left(\rho J_1^{-1}\right) A_0^{-1/2} I_{12} I_{21} A_0^{-1/2}. \quad (5.42)$$

For $\gamma \leq 0$ whence it follows that $L'(\gamma) \geq I_0 \gg 0$. For $\gamma > 0$, by virtue to the previous estimate for $\|I_{12}\|$ we obtain

$$L'(\gamma) \geq \left(1 - \frac{2\dfrac{\gamma\omega_0^2}{\nu^2}}{\left(1 + \dfrac{\gamma^2\omega_0^2}{\nu^2}\right)^2} \|A_0^{-1}\| \, q\right) I_0 \gg 0, \qquad 0 < \gamma \leq \varepsilon_2,$$

if the parameter ν is large enough.

Thus, on the chosen interval $[-\varepsilon_1, \varepsilon_2]$ the conditions (1.6.9) for the operator pencil $L(\gamma)$ are fulfilled, and from this, according to the statements of Section 1.6.10 it follows that the system of eigenelements $\{\boldsymbol{u}_k\}_{k=1}^{\infty}$ of problem (5.38) corresponding to eigenvalues $\gamma_k \in (-\varepsilon_1, \varepsilon_2)$, forms a Riesz basis and a p-basis with $p > 3/2$ in the space $\boldsymbol{J}_0(\Omega)$. In this, the eigenvalues $\lambda_k = \nu\gamma_k^{-1}$ satisfy the condition $\lambda_k > \nu\|A_0^{-1}\| = \nu\lambda_1(A_0)$.

The statement formulated at the beginning of Section 7.5.5 has been completely proved.

7.6 Problems on Fluids Flowing Through a Given Cavity

Here the problem on motion of a viscous fluid in a given region for a certain mode of flowing in and flowing out on parts of its boundary is considered in a linear statement.

7.6.1 The Basic Equations

Let us assume that the region Ω completely filled with a viscous incompressible fluid has two plane orifices Γ_1 and Γ_2 located in horizontal planes. Through the lower orifice Γ_1 the same fluid with a given small normal velocity $u_3|_{\Gamma_1} = \varphi(t,x),\ x \in \Gamma_1$, is flowing. Through the upper orifice Γ_2 the fluid flows out spontaneously.

Let us assume that on the upper orifice Γ_2 in the flowing out mode the tangent and normal stresses equal zero, and on the lower orifice Γ_1 the tangent stresses equal zero. Then, the problem on small movements of the fluid in the region Ω for the mentioned modes of flowing out and flowing in and a small field of external forces can be reduced to the following system of equations, boundary, and initial conditons:

$$\begin{aligned}
&\frac{\partial \boldsymbol{u}}{\partial t} = -\frac{1}{\rho}\nabla p + \nu\Delta\boldsymbol{u} + \boldsymbol{f}, \qquad \operatorname{div}\boldsymbol{u} = 0 \quad \text{in } \Omega,\\
&\boldsymbol{u} = \boldsymbol{0} \quad \text{on } S,\\
&\tilde{\tau}_{13}(\boldsymbol{u}) = \tilde{\tau}_{23}(\boldsymbol{u}) = \tilde{\tau}_{33}(\boldsymbol{u}) = 0 \quad \text{on } \Gamma_2,\\
&\tilde{\tau}_{13}(\boldsymbol{u}) = \tilde{\tau}_{23}(\boldsymbol{u}) = 0, \qquad u_3 = \varphi \quad \text{on } \Gamma_1,\\
&\boldsymbol{u}(0,x) = \boldsymbol{u}^0(x),\ \tilde{\tau}_{ij}(\boldsymbol{u}) := -p\delta_{ij} + \rho\nu\left(\frac{\partial u_i}{\partial x_j} + \frac{\partial u_j}{\partial x_i}\right).
\end{aligned} \tag{6.1}$$

Hence, a solution of the problem (6.1) is defined by the field $\boldsymbol{f} = \boldsymbol{f}(t,x)$, the function $\varphi = \varphi(t,x)$ and the initial velocity field $\boldsymbol{u}^0(x)$.

7.6.2 Application of the Abstract Scheme

A generalized solution $\boldsymbol{u}(t,x)$ of problem (6.1) can be naturally considered as a function of variable t with values in the space $\boldsymbol{J}^1_{0.S}(\Omega)$ of solenoidal fields from $\boldsymbol{H}^1(\Omega)$, which turn to zero on the solid boundary S. We consider two auxiliary boundary value problems within the abstract scheme from Section 1.8 and according to the method stated in Section 2.2.

Let us introduce the operator γ_n, acting from the space $\boldsymbol{J}^1_{0,S}(\Omega)$ to the space $L_2(\Gamma_1)$, $\gamma_n\boldsymbol{u} := u_3|_{\Gamma_1}$. The kernel $\operatorname{Ker}\gamma_n := \boldsymbol{N}_1(\Omega) \subset \boldsymbol{J}^1_{0,S}(\Omega)$ of this operator consists of solenoidal functions $\boldsymbol{u}$, for which $\boldsymbol{u} = \boldsymbol{0}$ on S and $u_n = 0$ on Γ_1. This set is dense in the space of functions $\boldsymbol{J}_{0.\partial\Omega\backslash\Gamma_2}(\Omega) \supset \boldsymbol{J}_0(\Omega)$, for which $\operatorname{div}\boldsymbol{u} = 0$ in Ω and $u_n = 0$ on $S \cup \Gamma_1 = \partial\Omega/\Gamma_2$. It makes it possible to consider the Hilbert pair $(\boldsymbol{N}_1(\Omega); \boldsymbol{J}_{0,\partial\Omega\backslash\Gamma_2}(\Omega))$. The generating operator L of this pair can be determined from the identity

$$E(\boldsymbol{u},\boldsymbol{v}) = \int_\Omega \boldsymbol{u}\cdot L\boldsymbol{v}\mathrm{d}\Omega, \qquad \boldsymbol{u}\in \boldsymbol{N}_1(\Omega),\ \boldsymbol{v}\in \mathcal{D}(L)\subset \boldsymbol{N}_1(\Omega). \tag{6.2}$$

For two times continuously differentiable fields $\boldsymbol{v}$ from $\boldsymbol{N}_1(\Omega)$ and $\boldsymbol{u}$ from $\boldsymbol{N}_1(\Omega)$, the Green formula has the form

$$\begin{aligned} E(\boldsymbol{u},\boldsymbol{v}) &= -\int_\Omega \boldsymbol{u}\cdot\Delta\boldsymbol{v}\mathrm{d}\Omega - \int_{\Gamma_1}\sum_{i=1}^{2} u_i\tau_{i3}(\boldsymbol{v})\mathrm{d}\Gamma_1 + \int_{\Gamma_2}\sum_{i=1}^{3} u_i\tau_{i3}(\boldsymbol{v})\mathrm{d}\Gamma_2,\\ \tau_{i3}(\boldsymbol{v}) &:= \frac{\partial v_i}{\partial x_3} + \frac{\partial v_3}{\partial x_i}. \end{aligned} \tag{6.3}$$

Then from (6.2) it follows that

$$\int_\Omega \boldsymbol{u}\cdot L\boldsymbol{v}\mathrm{d}\Omega = -\int_\Omega \boldsymbol{u}\cdot\Delta\boldsymbol{v}\mathrm{d}\Omega - \int_{\Gamma_1}\sum_{i=1}^{2} u_i\tau_{i3}(\boldsymbol{v})\mathrm{d}\Gamma_1 + \int_{\Gamma_2}\sum_{i=1}^{3} u_i\tau_{i3}(\boldsymbol{v})\mathrm{d}\Gamma_2.$$

Repeating the reasoning performed in Section 2.2.7 we come to the conclusion that the self-adjoint positive definite operator L, acting in $\boldsymbol{J}_{0,\partial\Omega\setminus\Gamma_2}(\Omega)$ and defined in this space on a dense set $\mathcal{D}(L)$, is given by the equality

$$L\boldsymbol{v} := -\Delta\boldsymbol{v} + \nabla q \tag{6.4}$$

for fields $\boldsymbol{v}\in\mathcal{D}(L)\subset \boldsymbol{J}^1_{0,S}(\Omega)$ that also satisfy the boundary conditions

$$\begin{aligned} \tau_{i3}(\boldsymbol{v}) &= 0 \quad \text{on } \Gamma_2, \qquad i=1,2,\\ -q+\tau_{33}(\boldsymbol{v}) &= 0 \quad \text{on } \Gamma_2,\\ \tau_{i3}(\boldsymbol{v}) &= 0 \quad \text{on } \Gamma_1, \qquad i=1,2,\\ v_3 &= 0 \quad \text{on } \Gamma_1. \end{aligned} \tag{6.5}$$

Now let us find out what is the orthogonal complement $\boldsymbol{M}_1(\Omega)$ in $\boldsymbol{J}^1_{0,S}$ of the kernel $\boldsymbol{N}_1(\Omega)$ of operator γ_n. From Green formula for $\boldsymbol{w}\in\boldsymbol{M}_1(\Omega)$ and $\boldsymbol{v}\in\boldsymbol{N}_1(\Omega)$ we have

$$0 = E(\boldsymbol{w},\boldsymbol{v}) = -\int_\Omega \Delta\boldsymbol{w}\cdot\boldsymbol{v}\mathrm{d}\Omega - \int_{\Gamma_1}\sum_{i=1}^{2}\tau_{i3}(\boldsymbol{w})v_i\mathrm{d}\Gamma_1 + \int_{\Gamma_2}\sum_{i=1}^{3}\tau_{i3}(\boldsymbol{w})v_i\mathrm{d}\Gamma_2.$$

Hence, we obtain that the smooth fields $\boldsymbol{w}$ from $\boldsymbol{M}_1(\Omega)$ satisfy the following equations and boundary conditions,

$$
\begin{aligned}
-\Delta \boldsymbol{w} + \nabla p = \mathbf{0}, &\qquad \operatorname{div} \boldsymbol{w} = 0 \quad \text{in } \Omega,\\
\boldsymbol{w} = \mathbf{0} &\quad \text{on } S,\\
\tau_{i3}(\boldsymbol{w}) = 0 &\quad \text{on } \Gamma_1, \quad i = 1, 2,\\
\tau_{i3}(\boldsymbol{w}) = 0 &\quad \text{on } \Gamma_2, \quad i = 1, 2,\\
-p + \tau_{33}(\boldsymbol{w}) = 0 &\quad \text{on } \Gamma_2.
\end{aligned}
$$

Since operator γ_n establishes a one-to-one correspondence between $\boldsymbol{M}_1(\Omega)$ and $H^{1/2}(\Gamma_1)$, then operator γ_n^{-1} gives the solution of the boundary value problem

$$
\begin{aligned}
&-\Delta \boldsymbol{w} + \nabla p = \mathbf{0}, \qquad \operatorname{div} \boldsymbol{w} = 0 \quad \text{in } \Omega,\\
&\boldsymbol{w} = \mathbf{0} \quad \text{on } S, \qquad \tau_{i3}(\boldsymbol{w}) = 0 \quad \text{on } \Gamma_1,\ i = 1, 2;\\
&w_3 = \psi \quad \text{on } \Gamma_1, \qquad \tau_{i3}(\boldsymbol{w}) = 0 \quad \text{on } \Gamma_2,\ i = 1, 2;\\
&-p + \tau_{33}(\boldsymbol{w}) = 0 \quad \text{on } \Gamma_2.
\end{aligned} \tag{6.6}
$$

7.6.3 Transition to Operator Equations in Orthogonal Subspaces

We consider problem (6.1). By a given function $\varphi(t, x)$ let us find a solution $\boldsymbol{w}(t, x)$ of the problem of the form (6.6), which belongs to the space $M_1(\Omega)$,

$$
\begin{aligned}
&-\nu \Delta \boldsymbol{w} + \frac{1}{\rho} \nabla p_1 = \mathbf{0}, \qquad \operatorname{div} \boldsymbol{w} = 0 \quad \text{in } \Omega,\\
&\boldsymbol{w} = \mathbf{0} \quad \text{on } S, \qquad \tilde{\tau}_{i3}(\boldsymbol{w}) = 0 \quad \text{on} \Gamma_1,\ i = 1, 2;\\
&w_3 = \varphi \quad \text{on } \Gamma_1, \qquad \tilde{\tau}_{i3}(\boldsymbol{w}) = 0 \quad \text{on } \Gamma_2,\\
&\tilde{\tau}_{33}(\boldsymbol{w}) \equiv -p_1 + \rho\nu\tau_{33}(\boldsymbol{w}) = 0 \quad \text{on } \Gamma_2
\end{aligned} \tag{6.7}
$$

Then

$$
\boldsymbol{w}(t, x) = \gamma_n^{-1} \varphi(t, x). \tag{6.8}
$$

We subtract the right and left sides of the obtained relations (6.7) from the corresponding sides of the equations and boundary conditions of problem (6.1); we have

$$
\begin{aligned}
\frac{\partial \boldsymbol{u}}{\partial t} &= -\frac{1}{\rho} \nabla \left(p - p_1\right) + \nu \Delta (\boldsymbol{u} - \boldsymbol{w}) + \boldsymbol{f},\\
\operatorname{div} (\boldsymbol{u} - \boldsymbol{w}) &= 0 \quad \text{in } \Omega, \qquad \boldsymbol{u} - \boldsymbol{w} = \mathbf{0} \quad \text{on } S,\\
\tilde{\tau}_{i3}(\boldsymbol{u} - \boldsymbol{w}) &= 0 \quad \text{on } \Gamma_1 \cup \Gamma_2,\ i = 1, 2;\\
-\left(p - p_1\right) + \rho\nu\tau_{33}(\boldsymbol{u} - \boldsymbol{w}) &= 0 \quad \text{on } \Gamma_2,\\
u_3 - w_3 &= 0 \quad \text{on } \Gamma_1.
\end{aligned} \tag{6.9}
$$

Let us introduce the notations $\boldsymbol{u} - \boldsymbol{w} =: \boldsymbol{v}$ and $p - p_1 =: p_2$. Then from (6.9), (6.4) and (6.5) it follows that $\boldsymbol{v}(t,x)$ can be considered as a function of t with values in $\boldsymbol{N}_1(\Omega)$ and the problem (6.9) can be considered as an abstract equation

$$\frac{\mathrm{d}\boldsymbol{v}}{\mathrm{d}t} = -\nu L\boldsymbol{v} + P_{0.\partial\Omega\setminus\Gamma_2}\left(\boldsymbol{f} - \frac{\mathrm{d}\boldsymbol{w}}{\mathrm{d}t}\right) \tag{6.10}$$

in the space $\boldsymbol{J}_{0,\partial\Omega\setminus\Gamma_2}(\Omega)$. Here, L is the operator of the Hilbert pair $(\boldsymbol{N}_1(\Omega);\ \boldsymbol{J}_{0,\partial\Omega\setminus\Gamma_2}(\Omega))$, $\boldsymbol{w} = \boldsymbol{w}(t)$ is the function (6.8), and $P_{0.\partial\Omega\setminus\Gamma_2}$ is the orthoprojector onto $\boldsymbol{J}_{0,\partial\Omega\setminus\Gamma_2}$.

Hence, problem (6.1) is reduced to the relation (6.8) and the Cauchy problem

$$\begin{aligned} \frac{\mathrm{d}\boldsymbol{v}}{\mathrm{d}t} &= -\nu L\boldsymbol{v} + P_{0.\partial\Omega\setminus\Gamma_2}\left(\boldsymbol{f} - \frac{\mathrm{d}\boldsymbol{w}}{\mathrm{d}t}\right), \\ \boldsymbol{v}(0) &= -\boldsymbol{u}(0) - \boldsymbol{w}(0) = \boldsymbol{u}^0 - \gamma_n^{-1}\varphi(0,x) \end{aligned} \tag{6.11}$$

in the subspace $\boldsymbol{J}_{0,\partial\Omega\setminus\Gamma_2}$.

7.6.4 Theorem on Existence of a Generalized Solution

For $t > 0$, a solution $\boldsymbol{v}$ of the equation (6.11) belongs to the domain of definition $\mathcal{D}(L) \subset \boldsymbol{N}_1(\Omega)$ of operator L and therefore, satisfies all boundary conditions (6.9). This solution exists, if the field $\boldsymbol{v}(t,x) - \mathrm{d}\boldsymbol{w}/\mathrm{d}t$ satisfies by t a Hölder condition in the norm of the space $\boldsymbol{L}_2(\Omega)$.

Let us assume that the initial and boundary conditions of the problem (6.1) are coordinated on Γ_1, that is,

$$u_3^0(x) = \varphi(0,x), \qquad x \in \Gamma_1. \tag{6.12}$$

Then in (6.11) we have $\boldsymbol{v}(0) \in \boldsymbol{J}_{0,\partial\Omega\setminus\Gamma_2}(\Omega)$.

For weaker limitations we obtain the following conclusion.

If $\varphi(t,x)$ is a continuous function of variable t with values in $H^{1/2}(\Gamma_1)$, the field $\boldsymbol{f}(t,x)$ is a continuous function of t with values in $\boldsymbol{L}_2(\Omega), \boldsymbol{u}^0 \in \boldsymbol{J}_{0,S}(\Omega)$, and the coordination condition (6.12) holds true, then problem (6.1) has the unique generalized solution $\boldsymbol{u}(t,x)$ that, in operator form, can be written down as

$$\begin{aligned} \boldsymbol{u}(t) &= \exp(-\nu t L)\,\boldsymbol{v}(0) + \int_0^t \exp(-\nu(t-s)L)\, P_{0.\partial\Omega\setminus\Gamma_2}\left(\boldsymbol{f}(s) - \frac{\mathrm{d}\boldsymbol{w}}{\mathrm{d}s}\right)\mathrm{d}s + \boldsymbol{w}(t), \\ \boldsymbol{w}(t) &= \gamma_n^{-1}\varphi(t), \\ \boldsymbol{v}(0) &= \boldsymbol{u}^0 - \gamma_n^{-1}\varphi(0). \end{aligned} \tag{6.13}$$

In conclusion, let us note that along with the boundary conditions on Γ_1 and Γ_2 taking place in (6.1), other boundary conditions are possible; for example, the output mode $u_3|_{\Gamma_2} = \psi$, or some tangent stresses on these boundaries are given. In this case, a modification of the abstract scheme will be required according to the general method stated in Section 1.8. Such variants can be considered independently by the reader.

7.7 Convective Movements of Fluids in a Closed Cavity

The methods developed in the previous sections of this chapter can be applied to the problem on small movements and normal oscillations of a nonuniformly heated fluid that completely fills a certain volume. Here, we consider two important cases that admit a state of mechanical equilibrium: the cases of fluid heating from above, and from below.

7.7.1 Equations of Thermal Convection

Let us assume that a nonuniformly heated viscous fluid fills some immovable container Ω, and is influenced by an external gravitational field with acceleration $\boldsymbol{g} = -g\boldsymbol{e}_3$. As it is usually assumed in such problems, the fluid is supposed to be incompressible, that is, the velocity field $\boldsymbol{u}(t,x)$ is solenoidal. Further, the density of the fluid $\tilde{\rho}(t,x)$ depends only on modifications of the temperature field, that is, the deviation $T(t,x)$ of the temperature from some mean value T_m =const. For small $T(t,x)$, this dependence is a linear function,

$$\tilde{\rho}(t,x) = \rho\,(1 - \beta T(t,x))\,, \tag{7.1}$$

where $\rho = \rho_m$ = const is the density corresponding to the temperature T_m, and $\beta > 0$ is the coefficient of thermal extension.

To describe convective movements in the nonuniformly heated fluid, usually the equations of fluid movement and equations of heat translation are used in the following form, which is called Boussinesq approximation,

$$\begin{aligned} &\frac{\partial \boldsymbol{u}}{\partial t} + (\boldsymbol{u}\cdot\nabla)\boldsymbol{u} = -\frac{1}{\rho}\nabla \boldsymbol{P} + \nu\Delta\boldsymbol{u} + g\beta T\boldsymbol{e}_3, \\ &\operatorname{div}\boldsymbol{u} = 0, \qquad \frac{\partial T}{\partial t} + \boldsymbol{u}\cdot\nabla T = \chi\Delta T \quad \text{in } \Omega. \end{aligned} \tag{7.2}$$

Here $\chi > 0$ is the coefficient of temperature conductivity, and $P(t,x)$ is the deviation of pressure from the hydrostatic pressure, corresponding to the constant temperature T_m.

While deducing equations (7.2) it has been taken into account that convection is sufficiently weak, that is, the deviations of fluid's density from its mean value ρ_m, caused by the fact that the temperature is not homogeneous, are so small, that they can be omitted in all equations except for the equations of motion, in which the mentioned deviation is taken into account only in the last term of the form $g\beta T_m$.

7.7.2 Conditions of Mechanical Equilibrium

There exist such special conditions of fluid heating, that the considered system can be in the state of mechanical equilibrium, that is, to have zero velocity field. But in this case there is no thermodynamic equilibrium in the fluid, because the space nonhomogeneity of temperature leads to the appearance of a thermal flow.

To find out the conditions of mechanical equilibrium, let us assume in (7.2) that $\boldsymbol{u}(t,x) \equiv \mathbf{0}$, $P = p_0(x)$, $T = T_0(x)$; we obtain

$$-\frac{1}{\rho}\nabla p_0 + g\beta T_0 \boldsymbol{e}_3 = \mathbf{0}, \quad \Delta T_0 = 0 \qquad \text{in } \Omega. \tag{7.3}$$

Hence, it follows that $\partial p_0/\partial x_i = 0$, $i = 1,2$ and therefore $p_0 = p_0(x_3)$. Then also $T_0 = T_0(x_3)$ and from the second equation one obtains $\mathrm{d}^2 T_0/\mathrm{d}x_3^2 = 0$, that is, T_0 is a linear function of the form

$$T_0 = -\alpha x_3 + \alpha_0, \tag{7.4}$$

where α and α_0 are some constants. As far as axis Ox_3 is directed upward, then $\alpha > 0$ corresponds to linear decreasing, and $\alpha < 0$ corresponds to linear increasing of temperature with height. Hence, for mechanical equilibrium, the gradient of temperature is vertical and has constant value: $\nabla T_0 = -\alpha \boldsymbol{e}_3$.

Now let us obtain boundary conditions on the boundary $S = \partial\Omega$ of the region Ω. As far as cavity Ω is immovable, then on S for the velocity field $\boldsymbol{u}(t,x)$ the ordinary stickiness condition should be valid: $\boldsymbol{u} = \mathbf{0}$ on S. As for boundary conditions for temperature, we assume that on the walls of cavity Ω there is an unchangeable equilibrium temperature distribution $T_0 = T_0(x_3)$, which is defined by the formula (7.4), and therefore the deviation $\theta = \theta(t,x)$ from the equilibrium temperature $T_0(x_3)$ on S disappears: $\theta = 0$ on S. Such a boundary condition corresponds to the case when the cavity Ω is surrounded by a rigid mass with a higher heat conductivity than in fluid.

7.7.3 Final Statement of the Problem

Now we give the full statement of the linearized initial boundary value problem on small convective movements of a fluid completely filling the cavity Ω. Let us assume that in the nonuniformly heated fluid mechanical equilibrium can exist; this state corresponds to temperature $T_0(x_3)$ and pressure $p_0(x_3)$, which can be determined by the formulas (7.4) and (7.3). Looking for solutions of the problem (7.2) in the form

$$\boldsymbol{u} = \boldsymbol{u}(t,x), \quad P = p_0(x_3) + p(t,x), \quad T = T_0(x_3) + \theta(t,x), \tag{7.5}$$

and assuming that the functions $\boldsymbol{u}, p$ are θ have small magnitudes of the first order, we obtain the following equations from (7.2),

$$\begin{aligned} &\frac{\partial \boldsymbol{u}}{\partial t} = -\frac{1}{\rho}\nabla p + \nu\Delta\boldsymbol{u} + g\beta\theta\boldsymbol{e}_3, \\ &\operatorname{div}\boldsymbol{u} = 0, \\ &\frac{\partial\theta}{\partial t} - \alpha\boldsymbol{u}\cdot\boldsymbol{e}_3 = \chi\Delta\theta \qquad \text{in } \Omega, \end{aligned} \tag{7.6}$$

and the following initial and boundary conditions

$$\boldsymbol{u} = \boldsymbol{0}, \qquad \theta = 0 \quad \text{on } S, \tag{7.7}$$

$$\boldsymbol{u}(0,x) = \boldsymbol{u}^0(x), \qquad \theta(0,x) = \theta^0(x). \tag{7.8}$$

Here the constant α, caused by the dependence (6.4), characterizes heating conditions for mechanical equilibrium: $\alpha > 0$ for heating from below, and $\alpha < 0$ for heating from above.

7.7.4 Transition to an Operator Equation

Equations (7.6) and the stickiness condition (7.7) for the field $\boldsymbol{u}(t,x)$ show that it can be considered as a function with values in the space $\boldsymbol{J}_0(\Omega)$. That is why, applying orthoprojector P_0 to the first equation (7.6), we obtain

$$\frac{d\boldsymbol{u}}{dt} + \nu A_0\boldsymbol{u} - g\beta P_0\left(\theta\boldsymbol{e}_3\right) = \boldsymbol{0}, \tag{7.9}$$

where A_0 is the Stokes operator.

Further, let us introduce the pair of Hilbert spaces $(H_0^1(\Omega); L_2(\Omega))$ and the corresponding generating operator A_1; the squared norm in $H_0^1(\Omega)$ is defined as a Dirichlet integral. As it follows from the considerations of Section 1.8, A_1 is the operator of the boundary value problem

$$A_1\theta := -\Delta\theta = \varphi \quad \text{in } \Omega, \qquad \theta = 0 \quad \text{on } S; \tag{7.10}$$

it is positive definite and self-adjoint in $L_2(\Omega)$, and $\mathcal{D}(A_1^{1/2}) = H_0^1(\Omega)$. Its eigenvalues $\{\lambda_k(A_1)\}_{k=1}^{\infty}$ form a discrete spectrum and have the following asymptotic behavior

$$\lambda_k\left(A_1\right)) = c_{A_1}^{-2/3}k^{2/3}[1 + o(1)], \qquad c_{A_1} = \operatorname{mes}\Omega/(6\pi^2) > 0; \tag{7.11}$$

from the latter it follows that the inverse operator A_1^{-1} is compact and belongs to the class $\mathfrak{S}_p$ for $p > 3/2$.

With regard to above mentioned facts, the Dirichlet condition (7.7) for θ, and (7.10), the second equations (7.6) can be written down in the form

$$\frac{\mathrm{d}\theta}{\mathrm{d}t} + \chi A_1\theta - \alpha\overline{\boldsymbol{u}}\cdot\boldsymbol{e}_3 = 0. \tag{7.12}$$

Further, let us assume that parameter α, which characterizes the temperature distribution in the state of mechanical equilibrium, does not equal zero. (For $\alpha = 0$, the initial problem splits into two independent problems for determining the field $\theta(t,x)$; the field $\boldsymbol{u}(t,x)$ is defined by $\theta(t,x)$.) It allows us to make the following substitution in (7.9) and (7.12),

$$\theta(t,x) = \left(\frac{|\alpha|}{\beta g}\right)^{1/2} v(t,x), \tag{7.13}$$

which leads to the system of equations

$$\frac{\mathrm{d}}{\mathrm{d}t}\begin{pmatrix}\boldsymbol{u}(t)\\ v(t)\end{pmatrix} + \begin{pmatrix}\nu A_0 & -\varepsilon C\\ \mp\varepsilon C^* & \chi A_1\end{pmatrix}\begin{pmatrix}\boldsymbol{u}(t)\\ v(t)\end{pmatrix} = \begin{pmatrix}\boldsymbol{0}\\ 0\end{pmatrix}, \tag{7.14}$$

where the transition to nondimensional variables has been already performed and the following notations are introduced,

$$\varepsilon = \sqrt{|\alpha| g\beta} > 0, \quad Cv := P_0\left(v\boldsymbol{e}_3\right), \quad C^*\boldsymbol{u} := \boldsymbol{u}\cdot\boldsymbol{e}_3; \tag{7.15}$$

to the case "minus" there corresponds heating from below ($\alpha > 0$), to the case "plus" —heating from above ($\alpha < 0$).

Relations (7.14) can be naturally considered as a differential equation of the form

$$\frac{\mathrm{d}y}{\mathrm{d}t} + \mathcal{A}_{\mp}y = 0, \qquad y = (\boldsymbol{u}; v)^{\mathrm{t}}, \tag{7.16}$$

in the space $\boldsymbol{J}_0(\Omega) \oplus L_2(\Omega)$, where operator $\mathcal{A}_{\mp}$ is defined by the matrix (7.14). Let us note that operators C and C^* from (7.15) are mutually adjoint: for any $\boldsymbol{u}$ from $\boldsymbol{J}_0(\Omega)$ and v from $L_2(\Omega)$,

$$\begin{aligned}(\boldsymbol{u}, Cv)_{\boldsymbol{J}_0(\Omega)} &= \int_\Omega \boldsymbol{u}\cdot P_0\left(v\boldsymbol{e}_3\right)\mathrm{d}\Omega = \int_\Omega P_0\boldsymbol{u}\cdot\left(v\boldsymbol{e}_3\right)\mathrm{d}\Omega\\ &= \int_\Omega \boldsymbol{u}\cdot\left(v\boldsymbol{e}_3\right)\mathrm{d}\Omega = \int_\Omega \left(\boldsymbol{u}\cdot\boldsymbol{e}_3\right)v\,\mathrm{d}\Omega = \left(C^*\boldsymbol{u}, v\right)_{L_2(\Omega)}.\end{aligned}$$

From the latter we obtain the inequalities

$$\left|(\boldsymbol{u}, Cv)_{\boldsymbol{J}_0(\Omega)}\right| = \left|(C^*\boldsymbol{u}, v)_{L_2(\Omega)}\right| \le \|\boldsymbol{u}\|_{\boldsymbol{J}_0(\Omega)}\cdot\|v\|_{L_2(\Omega)}, \tag{7.17}$$

which shows that these operators are bounded and

$$\|C\| = \|C^*\| \le 1. \tag{7.18}$$

7.7.5 Solvability of the Initial Boundary Value Problem

From the properties of Stokes operator A_0 and operator A_1 it follows that operator $\mathcal{A}_0 :=\text{diag}\,(\nu A_0; \chi A_1)$ is a generating operator of the analytic semigroup $\exp(-t\mathcal{A}_0)$. Since operators C and C^* from (7.14) are bounded, then operator $\mathcal{A}_{\mp}$ is also a generating operator of the analytic semigroup $\exp(-t\mathcal{A}_{\mp})$. That is why the solution of equation (7.16) is expressed by the formula

$$y(t) = \exp\,(-t\mathcal{A}_{\mp})\,y(0), \qquad y(0) = \left(\boldsymbol{u}^0; (g\beta/|\alpha|)^{1/2}\theta^0\right)^{\text{t}}. \tag{7.19}$$

Hence, it appears that if $\boldsymbol{u}^0 \in \boldsymbol{J}_0(\Omega)$, and $\theta^0(x) \in L_2(\Omega)$, then the initial boundary value problem (7.6)–(7.8) is univalently solvable, its weak solution can be found by formula (7.19), where for $t > 0$ the function $\boldsymbol{u}(t,x)$ belongs to the space $\boldsymbol{J}_0^1(\Omega)$ and $\theta(t,x)$ belongs to $H_0^1(\Omega)$.

7.7.6 Normal Movements of a System Heated from Below

Let us consider solutions of the problem (7.16) that depend on time according to the law $\exp(-\lambda t)$; we obtain

$$\mathcal{A}_{\mp} y = \lambda y, \tag{7.20}$$

where operator $\mathcal{A}_{\mp}$ is defined by the matrix (7.14). First, let us assume that heating from below is being performed, that is, $\alpha > 0$ and therefore, in (7.20) the sign "minus" takes place, that is,

$$\mathcal{A}_{-} = \begin{pmatrix} \nu A_0 & -\varepsilon C \\ -\varepsilon C^* & \chi A_1 \end{pmatrix}. \tag{7.21}$$

Since C, C^* are mutually adjoint bounded operators, and A_0, A_1 are unbounded positive definite operators acting in $\boldsymbol{J}_0(\Omega)$, $L_2(\Omega)$, respectively, and having compact inverse operators A_0^{-1}, A_1^{-1} from $\mathfrak{S}_p$ for $p > 3/2$, then operator $\mathcal{A}_{-}$ from (7.21) is an unbounded self-adjoint and bounded below operator acting in the space $\boldsymbol{J}_0(\Omega)\oplus L_2(\Omega)$. If $\mathcal{A}_{-}$ is invertible, then the inverse operator $(\mathcal{A}_{-})^{-1}$ is a compact self-adjoint operator from the class $\mathfrak{S}_p$ for $p > 3/2$.

From these properties of operator $\mathcal{A}_{-}$ it follows that the problem (7.20)–(7.21) has a real discrete spectrum $\{\lambda_k^-\}_{k=1}^{\infty}$, which is located on the positive halfaxis, except for a possible finite number of nonpositive eigenvalues, and has the limit point $\lambda = +\infty$. The corresponding eigenelements $\{y_k^-\}_{k=1}^{\infty}$, $y_k^- = (\boldsymbol{u}_k^-; v_k^-)^{\text{t}}$ form an orthonormal basis in the Hilbert space $\boldsymbol{J}_0(\Omega)\oplus L_2(\Omega)$ and the orthogonality conditions hold true,

$$(\mathcal{A}_{-} y_k^-, y_l^-) = \nu \left(A_0^{1/2}\boldsymbol{u}_k^-, A_0^{1/2}\boldsymbol{u}_l^-\right)_{\boldsymbol{J}_0(\Omega)} + \chi \left(A_1^{1/2} v_k^-, A_1^{1/2} v_l^-\right)_{L_2(\Omega)}$$

$$
\begin{aligned}
&- \varepsilon \left[\left(C v_k^-, \boldsymbol{u}_l^- \right)_{J_0(\Omega)} + \left(v_l^-, C^* \boldsymbol{u}_k^- \right)_{L_2(\Omega)} \right] \\
&= \nu \int_\Omega \operatorname{rot} \boldsymbol{u}_k^- \cdot \operatorname{rot} \boldsymbol{u}_l^- \mathrm{d}\Omega + \chi \int_\Omega \nabla v_k^- \cdot \nabla v_l \mathrm{d}\Omega - 2\varepsilon \mathrm{Re} \int_\Omega \left(\boldsymbol{u}_k^- \cdot \boldsymbol{e}_3 \right) v_l^- \mathrm{d}\Omega \\
&= \lambda_k^- \left(y_k^-, y_l^- \right) = \lambda_k \left(\int_\Omega \boldsymbol{u}_k^- \cdot \boldsymbol{u}_l^- \mathrm{d}\Omega + \int_\Omega v_k^- v_l^- \mathrm{d}\Omega \right) = \lambda_k^- \delta_{kl}.
\end{aligned} \tag{7.22}
$$

The eigenvalues $\{\lambda_k^-\}_{k=1}^\infty$ of problem (7.20) are consecutive minima of the variational ratio

$$
\frac{\nu \int_\Omega |\operatorname{rot} \boldsymbol{u}|^2 \mathrm{d}\Omega + \chi \int_\Omega |\nabla v|^2 \mathrm{d}\Omega - 2\varepsilon \mathrm{Re} \int_\Omega u_3 v \mathrm{d}\Omega}{\int_\Omega |\boldsymbol{u}|^2 \mathrm{d}\Omega + \int_\Omega |v|^2 \mathrm{d}\Omega}. \tag{7.23}
$$

Since

$$
\mathcal{A}_- = \mathcal{A}_0 - \varepsilon V, \quad \mathcal{A}_0 = \operatorname{diag}\left(\nu A_0; \chi A_1 \right), \quad V = \begin{pmatrix} 0 & C \\ C^* & 0 \end{pmatrix} = V^*, \tag{7.24}
$$

and conditions (7.18) hold true, then from (7.23), using the minimax principle for eigenvalues, we obtain that

$$
\lambda_k \left(\mathcal{A}_0 \right) - \varepsilon \le \lambda_k^- \le \lambda_k \left(\mathcal{A}_0 \right) + \varepsilon, \qquad k = 1, 2, \ldots. \tag{7.25}
$$

If the following condition is valid,

$$
\lambda_1 \left(\mathcal{A}_0 \right) = \min \left\{ \nu \lambda_1 \left(A_0 \right); \chi \lambda_1 \left(A_1 \right) \right\} > \varepsilon, \tag{7.26}
$$

then the minimal eigenvalue of problem (7.20) for $\alpha > 0$ is positive, and that is why all normal oscillations of the system are nonperiodically fading modes. For increasing ε (the value of gradient of temperature T_0 for heating from below) eigenvalues $\lambda_k^- = \lambda_k(\mathcal{A}_-)$ continuously change, but do not move from the real axis. Here, if $\lambda_1(\mathcal{A}_-) > 0$, then all perturbations monotonically fade with time; if $\lambda_1(\mathcal{A}_-) \le \cdots \le \lambda_\kappa(\mathcal{A}_-) < 0$ and $\lambda_{\kappa+1}(\mathcal{A}_-) > 0$, then the first κ normal movements of the system monotonically increase with time, and other movements monotonically fade.

Hence, for heating from below the "principle of perturbations monotony" holds true. In particular, if for some critical value $\varepsilon = \varepsilon_*$ the condition $\lambda_1(\mathcal{A}_-) = 0$ is valid, and for $\varepsilon > \varepsilon_*$ this eigenvalue moves to the left half-plane (and that is why a stable system becomes unstable), then it is said that the"principle of stability change (or stability inversion)" takes place.

From the inequalities (7.25) and asymptotic formulas for eigenvalues of operators A_0 and A_1 we obtain that

$$\lambda_k(\mathcal{A}_0) = C_{\mathcal{A}_0}^{-2/3} k^{2/3}[1+o(1)], \qquad k \to \infty,$$
$$C_{\mathcal{A}_0} = \nu^{2/3}\frac{\operatorname{mes}\Omega}{3\pi^2} + \chi^{2/3}\frac{\operatorname{mes}\Omega}{6\pi^2} > 0, \tag{7.27}$$

and this makes it possible to estimate the eigenvalues $\lambda_k^- = \lambda_k(\mathcal{A}_-)$ for large k (in virtue of (7.25)).

Hence, for heating from below, nonperiodic eigenmodes with the above mentioned properties appear in this system. These modes can be naturally called dissipatively thermal waves. If the condition (7.26) holds true, that is, the heating from below is sufficiently small, then the system remains stable. For large gradients of equilibrium temperature, when $\lambda_1(\mathcal{A}_-) < 0$, exponentially increasing eigenmodes appear in the system, and the initial state of mechanical equilibrium becomes unstable.

7.7.7 Normal Oscillations for Heating From Above

Now let us consider the problem (7.20) for $\alpha < 0$, that is, the case of heating from above. Then the following representation takes place

$$\mathcal{A}_+ = \mathcal{A}_0 - \varepsilon \mathcal{J} V, \qquad \mathcal{J} = \operatorname{diag}(I_0; -I_1), \tag{7.28}$$

where the matrix operator V was defined in (7.24). The representation

$$\mathcal{A}_+ = \mathcal{A}_0 + i\varepsilon \tilde{V}, \qquad \tilde{V} = \begin{pmatrix} 0 & \mathrm{i}C \\ -\mathrm{i}C^* & 0 \end{pmatrix} = \tilde{V}^* \tag{7.29}$$

is also valid. From (7.29) it follows that the eigenvalues λ of problem (7.20) are located in the halfband

$$\operatorname{Re}\lambda \geq \lambda_1(\mathcal{A}_0) > 0, \qquad |\operatorname{Im}\lambda| \leq \varepsilon. \tag{7.30}$$

Indeed, for normalized solutions of the problem (7.20) we have

$$(\mathcal{A}_0 y, y) + \mathrm{i}\varepsilon\left(\tilde{V}y, y\right) = \operatorname{Re}\lambda + \mathrm{i}\operatorname{Im}\lambda; \tag{7.31}$$

from (7.31) the inequalities (7.30) follow, if we take into account that $\|\tilde{V}\| \leq 1$ in virtue of the inequalities (7.18).

Now let us prove the discreteness of the spectrum of problem (7.20) for $\alpha < 0$ and the property on completeness of its eigen- and associated elements. In the equation

$$\mathcal{A}_+ y := \left(\mathcal{A}_0 + \mathrm{i}\varepsilon\tilde{V}\right) y = \lambda y \tag{7.32}$$

let us perform the substitution

$$\mathcal{A}_0 y = w. \tag{7.33}$$

Then for w we have the equation

$$\left(\mathcal{I} + \mathrm{i}\varepsilon\tilde{V}\mathcal{A}_0^{-1} - \lambda\mathcal{A}_0^{-1}\right) w = 0, \tag{7.34}$$

to which the general results stated in Section 1.6.4 are applicable. Indeed, $\mathcal{A}_0^{-1}$ is a compact positive operator from the class $\mathfrak{S}p$ for $p > 3/2$ and operator $\tilde{V}\mathcal{A}_0^{-1}$ is also compact. Besides, operator $I + \mathrm{i}\varepsilon\tilde{V}\mathcal{A}_0^{-1}$ is invertible. From the latter and the two Keldysh theorems it follows that the problem (7.34) and (7.32) have a discrete spectrum $\{\lambda_k^+\}_{k=1}^{\infty}$, which is located in the halfband (7.30), and a system of eigen- and associated elements $\{w_{k,q}\}_{k=1}^{\infty}$, complete in the space $\boldsymbol{J}_0(\Omega)\oplus L_2(\Omega)$. Hence it appears that the set $\{y_{k,q}^+\}_{k=1}^{\infty} = \{\boldsymbol{u}_{k,q}^+; v_{k.q}^+\}_{k=1}^{\infty} = \{\mathcal{A}_0^{-1}w_{k,q}\}_{k=1}^{\infty}$ of eigen- and associated elements of the problen (7.32) is complete in $\boldsymbol{J}_0(\Omega) \oplus L_2(\Omega)$, and moreover, it is complete in the norm

$$\|\mathcal{A}_0 y\|^2 = \|\nu A_0 \boldsymbol{u}\|^2_{\boldsymbol{J}_0(\Omega)} + \|\chi A_1 v\|^2_{L_2(\Omega)},$$

that is, in the graph norm of operator $\mathcal{A}_0$.

Further, from the asymptotic formula (7.27) and assertions from Section 1.6.8, it follows that for the eigenvalues of problem (7.34) the following asymptotic formula holds true,

$$\lambda_k^+ = \lambda_k(\mathcal{A}_0)\,[1 + \mathrm{o}(1)], \qquad k \to \infty. \tag{7.35}$$

Let us note that in virtue of the representation (7.28), operator $\mathcal{A}_+$ is $\mathcal{J}$-selfadjoint, because

$$\mathcal{J}\mathcal{A}_+ = \mathrm{diag}\,(\nu A_0; -\chi A_1) - \varepsilon V = (\mathcal{J}\mathcal{A}_+)^*.$$

Therefore, the spectrum of problem (7.32) is located symmetrically relatively to the real axis, and the eigenelements y that correspond to nonreal eigenvalues λ are neutral, that is, for them $(\mathcal{J}y, y) = 0$.

Hence, for heating from above the solutions of problem (7.32) on normal oscillations are dissipatively thermal waves, for which the complex fading decrements $\{\lambda_k^+\}_{k=1}^{\infty}$ are located in the band (7.30). Here, along with nonperiodically fading normal modes the oscillative fading modes with frequencies $|\mathrm{Im}\lambda_k^+|$, which are not bigger than ε, are possible. From the latter and conclusions of Section 7.7.6 it follows that the cases of heating from below ($\alpha > 0$) and heating from above ($\alpha < 0$) are completely different physically. In particular, for heating from above the system remains stable ($\mathrm{Re}\lambda_k^+ \geq \lambda_1(\mathcal{A}_0) > 0$), and for heating from below it can be unstable ($\lambda_1(\mathcal{A}_-) < 0$).

7.7.8 On Transition of Eigenvalues to the Left Half-Plane in the Case of Heating from Below

Let us consider in details the mechanism of transition of eigenvalues λ into the left complex half-plane in the problem from Section 7.7.6:

$$\begin{aligned} &\nu A_0 \boldsymbol{u} - \varepsilon C\theta = \lambda \boldsymbol{u}, &&\varepsilon = (\alpha g\beta)^{1/2} > 0, \\ &-\varepsilon C^* \boldsymbol{u} + \chi A_1 \theta = \lambda\theta, &&(\boldsymbol{u};\theta)^{\mathrm{t}} \in \mathcal{D}(A_0) \oplus \mathcal{D}(A_1). \end{aligned} \tag{7.36}$$

Performing the substitution

$$\nu^{1/2} A_0^{1/2} \boldsymbol{u} = \boldsymbol{\xi}, \quad \chi^{1/2} A_1^{1/2} \theta = \mu, \tag{7.37}$$

we can obtain the problem

$$\begin{aligned} &\left(\hat{I} - \tilde{\varepsilon}\hat{C}\right) y = \lambda \hat{A} y, \qquad y := (\boldsymbol{\xi};\mu)^{\mathrm{t}} \in \boldsymbol{J}_0(\Omega) \oplus L_2(\Omega), \\ &\qquad \hat{I} := \operatorname{diag}\left(I_0; I_1\right), \qquad \tilde{\varepsilon} = \varepsilon(\nu\chi)^{-1/2} > 0, \\ &\qquad \hat{A} = \operatorname{diag}\left(\nu^{-1}A_0^{-1}; \chi^{-1}A_1^{-1}\right), \\ &\qquad \hat{C} = \begin{pmatrix} 0 & \tilde{C} \\ \tilde{C}^* & 0 \end{pmatrix}, \qquad \tilde{C} = A_0^{-1/2} C A_1^{-1/2}, \ \tilde{C}^* = A_1^{-1/2} C^* A_0^{-1/2}. \end{aligned} \tag{7.38}$$

Since operator $\hat{A}$ is a compact positive operator, and operator $\hat{C}$ is compact and self-adjoint, then according to the assertions from Section 1.4.5, the problem (7.38) has a discrete real spectrum with the accumulation point $+\infty$, and the amount of negative eigenvalues coincides with the amount of negative eigenvalues of operator $\hat{I} - \tilde{\varepsilon}\hat{C}$. Hence, the following conclusions can be made. If the magnitude $\tilde{\varepsilon} > 0$ is sufficiently small, more precisely, $1 - \tilde{\varepsilon}\lambda^+_{\max}(\hat{C}) > 0$, then all eigenvalues of the problem (7.36) are located in the right half-plane, that is, each normal convective movement is stable. If the following condition holds true

$$\tilde{\varepsilon} > \left[\lambda^+_{\max}\left(\hat{C}\right)\right]^{-1} \tag{7.39}$$

then there is at least one eigenvalue in the left half-plane, and this eigenvalue is located on the negative halfaxis. To this eigenvalue there corresponds a mode of normal convective movements, which aperiodically increases with time.

Now let us note that in virtue of self-adjointness of the problem (7.38), its eigenvalues $\lambda_k = \lambda_\kappa(\tilde{\varepsilon})$ and eigenelements $y_k = y_k(\tilde{\varepsilon})$ are analytic functions of the parameter $\tilde{\varepsilon} \geq 0$. For any $\lambda(\tilde{\varepsilon})$ and $y(\tilde{\varepsilon})$, from (7.36), we obtain

$$\left(y(\tilde{\varepsilon}), y(\tilde{\varepsilon})\right) - \tilde{\varepsilon}\left(\hat{C}y(\tilde{\varepsilon}), y(\tilde{\varepsilon})\right) = \lambda(\tilde{\varepsilon})\left(\hat{A}y(\tilde{\varepsilon}), y(\tilde{\varepsilon})\right). \tag{7.40}$$

Differentiating this identity by $\tilde{\varepsilon}$ and using the fact that the pair $\{\lambda(\tilde{\varepsilon}), y(\tilde{\varepsilon})\}$ satisfies the equation (7.38), we obtain the formula

$$\frac{\mathrm{d}\lambda}{\mathrm{d}\tilde{\varepsilon}} = -\frac{\left(\hat{C}y(\tilde{\varepsilon}), y(\tilde{\varepsilon})\right)}{\left(\hat{A}y(\tilde{\varepsilon}), y(\tilde{\varepsilon})\right)}. \tag{7.41}$$

In particular, the following conclusions can be made.

1° The nonpositive eigenvalues of problem (7.38) move to the left if $\tilde{\varepsilon}$ increases, and

$$\frac{\mathrm{d}\lambda}{\mathrm{d}\tilde{\varepsilon}} \leq -\tilde{\varepsilon}^{-1}\lambda_{\min}\left(\hat{A}^{-1}\right). \tag{7.42}$$

Indeed, if $\lambda(\tilde{\varepsilon}) \leq 0$, then from (7.40) it follows that

$$(y(\tilde{\varepsilon}), y(\tilde{\varepsilon})) - \tilde{\varepsilon}(\hat{C}y(\tilde{\varepsilon}), y(\tilde{\varepsilon})) \leq 0,$$

and, therefore, from (7.41) we obtain

$$-\frac{\mathrm{d}\lambda}{\mathrm{d}\tilde{\varepsilon}} \geq -\frac{\tilde{\varepsilon}^{-1}\,(y(\tilde{\varepsilon}), y(\tilde{\varepsilon}))}{\left(\hat{A}y(\tilde{\varepsilon}), y(\tilde{\varepsilon})\right)} \geq \tilde{\varepsilon}^{-1}\left\|\hat{A}\right\|^{-1} = \tilde{\varepsilon}^{-1}\lambda_{\min}\left(\hat{A}^{-1}\right).$$

The formula (7.42) gives an estimate of the velocity of movement to the left of eigenvalues from the left half-plane.

For $\lambda(\tilde{\varepsilon}) > 0$, the eigenpair $\{\lambda(\tilde{\varepsilon}); y(\tilde{\varepsilon})\}$ for equation (7.38) is called a pair of the first type if for this pair $(\hat{C}y(\tilde{\varepsilon}), y(\tilde{\varepsilon})) > 0$; the same pair is called a pair of the second type, if $(\hat{C}y(\tilde{\varepsilon}), y(\tilde{\varepsilon})) < 0$.

2° The eigenvalue $\lambda(\tilde{\varepsilon})$ of a pair of the first type moves to the left if $\tilde{\varepsilon}$ increases, and $\lambda(\tilde{\varepsilon})$ of a pair of the second type moves to the right under the same conditions.

3° In the interval $0 < \lambda(\tilde{\varepsilon}) < \lambda_{\min}(\hat{A}^{-1})$ all eigenvalues correspond to pairs of the first type, and

$$\mathrm{d}\lambda/\mathrm{d}\tilde{\varepsilon} \leq -\tilde{\varepsilon}^{-1}\left[\lambda_{\min}\left(\hat{A}^{-1}\right) - \lambda\left(\tilde{\varepsilon}\right)\right] \;\; (< 0). \tag{7.43}$$

Indeed, from (7.40) and (7.42) we have

$$-\frac{\mathrm{d}\lambda}{\mathrm{d}\tilde{\varepsilon}} = \frac{\left(\hat{C}y(\tilde{\varepsilon}), y(\tilde{\varepsilon})\right)}{\left(\hat{A}y(\tilde{\varepsilon}), y(\tilde{\varepsilon})\right)}$$

$$= \frac{\tilde{\varepsilon}^{-1}\left[(y(\tilde{\varepsilon}), y(\tilde{\varepsilon})) - \lambda(\tilde{\varepsilon})\left(\hat{A}y(\tilde{\varepsilon}), y(\tilde{\varepsilon})\right)\right]}{\left(\hat{A}y(\tilde{\varepsilon}), y(\tilde{\varepsilon})\right)}$$

$$= \tilde{\varepsilon}^{-1}\left[\frac{(y(\tilde{\varepsilon}), y(\tilde{\varepsilon}))}{\left(\hat{A}y(\tilde{\varepsilon}), y(\tilde{\varepsilon})\right)} - \lambda(\tilde{\varepsilon})\right]$$

$$\geq \tilde{\varepsilon}^{-1}\left[\left\|\hat{A}\right\|^{-1} - \lambda(\tilde{\varepsilon})\right]$$

$$= \tilde{\varepsilon}^{-1}\left[\lambda_{\min}\left(\hat{A}^{-1}\right) - \lambda(\tilde{\varepsilon})\right] > 0.$$

From the same inequality it can be obtained that $(\hat{C}y(\tilde{\varepsilon}), y(\tilde{\varepsilon})) > 0$.

The following assertion is a corollary of the properties 2° and 3°.

4° For eigenpairs of the second type and for neutral eigenpairs, the eigenvalues satisfy $\lambda(\tilde{\varepsilon}) \geq \lambda_{\min}(\hat{A}^{-1})$.

Now let us consider the process, when the magnitude of heating $\tilde{\varepsilon}$ increases from zero to some small positive value, and show that there are such eigenvalues $\lambda(\tilde{\varepsilon})$ of the problems (7.36), (7.38), which satisfy the condition $\lambda(\tilde{\varepsilon}) < \lambda_{\min}(\hat{A}^{-1})$, and in virtue of the properties 2° and 3° to them there correspond eigenpairs of the first type. If $\tilde{\varepsilon}$ keeps increasing, then these $\lambda(\tilde{\varepsilon})$ move only to the left and pass to the left half-plane.

Instead of (7.38), let us consider the problem (7.36) and formally replace νA_0 by A_0 and χA_1 by A_1 (in order to simplify further reasoning.)

For $\varepsilon = 0$, the system of equations

$$\begin{pmatrix} A_0 & 0 \\ 0 & A_1 \end{pmatrix}\begin{pmatrix} \boldsymbol{u} \\ \theta \end{pmatrix} - \varepsilon\begin{pmatrix} 0 & C \\ C^* & 0 \end{pmatrix}\begin{pmatrix} \boldsymbol{u} \\ \theta \end{pmatrix} = \lambda\begin{pmatrix} \boldsymbol{u} \\ \theta \end{pmatrix} \tag{7.44}$$

has two obvious groups of solutions:

$$\text{(a)}\quad \boldsymbol{u} = \boldsymbol{u}_k(A_0), \quad \lambda = \lambda_k(A_0), \quad \theta_k = 0 \qquad k = 1, 2, \ldots,$$

$$\text{(b)}\quad \boldsymbol{u}_n = \boldsymbol{0}, \quad \theta = \theta_n(A_1), \quad \lambda_n = \lambda_n(A_1), \qquad n = 1, 2, \ldots, \tag{7.45}$$

which correspond to dissipative and heat waves in the fluid, respectively. When $\varepsilon > 0$, then their interaction takes place. Let us note that to the solution (7.45) there correspond neutral pairs.

Let us consider that in (7.44) the magnitude ε is a small parameter, and apply the perturbation theory using no more than the second approximation by ε. For the solutions close to the pair (a) from (7.45), we have

$$\begin{aligned} \boldsymbol{u}_k(\varepsilon) &= \boldsymbol{u}_k(A_0) + \varepsilon \boldsymbol{v}_k + \varepsilon^2 \boldsymbol{w}_k + \mathrm{O}(\varepsilon^3), \\ \theta_k(\varepsilon) &= 0 + \varepsilon \varphi_k + \varepsilon^2 \psi_k + \mathrm{O}(\varepsilon^3), \\ \lambda_k(\varepsilon) &= \lambda_k(A_0) + \varepsilon \mu_k + \varepsilon^2 \nu_k + \mathrm{O}(\varepsilon^3). \end{aligned} \tag{7.46}$$

The first approximation gives us the system of equations

$$\begin{aligned} A_0 \boldsymbol{v}_k - \lambda_k(A_0) \boldsymbol{v}_k &= \mu_k \boldsymbol{u}_k(A_0), \\ A_1 \varphi_k - \lambda_k(A_0) \varphi_k &= C^* \boldsymbol{u}_k(A_0). \end{aligned} \tag{7.47}$$

For simplicity, let us further assume that all eigenvalues $\lambda_k(A_0)$ of operator A_0 are one-multiple; besides, let us consider the case when $\lambda_k(A_0) \neq \lambda_m(A_1)$ for all k and m from $\mathbb{N}$ (absence of internal resonance).

Then from the first equation (7.47), after scalar multiplication (in $\boldsymbol{J}_0(\Omega)$) by $\boldsymbol{u}_k(A_0)$, we obtain that $\mu_k = 0$, $\boldsymbol{v}_k = c_k \boldsymbol{u}_k(A_0)$, where c_k is an arbitrary constant. From the second equation (7.47), in virtue of the invertibility of operator $A_1 - \lambda_k(A_0) I_1$, we have $\varphi_k = (A_1 - \lambda_k(A_0) I_1)^{-1} C^* \boldsymbol{u}_k(A_0)$.

The second approximation by ε leads us to the system

$$\begin{aligned} (A_0 - \lambda_k(A_0) I_0)\, \boldsymbol{w}_k &= C \left(A_1 - \lambda_k(A_0) I_1\right)^{-1} C^* \boldsymbol{u}_k(A_0) + \nu_k \boldsymbol{u}_k(A_0), \\ (A_1 - \lambda_k(A_0) I_1)\, \psi_k &= c_k C^* \boldsymbol{u}_k(A_0). \end{aligned} \tag{7.48}$$

Multiplying the first equation by $\boldsymbol{u}_k(A_0)$ again, we obtain

$$\nu_k = -\frac{\Big(\left(A_1 - \lambda_k(A_0) I_1\right)^{-1} C^* \boldsymbol{u}_k(A_0), C^* \boldsymbol{u}_k(A_0)\Big)_{L_2(\Omega)}}{\|\boldsymbol{u}_k(A_0)\|^2_{\boldsymbol{L}_2(\Omega)}}. \tag{7.49}$$

Since in virtue of (7.15) $C^* \boldsymbol{u}_k(A_0) = \boldsymbol{u}_k(A_0) \cdot \boldsymbol{e}_3$ and for eigenelements of operator A_0 there cannot be true that $\boldsymbol{u}_k(A_0) \cdot \boldsymbol{e}_3 \equiv 0$, then from (7.49) we obtain that $\nu_k < 0$ if $(A_1 - \lambda_k(A_0) I_1)^{-1} > 0$, and $\nu_k > 0$ if $(A_1 - \lambda_k(A_0) I_1)^{-1} < 0$.

For those solution of the problem (7.44) which are close to (b), from (7.45), similarly to (7.46) we have

$$\begin{aligned} \boldsymbol{u}_n(\varepsilon) &= 0 + \varepsilon \boldsymbol{v}_n + \varepsilon^2 \boldsymbol{w}_n + \mathrm{O}(\varepsilon^3), \\ \theta_n(\varepsilon) &= \theta_n(A_1) + \varepsilon \varphi_n + \varepsilon^2 \psi_n + \mathrm{O}(\varepsilon^3), \\ \lambda_n(\varepsilon) &= \lambda_n(A_1) + \varepsilon \mu_n = \varepsilon^2 \nu_n + \mathrm{O}(\varepsilon^3). \end{aligned} \tag{7.50}$$

From the equations of the first approximation by ε it can be obtained that $\mu_n = 0$, $\varphi_n = c_n\theta_n(A_1)$, and the second approximation gives us a formula that is similar to (7.49),

$$\nu_n = -\frac{\left((A_0 - \lambda_n(A_1)I_0)^{-1} C\theta_n(A_1), C\theta_n(A_1)\right)_{L_2(\Omega)}}{\|\theta_n(A_1)\|^2_{L_2(\Omega)}}. \qquad (7.51)$$

From (7.49), (7.51) the following conclusions can be made.

(a) The minimal eigenvalue of problem (7.36) corresponding to the case $\varepsilon = 0$ and to a neutral eigenpair, moves to the left when ε increases, and therefore, the given neutral pair becomes a pair of the first type. Further, according to Property 3°, it moves to the left with a velocity that satisfies the inequality (7.43).

Indeed, the minimal eigenvalue of the problem (7.36) for $\varepsilon = 0$ is either the minimal eigenvalue of operator νA_0 or the minimal eigenvalue of χA_1; in the notations of the problem (7.44), it is either $\lambda_1(A_0)$ or $\lambda_1(A_1)$, respectively. If $\min(\lambda_1(A_0); \lambda_1(A_1)) = \lambda_1(A_0) < \lambda_1(A_1)$, then $(A_1 - \lambda_1(A_0)I_1)^{-1} > 0$, because $A_1 - \lambda_1(A_0)I_1 \geq (\lambda_1(A_1) - \lambda_1(A_0))I_1 \gg 0$. Then from (7.49) we obtain that $\nu_1 < 0$ and therefore $\lambda_{\min}(\hat{A}^{-1}) - \lambda_1(\tilde{\varepsilon}) > 0$ for small $\tilde{\varepsilon} = \varepsilon(\nu\chi)^{-1/2}$. In the opposite case, when $\min(\lambda_1(A_0), \lambda_1(A_1)) = \lambda_1(A_1) < \lambda_1(A_0)$, the similar conclusion about the movement of the minimal eigenvalue follows from (7.51) and again from (7.43). Hence, when ε increases from zero, both the minimal eigenvalue corresponding to the heat wave and the minimal eigenvalue corresponding to the dissipative wave of normal oscillations can move to the left.

(b) As for eigenvalues of the problem (7.36) with other subscripts, according to the formulas (7.49) and (7.51) they can move both to the left and to the right when ε increases from zero. Let us remind, that for any $\varepsilon > 0$ the number of negative eigenvalues of the problem (7.38) exactly coincides with the number of negative eigenvalues of operator $\hat{I} - \tilde{\varepsilon}\hat{C}$ in the problem (7.38), and therefore, the number of eigenvalues, which always move to the left, increases according to the growth of ε.

Chapter 8

Motion of Viscous Fluids in Open Containers

In this chapter we study the motion of a viscous fluid in an open container. First we will prove the solvability of the initial boundary value problem. Then, in order to investigate the normal oscillations, we introduce the main operator pencil, whose spectrum is analyzed in details using the methods presented in Section 1.6.

The motions studied next are close to uniform rotations. We will find asymptotic solutions of the intial boundary value problem in the case of high viscousity. We also consider joint movements around a fixed point of the system "body + fluid".

At the end of this chapter we will address the problem of convection in a partially filled container. Various phenomena occurring in the cases of heating from above or below will be described.

8.1 Small Movements of Viscous Fluids in an Open Immovable Container

We begin by recalling the classical formulation of the boundary value problem for small movements of a viscous fluid in an open container. The equivalent first order differential equation with an operator that generates a semigroup is next obtained for this problem. Based on that equation, we will prove the theorem on the existence of a generalized solution for the initial problem and show that for this solution the full energy of the system and the velocity of its dissipation are continuous.

8.1.1 Classical Statement of the Problem

Let us assume that a heavy viscous fluid with density ρ fills partially an immovable container and in the equilibrium state occupies a region Ω that is bounded by the solid boundary S and the free surface Γ. Since for a heavy fluid the surface tension forces are not taken into account, then, in the state of equilibrium, surface Γ is a plane orthogonal to the acceleration $\boldsymbol{g}$ of the gravitation field. As usual, we choose the system of coordinates $Ox_1x_2x_3$ such that $\boldsymbol{g} = -g\boldsymbol{e}_3$ and its center O is located on the equilibrium surface Γ. Then the static pressure in the fluid equals

$$P_{st}(x_3) = p_a - \rho g x_3,$$

where p_a is the atmospheric pressure, which is assumed to be constant.

Let us now write down the equations and the boundary and initial conditions of the problem in the case when small movements of the fluid close to the immovable state are considered. The linearized Navier-Stokes equations in this case have the form

$$\frac{\partial \boldsymbol{u}}{\partial t} = -\frac{1}{\rho}\nabla p + \nu\Delta\boldsymbol{u} + \boldsymbol{f},$$
$$\operatorname{div}\boldsymbol{u} = 0 \qquad \text{in } \Omega, \tag{1.1}$$

where $\boldsymbol{u} = \boldsymbol{u}(t,x)$ is the velocity field, $p = p(t,x)$ is the dynamic pressure, and $\boldsymbol{f}(t,x)$ is a small field of external forces superimposed to the gravitation field $\boldsymbol{F}_0 = -\nabla(gx_3)$.

The stickiness condition for a viscous fluid holds true on the solid boundary S, that is,

$$\boldsymbol{u} = \boldsymbol{0} \qquad \text{on } S, \tag{1.2}$$

and the dynamic and kinematic conditions hold true on the free equilibrium surface Γ. If the equation of the free moving surface $\Gamma(t)$ we look for has the form

$$x_3 = \zeta(t, x_1, x_2) \qquad (x_1, x_2) \in \Gamma,$$

then the above-mentioned conditions become (see (1.33)–(1.35) in Section 3.1.5)

$$\tilde{\tau}_{i3}(\boldsymbol{u}) := \rho\nu\left(\frac{\partial u_i}{\partial x_3} + \frac{\partial u_3}{\partial x_i}\right) = 0, \qquad i = 1, 2,$$
$$\tilde{\tau}_{33}(\boldsymbol{u}) := -p + 2\rho\nu\frac{\partial u_3}{\partial x_3} = -\rho g\zeta,$$
$$\frac{\partial \zeta}{\partial t} = u_n := \boldsymbol{u}\cdot\boldsymbol{n} \qquad \text{on } \Gamma. \tag{1.3}$$

It is natural to assume that the distribution of the velocity field and the shape of the free surface at the initial moment are given:

$$\boldsymbol{u}(0,x) = \boldsymbol{u}^0(x), \quad \zeta(0,x_1,x_2) = \zeta^0(x_1,x_2). \tag{1.4}$$

Therefore, the problem on small movements for a heavy viscous fluid in an open immovable container can be reduced to integrating equations (1.1) with the boundary and initial conditions (1.2)–(1.4). This problem is more complicated than the problem on the motion of a fluid in a wholly filled immovable container presented in Section 7.1 because of the boundary conditions (1.3).

8.1.2 Auxiliary Boundary Value Problems

To switch from problem (1.1)–(1.4) to an evolution operator equation in a Hilbert space, we need to make some additional comments.

The spaces $\boldsymbol{J}_{0,S}(\Omega)$ and $\boldsymbol{J}^1_{0,S}(\Omega)$ that have been described in detail in Sections 2.1–2.2 form a natural pair of spaces to deal with problem (1.1)–(1.4). The space $\boldsymbol{J}_{0,S}(\Omega)$ is related to the kinetic energy of a fluid in an open container and the space $\boldsymbol{J}^1_{0,S}(\Omega)$, consisting of functions with a finite velocity of dissipation and equipped with the scalar product

$$(\boldsymbol{u},\boldsymbol{v})_{\boldsymbol{J}^1_{0,S}(\Omega)} = E(\boldsymbol{u},\boldsymbol{v}), \tag{1.5}$$

where $E(\boldsymbol{u},\boldsymbol{v})$ is defined by formula (2.3) in Section 2.2 is dense in $\boldsymbol{J}_{0,S}(\Omega)$.

By using the pair of spaces $(F;G) = (\boldsymbol{J}^1_{0,S}(\Omega);\boldsymbol{J}_{0,S}(\Omega))$, the relationship between the boundary value problems addressed in in Section 2.2.7 and the problem on small movements of a viscous fluid in an open container becomes evident. In the sequel, we still need to identify the operators A, T, and the trace operator γ_n for these problems.

Let us recall that the operator A generated by the pair of Hilbert spaces $\boldsymbol{J}_{0,S}(\Omega)$ and $\boldsymbol{J}^1_{0,S}(\Omega)$ can be obtained from the identity

$$E(\boldsymbol{u},\boldsymbol{v}) = \int_\Omega A\boldsymbol{u}\cdot\boldsymbol{v}\,d\Omega, \tag{1.6}$$

where $\boldsymbol{u}\in\mathcal{D}(A)\subset\boldsymbol{J}^1_{0,S}(\Omega)$ and $\boldsymbol{v}$ is an arbitrary element in $\boldsymbol{J}^1_{0,S}(\Omega)$. The inverse operator, A^{-1}, gives the generalized solution of the boundary value problem (2.30) in Section 2.2.7 by the formula $\boldsymbol{u} = A^{-1}\boldsymbol{f}$.

Let us now consider an auxiliary boundary value problem which will be called **Problem I** in what follows. The problem is formulated as follows: One needs to find the solution to the system of Stokes equations

$$-\nu\Delta\boldsymbol{s} + \nabla p_1 = \boldsymbol{f} \quad \operatorname{div}\boldsymbol{s} = 0, \qquad \text{in } \Omega, \tag{1.7}$$

that satisfies the boundary conditions

$$\begin{aligned} \boldsymbol{s} &= 0 \qquad \text{on } S, \\ \tilde{\tau}_{i3}(\boldsymbol{s}) &:= \nu\left(\frac{\partial s_i}{\partial x_3} + \frac{\partial s_3}{\partial x_i}\right) = 0, \qquad i = 1, 2, \\ \tilde{\tau}_{33}(\boldsymbol{s}) &:= -p_1 + 2\nu\frac{\partial s_3}{\partial x_3} = 0 \qquad \text{on } \Gamma. \end{aligned} \tag{1.8}$$

By (1.6), the generalized solution of Problem I is derived from the following identity for $\boldsymbol{f} \in \boldsymbol{J}_{0.S}(\Omega)$,

$$\nu E(\boldsymbol{s}, \boldsymbol{v}) = \int_\Omega \boldsymbol{f} \cdot \boldsymbol{v} d\Omega, \tag{1.9}$$

which should be valid for any $\boldsymbol{v} \in \boldsymbol{J}^1_{0.S}(\Omega)$. According to the general results from Sections 1.8 and the conclusions of Section 2.2.7, the boundary value problem I admits a unique generalized solution, $\boldsymbol{s} \in \boldsymbol{J}^1_{0.S}(\Omega)$, for any $\boldsymbol{f} \in \boldsymbol{J}_{0.S}(\Omega)$. If $\boldsymbol{f} \in (\boldsymbol{J}^1_{0.S}(\Omega))^* =: \boldsymbol{J}^{-1}_{0.S}(\Omega)$, then Problem I admits a unique weak solution $\boldsymbol{s}$ that belongs to $\boldsymbol{J}^1_{0.S}(\Omega)$, too. Let us recall that A as an operator acting on $\boldsymbol{J}_{0.S}(\Omega)$ is an unbounded self-adjoint positive definite operator, defined on the dense set $\mathcal{D}(A) \subset \boldsymbol{J}^1_{0.S}(\Omega) \subset \boldsymbol{J}_{0.S}(\Omega)$. Here, the domain of definition $\mathcal{D}(A^{1/2})$ coincides with the whole space $\boldsymbol{J}^1_{0.S}(\Omega)$ and

$$E(\boldsymbol{s}, \boldsymbol{v}) = (A^{1/2}\boldsymbol{s}, A^{1/2}\boldsymbol{v})_{\boldsymbol{L}_2(\Omega)} \tag{1.10}$$

for any $\boldsymbol{s}, \boldsymbol{v}$ in $\boldsymbol{J}^1_{0.S}(\Omega)$.

Let us also notice that the operator A has a discrete positive spectrum $\{\lambda_n(A)\}_{n=1}^\infty$ with $\lim_{n\to\infty} \lambda_n(A) = +\infty$. Its eigenvalues $\lambda_n(A)$ are consecutive minima of the variational ratio

$$\frac{\|A^{1/2}\boldsymbol{u}\|^2_{\boldsymbol{L}_2(\Omega)}}{\|\boldsymbol{u}\|^2_{\boldsymbol{L}_2(\Omega)}} = \frac{E(\boldsymbol{u}, \boldsymbol{u})}{(\boldsymbol{u}, \boldsymbol{u})_{\boldsymbol{L}_2(\Omega)}}, \tag{1.11}$$

that is considered for elements $\boldsymbol{u} \in \boldsymbol{J}^1_{0.S}(\Omega)$. The asymptotic behavior of the eigenvalues has the form

$$\lambda_n(A) = c_A^{-2/3} n^{2/3}[1 + o(1)], \qquad n \to \infty, \tag{1.12}$$

where $c_A = \operatorname{mes}\Omega/(3\pi^2)$. Hence, it appears that the inverse operator A^{-1} which is compact, belongs to the class $\mathfrak{S}_p$ for $p > 3/2$.

Therefore, the solution to the boundary value problem I is given by the formula $\nu\boldsymbol{s} = A^{-1}f$, where A is the operator with the above mentioned properties.

Let us consider now another auxiliary boundary value problem which will be called from now on **Problem II**. This problem is connected with the operator $T = \partial^{-1}$ in the abstract scheme of Section 1.8 and it requires to find a solution to the system of equations

$$-\nu\Delta \boldsymbol{w} + \nabla p_2 = \boldsymbol{f}, \quad \operatorname{div} \boldsymbol{w} = 0 \qquad \text{in } \Omega, \tag{1.13}$$

that satisfies the boundary conditions

$$\begin{aligned} &\boldsymbol{w} = 0 \qquad \text{on } S,\\ &\tilde{\tau}_{i3}(\boldsymbol{w}) := \nu\left(\frac{\partial w_i}{\partial x_3} + \frac{\partial w_3}{\partial x_i}\right) = 0, \qquad i = 1, 2,\\ &\tilde{\tau}_{33}(\boldsymbol{w}) := -p_2 + 2\nu\frac{\partial w_3}{\partial x_3} = \psi \qquad \text{on } \Gamma,\\ &\int_\Gamma \psi d\Gamma = 0. \end{aligned} \tag{1.14}$$

Let us recall that for the boundary value problem (2.35) in Section 2.2.7, which coincides with (1.13)–(1.14), the operator T for $\boldsymbol{u} = \nu\boldsymbol{w}$ can be obtained from the identity

$$E(T\psi, \boldsymbol{v}) = \int_\Gamma \psi\gamma_n \boldsymbol{v} d\Gamma, \qquad \psi \in H_\Gamma^{-1/2}, \tag{1.15}$$

which should be true for any $\boldsymbol{v} \in \boldsymbol{J}_{0.S}^1(\Omega)$ and $\gamma_n \boldsymbol{v} = v_n|_\Gamma = v_3|_\Gamma$. By the general considerations in Section 1.8 and the detailed explanation at the end of Section 2.2.7, we conclude that there exist only one generalized solution $\boldsymbol{w} \in \boldsymbol{J}_{0.S}^1(\Omega)$ to Problem II for any $\psi \in L_{2,\Gamma}$. Comparing it with problem (1.13)–(1.14) and identity (1.15), we get that for that generalized solution the following expression

$$\nu E(\boldsymbol{w}, \boldsymbol{v}) = \int_\Gamma \psi\gamma_n \boldsymbol{v} d\Gamma \tag{1.16}$$

is valid for any $\boldsymbol{v} \in \boldsymbol{J}_{0.S}^1(\Omega)$. If $\psi \in H_\Gamma^{-1/2}$, then there exists a unique weak solution, $\boldsymbol{w}$, to Problem II, which belongs to the space $\boldsymbol{J}_{0.S}^1(\Omega)$ as well. Hence, a solution to Problem II is defined by the formula $\nu\boldsymbol{w} = T\psi$, where T is an operator that acts isometrically from $H_\Gamma^{-1/2}$ to $\boldsymbol{J}_{0.S}^1(\Omega)$.

According to the general scheme in Section 1.8, the operator $C := \gamma_n T$ that maps isometrically the space $H_\Gamma^{-1/2}$ into $H_\Gamma^{1/2}$ can also be considered. Its restriction to $L_{2,\Gamma}$ is a bounded self-adjoint operator on $L_{2,\Gamma}$. Moreover, by the theorem on traces, the space $H_\Gamma^{1/2}$ is compactly embedded into $L_{2,\Gamma}$, therefore, operator C is compact.

8.1.3 Generalized Solutions of the Homogeneous Nonstationary Problem

Let $\boldsymbol{u}(x)$ be a smooth solenoidal field satisfying the stickiness condition (1.2) and the first two dynamic conditions (1.3), which mean that the tangent stresses equal zero on Γ. Further, let $\tilde{p}(x_1, x_2)$ be a smooth function on Γ, for which

$$\int_\Gamma \tilde{p} \mathrm{d}\Gamma = 0.$$

Let us denote some smooth extension of $\tilde{p} = \tilde{p}(x_1, x_2)$ on the whole region Ω by $p(x)$. For the chosen $\boldsymbol{u}(x)$, $p(x)$, and any field $\boldsymbol{v} \in \boldsymbol{J}^1_{0.S}(\Omega)$, from the Green formula (2.2.10) the equality below follows,

$$\int_\Omega \left(-\nu\Delta\boldsymbol{u} + \frac{1}{\rho}\nabla p\right)\cdot \boldsymbol{v}\mathrm{d}\Omega = \nu E(\boldsymbol{u},\boldsymbol{v}) - \int_\Gamma \left(-\frac{1}{\rho}p + 2\nu\frac{\partial u_3}{\partial x_3}\right) v_3 \mathrm{d}\Gamma. \tag{1.17}$$

Let us consider the solutions $\boldsymbol{s}$ and $\boldsymbol{w}$ of the boundary value problems I and II through the identities

$$\nu E(\boldsymbol{s},\boldsymbol{v}) = \int_\Omega \left(-\nu\Delta\boldsymbol{u} + \frac{1}{\rho}\nabla p\right)\cdot \boldsymbol{v}\mathrm{d}\Omega, \tag{1.18}$$

$$\nu E(\boldsymbol{w},\boldsymbol{v}) = \int_\Omega \left(-\frac{1}{\rho}p + 2\nu\frac{\partial u_3}{\partial x_3}\right) v_3 \mathrm{d}\Gamma, \tag{1.19}$$

which should be valid for any field $\boldsymbol{v} \in \boldsymbol{J}^1_{0.S}(\Omega)$. (Let us recall that according to the facts proved at the end of Section 2.2.6, the mean value on Γ of the function $\partial u_3/\partial x_3$ equals zero.) Then, the equality (1.17) takes the form

$$\nu E(\boldsymbol{s},\boldsymbol{v}) = \nu E(\boldsymbol{u},\boldsymbol{v}) - \nu E(\boldsymbol{w},\boldsymbol{v});$$

in virtue of the arbitrary choice of $\boldsymbol{v} \in \boldsymbol{J}^1_{0.S}(\Omega)$, from the latter it follows that $\boldsymbol{u} = \boldsymbol{s} + \boldsymbol{w}$.

Using the above mentioned methods, the fields $\boldsymbol{s}$ and $\boldsymbol{w}$ can be univalently defined by the field $\boldsymbol{u}$, and independently from the extension of the function $\tilde{p}(x_1, x_2)$ on the whole region Ω. Indeed, a substitution of an extension $p_1(x)$ by another extension $p_2(x)$ does not change the equation for $\boldsymbol{w}$ because in the identity (1.19) only the values of $p_1|_\Gamma = \tilde{p}(x_1, x_2)$ are involved; the equation for $\boldsymbol{s}$ derived from the identity (1.18) also does not change because the potential of the field $\nabla(p_1 - p_2) \in \boldsymbol{G}_{0,\Gamma}(\Omega)$ turns out to be zero on Γ and in virtue of the decomposition $\boldsymbol{L}_2(\Omega) = \boldsymbol{G}_{0,\Gamma}(\Omega) \oplus \boldsymbol{J}_{0,S}(\Omega)$, the field $\nabla(p_1 - p_2)$ is orthogonal to the field $\boldsymbol{v} \in \boldsymbol{J}^1_{0.S}(\Omega) \subset \boldsymbol{J}_{0.S}(\Omega)$.

Now, let the functions $\boldsymbol{u} = \boldsymbol{u}(t,x)$, $p = p(t,x)$, and $\zeta = \zeta(t, x_1, x_2)$ be a classical solution of the homogeneous problem (1.1)–(1.4) at some fixed moment of time $t > 0$. Then, according to the above mentioned facts, one may represent

$$\boldsymbol{u}(t,x) = \boldsymbol{s}(t,x) + \boldsymbol{w}(t,x), \quad \rho^{-1}p(t,x) = p_1(t,x) + p_2(t,x), \tag{1.20}$$

where $\boldsymbol{s}(t,x)$, $p_1(t,x)$ is a solution of the boundary value problem I when the right hand side equals $(-\partial\boldsymbol{u}/\partial t)$, and $\boldsymbol{w}(t,x)$, $p_2(t,x)$ is a solution of the boundary value problem II with the function $\psi(t,x_1,x_2) = -g\zeta(t,x_1,x_2)$. According to (1.9) and (1.16), for the functions $\boldsymbol{s}$ and $\boldsymbol{w}$ the following identities hold true,

$$\nu E(\boldsymbol{s},\boldsymbol{v}) = -\int_\Omega \frac{\partial\boldsymbol{u}}{\partial t}\cdot\boldsymbol{v}\mathrm{d}\Omega, \tag{1.21}$$

$$\nu E(\boldsymbol{w},\boldsymbol{v}) = -g\int_\Gamma \zeta v_3\mathrm{d}\Gamma, \tag{1.22}$$

where $\boldsymbol{v}$ is an arbitrary field in $\boldsymbol{J}^1_{0,S}(\Omega)$.

Taking into account the kinematic condition (1.3) and definitions (1.6), (1.15) of operators A, T, the following system of equations in operator form can be obtained from (1.21), (1.22),

$$\begin{aligned} \nu A\boldsymbol{s} &= -\frac{\mathrm{d}\boldsymbol{u}}{\mathrm{d}t}, & \frac{\mathrm{d}\zeta}{\mathrm{d}t} &= \gamma_n\boldsymbol{u}, \\ \nu\boldsymbol{w} &= -gT\zeta, & \boldsymbol{u} &= \boldsymbol{s}+\boldsymbol{w}; \end{aligned} \tag{1.23}$$

to these equations initial conditions (1.4) should be added.

Equations (1.23) make it possible to exclude the functions $\boldsymbol{u}(t,x)$, $\zeta(t,x_1,x_2)$ and finally we obtain the problem

$$\begin{aligned} \frac{\mathrm{d}\boldsymbol{s}}{\mathrm{d}t} &= -\nu A\boldsymbol{s} + g\nu^{-1}T\gamma_n(\boldsymbol{s}+\boldsymbol{w}), \\ \frac{\mathrm{d}\boldsymbol{w}}{\mathrm{d}t} &= -g\nu^{-1}T\gamma_n(\boldsymbol{s}+\boldsymbol{w}). \end{aligned} \tag{1.24}$$

As far as $\boldsymbol{w}(0) = -g\nu^{-1}T\zeta(0) = -g\nu^{-1}T\zeta^0$, $\boldsymbol{s}(0) = \boldsymbol{u}(0) - \boldsymbol{w}(0)$, then the following conditions

$$\boldsymbol{s}(0) = \boldsymbol{u}^0 + g\nu^{-1}T\zeta^0, \quad \boldsymbol{w}(0) = -g\nu^{-1}T\zeta^0 \tag{1.25}$$

should be taken as initial conditions for the problem (1.24). Hence, the classical solution of the problem (1.1)–(1.9) is a solution of the Cauchy problem (1.24), (1.25) in operator form.

The field $\boldsymbol{u}(t,x) = \boldsymbol{s}(t,x) + \boldsymbol{w}(t,x)$ is called a *generalized solution of the problem (1.1)–(1.4)*, if the functions $\boldsymbol{s}(t,x)$ and $\boldsymbol{w}(t,x)$ are generalized solutions of the Cauchy problem (1.24), (1.25) in the space $\boldsymbol{J}^1_{0,S}(\Omega)$.

To prove the solvability of the Cauchy problem (1.24), (1.25), let us transform it into an equivalent form. As long as the elements $\boldsymbol{u}$, $\boldsymbol{s}$, and $\boldsymbol{v}$ belong to the space $\boldsymbol{J}^1_{0,S}(\Omega) = \mathcal{D}(A^{1/2}) \subset \boldsymbol{J}_{0,S}(\Omega)$, then they can be represented as

$$\boldsymbol{u} = A^{-1/2}\boldsymbol{\xi}, \quad \boldsymbol{s} = A^{-1/2}\boldsymbol{\eta}, \quad \boldsymbol{w} = A^{-1/2}\boldsymbol{\delta}, \tag{1.26}$$

where $\boldsymbol{\xi}, \boldsymbol{\eta}, \boldsymbol{\delta}$ are elements in $\boldsymbol{J}_{0,S}(\Omega)$. Substituting (1.26) into (1.24) we get

$$\begin{aligned} A^{-1/2}\frac{d\boldsymbol{\eta}}{dt} &= -\nu A^{1/2}\boldsymbol{\eta} + g\nu^{-1}T\gamma_n A^{-1/2}(\boldsymbol{\eta}+\boldsymbol{\delta}), \\ A^{-1/2}\frac{d\boldsymbol{\delta}}{dt} &= -g\nu^{-1}T\gamma_n A^{-1/2}(\boldsymbol{\eta}+\boldsymbol{\delta}). \end{aligned} \tag{1.27}$$

The second formula (1.27) shows that its right side belongs to $\mathcal{D}(A^{1/2})$, therefore operator $A^{1/2}$ can be applied to both equations in (1.27). Finally, instead of (1.24), (1.25) we obtain the Cauchy problem

$$\begin{aligned} \frac{d\boldsymbol{\eta}}{dt} &= -\nu A\boldsymbol{\eta} + g\nu^{-1}B(\boldsymbol{\eta}+\boldsymbol{\delta}), \\ \frac{d\boldsymbol{\delta}}{dt} &= -g\nu^{-1}B(\boldsymbol{\eta}+\boldsymbol{\delta}), \qquad B := A^{1/2}T\gamma_n A^{-1/2}, \end{aligned} \tag{1.28}$$

$$\boldsymbol{\eta}(0) = \boldsymbol{\eta}^0 := A^{1/2}\boldsymbol{s}(0), \qquad \boldsymbol{\delta}(0) = \boldsymbol{\delta}^0 := A^{1/2}\boldsymbol{w}(0). \tag{1.29}$$

Let us now consider B as an operator acting on the space $\boldsymbol{J}_{0,S}(\Omega)$. The operator $A^{-1/2}$ maps $\boldsymbol{J}_{0,S}(\Omega)$ onto $\boldsymbol{J}^1_{0,S}(\Omega)$ and the operator γ_n (according to the theorem on traces) is compact as an operator from $\boldsymbol{J}^1_{0,S}(\Omega)$ to $L_{2,\Gamma}$. That is why the product $Q := \gamma_n A^{-1/2}$ is a compact operator from $\boldsymbol{J}_{0,S}(\Omega)$ to $L_{2,\Gamma}$. Further, let us perform the substitution $\boldsymbol{v} = A^{-1/2}\boldsymbol{w}$ in the identity (1.15) and use the property (1.10). Then for any $\psi \in L_{2,\Gamma}$ and arbitrary $\boldsymbol{w} \in \boldsymbol{J}_{0,S}(\Omega)$ the following relation holds true,

$$(A^{1/2}T\psi, \boldsymbol{w})_{\boldsymbol{L}_2(\Omega)} = (\psi, \gamma_n A^{-1/2}\boldsymbol{w})_{L_{2,\Gamma}}. \tag{1.30}$$

According to the definition of an adjoint operator, this equality means that $A^{1/2}T$ is adjoint to the operator $\gamma_n A^{-1/2} = Q$ (as an operator acting from $L_{2,\Gamma}$ to $\boldsymbol{J}_{0,S}(\Omega)$) and, therefore, it is compact too. Hence, $B = (A^{1/2}T)(\gamma_n A^{-1/2}) = Q^*Q$ is a compact and nonnegative operator acting on $\boldsymbol{J}_{0,S}(\Omega)$.

Let us note that the operator B has the infinite dimensional kernel and is infinite dimensional itself. Indeed, if we choose a smooth finite field $\boldsymbol{w}$ and assume $\boldsymbol{v} = A^{1/2}\boldsymbol{w}$, then $\gamma_n A^{-1/2}\boldsymbol{v} = \gamma_n \boldsymbol{w} = 0$ and therefore, $B\boldsymbol{w} = \boldsymbol{0}$. On the other hand, the set of functions $\{\gamma_n A^{-1/2}\boldsymbol{v}\}$ fills the whole space $H^{1/2}_\Gamma$ and is infinite dimensional for any $\boldsymbol{v} \in \boldsymbol{J}_{0,S}(\Omega)$, therefore, the operator B is infinite dimensional.

Let us list the eigenvalues of operator B in decreasing order with regard to their multiplicities. Obviously, they are consecutive maxima of the variational ratio

$$\frac{(B\boldsymbol{\xi},\boldsymbol{\xi})_{\boldsymbol{L}_2(\Omega)}}{(\boldsymbol{\xi},\boldsymbol{\xi})_{\boldsymbol{L}_2(\Omega)}} = \frac{\left(A^{1/2}T\gamma_n A^{-1/2}\boldsymbol{\xi},\boldsymbol{\xi}\right)_{\boldsymbol{L}_2(\Omega)}}{(\boldsymbol{\xi},\boldsymbol{\xi})_{\boldsymbol{L}_2(\Omega)}}, \tag{1.31}$$

which is considered for nonzero elements $\boldsymbol{\xi} \in \boldsymbol{J}_{0.S}(\Omega)$. In (1.31), let us perform the substitution $A^{-1/2}\boldsymbol{\xi} = \boldsymbol{v} \in \mathcal{D}(A^{1/2}) = \boldsymbol{J}^1_{0.S}(\Omega)$ and use the identities (1.10), (1.15); we obtain the following variational ratio instead of (1.31)

$$\frac{(\gamma_n\boldsymbol{v},\gamma_n\boldsymbol{v})_{L_{2.\Gamma}}}{(A^{1/2}\boldsymbol{v},A^{1/2}\boldsymbol{v})_{\boldsymbol{L}_2(\Omega)}} = \frac{\int_\Gamma |v_n|^2 d\Gamma}{E(\boldsymbol{v},\boldsymbol{v})}, \tag{1.32}$$

which is considered for nonzero elements $\boldsymbol{v}$ from $\boldsymbol{J}^1_{0.S}(\Omega)$. Hence the next asymptotic formula can be obtained,

$$\lambda_n(B) = c_B^{1/2} n^{-1/2}[1+\mathrm{o}(1)], \quad n\to\infty, \qquad c_B := \frac{\operatorname{mes}\Gamma}{16\pi} > 0. \tag{1.33}$$

From (1.33) it follows that B belongs to the class $\mathfrak{S}_p$ for $p > 2$.

Let us now consider the problem (1.28), (1.29); it can be investigated as the Cauchy problem for an ordinary differential equation of the first order in the doubled Hilbert space $\boldsymbol{E} = \boldsymbol{J}_{0,S}(\Omega) \oplus \boldsymbol{J}_{0.S}(\Omega)$:

$$\frac{dy}{dt} = -\nu\mathcal{A}y + g\nu^{-1}\mathcal{B}y, \quad y(0) = y^0, \tag{1.34}$$

where

$$\mathcal{A} = \operatorname{diag}(A;0), \quad y = (\boldsymbol{\eta};\boldsymbol{\delta})^{\mathrm{t}}, \quad \mathcal{B} = \begin{pmatrix} B & B \\ -B & -B\end{pmatrix}, \quad y^0 = \begin{pmatrix}\boldsymbol{\eta}^0 \\ \boldsymbol{\delta}^0\end{pmatrix}.$$

As far as $A \gg 0$, then operator $\mathcal{A}$ is self-adjoint and nonnegative in $\boldsymbol{E}$, and therefore, the operator $-\nu\mathcal{A}$ generates an analytic semigroup of operators. Since B is compact, $\mathcal{B}$ is a bounded and compact operator. Therefore, the operator $(-\nu\mathcal{A} + g\nu^{-1}\mathcal{B})$ generates the analytic semigroup $\mathcal{U}(t) := \exp(-t(\nu\mathcal{A} - g\nu^{-1}\mathcal{B}))$. Hence, it appears that the homogeneous Cauchy problem (1.34) for an arbitrary initial data $y(0) = y^0 \in \boldsymbol{E}$ has the unique generalized solution $y(t) = \mathcal{U}(t)y^0$, which is a continuous function in $\boldsymbol{E}$ on $[0,\infty)$ and analytic for $t > 0$.

We next consider the initial boundary value problem (1.1)–(1.4) for the field $\boldsymbol{f}(t,x) \equiv \boldsymbol{0}$. Let us assume that the initial functions satisfy the conditions

$$\boldsymbol{u}^0(x) \in \boldsymbol{J}^1_{0,S}(\Omega), \quad \zeta^0(x_1,x_2) \in L_{2,\Gamma}. \tag{1.35}$$

Then, according to (1.25) we have $\boldsymbol{w}(0) = -g\nu^{-1}T\zeta^0 \in \boldsymbol{J}^1_{0,S}(\Omega)$, $\boldsymbol{s}(0) = \boldsymbol{u}^0 - \boldsymbol{w}(0) \in \boldsymbol{J}^1_{0,S}(\Omega) = \mathcal{D}(A^{1/2})$. That is way from (1.29) we obtain $\boldsymbol{\eta}^0 = A^{1/2}\boldsymbol{s}(0) \in \boldsymbol{J}_{0,S}(\Omega)$, $\boldsymbol{\delta}^0 = A^{1/2}\boldsymbol{w}(0) \in \boldsymbol{J}_{0,S}(\Omega)$, i.e., $y^0 = (\boldsymbol{\eta}^0; \boldsymbol{\delta}^0)^{\mathrm{t}} \in \boldsymbol{E} = \boldsymbol{J}_{0,S}(\Omega) \oplus \boldsymbol{J}_{0,S}(\Omega)$. Hence, if the conditions in (1.35) hold true, then $y^0 \in \boldsymbol{E}$ and the generalized solution $y(t) = \mathcal{U}(t)y^0 = (\boldsymbol{\eta}(t); \boldsymbol{\delta}(t))^{\mathrm{t}}$ has the components $\boldsymbol{\eta}(t)$, $\boldsymbol{\delta}(t)$, which are continuous functions of t with values in $\boldsymbol{J}_{0,S}(\Omega)$ and analytic for $t > 0$. But in this case $\boldsymbol{s}(t) = A^{-1/2}\boldsymbol{\eta}(t)$, $\boldsymbol{w}(t) = A^{-1/2}\boldsymbol{\delta}(t)$ and $\boldsymbol{u}(t) = \boldsymbol{s}(t) + \boldsymbol{w}(t)$ are continuous functions of t with values in $\boldsymbol{J}^1_{0,S}(\Omega)$; these functions are analytic for $t > 0$. From this fact and the second relation (1.23) it can be obtained that $\mathrm{d}\zeta/\mathrm{d}t = \gamma_n\boldsymbol{u} = u_3|_\Gamma$ is a continuous function of t with values in $H^{1/2}_\Gamma$; from the latter and (1.35) it follows that $\zeta(t, x_1, x_2) = \zeta^0(x_1, x_2) + \int_0^t \gamma_n\boldsymbol{u}(s)\mathrm{d}s$ is a continuously differentiable function with values in $L_{2,\Gamma}$.

8.1.4 Motions with Small Mass Forces

Now let us consider the case of forced motions of a fluid in a container, i.e., when the small field of external forces with mass density $\boldsymbol{f}(t,x)$ acts on the system.

First of all, let us notice that the velocity field $\boldsymbol{u}(t,x)$ is influenced only by those components of the field $\boldsymbol{f}(t,x)$ that belong to the subspace $\boldsymbol{J}_{0,S}(\Omega)$. The projection of the field $\boldsymbol{f}(t,x)$ onto the potential subspace $\boldsymbol{G}_{0,\Gamma}(\Omega)$ that is orthogonal to $\boldsymbol{J}_{0,S}(\Omega)$ results only in modifications of the pressure in the fluid. Moreover, since the potential on Γ of all the elements of $\boldsymbol{G}_{0,\Gamma}(\Omega)$ is equal to zero, then the presence or absence of a potential component of the field $\boldsymbol{f}(t,x)$ from the subspace $\boldsymbol{G}_{0,\Gamma}(\Omega)$ does not change the boundary conditions on Γ for the problem (1.1)–(1.4). Taking into account the above mentioned facts, we will assume that $\boldsymbol{f}(t,x)$ is a function of t with values in $\boldsymbol{J}_{0,S}(\Omega)$.

According to the scheme developed in Section 8.1.3, let us perform the transition from the boundary value problem (1.1)–(1.4) to the system of equations (1.24), and then to the problems (1.28) and (1.34). In this case we obtain an additional term, $\boldsymbol{f}(t)$, in the right side of equation (1.24), the term $A^{1/2}\boldsymbol{f}(t)$ appears in the first equation (1.28), and problem (1.34) is no longer homogeneous having a new term $F(t) := (A^{1/2}\boldsymbol{f}(t); 0)^{\mathrm{t}}$ in the right hand side.

To explain the performed transformations, let us assume (in addition to conditions (1.35)) that the field of external forces $\boldsymbol{f}(t,x)$ is a continuous function of t with values in $\boldsymbol{J}^1_{0,S}(\Omega)$. Then $F(t)$ is a continuous function of t with values in $\boldsymbol{E}$ and the nonhomogeneous problem (1.34) has a unique generalized solution given by the formula

$$y(t) = \mathcal{U}(t)y^0 + \int_0^t \mathcal{U}(t-s)F(s)\mathrm{d}s,$$

where $\mathcal{U}(t) = \exp(-t(\nu\mathcal{A} - g\nu^{-1}\mathcal{B}))$ is an analytic semigroup. This solution is a continuous function of t with values in $\boldsymbol{E}$. As in Section 8.1.3, from the latter we obtain that under conditions (1.35) and for the above mentioned constraints for $\boldsymbol{f}(t,x)$ the nonhomogeneous problem (1.1)–(1.4) has the generalized solution $\{\boldsymbol{u}(t,x);\zeta(t,x_1,x_2)\}$ for which both the kinetic and potential energies of the system are continuous functions of t, and the dissipation velocity of the full energy is also continuous.

8.1.5 Equation of Energy Balance

In order to deduce this equation, let us write down the system (1.23) with regard to mass forces so that the above mentioned generalized solutions satisfy the following system:

$$\nu A^{1/2}\boldsymbol{s} + A^{-1/2}\frac{\mathrm{d}\boldsymbol{u}}{\mathrm{d}t} = A^{-1/2}\boldsymbol{f}, \quad \nu A^{1/2}\boldsymbol{w} + gA^{1/2}T\zeta = 0,$$
$$\frac{\mathrm{d}\zeta}{\mathrm{d}t} = \gamma_n\boldsymbol{u}, \quad \boldsymbol{u} = \boldsymbol{s} + \boldsymbol{w}. \tag{1.36}$$

Let us scalarly multiply the first and second equations by $A^{1/2}\boldsymbol{u}$ and add them up. Then we get

$$\nu(A^{1/2}\boldsymbol{u}, A^{1/2}\boldsymbol{u})_{\boldsymbol{L}_2(\Omega)} + \left(A^{-1/2}\frac{\mathrm{d}\boldsymbol{u}}{\mathrm{d}t}, A^{1/2}\boldsymbol{u}\right)_{\boldsymbol{L}_2(\Omega)} + g(A^{1/2}T\zeta, A^{1/2}\boldsymbol{u})_{\boldsymbol{L}_2(\Omega)}$$
$$= (A^{-1/2}\boldsymbol{f}, A^{1/2}\boldsymbol{u})_{\boldsymbol{L}_2(\Omega)}.$$

Using equations (1.10) and (1.30) we obtain

$$\nu E(\boldsymbol{u},\boldsymbol{u}) + \frac{1}{2}\frac{\mathrm{d}}{\mathrm{d}t}(\boldsymbol{u},\boldsymbol{u})_{\boldsymbol{L}_2(\Omega)} + g(\zeta,\gamma_n\boldsymbol{u})_{L_2(\Gamma)} = (\boldsymbol{f},\boldsymbol{u})_{\boldsymbol{L}_2(\Omega)}.$$

Taking into account the third equation in (1.36) and integrating by t, we can deduce the equation for full energy balance which is written down in an ordinary integral form

$$\frac{1}{2}\rho\int_\Omega |\boldsymbol{u}|^2\mathrm{d}\Omega + \frac{1}{2}g\rho\int_\Gamma |\zeta|^2\mathrm{d}\Gamma$$
$$= \frac{1}{2}\rho\int_\Omega |\boldsymbol{u}^0|^2\mathrm{d}\Omega + \frac{1}{2}g\rho\int_\Gamma |\zeta^0|^2\mathrm{d}\Gamma - \rho\nu\int_0^t E(\boldsymbol{u},\boldsymbol{u})(s)\mathrm{d}s$$
$$+ \rho\int_0^t \left(\int_\Omega \boldsymbol{f}\cdot\boldsymbol{u}\mathrm{d}\Omega\right)(s)\mathrm{d}s.$$

Let us consider once again the boundary conditions (1.35). It can be assumed that the initial value ζ^0 belongs to the space of generalized functions $H_\Gamma^{-1/2}$. Then $\boldsymbol{w}(0) = -g\nu^{-1}T\zeta^0 \in \boldsymbol{J}^1_{0,S}(\Omega)$, and if $\boldsymbol{u}^0(x) \in \boldsymbol{J}^1_{0,S}(\Omega)$, then the generalized solution $\boldsymbol{u}(t)$ of problem (1.1)–(1.4) exists. However, as it follows from the equation of energy balance, the solutions for which $\zeta^0 \notin L_{2,\Gamma}$ do not have a finite full energy and, therefore, they do not have any physical meaning.

8.1.6 Equation of Normal Oscillations

Let us consider normal oscillations of a viscous fluid in an open container, i.e., those solutions of the homogeneous problem (1.1)–(1.4) for which

$$\boldsymbol{u}(t,x) = \mathrm{e}^{-\lambda t}\boldsymbol{u}(x), \quad p(t,x) = \mathrm{e}^{-\lambda t}p(x), \quad \zeta(t,x_1,x_2) = \mathrm{e}^{-\lambda t}\zeta(x_1,x_2). \tag{1.37}$$

Performing the transition to the system of equations (1.27) as we did in Section 8.1.3, let us look for its solution in the form

$$\boldsymbol{\eta}(t,x) = \mathrm{e}^{-\lambda t}\boldsymbol{\eta}(x), \quad \boldsymbol{\delta}(t,x) = \mathrm{e}^{-\lambda t}\boldsymbol{\delta}(x), \quad \boldsymbol{\xi}(t,x) = \mathrm{e}^{-\lambda t}\boldsymbol{\xi}(x). \tag{1.38}$$

This gives the system of equations

$$\begin{aligned} -\lambda A^{-1/2}\boldsymbol{\eta} &= -\nu A^{1/2}\boldsymbol{\eta} + g\nu^{-1}T\gamma_n A^{-1/2}\boldsymbol{\xi}, \\ -\lambda A^{-1/2}\boldsymbol{\delta} &= -g\nu^{-1}T\gamma_n A^{-1/2}\boldsymbol{\xi}, \qquad \boldsymbol{\eta} + \boldsymbol{\delta} = \boldsymbol{\xi}. \end{aligned} \tag{1.39}$$

Adding up the first two equations and applying $A^{-1/2}$ to the sum, and then applying $A^{1/2}$ to the second equation, we obtain the following from (1.39),

$$\nu\boldsymbol{\eta} = \lambda A^{-1}\boldsymbol{\xi}, \quad \nu\boldsymbol{\delta} = g\lambda^{-1}B\boldsymbol{\xi}, \quad \boldsymbol{\eta} + \boldsymbol{\delta} = \boldsymbol{\xi}. \tag{1.40}$$

Now, if we add up the first two equations and use the third one, we obtain the following main spectral problem

$$\nu\boldsymbol{\xi} = \lambda A^{-1}\boldsymbol{\xi} + g\lambda^{-1}B\boldsymbol{\xi}, \tag{1.41}$$

which is considered in the Hilbert space $\boldsymbol{J}_{0,S}(\Omega)$. Here, A^{-1} is a positive and compact operator, and B is an infinite dimensional nonnegative compact operator with infinite dimensional kernel.

The next section is devoted to a detailed investigation of the spectrum structure and properties of eigen- and associated elements of the operator pencil which is generated by problem (1.41).

8.2 The Main Operator Pencil

Disregarding the notations in the previous subsection, let us consider the spectral problem

$$\varphi = \lambda A\varphi + \lambda^{-1}B\varphi, \qquad \lambda \neq 0, \tag{2.1}$$

in the separable Hilbert space E, where A is a compact positive operator and B is a compact nonnegative operator; both A and B act on E.

8.2.1 Structure of the Spectrum of the Problem

The pencil $L(\lambda) := I - \lambda A - \lambda^{-1}B$ is a Fredholm pencil on the whole complex plane $\mathbb{C}$ except for the point $\lambda = 0$. In virtue of the properties of operators A and B, $L(\lambda)$ is a positive definite operator and, therefore, invertible for negative λ. That is why, the assertion in Section 1.6.4 applies to this operator. According to that assertion, the spectrum of $L(\lambda)$ in $\mathbb{C} \setminus \{0\}$ consists of isolated points that are eigenvalues. These points can have only 0 and ∞ as accumulation points.

If $\lambda_0 \neq 0$ is an eigenvalue, then, in virtue of the compactness of operators A and B, the eigensubspace that corresponds to the number λ_0 is finite dimensional. Moreover, the resolvent $L^{-1}(\lambda) = (I - \lambda A - \lambda^{-1}B)^{-1}$ has a pole in λ_0 and the entire root subspace corresponding to the number λ_0 is finite dimensional.

Let λ_0 be an eigenvalue and φ_0 the eigenelement corresponding to it. We have

$$\varphi_0 = \lambda_0 A\varphi_0 + \lambda_0^{-1}B\varphi_0. \tag{2.2}$$

Multiplying both sides of this equation by φ_0 we obtain the quadratic equation satisfied by λ_0, that is,

$$\lambda_0^2(A\varphi_0, \varphi_0) - \lambda_0(\varphi_0, \varphi_0) + (B\varphi, \varphi_0) = 0, \tag{2.3}$$

hence

$$\lambda_0 = \frac{(\varphi_0, \varphi_0) \pm \sqrt{(\varphi_0, \varphi_0)^2 - 4(A\varphi_0, \varphi_0)(B\varphi_0, \varphi_0)}}{2(A\varphi_0, \varphi_0)}. \tag{2.4}$$

According to this formula, $\operatorname{Re}\lambda_0 > 0$. We exclude the case $(B\varphi_0, \varphi_0) = 0$ that leads to one of the values $\lambda_0 = 0$. Further, let $\operatorname{Im}\lambda_0 \neq 0$. That means that

$$(\varphi_0, \varphi_0)^2 < 4(A\varphi_0, \varphi_0)(B(\varphi_0, \varphi_0) \Longleftrightarrow \frac{(\varphi_0, \varphi_0)}{2(A\varphi_0, \varphi_0)} < \frac{2(B\varphi_0, \varphi_0)}{(\varphi_0, \varphi_0)}. \tag{2.5}$$

From (2.4) it follows that

$$|\lambda_0|^2 = \frac{(B\varphi_0, \varphi_0)}{(A\varphi_0, \varphi_0)} = \frac{2(B\varphi_0, \varphi_0)}{(\varphi_0, \varphi_0)} \cdot \frac{(\varphi_0, \varphi_0)}{2(A\varphi_0, \varphi_0)}.$$

Then from inequalities (2.5) we can obtain the following

$$\left(\frac{(\varphi_0,\varphi_0)}{2(A\varphi_0,\varphi_0)}\right)^2 < |\lambda_0|^2 < \left(\frac{2(B\varphi_0,\varphi_0)}{(\varphi_0,\varphi_0)}\right)^2. \tag{2.6}$$

Hence we get the estimates

$$(2\|A\|)^{-1} \le |\lambda_0| \le 2\|B\|.$$

Let us notice that if the following inequality

$$4(A\varphi,\varphi)(B\varphi,\varphi) < (\varphi,\varphi)^2 \tag{2.7}$$

holds true for all $\varphi \in E$, then the left side in inequality (2.6) is greater than the right side which means that there are no nonreal eigenvalues. Inequality (2.7) is called the *overdamped condition* for the pencil $L(\lambda)$.

If $\operatorname{Im}\lambda_0 \ne 0$, from (2.4) we get that

$$\operatorname{Re}\lambda_0 = \frac{(\varphi_0,\varphi_0)}{2(A\varphi_0,\varphi_0)} \ge (2\|A\|)^{-1}$$

holds true. Hence, all nonreal eigenvalues are located on the segment

$$\operatorname{Re}\lambda \ge (2\|A\|)^{-1}, \qquad |\lambda| \le 2\|B\|. \tag{2.8}$$

Let us note that if a nonreal number λ_0 is an eigenvalue of the pencil $L(\lambda)$, then the complex-adjoint number $\bar\lambda_0$ is an eigenvalue of $L(\lambda)$ as well. Indeed, since the pencil $L(\lambda)$ is self-adjoint, then $L(\lambda_0)$ is an invertible operator if and only if $(L(\lambda_0))^* = L(\bar\lambda_0)$ is an invertible operator.

Let us consider now the associated elements of φ_0 in the case $\operatorname{Im}\lambda_0 = 0$. The very first of them, φ_1, is a solution of the equation

$$(I - \lambda_0 A - \lambda_0^{-1}B)\varphi_1 + (-A + \lambda_0^{-2}B)\varphi_0 = 0.$$

Scalarly multiplying both sides of this equation by φ_0, using the fact that $L(\lambda_0) = I - \lambda_0 A - \lambda_0^{-1}B$ is a self-adjoint operator for $\lambda_0 \in \mathbb{R}$, and equation (2.2), we obtain

$$((-A + \lambda_0^{-2}B)\varphi_0,\varphi_0) = 0.$$

Considering this equation and equation (2.5) as a system of equations with respect to $(A\varphi_0,\varphi_0)$ and $(B\varphi_0,\varphi_0)$, we get that

$$2\lambda_0(A\varphi_0,\varphi_0) = (\varphi_0,\varphi_0), \quad 2\lambda_0^{-1}(B\varphi_0,\varphi_0) = (\varphi_0,\varphi_0).$$

We can rewrite these two equations as

$$\frac{(\varphi_0, \varphi_0)}{2(A\varphi_0, \varphi_0)} = \lambda_0 = \frac{2(B\varphi_0, \varphi_0)}{(\varphi_0, \varphi_0)},$$

a form that easily leads to the inequalities in (2.8). These equalities though do not hold true for the overdamped condition (2.7) in which case there are no associated elements.

Let us now formulate the final conclusions on the spectrum structure for problem (2.1). The spectrum of this problem consists of no more than a countable set of eigenvalues with finite algebraic multiplicities that are located in the right semiplane. The points $\lambda = 0$ and $\lambda = \infty$ are the only accumulation points of the spectrum. The nonreal eigenvalues are located symmetrically relatively to the real axis on the segment (2.8), and there are no more than a finite number of such eigenvalues. The corresponding eigenelements can have associated elements. If either the overdamped condition (2.7) or the more restrictive one,

$$4\|A\| \cdot \|B\| < 1, \tag{2.9}$$

hold true, then all the eigenvalues are real and there are no associated elements.

For further considerations let us notice that without loss of generality we can assume that $\lambda = 1$ does not belong to the spectrum of the pencil $L(\lambda)$. It is always valid due to the substitution $\lambda \mapsto a\lambda$, where $a > 0$ and does not belong to the spectrum of $L(\lambda)$. This substitution will leave all the properties of operators A and B unchanged.

8.2.2 Linearization of the Pencil

Let us perform the linearization of the pencil $L(\lambda)$ in the following special way. We will rewrite equation (2.1) in two equivalent forms, that is,

$$\varphi = (\lambda - \lambda^{-1})A\varphi + \lambda^{-1}(A+B)\varphi,$$
$$\lambda^{-1}\varphi = (A+B)\varphi - \lambda^{-1}(\lambda - \lambda^{-1})B\varphi.$$

Setting $\lambda - \lambda^{-1} = \mu$, $\lambda^{-1}\varphi = \psi$, we get the following system of equations

$$\begin{cases} \varphi = \mu A\varphi + (A+B)\psi, \\ \psi = (A+B)\varphi - \mu B\psi. \end{cases}$$

It will be convenient to consider this system of equations as an equation in the space $E^2 = E \oplus E$:

$$z = Rz + \mu Hz, \tag{2.10}$$

where

$$z = \begin{pmatrix} \varphi \\ \psi \end{pmatrix}, \quad R = \begin{pmatrix} 0 & A+B \\ A+B & 0 \end{pmatrix}, \quad H = \begin{pmatrix} A & 0 \\ 0 & -B \end{pmatrix}.$$

Let us consider the linear pencil $F(\mu) := I - R - \mu H$, which corresponds to problem (2.10). Going backwards, we can prove that for elements of the form

$$z = (\varphi; \lambda^{-1}\varphi)^{\mathrm{t}} \tag{2.11}$$

the next identity holds true,

$$F(\lambda - \lambda^{-1}) \begin{pmatrix} \varphi \\ \lambda^{-1}\varphi \end{pmatrix} = \begin{pmatrix} \varphi - \lambda A\varphi - \lambda^{-1}B\varphi \\ \lambda^{-1}\varphi - A\varphi - \lambda^{-2}B\varphi \end{pmatrix} = \begin{pmatrix} L(\lambda)\varphi \\ \lambda^{-1}L(\lambda)\varphi \end{pmatrix}. \tag{2.12}$$

Let us now find out the connections between the spectra of the pencils $L(\lambda)$ and $F(\mu)$. If λ_0 is an eigenvalue of the pencil $L(\lambda)$ and φ_0 is its corresponding eigenelement, then from (2.12) it follows that the element $z_0 = (\varphi_0; \lambda_0^{-1}\varphi_0)^{\mathrm{t}}$ is an eigenelement of the pencil $F(\mu)$ corresponding to the eigenvalue $\mu_0 = \lambda_0 - \lambda_0^{-1}$. Thus, each number $\mu_0 = \lambda_0 - \lambda_0^{-1}$, where λ_0 is a point in the spectrum of the pencil $L(\lambda)$, belongs to the spectrum of the pencil $F(\mu)$.

Now let μ_0 be an arbitrary complex number and $z_0 = (\varphi_0; \psi_0)^{\mathrm{t}}$ be an element in E^2 such that $F(\mu_0)z_0 = 0$. If the number λ_0 is a root of the equation $\lambda - \lambda^{-1} = \mu_0$, then the number $-\lambda_0^{-1}$ is a root of the same equation as well. Let us assume now that neither one of λ_0 and $-\lambda_0^{-1}$ belong to the spectrum of the pencil $L(\lambda)$. The auxiliary operator in E^2 defined as

$$D(\lambda) := \begin{pmatrix} \lambda^2 I & \lambda I \\ \lambda I & I \end{pmatrix}$$

commutes with the operator $F(\lambda - \lambda^{-1})$. Therefore,

$$F(\lambda_0 - \lambda_0^{-1})D(\lambda_0)z_0 = D(\lambda_0)F(\lambda_0 - \lambda_0^{-1})z_0 = 0.$$

Applying identity (2.12) to the element

$$D(\lambda_0)z_0 = \begin{pmatrix} \lambda_0^2\varphi_0 + \lambda_0\psi_0 \\ \lambda_0\varphi_0 + \psi_0 \end{pmatrix} \tag{2.13}$$

we obtain

$$F(\lambda_0 - \lambda_0^{-1})D(\lambda_0)z_0 = \begin{pmatrix} L(\lambda_0)(\lambda_0^2\varphi_0 + \lambda_0\psi_0) \\ \lambda_0^{-1}L(\lambda_0)(\lambda_0\varphi_0 + \psi_0) \end{pmatrix} = 0.$$

According to our assumption, λ_0 does not belong to the spectrum of $L(\lambda)$ and, therefore, $\lambda_0\varphi_0 + \psi_0 = 0$, or $\psi_0 = -\lambda_0\varphi_0$. Using now identity (2.12) for $\lambda = -\lambda_0^{-1}$ we get that

$$F(\lambda_0 - \lambda_0^{-1})z_0 = \begin{pmatrix} L(-\lambda_0^{-1})\varphi_0 \\ -\lambda_0 L(-\lambda_0^{-1})\varphi_0 \end{pmatrix} = 0. \tag{2.14}$$

Since the point $-\lambda_0^{-1}$ does not belong to the spectrum of $L(\lambda)$, $\varphi_0 = 0$ and, therefore, $\psi_0 = -\lambda_0\varphi_0 = 0$ and $z_0 = (\varphi_0; \psi_0)^t = 0$.

Finally, the spectrum of the pencil $F(\mu)$ is a univalent image of the pencil $L(\lambda)$ for the mapping $\mu = \lambda - \lambda^{-1}$. This univalence follows from the fact that if λ_0 belongs to the spectrum of the pencil $L(\lambda)$ and thus lies in the right half-plane, then $-\lambda_0^{-1}$ lies in the left half-plane and hence does not belong to the spectrum of the pencil $L(\lambda)$.

Let us note the following corollary:

Since, according to our assumptions, the point $\lambda = 1$ does not belong to the spectrum of the pencil $L(\lambda)$, then the number $\mu = 0$ does not belong to the spectrum of the pencil $F(\mu)$. Hence, it appears that the operator $F(0) = I - R$ is invertible.

8.2.3 Mutual Relationships between Eigen- and Associated Elements of the Two Pencils

If $z_0 = (\varphi_0; \psi_0)^t$ is an eigenelement of the pencil $F(\mu)$ corresponding to the eigenvalue μ_0 and λ_0 is a root of the equation $\lambda - \lambda^{-1} = \mu_0$ that belongs to the spectrum of $L(\lambda)$, then $D(\lambda_0)z_0 \neq 0$. In the opposite case, from (2.13) it follows that $\psi_0 = -\lambda_0\varphi_0$; but then from (2.14) it can be assumed that the number $-\lambda_0^{-1}$ belongs to the spectrum of $L(\lambda)$, which is not true. That is why, by choosing arbitrarily a basis in the eigensubspace of the pencil that corresponds to the eigenvalue μ_0, and by applying the operator $D(\lambda_0)$, with $\lambda_0 \in \sigma(L(\lambda_0))$, to each element in this basis, we obtain another basis whose elements have the form (2.11). The first components of these elements form a basis in the eigensubspace of the pencil $L(\lambda)$ that corresponds to the eigenvalue λ_0.

A similar reasoning can be applied to associated elements if we use the theory of root functions (see Section 1.6.3). Let $z(\mu)$ be a root function of the pencil $F(\mu)$ of order m relatively to the point μ_0. It means that $z(\mu_0) \neq 0$ and the function $F(\mu)z(\mu)$ has a zero of the order $m+1$ at the point $\mu = \mu_0$. Let λ_0 be a root of the equation $\lambda - \lambda^{-1} = \mu_0$, located in the spectrum of $L(\lambda)$. Since $\mu - \mu_0 = \lambda^{-1}(\lambda - \lambda_0)(\lambda + \lambda_0^{-1})$, then the function $F(\lambda - \lambda^{-1})z(\lambda - \lambda^{-1})$ also has a zero of the order $m+1$ at the point $\lambda = \lambda_0$. The function $D(\lambda)F(\lambda - \lambda^{-1})z(\lambda - \lambda^{-1}) = F(\lambda - \lambda^{-1})D(\lambda)z(\lambda - \lambda^{-1})$ has a zero of order not less than $m+1$ at the same point. According to the above proved facts, $D(\lambda_0)z(\lambda_0 - \lambda_0^{-1}) \neq 0$ and that is way the function $D(\lambda)z(\lambda - \lambda^{-1})$ is a root function of order not less than m of the pencil $F(\lambda - \lambda^{-1})$ relatively to the point $\lambda = \lambda_0$.

Let us denote

$$D(\lambda)z(\lambda - \lambda^{-1}) =: \begin{pmatrix} \tilde{\varphi}(\lambda) \\ \lambda^{-1}\tilde{\varphi}(\lambda) \end{pmatrix}. \tag{2.15}$$

In virtue of (2.12), the function $L(\lambda)\tilde{\varphi}(\lambda)$ has a zero of order not less than $m+1$ for $\lambda = \lambda_0$, and therefore, $\tilde{\varphi}(\lambda)$ is a root function of the pencil $L(\lambda)$ of order not less than m relatively to λ_0. Let

$$\tilde{\varphi}(\lambda) = \varphi_0 + (\lambda - \lambda_0)\varphi_1 + \cdots + (\lambda - \lambda_0)^m \varphi_m + \cdots. \tag{2.16}$$

Then φ_0 is an eigenelement, and $\varphi_1, \ldots, \varphi_m$ are associated elements to φ_0 of the pencil $L(\lambda)$.

If the holomorphic branch of solutions of the equation $\lambda - \lambda^{-1} = \mu$ for which $\lambda(\mu_0) = \lambda_0$ is denoted by $\lambda(\mu)$, then the function $D(\lambda(\mu))z(\mu)$ is a root function of order not less than m of the pencil $F(\mu)$ relatively to μ_0. Let us find the corresponding eigenelement and the associated elements of the pencil $F(\mu)$. We have

$$\begin{aligned} z_0 &= D(\lambda_0)z(\mu_0), \\ z_k &= \frac{1}{k!}\frac{d^k}{d\mu^k}(D(\lambda(\mu))z(\mu))\Big|_{\mu=\mu_0}, \qquad k = 1, \ldots, m. \end{aligned}$$

Calculating derivatives of the composite function, we obtain

$$\begin{aligned} z_k &= \frac{1}{k!}\sum_{j=1}^{k}\frac{\mathrm{d}^j}{\mathrm{d}\lambda^j}(D(\lambda)z(\lambda - \lambda^{-1}))\Big|_{\lambda=\lambda_0} \gamma_j(\mu_0) \\ &= \frac{1}{k!}\sum_{j=1}^{k}\frac{\mathrm{d}^j}{\mathrm{d}\lambda^j}\begin{pmatrix} \tilde{\varphi}(\lambda) \\ \lambda^{-1}\tilde{\varphi}(\lambda)\end{pmatrix}\Big|_{\lambda=\lambda_0} \gamma_j(\mu_0), \end{aligned}$$

where $\gamma_j(\mu)$ are coefficients that are expressed by derivatives of the function $\lambda(\mu)$. Using (2.16) we finally have

$$z_k = \frac{j!}{k!}\sum_{j=1}^{k}\left(\varphi_j; \sum_{i=0}^{j}(-1)^i \lambda_0^{-(i+1)}\varphi_{j-i}\right)^{\mathrm{t}} \gamma_j(\mu_0).$$

Elements of the form $(\varphi_j; \sum_{i=0}^{j}(-1)^i\lambda_0^{-(i+1)}\varphi_{j-i})^{\mathrm{t}}$, where φ_0 is eigenelement and φ_k are associated elements of the pencil $L(\lambda)$, are called the *elements of special form*.

Hence, the following assertion is proved:

In each root subspace of the pencil $F(\mu)$ a basis may be chosen that consists of elements of the special form.

8.2.4 Transformation to a Nondegenerate Pencil

The pencil $F(\mu)$ is degenerate in the following sense: The operator H in (2.10) has a nonzero kernel, $\operatorname{Ker} H \neq \{0\}$, that consists of elements of the form $(0;\psi)^{\mathrm{t}}$, where $\psi \in \operatorname{Ker} B \neq \{0\}$. Let Π_0 be the orthoprojector onto $\operatorname{Ker} H$ and $\Pi_1 = I - \Pi_0$ is the orthoprojector onto the closure of the range of operator H. Each element $z \in E^2$ admits the decomposition $z = \Pi_0 z + \Pi_1 z =: z_0 + z_1$. Obviously, operator Π_0 has the form $\operatorname{diag}(0; Q_0)$, where Q_0 is the orthoprojector onto the kernel of operator B. That is why

$$\Pi_0 R \Pi_0 = 0, \quad H\Pi_0 = 0. \tag{2.17}$$

Substituting the decomposition of z into equation (2.10) and taking into account (2.17), we obtain

$$z_0 + z_1 = Rz_0 + Rz_1 + \mu H z_1.$$

Applying the operators Π_0 and Π_1 leads to the following equations:

$$z_0 = \Pi_0 R z_1,$$
$$z_1 = \Pi_1 R z_0 + \Pi_1 R z_1 + \mu H z_1.$$

Excluding z_0, we come to the equation

$$z_1 = \Pi_1 R \Pi_0 R \Pi_1 z_1 + \Pi_1 R \Pi_1 z_1 + \mu H z_1.$$

If this equation is written down in the matrix form, then it looks like the following

$$z_1 = \begin{pmatrix} AQ_0A & (A+B)Q_1 \\ Q_1(A+B) & 0 \end{pmatrix} z_1 + \mu \begin{pmatrix} A & 0 \\ 0 & -B \end{pmatrix} z_1 =: Tz_1 + \mu G z_1, \tag{2.18}$$

where $z_1 = (\varphi;\psi)^{\mathrm{t}}$, $\varphi \in E$, $\psi \in E_1 := Q_1 E$, $Q_1 = I - Q_0$.

Let z_1 be a nonzero solution of the equation (2.18). The inverse transition to nonzero solutions of the equation (2.10), that is, to the eigenelements of the pencil $F(\mu)$, can be performed by the formula

$$z = z_0 + z_1 = \Pi_0 R z_1 + z_1.$$

Similarly, the equations for associated elements can be transformed.

8.2.5 The Property of Two-Multiple Basicity

Since the point $\mu = 0$ does not belong to the spectrum of the pencil $F(\mu)$, then it does not belong to the spectrum of problem (2.18). Hence, it appears that the operator $I - T$ is invertible. Here the following two cases are possible:

1) The eigenvalues of the operator T are less than 1;

2) There is a finite number (let us denote it by κ) of eigenvalues of the operator T that are greater than 1.

In the first case, the operator $I - T$ is positive. In the second case, $I - T$ does not have this property, but it can be represented as $I - T = J|I - T|$, with $J = P_+ - P_-$, where P_+ is an orthoprojector onto the invariant subspace corresponding to the positive part of the spectrum of the operator $I - T$ and $P_- = I - P_+$. The operator P_- is κ-dimensional.

In both cases, let us perform the substitution $|I - T|^{1/2} z_1 = y_1$ in equation (2.18) and apply the operator $|I - T|^{1/2}$ to both sides of this equation. Then equation (2.18) becomes

$$y_1 = \mu K y_1, \tag{2.19}$$

where $K := (I - T)^{-1/2} G (I - T)^{-1/2}$ in the first case, and $K := J|I - T|^{-1/2} G |I - T|^{-1/2}$ in the second case.

In the first case, K is compact and self-adjoint. Here there exists an orthonormal basis in the space $E \oplus E_1$. This basis consists of the eigenelements y_{1n}, $n = 1, 2, \ldots$, of operator K that correspond to the eigenvalues $\lambda_n = \lambda_n(K)$. The elements $z_{1n} = (I - T)^{-1/2} y_{1n}$ are solutions of the equation (2.18) for $\mu = \mu_n = 1/\lambda_n(K)$, and they form a Riesz basis in the space $E \oplus E_1$ and satisfy the orthogonality conditions

$$((I - T)z_{1j}, z_{1k})_{E\oplus E_1} = \left((I - T)^{1/2} z_{1j}, (I - T)^{1/2} z_{1k}\right)_{E\oplus E_1} = (y_{1j}, y_{1k})_{E\oplus E_1} = \delta_{jk}. \tag{2.20}$$

In the second case, let us introduce on the space $E \oplus E_1$ the indefinite scalar product

$$[u_1, y_1] := (J u_1, y_1)_{E\oplus E_1}, \tag{2.21}$$

that transforms this space into a space Π_κ. The operator K is compact and J-self-adjoint, that is,

$$\begin{aligned} [K v_1, y_1] &= (J K v_1, y_1)_{E\oplus E_1} = \left(|I - T|^{-1/2} G |I - T|^{-1/2} v_1, y_1\right)_{E\oplus E_1} \\ &= \left(v_1, |I - T|^{-1/2} G |I - T|^{-1/2} y_1\right)_{E\oplus E_1} = (v_1, J K y_1)_{E\oplus E_1} = [v_1, K y_1]. \end{aligned} \tag{2.22}$$

That is why the whole space $E \oplus E_1$ can be decomposed into the J-orthogonal sum $E \oplus E_1 = E_0[+]E_+$ whose terms are subspaces invariant relatively to K (see Section 1.3.9). The dimension of the subspace E_0 does not exceed 2κ, the subspace E_+ is infinite dimensional and the form $[y, y]$ is positive definite in E_+. In E_+, there exists a J-orthogonal basis consisting of eigenelements y_{1n}, $n = 1, 2, \ldots$, and corresponding to the eigenvalues $\lambda_n(K)$ of the operator K. The elements $z_{1n} = |I - T|^{-1/2} y_{1n}$ satisfy the equation (2.18) for $\mu_n = 1/\lambda_n(K)$ and form a Riesz basis in the subspace $|I - T|^{-1/2} E_+$.

From the conditions of J-orthonormality it follows that

$$\begin{aligned} ((I-T)z_{1j}, z_{1k})_{E\oplus E_1} &= (J|I-T|z_{1j}, z_{1k})_{E\oplus E_1} \\ &= \left(J|I-T|^{1/2} z_{1j}, |I-T|^{1/2} z_{1k}\right)_{E\oplus E_1} \\ &= (Jy_{1j}, y_{1k})_{E\oplus E_1} = \delta_{jk}. \end{aligned} \tag{2.23}$$

The subspace E_0 is finite-dimensional, therefore, there exists the basis of eigen- and associated elements of the pencil $I - \mu K$.

For the mapping $|I - T|^{-1/2}$, the images of these elements form a basis in the space $|I - T|^{-1/2} E_0$. This basis consists of the eigen- and associated elements of problem (2.18). The union of this particular basis and the previously mentioned one forms a Riesz basis in the entire space $E \oplus E_1$.

Considering the pencil $F(\mu)$ and then $L(\lambda)$, we obtain the following result that characterizes the property of two-multiple basicity of the pencil $L(\lambda)$. The system of eigen- and associated elements of the pencil $L(\lambda)$ that correspond to its spectrum $\{\lambda_n\}$, has the following property. For each pair of elements φ and ψ from E there exists a unique system of coefficients $\{c_{nk}\}$ for which

$$\varphi = \sum_{n,k} c_{nk} \varphi_{k,n},$$

$$Q_1 \psi = \sum_{n,k} c_{nk} \sum_{j=0}^{k} (-1)^j \lambda_n^{-(j+1)} Q_1 \varphi_{k-j,n}.$$

Here the following series converge

$$\sum_{n,k} |c_{nk}|^2 \|\varphi_{k,n}\|_E^2 < \infty,$$

$$\sum_{n,k} |c_{nk}|^2 \sum_{j=0}^{k} \|Q_1 \varphi_{k-j,n}\|_E^2 |\lambda_n|^{-2(j+1)} < \infty. \tag{2.24}$$

8.2.6 Spectral Factorization of the Overdamped Pencil. Separate Basicity

In Section 1, it was shown that if the overdamped condition (2.7) holds true, then problem (2.1) does not have nonreal eigenvalues and the eigenelements corresponding to the real eigenvalues do not have associate elements.

Let us consider the quadratic pencil $M(\lambda) := \lambda L(\lambda)$. The totality of all λ is called a *root region* of the pencil if for each λ there exists an element $\varphi \neq 0$, such that $(M(\lambda)\varphi, \varphi) = 0$. Let us denote by $p_\pm(\varphi)$ the roots of the latter quadratic equation with respect to λ:

$$p_\pm(\varphi) := \frac{(\varphi, \varphi) \pm \sqrt{(\varphi, \varphi)^2 - 4(A\varphi, \varphi)(B\varphi, \varphi)}}{2(A\varphi, \varphi)}. \tag{2.25}$$

If the condition (2.7) holds true, then $p_\pm(\varphi)$ are positive and distinct. By the homogeneity of these functionals, $p_\pm(\varphi)$ can be investigated on the unit sphere S of the space E. Let us divide the root region of the pencil $M(\lambda)$ into two root zones, Δ_+ and Δ_-. Δ_+ (Δ_-) consits of all values taken by the functional $p_+(\varphi)$ $(p_-(\varphi))$ on the sphere S. If the conditions (2.7) hold true, then the zones are nonempty connected subsets of the nonnegative semiaxis.

It is easy to prove that $\inf_{\varphi \in S} \Delta_- = 0$ and $\sup_{\varphi \in S} \Delta_+ = +\infty$. Indeed, if $\|\varphi_0\| = 1$ and $B\varphi_0 = 0$, that is, $\varphi_0 \in \operatorname{Ker} B \neq \{0\}$, then $p_-(\varphi_0) = 0$. If $B > 0$, then let us choose a sequence of eigenvalues $\lambda_n(B)$ of the operator B and the corresponding sequence of orthonormal eigenelements $\varphi_n(B)$. Then

$$\begin{aligned} p_-(\varphi_n(B)) &= \frac{1 - \sqrt{1 - 4(A\varphi_n, \varphi_n)\lambda_n(B)}}{2(A\varphi_n, \varphi_n)} \\ &= \frac{2\lambda_n(B)}{1 + \sqrt{1 - 4(A\varphi_n, \varphi_n)\lambda_n(B)}} \sim \lambda_n(B) \to 0, \qquad n \to \infty. \end{aligned}$$

Further, if $\{\lambda_n(A)\}_{n=1}^\infty$ is a sequence of eigenvalues of the operator A and $\{\varphi_n(A)\}_{n=1}^\infty$ is a sequence of its orthonormal eigenelements, then for the elements of this sequence we have

$$p_+(\varphi_n(A)) = \frac{1 + \sqrt{1 - 4\lambda_n(A)(B\varphi_n, \varphi_n)}}{2\lambda_n(A)} \sim [\lambda_n(A)]^{-1} \to +\infty, \qquad n \to \infty.$$

Let us introduce the notations

$$\alpha_- := \sup_{\varphi \in S} \Delta_-, \qquad \alpha_+ := \inf_{\varphi \in S} \Delta_+.$$

It turns out that, under the overdamped condition, the root zones Δ_- and Δ_+ are separated from each other, that is, the following inequality holds true,

$$\alpha_+ = \inf_{\varphi \in S} \Delta_+ > \alpha_- = \sup_{\varphi \in S} \Delta_-. \tag{2.26}$$

If condition (2.9)—called the *rough overdamped condition*—holds true, then the previous fact can be easily proved. Assuming (2.9) is true, let us introduce the quantities

$$r_\pm = \frac{1 \pm \sqrt{1 - 4\|A\| \cdot \|B\|}}{2\|A\|}.$$

Obviously, $p_-(\varphi) \leq r_-$ and $p_+(\varphi) \geq r_+$. That is why

$$\alpha_+ - \alpha_- \geq r_+ - r_- = \frac{\sqrt{1 - 4\|A\| \cdot \|B\|}}{\|A\|} > 0.$$

If only the condition (2.7) is valid, then the proof of (2.26) is rather complicated. This proof is being omitted here.

Now let us check that condition (2.7) is sufficient for the factorization of the pencil $M(\lambda) = \lambda L(\lambda) = \lambda I - B - \lambda^2 A$ relatively to the interval $[a, b] = [-r, r]$, where $r \in (\alpha_-, \alpha_+)$. For this matther let us use the conditions of spectral factorization of the operator-valued function $M(\lambda)$; these conditions have been formulated in the form 1° in Section 1.6.9, namely, if

$$M(-r) \ll 0, \quad M(r) \gg 0 \tag{2.27}$$

and the function $(M(\lambda)\varphi, \varphi) = 0$ has exactly one root in some neighbourhood $\mathcal{U}$ of the interval $[-r, r]$ for any $\varphi \neq 0$, then $M(\lambda)$ admits factorization.

Existence of unique root of the function $(M(\lambda)\varphi, \varphi)$ on the interval $[-r, r]$ is obvious in virtue of the fact that zone $\Delta_- = [0, \alpha_-] \subset [-r, r]$ and $r \in (\alpha_-, \alpha_+)$. Further, as far as $A > 0$, $B \geq 0$, then $M(-r) = -rI - B - r^2 A \leq -rI \ll 0$. Now let us prove the property $M(r) \gg 0$.

In virtue of the choice of r, this point does not belong to the closure $\overline{R(M(\lambda))}$ of the root region $R(M(\lambda))$. Let us show that $0 \notin \overline{W(M(r))}$, where $W(M(r))$ is the numerical range of operator $M(r)$. Let us assume the opposite case: $0 \in \overline{W(M(r))}$. Then there is a normalized sequence $\{\varphi_n\}_{n=1}^\infty$, such that $(M(r)\varphi_n, \varphi_n) \to 0$ as $n \to \infty$. The sequence of quadratic polynomials $(M(\lambda)\varphi_n, \varphi_n) =: p_n(\lambda)$ contains a subsequence that converges to the quadratic polynomial $p(\lambda)$ and $p(\lambda) \not\equiv 0$, because the coefficient of the first power of λ is 1. Since $p(r) = 0$, then, according to the Hurwitz Theorem, in any neighborhood of the point $\lambda = r$ there is at least one root of a quadratic polynomial with a large enough subscript from the mentioned subsequence. This contradicts the condition $r \notin \overline{R(M(\lambda))}$.

Hence, $M(r)$ is a positive definite ($\gg 0$) or negative definite ($\ll 0$) operator. Let us show that $M(r) \gg 0$. Indeed, for any $\varphi \neq 0$, the following decomposition to multipliers is valid

$$(M(\lambda)\varphi,\varphi) = ((\lambda I - B - \lambda^2 A)\varphi,\varphi) = -(A\varphi,\varphi)(\lambda - p_-(\varphi))(\lambda - p_+(\varphi)).$$

If $\lambda = r \in (\alpha_-, \alpha_+)$, then $\lambda - p_+(\varphi) < 0$, $\lambda - p_-(\varphi) > 0$, and since $(A\varphi,\varphi) > 0$, then $(M(r)\varphi,\varphi) > 0$.

Hence, it is proved that for any $r \in (\alpha_-, \alpha_+)$,

$$(M(r)\varphi,\varphi) \geq \delta(\varphi,\varphi), \qquad \delta > 0, \tag{2.28}$$

i.e., the second property (2.27) is valid too. Therefore, the operator quadratic polynomial $M(\lambda) = \lambda L(\lambda)$ admits the spectral factorization

$$M(\lambda) = \lambda I - B - \lambda^2 A = M_+(\lambda)(\lambda I - Z), \tag{2.29}$$

where $\sigma(Z) \subset [-r, r]$, and $M_+(\lambda)$ is a holomorphic and holomorphically invertible operator-valued function in the neighborhood of the interval $[-r, r]$, $r \in (\alpha_-, \alpha_+)$. Moreover, since the spectrum of $M(\lambda)$ belongs to the union of the sets $\bar{\Delta}_-$ and $\bar{\Delta}_+$, then $\sigma(Z) \subset [0, \alpha_-]$, and $M_+(\lambda)$ is holomorphic and holomorphically invertible in the circle of any radius $r \in (\alpha_-, \alpha_+)$.

For the operator Z there exists a positive definite symmetrizator and that is why Z is similar to a self-adjoint operator. From the equality (2.29) for $\lambda = 0$ it follows that $Z = M_+^{-1}(0)B$, and therefore, Z is a compact operator. From the latter equality it also follows that $\operatorname{Ker} Z = \operatorname{Ker} B \neq \{0\}$. The last circumstance makes the situation more complicated, compared to the one considered in Section 6.5.

Let us consider the spectral problem

$$Z\varphi = \lambda\varphi, \qquad \varphi \in E. \tag{2.30}$$

We represent the space E as an orthogonal sum $E = E_0 \oplus E_1$, where $E_0 := \operatorname{Ker} B$, and introduce operators of orthogonal projection P_i onto the subspaces E_i, $i = 0, 1$. For $\varphi \in E$ let us assume $\varphi_0 = P_0\varphi$ and $\varphi_1 = P_1\varphi$, and then $\varphi = \varphi_0 + \varphi_1$. Substituting this expression for φ into the equation (2.30) and projecting it onto E_1 and E_0 we obtain the following two equations,

$$P_1 Z P_1 \varphi_1 = \lambda\varphi_1, \quad \varphi_0 = \lambda^{-1} P_0 Z P_1 \varphi_1. \tag{2.31}$$

The first equation contains only φ_1. After finding its solution, we determine φ_0 from the second equation.

Let us denote by F the positive definite symmetrizator of operator Z and define

$$F_1 := P_1 F P_1 = F_1^*,$$
$$K_1 := P_1 Z F P_1 = K_1^*.$$

Since $Z\varphi = ZP_1\varphi$, $\varphi \in E$, then

$$K_1 = P_1 Z P_1 \cdot P_1 F P_1 =: P_1 Z P_1 \cdot F_1. \tag{2.32}$$

From the positive definiteness of F in E it follows that the operator F_1 is positive definite in E_1. In particular, it has an inverse operator in E_1. Then $P_1 Z P_1 = K_1 \cdot F_1^{-1}$ and the first equation (2.31) takes the form

$$K_1 F_1^{-1} \varphi_1 = \lambda \varphi_1. \tag{2.33}$$

We next make the substitution

$$F_1^{-1/2} \varphi_1 = \psi_1 \in E_1$$

and apply the operator $F_1^{-1/2}$ from the left to both sides of (2.33). Then we obtain

$$N\psi_1 := F_1^{-1/2} K_1 F_1^{-1/2} \psi_1 = \lambda \psi_1, \qquad \psi_1 \in E_1. \tag{2.34}$$

Operator N is a compact self-adjoint operator in E_1 and $\operatorname{Ker} N = \{0\}$. Indeed, if $N\psi_1 = 0$, then $K_1\varphi_1 = 0$, where $\varphi_1 = F_1^{-1/2}\psi_1 \in E_1$. Then from (2.32) it follows that $P_1 Z P_1 \cdot F_1 \varphi_1 = 0$ and therefore $F_1\varphi_1 = 0$ and $\psi_1 = F_1^{-1/2} F_1 \varphi_1 = 0$.

According to the Hilbert-Schmidt Theorem, the problem (2.34) has a discrete spectrum $\{\lambda_n^-\}_{n=1}^{\infty}$, $\lambda_n^- = \lambda_n(N) \neq 0$, which is located in the interval $[0, \alpha_-]$ and has the limit point $\lambda = 0$. The spectrum of operator Z coincides with the spectrum of operator N. The eigenelements $\{\psi_{1n}\}_{n=1}^{\infty}$ of operator N corresponding to the eigenvalues λ_n^- form an orthogonal basis in E_1. For solutions φ_n of the problem (2.30), the components $\varphi_{1n} = P_1\varphi_n$ can be obtained from solutions ψ_{1n} of the problem (2.34) by means of operator $F_1^{-1/2}$, which is bounded and invertible in E_1; therefore, these components form a Riesz basis in the space E_1.

Until now we have considered the spectral properties of the operator pencil in the zone $\Delta_- = [0, \alpha_-]$ which is bounded in $\mathbb{R}$ and adjacent to the coordinate center. To consider the similar issues for the root zone $\Delta_+ = [\alpha_+, \infty)$, let us perform a transformation of parameter in the pencil $L(\lambda)$ by setting $\lambda = 1/\mu$. Then instead of $L(\lambda)$ we have the following:

$$\tilde{L}(\mu) := L\left(\frac{1}{\mu}\right) = I - \mu B - \mu^{-1} A. \tag{2.35}$$

The new pencil $\tilde{L}(\mu)$ differs from $L(\lambda)$ only by notation, that is, using the formal substitution $\mu \mapsto \lambda$, $B \mapsto A$ and $A \mapsto B$ it can be converted into $L(\lambda)$. That is why all the reasoning performed above can be applied to $\tilde{L}(\mu)$.

Here, let us note that the initial root zone $\Delta_+ = [\alpha_+, \infty)$ is transformed to the zone $\tilde{\Delta}_- = [0, \tilde{\alpha}_-]$, $\tilde{\alpha}_- = 1/\alpha_+$, under the substitution $\mu = 1/\lambda$. It is still separated from the zone $\tilde{\Delta}_+ = [\tilde{\alpha}_+, \infty]$, $\tilde{\alpha}_+ = 1/\alpha_-$, and the new factorization of the pencil $\tilde{M}(\mu) := \mu\tilde{L}(\mu)$ has the following form, instead of (2.29),

$$\tilde{M}(\mu) = \tilde{M}_+(\mu)(\mu I - \tilde{Z}). \tag{2.36}$$

Further we once again derive conclusions concerning the factor $\tilde{Z} = \tilde{M}_+^{-1}(0)A$ from (2.36), and performed a transition to a spectral problem of the form (2.30),

$$\tilde{Z}\varphi = \mu\varphi, \qquad \varphi \in E,\ \mu = \lambda^{-1}, \tag{2.37}$$

which is being considered in the zone $\tilde{\Delta}_-$. Since operator A has a zero kernel, i.e., $E_0 = \{0\}$, then this problem is simpler than problem (2.30). That is why it is not required to perform any transition to a problem of the form (2.31).

Let us now formulate final conclusions that are based on considering the problem (2.37).

1° In the zone $\tilde{\Delta}_-$, the problem (2.37) has a discrete spectrum $\{\mu_n^-\}_{n=1}^{\infty}$ consisting of positive eigenvalues with finite multiplicities and the only limit point $\mu = 0$; to this spectrum, there corresponds a discrete spectrum $\{\lambda_n^+\}_{n=1}^{\infty}$, $\lambda_n^+ = 1/\mu_n^-$, of the pencil $L(\lambda)$ which is located in the zone $\Delta_+ = [\alpha_+, \infty)$.

2° Eigenelements of the problem (2.37) or, similarly, eigenelements of the pencil $L(\lambda)$ corresponding to the eigenvalues λ_n^+ from Δ_+ form a Riesz basis in the space E.

Hence, if the overdamped condition (2.7) holds true, then the spectrum of the pencil $L(\lambda)$ is located in the zones Δ_-, Δ_+ and has two branches $\{\lambda_n^-\}_{n=1}^{\infty}$, $\lambda_n^- \to 0$, $n \to \infty$, and $\{\lambda_n^+\}_{n=1}^{\infty}$, $\lambda_n^+ \to \infty$, $n \to \infty$; the eigenelements $\{\varphi_n^-\}_{n=1}^{\infty}$ and $\{\varphi_n^+\}_{n=1}^{\infty}$ that correspond to the two mentioned branches of the spectrum, have the above-formulated properties of basicity in E_1 and E, respectively.

8.2.7 Double-Sided Inequalities for the Two Branches of Eigenvalues

To obtain double-sided estimates for the numbers $\{\lambda_n^-\}_{n=1}^{\infty}$ and $\{\lambda_n^+\}_{n=1}^{\infty}$, let us use the variational principles stated in Section 1.6.11.

We assume that the overdamped condition (2.7) holds true. Then, in the zone $\Delta_- = [0, \alpha_-]$, the eigenvalues $\{\lambda_n^-\}_{n=1}^{\infty}$ of the pencil $M(\lambda) = \lambda L(\lambda)$ form a nonincreasing sequence $\lambda_n^- \searrow 0$, $n \to \infty$. Further, for the pencil $M(\lambda) = \lambda I - B - \lambda^2 A$ (as it has already been proved in Section 8.2.6), conditions 1° in Section 1.6.10 hold true. These conditions are sufficient for its spectral factorization. Besides, for $\lambda = 0$, we have $M(0) = -B \in \mathfrak{S}_\infty$. That is why, according to assertion 9° in Section 1.6.11, the following variational principles are valid for the eigenvalues $\{\lambda_n^-\}_{n=1}^\infty$:

$$\lambda_n^- = \min_{\dim N^\perp = n-1} \max_{0 \neq \varphi \in N} p_-(\varphi), \tag{2.38}$$

$$\lambda_n^- = \max_{\dim M = n} \min_{0 \neq \varphi \in M} p_-(\varphi). \tag{2.39}$$

We recall that here M is an arbitrary n-dimensional subspace of E, and N is an arbitrary subspace of E with codimension $n-1$.

Let us make a formal substitution $\mu = 1/\lambda$ in the pencil $M(\lambda)$. Then after this substitution, the eigenvalues λ_n^- are converted into the numbers $\mu_n^+ = 1/\lambda_n^-$ located in the unbounded zone $[\tilde{\alpha}_+, \infty)$, $\tilde{\alpha}_+ = 1/\alpha_-$, and for them the following variational principles are valid, instead on (2.38)–(2.39):

$$\mu_n^+ = \frac{1}{\lambda_n^-} = \max_{\dim N^\perp = n-1} \min_{0 \neq \varphi \in N} \frac{1}{p_-(\varphi)} =: \max_{\dim N^\perp = n-1} \min_{0 \neq \varphi \in N} \tilde{p}_+(\varphi), \tag{2.40}$$

$$\mu_n^+ = \frac{1}{\lambda_n^-} = \min_{\dim M = n} \max_{0 \neq \varphi \in M} \tilde{p}_+(\varphi). \tag{2.41}$$

We next use the obvious inequalities

$$\begin{aligned}
0 &\leq \frac{(\varphi,\varphi)}{(B\varphi,\varphi)} - \tilde{p}_+(\varphi) \\
&= \frac{(\varphi,\varphi)}{(B\varphi,\varphi)} - \frac{(\varphi,\varphi) + \sqrt{(\varphi,\varphi)^2 - 4(A\varphi,\varphi)(B\varphi,\varphi)}}{2(B\varphi,\varphi)} \\
&= \frac{(\varphi,\varphi) - \sqrt{(\varphi,\varphi)^2 - 4(A\varphi,\varphi)(B\varphi,\varphi)}}{2(B\varphi,\varphi)} \\
&= \frac{2(A\varphi,\varphi)}{(\varphi,\varphi) + \sqrt{(\varphi,\varphi)^2 - 4(A\varphi,\varphi)(B\varphi,\varphi)}} \\
&\leq 2\frac{(A\varphi,\varphi)}{(\varphi,\varphi)} \leq 2\|A\|.
\end{aligned}$$

Then,

$$\frac{(\varphi,\varphi)}{(B\varphi,\varphi)} - 2\|A\| \leq \tilde{p}_+(\varphi) \leq \frac{(\varphi,\varphi)}{(B\varphi,\varphi)}. \tag{2.42}$$

Now, let us use the principle (2.40) and the following obvious principle,

$$\max_{\dim N^\perp = n-1} \min_{0 \neq \varphi \in N} \frac{(\varphi,\varphi)}{(B\varphi,\varphi)} = \frac{1}{\min\limits_{\dim N^\perp = n-1} \max\limits_{0 \neq \varphi \in N} \dfrac{(B\varphi,\varphi)}{(\varphi,\varphi)}} = \frac{1}{\lambda_n(B)}.$$

Then, from the inequalities (2.42) it follows that

$$\frac{1}{\lambda_n(B)} - 2\|A\| \le \frac{1}{\lambda_n^-} \le \frac{1}{\lambda_n(B)}, \tag{2.43}$$

and hence we finally obtain the double-sided estimates,

$$\lambda_n(B) \le \lambda_n^- \le \frac{\lambda_n(B)}{1 - 2\lambda_n(B)\|A\|}, \qquad n = 1, 2, \dots. \tag{2.44}$$

Inequalities (2.44) show that the branch $\{\lambda_n^-\}_{n=1}^{\infty}$ of eigenvalues moves to the right relatively to the eigenvalues of operator B and formula (2.43) shows that for the inverse magnitudes the corresponding displacement to the left does not exceed $2\|A\|$.

Similar double-sided inequalities can be also obtained for the second branch $\{\lambda_n^+\}_{n=1}^{\infty}$ of the pencil $L(\lambda) = I - \lambda A - \lambda^{-1}B$, which is located in the zone $[\alpha_+, \infty)$. Without adducing the corresponding reasoning, let us note that the result can be obtained by the formal substitution $\lambda_n^- \mapsto 1/\lambda_n^+$, $A \mapsto B$, $B \mapsto A$. Then instead of (2.43) we have

$$\frac{1}{\lambda_n(A)} - 2\|B\| \le \lambda_n^+ \le \frac{1}{\lambda_n(A)}, \qquad n = 1, 2, \dots. \tag{2.45}$$

These formulas show that the branch $\{\lambda_n^+\}_{n=1}^{\infty}$ moves to the left compared to the eigenvalues $\lambda_n(A^{-1}) = 1/\lambda_n(A)$ of the (unbounded positive definite) operator A^{-1}. The magnitude of this displacement to the left for any number λ_n^+ compared to $\lambda_n(A^{-1})$ does not excced $2\|B\|$. The following asymptotic formulas are corollaries of the inequalities (2.44) and (2.45):

$$\lambda_n^- = \lambda_n(B)[1 + \mathrm{o}(1)], \qquad n \to \infty, \tag{2.46}$$

$$\lambda_n^+ = \frac{1}{\lambda_n(A)} + \mathrm{O}(1), \qquad n \to \infty. \tag{2.47}$$

Summing up the result of the previous analysis of the spectral problem for the pencil $L(\lambda) = I - \lambda A - \lambda^{-1}B$ under the overdamped condition, let us note the following. The given pencil can be considered as a perturbation of the linear pencil $L_1(\lambda) = I - \lambda A$ and at the same time of the pencil $L_2(\lambda) = I - \lambda^{-1}B$, which can be reduced to a linear pencil. Obviously, for the pencil $L_1(\lambda)$ the spectrum consists of isolated eigenvalues $\lambda_n^{(1)} = 1/\lambda_n(A) \to \infty$, $n \to \infty$, and the eigenelements form an orthonormal basis in E. In the second case, the spectrum of the pencil $L_2(\lambda)$ consists of eigenvalues $\lambda_n^{(2)} = \lambda_n(B) \to 0$, $n \to \infty$, and the eigenelements corresponding to nonzero eigenvalues form an orthonormal basis in $E_1 \subset E$. For the mentioned perturbation of the pencils $L_1(\lambda)$ and $L_2(\lambda)$, i.e., the transition to the pencil $L(\lambda)$, the spectra of both nonperturbed problems move to the left with no more than $2\|B\|$ in the case of $L_1(\lambda)$, and to the right with no more than $2\|A\|$ in the case of $L_2(\lambda)$ for the inverse values $1/\lambda_n^{(2)}$. Further, the eigenelements of the pencils $L_1(\lambda)$ and $L_2(\lambda)$ are orthonormal bases in E and E_1 respectively; in the perturbed case, they become Riesz bases in E and (after projecting) in E_1.

8.2.8 The General Case

Before considering the general situation, i.e., when the overdamped condition (2.7) is not necessarily valid for the pencil $L(\lambda)$, we give some definitions.

An eigenelement φ_0 of the pencil $M(\lambda) = \lambda L(\lambda) = \lambda I - B - \lambda^2 A$ corresponding to the real eigenvalue λ_0 is called *positive* (*negative*) provided $-(M'(\lambda_0)\varphi_0, \varphi_0) > 0$ (< 0). The eigenvalue $\lambda_0 \in \mathbb{R}$ is called *positive* (*negative*) if all the corresponding eigenelements are positive (negative).

Let us note that (as it has been mentioned at the beginning of Section 1.6.10) there are no associated elements corresponding to positive or negative eigenelements because in this case $(M'(\lambda_0)\varphi_0, \varphi_0) \neq 0$.

If the overdamped condition holds true, then for any $\varphi \neq 0$,

$$(M'(p_\pm(\varphi))\varphi, \varphi) = -((I - 2p_\pm(\varphi))\varphi, \varphi) = \pm\sqrt{(\varphi,\varphi)^2 - 4(A\varphi,\varphi)(B\varphi,\varphi)} \neq 0. \tag{2.48}$$

Hence it appears that all the nonzero eigenvalues of the pencil $M(\lambda)$ (or, which is the same, eigenvalues of the pencil $L(\lambda)$) located in the zone Δ_+ are positive, and the eigenvalues from the zone Δ_- are negative. It is obvious that in this case all eigenvalues of $M(\lambda)$ are positive or negative.

If condition (2.7) is not valid and $4\|A\| \cdot \|B\| > 1$, then the eigenvalues $\lambda_0 \in \mathbb{R}$ that are neither positive nor negative (let us call them *degenerate*), are located in the interval $[(2\|A\|)^{-1}, 2\|B\|]$. Here, as it follows from (2.48), the negative eigenvalues $\{\lambda_n^-\}_{n=1}^\infty$ are located in the interval $(0, (2\|A\|)^{-1})$ and the positive eigenvalues $\{\lambda_n^+\}_{n=1}^\infty$ are located in the interval $(2\|B\|, +\infty)$.

Now let us consider the issues on basicity of eigenelements corresponding to $\{\lambda_n^+\}_{n=1}^\infty$ and $\{\lambda_n^-\}_{n=1}^\infty$, the positive and negative branches of eigenvalues. We show that here the properties of Riesz basicity in E and E_1, that should take place in the overdamped case, are replaced by the property of Riesz basicity with finite defect accuracy.

Let us consider the equation

$$M(\lambda)\varphi := (\lambda I - B - \lambda^2 A)\varphi = 0, \qquad \varphi \in E. \tag{2.49}$$

Since the operator B equals zero on the subspace $E_0 = \operatorname{Ker} B \neq \{0\}$, then the vector-matrix form of equation (2.49) is the following,

$$\lambda \begin{pmatrix} \varphi_1 \\ \varphi_0 \end{pmatrix} = \begin{pmatrix} B_1 & 0 \\ 0 & 0 \end{pmatrix} \begin{pmatrix} \varphi_1 \\ \varphi_0 \end{pmatrix} + \lambda^2 \begin{pmatrix} P_1AP_1 & P_1AP_0 \\ P_0AP_1 & P_0AP_0 \end{pmatrix} \begin{pmatrix} \varphi_1 \\ \varphi_0 \end{pmatrix}, \tag{2.50}$$

where $\varphi_0 \in E_0$, $\varphi_1 \in E_1$, and P_0, P_1 are orthoprojectors onto the subspaces E_0, E_1 corresponding to the decomposition $E = E_0 \oplus E_1$, $B_1 := P_1 B P_1$.

From the second equation (2.50) we have (for $\lambda \neq 0$)

$$(I_0 - \lambda P_0 A P_0)\varphi_0 = \lambda P_0 A P_1 \varphi_1. \tag{2.51}$$

Let us assume that λ does not belong to the spectrum of the linear pencil $I_0 - \lambda P_0 A P_0$. For example, suppose that λ is sufficiently small. Then, determining φ_0 from (2.51) and substituting it into the first equation (2.50), we obtain the spectral problem in the subspace E_1,

$$M_1(\lambda)\varphi_1 := \left(\lambda I_1 - B_1 - \lambda^2 P_1 A P_1 - \lambda^3 P_1 A P_0 (I_0 - \lambda P_0 A P_0)^{-1} P_0 A P_1\right)\varphi_1 = 0. \tag{2.52}$$

It is easy to notice that $M_1(\lambda)$ is a self-adjoint operator-valued function, which is analytic at zero, and $\operatorname{Ker} B_1 = \{0\}$. Moreover, for $M_1(\lambda)$ the following conditions hold true,

$$\begin{aligned} M_1'(0) &= I_1 \gg 0 \qquad \text{in } E_1, \\ M_1(0) &= -B_1 \in \mathfrak{S}_\infty. \end{aligned} \tag{2.53}$$

That is why, according to assertion 4° in Section 1.6.10, we get that for any sufficiently small $\varepsilon > 0$ the eigenelements of the pencil $M_1(\lambda)$ corresponding to (negative) eigenvalues λ_n^- from the interval $(0, \varepsilon)$ form a Riesz basis in E_1 with accuracy to a finite defect.

Since there is no more than a finite number of negative eigenvalues outside the interval $(0, \varepsilon)$, we can conclude that all negative eigenelements of problem (2.49) corresponding to the negative eigenvalues $\{\lambda_n^-\}_{n=1}^\infty$ form a Riesz basis in E_1 with accuracy to a finite defect (after projecting onto E_1).

In order to find out the similar properties for eigenvalues $\{\lambda_n^+\}_{n=1}^\infty$ and their corresponding eigenelements, let us perform the substitution $\lambda = 1/\mu$ in the pencil $L(\lambda)$ and consider the following pencil in the space E,

$$\tilde{M}(\mu) := \mu L\left(\frac{1}{\mu}\right) = \mu I - A - \mu^2 B; \tag{2.54}$$

this pencil differs from $M(\lambda)$ only in notation. Since in this case

$$\operatorname{Ker} A = \{0\}, \quad \tilde{M}'(0) = I \gg 0, \quad \tilde{M}(0) = -A \in \mathfrak{S}_\infty, \tag{2.55}$$

then again we can use conclusion 4° in Section 1.6.10 on basicity with accuracy to a finite defect in E of eigenelements of the pencil $\tilde{M}(\mu)$ corresponding to negative eigenvalues from the interval $(0, \varepsilon)$. It can be easily checked that to these numbers there correspond positive eigenvalues of the pencil $M(\lambda)$; hence we have the following conclusion: all the positive eigenelements of the problem (2.49) corresponding to the positive eigenvalues $\{\lambda_n^+\}_{n=1}^{\infty}$ form a Riesz basis in a subspace that has finite defect in E.

The following natural question arises. How many and what eigen- and associated elements of the pencil $L(\lambda)$ should be added to the totality of positive eigenelements corresponding to the positive eigenvalues $\{\lambda_n^+\}_{n=1}^{\infty}$ in order to obtain a Riesz basis in E? A similar question for the set of negative eigenelements corresponding to the negative eigenvalues $\{\lambda_n^-\}_{n=1}^{\infty}$ can be asked: after extending and projecting it onto E_1 will this set become a Riesz basis in E_1?

It turns out that we can add (following a certain law) a finite number of eigen- and associated elements corresponding to nonreal eigenvalues and also to the real ones that are neither positive nor negative to the set of positive eigenelements corresponding to the positive eigenvalues of the pencil $M(\lambda) = \lambda L(\lambda)$. Extended in this way, the set becomes a Riesz basis in E. Similarly, there exists a certain rule by which we can add eigen- and associated elements (possibly different ones) to the set of negative eigenelements to tranform it into a Riesz basis in E_1 after projecting it onto E_1.

Hence, the properties of separate Riesz basicity of two "one-halves" of the system of eigen- and associated elements of the pencil $L(\lambda)$ are valid, together with the properties of the two-multiple Riesz basicity of the system of eigen- and associated elements of the pencil $L(\lambda)$ that were proved in Section 8.2.5 and mean Riesz basicity of the system of eigen- and associated elements in the space $E \oplus E_1$ for the pencil (2.18). Let us note that the first "half" meant is in E (the extended system of positive elements) and the second one in E_1 after projecting onto E_1 (the extended system of negative elements).

8.3 Normal Oscillations and the Spectrum of the Hydrodynamics Problem

Based on the general conclusions of the previous section, we formulate here the properties of normal oscillations of the problem stated in Section 8.1.

8.3.1 General Properties of the Spectrum

Let us consider the problem (1.41) on normal oscillations of a heavy viscous fluid in a partially filled container:

$$L_\nu(\lambda)\boldsymbol{\xi} := [I - \nu^{-1}(\lambda A^{-1} + g\lambda^{-1}B)]\boldsymbol{\xi} = \mathbf{0}, \qquad \boldsymbol{\xi} \in \boldsymbol{J}_{0,S}(\Omega). \tag{3.1}$$

According to the conclusions of Section 8.1, the operator A^{-1} acting in the space $\boldsymbol{J}_{0,S}(\Omega)$ is compact and positive and B is a compact nonnegative operator. In addition, the physical constants g (the acceleration of the gravity force) and ν (the coefficient of kinematic viscousity) are positive too.

Hence, it appears that (3.1) is a special case of an operator pencil corresponding to problem (2.1). It can be obtained from (2.1) by the formal substitutions $A \mapsto A^{-1}/\nu$, $B \mapsto gB/\nu$. That is why, using the conclusions of Section 8.2, the following general properties of solutions of the problem (3.1) can be formulated.

1° The spectrum of the problem (3.1) is discrete and consists of a countable set of eigenvalues with finite algebraic multiplicity that are located in the right half-plane and have the points $\lambda = 0$, $\lambda = \infty$ as accumulation points.

2° Depending on the value of viscousity ν, the spectrum of problem (3.1) can be naturally divided into three, or two subsets. That is, if the following condition holds true,

$$\nu^2 \leq 4g\|A^{-1}\| \cdot \|B\|, \tag{3.2}$$

then the branch $\{\lambda_k^+\}_{k=1}^{\infty}$ of positive eigenvalues with the limit point $\lambda = +\infty$ is located in the interval $\Delta_+ = (2g\nu^{-1}\|B\|; +\infty)$; the corresponding eigenelements $\{\boldsymbol{\xi}_k^+\}_{k=1}^{\infty}$ do not have associated elements and the following conditions are valid,

$$-\left((\lambda L_\nu(\lambda))'_{\lambda=\lambda_k^+}\boldsymbol{\xi}_k^+, \boldsymbol{\xi}_k^+\right)_{\boldsymbol{L}_2(\Omega)} > 0, \qquad k = 1, 2, \ldots. \tag{3.3}$$

The branch $\{\lambda_k^-\}_{k=1}^{\infty}$ of negative eigenvalues, $\lambda_k^- \to 0$, $k \to \infty$, is located in the interval $\Delta_- = (0; \nu(2\|A^{-1}\|)^{-1}$. The corresponding eigenelements $\{\boldsymbol{\xi}_k^-\}_{k=1}^{\infty}$ do not have associated elements and

$$-\left((\lambda L_\nu(\lambda))'_{\lambda=\lambda_k^-}\boldsymbol{\xi}_k^-, \boldsymbol{\xi}_k^-\right)_{\boldsymbol{L}_2(\Omega)} < 0, \qquad k = 1, 2, \ldots. \tag{3.4}$$

Finally, in the segment

$$\Delta_0 = \left\{\lambda : \operatorname{Re}\lambda \geq \nu(2\|A^{-1}\|)^{-1}, |\lambda| \leq 2g\nu^{-1}\|B\|\right\} \tag{3.5}$$

there are only a finite number of nonreal eigenvalues that are located symmetrically relative to the real axis, and real eigenvalues for which the corresponding eigenelements have associated elements. Let us call this finite (or empty) set of eigenvalues for problem (3.1) from the segment Δ_0 the set of *intermediate eigenvalues*.

3° If the rough overdamped condition

$$\nu^2 > 4g\|A^{-1}\| \cdot \|B\| \tag{3.6}$$

holds true, then there are no intermediate eigenvalues. In this case, the branch of positive eigenvalues is located in the interval $(r_+, +\infty)$ and the branch of negative eigenvalues is situated in the interval $(0, r_-)$,

$$r_\pm = \frac{\nu \pm \sqrt{\nu^2 - 4g\|A^{-1}\| \cdot \|B\|}}{2\|A^{-1}\|}, \qquad 0 < r_- < r_+ < \infty. \tag{3.7}$$

4° To positive eigenvalues $\{\lambda_k^+\}$ with large numbers k there correspond quickly fading nonperiodic modes of normal movements of the viscous fluid. Indeed, in this case, the dependence of the normal solutions on time has the form $\exp(-t\lambda_k^+)$, where the (logarithmic) fading decrements λ_k^+ are big. Such movements, which are similar to the corresponding normal oscillations of a viscous fluid in a wholly filled container described in Section 7.1, can be naturally called the *internal dissipative waves.*

5° To negative eigenvalues $\{\lambda_k^-\}$ there corresponds a new form of normal movements that does not have an analogue in a wholly filled container: for large numbers k, these movements are modes with arbitrarily small fading decrements λ_k^-. Their appearence is caused by the presence of the free surface of the fluid and by the influence of the gravitation field. Such movements are called the *surface gravitation waves.*

6° Finally, if condition (3.2) holds true, then the finite number of fading decrements λ located in segment (3.5) is possible in the system. The corresponding normal oscillations are called the *intermediate waves.* For this type of movements the dependence of time for real fading decrements λ can be described not only by the law $\exp(-t\lambda)$, but also by $t^m \exp(-t\lambda)$ with some integer m, which correspnds to the existence not only of eigenelements, but also of associated ones. Moreover, to nonreal fading decrements λ that are located in the mentioned segment, there correspond normal movements taking place according to the law $\exp(-t\operatorname{Re}\lambda)\exp(-it\operatorname{Im}\lambda)$. Hence, it appears that the presence of the free surface of a fluid generates the appearance of oscillating time fading modes of normal oscillations (in the case of small viscousity).

7° For the decrements λ_k^+ of internal dissipative waves the following double-sided inequalities hold true,

$$\nu\lambda_k(A) - 2g\nu^{-1}\|B\| \le \lambda_k^+ \le \nu\lambda_k(A), \qquad k = 1, 2, \ldots; \tag{3.8}$$

from (3.8) the asymptotic formula below follows for $k \to \infty$,

$$\lambda_k^+ = \nu\lambda_k(A)[1 + \mathrm{o}(1)] = \nu\left(\frac{\operatorname{mes}\Omega}{3\pi^2}\right)^{-2/3} k^{2/3}[1 + \mathrm{o}(1)]. \tag{3.9}$$

Respectively, for the decrements λ_k^- of surface gravitation waves the following inequalities,

$$g\nu^{-1}\lambda_k(B) \le \lambda_k^- \le g\nu^{-1}\frac{\lambda_k(B)}{1-2g\lambda_k(B)\|A^{-1}\|\nu^{-2}}, \tag{3.10}$$

and the asymptotic formula

$$\lambda_k^- = g\nu^{-1}\lambda_k(B)[1+\mathrm{o}(1)] = \frac{g}{\nu}\left(\frac{\operatorname{mes}\Gamma}{16\pi}\right)^{1/2} k^{-1/2}[1+\mathrm{o}(1)] \tag{3.11}$$

take place.

8.3.2 Influence of Fluid Viscosity on the Structure of the Spectrum of the Problem

The conclusions of Section 8.3.1 show that the spectrum structure of problem (3.1) depends on the value of the fluid viscousity ν. Let us consider the quantitative changes of spectrum in the process when all parameters of the physical system remain unchanged, fluid viscousity ν changes from sufficiently large to sufficiently small values and turns zero after transition to the limit.

Obviously, for big ν the condition (3.6) holds true and, therefore, only internal dissipative and surface gravitation waves take place in this system. The distance between the branches $\{\lambda_n^+\}_{n=1}^\infty$ and $\{\lambda_n^-\}_{n=1}^\infty$ of these waves is not less than the magnitude

$$r_+ - r_- = \frac{\sqrt{\nu^2 - 4g\|A^{-1}\|\cdot\|B\|}}{\|A^{-1}\|} \sim \frac{\nu}{\|A^{-1}\|}, \qquad \nu\to\infty.$$

For decreasing ν this difference lessens and turns to zero for $\nu=\nu_*$, where $\nu_*^2 = 4g\|A^{-1}\|\cdot\|B\| > 0$. Here, during the process of decreasing ν, the positive eigenvalues λ_k^+ (as shown by (3.8)) move to the left and the negative eigenvalues move to the right according to (3.10).

Further, the following phenomenon takes place. If for some $\nu < \nu_*$ a collision of any positive and any negative eigenvalues occurs, then a qualitative modification of spectrum structure takes place. The essence of this modification is the following: these eigenvalues move from the real axis and become a complex-conjugate pair of eigenvalues of the problem (3.1): this pair is located in the segment (3.5). If the parameter ν decreases continuously from one positive value to another, then there is no more than a finite number of such collisions. This fact explains the finiteness of the number of eigenvalues of the problem (3.1) for any $\nu > 0$.

If ν decreases from $\nu=\nu_*$ to $\nu=0$, then the segment (3.5) extends and after transition to the limit moves to the right complex half-plane. In this case, the number of intermediate eigenvalues increases to infinity and the complex-conjugate pairs $(\lambda_0)^{\pm}$ come close to the imaginary axis for $\nu \to 0$ (the smaller the number of pairs, the closer they come to the imaginary axis). Such a qualitative situation can be easily explained physically because for $\nu = 0$ there should be a countable set of intermediate eigenvalues $\{(\lambda_k^0)^{\pm}\}_{k=1}^{\infty}$, they should be located on the imaginary axis and generate oscillation frequencies of a heavy ideal fluid in a partially filled container (see Section 3.3).

For small values of ν, the eigenvalues $(\lambda_k^0)^{\pm}$ determine oscillation frequency and fading decrement and are expressed by the asymptotic formula

$$(\lambda_k^0)^{\pm} = \mathrm{i}\omega_k^{\pm} + \frac{1+\mathrm{i}}{2\sqrt{2}}\nu^{1/2} g \left|\omega_k^{\pm}\right|^{3/2} a_k + \mathrm{O}(\nu), \qquad \nu \to 0, \tag{3.12}$$

with

$$a_k = \frac{\displaystyle\int_S |\nabla \Phi_k|^2 dS}{\displaystyle\int_\Gamma |\Phi_k|^2 d\Gamma},$$

where $\omega_k^{\pm}$ are oscillation frequencies of an ideal heavy fluid in a container which are defined by (3.3.21) and (3.3.25), and $\Phi_k = \Phi_k(x)$ is a potential of the fluid velocity corresponding to the frequency $\omega = \omega_k^{\pm}$.

Another limit case, that is, when $\nu \to \infty$, will be considered in detail in Section 8.5.

8.3.3 Properties of Surface and Internal Waves

To specify the properties of internal and surface waves, we perform the transition from (3.1), that is, from the equation

$$\nu \boldsymbol{\xi} = \lambda A^{-1} \boldsymbol{\xi} + g \lambda^{-1} B \boldsymbol{\xi} \tag{3.13}$$

to the system of equations

$$\begin{aligned} \nu \boldsymbol{\eta} &= \lambda A^{-1} \boldsymbol{\xi}, \\ \nu \boldsymbol{\delta} &= g \lambda^{-1} B \boldsymbol{\xi}, \\ \boldsymbol{\xi} &= \boldsymbol{\eta} + \boldsymbol{\delta}. \end{aligned} \tag{3.14}$$

Performing substitutions inverse to (1.26) and introducing the deviation ζ of the free surface Γ, we obtain the following relations instead of (1.23) in the case of normal oscillations,

$$\begin{aligned} \nu A\boldsymbol{s} &= \lambda \boldsymbol{u}, \\ -\lambda\zeta &= \gamma_n \boldsymbol{u}, \\ \nu \boldsymbol{w} &= -gT\zeta, \\ \boldsymbol{u} &= \boldsymbol{s} + \boldsymbol{w}. \end{aligned} \tag{3.15}$$

Let us note, that in the latter sum the fields $\boldsymbol{s}$ and $\boldsymbol{w}$ can be univalently determined by the given field $\boldsymbol{u}$; this fact is also true for the last sum in (3.14).

We normalize the solutions of problem (3.13) by the condition

$$\|\boldsymbol{\xi}\|_{\boldsymbol{J}_{0,S}(\Omega)} = \|\boldsymbol{\eta} + \boldsymbol{\delta}\|_{\boldsymbol{J}_{0,S}(\Omega)} = 1.$$

Then, for the solutions of problem (3.15) we have the following,

$$\|\boldsymbol{u}\|_{\boldsymbol{J}^1_{0,S}(\Omega)} = \|A^{1/2}\boldsymbol{u}\|_{\boldsymbol{J}_{0,S}(\Omega)} = \|\boldsymbol{\xi}\|_{\boldsymbol{J}_{0,S}(\Omega)} = 1 = \|\boldsymbol{s} + \boldsymbol{w}\|_{\boldsymbol{J}^1_{0,S}(\Omega)}. \tag{3.16}$$

At first let us consider the properties of the solutions of either problem (3.13), or of the equivalent problems (3.14) or (3.15), for the case of dissipative waves. Since in this case the eigenvalues $\lambda = \lambda_k^+ \to \infty$ as $k \to \infty$, then from the second equality in (3.14) it can be obtained that $\boldsymbol{\delta} = \boldsymbol{\delta}_k^+ \to \boldsymbol{0}$ (in $\boldsymbol{J}_{0,S}(\Omega)$). That is why $\boldsymbol{w} = \boldsymbol{w}_k^+ = A^{-1/2}\boldsymbol{\delta}_k^+ \to \boldsymbol{0}$ (in $\boldsymbol{J}^1_{0,S}(\Omega)$). Further, since the operator γ_n acts continuously from $\boldsymbol{J}^1_{0,S}(\Omega)$ to $H^{1/2}_\Gamma$, then from the second equality (3.15) we obtain $\zeta_k^+ = -(\lambda_k^+)^{-1}\gamma_n \boldsymbol{u}_k^+ \to 0$ (in $H^{1/2}_\Gamma$). Finally, from the first and last equations (3.15) it follows that

$$\nu\left(\lambda_k^+\right)^{-1} A\boldsymbol{s}_k^+ - \boldsymbol{s}_k^+ = \boldsymbol{w}_k^+ \to \boldsymbol{0} \qquad \text{in } \boldsymbol{J}^1_{0,S}(\Omega).$$

Summing up the results of this reasoning, let us formulate them in the form of the following conclusions.

1° Any field $\boldsymbol{u}$ corresponding to some normal oscillation of a viscous fluid can be represented in the form of two summands $\boldsymbol{s}$ and $\boldsymbol{w}$ that can be univalently determined by $\boldsymbol{u}$ and are of different physical nature. Specifically, the component $\boldsymbol{w} = -g\nu^{-1}T\zeta$ is univalently determined by the deviation ζ of the free surface of the fluid from its equilibrium position; this component can be naturally called the *surface component of the solution* $\boldsymbol{u}$. The other summand $\boldsymbol{s}$ in the sum for $\boldsymbol{u}$ is called the *internal component of the solution* $\boldsymbol{u}$.

2° For the positive eigenvalues $\lambda = \lambda_k^+$, the following properties are valid as

$k \to \infty$:

$$\begin{aligned} \boldsymbol{w}_k^+ &\to \mathbf{0} \qquad \text{in } \boldsymbol{J}_{0.S}^1(\Omega), \\ \zeta_k^+ &\to 0 \qquad \text{in } H_\Gamma^{1/2}, \\ \|\boldsymbol{s}_k^+\|_{\boldsymbol{J}_{0.S}^1(\Omega)} &\to 1, \\ \nu\left(\lambda_k^+\right)^{-1} A\boldsymbol{s}_k^+ - \boldsymbol{s}_k^+ &\to \mathbf{0} \qquad \text{in } \boldsymbol{J}_{0.S}^1(\Omega). \end{aligned} \tag{3.17}$$

These relations show that the solutions corresponding to the numbers λ_k^+ have the properties of internal waves: asymptotically, for such solutions, the deviations of the free surface of the fluid are very small in the process of oscillations for $k \to \infty$, and the main problem (3.15) is converted into the spectral problem $\nu As = \lambda s$ which is close to the problem in Section 7.1 on oscilations of a viscous fluid in a wholly filled container. Let us note that both the mentioned simple problems have similar asymptotics of eigenvalues for $k \to \infty$.

Now let us consider the solutions of problem (3.13) that correspond to the negative eigenvalues $\lambda = \lambda_k^-$, $\lambda_k^- \to 0$ as $k \to \infty$. If for the fields $\boldsymbol{u} = \boldsymbol{u}_k^-$ the norm $\|\boldsymbol{u}_k^-\|_{\boldsymbol{J}_{0,S}^1(\Omega)} = 1$, then from the first equation (3.15) it follows that $\nu A\boldsymbol{s}_k^+ = \lambda_k^- \boldsymbol{u}_k^- \to \mathbf{0}$ in $\boldsymbol{J}_{0,S}^1(\Omega)$; from this fact it also follows that $\boldsymbol{s}_k^- \to \mathbf{0}$ in $\boldsymbol{J}_{0,S}^1(\Omega)$. Here, $\|\boldsymbol{w}_k^-\|_{\boldsymbol{J}_{0,S}^1(\Omega)} = \|\boldsymbol{u}_k^- - \boldsymbol{s}_k^-\|_{\boldsymbol{J}_{0,S}^1(\Omega)} \to 1$ for $k \to \infty$. From the first and third equations in (3.15) we have

$$\nu\lambda_k^- g^{-1}\boldsymbol{w}_k^- - T\gamma_n \boldsymbol{w}_k^- = T\gamma_n \boldsymbol{s}_k^- .$$

Now let us note that the operator $T\gamma_n$ acts boundedly from $\boldsymbol{J}_{0.S}^1(\Omega)$ to $\boldsymbol{J}_{0,S}^1(\Omega)$. That is why, in virtue of the property $\boldsymbol{s}_k^- \to \mathbf{0}$, $k \to \infty$, we have

$$\nu\lambda_k^- g^{-1}\boldsymbol{w}_k^- - T\gamma_n \boldsymbol{w}_k^- \to \mathbf{0} \qquad \text{in } \boldsymbol{J}_{0.S}^1(\Omega). \tag{3.18}$$

Hence, the component $\boldsymbol{w}_k^-$ of the solution $\boldsymbol{u}_k^-$ satisfies the equation

$$\nu\lambda g^{-1}\boldsymbol{w} - T\gamma_n \boldsymbol{w} = \mathbf{0}, \qquad \|\boldsymbol{w}\|_{\boldsymbol{J}_{0.S}^1(\Omega)} = 1, \tag{3.19}$$

after transition to the limit. Equation (3.19) can be called the *auxiliary equation of gravitation waves in a viscous fluid.* Indeed, on one hand, problem (3.19) is equivalent to the eigenvalue problem for the operator $g\nu^{-1}B$,

$$g\nu^{-1}B\boldsymbol{\delta} := g\nu^{-1}A^{1/2}T\gamma_n A^{-1/2}\boldsymbol{\delta} = \lambda\boldsymbol{\delta}, \qquad \boldsymbol{\delta} = A^{1/2}\boldsymbol{w}, \tag{3.20}$$

and an the other hand it is equivalent to the problem

$$g\nu^{-1}C\zeta := g\nu^{-1}\gamma_n T\zeta = \lambda\zeta, \qquad \boldsymbol{w} = -g\nu^{-1}T\zeta, \tag{3.21}$$

which is an eigenvalue problem for the operator $g\nu^{-1}C$, where $C = \gamma_n T$ is the operator appearing in the abstract scheme of Section 1.8 and $\zeta \in H_\Gamma^{-1/2}$ is the deviation of the free surface. After the substitution $\zeta = C^{-1/2}v$, we come to the following equation,

$$g\nu^{-1}Cv = \lambda v, \qquad v \in L_{2,\Gamma}. \tag{3.22}$$

Hence, we conclude that the problems (3.19)–(3.22) have the same spectrum, and $v = V\boldsymbol{\delta}$, where $V := C^{-1/2}\gamma_n A^{-1/2}$ is an isometric operator acting from $\boldsymbol{J}_{0,S}(\Omega)$ into $L_{2,\Gamma}$.

The above-mentioned facts allow us to formulate the following assertion in addition to properties 1°–2° of the internal and surface waves.

3° For negative eigenvalues $\lambda = \lambda_k^-$ the following properties are valid for $k \to \infty$:

$$\boldsymbol{s}_k^- \to \boldsymbol{0} \qquad \text{in } \boldsymbol{J}_{0.S}^1(\Omega),$$
$$\|\boldsymbol{w}_k^-\|_{\boldsymbol{J}_{0.S}^1(\Omega)} \to 1,$$

and relation (3.18) holds true. Hence, asymptotically, the internal component $\boldsymbol{s}_k^-$ of the field $\boldsymbol{u}_k^-$ equals zero for $k \to \infty$, and the main problem (3.15) is transformed into either the spectral problem (3.19) or the equivalent problems (3.20)–(3.22), which are eigenvalue problems for the operators B and C related only to the surface oscillations of the system. Here, as already mentioned above, the asymptotics of the negative eigenvalues λ_k^- is determined by the eigenvalues of operator B (or by the ones of operator C which equal the ones for B).

8.3.4 On the Basicity of Modes of Normal Oscillations

The general results of Section 8.2 enable us to find the properties of basicity for the set of all the modes of normal oscillations and for its separate subsets in problem (3.13) and the corresponding problem (3.15). At first let us consider some issues related to the decompositioin of spaces $\boldsymbol{J}_{0,S}^1(\Omega)$ and $\boldsymbol{J}_{0.S}(\Omega)$ into some orthogonal subspaces that are connected to the operators A, γ_n, T, and $C = \gamma_n T$.

We recall that, according to the general scheme of Section 1.8 as well as the conclusions of Sections 1.3 and 2.2, the space $\boldsymbol{J}_{0,S}^1(\Omega)$ admits the orthogonal decomposition

$$\boldsymbol{J}_{0.S}^1(\Omega) = \boldsymbol{M}_1(\Omega) \oplus \boldsymbol{N}_1(\Omega), \tag{3.23}$$

where $\boldsymbol{M}_1(\Omega)$ is the set of solutions $\boldsymbol{w} = T\psi$ of the auxiliary boundary value problem II (see (1.13) and (1.14)) corresponding to all elements $\psi \in H_\Gamma^{-1/2} = \mathcal{R}(C^{-1/2})$, and $\boldsymbol{N}_1(\Omega) := \{\boldsymbol{u} \in \boldsymbol{J}_{0,S}^1(\Omega) : \gamma_n \boldsymbol{u} = 0\}$ is the kernel of operator γ_n. Since $\boldsymbol{J}_{0,S}^1(\Omega) = \mathcal{D}(A^{1/2}) \subset \boldsymbol{J}_{0,S}(\Omega)$, then the following decomposition corresponds to the decomposition (3.23),

$$\boldsymbol{J}_{0,S}(\Omega) = \boldsymbol{M}_0(\Omega) \oplus \boldsymbol{N}_0(\Omega),$$
$$\boldsymbol{M}_0(\Omega) := A^{1/2}\boldsymbol{M}_1(\Omega), \ \boldsymbol{N}_0(\Omega) := A^{1/2}\boldsymbol{N}_1(\Omega). \tag{3.24}$$

It turns out that the subspace $\boldsymbol{N}_0(\Omega)$ coincides with the kernel $\operatorname{Ker} B$ of operator B. Indeed, if $B\boldsymbol{\xi} := A^{1/2}T\gamma_n A^{-1/2}\boldsymbol{\xi} = \mathbf{0}$, then $\|\gamma_n A^{-1/2}\boldsymbol{\xi}\|_{L_{2,\Gamma}} = 0$, because $(A^{1/2}T)^* = \gamma_n A^{-1/2}$. Hence we obtain that $A^{-1/2}\boldsymbol{\xi} \in \boldsymbol{N}_1(\Omega)$ and therefore $\boldsymbol{\xi} \in \boldsymbol{N}_0(\Omega)$. If $\boldsymbol{\xi} \in \boldsymbol{N}_0(\Omega)$, then $\boldsymbol{\xi} = A^{1/2}\boldsymbol{s}$, $\boldsymbol{s} \in \boldsymbol{N}_1(\Omega)$, $\gamma_n \boldsymbol{s} = 0$ and, therefore, $B\boldsymbol{\xi} = A^{1/2}T\gamma_n\boldsymbol{s} = \mathbf{0}$.

Let us now introduce the operators

$$V := C^{-1/2}\gamma_n A^{-1/2},$$
$$V^* := A^{1/2}TC^{-1/2}, \tag{3.25}$$

where $C = \gamma_n T$, acting from $\boldsymbol{J}_{0,S}(\Omega)$ to $H_\Gamma = L_{2,\Gamma}$ and from H_Γ to $\boldsymbol{J}_{0,S}(\Omega)$, respectively. By the equality $(A^{1/2}T)^* = \gamma_n A^{-1/2}$ and the fact that C is self-adjoint, V and V^* are mutually adjoint. Moreover, according to the general considerations of Section 1.8, V^* is an isometric operator acting from H_Γ to $\boldsymbol{M}_0(\Omega)$; as it was mentioned at the end of Section 8.3.3, the operator V has the isometry property as well.

We prove next that the following properties hold true for V and V^*:

$$VV^* = I \qquad \text{in } H_\Gamma,$$
$$V^*V = Q_{\boldsymbol{M}_0},$$

where $Q_{\boldsymbol{M}_0}$ is an orthoprojector onto $\boldsymbol{M}_0(\Omega) \subset \boldsymbol{J}_{0,S}(\Omega)$.

Indeed, since $C = \gamma_n T$,

$$VV^* = \left(A^{-1/2}\gamma_n A^{-1/2}\right)\left(A^{1/2}TC^{-1/2}\right) = C^{-1/2}(\gamma_n T)C^{-1/2} = I \qquad \text{in } H_\Gamma.$$

Further,

$$V^*V = \left(A^{1/2}TC^{-1/2}\right)\left(C^{-1/2}\gamma_n A^{-1/2}\right) = A^{1/2}TC^{-1}\gamma_n A^{-1/2}.$$

If $\boldsymbol{\delta} \in \boldsymbol{M}_0(\Omega)$, then $\boldsymbol{\delta} = A^{1/2}\boldsymbol{w}$, $\boldsymbol{w} \in \boldsymbol{M}_1(\Omega)$ and, therefore,

$$V^*V\boldsymbol{\delta} = A^{1/2}TC^{-1}\gamma_n\boldsymbol{w} = A^{1/2}T\left(T^{-1}\gamma_n^{-1}\right)\gamma_n\boldsymbol{w} = A^{1/2}\boldsymbol{w} = \boldsymbol{\delta}.$$

If $\boldsymbol{\eta} \in \boldsymbol{N}_0(\Omega)$, then $\boldsymbol{\eta} = A^{1/2}\boldsymbol{s}$, $\boldsymbol{s} \in \boldsymbol{N}_1(\Omega)$ and, therefore, $V^*V\boldsymbol{\eta} = A^{1/2}TC^{-1}\gamma_n\boldsymbol{s} = \mathbf{0}$, because $\gamma_n\boldsymbol{s} = 0$.

Using these proved facts, let us describe the properties of basicity of modes of normal oscillations for a viscous fluid. Since the pencil (3.1) is a special case of the pencil (2.1), the general conclusions of Sections 8.2.5 and 8.2.6 on the two-fold basicity of the system of eigen- and associated elements of the pencil (2.1) are applicable to (3.1). Namely, let $\{\lambda_j\}_{j=1}^{\infty}$ be the set of eigenvalues of problem (3.13) counted with regard to their multiplicities and $\{\boldsymbol{\xi}_{j,q}\}_{j=1}^{\infty}$ is the corresponding set of eigen- and associated elements $q = 0, \ldots, m_j - 1$, where m_j is the multiplicity of the eigenelement $\boldsymbol{\xi}_{j,0}$. Let us recall that only the intermediate eigenvalues can have associated elements and that is why there is no more than a finite number of intermediate eigenvalues.

We introduce the following elements of a special form

$$\left\{\boldsymbol{\xi}_{j,q}; \sum_{i=0}^{q}(-1)^i(\lambda_j)^{-i-1}Q_{\boldsymbol{M}_0}\boldsymbol{\xi}_{j,q-i}\right\}^{\mathrm{t}} \in \boldsymbol{J}_{0,S}(\Omega) \oplus \boldsymbol{M}_0(\Omega). \tag{3.26}$$

As it was proved in Section 8.2.5, the set of these elements forms a Riesz basis in the space $\boldsymbol{J}_{0,S}(\Omega) \oplus \boldsymbol{M}_0(\Omega)$.

Let us introduce the elements $\boldsymbol{u}_{j,q} = A^{-1/2}\boldsymbol{\xi}_{j,q} \in \boldsymbol{J}^1_{0,S}(\Omega)$ corresponding to the solutions $\boldsymbol{\xi}_{j,q} \in \boldsymbol{J}_{0,S}(\Omega)$ and then, according to (3.15), let us introduce the elements $\zeta_{j,q} = -(\lambda_j)^{-1}\gamma_n\boldsymbol{u}_{j,q} \in H_\Gamma^{-1/2}$. Since $Q_{\boldsymbol{M}_0} = V^*V = A^{1/2}TC^{-1}\gamma_n A^{-1/2}$, then the set of elements of the form

$$\left\{\boldsymbol{u}_{j,q}; \sum_{i=0}^{q}(-1)^i(\lambda_j)^{-i-1}TC^{-1}\gamma_n\boldsymbol{u}_{j,q-i}\right\}^{\mathrm{t}}, \qquad j = 1, 2, \ldots, \tag{3.27}$$

forms a Riesz basis in the space $\boldsymbol{J}^1_{0,S}(\Omega) \oplus \boldsymbol{M}_1(\Omega)$. From the latter and the isometry properties of operators T (from $H_\Gamma^{-1/2}$ to $\boldsymbol{M}_1(\Omega)$) and C^{-1} (from $H_\Gamma^{1/2}$ to $H_\Gamma^{-1/2}$) it follows that the set of elements of the form

$$\left\{\boldsymbol{u}_{j,q}; \sum_{i=0}^{q}(-1)^{i+1}(\lambda_j)^{-i}\zeta_{j,q-i}\right\}^{\mathrm{t}} \tag{3.28}$$

is a Riesz basis in the space $\boldsymbol{J}^1_{0,S}(\Omega) \oplus H_\Gamma^{1/2}$.

Hence, from the properties of two-fold basicity for the solutions $\{\boldsymbol{\xi}_{j,q}\}$ of the problem (3.13) we get the corresponding properties of two-fold basicity for the solutions $\{\boldsymbol{u}_{j,q}\}$ and $\{\zeta_{j,q}\}$ of the problem (3.15).

Let us consider now the basicity of the sets of eigenelements for problem (3.15) that correspond to positive and negative eigenvalues.

First let us assume that the rough overdamped condition (3.6) holds true, that is, when the spectrum of problem (3.13) consists only of branches of positive and negative eigenvalues. In this case, the positive eigenelements $\{\boldsymbol{\xi}_k^+\}_{k=1}^{\infty}$ of problem (3.13) form a Riesz basis in $\boldsymbol{J}_{0,S}(\Omega)$.

Further, for the negative eigenvalues $\{\lambda_k^-\}_{k=1}^{\infty}$, the corresponding negative eigenelements $\{\boldsymbol{\xi}_k^-\}_{k=1}^{\infty}$ form a Riesz basis in $\boldsymbol{M}_0(\Omega)$ after projecting onto $\boldsymbol{M}_0(\Omega)$. Let us introduce the elements $\boldsymbol{u}_k^- = A^{-1/2}\boldsymbol{\xi}_k^-$ along with the elements of the basis $\{Q_{\boldsymbol{M}_0}\boldsymbol{\xi}_k^-\}_{k=1}^{\infty} = \{A^{1/2}TC^{-1}\gamma_n A^{-1/2}\boldsymbol{\xi}_k^-\}_{k=1}^{\infty}$. Then, the set $\{TC^{-1}\gamma_n \boldsymbol{u}_k^-\}_{k=1}^{\infty}$ is a Riesz basis in $\boldsymbol{M}_1(\Omega) \subset \boldsymbol{J}_{0,S}^1(\Omega)$. Now if we introduce the displacements $\zeta_k^- = -(\lambda_k^-)^{-1}\gamma_n \boldsymbol{u}_k^-$, then, according to the properties of operators T and C^{-1}, we obtain that the set $\{\lambda_k^-\zeta_k^-\}_{k=1}^{\infty}$ forms a Riesz basis in $H_\Gamma^{1/2}$. Hence, it appears that the elements $\{\lambda_k^- C^{-1/2}\zeta_k^-\}_{k=1}^{\infty}$ form a Riesz basis in H_Γ and the elements $\{\lambda_k^- C^{-1}\zeta_k^-\}_{k=1}^{\infty}$ form a Riesz basis in $H_\Gamma^{-1/2}$.

If condition (3.6) is not valid, then the spectrum of problem (3.13) can contain, besides positive and negative eigenvalues, intermediate ones as well. In this case, to the positive eigenvalues $\{\lambda_k^+\}_{k=1}^{\infty}$ there corresponds a set of positive eigenelements $\{\boldsymbol{\xi}_k^+\}_{k=1}^{\infty}$ that form a Riesz basis in $\boldsymbol{J}_{0,S}(\Omega)$ up to a finite defect. Hence, we obtain that the elements $\{\boldsymbol{u}_k^+\}_{k=1}^{\infty}$ with $\boldsymbol{u}_k^+ = A^{-1/2}\boldsymbol{\xi}_k^+$ form a Riesz basis in $\boldsymbol{J}_{0,S}^1(\Omega)$ up to a finite defect. For negative eigenvalues the set of corresponding negative eigenelements, that is, the set $\{Q_{\boldsymbol{M}_0}\boldsymbol{\xi}_k^-\}_{k=1}^{\infty}$, forms a Riesz basis in $\boldsymbol{M}_0(\Omega)$ up to a finite defect, after projecting it onto $\boldsymbol{M}_0(\Omega)$. As it was argued previously, from the latter fact we can find the following: the sets $\{TC^{-1}\gamma_n \boldsymbol{u}_k^-\}_{k=1}^{\infty}$ and $\{\lambda_k^-\zeta_k^-\}_{k=1}^{\infty}$ form Riesz bases in $\boldsymbol{M}_1(\Omega)$ and $H_\Gamma^{1/2}$ up to a finite defect.

According to a certain rule, eigen- and associated elements of problem (3.18) corresponding to intermediate eigenvalues can be added to the above-mentioned basis elements from $\boldsymbol{J}_{0,S}^1(\Omega)$, $\boldsymbol{M}_1(\Omega)$, and $H_\Gamma^{1/2}$. This is done so that the sets of these elements form Riesz bases in the corresponding subspaces after extension.

8.4 Oscillations of a Heavy Rotating Fluid

In this section we consider a problem, which, on one hand, is a generalization of the problem in the previous section and is connected with uniform rotation of the whole system and Coriolis forces, and, on the other hand, generalizes the problem in Sections 6.3–6.6 for the case of small movements of a viscous fluid.

8.4.1 Statement of the Problem

As in Section 8.1, let us assume that a heavy viscous fluid with density ρ partially fills a container and rotates around the vertical axis Ox_3 with the angular velocity

ω_0. In the state of relative equilibrium, the fluid occupies the region $\Omega \subset \mathbb{R}^3$ bounded by the solid boundary S and the equilibrium surface Γ. The equilibrium pressure of the fluid is given by

$$P_{\text{equil}}(x) = p_{\text{a}} - \rho g x_3 + \frac{1}{2}\rho\omega_0^2\left(x_1^2 + x_2^2\right),$$

where p_{a} is a constant atmospheric pressure.

Let us consider those movements of the fluid that are close to a uniform rotation with an angular velocity $\omega_0 > 0$. In the coordinate system $Ox_1x_2x_3$ rigidly connected to the container, we can obtain the following linear equations that are similar to the ones in Sections 8.1 and 6.3,

$$\frac{\partial \boldsymbol{u}}{\partial t} - 2\omega_0 \boldsymbol{u} \times \boldsymbol{e}_3 = -\frac{1}{\rho}\nabla p + \nu\Delta\boldsymbol{u} + \boldsymbol{f}, \quad \operatorname{div}\boldsymbol{u} = 0 \qquad \text{in } \Omega, \tag{4.1}$$

and the boundary stickiness condition

$$\boldsymbol{u} = \boldsymbol{0} \qquad \text{on } S. \tag{4.2}$$

Kinematic and dynamic conditions on the surface Γ can be written down in the curvilinear coordinate system $\tilde{O}\xi^1\xi^2\xi^3$ (see Section 4.1.2). In this coordinate system, the equation for Γ has the form $\xi^3 = 0$, the Lame coefficient is $h_3|_\Gamma = 1$ and the normal $\boldsymbol{n}$ on Γ is directed outward Ω. The following kinematic condition can be obtained,

$$\frac{\partial \zeta}{\partial t} = u_n \qquad \text{on } \Gamma, \tag{4.3}$$

if we assume that the free moving surface has the equation $\xi^3 = \zeta(t, \xi^1, \xi^2)$. The dynamic conditions on Γ are the following: the tangent stresses equal zero and the normal stress equals the leap of pressure caused by the centrifugal and gravitation forces. If the covariant derivative of the covariant vector u_i by the variable ξ^k is denoted by $u_{i,k}$, then the above-mentioned conditions on Γ have the following form:

$$\begin{aligned} \rho\nu(u_{i,3} + u_{3,i}) &= 0, \qquad i = 1, 2, \\ -p + 2\rho\nu u_{3,3} &= a(\hat{\xi})\zeta, \\ a(\hat{\xi}) &:= (\nabla P_{\text{equil}} \cdot \boldsymbol{n})_\Gamma, \qquad \hat{\xi} := (\xi^1; \xi^2). \end{aligned} \tag{4.4}$$

As in Section 8.1, the initial conditions are given by

$$u(0, x) = u^0(x), \quad \zeta(0, \hat{\xi}) = \zeta^0(\hat{\xi}). \tag{4.5}$$

8.4.2 Transition to a System of Operator Equations

In this section we investigate the initial boundary value problem (4.1)–(4.5) according to the scheme stated in Section 8.1. We should note here that, for the mentioned geometry of the region Ω, the Green formula (2.2.27) remains unchanged by substituting the ordinary derivatives of a vector field with its covariant ones. It makes it possible to apply the reasoning carried out in Sections 8.1.2–8.1.4 to the problem (4.1)–(4.5) and to obtain a system of operator equations similar to (1.23), that is,

$$\begin{aligned} \boldsymbol{u}(t) &= \boldsymbol{s}(t) + \boldsymbol{w}(t), \\ \frac{\mathrm{d}\zeta}{\mathrm{d}t} &= \gamma_n \boldsymbol{u}, \\ \nu \boldsymbol{w} + T B_0 \zeta &= 0, \\ \nu A^{1/2} \boldsymbol{s} + A^{-1/2}\left(\frac{\mathrm{d}\boldsymbol{u}}{\mathrm{d}t} - 2\mathrm{i}\omega_0 S_0 \boldsymbol{u} - \boldsymbol{f}(t)\right) &= 0. \end{aligned} \tag{4.6}$$

Here, $\boldsymbol{u}(t)$, $\boldsymbol{s}(t)$, and $\boldsymbol{w}(t)$ are functions with values in $\boldsymbol{J}^1_{0,S}(\Omega)$, γ_n is the normal trace operator acting from $\boldsymbol{J}^1_{0,S}(\Omega)$ to $H^{1/2}_\Gamma$, A is the operator of the boundary value problem I in Section 8.1.2 (where the ordinary derivatives were substituted with the covariant ones on Γ), and T is the operator of the boundary value problem II. As in Section 8.1.4, it is assumed that the field of external forces $\boldsymbol{f}(t)$ is a function with values in $\boldsymbol{J}_{0,S}(\Omega)$. Further, the operator S_0 is defined by $S_0\boldsymbol{u} := -\mathrm{i}P_{0,S}(\boldsymbol{u}\times\boldsymbol{e}_3)$ for any $\boldsymbol{u}\in\boldsymbol{J}_{0,S}(\Omega)$, where $P_{0,S}$ is an orthoprojector onto $\boldsymbol{J}_{0,S}(\Omega)$. This operator—naturally called the *gyroscopic operator*—has the following properties:

$$S_0 = S_0^*, \quad \|S_0\| = 1. \tag{4.7}$$

Let us denote by B_0 an operator defined for any $\zeta\in H_\Gamma$ by

$$B_0\zeta := P_\Gamma a(\hat{\xi}) P_\Gamma. \tag{4.8}$$

Here, $a(\hat{\xi})$ is a function on Γ defined by (4.4) and P_Γ is the orthoprojector onto $H_\Gamma = L_{2,\Gamma}$. Since $a(\hat{\xi})$ is a continuous function on Γ, B_0 is a bounded self-adjoint operator acting in H_Γ.

8.4.3 Solvability of the Nonstationary Problem

Using equations (4.6), let us prove the univalent solvability of the initial boundary value problem (4.1)–(4.5). From the second equation in (4.6) we have

$$\zeta(t) = \zeta^0 + \int_0^t \gamma_n(\boldsymbol{s}+\boldsymbol{w})(\tau)\mathrm{d}\tau, \tag{4.9}$$

and, therefore, the third equation leads to

$$\nu \boldsymbol{w}(t) = -TB_0\zeta^0 - \int_0^t TB_0\gamma_n(\boldsymbol{s}+\boldsymbol{w})(\tau)\mathrm{d}\tau. \tag{4.10}$$

Then, from the first and last equations in (4.6) we obtain

$$A^{-1/2}\frac{\mathrm{d}\boldsymbol{s}}{\mathrm{d}t} + \nu A^{1/2}\boldsymbol{s} = \nu^{-1}A^{-1/2}TB_0\gamma_n(\boldsymbol{s}+\boldsymbol{w}) + A^{-1/2}(2\mathrm{i}\omega_0 S_0(\boldsymbol{s}+\boldsymbol{w}) + \boldsymbol{f}). \tag{4.11}$$

Performing the substitution (1.26) in (4.10) and (4.11) and applying the operator $A^{1/2}$ from the left (further considerations will make it clear that this operator is applicable), we obtain

$$\boldsymbol{\delta}(t) = -\nu^{-1}A^{1/2}TB_0\zeta^0 - \nu^{-1}\int_0^t B(\boldsymbol{\eta}+\boldsymbol{\delta})(\tau)\mathrm{d}\tau, \tag{4.12}$$

$$A^{-1/2}\frac{\mathrm{d}\boldsymbol{\eta}}{\mathrm{d}t} + \nu A^{1/2}\boldsymbol{\eta} = \nu^{-1}A^{-1/2}B(\boldsymbol{\eta}+\boldsymbol{\delta}) + 2\mathrm{i}\omega_0 S_0 A^{-1/2}(\boldsymbol{\eta}+\boldsymbol{\delta}) + \boldsymbol{f},$$
$$B := A^{1/2}TB_0\gamma_n A^{-1/2}. \tag{4.13}$$

Further, let us assume that the state of relative equilibrium of the rotating fluid in the container is statistically stable in linear approximation, that is, the operator B_0 is positive definite in $H_\Gamma = L_{2,\Gamma}$. This condition is equivalent to the following:

$$a(\hat{\xi}) \geq a_0 > 0, \qquad \hat{\xi} \in \Gamma. \tag{4.14}$$

It has been already proven previously that the operators $\gamma_n A^{-1/2}$ and $A^{1/2}T = (\gamma_n A^{-1/2})^*$ are compact and B_0 is a bounded and positive operator by virtue of (4.14). Then the operator B in (4.13) is compact and nonnegative. As before, its kernel Ker B and the closure of its range are infinite dimensional.

To make the transition from (4.13) to an integral equation, let us write down (4.13) for $t = \tau$, apply the bounded operator $A^{1/2}\exp[-\nu(t-\tau)A]$ from the left, and integrate between the limits 0 and t. Thus we obtain

$$\begin{aligned}\boldsymbol{\eta}(t) &= \exp(-\nu tA)\boldsymbol{\eta}(0) + \nu^{-1}\int_0^t \exp[-\nu(t-\tau)A]B(\boldsymbol{\eta}+\boldsymbol{\delta})(\tau)\mathrm{d}\tau \\ &+ 2\mathrm{i}\omega_0\int_0^t A^{1/2}\exp[-\nu(t-\tau)A]S_0A^{-1/2}(\boldsymbol{\eta}+\boldsymbol{\delta})(\tau)\mathrm{d}\tau \\ &+ \int_0^t \exp[-\nu(t-\tau)A]A^{1/2}\boldsymbol{f}(\tau)\mathrm{d}\tau.\end{aligned} \tag{4.15}$$

Let us consider both the integral equations (4.12) and (4.15). They represent a system of Volterra type integral equations with respect to the functions $\boldsymbol{\eta}(t)$ and $\boldsymbol{\delta}(t)$ with values in $\boldsymbol{J}_{0,S}(\Omega)$. In this system, all the kernels are continuous operators except for the one that has the weak singularity

$$\left\|A^{1/2}\exp[-\nu(t-\tau)A]AS_0A^{-1/2}\right\| \leq \frac{\text{const}}{|t-\tau|^{1/2}}\|A^{-1/2}\|, \qquad \|S_0\|=1.$$

That is why the solution of the system (4.12), (4.15) can be obtaind by the method of successive approximations and it exists for any

$$\boldsymbol{\eta}(0) \in \boldsymbol{J}_{0,S}(\Omega), \quad A^{1/2}TB_0\zeta^0 \in \boldsymbol{J}_{0,S}(\Omega) \tag{4.16}$$

if $\boldsymbol{f}(t)$ is a continuous function with values in $\boldsymbol{J}^1_{0,S}(\Omega)$.

Thus, we obtain the following conclusion:

If in the initial boundary value problem (4.1)–(4.5)

$$\boldsymbol{u}^0(x) \in \boldsymbol{J}^1_{0,S}(\Omega), \quad \zeta^0(\hat{\xi}) \in H_\Gamma, \tag{4.17}$$

and the field of external forces $\boldsymbol{f}(t,x)$ is a continuous function of t with values in $\boldsymbol{J}^1_{0,S}(\Omega)$, then the problem (4.2)–(4.5) *has a unique generalized solution for which $\boldsymbol{u}(t,x)$ is a continuous function of t with values in $\boldsymbol{J}^1_{0,S}(\Omega)$ and $\zeta(t,\hat{\xi})$ is a continuous function of t with values in H_Γ. Here the kinetic and potential energies of the system are continuous in t for the above-mentioned solution.*

Indeed, if conditions (4.17) hold true, then

$$\boldsymbol{w}(0) = -\nu^{-1}TB_0\zeta^0 \in \boldsymbol{J}^1_{0,S}(\Omega),$$
$$\boldsymbol{\delta}(0) = A^{1/2}\boldsymbol{w}(0) \in \boldsymbol{J}_{0,S}(\Omega),$$

therefore, $\boldsymbol{\eta}(0) = A^{1/2}(\boldsymbol{u}^0 - \boldsymbol{w}(0)) \in \boldsymbol{J}_{0,S}(\Omega)$. Moreover, $A^{1/2}\boldsymbol{f}(t,x)$ is a continuous function with values in $\boldsymbol{J}_{0,S}(\Omega)$.

Since the Coriolis forces do not perform any work, then the law of full energy balance—in the form mentioned in Section 8.1.5—holds true for the generalized solution of problem (4.2)–(4.5). The remark at the and of that section applies to this solution too.

8.4.4 Equations of Normal Oscillations

Let us assume in (4.6) that $\boldsymbol{f}(t,x) \equiv 0$ and consider the solutions that depend on t according to the law $\exp(-\lambda t)$. Thus we obtain

$$\begin{aligned} \boldsymbol{u} &= \boldsymbol{s} + \boldsymbol{w}, \\ -\lambda\zeta &= \gamma_n \boldsymbol{u}, \\ \nu \boldsymbol{w} + TB_0\zeta &= 0, \\ \nu A^{1/2}\boldsymbol{s} + A^{-1/2}(-\lambda\boldsymbol{u} - 2\mathrm{i}\omega_0 S_0 \boldsymbol{u}) &= 0. \end{aligned} \tag{4.18}$$

From the latter, after the substitutions (1.26), we obtain the following system of equations:

$$\begin{aligned} \boldsymbol{\xi} &= \boldsymbol{\eta} + \boldsymbol{\delta}, \\ \nu\boldsymbol{\delta} &= \lambda^{-1}B\boldsymbol{\xi}, \\ \zeta &= -\lambda^{-1}\gamma_n A^{-1/2}\boldsymbol{\xi}, \\ \nu\boldsymbol{\eta} &= \lambda A^{-1}\boldsymbol{\xi} + 2\mathrm{i}\omega_0 S\boldsymbol{\xi}, \\ S &:= A^{-1/2}S_0A^{-1/2}. \end{aligned} \tag{4.19}$$

From (4.19) for $\boldsymbol{\xi} = \boldsymbol{\eta} + \boldsymbol{\delta}$ we have the following,

$$\nu\boldsymbol{\xi} = \lambda A^{-1}\boldsymbol{\xi} + \lambda^{-1}B\boldsymbol{\xi} + 2\mathrm{i}\omega_0 S\boldsymbol{\xi}. \tag{4.20}$$

This equation is a generalization of the equation (3.13) for the case of a rotating fluid. For $\omega_0 \to \infty$, the last summand in the right hand side of (4.20) disappears, Ω is continuously converted into a region with a horizontal equilibrium surface Γ, the operators A^{-1} and B change continuously also and coincide with the operators A^{-1} and gB of problem (3.13) after transition to the limit (in particular, the latter fact follows from the representation (4.8) of operator B_0 and formulas (4.4) and (4.1) for $a(\hat{\xi})$ and $P_{\mathrm{equil}}(x)$).

Let us point out the asymptotic property of the eigenvalues of the operator $B = A^{1/2}TB_0\gamma_n A^{-1/2}$:

$$\begin{aligned} \lambda_k(B) &= (c_B)^{1/2}k^{-1/2}[1 + \mathrm{o}(1)], \qquad k \to \infty, \\ c_B &= \frac{1}{16\pi}\frac{\displaystyle\int_\Gamma |a(\hat{\xi})|^2 \mathrm{d}\Gamma}{\rho^2}. \end{aligned} \tag{4.21}$$

In particular, from (4.21) it follows that $B \in \mathfrak{S}_p$ for $p > 2$.

8.4.5 Investigation of the Spectral Problem

Since $S = S^*$, a transition can be made from (4.20) to the equivalent equation

$$L_{\nu,\omega_0}(\lambda)\boldsymbol{\xi} := \left[I - \nu^{-1}I_0(\lambda A^{-1} + \lambda^{-1}B)\right]\boldsymbol{\xi} = \mathbf{0},$$
$$I_0 := \left(I - 2\mathrm{i}\omega_0\nu^{-1}S\right)^{-1}. \tag{4.22}$$

Though at a first sight this equation is only a little bit different from equation (3.1), problem (4.22) looses many of the properties found out for the solutions of problem (3.1). In particular, since $I_0 \neq I$ for $\omega_0 \neq 0$, the property of self-adjointness of the pencil $L_{\nu,\omega_0}(\lambda)$ is no longer valid. Nevertheless, many spectral properties of the pencils $L_\nu(\lambda)$ in (3.1) and $L_{\nu,\omega_0}(\lambda)$ in (4.22) coincide.

First let us note that for the operator I_0, the following properties hold true:

$$(1) \qquad I_0 = I + \Phi_0, \qquad \Phi_0 \in \mathfrak{S}_\infty$$
$$(2) \qquad \|I_0\| = 1. \tag{4.23}$$

The first property is easy to prove and the second one can be proved as follows: If $\lambda_k = \lambda_k(S)$, $k = 1, 2, \ldots$, is the sequence of eigenvalues of the operator $S = S^* \in \mathfrak{S}_\infty$ corresponding to the orthonormal elements $\{\varphi_k\}_{k=1}^\infty$, then

$$\left(I_0^{-1}\varphi_k, \varphi_k\right)_{\boldsymbol{J}_{0,S}(\Omega)} = 1 - 2\mathrm{i}\omega_0\lambda_k(S) \to 1, \qquad \text{as } k \to \infty.$$

On the other hand, for any φ from $\boldsymbol{J}_{0,S}(\Omega)$

$$\operatorname{Re}\left(I_0^{-1}\varphi, \varphi\right)_{\boldsymbol{J}_{0,S}(\Omega)} = \|\varphi\|^2_{\boldsymbol{J}_{0,S}(\Omega)},$$

and therefore, $\|I_0\| \leq 1$. Both properties lead to the second relation in (4.23).

Let us find out some of the properties that the operator pencil $L_{\nu,\omega_0}(\lambda)$ has.

1° The spectrum of either problem (4.20) or (which is the same) problem (4.22) is no more than countable with possible limit points $\lambda = 0$ and $\lambda = \infty$. All the other points in the spectrum that are different from 0 and ∞ are eigenvalues with finite algebraic multiplicity.

Indeed, since the operators A^{-1}, B, and S are compact, there is a Fredholm holomorphic pencil $\nu I - A(\lambda)$, where $A(\lambda) := \lambda A^{-1} + \lambda^{-1}B + 2\mathrm{i}\omega_0 S$, that corresponds to equation (4.20), invertible on the negative semiaxis because

$$\operatorname{Re}((I - A(\lambda))\boldsymbol{\xi}, \boldsymbol{\xi})_{\boldsymbol{J}_0(\Omega)} = \|\boldsymbol{\xi}\|^2_{\boldsymbol{J}_{0,S}(\Omega)} - \lambda\|A^{-1/2}\boldsymbol{\xi}\|^2_{\boldsymbol{J}_{0,S}(\Omega)} - \lambda^{-1}\|B^{1/2}\boldsymbol{\xi}\|^2_{\boldsymbol{J}_{0,S}(\Omega)}$$
$$\geq \|\boldsymbol{\xi}\|^2_{\boldsymbol{J}_{0,S}(\Omega)}, \qquad \lambda < 0.$$

2° All eigenvalues of problem (4.20) are located in the right half-plane. Indeed, for the eigenelements $\boldsymbol{\xi}$ of problem (4.20) we have

$$\operatorname{Re}\lambda = \nu\|\boldsymbol{\xi}\|^2_{\boldsymbol{J}_{0,S}(\Omega)}|\lambda|^2\left(|\lambda|^2\|A^{-1/2}\boldsymbol{\xi}\|^2_{\boldsymbol{J}_{0,S}(\Omega)} + \|B^{1/2}\boldsymbol{\xi}\|^2_{\boldsymbol{J}_{0,S}(\Omega)}\right)^{-1} > 0.$$

The next step we are going to take in the investigation of problem (4.20) is based on posibility of reducing it to the Keldysh pencil. This reduction is similar to the transition from equation (2.1) to (2.10) and then to the degenerate pencil (2.18) in Section 8.2. Specifically, we perform the following substitutions,

$$\begin{gathered}\lambda - \lambda^{-1} = \mu, \\ \lambda^{-1}Q_{\boldsymbol{M}_0}\boldsymbol{\xi} = \boldsymbol{\psi} \in \boldsymbol{M}_0(\Omega),\end{gathered} \tag{4.24}$$

where $Q_{\boldsymbol{M}_0}$ is the orthoprojector onto the subspace $\boldsymbol{M}_0(\Omega)$ from the orthogonal decomposition (3.24). As in Section 8.3, $\boldsymbol{M}_0(\Omega)$ is the closure of the range of operator B in the norm of $\boldsymbol{J}_{0,S}(\Omega)$.

Let us next apply the operator $Q_{\boldsymbol{N}_0} = I - Q_{\boldsymbol{M}_0}$ to the both sides of (4.20) and use the property $Q_{\boldsymbol{N}_0}B = 0$. From the thus obtained equation, we find the following,

$$Q_{\boldsymbol{N}_0}\boldsymbol{\xi} = \lambda\nu^{-1}I_{0N}\left(Q_{\boldsymbol{N}_0}A^{-1}\boldsymbol{\xi} + 2\mathrm{i}\omega_0 Q_{\boldsymbol{N}_0}SQ_{\boldsymbol{M}_0}\left(\lambda^{-1}Q_{\boldsymbol{M}_0}\boldsymbol{\xi}\right)\right),$$

where $I_{0N} := Q_{\boldsymbol{N}_0}(I - 2\mathrm{i}\omega_0\nu^{-1}Q_{\boldsymbol{N}_0}SQ_{\boldsymbol{N}_0})^{-1}Q_{\boldsymbol{N}_0}$. Applying now the operator $\lambda^{-1}Q_{\boldsymbol{M}_0}$ to the both sides of (4.20) we obtain a relation, which together with the initial equation and the just obtained expression for $Q_{\boldsymbol{N}_0}\boldsymbol{\xi}$ can be written down as a vector-matrix equation in the space $\boldsymbol{J}_{0,S}(\Omega) \oplus \boldsymbol{M}_0(\Omega)$:

$$\nu z = \mu G z + T_0 z, \tag{4.25}$$

with

$$\begin{aligned} z &= (\boldsymbol{\xi}, \boldsymbol{\psi})^{\mathrm{t}}, \\ G &= \operatorname{diag}(A^{-1}; -B), \\ T_0 &= \begin{pmatrix} 2\mathrm{i}\omega_0 S + \nu^{-1}A^{-1}I_{0N}A^{-1} & CQ_{\boldsymbol{M}_0} + 2\mathrm{i}\omega_0\nu^{-1}A^{-1}I_{0N}SQ_{\boldsymbol{M}_0} \\ Q_{\boldsymbol{M}_0}C + 2\mathrm{i}\omega_0\nu^{-1}Q_{\boldsymbol{M}_0}SI_{0N}A^{-1} & 2\mathrm{i}\omega_0 Q_{\boldsymbol{M}_0}SQ_{\boldsymbol{M}_0} - 4\omega_0^2\nu^{-1}Q_{\boldsymbol{M}_0}SI_{0N}SQ_{\boldsymbol{M}_0}\end{pmatrix},\end{aligned}$$

where $C = B + A^{-1}$.

Since $A^{-1} \in \mathfrak{S}_p$ for $p > 3/2$ and $B \in \mathfrak{S}_p$ for $p > 2$, then $G \in \mathfrak{S}_p$ for $p > 2$. Moreover, the operator T_0 in (4.25) is compact because the operators S, A^{-1}, and B are compact. Let us point out without a proof that, by the substitution $\lambda \mapsto a\lambda$ with some $a > 0$, we can always obtain the invertibility of operator $I - \nu^{-1}T_0$. For $\nu > \|T_0\|$, the invertibility of the operator is obvious. Therefore, it can be further assumed that the bounded inverse operator exists and

$$\left(I - \nu^{-1}T_0\right)^{-1} =: I + F_0, \qquad F_0 \in \mathfrak{S}_\infty. \tag{4.26}$$

In virtue of (4.26), problem (4.25) is equivalent to the eigenvalue problem

$$(I + F_0)Gz = \mu^{-1}z \tag{4.27}$$

for the weakly perturbed self-adjoint operator $(I + F_0)G$. Since $F_0 \in \mathfrak{S}_\infty$, $I + F_0$ is invertible, and G is a complete self-adjoint operator from the class $\mathfrak{S}_p$ for $p > 2$, then, according to Keldysh theorem, we can infer that problem (4.27) has a discrete spectrum $\{\mu_k^{-1}\}_{k=1}^\infty$ with the limit point zero and all the eigenvalues μ_k^{-1}, except for maybe a finite number of them, are located in the sectors $|\arg \mu_k^{-1}| < \varepsilon$ and $|\pi - \arg \mu_k^{-1}| < \varepsilon$ for any $\varepsilon > 0$. Moreover, the system of eigen- and associated elements of problem (4.27) corresponding to the eigenvalues $\{\mu_k^{-1}\}_{k=1}^\infty$ is complete in the space $\boldsymbol{J}_{0,S}(\Omega) \oplus \boldsymbol{M}_0(\Omega)$. Thus, in addition to Properties 1° and 2°, we obtain the following properties of the problem (4.20).

3° The spectrum of problem (4.20) is countable and consists of two branches of eigenvalues $\{\lambda_k^+\}_{k=1}^\infty$ and $\{\lambda_k^-\}_{k=1}^\infty$ with the limit points ∞ and 0, respectively. For any $\varepsilon > 0$ all eigenvalues λ_k^+ and λ_k^-, except for maybe a finite number, are located in the angle $|\arg \lambda| < \varepsilon$.

Indeed, let us denote by $\sqrt{z}$ the branch of the root for which $\operatorname{Re}\sqrt{z} \geq 0$. Then, by the connection $\lambda - \lambda^{-1} = \mu$, the eigenvalues $\lambda = \lambda(\mu)$ can be obtained by the formula $\lambda(\mu) = (\mu + \sqrt{\mu^2 + 4})/2$. Since the eigenvalues $\mu^{-1} = \mu_k^{-1}$ of the operator $(I + F_0)G$ are located in the right and left half-planes, all the eigenvalues λ can be devided into two branches assuming that $\lambda = \lambda_k^+$ if $\lambda = (\mu_k + \sqrt{\mu_k^2 + 4})/2$, $\operatorname{Re}\mu_k > 0$, and $\lambda = \lambda_j^-$ if $\lambda = (\mu_j + \sqrt{\mu_j^2 + 4})/2$, $\operatorname{Re}\mu_j < 0$. Here the arguments of all eigenvalues λ, except for maybe a finite number, are located in the sector $|\arg \lambda| < \varepsilon$ for an arbitrarily small $\varepsilon > 0$.

4° The system of eigen- and associated elements of problem (4.20) is two-multiply complete in the space $\boldsymbol{J}_{0,S}(\Omega)$, namely, the system of eigen- and associated elements of problem (4.27) is complete in $\boldsymbol{J}_{0,S}(\Omega) \oplus \boldsymbol{M}_0(\Omega)$.

5° For any $\delta > 0$ the following inequality holds true

$$|\operatorname{Im}\lambda_k| \leq 2\omega_0(1 + \delta), \tag{4.28}$$

for all the eigenvalues of the problem (4.20), except for maybe a finite number of them.

To prove property (4.28), it is sufficient to consider the branch $\{\lambda_k^+\}_{k=1}^\infty$ because for the branch $\{\lambda_k^-\}_{k=1}^\infty$ we have $|\arg \lambda_k^-| < \varepsilon$ for any $\varepsilon > 0$. Let $\{\xi_k^+\}_{k=1}^\infty$ be the normalized eigenelements of problem (4.20) corresponding to the eigenvalues $\lambda = \lambda_k^+ \to \infty$, $k \to \infty$. Then by the definition of the operator S we obtain from (4.20) the following

$$\nu - 2\mathrm{i}\omega_0 \left(S_0 A^{-1/2}\boldsymbol{\xi}_k^+, A^{-1/2}\boldsymbol{\xi}_k^+\right)$$

$$= \left(\operatorname{Re}\lambda_k^+ + \mathrm{i}\operatorname{Im}\lambda_k^+\right)\left\|A^{-1/2}\boldsymbol{\xi}_k^+\right\|^2 + \left|\lambda_k^+\right|^{-2}\left(\operatorname{Re}\lambda_k^+ - \mathrm{i}\operatorname{Im}\lambda_k^+\right)\left\|B^{1/2}\boldsymbol{\xi}_k^+\right\|^2.$$

Hence, we obtain

$$\begin{aligned} \nu &= \operatorname{Re}\lambda_k^+\left(\left\|A^{-1/2}\boldsymbol{\xi}_k^+\right\|^2 + \left|\lambda_k^+\right|^{-2}\left\|B^{1/2}\boldsymbol{\xi}_k^+\right\|^2\right), \\ -2\omega_0\left(S_0A^{-1/2}\boldsymbol{\xi}_k^+, A^{-1/2}\boldsymbol{\xi}_k^+\right) &= \operatorname{Im}\lambda_k^+\left(\left\|A^{-1/2}\boldsymbol{\xi}_k^+\right\|^2 - \left|\lambda_k^+\right|^{-2}\left\|B^{1/2}\boldsymbol{\xi}_k^+\right\|^2\right). \end{aligned} \tag{4.29}$$

(Here, the norm and the scalar product are to be calculated in $\boldsymbol{J}_{0,S}(\Omega)$.)

Since $\|S_0\| \le 1$ and taking into account the second equality in (4.29) we obtain that

$$\begin{aligned} \left|\operatorname{Im}\lambda_k^+\right| &\le 2\omega_0\frac{\left|\left(S_0A^{-1/2}\boldsymbol{\xi}_k^+, A^{-1/2}\boldsymbol{\xi}_k^+,\right)\right|}{\|A^{-1/2}\boldsymbol{\xi}_k^+\|^2}\cdot\frac{\|A^{-1/2}\boldsymbol{\xi}_k^+\|^2}{\left|\|A^{-1/2}\boldsymbol{\xi}_k^+\|^2 - \left|\lambda_k^+\right|^{-2}\|B^{1/2}\boldsymbol{\xi}_k^+\|^2\right|} \\ &\le 2\omega_0 M_k, \end{aligned} \tag{4.30}$$

with

$$M_k := \frac{\|A^{-1/2}\boldsymbol{\xi}_k^+\|^2}{\left|\|A^{-1/2}\boldsymbol{\xi}_k^+\|^2 - \left|\lambda_k^+\right|^{-2}\|B^{1/2}\boldsymbol{\xi}_k^+\|^2\right|} =: |1-\alpha_k|^{-1}.$$

Let us prove that $\alpha_k \to 0$ as $k \to \infty$. Indeed, from the first equality in (4.29) we have

$$\left\|A^{-1/2}\boldsymbol{\xi}_k^+\right\|^2 = \nu\left(\operatorname{Re}\lambda_k^+\right)^{-1} - \left|\lambda_k^+\right|^{-2}\left\|B^{1/2}\boldsymbol{\xi}_k^+\right\|^2,$$

therefore,

$$\alpha_k = \frac{\|B^{1/2}\boldsymbol{\xi}_k^+\|^2}{\left|\nu\dfrac{|\lambda_k^+|}{\operatorname{Re}\lambda_k^+} - \dfrac{1}{\lambda_k^+}\left\|B^{1/2}\boldsymbol{\xi}_k^+\right\|^2\right|}\cdot\frac{1}{|\lambda_k^+|}.$$

Since $|\lambda_k^+|^{-1}\|B^{1/2}\boldsymbol{\xi}_k^+\|^2 \le |\lambda_k^+|^{-1}\|B\| \to 0$ as $k \to \infty$ and for any sufficiently large k we have the following inequality

$$1 \le \frac{|\lambda_k^+|}{\operatorname{Re}\lambda_k^+} < (\cos\varepsilon)^{-1},$$

where ε is an arbitrarily small positive number providing the inequality $|\arg\lambda_k^+| < \varepsilon$, $k \ge N(\varepsilon)$, then the term within the square brackets in the expression for α_k is close to $1\cdot\nu$ for large k's. From the latter and the definition of α_k, it follows that $\alpha_k \to 0$ as $k \to \infty$.

Now from (4.30) we obtain that $M_k \to 1$ as $k \to \infty$ and assertion (4.28) is proved for the branch $\{\lambda_k^+\}_{k=1}^\infty$.

The following assertion is a corollary of this fact:

All the eigenvalues $\{\lambda_j\}$ of problem (4.20) with the exception of a finite number of them are located in the region

$$D_{\varepsilon,\delta} = \{\lambda \in \mathbb{C} : |\arg \lambda| < \varepsilon, \ |\operatorname{Im} \lambda| \leq 2\omega_0(1+\delta)\} \tag{4.31}$$

for any positive ε and δ.

6° The asymptotic behavior of the two branches of eigenvalues is the following:

$$\begin{aligned} \lambda_k^+ &= \nu\lambda_k(A)[1+\mathrm{o}(1)], \qquad k \to \infty, \\ \lambda_k^- &= \nu^{-1}\lambda_k(B)[1+\mathrm{o}(1)], \qquad k \to \infty. \end{aligned} \tag{4.32}$$

Both the equalities in (4.32) follow from the assertions in Section 1.6.11 regarding the asymptotic of eigenvalues of an operator pencil. The pencil of problem (4.20) is a special case of the pencil considered in Section 1.6.11. Here it is important that the operators A^{-1} and B—as it has been previously established—have power asymptotics of eigenvalues. Let us notice that the second formula in (4.32) can be obtain from the results in Section 1.6.11, after the formal substitution $\lambda = 1/\tilde{\lambda}$ is performed in problem (4.20).

Formulas (4.32) show that the fading decrements $\lambda_k^\pm$ of the considered problem depend asymptotically (for $k \to \infty$) on gravitation, centrifugal and dissipative forces acting in the system, and do not depend on the Coriolis forces at all.

7° As for the problem in Section 8.3, let us consider valid the rough over-damped condition

$$\nu^2 > 4\|A^{-1}\| \cdot \|B\|. \tag{4.33}$$

Then the eigenvalues λ_k^- are located in the semicircle $D_- := \{\lambda \in \mathbb{C} : |\lambda| \leq r_-, \ \operatorname{Re}\lambda > 0\}$ and the eigenvalues λ_k^+ are located in the region $D_+ := \{\lambda \in \mathbb{C} : |\lambda| \geq r_+, \ \operatorname{Re}\lambda > 0\}$, where $r_\pm = (\nu \pm \sqrt{\nu^2 - 4\|A^{-1}\| \cdot \|B\|})(2\|A^{-1}\|)^{-1}$. Here, to the eigenvalues $\{\lambda_k^+\}_{k=1}^\infty$ there correspond the eigen- and associated elements of problem (4.20). These elements form a complete system in the space $\boldsymbol{J}_{0,S}(\Omega)$. Respectively, to eigenvalues $\{\lambda_k^-\}_{k=1}^\infty$ there correspond eigen- and associated elements that form a complete system of elements in $\boldsymbol{M}_0(\Omega)$ after projecting them onto $\boldsymbol{M}_0(\Omega)$.

Let us prove the above-mentioned statements. If (4.33) holds true, then the pencil $L_{\nu,\omega_0}(\lambda)$ in problem (4.22) admits a spectral factorization in two forms. Indeed, introducing the operator-valued function

$$M_{\nu,\omega_0}(\lambda) := -\lambda L_{\nu,\omega_0}(\lambda) = -\lambda I + \nu^{-1} I_0 B + \lambda^2 \nu^{-1} I_0 A^{-1},$$

we can check that, for $r \in (r_-, r_+)$, the following condition

$$\nu^{-1}\left(\|I_0 B\| r^{-1} + \|I_0 A^{-1}\| r\right) \le \left(\|B\| r^{-1} + \|A^{-1}\| r\right) < 1, \qquad \|I_0\| = 1,$$

holds true for this function and this is sufficient for the spectral factorization relatively to the circle of radius r. Therefore, for $M_{\nu,\omega_0}(\lambda)$ we have a decomposition into linear multipliers and for $L_{\nu,\omega_0}(\lambda)$ the decomposition has the form

$$\begin{aligned} L_{\nu,\omega_0}(\lambda) &= I_0 Y^{-1}(I - \lambda\nu^{-1} Y A^{-1})(I - (\lambda\nu)^{-1} Y B), \\ Y &= I_0(I + \nu^{-2} A^{-1} Y B Y), \end{aligned} \tag{4.34}$$

where the expression in the first set of parantheses is a holomorphic and holomorphically invertible operator-valued function for $|\lambda| \le r \in (r_-, r_+)$ and the spectrum $\{\lambda_k^-\}_{k=1}^\infty$ of the pencil $I - (\lambda\nu)^{-1} Y B$ coincides with the spectrum of $L_{\nu,\omega_0}(\lambda)$ for $|\lambda| \le r$.

Hence, for $|\lambda| \le r_-$, it is sufficient to consider the spectral problem

$$Y B \boldsymbol{\xi} = \lambda\nu\boldsymbol{\xi}, \qquad \boldsymbol{\xi} \in \boldsymbol{J}_{0,S}(\Omega), \tag{4.35}$$

where $Y = I + \Phi$, $\Phi \in \mathfrak{S}_\infty$, $Y^{-1} \in \mathcal{L}(\boldsymbol{J}_{0.S}(\Omega))$. Here B is a nononegative compact operator in the class $\mathfrak{S}_p$ for $p > 2$. Its kernel coincides with the subspace $\boldsymbol{N}_0(\Omega)$ and the closure of its range coincides with the subspace $\boldsymbol{M}_0(\Omega)$.

By performing the substitution $\tilde{\lambda} = 1/\lambda$ and by applying the operator $Y^{-1} = I + \tilde{\Phi}$, $\tilde{\Phi} \in \mathfrak{S}_\infty$, to (4.35), we can reduce (4.35) to the following problem with a degenerate Keldysh pencil:

$$\left(\tilde{\lambda}\nu^{-1} B - I - \tilde{\Phi}\right)\boldsymbol{\xi} = \boldsymbol{0}. \tag{4.36}$$

Since in this case the operator $I + Q_{\boldsymbol{N}_0}\tilde{\Phi}Q_{\boldsymbol{N}_0} = Q_{\boldsymbol{N}_0}(I + \tilde{\Phi})Q_{\boldsymbol{N}_0} = Q_{\boldsymbol{N}_0} Y^{-1} Q_{\boldsymbol{N}_0}$ is invertible on the kernel $\boldsymbol{N}_0$ of operator B, the conclusions in Section 1.6.8 apply to equation (4.36). According to these conclusions, the system of eigen- and associated elements of the problem (4.36) forms a complete system of elements in the space $\boldsymbol{M}_0(\Omega) = \boldsymbol{J}_{0,S}(\Omega) \ominus \boldsymbol{N}_0(\Omega)$ after projecting them onto $\boldsymbol{M}_0(\Omega)$. Hence, we proved that the branch $\{\lambda_k^-\}_{k=1}^\infty$ is located in the semicircle $|\lambda| \le r_-$, $\operatorname{Re}\lambda > 0$, and the property of completeness for the system of eigen- and associated elements.

The second part of 7° regarding the branch $\{\lambda_k^+\}_{k=1}^\infty$ of eigenvalues of the problem (4.20) can be proved in a similar way. For condition (4.33), we use here the existence of the following factorization relatively to the circle $|\lambda| = r \in (r_-, r_+)$ for $L_{\nu,\omega_0}(\lambda)$,

$$\begin{aligned} L_{\nu,\omega_0}(\lambda) &= I_0 X^{-1}(I - (\lambda\nu)^{-1} X B)(I - \lambda\nu^{-1} X A^{-1}), \\ X &= I_0(I + \nu^{-2} B X A X), \end{aligned} \tag{4.37}$$

as well as the property of invertibility of operator X, its structure as indicated in (4.37), the property $A^{-1} \in \mathfrak{S}_p$ for $p > 3/2$, and the fact that $\operatorname{Ker} A^{-1} = \{0\}$. Since we obtain thus a nondegenerate Keldysh pencil, then to the eigenvalues $\{\lambda_k^+\}_{k=1}^{\infty}$, with $|\lambda_k^+| \geq r_+$, there correspond a system of eigen- and associated elements that is complete in $\boldsymbol{J}_{0,S}(\Omega)$.

8° If condition (4.33) is not valid, then instead of the above-proved properties of completeness of subsystems of eigen- and associated elements of problem (4.20), similar properties of completeness with finite defect hold true. Specifically, the following result is valid. For arbitrary viscousity $\nu > 0$ and any $r > 0$, the system of eigen- and associated elements of problem (4.20) corresponding to eigenvalues from the semicircle $|\lambda| < r$, $\operatorname{Re}\lambda > 0$, has no more than a finite defect in $\boldsymbol{M}_0(\Omega)$ after projecting onto $\boldsymbol{M}_0(\Omega)$, and the system of eigen- and associated elements of problem (4.20) corresponding to eigenvalues in the region $|\lambda| \geq r$, $\operatorname{Re}\lambda > 0$, has no more than a finite defect in the space $\boldsymbol{J}_{0,S}(\Omega)$.

Let us prove only the first assertion, because the proof of the second one is easier and does not require additional projecting onto the subspace $\boldsymbol{M}_0(\Omega) \subset \boldsymbol{J}_{0,S}(\Omega)$.

First, let $|\lambda| < r_1 := \nu\|A^{-1}\|^{-1}$. Since $\|I_0\| = 1$, the operator $I - \nu^{-1}\lambda I_0 A^{-1}$ is invertible for $|\lambda| < r_1$, and the pencil $L_{\nu,\omega_0}(\lambda)$ can be represented as

$$\begin{aligned} L_{\nu,\omega_0}(\lambda) &= I - \frac{1}{\nu} I_0(\lambda A^{-1} + \lambda^{-1} B) \\ &= \left(I - \lambda\nu^{-1} I_0 A^{-1}\right)\left(I - (\nu\lambda)^{-1} \sum_{k=0}^{\infty} \left(\lambda\nu^{-1} I_0 A^{-1}\right)^k I_0 B\right) \\ &:= \left(I - \lambda\nu^{-1} I_0 A^{-1}\right)\left(I - (\nu\lambda)^{-1} I_0 B - G_1(\lambda) B\right), \\ G_1(\lambda) &:= \sum_{k=1}^{\infty} \lambda^{k-1}\nu^{-k-1}(I_0 A^{-1})^k I_0, \end{aligned} \tag{4.38}$$

where $G_1(\lambda)$ is an operator-valued function which is analytic in the circle $|\lambda| < r_1$ and take compact values. Since the first multiplier in (4.38) is invertible for $|\lambda| < r_1$, the pencils $L_{\nu,\omega_0}(\lambda)$ and $F_1(\lambda) := I - (\nu\lambda)^{-1} I_0 B - G_1(\lambda) B$ have the same eigenvalues and systems of eigen- and associated elements for $|\lambda| < r_1$.

Let us show now that the system of eigen- and associated elements of the pencil $F_1(\lambda)$ has no more than a finite defect in $\boldsymbol{M}_0(\Omega)$ after projecting onto $\boldsymbol{M}_0(\Omega)$. We apply the orthoprojectors $Q_{\boldsymbol{M}_0}$ and $Q_{\boldsymbol{N}_0}$ to the left and right hand sides of the equation $F_1(\lambda)\boldsymbol{\xi} = \boldsymbol{0}$, respectively. Considering the equalities $BQ_{\boldsymbol{M}_0} = Q_{\boldsymbol{M}_0} B$ and $BQ_{\boldsymbol{N}_0} = 0$, we obtain

$$\begin{aligned} Q_{\boldsymbol{M}_0}\boldsymbol{\xi} &= (\nu\lambda)^{-1}(Q_{\boldsymbol{M}_0} I_0 Q_{\boldsymbol{M}_0}) B(Q_{\boldsymbol{M}_0}\boldsymbol{\xi}) + (Q_{\boldsymbol{M}_0} G_1(\lambda) Q_{\boldsymbol{M}_0}) B(Q_{\boldsymbol{M}_0}\boldsymbol{\xi}), \\ Q_{\boldsymbol{N}_0}\boldsymbol{\xi} &= (\nu\lambda)^{-1}(Q_{\boldsymbol{N}_0} I_0 Q_{\boldsymbol{M}_0}) B(Q_{\boldsymbol{M}_0}\boldsymbol{\xi}) + (Q_{\boldsymbol{N}_0} G_1(\lambda) Q_{\boldsymbol{M}_0}) B(Q_{\boldsymbol{M}_0}\boldsymbol{\xi}). \end{aligned} \tag{4.39}$$

If $Q_{M_0}\boldsymbol{\xi}$ is known, then $Q_{N_0}\boldsymbol{\xi}$ can be determined from the second equation in (4.39). The first equation contains only $Q_{M_0}\boldsymbol{\xi}$; to this equation it corresponds the following operator-valued function,

$$\tilde{F}_1(\lambda) := -\lambda I_{M_0} + \nu^{-1}(Q_{M_0} I_0 Q_{M_0})B + \lambda(Q_{M_0} G_1(\lambda) Q_{M_0})B, \qquad (4.40)$$

that acts in $\boldsymbol{M}_0(\Omega)$.

Since $Q_{M_0} I_0 Q_{M_0} = I_{M_0} + \Phi_1$, with $\Phi_1 \in \mathfrak{S}_\infty$ (I_{M_0} is the identity operator in $\boldsymbol{M}_0(\Omega)$), the operator B is complete in $\boldsymbol{M}_0(\Omega)$, and the last term in the right hand side of (4.40) is an operator-valued function holomorphic in the circle $|\lambda| < r_1$, then, according to the theorem in Section 1.6.10, the system of eigen- and associated elements of the first equation in (4.39) has no more than a finite defect in $\boldsymbol{M}_0(\Omega)$.

If we consider the semicircle $|\lambda| < r$, $\operatorname{Re}\lambda > 0$, with an arbitrary $r > 0$, then in the semiring of the right complex half-plane enclosed between the circles $|\lambda| = r$ and $|\lambda| = r_1$ there are no more than a finite number of eigenvalues of the problem (4.20). Each of these eigenvalues has a finite algebraic multiplicity. Hence, it appears that the proved statement on completeness with accuracy to a finite defect takes place not only for eigenvalues λ with $|\lambda| < r_1$ but also for λ with $|\lambda| < r$ for any r, $0 < r < \infty$.

This proof shows that the second part of assertion 7° can be proved similarly by performing the substitution $\lambda = 1/\tilde{\lambda}$ without projecting onto $\boldsymbol{M}_0(\Omega)$.

Summing up the results in this section, let us compare them with the similar ones for the problem in Section 8.3. on normal oscillations of a nonrotating fluid. In short, we can say that if the fluid rotates, then the spectrum of the fluid moves out of the real axis and all the eigenvalues, with the exception of maybe a finite number, are located in the region $D_{\varepsilon,\delta}$ defined in (4.31). Moreover, the property of Riesz basicity of the corresponding systems of eigen- and associated elements can be replaced in this case by the property of completeness (for a fluid with sufficiently high viscousity), and the property of defective basicity (for arbitrary viscousity) can be replaced by the property of defective completeness in the same space.

8.4.6 On the Completeness of Systems of Modes of Normal Oscillations of the Initial Problem

The conclusions of the previous section on the completeness of the systems of eigen- and associated elements of problem (4.20) make it possible to prove (as we did in Section 8.3.4) the properties of completeness for solutions of the initial problem on normal oscillations or, which is the same, for eigen- and associated elements of the problem (4.18). Here one can repeat the conclusions of Section 8.3.4, in which the word "basicity" is replaced by the word "completeness." The reader can do this independently; we would like to formulate the final properties for the solutions of problem (4.18).

1° All the elements of the special form

$$\left\{ \boldsymbol{u}_{j,q}; \sum_{i=0}^{q} (-1)^{i+1} (\lambda_j)^{-i} \zeta_{j.q-i} \right\}^{\mathrm{t}}, \qquad j = 1, 2, \ldots, \tag{4.41}$$

where $\zeta_{j,q} = (-\lambda_j)^{-1} \gamma_n \boldsymbol{u}_{j,q}$, that correspond to the eigenvalues λ_j of problem (4.18), form a complete system of elements in the space $\boldsymbol{J}^1_{0.S}(\Omega) \oplus H^{1/2}_\Gamma$.

2° If the condition (4.33) holds true, then the eigen- and associated elements $\{\boldsymbol{u}^+_{k,q}\}$ corresponding to the eigenvalues $\{\lambda^+_k\}^\infty_{k=1}$, $|\lambda^+_k| \geq r_+$, $\operatorname{Re} \lambda^+_k > 0$, form a complete system in the space $\boldsymbol{J}^1_{0.S}(\Omega)$. Similarly, the elements of the form $\zeta^-_{k,q} = -(\lambda^-_k)^{-1} \gamma_n \boldsymbol{u}^-_{k,q}$, where $\boldsymbol{u}^-_{k.q}$ are the eigen- and the associated elements of problem (4.18) corresponding to the eigenvalues $\{\lambda^-_k\}^\infty_{k=1}$, $|\lambda^-_k| \leq r_-$, $\operatorname{Re} \lambda^-_k > 0$, form a complete system in the space $H^{1/2}_\Gamma$.

3° For arbitrary viscousity $\nu > 0$, the above-mentioned elements $\{\boldsymbol{u}^+_{k.q}\}^\infty_{k=1}$ corresponding to the eigenvalues $\{\lambda^+_k\}^\infty_{k=1}$ located in the region $|\lambda| \geq r$, $\operatorname{Re} \lambda > 0$, for any $r > 0$, form a system that has no more than a finite defect in $\boldsymbol{J}^1_{0.S}(\Omega)$. Similarly, the elements $\{\zeta^-_{k,q}\}^\infty_{k=1}$ corresponding to the eigenvalues $\{\lambda^-_k\}^\infty_{k=1}$ located in the region $|\lambda| \leq r$, $\operatorname{Re} \lambda > 0$, form a system that has no more than a finite defect in $H^{1/2}_\Gamma$.

8.5 Asymptotic Solutions for High Viscosity

In the case of high viscousity, ν, we can apply the asymptotic method stated in Section 1.7 to the problems considered in the previous sections.

8.5.1 The Cauchy Problem

First, let us investigate the problem on free oscillations of a fluid in an open immovable container. We can write down the equations (1.34) as

$$\nu^{-1} \frac{\mathrm{d}y}{\mathrm{d}t} = -\mathcal{A} y + g \nu^{-2} \mathcal{B} y, \qquad y(0) = y^0, \tag{5.1}$$

where

$$y = (\boldsymbol{\eta}; \boldsymbol{\delta})^{\mathrm{t}} \in \boldsymbol{J}_{0.S}(\Omega) \oplus \boldsymbol{J}_{0.S}(\Omega), \qquad y^0 = (\boldsymbol{\eta}^0; \boldsymbol{\delta}^0)^{\mathrm{t}},$$

$$\mathcal{A} = \begin{pmatrix} A & 0 \\ 0 & 0 \end{pmatrix},$$

$$\mathcal{B} = \begin{pmatrix} B & B \\ -B & -B \end{pmatrix}.$$

Unlike the case when the fluid fills completely the cavity (see Section 7.4), here the operator coefficient $\mathcal{A}$ does not have an inverse one in the space $\boldsymbol{E} = \boldsymbol{J}_{0,S}(\Omega) \oplus \boldsymbol{J}_{0,S}(\Omega)$. This space is naturally decomposed into the orthogonal sum of subspaces $\boldsymbol{E}_1$ and $\boldsymbol{E}_2$, which consists of elements of the form $(\boldsymbol{\eta}; \mathbf{0})^{\mathrm{t}}$ and $(\mathbf{0}; \boldsymbol{\delta})^{\mathrm{t}}$, respectively. The restriction $\mathcal{A}_1$ of the operator $\mathcal{A}$ to $\boldsymbol{E}_1$ coincides with the operator A and the restriction $\mathcal{A}_2$ of the operator $\mathcal{A}$ to $\boldsymbol{E}_2$ equals the zero operator. Let us recall that A is a positive definite self-adjoint operator in $\boldsymbol{E}_1$ and has a compact inverse operator $A^{-1} > 0$.

Assuming that $\varepsilon = \nu^{-1}$ is a small parameter, we can use the methods in Section 1.7 to construct the third approximation to the solution of the Cauchy problem (5.1). Comparing equation (5.1) with (1.7.7), we realize that in (5.1) $-\mathcal{A}$ plays the role of A in (1.7.7) and $g\nu^{-1}\mathcal{B}$ plays the role of L. With this in mind, the solution of problem (5.1) should look like (1.7.8), that is,

$$y(t) = (I + \nu^{-1} Y) v(t), \tag{5.2}$$

where Y is an interlacing operator that does no depend on t. Here, the new unknown function, $v(t)$, can be found as a solution of equation (1.7.9), that is,

$$\nu^{-1} \frac{dv}{dt} = -\mathcal{A} v + \nu^{-1} \mathcal{C} v, \tag{5.3}$$

where the linear operator $\mathcal{C} = \mathcal{C}(\nu^{-1})$ keeps the subspaces $\boldsymbol{E}_1$ and $\boldsymbol{E}_2$ invariant.

In this case the decomposition (1.7.27) for $L = \sum_{k=0}^{\infty} \varepsilon^k L^{(k)}$ is the following,

$$L^{(0)} = 0, \quad L^{(1)} = g\mathcal{B}, \quad L^{(k)} = 0, \qquad k = 2, 3, \ldots.$$

Therefore, by using (1.7.40) we get that

$$\begin{aligned} Y_{12}^{(0)} &= 0, \quad Y_{12}^{(1)} = gA^{-1}B, \quad Y_{12}^{(2)} = 0, \\ Y_{21}^{(0)} &= 0, \quad Y_{21}^{(1)} = gBA^{-1}, \quad Y_{21}^{(2)} = 0. \end{aligned}$$

Hence, for the partial sums $Y_{ij(N)}$ we have

$$Y_{12(2)} = g\nu^{-1} A^{-1} B, \qquad Y_{21(2)} = g\nu^{-1} B A^{-1}. \tag{5.4}$$

Further, by using (1.7.30) we get that

$$\begin{aligned} S_{11}^{(0)} &= 0, \quad S_{11}^{(1)} = gB, \quad S_{11}^{(2)} = 0, \\ S_{22}^{(0)} &= 0, \quad S_{22}^{(1)} = -gB, \quad S_{22}^{(2)} = 0. \end{aligned}$$

Then

$$S_{11(2)} = \nu^{-1} gB, \qquad S_{22(2)} = -\nu^{-1} gB. \tag{5.5}$$

Therefore, according to (1.7.31) for $N = 3$, equation (5.3) becames

$$\nu^{-1} \frac{d\boldsymbol{v}_1^{(3)}}{dt} = -A\boldsymbol{v}_1^{(3)} + g\nu^{-2} B\boldsymbol{v}_1^{(3)},$$
$$\nu^{-1} \frac{d\boldsymbol{v}_2^{(3)}}{dt} = -g\nu^{-2} B\boldsymbol{v}_2^{(3)}.$$

Hence

$$\boldsymbol{v}_1^{(3)}(t) = \exp(-t(\nu A - g\nu^{-1}B))\boldsymbol{v}_1^{(3)}(0),$$
$$\boldsymbol{v}_2^{(3)}(t) = \exp(-tg\nu^{-1}B)\boldsymbol{v}_2^{(3)}(0). \tag{5.6}$$

By (1.7.33) we obtain that

$$y^{(3)}(t) = (I + \nu^{-1} Y_2) v^{(3)}(t),$$

and in virtue of (5.4) and (5.6) we have

$$\begin{aligned}
\boldsymbol{\eta}^{(3)}(t) &= \boldsymbol{v}_1^{(3)}(t) + \nu^{-1} Y_{12(2)} \boldsymbol{v}_2^{(3)}(t) \\
&= \exp(-t(\nu A - g\nu^{-1}B))\boldsymbol{v}_1^{(3)}(0) + g\nu^{-2} A^{-1} B \exp(-tg\nu^{-1}B)\boldsymbol{v}_2^{(3)}(0), \\
\boldsymbol{\delta}^{(3)}(t) &= \boldsymbol{v}_2^{(3)}(t) + \nu^{-1} Y_{21(2)} \boldsymbol{v}_1^{(3)}(t) \\
&= \exp(-tg\nu^{-1}B))\boldsymbol{v}_2^{(3)}(0) + g\nu^{-2} BA^{-1} \exp(-t(\nu A - g\nu^{-1}B))\boldsymbol{v}_1^{(3)}(0).
\end{aligned}$$

The initial values $\boldsymbol{v}_1^{(3)}(0)$ and $\boldsymbol{v}_2^{(3)}(0)$ can be determined using equations of the form (1.7.25), that is,

$$\boldsymbol{v}_1^{(3)}(0) + g\nu^{-2} A^{-1} B\boldsymbol{v}_2^{(3)}(0) = \boldsymbol{\eta}^0,$$
$$\boldsymbol{v}_2^{(3)}(0) + g\nu^{-2} BA^{-1}\boldsymbol{v}_1^{(3)}(0) = \boldsymbol{\delta}^0.$$

Hence, with accuracy up to terms of order not less than ν^{-4}, we obtain the following:

$$\boldsymbol{v}_1^{(3)}(0) = \boldsymbol{\eta}^0 - g\nu^{-2} A^{-1} B\boldsymbol{\delta}^0,$$
$$\boldsymbol{v}_2^{(3)}(0) = \boldsymbol{\delta}^0 - g\nu^{-2} BA^{-1}\boldsymbol{\eta}^0.$$

Finally we get

$$\boldsymbol{\eta}^{(3)}(t) = \exp(-t\nu(A - g\nu^{-2}B))(\boldsymbol{\eta}^0 - g\nu^{-2} A^{-1} B\boldsymbol{\delta}^0) + g\nu^{-2} A^{-1} B \exp(-gt\nu^{-1}B)\boldsymbol{\delta}^0,$$
$$\boldsymbol{\delta}^{(3)}(t) = \exp(-gt\nu^{-1}B)(\boldsymbol{\delta}^0 - g\nu^{-2} BA^{-1}\boldsymbol{\eta}^0) + g\nu^{-2} BA^{-1} \exp(-t\nu(A - g\nu^{-2}B))\boldsymbol{\eta}^0.$$

Let us recall that the velocity field of a fluid in a container is $\boldsymbol{u}(t) = \boldsymbol{s}(t)+\boldsymbol{w}(t)$, where $\boldsymbol{s} = A^{-1/2}\boldsymbol{\eta}$, $\boldsymbol{w} = A^{-1/2}\boldsymbol{\delta}$. Therefore,

$$\begin{aligned}\boldsymbol{u}^{(3)}(t) &= A^{-1/2}\exp(-t\nu(A-g\nu^{-2}B))(A^{1/2}\boldsymbol{s}(0) - g\nu^{-2}A^{-1}BA^{1/2}\boldsymbol{w}(0))\\ &+ A^{-1/2}\exp(-gt\nu^{-1}B)(A^{1/2}\boldsymbol{w}(0) - g\nu^{-2}BA^{-1/2}\boldsymbol{s}(0))\\ &+ g\nu^{-2}A^{-3/2}B\exp(-gt\nu^{-1}B)A^{1/2}\boldsymbol{w}(0)\\ &+ g\nu^{-2}A^{-1/2}BA^{-1}\exp(-t\nu(A-g\nu^{-2}B))A^{1/2}\boldsymbol{s}(0).\end{aligned}$$

If in this expression the terms containing ν^{-2} are omitted, then we obtain the main members of the decomposition of the velocity field by powers of the parameter ν^{-1}:

$$\boldsymbol{u}^{(1)}(t) = \exp(-t\nu A)\boldsymbol{s}(0) + A^{-1/2}\exp(-gt\nu^{-1}B)\boldsymbol{w}(0). \tag{5.7}$$

Now let us point out that, according to (1.25),

$$\begin{aligned}\boldsymbol{w}(0) &= -g\nu^{-1}T\zeta^0,\\ \boldsymbol{s}(0) &= \boldsymbol{u}^0 - \boldsymbol{w}(0) = \boldsymbol{u}^0 + g\nu^{-1}T\zeta^0,\end{aligned} \tag{5.8}$$

where $\boldsymbol{u}^0(x)$ is the initial velocity field and $\zeta^0(x_1,x_2)$ is the initial deviation of the free surface. Therefore,

$$\boldsymbol{u}^{(1)}(t) = \exp(-t\nu A)(\boldsymbol{u}^0 + g\nu^{-1}T\zeta^0) - g\nu^{-1}A^{-1/2}\exp(-gt\nu^{-1}B)A^{1/2}T\zeta^0. \tag{5.9}$$

For high viscousity ν, the first term in (5.9) decreases quickly enough with time (it decreases not slower than $\exp(-t\nu\lambda_1(A))$ with $\lambda_1(A) > 0$) and the second term for $\zeta^0 \neq 0$ decreases slow enough (because $\lambda_k(B) \to 0$ for $k \to \infty$). That is why, for large t, the first term in (5.9) can be omitted and we may assume that

$$\boldsymbol{u}^{(1)}(t) \approx -g\nu^{-1}A^{-1/2}\exp(-gt\nu^{-1}B)A^{1/2}T\zeta^0. \tag{5.10}$$

Hence, the approximate solution $\boldsymbol{u}^{(1)}(t)$ does not depend on the initial field $\boldsymbol{u}^0(x)$ in the fluid, but depends only on the initial shape $\zeta^0(x_1,x_2)$ of the free surface.

8.5.2 Normal Oscillations

In order to find modes of normal oscillations of a fluid in the form

$$y(t) = \exp(-\lambda t)z, \qquad z = (\boldsymbol{\eta};\boldsymbol{\delta})^t, \tag{5.11}$$

where z does not depend on t, one should obtain the following equation from (5.1),

$$-\lambda\nu^{-1}z = -\mathcal{A}z + g\nu^{-2}\mathcal{B}z. \tag{5.12}$$

After splitting it according to the scheme in Section 1.7, we obtain the system of equations of type (1.7.63) and (1.7.64),

$$-A\boldsymbol{v}_1 + \nu^{-1}S_{11}\boldsymbol{v}_1 = -\lambda\nu^{-1}\boldsymbol{v}_1,$$
$$S_{22}\boldsymbol{v}_2 = -\lambda\boldsymbol{v}_2,$$

where $v = (\boldsymbol{v}_1; \boldsymbol{v}_2)^{\mathrm{t}}$ and z in (5.12) are connected by (5.2). Taking into account only the values (5.5) for $\mathcal{C} = \mathcal{C}(\nu^{-1})$, the following approximate equations can be obtained,

$$A\boldsymbol{v}_1 - g\nu^{-2}B\boldsymbol{v}_1 = \lambda\nu^{-1}\boldsymbol{v}_1, \tag{5.13}$$
$$g\nu^{-1}B\boldsymbol{v}_2 = \lambda\boldsymbol{v}_2. \tag{5.14}$$

Hence, we have two independent eigenvalue problems for the self-adjoint operators $\nu A - g\nu^{-1}B$ and $g\nu^{-1}B$. For a sufficiently large ν, all the eigenvalues $\lambda = \lambda_n^+$ of the first problem are positive and have a limit point $\lambda = +\infty$. The eigenvalues $\lambda = \lambda_n^-$ of the second problem are nonnegative and have a limit point $\lambda = 0$, which is also an eigenvalue with infinite multiplicity. However, as it follows from Section 8.3 and can be directly deduced from (5.12), the number $\lambda = 0$ is not an eigenvalue of the problem (5.12) and this number should not be considered.

Let us look now for the solutions of problem (5.13) in the form of asymptotic series by the powers ν^{-2} taking into account the first two terms only:

$$\boldsymbol{v}_1 = \boldsymbol{v}_1^{(0)} + \nu^{-2}\boldsymbol{v}_1^{(1)}, \qquad \lambda/\nu = \mu^{(0)} + \nu^{-2}\mu^{(1)}. \tag{5.15}$$

Substituting them into (5.12) leads to the relations

$$A\boldsymbol{v}_1^{(0)} - \mu^{(0)}\boldsymbol{v}_1^{(0)} = \mathbf{0},$$
$$A\boldsymbol{v}_1^{(1)} - \mu^{(0)}\boldsymbol{v}_1^{(1)} = gB\boldsymbol{v}_1^{(0)} + \mu^{(1)}\boldsymbol{v}_1^{(0)}. \tag{5.16}$$

From the first equation we obtain the the numbers $\mu^{(0)}$ should coincide with the eigenvalues $\lambda_k(A)$ of the operator A, and the elements $\boldsymbol{v}_1^{(0)}$ should coincide with its eigenelements $\boldsymbol{\xi}_k(A)$. To obtain further corrections for (5.15), let us assume that the eigenvalues $\lambda_k(A)$ are simple. Then, using the orthogonality condition of the right hand side of the second equation in (5.16) to the normalized solution $\boldsymbol{\xi}_k(A)$ of the homogeneous equation, we obtain

$$\mu^{(1)} = \mu_k^{(1)} = -g\left\|B^{1/2}\boldsymbol{\xi}_k(A)\right\|^2_{\boldsymbol{J}_{0,S}(\Omega)}$$
$$= -g\left\|\gamma_n A^{-1/2}\boldsymbol{\xi}_k(A)\right\|^2_{H_\Gamma} = -g\|\gamma_n \boldsymbol{s}_k(A)\|^2_{H_\Gamma}. \tag{5.17}$$

Here $\boldsymbol{s}_k(A)$ are eigenelements, which are normalized in $\boldsymbol{J}^1_{0,S}(\Omega)$, of the boundary value problem

$$-\Delta \boldsymbol{s} + \nabla p = \mu^{(0)} \boldsymbol{s}, \qquad \operatorname{div} \boldsymbol{s} = 0 \text{ in } \Omega, \ \boldsymbol{s} = \boldsymbol{0} \text{ on } S,$$
$$\frac{\partial s_i}{\partial x_3} + \frac{\partial s_3}{\partial x_i} = 0, \quad i = 1,2, \qquad -p + 2\frac{\partial s_3}{\partial x_3} = 0 \text{ on } \Gamma. \tag{5.18}$$

Finally we obtain

$$\lambda = \lambda_k^+(\nu) = \nu\lambda_k(A) - g\nu^{-1}\|\gamma_n \boldsymbol{s}_k(A)\|^2_{H_\Gamma}[1 + \mathrm{o}(\nu^{-2})], \qquad k = 1,2,\ldots; \ \nu \to \infty. \tag{5.19}$$

This formula shows that if ν increases, then the eigenvalues λ_k^+ move to the right (this fact has been pointed out already) and are located to the left from the numbers $\nu\lambda_k(A)$. For any k, the difference between these numbers is not greater asymptotically than the magnitude $g\nu^{-1}\|B\|$ and, therefore, this difference converges to zero as $\nu \to \infty$.

As for the solutions of problem (5.14), it is obvious that for its eigenvalues we have the following asymptotics,

$$\lambda = \lambda_k^-(\nu) = g\nu^{-1}\lambda_k(B)[1 + \mathrm{o}(1)], \qquad k = 1,2,\ldots, \ \nu \to \infty, \tag{5.20}$$

and the eigenelements can be determined by the formula

$$\boldsymbol{v}_2 = \boldsymbol{v}_{2k} = \boldsymbol{\xi}_k(B), \qquad k = 1,2,\ldots. \tag{5.21}$$

From (5.20) it follows that, for increasing ν, the eigenvalue $\lambda_k^-(\nu)$ moves to the left and after transition to the limit for $\nu \to \infty$ it coincides with $\lambda_0 = 0$.

Let us note that, in classical terms, the problem on determining the eigenvalues $\lambda_k^-(\nu)$ is equivalent to the boundary value spectral problem below,

$$-\Delta \boldsymbol{w} + \nabla p = \boldsymbol{0}, \qquad \operatorname{div} \boldsymbol{w} = 0 \text{ in } \Omega, \ \boldsymbol{w} = \boldsymbol{0} \text{ on } S,$$
$$-p + 2\frac{\partial w_3}{\partial x_3} = \mu w_3, \qquad \mu = \lambda\nu/g,$$
$$\frac{\partial w_i}{\partial x_3} + \frac{\partial w_3}{\partial x_i} = 0, \qquad i = 1,2; \text{ on } \Gamma, \tag{5.22}$$

due to the relations $B = A^{1/2}T\gamma_n A^{-1/2}$, and $\boldsymbol{v}_2 = A^{1/2}\boldsymbol{w}$.

Let us deduce the final formulas that characterize the two types of normal movements of a fluid with high viscousity in an open container. For quickly fading movements, that is, when the fading decrements equal $\lambda = \lambda_k^+(\nu)$ and can be calculated using (5.19), the corresponding normal movements $\boldsymbol{u} = \boldsymbol{u}_k^+(t,x)$ have the form

$$\begin{aligned}\boldsymbol{u}_k^+(t,x) &= \exp\left(-\lambda_k^+(\nu)t\right) A^{-1/2}\left(\boldsymbol{\eta}_k^+ + \boldsymbol{\delta}_k^+\right)\\ &= \exp\left(-\lambda_k^+(\nu)t\right) A^{-1/2}\left(I + \nu^{-2}Y_{21}^{(1)}\right)\boldsymbol{\xi}_k(A)\\ &= (\lambda_k(A))^{-1/2}\exp\left(-\lambda_k^+(\nu)t\right)\left(\boldsymbol{\xi}_k(A) + \nu^{-2}gT\gamma_n\boldsymbol{\xi}_k(A)\right).\end{aligned} \tag{5.23}$$

Here we used (1.7.66) and the fact that $\boldsymbol{\xi}_k(A) = \boldsymbol{s}$ and $\lambda_k(A) = \mu^{(0)}$ are the solutions of problem (5.18).

For the second class of movements, some similar formulas can be obtained by using (1.7.65) and (5.4). Finally we have:

$$\begin{aligned}
\boldsymbol{u}_k^-(t,x) &= \exp\left(-\lambda_k^-(\nu)t\right) A^{-1/2}\left(\boldsymbol{\eta}_k^- + \boldsymbol{\delta}_k^-\right) \\
&= \exp\left(-\lambda_k^-(\nu)t\right) A^{-1/2}\left(I + \nu^{-2}Y_{12}^{(1)}\right)\boldsymbol{\xi}_k(B) \\
&\approx \exp\left(-g\nu^{-1}\lambda_k(B)t\right)\left(\boldsymbol{w}_k(B) + g\nu^{-2}A^{-1}T\gamma_n\boldsymbol{w}_k(B)\right),
\end{aligned} \tag{5.24}$$

where $\boldsymbol{w}_k(B) := A^{-1/2}\boldsymbol{\xi}_k(B)$ are the eigenfunctions of the boundary value problem (5.22).

Formulas (5.20) and (5.24) show that for the second class of movements which can be considered as slowly fading normal movements, a logarithmic fading decrement corresponding to the kth movement decreases with the increasing of ν.

Let us note that, according to Section 8.3, all the eigenvalues $\lambda = \lambda_k^-(\nu)$ are located in the interval $[0, r_-]$ and the eigenvalues $\lambda = \lambda_k^+(\nu)$ are located in the interval $[r_+, \infty)$, where

$$r_\pm(\nu) = \frac{\nu \pm \sqrt{\nu^2 - 4g\|A^{-1}\| \cdot \|B\|}}{2\|A^{-1}\|},$$

and therefore,

$$\begin{aligned}
r_-(\nu) &\sim g\nu^{-1} \to 0, \\
r_+(\nu) &\sim \frac{\nu}{g\|A^{-1}\| \cdot \|B\|} \to \infty, \qquad \nu \to \infty.
\end{aligned}$$

Hence, it appears that the following holds true uniformly for all $k = 1, 2, \ldots$,

$$\lambda_k^-(\nu) \to 0, \quad \lambda_k^+(\nu) \to \infty, \qquad \nu \to \infty.$$

8.5.3 Motions Under Mass Forces

Let us consider again the evolution problem of the motion of a fluid in a container, but assuming now that the field of external forces $\boldsymbol{f}(t,x)$ influences the system. Then, according to Section 8.1.4, we obtain that in the right hand side of (5.1) we have an additional term, $\nu^{-1}\mathcal{F}$, where $\mathcal{F} = (A^{1/2}\boldsymbol{f}; \boldsymbol{0})^t$:

$$\nu^{-1}\frac{\mathrm{d}y}{\mathrm{d}t} = -\mathcal{A}y + g\nu^{-2}\mathcal{B}y + \nu^{-1}\mathcal{F}, \qquad y(0) = y^0. \tag{5.25}$$

According to the general theory of asymptotic solutions for equations of a similar type stated in Section 1.7.6, the Nth approximation $y^{(N)} := y_{\text{in}}^{(N)}(t)$ to some particular solution $y_{\text{in}}(t)$ of the inhomogeneous problem (5.25) can be determined from the following equations:

$$y^{(N)} = \left(I + \nu^{-1} Y_{(N-1)}\right) v^{(N)} + h_{(N)}, \tag{5.26}$$

$$\nu^{-1} \frac{\mathrm{d}v^{(N)}}{\mathrm{d}t} = -\mathcal{A} v^{(N)} + \nu^{-1} \mathcal{C}_{(N-1)} v^{(N)} + \nu^{-1} \varphi_{(N-1)}, \tag{5.27}$$

$$\left(I + \nu^{-1} Y_{(N-1)}\right) \varphi_{(N-1)} = \nu^{-1} g \mathcal{B} h_{(N-1)}, \tag{5.28}$$

$$\nu^{-1} \frac{\mathrm{d}h_{(N)}}{\mathrm{d}t} = -\mathcal{A} h_{(N)} + \nu^{-1} \mathcal{F}_{(N)}, \tag{5.29}$$

where $Y = Y(\nu)$ and $\mathcal{C} = \mathcal{C}(\nu)$ are the same operators as in Section 8.5.1.

It is natural to solve the system of equations (5.26)–(5.29) in reverse order. We will do this for the case $N = 3$.

Substituting the expressions for $\mathcal{A}$ and $\mathcal{F}$ into (5.26), we obtain the system

$$\begin{aligned} \nu^{-1} \frac{\mathrm{d}\boldsymbol{h}_{1(3)}}{\mathrm{d}t} &= -A \boldsymbol{h}_{1(3)} + \nu^{-1} A^{1/2} \boldsymbol{f}, \\ \nu^{-1} \frac{\mathrm{d}\boldsymbol{h}_{2(3)}}{\mathrm{d}t} &= 0. \end{aligned} \tag{5.30}$$

The field $\boldsymbol{h}_{2(3)} = 0$ satisfies the second equation in (5.30).

Looking for $\boldsymbol{h}_{1(3)}$ in the form of an asymptotic series by powers of ν^{-1}, substituting this solution into (5.30) and equating the coefficients by the same powers of ν^{-1}, we obtain

$$\begin{aligned} A \boldsymbol{h}_1^{(0)} &= \mathbf{0}, \\ \frac{\mathrm{d}\boldsymbol{h}^{(0)}}{\mathrm{d}t} &= -A \boldsymbol{h}_1^{(1)} + A^{1/2} \boldsymbol{f}, \\ \frac{\mathrm{d}\boldsymbol{h}^{(k)}}{\mathrm{d}t} &= -A \boldsymbol{h}_1^{(k+1)}, \qquad k = 2, 3, \cdots. \end{aligned}$$

Hence

$$\begin{aligned} \boldsymbol{h}_1^{(0)} &\equiv \mathbf{0}, \\ \boldsymbol{h}_1^{(1)}(t) &= A^{-1/2} \boldsymbol{f}(t), \\ \boldsymbol{h}_1^{(2)}(t) &= -A^{-3/2} \frac{\mathrm{d}\boldsymbol{f}}{\mathrm{d}t}, \\ \boldsymbol{h}_1^{(3)}(t) &= A^{-5/2} \frac{\mathrm{d}^2 \boldsymbol{f}}{\mathrm{d}t^2}, \end{aligned}$$

therefore,

$$\boldsymbol{h}_{1(3)}(t) = \nu^{-1}A^{-1/2}\boldsymbol{f} - \nu^{-2}A^{-3/2}\frac{\mathrm{d}\boldsymbol{f}}{\mathrm{d}t} + \nu^{-3}A^{-5/2}\frac{\mathrm{d}^2\boldsymbol{f}}{\mathrm{d}t^2}. \tag{5.31}$$

Now let us determine $\varphi_{(2)}$ from (5.20) or, which is more concrete, from the system

$$\begin{aligned} \boldsymbol{\varphi}_{1(2)} + \nu^{-1}Y_{12(2)}\boldsymbol{\varphi}_{2(2)} &= \nu^{-1}gB\boldsymbol{h}_{1(2)}, \\ \boldsymbol{\varphi}_{2(2)} + \nu^{-1}Y_{21(2)}\boldsymbol{\varphi}_{1(2)} &= -\nu^{-1}gB\boldsymbol{h}_{1(2)}. \end{aligned} \tag{5.32}$$

Equating the coefficients by the equal powers of ν^{-1} we obtain

$$\begin{aligned} \boldsymbol{\varphi}_1^{(0)} &= \boldsymbol{\varphi}_2^{(0)} = \boldsymbol{\varphi}_1^{(1)} = \boldsymbol{\varphi}_2^{(1)} = 0, \\ \boldsymbol{\varphi}_1^{(2)} &= gBA^{-1/2}\boldsymbol{f}, \\ \boldsymbol{\varphi}_2^{(2)} &= -gBA^{-1/2}\boldsymbol{f}, \end{aligned}$$

hence, it appears that

$$\begin{aligned} \boldsymbol{\varphi}_{1(2)} &= \nu^{-2}gBA^{-1/2}\boldsymbol{f}, \\ \boldsymbol{\varphi}_{2(2)} &= -\nu^{-2}gBA^{-1/2}\boldsymbol{f}. \end{aligned} \tag{5.33}$$

Further, the function $v^{(3)}(t)$ is a solution of equation (5.27) for $N = 3$, that is, it is a solution of the following system of equations,

$$\begin{aligned} \nu^{-1}\frac{\mathrm{d}\boldsymbol{v}_1^{(3)}}{\mathrm{d}t} &= -A\boldsymbol{v}_1^{(3)} + \nu^{-2}gB\boldsymbol{v}_1^{(3)} + \nu^{-3}gBA^{-1/2}\boldsymbol{f}, \\ \nu^{-1}\frac{\mathrm{d}\boldsymbol{v}_2^{(3)}}{\mathrm{d}t} &= -\nu^{-2}gB\boldsymbol{v}_2^{(3)} - \nu^{-3}gBA^{-1/2}\boldsymbol{f}. \end{aligned}$$

Hence we obtain

$$\begin{aligned} \boldsymbol{v}_1^{(3)}(t) &= \exp(-t\nu(A - g\nu^{-2}B))\boldsymbol{v}_1^{(3)}(0) \\ &\quad + g\nu^{-2}\int_0^t \exp(-\nu(t-\tau)(A - g\nu^{-2}B))BA^{-1/2}\boldsymbol{f}(\tau)\mathrm{d}\tau, \\ \boldsymbol{v}_2^{(3)}(t) &= \exp(-gt\nu^{-1}B)\boldsymbol{v}_2^{(3)}(0) - g\nu^{-2}\int_0^t \exp(-(t-\tau)g\nu^{-1}B)BA^{-1/2}\boldsymbol{f}(\tau)\mathrm{d}\tau. \end{aligned} \tag{5.34}$$

Since we are looking for any particular solution of the initial problem (5.25), it can be assumed that $\boldsymbol{v}_1^{(3)}(0) = \boldsymbol{v}_2^{(3)}(0) = \mathbf{0}$.

By (5.26) and taking into account (5.4), (5.34), and (5.31), the asymptotic particular solution of the nonhomogeneous problem (5.25) can be obtain as follows:

$$\begin{aligned}\boldsymbol{\eta}^{(3)}(t) &= g\nu^{-2}\int_0^t \exp(-\nu(t-\tau)(A-g\nu^{-2}B))(BA^{-1/2})\boldsymbol{f}(\tau)\mathrm{d}\tau \\ &\quad + \nu^{-1}A^{-1/2}\boldsymbol{f} - \nu^{-2}A^{-3/2}\frac{\mathrm{d}\boldsymbol{f}}{\mathrm{d}t}, \\ \boldsymbol{\delta}^{(3)}(t) &= -g\nu^{-2}\int_0^t \exp(-\nu(t-\tau)gB)(BA^{-1/2})\boldsymbol{f}(\tau)\mathrm{d}\tau. \end{aligned} \tag{5.35}$$

If in (5.35) the terms of order ν^{-2} are omitted, then we can obtain that

$$\begin{aligned}\boldsymbol{\eta}^{(1)}(t) &= \nu^{-1}A^{-1/2}\boldsymbol{f}, \\ \boldsymbol{\delta}^{(1)}(t) &\equiv 0.\end{aligned}$$

Therefore, the main member of the asymptotic decomposition by powers of ν^{-1} of the velocity field of forced motion of a fluid has the form

$$\boldsymbol{u}_{\text{for}}^{(1)}(t) = A^{-1/2}(\boldsymbol{\eta}^{(1)}(t) + \boldsymbol{\delta}^{(1)}(t)) = \nu^{-1}A^{-1}\boldsymbol{f}(t). \tag{5.36}$$

In classical terms, formula (5.36) states that the field $\boldsymbol{u} = \boldsymbol{u}_{\text{for}}^{(1)}(t,x)$ is a solution, for each t, of the following problem,

$$\begin{aligned} &-\nu\Delta\boldsymbol{u} + \nabla p = \boldsymbol{f}(t,x), \qquad \operatorname{div}\bar{\boldsymbol{u}} = 0 \text{ in } \Omega, \ \boldsymbol{u} = \boldsymbol{0} \text{ on } S, \\ &\frac{\partial u_i}{\partial x_3} + \frac{\partial u_3}{\partial x_i} = 0, \qquad i = 1,2, \\ &-p + 2\nu\frac{\partial u_3}{\partial x_3} = \boldsymbol{0} \text{ on } \Gamma. \end{aligned}$$

8.5.4 The Case of Rotating Viscous Fluids

Let us now obtain the asymptotic solution of the problem on free movements of a viscous fluid in a rotating partially filled container. If in the system of equations (4.6) we perform the change of variables $\boldsymbol{\eta}(t) = A^{1/2}\boldsymbol{s}(t)$ and $\boldsymbol{\delta}(t) = A^{1/2}\boldsymbol{w}(t)$, the system can be rewritten as

$$\nu^{-1}\frac{\mathrm{d}y}{\mathrm{d}t} = -\mathcal{A}y + 2\mathrm{i}\omega_0\nu^{-1}\mathcal{A}^{1/2}\mathcal{D}y + \nu^{-2}\mathcal{B}y, \qquad y = (\boldsymbol{\eta};\boldsymbol{\delta})^{\mathrm{t}}, \tag{5.37}$$

where the operator matrices $\mathcal{A}$ and $\mathcal{B}$ are as in (5.1),

$$\mathcal{D} = \begin{pmatrix} D & D \\ 0 & 0 \end{pmatrix}, \qquad D = S_0A^{-1/2} \in \mathfrak{S}_\infty, \ S_0\boldsymbol{u} := \mathrm{i}P_{0,S}(\boldsymbol{u}\times\boldsymbol{e}_3), \tag{5.38}$$

and $B = A^{1/2}TB_0\gamma_nA^{-1/2}$ (see Section 8.4).

As in Section 8.5.1, we apply the asymptotic method stated in detail in Section 1.7 to problem (5.37). Without going into too much detail, let us write down the formulas to be used in the sequel,

$$\begin{aligned} Y_{12(1)} &= 2\mathrm{i}\omega_0 V + \nu^{-1}\left(A^{-1}B - 4\omega_0^2 V^2\right), \\ V &:= A^{-1/2}S_0A^{-1/2} = V^*, \\ Y_{21(1)} &= \nu^{-1}BA^{-1}, \\ S_{11(1)} &= \nu^{-1}B + 2\mathrm{i}\omega_0 AV, \\ S_{22(1)} &= -\nu^{-1}B. \end{aligned} \tag{5.39}$$

The splitted system of equations for the second approximations has the form

$$\begin{aligned} \nu^{-1}\frac{d\boldsymbol{v}_1^{(2)}}{dt} &= -A\boldsymbol{v}_1^{(2)} + 2\mathrm{i}\omega_0 A^{1/2}D\boldsymbol{v}_1^{(2)} + \nu^{-2}B\boldsymbol{v}_1^{(2)}, \\ \nu^{-1}\frac{d\boldsymbol{v}_2^{(2)}}{dt} &= -\nu^{-2}B\boldsymbol{v}_2^{(2)}. \end{aligned} \tag{5.40}$$

Since $D = S_0A^{-1/2} \in \mathfrak{S}_\infty$, the operator $A^{1/2}D$ is quite subordinated to the operator $A^{1/2}$. That is why the operator in the right hand side of the first equation in (5.40) multiplied by ν generates an analytic semigroup $\mathcal{U}(t)$. Then,

$$\begin{aligned} \boldsymbol{v}_1^{(2)}(t) &= \mathcal{U}(t)\boldsymbol{v}_1^{(2)}(0), \\ \boldsymbol{v}_2^{(2)}(t) &= \exp(-\nu^{-1}tB)\boldsymbol{v}_2^{(2)}(0). \end{aligned} \tag{5.41}$$

From (5.39), (5.41), and the following formula

$$y^{(2)}(t) = \left(I + \nu^{-1}Y_{(1)}(\nu)\right)v^{(2)}(t),$$

we obtain the following, with accuracy up to terms of order ν^{-2},

$$\begin{aligned} \boldsymbol{\eta}^{(2)}(t) &= \boldsymbol{v}_1^{(2)}(t) + \nu^{-1}Y_{12(1)}\boldsymbol{v}_2^{(2)}(t) = \mathcal{U}(t)\boldsymbol{v}_1^{(2)}(0) + 2\mathrm{i}\omega_0\nu^{-1}V\exp(-\nu^{-1}tB)\boldsymbol{v}_2^{(2)}(0), \\ \boldsymbol{\delta}^{(2)}(t) &= \boldsymbol{v}_2^{(2)}(t) + \nu^{-1}Y_{21(1)}\boldsymbol{v}_1^{(2)}(t) = \exp(-\nu^{-1}tB)\boldsymbol{v}_2^{(2)}(0). \end{aligned}$$

For the initial data we have the system

$$\begin{aligned} \boldsymbol{v}_1^{(2)}(0) + 2\mathrm{i}\omega_0\nu^{-1}V\boldsymbol{v}_2^{(2)}(0) &= \boldsymbol{\eta}^0, \\ \boldsymbol{v}_2^{(2)}(0) &= \boldsymbol{\delta}^0. \end{aligned}$$

Hence it can finally be obtained that

$$\begin{aligned} \boldsymbol{\eta}^{(2)}(t) &= \mathcal{U}(t)\left(\boldsymbol{\eta}^0 - 2\mathrm{i}\omega\nu^{-1}V\boldsymbol{\delta}^0\right) + 2\mathrm{i}\omega_0\nu^{-1}V\exp(-\nu^{-1}tB)\boldsymbol{\delta}^0, \\ \boldsymbol{\delta}^{(2)}(t) &= \exp(-\nu^{-1}tB)\boldsymbol{\delta}^0. \end{aligned}$$

Therefore, for the velocity field $\boldsymbol{u} = A^{-1/2}(\boldsymbol{\eta} + \boldsymbol{\delta})$, the asymptotic solution equals—with accuracy up to terms of order ν^{-2}—the following

$$\boldsymbol{u}^{(2)}(t) = A^{-1/2}\left(I + 2\mathrm{i}\omega_0\nu^{-1}V\right)\exp(-\nu^{-1}tB)\boldsymbol{\delta}^0 + A^{-1/2}\mathcal{U}(t)\left(\boldsymbol{\eta}^0 - 2\mathrm{i}\omega\nu^{-1}V\boldsymbol{\delta}^0\right), \tag{5.42}$$

where $\mathcal{U}(t)$ is the semigroup generated by the operator

$$-\nu A + 2\mathrm{i}\omega_0 A^{1/2}D + \nu^{-1}B.$$

From (5.42) it follows that the main member of the asymptotic decomposition of a solution can be determined by the formula

$$\boldsymbol{u}(t) \approx A^{-1/2}\mathcal{U}(t)A^{1/2}\boldsymbol{s}(0) + A^{-1/2}\exp(-\nu^{-1}tB)A^{1/2}\boldsymbol{w}(0). \tag{5.43}$$

Here, the first term is the quickly fading with time part of the solution and the second one is the slowly fading part. Formula (5.43) shows that the slowly fading motion of the system does not depend explicitly on ω_0. Yet it obviously depends on ω_0 implicitly, because the shape of the region Ω that is filled with fluid in the state of uniform rotation depends on ω_0. Therefore, the operators A and B depend on ω_0, too. Hence, the slowly fading part does not depend on the Coriolis forces influencing the system.

Summing up the results related to the evolution problem, let us note that for the forced movements of a rotating fluid we can obtain an asymptotic solution in a similar way as in Section 8.5.3. The reader can do it independently.

Let us consider normal movements in problem (5.37) while assuming $y(t) = \exp(-\lambda t)z$. Then

$$-\lambda\nu^{-1}z = -\mathcal{A}z + 2\mathrm{i}\omega_0\nu^{-1}\mathcal{A}^{1/2}\mathcal{D}z + \nu^{-2}\mathcal{B}z,$$

and, according to (5.40), the splitted system takes the following form:

$$\begin{aligned} -\nu^{-1}\lambda\boldsymbol{v}_1 &= -A\boldsymbol{v}_1 + 2\mathrm{i}\omega_0\nu^{-1}A^{1/2}D\boldsymbol{v}_1 + \nu^{-2}B\boldsymbol{v}_1, \\ -\nu^{-1}\lambda\boldsymbol{v}_2 &= -\nu^{-2}B\boldsymbol{v}_2. \end{aligned} \tag{5.44}$$

From the second equation in (5.44) we obtain that

$$\lambda = \lambda_k^- = \nu^{-1}\lambda_k(B), \quad \boldsymbol{v}_2 = \boldsymbol{\xi}_k(B), \qquad k = 1, 2, \ldots, \tag{5.45}$$

where $\boldsymbol{\xi}_k(B) = A^{1/2}\boldsymbol{w}_k(B)$ are the eigenelements of the operator B, that is, they are nontrivial solutions of the problem

$$\begin{aligned} &-\Delta \boldsymbol{w} + \nabla p = \boldsymbol{0}, \qquad \operatorname{div} \boldsymbol{w} = 0 \text{ in } \Omega, \;\; \boldsymbol{w} = \boldsymbol{0} \text{ on } S, \\ &w_{i,3} + w_{3,i} = 0, \qquad i = 1, 2, \\ &-p + 2w_{3,3} = \lambda^{-1} B_0 w_n \text{ on } \Gamma, \end{aligned}$$

where B_0 is the operator acting in H_Γ that has been defined in Section 8.4.2.

Expanding the solutions $\boldsymbol{v}_1$ and λ/ν of the first equation in (5.44) into series by powers of ν^{-1} and taking into account the first and zero approximations only, we obtain the first equation (5.16) and the following equation

$$A\boldsymbol{v}_1^{(1)} - \mu^{(0)}\boldsymbol{v}_1^{(1)} = \mu^{(1)}\boldsymbol{v}_1^{(0)} + 2\mathrm{i}\omega_0 A^{1/2} S_0 A^{-1/2}\boldsymbol{v}_1^{(0)}, \qquad \mu^{(0)} = \nu\lambda_k(A). \tag{5.47}$$

Assuming that the eigenvalues $\mu^{(0)}$ are of multiplicity one and using the solvability conditions for equation (5.47), the following formula for $\mu^{(1)}$ can be obtained,

$$\begin{aligned} \mu^{(1)'} &= -2\mathrm{i}\omega_0 \left(A^{1/2} S_0 A^{-1/2}\boldsymbol{\xi}_k(A), \boldsymbol{\xi}_k(A)\right)_{\boldsymbol{J}_{0,S}(\Omega)} \\ &= -2\mathrm{i}\omega_0 \left(S_0\boldsymbol{\xi}_k(A), \boldsymbol{\xi}_k(A)\right)_{\boldsymbol{J}_{0,S}(\Omega)}. \end{aligned}$$

Hence,

$$\lambda = \lambda_k^+(\nu) = \nu\lambda_k(A) - 2\mathrm{i}\omega_0 \left(S_0\boldsymbol{\xi}_k(A), \boldsymbol{\xi}_k(A)\right)_{\boldsymbol{J}_{0,S}(\Omega)} + O(\nu^{-1}), \quad k = 1, 2 \ldots, \; \nu \to \infty. \tag{5.48}$$

Since $S_0 = S_0^*$ and $\|S_0\| = 1$, then (5.48) leads to the following conclusions. For large ν, the fading decrements of quickly fading normal movements equal $\nu\lambda_k(A)$ with accuracy up to magnitudes of order ν^{-1}, and the frequencies of their oscillations equal $2\omega_0(S_0\boldsymbol{\xi}_k(A), \boldsymbol{\xi}_k(A))_{\boldsymbol{J}_{0,S}(\Omega)}$, where $\boldsymbol{\xi}_k(A)$ are the eigenelements of the operator A normalized in $\boldsymbol{J}_{0,S}(\Omega)$. The absolute values of these frequencies do not exceed $2\omega_0$ and do not depend on fluid's viscousity.

8.6 Oscillations of a System of Nonmixing Fluids

In this section we review the case when a cavity in a body is filled with a system of homogeneous incompressible fluids with different densities.It will be shown that, according to Section 3.6, Chaper 4 and Section 6.3, this class of problems can be reduced to the same type of operator equations as in the case of one fluid partially filling the container.

8.6.1 Statement of the Problem on Small Oscillations

Let us assume that a container Ω is filled with a system of m unmixing viscous fluids with the densities $\rho_1 > \rho_2 > \cdots > \rho_m > 0$. These fluids in an unperturbed state are rotated uniformly together with the container with an angular velocity $\boldsymbol{\omega}_0 = \omega_0 \boldsymbol{e}_3$. In this case, if the gravitational field is directed against the rotation axis Ox_3, the pressures $P_{0k}(x)$ in fluids are distributed according to the law (6.3.41). The equations of the separation boundaries Γ_i between the fluids are described by (6.3.42).

We denote by μ_k, $k = 1, 2, \ldots, m$, the dynamic fluid viscosities and put $\mu_k = \nu \rho_k^0$, where $\nu > 0$ is the average kinematic viscosity of the system and also a parameter of the problem, and ρ_k^0 are positive constants with density dimensionality. From now on, whenever we will mention a change in the system's viscosity, we will mean a change of the parameter ν alone.

Consider now small motions of the system of fluids close to the uniform rotation. As in (4.1), (4.2), and (6.3.46), for the velocity fields $\boldsymbol{u}^k(t,x)$ and the dynamic presssures $p_k(t,x)$ we get

$$\frac{\partial \boldsymbol{u}^k}{\partial t} - 2\omega_0 \boldsymbol{u}^k \times \boldsymbol{e}_3 = -\frac{1}{\rho_k}\nabla p_k + \nu_k \Delta \boldsymbol{u}^k + \boldsymbol{f},$$
$$\operatorname{div} \boldsymbol{u}^k = 0 \text{ in } \Omega_k, \ \boldsymbol{u}^k = \boldsymbol{0} \text{ on } S_k, \qquad k = 1, 2, \ldots, m. \tag{6.1}$$

Here $\nu_k = \mu_k/\rho_k$ are the kinematic fluid viscosities, Ω_k is the domain occupied by the fluid k in an unperturbed state, $S_k = S \cap \bar{\Omega}_k$ is the corresponding part of the rigid wall S, and $\boldsymbol{f} = \boldsymbol{f}(t,x)$ is a small field of external forces.

To write the boundary conditions on the equilibrium separation boundaries Γ_i, $i = 1, 2, \ldots, m-1$, which in this case are the paraboloids of rotation (6.3.42), we proceed similarly to Section 4.1.9. In the neighborhood of each surface Γ_i, we introduce a curvilinear coordinate system $\tilde{O}_i \xi^1 \xi^2 \xi^3$ just like in the case of the system of capillary fluids. Then the dynamic and kinematic conditions on Γ_i, consisting of a continuity of velocities and stresses on the moving surface $\tilde{\Gamma}_i(t)$, take the form

$$\mu_i \left(u^i_{j,3} + u^i_{3,j}\right) = \mu_{i+1}\left(u^{i+1}_{j.3} + u^{i+1}_{3.j}\right) \text{ on } \Gamma_i, \ j = 1, 2, \ i = 1, 2, \ldots, m-1, \tag{6.2}$$
$$\left(-p_i + 2\mu_i u^i_{3,3}\right) - \left(-p_{i+1} + 2\mu_{i+1} u^{i+1}_{3.3}\right) = -a_i(\hat{\xi}_i)\zeta_i,$$
$$a_i(\hat{\xi}_i) := (\nabla P_{0,i} - \nabla P_{0.i+1}) \cdot \boldsymbol{n}_i \text{ on } \Gamma_i, \ i = 1, 2, \ldots, m-1, \tag{6.3}$$
$$\frac{\partial \zeta_i}{\partial t} = u^i_{n_i}, \ \boldsymbol{u}^i = \boldsymbol{u}^{i+1} \text{ on } \Gamma_i, \ i = 1, 2, \ldots, m-1. \tag{6.4}$$

Here $\zeta_i = \zeta_i(t, \hat{\xi}_i)$ is the deviation along the normal $\boldsymbol{n}_i$ to Γ_i of the moving free surface $\tilde{\Gamma}_i(t)$ from Γ_i and $\hat{\xi}_i := (\xi_i^1, \xi_i^2)$. Because the equilibrium pressures are equal,

$$P_{0k}(x) = -\rho_k g x_3 + \frac{1}{2}\rho_k \omega_0^2 \left(x_1^2 + x_2^2\right) + c_k, \qquad k = 1, 2, \ldots, m,$$

then

$$a_i(\hat{\xi}_i) = (\rho_i - \rho_{i+1}) \left[g\cos(\widehat{\boldsymbol{n}_i, \boldsymbol{e}_3}) - \omega_0^2 r \cos(\widehat{\boldsymbol{n}_i, \boldsymbol{e}_r}) \right], \quad r = \sqrt{x_1^2 + x_2^2},\ i = 1, 2, \dots, m.$$

The initial conditions in the considered initial boundary value problem have the form

$$\begin{aligned} \boldsymbol{u}^k(0, x) &= \boldsymbol{u}^{k,0}(x), \qquad k = 1, 2, \dots, m; \\ \zeta_i(0, \hat{\xi}_i) &= \zeta_i^0(\hat{\xi}_i), \qquad i = 1, 2, \dots, m-1. \end{aligned} \tag{6.5}$$

8.6.2 Transition to a System of Operator Equations

Let us now introduce, as we did in Section 2.1, the Hilbert space $\hat{\boldsymbol{L}}_2(\Omega)$ of field sets $\hat{\boldsymbol{u}} := \{\boldsymbol{u}^k\}_{k=1}^m$ with the norm squared

$$\|\hat{\boldsymbol{u}}\|^2_{\hat{\boldsymbol{L}}_2(\Omega)} := \sum_{k=1}^m \rho_k \int_{\Omega_k} |\boldsymbol{u}^k|^2 d\Omega_k,$$

and think of the solution $\hat{\boldsymbol{u}}(t) = \{\boldsymbol{u}^k(t,x)\}_{k=1}^m$ of the problem (6.1)–(6.5) for every t as an element of the space $\hat{\boldsymbol{L}}_2(\Omega)$. Then from (6.1) it follows that $\hat{\boldsymbol{u}} \in \boldsymbol{E} := \hat{\boldsymbol{J}}_{0,S}(\Omega)$, where $\hat{\boldsymbol{J}}_{0.S}(\Omega)$ is a subspace of $\hat{\boldsymbol{L}}_2(\Omega)$ associated with it by the orthogonal expansions (2.1.42)–(2.1.44).

Consider further the space $\boldsymbol{F} := \hat{\boldsymbol{J}}^1_{0.S}(\Omega)$ densely embedded in $\boldsymbol{E} = \hat{\boldsymbol{J}}_{0,S}(\Omega)$ and characterizing the energy dissipation of the system. The properties of the elements $\hat{\boldsymbol{u}} = \{\boldsymbol{u}^k\}_{k=1}^m$ from $\hat{\boldsymbol{J}}^1_{0.S}(\Omega)$ and of the two basic boundary value problems (2.2.44) and (2.2.45) related to $\boldsymbol{F}$ are described explicitly in Section 2.2.8.

From (6.1) and (6.2) we get that for every t, $\hat{\boldsymbol{u}}(t) = \{\boldsymbol{u}^k(t,x)\}_{k=1}^m$ can be considered as an element not only in $\hat{\boldsymbol{J}}_{0.S}(\Omega)$, but in its dense subset $\hat{\boldsymbol{J}}^1_{0.S}(\Omega)$ as well.

Let us next represent the solution $\hat{\boldsymbol{u}}(t)$ of the problem (6.1)–(6.5) as a sum $\hat{\boldsymbol{u}}(t) = \hat{\boldsymbol{s}}(t) + \hat{\boldsymbol{w}}(t)$, where $\hat{\boldsymbol{s}}(t) = \{\boldsymbol{s}^k(t,x)\}_{k=1}^m$ is the solution of the following boundary value problem I of type (2.2.44),

$$\begin{aligned} &-\nu \frac{\rho_k^0}{\rho_k} \Delta \boldsymbol{s}^k + \frac{1}{\rho_k} \nabla p_k^{(1)} = -\frac{\partial \boldsymbol{u}^k}{\partial t} + 2\omega_0 (\boldsymbol{u}^k \times \boldsymbol{e}_3) + \boldsymbol{f}, \\ &\operatorname{div} \boldsymbol{u}^k = 0 \text{ in } \Omega_k, \quad \boldsymbol{u}^k = \boldsymbol{0} \text{ on } S_k, \qquad k = 1, 2, \dots, m; \\ &\boldsymbol{s}^i = \boldsymbol{s}^{i+1}, \quad \nu\rho_i^0 \left(s^i_{j,3} + s^i_{3.j}\right) - \nu\rho_{i+1}^0 \left(s^{i+1}_{j.3} + s^{i+1}_{3.j}\right) = 0, \quad j = 1, 2, \\ &\left(-p_i^{(1)} + 2\nu\rho_i^0 s^i_{3,3}\right) - \left(-p_{i+1}^{(1)} + 2\nu\rho_{i+1}^0 s^{i+1}_{3.3}\right) = 0 \text{ on } \Gamma_i, \quad i = 1, 2, \dots, m-1, \end{aligned} \tag{6.6}$$

and $\hat{\boldsymbol{w}}(t) = \{\boldsymbol{w}^k(t,x)\}_{k=1}^m$ is the solution of the following boundary value problem II of type (2.2.45),

$$
\begin{aligned}
&-\nu\rho_k^0\Delta\boldsymbol{w}^k + \nabla p_k^{(2)} = \boldsymbol{0},\\
&\operatorname{div}\boldsymbol{w}^k = 0 \text{ in } \Omega_k, \quad \boldsymbol{w}^k = \boldsymbol{0} \text{ on } S_k, \qquad k = 1,2,\ldots,m;\\
&\boldsymbol{w}^i = \boldsymbol{w}^{i+1}, \quad \nu\rho_i^0\left(w_{j.3}^i + w_{3.j}^i\right) - \nu\rho_{i+1}^0\left(w_{j.3}^{i+1} + w_{3.j}^{i+1}\right) = 0, \quad j = 1,2,\\
&\left(-p_i^{(2)} + 2\nu\rho_i^0 w_{3.3}^i\right) - \left(-p_{i+1}^{(2)} + 2\nu\rho_{i+1}^0 w_{3.3}^{i+1}\right) = -a_i\left(\hat{\xi}_i\right)\zeta_i \text{ on } \Gamma_i,\\
&\qquad\qquad i = 1,2,\ldots,m-1.
\end{aligned}
\tag{6.7}
$$

We choose the norm squared in $\hat{\boldsymbol{J}}_{0.S}^1(\Omega)$ not to be defined by a scalar product as in (2.2.39), but rather in the following form,

$$
\|\hat{\boldsymbol{u}}\|^2_{\hat{\boldsymbol{J}}_{0.S}^1(\Omega)} := \sum_{k=1}^m \rho_k^0 E_k(\boldsymbol{u}^k, \boldsymbol{u}^k), \tag{6.8}
$$

where $E_k(\boldsymbol{u}^k, \boldsymbol{u}^k)$ is a corresponding bilinear form of the kind (2.2.3) for the domain Ω_k. Then, in short, a solution of problem I can be written as

$$
\nu\boldsymbol{s} = \hat{A}^{-1}\left(-\frac{d\hat{\boldsymbol{u}}}{dt} + 2i\omega_0\hat{S}_0\hat{\boldsymbol{u}} + \hat{\boldsymbol{f}}\right), \tag{6.9}
$$

where $\hat{A}$ is the operator generated by the Hilbert pair $(\boldsymbol{F};\boldsymbol{E}) = \left(\hat{\boldsymbol{J}}_{0,S}^1(\Omega); \hat{\boldsymbol{J}}_{0,S}(\Omega)\right)$, $\hat{\boldsymbol{f}} = \hat{P}_{0,S}\{\boldsymbol{f}|_{\Omega_k}\}_{k=1}^m$,

$$
\hat{S}_0\hat{\boldsymbol{u}} := i\hat{P}_{0.S}\{\boldsymbol{u}^k \times \boldsymbol{e}_3\}_{k=1}^m, \tag{6.10}
$$

and $\hat{P}_{0,S}$ is the orthoprojector onto $\hat{\boldsymbol{J}}_{0.S}(\Omega)$.

Following the abstract scheme in Section 1.8, we get that $\hat{A}$ is positive definite and unbounded in $\hat{\boldsymbol{J}}_{0.S}(\Omega)$ and that $\mathcal{D}(\hat{A}^{1/2}) = \hat{\boldsymbol{J}}_{0.S}^1(\Omega)$, $0 < \hat{A}^{-1} \in \mathfrak{S}_\infty$. Its eigenvalues $\lambda_k(\hat{A})$ are obtained as successive minima of the variational ratio $\|\hat{\boldsymbol{u}}\|^2_{\hat{\boldsymbol{J}}_{0.S}^1(\Omega)}/\|\hat{\boldsymbol{u}}\|^2_{\hat{\boldsymbol{J}}_{0.S}(\Omega)}$ considered on nonzero elements $\hat{\boldsymbol{u}}$ in $\hat{\boldsymbol{J}}_{0.S}^1(\Omega)$. The asymptotic behavior of the eigenvalues $\lambda_k(\hat{A})$ as $k \to \infty$ is the following,

$$
\lambda_k(\hat{A}) = (c_{\hat{A}})^{-2/3}k^{2/3}[1 + o(1)], \qquad c_{\hat{A}} := \frac{1}{3\pi^2}\sum_{j=1}^m (\rho_j^0/\rho_j)^{3/2} \operatorname{mes}\Omega_j > 0. \tag{6.11}
$$

Whence it follows that $\hat{A}^{-1} \in \mathfrak{S}_p$ for $p > 3/2$.

It can also be verified that the operator $\hat{S}_0$ defined by (6.10) is bounded and self-adjoint and that its spectrum is $\sigma(S_0) = [-1, 1]$.

Let us denote by $\hat{T}$ the operator solving the boundary value problem (6.7), that is,

$$\nu\hat{\boldsymbol{w}} = -\hat{T}\hat{B}_0\hat{\zeta}, \qquad \hat{\zeta} := \{\zeta_j\}_{j=1}^{m-1}, \tag{6.12}$$

$$\hat{B}_0\hat{\zeta} := \hat{P}_\Gamma\{a_j(\hat{\xi}_j)\zeta_j\}_{j=1}^{m-1}. \tag{6.13}$$

In (6.12) and (6.13) it is natural to assume that $\hat{\zeta}(t) = \{\zeta_j\}_{j=1}^{m-1}$ is a function with values in the space $\hat{L}_{2,\Gamma} := \bigoplus_{j=1}^{m-1} L_{2,\Gamma_j}$, where $L_{2,\Gamma_j} := L_2(\Gamma_j) \ominus \{1_j\}$. The orthoprojector on $\hat{L}_{2,\Gamma}$ is denoted by $\hat{P}_\Gamma$ in (6.13). Just as in Section 8.4, we assume that all the functions $a_j(\hat{\xi}_j)$, $j = 1, 2, \ldots, m-1$ are continuous on Γ_j. Then (6.13) defines a self-adjoint bounded operator $\hat{B}_0$ acting in $\hat{L}_{2,\Gamma}$.

According to the general scheme in Section 1.8, it follows that the operator $\hat{T}$ acts boundedly from $\hat{H}_\Gamma^{-1/2} := (\hat{H}_\Gamma^{1/2})^*$ in $\hat{\boldsymbol{J}}_{0.S}^1(\Omega)$.

All the above-mentioned results point out the fact that problem (6.1)–(6.4) is equivalent to the system of equations (6.9), (6.12) and to the following relationships

$$\hat{\boldsymbol{u}}(t) = \hat{\boldsymbol{s}}(t) + \hat{\boldsymbol{w}}(t), \qquad \frac{\mathrm{d}\hat{\zeta}}{\mathrm{d}t} = \hat{\gamma}_n\hat{\boldsymbol{u}}, \tag{6, 14}$$

where the trace operator $\hat{\gamma}_n$ is defined for elements from $\hat{\boldsymbol{J}}_{0.S}^1(\Omega)$ by

$$\hat{\gamma}_n\hat{\boldsymbol{u}} := \left\{(\boldsymbol{u}^j \cdot \boldsymbol{n}_j)_{\Gamma_j}\right\}_{j=1}^{m-1}. \tag{6.15}$$

As it follows from the embedding theorems and the abstract scheme in Section 1.8, the operator $\hat{\gamma}_n$ is acting continuously from $\hat{\boldsymbol{J}}_{0.S}^1(\Omega)$ in $\hat{H}_\Gamma^{1/2}$.

We need to add to equations (6.9), (6.12), and (6.14) the initial conditions arising from (6.5), that is,

$$\hat{\boldsymbol{u}}(0) = \hat{\boldsymbol{u}}^0 := \left\{\boldsymbol{u}^{k.0}(x)\right\}_{k=1}^{m},$$

$$\hat{\zeta}(0) = \hat{\zeta}^0 := \left\{\zeta^{j,0}(\hat{\xi}_j)\right\}_{j=1}^{m-1}. \tag{6.16}$$

8.6.3 The Theorem on Existence of a Generalized Solution

Some further considerations of the initial problem (6.1)–(6.5) based on the problem (6.9), (6.12), (6.14), and (6.15) may be done similarly to the scheme developed in Section 8.4.3 for the problem on small motions of a rotating viscous fluid. It is obvious that these considerations lead to formulas similar to (4.9)–(4.16) where we have to add the $\widehat{\ }$ sign to all the elements, operators and spaces.

In particular, the state of relative equilibrium of the fluid system in a container will be called *statically stable* if the following condition is satisfied,

$$\min_{\hat{\xi}_j \in \Gamma_j,\ 1 \leq j \leq m-1} a_j(\hat{\xi}_j) \geq a_0 > 0. \tag{6.17}$$

In this case, the operator $\hat{B} := \hat{A}^{1/2}\hat{T}\hat{B}_0\hat{\gamma}_n\hat{A}^{-1/2}$ is nonnegative and compact with an infinite-dimensional kernel. If condition (6.17) is satisfied, then the eigenvalues of $\hat{B}$ have an asymptotic behavior as $k \to \infty$,

$$\lambda_k(\hat{B}) = (c_{\hat{B}})^{1/2} k^{-1/2}[1 + o(1)],$$
$$c_{\hat{B}} := \frac{1}{16\pi^2} \sum_{j=1}^{m-1} \int_{\Gamma_j} |a_j|^2 d\Gamma_j / (\rho_j^0 + \rho_{j+1}^0)^2 > 0, \tag{6.18}$$

whence we get that $\hat{B} \in \mathfrak{S}_p$ for $p > 2$.

As in Section 8.4.3, we come to the following conclusion. Let the conditions

$$\hat{\boldsymbol{u}}^0 \in \hat{\boldsymbol{J}}^1_{0,S}(\Omega), \quad \hat{\zeta}^0 \in \hat{L}_{2,\Gamma} \tag{6.19}$$

be satisfied and let the field of external forces $\boldsymbol{f}(t,x)$ be a continuous function of t with values in $\hat{\boldsymbol{J}}^1_{0,S_k}(\Omega_k)$ for every $k = 1, 2, \ldots, m$. Then problem (6.1)–(6.5) has a unique generalized solution

$$\hat{\boldsymbol{u}}(t) = \{\boldsymbol{u}^k(t,x)\}_{k=1}^m; \quad \hat{\zeta}(t) = \{\zeta_j(t, \hat{\xi}_j)\}_{j=1}^{m-1},$$

for which $\hat{\boldsymbol{u}}(t)$ is a continuous function with values in $\hat{\boldsymbol{J}}^1_{0,S}(\Omega)$, and $\hat{\zeta}(t)$ is a continuous function with values in $\hat{L}_{2,\Gamma}$. For this generalized solution, both the kinetic and the potential energy of the system are continuous in t and the law of balance of full energy is satisfied similarly to the one presented in Section 8.1.5, that is,

$$\frac{1}{2}\left\{\|\hat{\boldsymbol{u}}(t)\|^2_{\hat{\boldsymbol{L}}_2(\Omega)} + \|\hat{\zeta}(t)\|^2_{\hat{L}_{2,\Gamma}}\right\}$$
$$= \frac{1}{2}\left\{\|\hat{\boldsymbol{u}}^0\|^2_{\hat{\boldsymbol{L}}_2(\Omega)} + \|\hat{\zeta}^0\|^2_{\hat{L}_{2,\Gamma}}\right\} - \nu \int_0^t \|\hat{\boldsymbol{u}}(\tau)\|^2_{\hat{\boldsymbol{J}}^1_{0,S}(\Omega)} d\tau + \int_0^t (\hat{\boldsymbol{f}}(\tau), \hat{\boldsymbol{u}}(\tau))_{\hat{\boldsymbol{L}}_2(\Omega)} d\tau. \tag{6.20}$$

8.6.4 Normal Oscillations of a System of Fluids

We consider now the solutions of the problem on free oscillations that depend on time by the law $\exp(-\lambda t)$. As in Section 8.4.4, we obtain equations similar to (4.18)–(4.20) by a notation change and a new meaning for the parameter ν, which in this case will denote the average kinematic viscosity of the system of fluids.

Therefore, all the general results obtained in Sections 8.4.5 and 8.4.6 for one fluid that rotates in a partially filled container apply to the problem on normal oscillations of a system of rotating fluids as well. In particular, Properties 1°–8° in Section 8.4.5, and the asymptotic formulas (6.11) and (6.18) for eigenvalues of the operators $\hat{A}$ and $\hat{B}$ are preserved.

If the system of fluids and the container Ω do not rotate ($\omega_0 = 0$), then at rest the equilibrium surfaces Γ_i are horizontal and, therefore, $\boldsymbol{n}_i = \boldsymbol{e}_3$ for $i = 1, 2, \ldots, m-1$. The corresponding evolution equations for such a problem are obtained by performing a formal transition $\omega_0 \to 0$ in the system of equations (6.9), (6.12), (6.14), and (6.16). In this case, the asymptotic formula (6.11) stays the same (with a new meaning for Ω_j) and the functions $a_i(\hat{\xi}_i)$ equal $(\rho_i - \rho_{i+1})g$, that is, they become contants. At the same time, in the asymptotic formula (6.18), the constant $c_{\hat{B}}$ becomes

$$c_{\hat{B}}|_{\omega_0=0} = \frac{g^2}{16\pi} \sum_{j=1}^{m-1} \frac{(\rho_j - \rho_{j+1})^2}{(\rho_j^0 + \rho_{j+1}^0)^2} \operatorname{mes} \Gamma_j > 0. \tag{6.21}$$

The basic equation for normal oscillations takes the form

$$\nu \hat{\boldsymbol{\eta}} = \lambda \hat{A}^{-1} \hat{\boldsymbol{\eta}} + \lambda^{-1} \hat{B} \hat{\boldsymbol{\eta}}, \qquad \hat{\boldsymbol{\eta}} \in \hat{\boldsymbol{J}}_{0,S}(\Omega), \tag{6.22}$$

which was previously studied in detail in Section 8.3 for the problem on normal oscillations of one fluid partially filling a stationary container. Therefore, all the results in Section 8.3, namely, the structure and the asymptotics of the spectrum of the problem, the basicity and the double basicity of the eigenelements, the presence of the dissipative, surface, and intermediate waves, etc., hold true for equation (6.22) as well. The reader can independently follow these properties of the solutions.

8.6.5 On the Stability of the Relative Equilibrium State

Let us assume now that condition (6.17) of static stability in linear approximation is not satisfied, and at least one of the functions $a_i(\hat{\xi}_i)$ turns negative at some point $\hat{\xi}_j^0 \in \Gamma_j$. Then the considered system of rotating fluids is not dynamically stable: there is at least one normal motion of the system for which the eigenvalue λ is located in the left complex half-plane.

To prove this theorem on instability, we note that in virtuc of its continuity that was previously mentioned, the function $a_j(\hat{\xi}_j)$ will take negative values in a neighborhood of the point $\hat{\xi}_j^0 \in \Gamma_j$. In this case, one can easily notice that the form $(B_0\hat{\zeta}, \hat{\zeta})_{\hat{L}_{2,\Gamma}}$ takes negative values and, therefore, the form $(B_0\boldsymbol{\eta}, \boldsymbol{\eta})_{\hat{\boldsymbol{J}}_{0,S}(\Omega)}$ is also negative for some $\boldsymbol{\eta} \in \hat{\boldsymbol{J}}_{0,S}(\Omega)$. Hence it follows that the operator $\hat{B}$ has besides the positive eigenvalues $\lambda = \lambda_k^+(\hat{B})$ a branch of negative eigenvalues $\{\lambda_k^-(\hat{B})\}_{k=1}^{\infty}$, where $\lim_{k\to\infty} \lambda_k^-(\hat{B}) = 0$. The asymptotic behavior of these two branches of eigenvalues is defined by the following formula, rather than formula (6.18):

$$\lambda_k^{\pm}(\hat{B}) = \pm\left(c_{\hat{B}}^{\pm}\right)^{1/2} k^{-1/2}[1+\mathrm{o}(1)], \qquad k\to\infty,$$

$$c_{\hat{B}}^{\pm} = \frac{1}{16\pi}\sum_{j=1}^{m-1}\int_{\Gamma_j}\frac{\left|a_j^{\pm}\right|^2 \mathrm{d}\Gamma_j}{(\rho_j^0+\rho_{j+1}^0)^2} > 0, \qquad a_j^{\pm} := \frac{|a_j|\pm a_j}{2}. \tag{6.23}$$

Since the branches of positive and negative eigenvalues $\{\lambda_k^{\pm}(\hat{B})\}_{k=1}^{\infty}$ have power asymptotics according to (6.23), then from the conclusions in Section 1.8.11 we can—as we did in the proof of Property 6° in Section 8.4.5—single out two branches $\{\tilde{\lambda}_k^{\pm}\}_{k=1}^{\infty}$ of eigenvalues of the considered spectal problem tending to zero. The asymptotic behavior of these two branches is given by

$$\tilde{\lambda}_k^{\pm} = \nu^{-1}\lambda_k^{\pm}(\hat{B})[1+\mathrm{o}(1)], \qquad k\to\infty. \tag{6.24}$$

In particular, in virtue of (6.23) the branch $\{\lambda_k^{-}\}_{k=1}^{\infty}$ contains a countable set of eigenvalues in the left half-plane, that is, a countable set of unstable normal oscillating modes. Thus we proved the theorem on instability.

We note that this theorem is valid for the problem on oscillations of one rotating fluid if in condition (4.4) the function $a(\hat{\xi})$ takes a negative value at some point $\hat{\xi}^0 \in \Gamma$. Moreover, the oscillations of the system of nonmixing fluids in a stationary container described by problem (6.22) are unstable if instead of the inequalities $\rho_1 > \rho_2 > \cdots > \rho_m > 0$ we have $\rho_k < \rho_{k+1}$ for at least one pair of the numbers ρ_k and ρ_{k+1}.

The latter situation occurs when in a state of equilibrium a fluid with greater density is situated above a fluid with lesser density. Therefore, the resulting motion of the fluids will be nonlinear, that is, not small, and it leads in the end to the mixing and reconstructing of the complete system of fluids to such a state when the conditions $\rho_1 > \rho_2 > \cdots > \rho_m > 0$ are satisfied.

8.7 Small Motions Around a Fixed Point of a Body with a Cavity Partially Filled with Fluid

In this section we consider an evolution problem similar to the one in Section 3.5 for a viscous fluid.

8.7.1 Basic Equations

As in Section 3.5, let us assume that, in the nonperturbed state, the body, whose cavity is partially filled with a viscous fluid, is immovable and it is fixed at the pole O, which, generally speaking, does not coincide with the mass center C in the immovable state. We consider some small movements of the fluid, which are close to the nonperturbed state. Taking into account the influence of the homogeneous gravitational field and the smal field of external forces, using the previous notations, we obtain the following equations and boundary and initial conditions:

$$\boldsymbol{J}\boldsymbol{\varepsilon} + \rho \int_\Omega \left(\boldsymbol{r} \times \frac{\partial \boldsymbol{u}}{\partial t} \right) \mathrm{d}\Omega + mga(\delta_1 \boldsymbol{e}_1 + \delta_2 \boldsymbol{e}_2) - \rho g \int_\Gamma (\boldsymbol{e}_3 \times \boldsymbol{r}) \zeta \mathrm{d}\Gamma = \boldsymbol{M}(t), \tag{7.1}$$

$$\frac{\partial \boldsymbol{u}}{\partial t} + \boldsymbol{\varepsilon} \times \boldsymbol{r} = -\frac{1}{\rho} \nabla p + \nu \Delta \boldsymbol{u} + \boldsymbol{f},$$

$$\operatorname{div} \boldsymbol{u} = 0 \text{ in } \Omega, \quad \boldsymbol{u} = \boldsymbol{0} \text{ on } S, \quad \frac{\partial \zeta}{\partial t} = u_n \text{ on } \Gamma, \tag{7.2}$$

$$\rho\nu \left(\frac{\partial u_i}{\partial x_3} + \frac{\partial u_3}{\partial x_i} \right) = 0 \text{ on } \Gamma, \qquad i = 1, 2, \tag{7.3}$$

$$p - 2\rho\nu \frac{\partial u_3}{\partial x_3} = \rho g (\zeta - \delta_2 x_1 + \delta_1 x_2) \text{ on } \Gamma,$$

$$\boldsymbol{u}(0, x) = \boldsymbol{u}^0(x), \quad \zeta(0, x_1, x_2) = \zeta^0(x_1, x_2), \quad \boldsymbol{\delta}(0) = \boldsymbol{\delta}^0, \quad \boldsymbol{\omega}(0) = \boldsymbol{\omega}^0. \tag{7.4}$$

The law of full energy balance holds true for the classical solutions of the problem (7.1)–(7.4):

$$\frac{1}{2} \frac{\mathrm{d}}{\mathrm{d}t} \left\{ \rho \int_\Omega |\boldsymbol{u}|^2 \mathrm{d}\Omega + \boldsymbol{J}\boldsymbol{\omega} \cdot \boldsymbol{\omega} + 2\rho \int_\Omega (\boldsymbol{\omega} \times \boldsymbol{r}) \cdot \boldsymbol{u} \mathrm{d}\Omega \right\}$$
$$+ \frac{1}{2} \frac{\mathrm{d}}{\mathrm{d}t} \left\{ \rho g \int_\Gamma |\zeta|^2 \mathrm{d}\Gamma + mga \left(\delta_1^2 + \delta_2^2 \right) + 2\rho g \int_\Gamma (-\delta_2 x_1 + \delta_1 x_2) \zeta \mathrm{d}\Gamma \right\}$$
$$= -\rho\nu E(\boldsymbol{u}, \boldsymbol{u}) + \rho \int_\Omega \boldsymbol{f} \cdot \boldsymbol{u} \mathrm{d}\Omega + \boldsymbol{M} \cdot \boldsymbol{\omega}. \tag{7.5}$$

The deduction of relation (7.5) is similar to the one of the law (3.5.6) for an ideal fluid with the additional usage of the Green formula (2.2.10) and the dynamic boundary conditions (7.3) that contain the coefficient of kinematic viscousity.

8.7.2 Transition to a System of Operator Equations

The general method that have been used in Section 8.1 for the case of an immovable rigid body, can be applied to problem (7.1)–(7.4), too. In particular, the system of equations and boundary conditions (7.2) and (7.3) can be obtained from the corresponding equations in Section 8.1 if in the Navier-Stokes equation (1.1) $\boldsymbol{f}$ is formally replaced by $\boldsymbol{f} - \varepsilon \times \boldsymbol{r}$ and in the dynamic boundary condition ζ is replaced by $\zeta - \delta_2 x_1 + \delta_1 x_2$. That is why, in the considered problem, we obtain the following equations instead of (1.36),

$$\begin{aligned}
&\boldsymbol{u} = \boldsymbol{s} + \boldsymbol{w}, \\
&\nu A^{1/2}\boldsymbol{s} = A^{-1/2}\left(-\frac{\mathrm{d}\boldsymbol{u}}{\mathrm{d}t} - P_{0,S}(\boldsymbol{\varepsilon} \times \boldsymbol{r}) + P_{0,S}\boldsymbol{f}\right), \\
&\nu A^{1/2}\boldsymbol{w} + gA^{1/2}T(\zeta - \delta_2 x_1 + \delta_1 x_2) = 0, \\
&\frac{\mathrm{d}\zeta}{\mathrm{d}t} = \gamma_n \boldsymbol{u},
\end{aligned} \tag{7.6}$$

where $P_{0,S}$ is the orthoprojector onto $\boldsymbol{J}_{0,S}(\Omega)$, and A and T are the operators of the boundary value problems I and II in Section 8.1.

Hence, the initial problem (7.1)–(7.4) is reduced to the evolution problem (7.1), (7.6), which is condidered for the initial data (7.4).

Let us now transform this problem by reducing it to a system of integral Volterra equations with respect to functions with values in the space $\boldsymbol{J}_{0,S}(\Omega)$ and the angular velocity $\boldsymbol{\omega}(t)$.

To this end, let us apply the tensor $\boldsymbol{J}^{-1}$ to (7.1), determine the angular acceleration, $\boldsymbol{\varepsilon} = \mathrm{d}\boldsymbol{\omega}/\mathrm{d}t$, and substitute it into the second equation in (7.6). Thus we obtain

$$(I - \Pi)\frac{\mathrm{d}\boldsymbol{u}}{\mathrm{d}t} + \nu A\boldsymbol{s} = \boldsymbol{L}(t) + \int_0^t \boldsymbol{F}_1(\boldsymbol{\omega}, \tau)\mathrm{d}\tau - \int_0^t \boldsymbol{F}_2(\boldsymbol{r}, \gamma_n\boldsymbol{u})(\tau)\mathrm{d}\tau, \tag{7.7}$$

where

$$\begin{aligned}
\boldsymbol{L}(t) := {}& P_{0,S}\boldsymbol{f} - P_{0,S}((\boldsymbol{J}^{-1}\boldsymbol{M}) \times \boldsymbol{r}) - \rho g P_{0,S}\left(\left(\boldsymbol{J}^{-1}\int_\Gamma (\boldsymbol{e}_3 \times \boldsymbol{r})\zeta^0 \mathrm{d}\Gamma\right) \times \boldsymbol{r}\right) \\
&+ mgaP_{0,S}\left((\boldsymbol{J}^{-1}(\delta_1^0\boldsymbol{e}_1 + \delta_2^0\boldsymbol{e}_2)) \times \boldsymbol{r}\right),
\end{aligned} \tag{7.8}$$

$$\boldsymbol{F}_1(\boldsymbol{\omega}, t) := mgaP_{0,S}\left(\left(\boldsymbol{J}^{-1}\int_0^t (\omega_1(\tau)\boldsymbol{e}_1 + \omega_2(\tau)\boldsymbol{e}_2)\mathrm{d}\tau\right) \times \boldsymbol{r}\right), \tag{7.9}$$

$$\boldsymbol{F}_2(\boldsymbol{r}, \gamma_n\boldsymbol{u}) := \rho g P_{0,S}\left(\left(\boldsymbol{J}^{-1}\int_\Gamma (\boldsymbol{e}_3 \times \boldsymbol{r})\int_0^t \gamma_n\boldsymbol{u}(\tau)\mathrm{d}\tau\mathrm{d}\Gamma\right) \times \boldsymbol{r}\right), \tag{7.10}$$

$$\Pi\boldsymbol{v} := \rho P_{0,S}\left(\left(\boldsymbol{J}^{-1}\int_\Omega (\boldsymbol{r} \times \boldsymbol{v})\mathrm{d}\Omega\right) \times \boldsymbol{r}\right), \qquad \boldsymbol{v} \in \boldsymbol{J}_{0,S}(\Omega). \tag{7.11}$$

In deducing (7.7)–(7.11), we used the following relation,

$$\zeta(t, x_1, x_2) = \zeta^0(x_1, x_2) + \int_0^t (\gamma_n\boldsymbol{u})(\tau)\mathrm{d}\tau, \tag{7.12}$$

that follows from the kinematic condition (7.6), the formula

$$\delta(t) = \delta^0 + \int_0^t \omega(\tau)\mathrm{d}\tau, \tag{7.13}$$

and the initial conditions (7.4).

The operator Π defined by (7.11) is naturally called—similarly to Section 7.2—a *transition operator*. A comparison with formula (7.2.8) shows that it extends the transition operator employed there from the space $\boldsymbol{J}_0(\Omega)$ to the space $\boldsymbol{J}_{0,S}(\Omega)$. Reasoning according to the scheme in Section 7.2.3, we deduce that Π is a finite-dimensional self-adjoint operator acting on $\boldsymbol{J}_{0,S}(\Omega)$ and that $0 \leq \Pi < I$. Thus, $0 < I - \Pi \leq I$.

Let us apply now the bounded operator $(I-\Pi)^{-1}$ to the left part of (7.7) and make the following substitution in the obtained equation,

$$\boldsymbol{u} = A^{-1/2}\boldsymbol{\xi}, \qquad \boldsymbol{s} = A^{-1/2}\boldsymbol{\eta}, \qquad \boldsymbol{w} = A^{-1/2}\boldsymbol{\varphi}. \tag{7.14}$$

We make similar substitutions in the first and the third equations in (7.6) and (7.12) and then substitute the obtained expression for $\mathrm{d}\boldsymbol{\varphi}/\mathrm{d}t$ into the modified equation (7.7). Thus we obtain

$$\begin{aligned} A^{-1/2}&\frac{\mathrm{d}\boldsymbol{\eta}}{\mathrm{d}t} + \nu(I-\Pi)^{-1}A^{1/2}\boldsymbol{\eta} \\ &= (I-\Pi)^{-1}\boldsymbol{L}(t) \\ &+ (I-\Pi)^{-1}\int_0^t \boldsymbol{F}_1(\boldsymbol{\omega},\tau)\mathrm{d}\tau - (I-\Pi)^{-1}\int_0^t \boldsymbol{F}_2\left(\boldsymbol{r},\gamma_n A^{-1/2}(\boldsymbol{\eta}+\boldsymbol{\varphi})(\tau)\right)\mathrm{d}\tau \\ &+ g\nu^{-1}A^{-1/2}B(\boldsymbol{\eta}+\boldsymbol{\varphi})(t) + \boldsymbol{F}_3(\boldsymbol{\omega},t), \end{aligned} \tag{7.15}$$

where $B = A^{1/2}T\gamma_n A^{-1/2}$ is the nonnegative compact operator that we have investigated in Section 8.1,

$$\boldsymbol{F}_3(\boldsymbol{\omega},t) := g\nu^{-1}A^{1/2}TP_\Gamma(-\omega_2(t)x_1 + \omega_1(t)x_2), \tag{7.16}$$

and P_Γ is the orthoprojector onto $H_\Gamma = L_{2,\Gamma}$.

8.7.3 Auxiliary Results

The main special feature of equation (7.15) is that the unbounded operator $A^{1/2}$ is not applicable from the left to all the terms in that equation. This has to do with the fact that, generally speaking, the operator $(I-\Pi)^{-1}$ does not act on the space $\boldsymbol{J}^1_{0,S}(\Omega) = \mathcal{D}(A^{1/2})$.

In order to overcome this difficulty, let us consider the compact positive operator $A^{-1/2}(I-\Pi)A^{-1/2}$ and its inverse positive definite operator

$$A_\Pi := A^{1/2}(I-\Pi)^{-1}A^{1/2} \tag{7.17}$$

with the natural domain of definition $\mathcal{D}(A_\Pi)$, which is dense in $\boldsymbol{J}_{0,S}(\Omega)$. There is an analytic contractive semigroup of operators, $\exp(-\nu t A_\Pi)$, that corresponds to the operator νA_Π and for which the following estimate holds true for $t>0$:

$$\left\| A_\Pi^{1/2} \exp(-\nu t A_\Pi) \right\| \le \frac{c_1}{\sqrt{t}}. \tag{7.18}$$

Using the properties of the operator $I-\Pi$ we get that the following inequality is valid for some $c_2>0$,

$$\begin{aligned} c_2^2(A^{1/2}\boldsymbol{v}, A^{1/2}\boldsymbol{v})_{\boldsymbol{J}_{0,S}(\Omega)} &\le ((I-\Pi)A^{1/2}\boldsymbol{v}, A^{1/2}\boldsymbol{v})_{\boldsymbol{J}_{0,S}(\Omega)} \le \left(A_\Pi^{-1}\boldsymbol{v}, \boldsymbol{v}\right)_{\boldsymbol{J}_{0,S}(\Omega)} \\ &= \left(A_\Pi^{-1/2}\boldsymbol{v}, A_\Pi^{-1/2}\boldsymbol{v}\right)_{\boldsymbol{J}_{0,S}(\Omega)}, \end{aligned}$$

or, equivalently, the inequality below holds true,

$$c_2\|A^{-1/2}\boldsymbol{v}\|_{\boldsymbol{J}_{0,S}(\Omega)} \le \left\|A_\Pi^{-1/2}\boldsymbol{v}\right\|_{\boldsymbol{J}_{0,S}(\Omega)}.$$

If we assume that $A_\Pi^{-1/2}\boldsymbol{v} = \boldsymbol{w} \in \mathcal{D}(A_\Pi^{-1/2})$, then

$$c_2\left\|A^{-1/2}A_\Pi^{-1/2}\boldsymbol{w}\right\|_{\boldsymbol{J}_{0,S}(\Omega)} \le \|\boldsymbol{w}\|_{\boldsymbol{J}_{0,S}(\Omega)}. \tag{7.19}$$

Let us estimate (with regard to (7.19)) the scalar product

$$\begin{aligned} \left| c_2\left(A_\Pi^{1/2}\boldsymbol{w}, A_\Pi^{-1/2}\boldsymbol{z}\right)_{\boldsymbol{J}_{0,S}(\Omega)} \right| &= c_2\left| \left(A^{-1/2}A_\Pi^{1/2}\boldsymbol{w}, \boldsymbol{z}\right)_{\boldsymbol{J}_{0,S}(\Omega)} \right| \\ &\le c_2\left\|A^{-1/2}A_\Pi^{1/2}\boldsymbol{w}\right\|_{\boldsymbol{J}_{0,S}(\Omega)} \cdot \|\boldsymbol{z}\|_{\boldsymbol{J}_{0,S}(\Omega)} \\ &\le \|\boldsymbol{w}\|_{\boldsymbol{J}_{0,S}(\Omega)} \cdot \|\boldsymbol{z}\|_{\boldsymbol{J}_{0,S}(\Omega)}, \end{aligned}$$

for any $z \in \boldsymbol{J}_{0,S}(\Omega)$. From this inequality it follows that $A^{-1/2}\boldsymbol{z} \in \mathcal{D}(A_\Pi^{1/2})$ and

$$c_2\left\|A_\Pi^{1/2}A^{-1/2}\boldsymbol{z}\right\|_{\boldsymbol{J}_{0,S}(\Omega)} \le \|\boldsymbol{z}\|_{\boldsymbol{J}_{0,S}(\Omega)}.$$

Therefore, operator $A_\Pi^{1/2}A^{-1/2}$ is bounded on $\boldsymbol{J}_{0,S}(\Omega)$.

We now consider the following operator for $t > 0$:

$$\begin{aligned}\exp(-\nu t A_\Pi) A^{1/2}(I - \Pi)^{-1} &= \exp(-\nu t A_\Pi) A_\Pi A^{-1/2} \\ &= A_\Pi^{1/2} \exp(-\nu t A_\Pi) \left(A_\Pi^{1/2} A^{-1/2}\right).\end{aligned}$$

From this equality it follows that the operator in left hand side is defined in the set $(I - \Pi)\mathcal{D}(A^{1/2})$, which is dense in $\boldsymbol{J}_{0,S}(\Omega)$, and coincides with the product of two bounded operators in this space. Therefore, this operator admits a closure to the bounded operator

$$G(t) := \overline{\exp(-\nu t A_\Pi) A^{1/2}(I - \Pi)^{-1}} = \left(A_\Pi^{1/2} \exp(-\nu t A_\Pi)\right) \left(A_\Pi^{1/2} A^{-1/2}\right), \tag{7.20}$$

which is defined in the whole space $\boldsymbol{J}_{0,S}(\Omega)$. From (7.18) and (7.19) it follows that

$$\|G(t)\| = \left\|\exp(-\nu t A_\Pi) A^{1/2}(I - \Pi)\right\| \leq \frac{c_1}{c_2\sqrt{t}}. \tag{7.21}$$

8.7.4 Existence of Solution of the Boundary Value Problem

Using the just proved facts, let us replace t by τ in (7.15). By assuming $0 < \tau < t$, we can apply the operator $\exp(-\nu(t-\tau)A_\Pi)A^{1/2}$ to both sides of (7.15). Then in the left side we obtain

$$\exp(-\nu(t-\tau)A_\Pi)\frac{\mathrm{d}\boldsymbol{\eta}}{\mathrm{d}t} + \nu \exp(-\nu(t-\tau)A_\Pi)A_\Pi \boldsymbol{\eta} = \frac{\mathrm{d}}{\mathrm{d}\tau}(\exp(-\nu(t-\tau)A_\Pi)\boldsymbol{\eta}(\tau)).$$

Integrating this expression between the limits 0 and t, we get $\boldsymbol{\eta}(t) - \exp(-\nu t A_\Pi)\boldsymbol{\eta}(0)$. Therefore, the modified equation (7.15) with regard to the third relation (7.6) for $t = 0$ takes the following form

$$\begin{aligned}\boldsymbol{\eta}(t) &= \exp(-\nu t A_\Pi)\boldsymbol{\eta}(0) + \int_0^t G(t-\tau)\boldsymbol{L}(\tau)\mathrm{d}\tau + \int_0^t G(t-\tau)\mathrm{d}\tau \int_0^\tau \boldsymbol{F}_1(\boldsymbol{\omega}, s)\mathrm{d}s \\ &- \int_0^t G(t-\tau)\mathrm{d}\tau \int_0^\tau \boldsymbol{F}_2\left(\boldsymbol{r}, \gamma_n A^{-1/2}(\boldsymbol{\eta} + \boldsymbol{\varphi})(s)\right)\mathrm{d}s \\ &+ g\nu^{-1} \int_0^t \exp(-\nu(t-\tau)A_\Pi)B(\boldsymbol{\eta} + \boldsymbol{\varphi})(\tau)\mathrm{d}\tau \\ &+ \int_0^t \exp(-\nu(t-\tau)A_\Pi)A^{1/2}\boldsymbol{F}_3(\boldsymbol{\omega}, \tau)\mathrm{d}\tau,\end{aligned} \tag{7.22}$$

$$\begin{aligned}\boldsymbol{\eta}(0) &= A^{1/2}(\boldsymbol{u}(0) - \boldsymbol{w}(0)) = A^{1/2}\boldsymbol{u}^0 - \boldsymbol{\varphi}(0), \\ \boldsymbol{\varphi}(0) &= -g\nu^{-1}A^{1/2}TP_\Gamma\left(\zeta^0 - \delta_2^0 x_1 + \delta_1^0 x_2\right).\end{aligned} \tag{7.23}$$

According to the estimate (7.21), the function in the first integral in the right hand side of (7.22) has a weak singularity for $\tau = t$, therefore, integrating this function between the limits 0 and t yields a continuous function in $[0, t]$ provided $\boldsymbol{L}(t)$ is continuous. Further, since—according to the previously proved facts—the operator-valued function $\exp(-\nu(t-\tau)A_\Pi)A^{1/2}$ is bounded, and if the functions $\boldsymbol{\eta}(t)$, $\boldsymbol{\varphi}(t)$ (from $\boldsymbol{J}_{0,S}(\Omega)$), and $\boldsymbol{\omega}(t)$ are continuous, then the last two integrals in (7.22) are also continuous functions of t with values in $\boldsymbol{J}_{0,S}(\Omega)$. As for the second and the third integrals in the right hand side of (7.22), each of them is an expression of the form

$$\boldsymbol{R}(t) = \int_0^t G(t-\tau)\mathrm{d}\tau \int_0^\tau \boldsymbol{\beta}(s)\mathrm{d}s.$$

If $\boldsymbol{\beta}(s)$ is continuous, with regard to (7.20) we obtain the following integral

$$\begin{aligned}\boldsymbol{R}(t) &= \int_0^t \left(\int_s^t A_\Pi \exp(-\nu(t-\tau)A_\Pi)\mathrm{d}\tau\right) A^{-1/2}\boldsymbol{\beta}(s)\mathrm{d}s \\ &= \frac{1}{\nu}\int_0^t (I - \exp(-\nu(t-\tau)A_\Pi))A^{-1/2}\boldsymbol{\beta}(s)\mathrm{d}s\end{aligned}$$

with a continuous integral function.

Hence, for continuous functions $\boldsymbol{\eta}(t)$, $\boldsymbol{\varphi}(t)$ from $\boldsymbol{J}_{0.S}(\Omega)$, and $\boldsymbol{\omega}(t)$ from $\mathbb{R}^3$, the right hand side of (7.22) is a continuous function of t with values in $\boldsymbol{J}_{0,S}(\Omega)$ for $\boldsymbol{\eta}(0) \in \boldsymbol{J}_{0,S}(\Omega)$, and the integral operators satisfy the conditions required for applying the methods of successive approximations.

Now, taking into account the substitutions (7.14) and relations (7.12) and (7.13), let us transform equation (7.1) and the third condition in (7.6). Since $\omega(t) = \omega_0 + \int_0^t \varepsilon(\tau)\mathrm{d}\tau$, we obtain the following two equations

$$\begin{aligned}\boldsymbol{\varphi}(t) = \boldsymbol{\varphi}(0) &- g\nu^{-1}\int_0^t B(\boldsymbol{\eta}+\boldsymbol{\varphi})(\tau)\mathrm{d}\tau \\ &- g\nu^{-1}\int_0^t A^{1/2}TP_\Gamma(-x_1\omega_2(\tau) + x_2\omega_1(\tau))\mathrm{d}\tau, \qquad (7.24)\end{aligned}$$

$$\begin{aligned}\boldsymbol{\omega}(t) + KA^{-1/2}(\boldsymbol{\eta}(t)+\boldsymbol{\varphi}(t)) = \boldsymbol{\psi}(t) &- mga\boldsymbol{J}^{-1}\int_0^t (t-\tau)(\omega_1(\tau)\boldsymbol{e}_1 + \omega_2(\tau)\boldsymbol{e}_2)\mathrm{d}\tau \\ &- \rho g\boldsymbol{J}^{-1}\int_0^t (t-\tau)F_4(\boldsymbol{\eta}+\boldsymbol{\varphi})(\tau)\mathrm{d}\tau, \qquad (7.25)\end{aligned}$$

where

$$\boldsymbol{\psi}(t) := \boldsymbol{\omega}^0 - K\boldsymbol{u}^0 - t\left(mga\boldsymbol{J}^{-1}\left(\delta_1^0\boldsymbol{e}_1 + \delta_2^0\boldsymbol{e}_2\right) + \rho g\boldsymbol{J}^{-1}\left(\left(\int_\Gamma(\boldsymbol{e}_3\times\boldsymbol{r})\zeta^0\mathrm{d}\Gamma\right)\right)\right)$$

$$+\boldsymbol{J}^{-1}\int_0^t \boldsymbol{M}(\tau)\mathrm{d}\tau,$$

$$F_4\boldsymbol{v} := \int_\Gamma (\boldsymbol{e}_3 \times \boldsymbol{r})\gamma_n A^{-1/2}\boldsymbol{v}\mathrm{d}\Gamma,$$

$$K\boldsymbol{v} := \rho\boldsymbol{J}^{-1}\int_\Omega (\boldsymbol{r} \times \boldsymbol{v})\mathrm{d}\Omega, \tag{7.26}$$

and $\boldsymbol{\varphi}(0)$ is determined by formula (7.23).

In order to get rid of the second term in the left hand side of (7.25), let us multiply both (7.22) and (7.24) by the bounded operator $KA^{-1/2}$, add up the left sides, and subtract the result from (7.25). Thus, the system of equations (7.22) and (7.24) and the modified equation (7.25) turn out to be a system of integral Volterra equations with respect to the functions $\boldsymbol{\eta}(t)$, $\boldsymbol{\varphi}(t)$ with values in $\boldsymbol{J}_{0,S}(\Omega)$, and the function $\boldsymbol{\omega}(t)$ with values in $\mathbb{R}^3$. These equations contain integral operators whose kernels are continuous in t with the exception of one of them, which has a weak singularity. This singularity admits the estimate (7.21). Therefore, we can apply the method of successive approximations to the above mentioned system.

Let us note now that all the initial conditions (7.4) of the initial boundary value problem (7.1)–(7.4) are included in the right hand sides of the just obtained Volterra equations and also in the relations (7.12) and (7.13). Suppose the following conditions are true:

1°. $\boldsymbol{u}^0(x) \in \boldsymbol{J}^1_{0,S}(\Omega)$, $\zeta^0(x_1,x_2) \in H_\Gamma = L_{2,\Gamma}$, $\boldsymbol{\delta}^0 \in \mathbb{R}^3$, $\boldsymbol{\omega} \in \mathbb{R}^3$;

2°. $\boldsymbol{M}(t)$ is a continuous function with values in $\mathbb{R}^3$ and $\boldsymbol{f}(t,x)$ in (7.4) is a continuous function in t that has values in $\boldsymbol{L}_2(\Omega)$.

Then $\boldsymbol{L}(t)$ in (7.8) is a continuous function with values in $\boldsymbol{J}_{0,S}(\Omega)$, $\boldsymbol{\psi}(t)$ in (7.26) is a continuously differentiable function with values in $\mathbb{R}^3$ (with regard to the boundedness of the operator K), and $\boldsymbol{\eta}(0)$ and $\boldsymbol{\varphi}(0)$—according to (7.23)—belong to the space $\boldsymbol{J}_{0,S}(\Omega)$. Therefore, the system of integral Volterra equations has a unique generalized solution $\{\boldsymbol{\eta}(t);\boldsymbol{\varphi}(t);\boldsymbol{\omega}(t)\}$, for which the functions $\boldsymbol{\eta}(t)$, $\boldsymbol{\varphi}(t)$, and $\boldsymbol{\xi}(t) = \boldsymbol{\eta}(t) + \boldsymbol{\varphi}(t) = A^{1/2}\boldsymbol{u}(t)$ are continuous and take values in $\boldsymbol{J}_{0,S}(\Omega)$, and $\boldsymbol{\omega}(t)$ is a continous function with values in $\mathbb{R}^3$.

Thus, if conditions 1°–2° are satisfied, the initial boundary value problem is uninvalently solvable and has a generalized solution $\{\boldsymbol{u}(t);\boldsymbol{\omega}(t)\}$, where $\boldsymbol{u}(t)$ is a continuous function with values in $\boldsymbol{J}^1_{0,S}(\Omega)$ and $\boldsymbol{\omega}(t)$ is continuous in $\mathbb{R}^3$. According to (7.5), for this generalized solution the law of full energy balance holds true.

8.8 Normal Oscillations of a Pendulum Partially Filled with a Fluid (The Plane Problem)

In this section we consider a problem that, on one hand, generalizes the plane problem studied in Section 7.5 in the case of a cavity in a pendulum partially filled with a viscous fluid, and, on the other hand, is a particular case of the problem in Section 8.7, connected with a transition from a three-dimensional problem to a plane one.

Since the evolution (initial-boundary value) problem in the three-dimensional case has been studied already in Section 8.7, we are going to turn our attention here to the spectral problem and study it through a specific approach, which although different from the one in Section 8.7, allows us to study an initial-boundary value problem.

8.8.1 Statement of the Problem

We consider the plane analog to the problem (7.1)–(7.4), that is, let us suppose that a body with a cavity partially filled with a viscous homogeneous fluid has small oscillations with respect to a fixed point O. At rest, the fluid occupies the region Ω limited by a wall of the cavity S and the free horizontal surface Γ.

If $\Omega \subset Ox_2x_3$ and the vector of the angular displacement of the body is $\boldsymbol{\delta} = \boldsymbol{\delta}_1 \boldsymbol{e}_1$, for the displacement field $\boldsymbol{w}(t,x)$, the dynamic pressure $p(t,x)$, $x = (x_2; x_3) \in \Omega$, and the deviation function $x_3 = \zeta(t, x_2)$ of the moving free surface of the fluid from the equilibrium surface Γ, instead of (7.1)–(7.4) we get the following equations, boundary and initial conditions:

$$\rho \frac{\partial^2 \boldsymbol{w}}{\partial t^2} + \rho \left(\frac{\mathrm{d}^2 \boldsymbol{\delta}}{\mathrm{d}t^2} \times \boldsymbol{r} \right) = -\nabla p + \mu \Delta \frac{\partial \boldsymbol{w}}{\partial t}, \quad \operatorname{div} \boldsymbol{w} = 0 \text{ in } \Omega, \tag{8.1}$$

$$\rho \frac{\mathrm{d}^2}{\mathrm{d}t^2} \int_\Omega (\boldsymbol{r} \times \boldsymbol{w}) \mathrm{d}\Omega + J_1 \frac{\mathrm{d}^2 \boldsymbol{\delta}}{\mathrm{d}t^2} + mgl\boldsymbol{\delta} - \rho g \int_\Gamma (\boldsymbol{e}_3 \times \boldsymbol{r}) \zeta \mathrm{d}\Gamma = \boldsymbol{0},$$

$$\boldsymbol{r} = x_2 \boldsymbol{e}_2 + x_3 \boldsymbol{e}_3, \tag{8.2}$$

$$\boldsymbol{w} = \boldsymbol{0} \text{ on } S, \quad \int_\Gamma \zeta \mathrm{d}\Gamma = 0, \qquad \zeta := \gamma_n \boldsymbol{w}, \tag{8.3}$$

$$\mu \frac{\partial}{\partial t} \left(\frac{\partial w_2}{\partial x_3} + \frac{\partial w_3}{\partial x_2} \right) = 0 \text{ on } \Gamma, \tag{8.4}$$

$$p - 2\mu \frac{\partial}{\partial t} \frac{\partial w_3}{\partial x_3} = \rho g (\zeta + \delta_1 x_2) \text{ on } \Gamma, \tag{8.5}$$

$$\boldsymbol{w}(0,x) = \boldsymbol{w}^0(x), \quad \frac{\partial \boldsymbol{w}}{\partial t}(0,x) = \boldsymbol{u}^0(x), \quad \boldsymbol{\delta}(0) = \boldsymbol{\delta}^0, \quad \frac{\mathrm{d}\boldsymbol{\delta}(0)}{\mathrm{d}t} = \boldsymbol{\omega}^0. \tag{8.6}$$

8.8.2 Transition to a Differential Equation in a Hilbert Space

Let us assume that $\boldsymbol{w}(t,x)$ and $\nabla p(t,x)$ are functions of t with values in the Hilbert space $\boldsymbol{L}_2(\Omega)$. From (8.1) and (8.3) we get that $\boldsymbol{w}(t,x) \in \boldsymbol{J}_{0,S}(\Omega)$ and $\nabla p(t,x) \in \boldsymbol{G}(\Omega)$. Applying the orthoprojector $P_{0,S}$ on the subspace $\boldsymbol{J}_{0,S}(\Omega)$ to both sides of (8.1), we have the equation

$$\rho\frac{\partial^2 \boldsymbol{w}}{\partial t^2} + \rho P_{0,S}\left(\frac{\mathrm{d}^2\boldsymbol{\delta}}{\mathrm{d}t^2}\times r\right) = -\nabla\tilde{p} + \mu P_{0,S}\Delta\frac{\partial \boldsymbol{w}}{\partial t}, \qquad \nabla\tilde{p} = P_{0,S}\nabla p, \tag{8.7}$$

where $\nabla\tilde{p} \in \boldsymbol{G}_{\mathrm{h},S}(\Omega)$. Since $\nabla p - \nabla\tilde{p} \in \boldsymbol{G}_{0,\Gamma}(\Omega)$ and thus $p - \tilde{p} = 0$ on Γ, then in (8.5) we can change $p|_\Gamma$ by $\tilde{p}|_\Gamma$. Moreover, since $\int_\Gamma \partial w_3/\partial x_3 \mathrm{d}\Gamma = 0$ (see equation (2.2.25)) and we may require $\int_\Gamma p\mathrm{d}\Gamma = 0$ for p, then, by virtue of the second condition (8.3), the boundary condition (8.5) becomes:

$$\tilde{p} - 2\mu\frac{\partial}{\partial t}\frac{\partial w_3}{\partial x_3} = \rho g(\zeta + \delta_1\theta x_2), \tag{8.8}$$

where θ is the orthoprojector onto the subspace $L_2(\Gamma)\ominus\{1\} =: L_{2,\Gamma}$.

Let us now find a solution of the changed problem (8.1)–(8.6) by means of two additional boundary value problems. Let $\nabla\tilde{p} = \nabla p_1 + \nabla p_2$, and the fields $\boldsymbol{u} = \partial\boldsymbol{w}/\partial t$ and ∇p_i satisfy the following equations and boundary conditions.

Problem 1.

$$\begin{aligned}
&-\mu P_{0,S}\Delta\frac{\partial \boldsymbol{w}}{\partial t} + \nabla p_1 =: \mu A\frac{\partial \boldsymbol{w}}{\partial t} = \boldsymbol{\varphi} := \left(-\rho\frac{\partial^2 \boldsymbol{w}}{\partial t^2} - \rho P_{0,S}\left(\frac{\partial^2\boldsymbol{\delta}}{\partial t^2}\times \boldsymbol{r}\right) - \nabla p_2\right),\\
&\boldsymbol{u} := \frac{\partial \boldsymbol{w}}{\partial t} = \mathbf{0} \text{ on } S,\\
&\mu\left(\frac{\partial u_2}{\partial x_3} + \frac{\partial u_3}{\partial x_2}\right) = 0 \text{ on } \Gamma,\\
&p_1 - 2\mu\frac{\partial u_3}{\partial x_3} = 0 \text{ on } \Gamma.
\end{aligned} \tag{8.9}$$

Problem 2.

$$\begin{aligned}
&\Delta p_2 = 0 \text{ in } \Omega,\\
&\frac{\partial p_2}{\partial n} = 0 \text{ on } S,\\
&p_2 = \psi := \rho g(\zeta + \delta_1\theta x_2) \text{ on } \Gamma.
\end{aligned} \tag{8.10}$$

It is obvious that all the equations and boundary conditions (8.7), (8.3), (8.4), (8.8) hold true for the functions $\boldsymbol{w}$ and $\nabla\tilde{p} = \nabla p_1 + \nabla p_2$. If $\boldsymbol{\varphi} \in \boldsymbol{J}_{0,S}(\Omega)$, then the resolving operator A, as in the space problem, has all the well-known properties studied earlier. As for the problem (8.10), if $\psi \in H_\Gamma^{1/2}$, it has a unique solution that belongs to the subspace $H^1_\Gamma(\Omega)$. Therefore,

$$\nabla p_2 = G\psi = \rho g G(\zeta + \delta_1 \theta x_2), \tag{8.11}$$

where G is a bounded operator from $H^{1/2}_\Gamma$ in $\boldsymbol{G}_{\mathrm{h},S}(\Omega)$.

Substituting (8.11) into the first equation (8.9) we obtain

$$\rho\frac{\mathrm{d}^2\boldsymbol{w}}{\mathrm{d}t^2} + \rho P_{0,S}\left(\frac{\mathrm{d}^2\boldsymbol{\delta}}{\mathrm{d}t^2}\times\boldsymbol{r}\right) + \mu A\frac{\mathrm{d}\boldsymbol{w}}{\mathrm{d}t} + \rho g G\left(\gamma_n\boldsymbol{w} + (\boldsymbol{\delta}\cdot\boldsymbol{e}_1)\theta x_2\right) = \boldsymbol{0} \tag{8.12}$$

in the subspace $\boldsymbol{J}_{0,S}(\Omega)$. This equation together with (8.2) written as

$$\rho\frac{\mathrm{d}^2}{\mathrm{d}t^2}\int_\Omega(\boldsymbol{r}\times\boldsymbol{w})\mathrm{d}\Omega + J_1\frac{\mathrm{d}^2\boldsymbol{\delta}}{\mathrm{d}t^2} + mgl\boldsymbol{\delta} - \rho g\int_\Gamma(\boldsymbol{e}_3\times\boldsymbol{r})\gamma_n\boldsymbol{w}\mathrm{d}\Gamma = \boldsymbol{0}, \tag{8.13}$$

can be considered as a system of two evolution equations with two unknown functions $\boldsymbol{w} = \boldsymbol{w}(t)$ and $\boldsymbol{\delta} = \delta_1(t)\boldsymbol{e}_1$.

Introducing the following notation

$$\begin{aligned} I_{11}\boldsymbol{w} &:= \rho\boldsymbol{w}, \\ I_{12}\boldsymbol{\delta} &:= \rho P_{0.S}(\boldsymbol{\delta}\times\boldsymbol{r}), \\ I_{21}\boldsymbol{w} &:= \rho\int_\Omega(\boldsymbol{r}\times\boldsymbol{w})\mathrm{d}\Omega, \\ I_{22}\boldsymbol{\delta} &:= J_1\boldsymbol{\delta}, \end{aligned} \tag{8.14}$$

and

$$\begin{aligned} B_{11}\boldsymbol{w} &:= \rho g G\gamma_n\boldsymbol{w}, \\ B_{12}\boldsymbol{\delta} &:= \rho g(\boldsymbol{\delta}\cdot\boldsymbol{e}_1)G(\theta x_2), \\ B_{21}\boldsymbol{w} &:= -\rho g\int_\Gamma(\boldsymbol{e}_3\times\boldsymbol{r})\gamma_n\boldsymbol{w}\mathrm{d}\Gamma, \\ B_{22}\boldsymbol{\delta} &:= mgl\boldsymbol{\delta}, \end{aligned} \tag{8.15}$$

equations (8.12) and (8.13) together with the initial conditions (8.6) turn out to be a Cauchy problem for a second order differential equation in the Hilbert space $\boldsymbol{H} := \boldsymbol{J}_{0,S}(\Omega)\oplus\mathbb{C}$, namely,

$$\tilde{I}\frac{\mathrm{d}^2\boldsymbol{y}}{\mathrm{d}t^2} + \mu\tilde{A}\frac{\mathrm{d}\boldsymbol{y}}{\mathrm{d}t} + \tilde{B}\boldsymbol{y} = \boldsymbol{0}, \qquad \boldsymbol{y}(0) = \boldsymbol{y}^0,\ \boldsymbol{y}'(0) = \boldsymbol{y}^1, \tag{8.16}$$

with

$$\begin{aligned} &\tilde{I} := (I_{ik})^2_{i,k=1}, && \tilde{B} := (B_{ik})^2_{i.k=1}, && \tilde{A} := \mathrm{diag}(A;0), \\ &\boldsymbol{y} := (\boldsymbol{w},\boldsymbol{\delta})^{\mathrm{t}}, && \boldsymbol{y}^0 = (\boldsymbol{w}^0;\boldsymbol{\delta}^0)^{\mathrm{t}}, && \boldsymbol{y}^1 = (\boldsymbol{u}^0;\boldsymbol{\omega}^0)^{\mathrm{t}}. \end{aligned} \tag{8.17}$$

Here, $\tilde{I}$ is the operator of kinetic energy, $\tilde{A}$ is the dissipation operator, and $\tilde{B}$ is the operator of potential energy of the system.

8.8.3 Properties of Operator Coefficients of the Evolution Equation

We next analyze the properties of the operators $\tilde{I}$, $\tilde{A}$, and $\tilde{B}$ in (8.16) and (8.17). Since $A \gg 0$ and $A^{-1} \in \mathfrak{S}_\infty$, $\tilde{A}$ is an unbounded, nonnegative operator, whose spectrum consists of the eigenvalues with finite multiplicities $\lambda_k(A)$ of the operator A, and the number $\lambda_0 = 0$, which—in the case of the plane problem—is a simple eigenvalue.

Furthermore, the operator of kinetic energy, $\tilde{I}$, is bounded and positive definite in $\boldsymbol{H}$. Instead of proving these properties, it is enough to compare the definitions of its matrix elements, I_{ik}, given in (8.14) with the corresponding formulas (7.5.4) in Section 7.5 that describe the matrix elements of the operator of kinetic energy for the problem on the oscillations of a pendulum with a cavity entirely filled with fluid. (See also Sections 6.2.2, 6.2.3, 7.3.2, and 7.3.3.)

Let us study now in more detail the properties of the operator $\tilde{B}$. First, we note that $B_{22}\boldsymbol{\delta} = mgl\boldsymbol{\delta} = J_1\omega_0^2\boldsymbol{\delta}$, where $\omega_0 > 0$ is the frequency of the oscillations of a pendulum filled with a hardened fluid. Therefore, B_{22} is a bounded, positive definite operator acting in the one-dimensional space $\mathbb{C}$. Next we show that G and γ_n are mutually adjoint unbounded operators acting from $\mathcal{D}(G) = H_\Gamma^{1/2} \subset L_{2,\Gamma}$ into $\boldsymbol{G}_{\mathrm{h},S}(\Omega) \subset \boldsymbol{J}_{0,S}(\Omega)$, and from $\mathcal{D}(\gamma_n) \subset \boldsymbol{J}_{0,S}(\Omega)$ into $L_{2,\Gamma}$, respectively. In particular, if $\boldsymbol{u} \in \boldsymbol{J}_{0,S}^1(\Omega)$, then $\gamma_n\boldsymbol{u} \in H_\Gamma^{1/2}$. Thus, each of the observed operators is defined on a dense set and acts on the second space.

Let $p(x)$ be a smooth scalar field for which $p|_\Gamma \in H_\Gamma^{1/2}$ and let $\boldsymbol{w} \in \boldsymbol{J}_{0,S}^1(\Omega)$. Then $\gamma_n\boldsymbol{w} \in H_\Gamma^{1/2}$ and

$$\begin{aligned}(p|_\Gamma, \gamma_n\boldsymbol{w})_{\boldsymbol{L}_2(\Gamma)} &= \int_\Gamma p\gamma_n\boldsymbol{w}\mathrm{d}\Gamma = \int_{\partial\Omega} pw_n\mathrm{d}S = \int_\Omega \operatorname{div}(p\boldsymbol{w})\mathrm{d}\Omega \\ &= \int_\Omega p\operatorname{div}\boldsymbol{w}\mathrm{d}\Omega + \int_\Omega \nabla p\cdot\boldsymbol{w}\mathrm{d}\Omega = (G(p|_\Gamma), \boldsymbol{w})_{\boldsymbol{L}_2(\Omega)}\,. \qquad (8.18)\end{aligned}$$

Hence it follows that after extending γ_n from $\boldsymbol{J}_{0,S}^1(\Omega)$ to $\mathcal{D}(\gamma_n)$, the operators G and γ_n are mutually adjoint.

The trace operator γ_n may be further extended to the entire space $\boldsymbol{J}_{0,S}(\Omega)$ if it acts from $\boldsymbol{J}_{0,S}(\Omega)$ into $H_\Gamma^{-1/2}$. If $\boldsymbol{w} \in \boldsymbol{J}_{0,S}(\Omega)$, then (8.18) still holds true and the left hand side turns out to be a functional defined on the elements $p|_\Gamma \in H_\Gamma^{1/2}$ and $\gamma_n\boldsymbol{w} \in H_\Gamma^{-1/2}$.

The facts proved above ensure that B_{11} in (8.15) is an unbounded nonnegative symmetric operator in $\boldsymbol{J}_{0,S}(\Omega)$. If we assume that $\mathcal{D}(B_{11}) = \boldsymbol{J}_{0,S}^1(\Omega) \subset \boldsymbol{J}_{0,S}(\Omega)$, then the the entire space $H_\Gamma^{1/2} = \mathcal{D}(G)$ is included in the set $\{\gamma_n\boldsymbol{w}\}$ with all $\boldsymbol{w} \in \boldsymbol{J}_{0,S}^1(\Omega)$ and the entire space $\boldsymbol{G}_{\mathrm{h},S}(\Omega) \subset \boldsymbol{J}_{0,S}(\Omega)$ is included in the set $\{G\gamma_n\boldsymbol{w}\}$. Since $\operatorname{Ker} G\gamma_n = \boldsymbol{J}_0(\Omega) = \boldsymbol{J}_{0,S}(\Omega) \ominus \boldsymbol{G}_{\mathrm{h},S}(\Omega)$, we can assume that the operator $G\gamma_n$ is defined on $\mathcal{D}(G\gamma_n) = \boldsymbol{J}^1_{0,S}(\Omega) \cap \boldsymbol{G}_{\mathrm{h},S}(\Omega)$ and acts in $\boldsymbol{G}_{\mathrm{h},S}(\Omega)$. Moreover, we can extend $G\gamma_n$ from $\mathcal{D}(G\gamma_n)$ to the entire $\boldsymbol{J}^1_{0,S}(\Omega)$ by setting it to be the zero operator on $\boldsymbol{J}_0(\Omega)$.

To establish the properties of the operator $G\gamma_n$, let us consider the following problem on eigenvalues,

$$G\gamma_n \boldsymbol{u} = \lambda \boldsymbol{u}, \qquad \boldsymbol{u} \in \mathcal{D}(G\gamma_n) \subset \boldsymbol{J}_{0,S}(\Omega). \tag{8.19}$$

Setting $\boldsymbol{u} = \nabla p + \boldsymbol{w}$, $\boldsymbol{w} \in \boldsymbol{J}_0(\Omega)$, $\nabla p \in \boldsymbol{G}_{\mathrm{h},S}(\Omega) \cap \mathcal{D}(G\gamma_n)$, we have $\gamma_n \boldsymbol{w} = 0$ and, therefore, (8.19) yields the following two relations:

$$\lambda \boldsymbol{w} = \boldsymbol{0}, \qquad G\gamma_n \nabla p = \lambda \nabla p = \lambda G(p|_\Gamma). \tag{8.20}$$

From the first equation we get that problem (8.19) has a solution $\lambda = 0$, for all $\boldsymbol{w} \in \boldsymbol{J}_0(\Omega)$, that is, the zero eigenvalue has infinite multiplicity. The second equation in (8.20) is equivalent to the spectral boundary value problem

$$\begin{aligned} &\Delta p = 0 \text{ in } \Omega, \\ &\frac{\partial p}{\partial n} = 0 \text{ on } S, \\ &\frac{\partial p}{\partial n} = \lambda p \text{ on } \Gamma, \end{aligned} \tag{8.21}$$

called the *Steklov problem*. In Section 3.3, we dealt with a slightly different version of this problem, namely, the problem of eigen oscillations of a heavy ideal fluid in an open container (see (3.3.19), (3.3.22), and (3.3.23)). By introducing the operator $C := \gamma_\Gamma T$, applying the operator γ_n to the second equation in (8.20), and denoting $\gamma_n \nabla p = \partial p/\partial n =: v$, we get the following problem

$$C^{-1} v = \gamma_n G v = \lambda v, \qquad v \in L_{2,\Gamma} \tag{8.22}$$

on eigenvalues for the unbounded positive operator C^{-1}, with C a positive compact operator. Whence it follows that operator $G\gamma_n$ acting on $\boldsymbol{G}_{\mathrm{h},S}(\Omega)$ is an unbounded positive definte operator, with a positive compact inverse operator, and its eigenvalues $\lambda_k(G\gamma_n)$ coincide with the eigenvalues $\lambda_k^{-1}(C)$. The asymptotic behavior of $\lambda_k(C)$ in the three-dimensional space problem is represented by formula (3.3.24).

We check now that the operators B_{12} and B_{21} in (8.15) are mutually adjoint. For every $\boldsymbol{\delta} \in \mathbb{C}$ and every $\boldsymbol{w} \in \boldsymbol{J}_{0,S}(\Omega)$, by using (8.18) we have,

$$\begin{aligned} (\rho g)^{-1}(B_{12}\boldsymbol{\delta}, \boldsymbol{w})_{\boldsymbol{J}_{0,S}(\Omega)} &= \delta_1 \int_\Omega G(\theta x_2) \cdot \bar{\boldsymbol{w}} \, d\Omega = \delta_1 \int_\Gamma (\theta x_2) \overline{\gamma_n \boldsymbol{w}} \, d\Gamma, \\ (\rho g)^{-1} \boldsymbol{\delta} \cdot B_{21} \boldsymbol{w} &= -\delta_1 \int_\Gamma \boldsymbol{e}_1 \cdot (\boldsymbol{e}_3 \times \boldsymbol{r}) \overline{\gamma_n \boldsymbol{w}} \, d\Gamma = \delta_1 \int_\Gamma x_2 \overline{\gamma_n \boldsymbol{w}} \, d\Gamma \\ &= \delta_1 \int_\Gamma (\theta x_2) \overline{\gamma_n \boldsymbol{w}} \, d\Gamma = (\rho g)^{-1} (B_{12} \boldsymbol{\delta}, \boldsymbol{w})_{\boldsymbol{J}_{0,S}(\Omega)}, \end{aligned}$$

which proves the statement. Thus, we came to the conclusion that the operator of potential energy, $\tilde{B}$, is unbounded, self-adjoint, and defined on $\mathcal{D}(\tilde{B}) = \boldsymbol{J}_{0,S}^1(\Omega) \oplus \mathbb{C} \subset \boldsymbol{J}_{0,S}(\Omega) \oplus \mathbb{C} =: \boldsymbol{H}$. Its quadratic form is

$$
\begin{aligned}
(\tilde{B}\boldsymbol{y}, \boldsymbol{y})_{\boldsymbol{H}} &:= \rho g \int_\Omega G\gamma_n \boldsymbol{w} \cdot \bar{\boldsymbol{w}} \mathrm{d}\Omega + \rho g \delta_1 \int_\Omega G(\theta x_2) \cdot \bar{\boldsymbol{w}} \mathrm{d}\Omega \\
&\quad - \rho g \int_\Gamma (\boldsymbol{e}_3 \times \boldsymbol{r}) \gamma_n \boldsymbol{w} \mathrm{d}\Gamma \cdot \bar{\boldsymbol{\delta}} + J_1 \omega_0^2 |\boldsymbol{\delta}|^2 \\
&= \rho g \Big\{ \int_\Gamma |\gamma_n \boldsymbol{w}|^2 \mathrm{d}\Gamma + \delta_1 \int_\Gamma (\theta x_2) \overline{\gamma_n \boldsymbol{w}} \mathrm{d}\Gamma + \bar{\delta}_1 \int_\Gamma (\theta x_2) \gamma_n \boldsymbol{w} \mathrm{d}\Gamma \\
&\qquad + |\delta_1|^2 \int_\Gamma |\theta x_2|^2 \mathrm{d}\Gamma - |\delta_1|^2 \int_\Gamma |\theta x_2|^2 \mathrm{d}\Gamma \Big\} + J_1 \omega_0^2 |\boldsymbol{\delta}|^2 \\
&= \rho g \int_\Gamma |\gamma_n \boldsymbol{w} + \delta_1 (\theta x_2)|^2 \mathrm{d}\Gamma + |\delta_1|^2 \left(J_1 \omega_0^2 - \rho g \alpha_{22} \right), \qquad (8.23)
\end{aligned}
$$

with

$$
\alpha_{22} := \int_\Gamma |\theta x_2|^2 \mathrm{d}\Gamma = \int_\Gamma (\theta x_2) x_2 \mathrm{d}\Gamma > 0.
$$

We will say that for the considered hydrodynamic system the condition of static stability in linear approximation is fulfilled if

$$
mgl = J_1 \omega_0^2 > \rho g \alpha_{22}. \qquad (8.24)
$$

This condition takes place evidently if the distance l from the fixed point O to the mass center C of the system in equilibrium state is large enough, or the mass of the entire system $m = m_{\mathrm{b}} + m_{\mathrm{f}}$ is sufficiently large.

If condition (8.24) holds true, the operator $\tilde{B}$ is nonnegative on $\mathcal{D}(\tilde{B}) = \boldsymbol{J}_{0,S}^1(\Omega) \oplus \mathbb{C}$; if $(\boldsymbol{w}; \boldsymbol{\delta})^{\mathrm{t}} \in \boldsymbol{J}_{0,S}^1(\Omega) \cap \boldsymbol{G}_{\mathrm{h},S} \oplus \mathbb{C}$, then on this set the operator $\tilde{B}$ has the positiveness property. Indeed, if the property $(\tilde{B}\boldsymbol{y}, \boldsymbol{y})_{\boldsymbol{H}} = 0$ is satisfied for such elements $\boldsymbol{y} = (\boldsymbol{w}; \boldsymbol{\delta})^{\mathrm{t}}$, then from (8.23) and (8.24) we obtain first that $\delta_1 = 0$ and, therefore, $\boldsymbol{\delta} = \delta_1 \boldsymbol{e}_1 = \boldsymbol{0}$, and then we have

$$
(\tilde{B}\boldsymbol{y}, \boldsymbol{y})_{\boldsymbol{H}} = \rho g \int_\Gamma |\gamma_n \boldsymbol{w}|^2 \mathrm{d}\Gamma = 0,
$$

whence it follows that $\gamma_n \boldsymbol{w} = 0$ and, therefore, $\boldsymbol{w} = \boldsymbol{0}$. Next we show that in reality on the above mentioned set, under condition (8.24), the operator $\tilde{B}$ is positively defined. From (8.23) we have

$$
(\tilde{B}\boldsymbol{y}, \boldsymbol{y})_{\boldsymbol{H}} \geq \rho g \left(\int_\Gamma |\gamma_n \boldsymbol{w}|^2 \mathrm{d}\Gamma - 2 \int_\Gamma |\gamma_n \boldsymbol{w}| \cdot |\delta_1 (\theta x_2)| \mathrm{d}\Gamma + |\delta_1|^2 \alpha_{22} \right) + c|\delta_1|^2,
$$

where $c = J_1\omega_0^2 - \rho g\alpha_{22} > 0$. Using the inequality $2ab \leq \varepsilon a^2 + \varepsilon^{-1}b^2$, $\varepsilon > 0$, we get

$$(\tilde{B}\boldsymbol{y}, \boldsymbol{y})_{\boldsymbol{H}} \geq \rho g\left((1-\varepsilon)\int_\Gamma |\gamma_n \boldsymbol{w}|^2 d\Gamma + \alpha_{22}(1-\varepsilon^{-1})|\delta_1|^2\right) + c|\delta_1|^2. \tag{8.25}$$

We use next the inequality

$$\int_\Gamma |\gamma_n \boldsymbol{w}|^2 d\Gamma \geq c_1 \int_\Omega |\boldsymbol{w}|^2 d\Omega, \qquad \boldsymbol{w} = \nabla\varphi \in \boldsymbol{G}_{\mathrm{h},S}(\Omega), \tag{8.26}$$

which follows from the properties of the above mentioned operator C and its inverse, C^{-1}. Since $C^{-1} \gg 0$, then $c_1 := \lambda_{\min}(C^{-1}) > 0$, and (8.26) proves the fact that C^{-1} is a positively defined operator.

Using inequality (8.26) in (8.25) and choosing $\varepsilon > 0$ to satisfy the condition $\varepsilon < 1/(1 + c(\rho g\alpha_{22})^{-1}) < 1$, we get

$$(\tilde{B}\boldsymbol{y}, \boldsymbol{y})_{\boldsymbol{H}} \geq \min\left[\rho g(1-\varepsilon)c_1; \alpha_{22}(1-\varepsilon^{-1})\rho g + c\right]\left(\int_\Omega |\boldsymbol{w}|^2 d\Omega + |\delta_1|^2\right)$$
$$=: c_2\|y\|^2_{\boldsymbol{H}}, \qquad c_2 > 0.$$

If condition (8.24) is not satisfied and

$$c = J_1\omega_0^2 - \rho g\alpha_{22} < 0, \tag{8.27}$$

then the form $(\tilde{B}\boldsymbol{y}, \boldsymbol{y})_{\boldsymbol{H}}$ takes negative values on the one-dimensional subspace $\boldsymbol{E}_0$ of the elements

$$(\boldsymbol{w}_0; \boldsymbol{\delta})^{\mathrm{t}} \in \boldsymbol{H} : \gamma_n \boldsymbol{w}_0 = -\delta_1(\theta x_2), \qquad \boldsymbol{\delta} = \delta_1 \boldsymbol{e}_1,\ \forall\, \delta_1 \in \mathbb{C}. \tag{8.28}$$

Therefore, in this case, the operator $\tilde{B}$ has a negative eigenvalue that has but one eigenelement.

8.8.4 Normal Oscillations. Properties of the Spectral Problem

Taking into account the properties of the operator coefficients in problem (8.16), we consider in this section the normal oscillation corresponding to this evolution equation. Since such a solution depends on t according to the law $\exp(-\lambda t)$, for amplitude elements $\boldsymbol{y} \in \boldsymbol{H}$ we arrive to the spectral problem

$$\lambda^2 \tilde{I}\boldsymbol{y} - \lambda\mu\tilde{A}\boldsymbol{y} + \tilde{B}\boldsymbol{y} = \boldsymbol{0}, \qquad \boldsymbol{y} = (\boldsymbol{w}; \boldsymbol{\delta})^{\mathrm{t}}. \tag{8.29}$$

Let us now list the general properties of the solutions of problem (8.29).

1° The number $\lambda = 0$ is an eigenvalue with infinite multiplicity that corresponds—according to condition (8.24)—to the set of eigenelements of the form $\boldsymbol{y}_0 = (\boldsymbol{w}_0; \boldsymbol{0})^{\mathrm{t}}$, for all $\boldsymbol{w}_0 \in \boldsymbol{J}_0(\Omega)$.

2° There are no eigenvalues of problem (8.29) situated on the imaginary axis beyond zero.

In fact, if $\lambda = \mathrm{i}\gamma$, $0 \neq \gamma \in \mathbb{R}$, then from (8.29) we get

$$\mathrm{i}\gamma(\tilde{I}\boldsymbol{y}, \boldsymbol{y})_{\boldsymbol{H}} - \mu(\tilde{A}\boldsymbol{y}, \boldsymbol{y})_{\boldsymbol{H}} + (\mathrm{i}\gamma)^{-1}(\tilde{B}\boldsymbol{y}, \boldsymbol{y})_{\boldsymbol{H}} = 0,$$

whence it follows that $\|A^{1/2}\boldsymbol{w}\|^2_{\boldsymbol{L}_2(\Omega)} = 0$ and, therefore, $\boldsymbol{w} = 0$. Moreover, taking also the imaginary part, from here we get the relation $|\delta_1|^2(J_1 - \gamma^{-2}\omega_0^2) = 0$. It is possible either with $\delta_1 = 0$ and then the problem's solution is trivial, or with $\gamma = \pm\omega_0$. In the latter case, problem (8.29) has a trivial solution.

Indeed, let $\gamma = \pm\omega_0$. Then, from (8.29) with $\boldsymbol{w} = \boldsymbol{0}$, we have $-\omega_0^2 I_{12}\boldsymbol{\delta} + B_{12}\boldsymbol{\delta} = \boldsymbol{0}$ and $-\omega_0^2 I_{22}\boldsymbol{\delta} + B_{22}\boldsymbol{\delta} = \boldsymbol{0}$. The second equality is satisfied in our case. Using the notations (8.14) and (8.15), the first equality yields

$$\delta_1\left(-\omega_0^2 P_{0.S}(\boldsymbol{e}_1 \times \boldsymbol{r}) + gG(\theta x_2)\right) = \boldsymbol{0}.$$

We can prove that the expression in paranthesis is not equal to the zero vector. If it were, then by using the orthoprojector P_0 we would get $-\omega_0^2 P_0(\boldsymbol{e}_1 \times \boldsymbol{r}) = \boldsymbol{0}$. Hence, $\boldsymbol{e}_1 \times \boldsymbol{r} = \nabla\varphi \in \boldsymbol{G}(\Omega)$, that is, $-\boldsymbol{e}_2 x_3 + \boldsymbol{e}_3 x_2 = \nabla\varphi = (\partial\varphi/\partial x_2)\boldsymbol{e}_2 + (\partial\varphi/\partial x_3)\boldsymbol{e}_3$. Whence it follows that $\partial\varphi/\partial x_3 = -x_3$ and we arrive to a contradiction because $\partial^2\varphi/\partial x_2\partial x_3 \neq \partial^2\varphi/\partial x_3\partial x_2$. Statement 2° is proved.

3° As a rather important consequence of Property 2° we list the *principle of stability change*, which states that the eigenvalues of a hydrodynamic system that loses its stability, can shift from the right half-plane to the left one only by passing through zero.

4° Problem (8.29) has a discrete spectrum with possible limiting points $\lambda = 0$ and $\lambda = \infty$.

To prove this property we will write (8.29) as a system of two equations, that is,

$$\begin{aligned} \lambda^2(I_{11}\boldsymbol{w} + I_{12}\boldsymbol{\delta}) - \lambda\mu A\boldsymbol{w} + B_{11}\boldsymbol{w} + B_{12}\boldsymbol{\delta} &= \boldsymbol{0} \\ \lambda^2(I_{21}\boldsymbol{w} + I_{22}\boldsymbol{\delta}) + B_{21}\boldsymbol{w} + B_{22}\boldsymbol{\delta} &= \boldsymbol{0}, \end{aligned} \tag{8.30}$$

where I_{ik} and B_{ik} are defined by (8.14) and (8.15). From the second equation we have

$$\left(\lambda^2 + \omega_0^2\right) J_1\boldsymbol{\delta} = -\left(B_{21} + \lambda^2 I_{21}\right)\boldsymbol{w}. \tag{8.31}$$

According to Property 2°, $\lambda \neq \pm i\omega_0$ for the solution of problem (8.30). Then finding $\boldsymbol{\delta}$ from (8.31) and using it in the first equation (8.30) we get the following spectral problem for $\lambda \neq 0$,

$$\left(\nu A - \lambda I - g\lambda^{-1}G\gamma_n + \left(\rho\lambda J_1\left(\omega_0^2 + \lambda^2\right)\right)^{-1}\left(\lambda^2 I_{12} + B_{12}\right)\left(B_{21} + \lambda^2 I_{21}\right)\right)\boldsymbol{w} = \boldsymbol{0}, \tag{8.32}$$

where $\boldsymbol{w} \in \mathcal{D}(A) \subset \boldsymbol{J}_{0,S}^1(\Omega)$.

Performing now the substitution

$$A^{1/2}\boldsymbol{w} = \boldsymbol{v} \in \boldsymbol{J}_{0.S}(\Omega) \tag{8.33}$$

and using the operator $A^{-1/2}$ in the left hand side, we come to the following form of the same problem,

$$\begin{aligned}\mathcal{L}(\lambda)\boldsymbol{v} := \Big(I - \lambda\nu^{-1}A^{-1} - g(\nu\lambda)^{-1}B \\ - \left(\rho\nu J_1\lambda\left(\omega_0^2 + \lambda^2\right)\right)^{-1} A^{-1/2}\left(\lambda^2 I_{12} + B_{12}\right)\left(B_{21} + \lambda^2 I_{21}\right)A^{-1/2}\Big)\boldsymbol{v} \\ = \boldsymbol{0},\end{aligned} \tag{8.34}$$

where $B := (A^{-1/2}G)(\gamma_n A^{-1/2})$.

As stated in Section 8.1, the operator $\gamma_n A^{-1/2} =: Q$ acts compactly from $\boldsymbol{J}_{0,S}(\Omega)$ to $L_{2,\Gamma}$ and, therefore, $A^{-1/2}G = (\gamma_n A^{-1/2})^* = Q^*$ is also compact. Whence it follows that (8.34) is actually a problem on the eigenvalues of a Fredholm operator pencil of the form $I + T(\lambda)$, where $T(\lambda)$ is a holomorphic operator-function that takes compact values on the entire complex plane $\mathbb{C}$ except the points 0, ∞, and $\pm i\omega_0$. As it was stated in Property 2°, for the points λ situated on the imaginary axis, problem (8.30) as well as problem (8.34) have a trivial solution, that is, the pencil $\mathcal{L}(i\gamma)$ with $0 \neq \gamma \in \mathbb{R}$ is reversible. Therefore, according to Section 1.6.3, the pencil $\mathcal{L}(\lambda)$ is regular and its spectrum consists of finitely multiple eigenvalues with possible limiting points $\lambda = 0$ and $\lambda = \infty$. It will be shown later on that the points $\lambda = \pm i\omega_0$ are not limiting points of the eigenvalues.

Here we note that equation (8.34) is a generalization of a corresponding equation in Section 8.1 for the case of a moving container (a pendulum with a partially filled cavity). If in (8.34) we perform a transition to the limit $J_1 \to +\infty$, then we retrieve equation (1.41) that desribes the process of normal oscillations of a viscous fluid in an open immovable container (the two-dimensional problem).

5° If the condition of static stability (8.24) is satisfied, then all nonzero eigenvalues of problem (8.29) are situated in the right half-plane.

Indeed, if $\lambda \neq 0$, from (8.29) we have

$$\lambda(\tilde{I}\boldsymbol{y}, \boldsymbol{y})_{\boldsymbol{H}} - \mu(\tilde{A}\boldsymbol{y}, \boldsymbol{y})_{\boldsymbol{H}} + \lambda^{-1}(B\boldsymbol{y}, \boldsymbol{y})_{\boldsymbol{H}} = 0,$$

whence it follows that

$$\operatorname{Re}\lambda = \frac{\mu(\tilde{A}\boldsymbol{y}, \boldsymbol{y})_{\boldsymbol{H}}}{(\tilde{I}\boldsymbol{y}, \boldsymbol{y})_{\boldsymbol{H}} + |\lambda|^{-2}(\tilde{B}\boldsymbol{y}, \boldsymbol{y})_{\boldsymbol{H}}} \geq 0$$

because the operator $\tilde{I}$ is positive, and the operators $\tilde{A}$ and $\tilde{B}$ are nonnegative. However, there are no eigenvalues of problem (8.29) on the imaginary axis and, therefore, $\operatorname{Re}\lambda > 0$.

6° In problem (8.34) there are two branches of eigenvalues $\{\lambda_{\infty n}\}_{n=1}^{\infty}$ and $\{\lambda_{0n}\}_{n=1}^{\infty}$ with limiting points $\lambda = \infty$ and $\lambda = 0$, respectively. The branch $\{\lambda_{\infty n}\}_{n=1}^{\infty}$ consists of positive eigenvalues and has the asymptotic behavior

$$\lambda_{\infty n} = \nu\lambda_n(A)[1 + \mathrm{o}(1)], \qquad n \to \infty, \tag{8.35}$$

and the branch $\{\lambda_{0n}\}_{n=1}^{\infty}$ consists of positive eigenvalues as well and has the asymptotic behavior

$$\lambda_{0n} = g\nu^{-1}\lambda_n(B)[1 + \mathrm{o}(1)], \qquad n \to \infty. \tag{8.36}$$

We are not going to prove formulas (8.35) and (8.36) but we will note here that $\mathcal{L}(\lambda)$ in (8.34) can be written as an operator-function, namely,

$$I - \lambda\nu^{-1}A^{-1} + \lambda(\rho\nu J_1)^{-1}A^{-1/2}I_{12}I_{21}A^{-1/2} + T(\lambda), \tag{8.37}$$

where $T(\lambda) \to 0$ as $\lambda \to \infty$ and A^{-1} has a power asymptotic just as in the space problem (see (8.1.12)). Hence, according to the statements in Section 1.6.8, it follows that the pencil (8.37) has a branch of eigenvalues with an asymptotic behavior similar to the one of the shortened pencil $I - \lambda\nu^{-1}A^{-1/2}(I - I_{12}I_{21}/(\rho J_1))A^{-1/2}$. Here again a one-dimensional disturbance does not influence the asymptotics and as a result we get formula (8.35). Similarly, after performing the substitution $\lambda = \tilde{\lambda}^{-1}$, we find formula (8.36) on the base of power asymptotics in the plane case for the numbers $\lambda_n(B)$ of the operator B (in the three-dimensional problem, a similar formula is presented in (8.1.33)). At last, we note that the eigenvalues of the series $\{\lambda_{\infty n}\}$ and $\{\lambda_{0n}\}$ are positive because the pencil $\mathcal{L}(\lambda)$ is self-adjoint, and the eigenvalues $\lambda_n(A)$ and $\lambda_n(B)$ are positive.

7° The eigenvalues λ of problem (8.30) are the solutions of the following characteristic equation,

$$\lambda^2 + \omega_0^2 + f_\nu(\lambda) = 0, \tag{8.38}$$

where

$$f_\nu(\lambda) := (\rho\nu J_1\lambda)^{-1}\left(\left(\lambda^2 I_{21} + B_{21}\right)A^{-1/2}L_\nu^{-1}(\lambda)A^{-1/2}\left(\lambda^2 I_{12} + B_{12}\right)\boldsymbol{e}_1, \boldsymbol{e}_1\right)_{\boldsymbol{L}_2(\Omega)},$$

$$L_\nu(\lambda) := I - \lambda\nu^{-1}A^{-1} - g(\lambda\nu)^{-1}B. \tag{8.39}$$

Equation (8.38) follows immediately from the system of equations (8.30) if from the first equation $\boldsymbol{w}$ is expressed in terms of $\boldsymbol{\delta}$ and then substituted in the second one. Let us notice here that the characteristic equation (8.38) is a generalization of another characteristic equation of the problem on normal oscilations of a plane pendulum with a cavity fully filled with a viscous fluid presented in Section 7.5. If in (8.38) and (8.39) we let $B = B_{12} = B_{21} = 0$ and replace $P_{0,S}$ by P_0 and A by A_0, then we arrive at equation (7.5.24).

8° For large enough values of viscosity, ν, problem (8.38)–(8.39) has a complex conjugate pair of eigenvalues $\lambda_0^\pm(\nu)$ situated in the neighborhood of the points $\pm i\omega_0$, that is,

$$\lambda_0^\pm(\nu) = \pm i\omega_0 + o(1), \qquad \nu \to \infty. \tag{8.40}$$

The proof of this property uses the Rouché theorem applied to equation (8.38). Since it is almost similar—with only a few difficult steps—as the one employed to check Property 4° in Section 7.5, we are omitting it here.

Thus, equally with two branches of positive eigenvalues $\{\lambda_{0n}\}$ and $\{\lambda_{\infty n}\}$ the problem (8.30) has a pair of nonreal complex conjugate eigenvalues, if the viscosity ν is large. It seems the fact takes place for any value $\nu > 0$.

The physical meaning of the solutions $\{\lambda_{\infty n}\}$, $\{\lambda_{0n}\}$, and $\lambda_0^\pm(\nu)$ is the following. The branch $\{\lambda_{\infty n}\}$—just as in the case of an immovable container—corresponds to the internal dissipative waves with decrements of the oscillation fading as large as possible; the branch $\{\lambda_{0n}\}$ corresponds to the surface waves with decrements of the oscillation fading as small as possible; and the numbers $\lambda_0^\pm(\nu)$ give the oscillation frequency and the decrements of fading of a pendulum with a fluid filling as a whole. Besides these types of normal motions, the system can also have—like in the case of an immovable container—boundary waves that correspond to the nonreal eigenvalues of problem (8.30). For large viscosity ν, these nonreal eigenvalues are probably absent (except for the pair of values $\lambda_0^\pm(\nu)$).

Let us also note that the eigenelements $\{\boldsymbol{w}_{\infty n}\}$ and $\{\boldsymbol{w}_{0n}\}$ corresponding to the two branches of eigenvalues $\{\lambda_{\infty n}\}$ and $\{\lambda_{0n}\}$, respectively, have the properties of Riesz and p-basicity as described in Section 8.3. We are not presenting these statements here.

8.8.5 Theorem on Instability

We consider now the case when condition (8.24) of static stability in linear approximation is not satisfied and inequality (8.27) is valid. Then we have the following statement, which in mechanics is called the *reversal of Lagrange theorem on a system stabilty*. If condition (8.27) is satisfied and, therefore, the form $(\tilde{B}\boldsymbol{y}, \boldsymbol{y})_{\boldsymbol{H}}$ of potential energy of the hydrodynamic system takes negative values, then the spectral problem (8.30) has only one real eigenvalue in the left complex half-plane. In the evolution problem (8.16), this corresponds to a certain type of normal oscillations that are increasing in time exponentially.

We will prove the theorem on instability in six steps.

1° In problem (8.30), we replace $A^{1/2}\boldsymbol{w}$ by $\boldsymbol{v}$ and use the operator $A^{-1/2}$ in the left hand side of the first equation to obtain

$$\begin{aligned}\lambda^2(A^{-1/2}I_{11}A^{-1/2}\boldsymbol{v} + A^{-1/2}I_{12}\boldsymbol{\delta}) \\ - \lambda\mu\boldsymbol{v} + (A^{-1/2}B_{11}A^{-1/2}\boldsymbol{v} + A^{-1/2}B_{12}\boldsymbol{\delta}) = \mathbf{0}, \\ \lambda^2(I_{21}A^{-1/2}\boldsymbol{v} + I_{22}\boldsymbol{\delta}) + (B_{21}A^{-1/2}\boldsymbol{v} + B_{22}\boldsymbol{\delta}) = \mathbf{0}.\end{aligned} \tag{8.41}$$

In short, the system can be written down as

$$\lambda^2\hat{A}\boldsymbol{z} - \lambda\mu\hat{P}\boldsymbol{z} + \hat{B}\boldsymbol{z} = \mathbf{0}, \qquad \boldsymbol{z} = (\boldsymbol{v};\boldsymbol{\delta})^{\mathrm{t}} \in \boldsymbol{J}_{0,S}(\Omega) \oplus \mathbb{C} =: \boldsymbol{H}, \tag{8.42}$$

where $\hat{P} := \operatorname{diag}(I;0)$, the operator matrices $\hat{A}$ and $\hat{B}$ are compact and self-adjoint, and $\hat{A} > 0$ because the matrix $\tilde{I} = (I_{ik})_{i,k=1}^2$ in (8.30) is bounded and positive definite and $A^{-1/2}$ is a positive compact operator. Since $\gamma_n A^{-1/2}$ and $A^{-1/2}G = (\gamma_n A^{-1/2})^*$ are compact operators as well, the elements of the matrix $\hat{B}$ (see (8.41) and (8.15)) are compact operators too and, therefore, the operator $\hat{B}$ is compact and self-adjoint in $\boldsymbol{H}$.

2° We introduce now a new parameter $\eta = -\lambda$, and for $\eta > 0$ we consider the operator $D(\eta) := \eta^2\hat{A} + \eta\mu\hat{P}$. It is obvious that this operator is bounded and positively definite in $\boldsymbol{H}$. This follows easily from the the relations $(\tilde{I}\boldsymbol{y}, \boldsymbol{y})_{\boldsymbol{H}} \geq a^2\|y\|^2_{\boldsymbol{H}}$, $(\hat{P}\boldsymbol{z}, \boldsymbol{z})_{\boldsymbol{H}} = \|\boldsymbol{v}\|^2_{\boldsymbol{L}_2(\Omega)}$, and $\boldsymbol{y} = (\boldsymbol{w};\boldsymbol{\delta})^{\mathrm{t}}$ after the substitution $\boldsymbol{w} = A^{-1/2}\boldsymbol{v}$. We note also that if the quadratic form of the potential energy operator $\tilde{B}$ in (8.29) and (8.30) takes negative values, then the same happens to the quadratic form of the operator $\hat{B}$ in (8.41) and (8.42).

We consider now the equation

$$D(\eta)\boldsymbol{z} = -\hat{B}\boldsymbol{z}. \tag{8.43}$$

If this equation has a nontrivial solution for some $\eta > 0$, then problem (8.42) will apparently have a nontrivial solution for $\lambda = -\eta < 0$, that is, it will verify the above formulated statement on the system instability.

3° To generalize problem (8.43), let us consider the following problem on eigenvalues,

$$-\hat{B}\boldsymbol{z} = \beta D(\eta)\boldsymbol{z}, \tag{8.44}$$

where $\beta = \beta(\eta)$ is a new spectral parameter. Since $D(\eta) \gg 0$ for $\eta > 0$, then there exists a bounded inverse operator $D^{-1}(\eta)$. Performing the following substitution

$$D^{1/2}(\eta)\boldsymbol{z} = \boldsymbol{\xi} \tag{8.45}$$

in (8.44), we come to the next problem on eigenvalues for a compact self-adjoint operator $K(\eta)$,

$$K(\eta)\boldsymbol{\xi} := -D^{1/2}(\eta)\hat{B}D^{1/2}(\eta)\boldsymbol{\xi} = \beta\boldsymbol{\xi}. \tag{8.46}$$

As it easily follows, the quadratic form of this operator takes positive values.

Let us denote by $\beta_+(\eta)$ the maximum positive eigenvalue of the operator $K(\eta)$. For this eigenvalue the maximum principle is satisfied, that is,

$$\beta_+(\eta) = \max_{\boldsymbol{\xi}\neq \boldsymbol{0}} \frac{(-D^{1/2}(\eta)\hat{B}D^{1/2}(\eta)\boldsymbol{\xi}, \boldsymbol{\xi})_{\boldsymbol{H}}}{(\boldsymbol{\xi}, \boldsymbol{\xi})_{\boldsymbol{H}}} = \max_{\boldsymbol{z}\neq \boldsymbol{0}} \frac{(-\hat{B}\boldsymbol{z}, \boldsymbol{z})_{\boldsymbol{H}}}{(D(\eta)\boldsymbol{z}, \boldsymbol{z})_{\boldsymbol{H}}}. \tag{8.47}$$

If $\beta_+(\eta) = 1$ for a certain $\eta > 0$, then equation (8.43) has a nontrivial solution and the theorem is proved.

4° We next notice that the following inequalities are satisfied,

$$\varphi_-(\eta)(\hat{A} + \mu\hat{P}) \leq D(\eta) \leq \varphi_+(\eta)(\hat{A} + \mu\hat{P}), \tag{8.48}$$

where $\varphi_-(\eta) := \min\{\eta, \eta^2\}$ and $\varphi_+(\eta) := \max\{\eta, \eta^2\}$, for $\eta > 0$. Hence, from (8.47) we get

$$\max_{\boldsymbol{z}\neq \boldsymbol{0}} \frac{(-\hat{B}\boldsymbol{z}, \boldsymbol{z})_{\boldsymbol{H}}}{((\hat{A} + \mu\hat{P})\boldsymbol{z}, \boldsymbol{z})_{\boldsymbol{H}}} \cdot \frac{1}{\varphi_+(\eta)} \leq \max_{\boldsymbol{z}\neq \boldsymbol{0}} \frac{(-\hat{B}\boldsymbol{z}, \boldsymbol{z})_{\boldsymbol{H}}}{(D(\eta)\boldsymbol{z}, \boldsymbol{z})_{\boldsymbol{H}}} \leq \max_{\boldsymbol{z}\neq \boldsymbol{0}} \frac{(-\hat{B}\boldsymbol{z}, \boldsymbol{z})_{\boldsymbol{H}}}{((\hat{A} + \mu\hat{P})\boldsymbol{z}, \boldsymbol{z})_{\boldsymbol{H}}} \cdot \frac{1}{\varphi_-(\eta)}. \tag{8.49}$$

If we denote the left and right maximum in (8.49) by γ_+, it is obvious that γ_+ equals the only positive eigenvalue of the problem

$$-\hat{B}\boldsymbol{z} = \gamma(\hat{A} + \mu\hat{P})\boldsymbol{z}, \tag{8.50}$$

which exists due to the fact that the quadratic form of the operator $-\hat{B}$ takes positive values on an one-dimensional subspace.

From (8.49) we obtain the two-sided estimates

$$\frac{\gamma_+}{\varphi_+(\eta)} \le \beta_+(\eta) \le \frac{\gamma_+}{\varphi_-(\eta)}. \tag{8.51}$$

5° Using the estimates (8.51), we solve equation (8.43) graphically. We construct the graphs of the functions $\gamma_+/\varphi_+(\eta)$, $\beta_+(\eta)$, $\gamma_+/\varphi_-(\eta)$, and the unit function. There exist $\eta_1 > 0$ and $\eta_2 > 0$ such that $\gamma_+/\varphi_+(\eta_1) = 1$, $\gamma_+/\varphi_-(\eta_1) = 1$, and $\eta_1 \le \eta_2$. Then, using inequalities (8.51), there exists at least one number $\eta_0 > 0$ in the interval $[\eta_1, \eta_2]$ such that $\beta_+(\eta_0) = 1$. Thus, the number $\lambda = \lambda_0 = -\eta_0 < 0$ gives an unstable solution of the evolution problem (8.16) that depends on t by the law $\exp(\eta_0 t)$.

6° We show now that there is only one negative eigenvalue of the spectral problem (8.29). From Property 5° in Section 8.8.4, it follows that

$$\operatorname{Re}\lambda = \frac{\mu(\tilde{A}\boldsymbol{y}, \boldsymbol{y})_{\boldsymbol{H}}}{(\tilde{I}\boldsymbol{y}, \boldsymbol{y})_{\boldsymbol{H}} + |\lambda|^{-2}(\tilde{B}\boldsymbol{y}, \boldsymbol{y})_{\boldsymbol{H}}}.$$

Here, the quadratic form in the denominator is nonnegative, the quadratic form $(\tilde{I}\boldsymbol{y}, \boldsymbol{y})_{\boldsymbol{H}}$ is positive, and the quadratic form $(\tilde{B}\boldsymbol{y}, \boldsymbol{y})_{\boldsymbol{H}}$ may take negative values only on an one-dimensional subspace.

This concludes the proof of the theorem on instability.

8.8.6 On the Solvability of the Initial Boundary Value Problem

Let us return now to problem (8.16) and assume that condition (8.24) on the static stability of the system in linear approximations is satisfied.

In (8.16) we make the change

$$\mathrm{i}\tilde{B}^{1/2}\boldsymbol{y} = \mathrm{d}\boldsymbol{u}/\mathrm{d}t, \qquad \boldsymbol{u}(0) = \boldsymbol{0} \tag{8.52}$$

to obtain the following relation

$$\tilde{I}\frac{\mathrm{d}^2\boldsymbol{y}}{\mathrm{d}t^2} + \mu\tilde{A}\frac{\mathrm{d}\boldsymbol{y}}{\mathrm{d}t} - \mathrm{i}\tilde{B}^{1/2}\frac{\mathrm{d}\boldsymbol{u}}{\mathrm{d}t} = \boldsymbol{0}$$

that can be integrated on t if we take into account the initial conditions (8.16). Thus, we obtain

$$\tilde{I}\frac{\mathrm{d}\boldsymbol{y}}{\mathrm{d}t} + \mu A\boldsymbol{y} - \mathrm{i}\tilde{B}^{1/2}\boldsymbol{u} = \tilde{I}\boldsymbol{y}^1 + \mu\tilde{A}\boldsymbol{y}^0 = \text{const.} \tag{8.53}$$

The system of equations (8.53) and (8.52) leads to the following Cauchy problem

$$\begin{pmatrix} \tilde{I} & 0 \\ 0 & I \end{pmatrix} \frac{\mathrm{d}}{\mathrm{d}t} \begin{pmatrix} \boldsymbol{y} \\ \boldsymbol{u} \end{pmatrix} + \begin{pmatrix} \mu\tilde{A} & -\mathrm{i}\tilde{B}^{1/2} \\ -\mathrm{i}\tilde{B}^{1/2} & 0 \end{pmatrix} \begin{pmatrix} \boldsymbol{y} \\ \boldsymbol{u} \end{pmatrix} = \begin{pmatrix} \tilde{I}\boldsymbol{y}^1 + \mu\tilde{A}\boldsymbol{y}^0 \\ \boldsymbol{0} \end{pmatrix},$$
$$(\boldsymbol{y}(0); \boldsymbol{u}(0))^{\mathrm{t}} = (\boldsymbol{y}^0; \boldsymbol{0})^{\mathrm{t}}. \tag{8.54}$$

Here the operator $\mathrm{diag}(\tilde{I}; I)$ is apparently bounded and positive definite in $\boldsymbol{H} \oplus \boldsymbol{H}$, and the second matrix operator, $\mathcal{A}$, satisfies the following property

$$\operatorname{Re}\mathcal{A} = \mu\,\mathrm{diag}(\tilde{A}; 0) \geq 0. \tag{8.55}$$

Therefore, applying the operator $\check{I} := \mathrm{diag}(\tilde{I}^{-1}; I)$ to both sides of (8.54) and introducing on $\boldsymbol{H} \oplus \boldsymbol{H}$ a new scalar product associated with the bilinear form for the operator $\check{I}^{-1}$ we obtain an evolution problem with a dissipative operator (see Section 1.5.4). The reader is invited to formulate a theorem on the solvability of problem (8.16) and the initial-boundary problem (8.1)–(8.6), to obtain the stability of its solutions, and to formulate conditions for the initial data (8.6), by using all the information available so far.

8.9 Convection in a Partially Filled Container

The problem considered in this section is closely related to both the problem in Section 7.7 and the one in Sections 8.1–8.4.

8.9.1 Statement of the Problem

Let us assume that a nonuniformly heated fluid that fills partially an arbitrary container occupies in the state of mechanical equilibrium the region Ω which is bounded by the solid boundary S and a free surface Γ. Using the notations in Section 7.7, the pressure field $P_0 = P_0(x_3)$ takes the following form in the immovable state,

$$\begin{aligned} P_0(x_3) &= p_{\mathrm{a}} - \rho g x_3 + p_0(x_3), \\ \nabla p_0 &= \rho g \beta T_0(x_3) \boldsymbol{e}_3, \\ T_0(x_3) &= -\alpha x_3 + \alpha_0, \\ p_0(x_3) &= \rho g \beta \left(-\alpha \frac{x_3^2}{2} + \alpha_0 x_3 \right), \end{aligned} \tag{9.1}$$

where p_{a} is the atmospheric pressure, $p_0(x_3)$ is the pressure of the fluid caused by the equilibrium gradient of the temperature $\nabla T_0 = -\alpha \boldsymbol{e}_3$, $\alpha \neq 0$, in the state of mechanical equilibrium, and β is the coefficient of thermal extension.

As in Section 7.7, if we consider small convective movements of the fluid in the container, we obtain the following equations and boundary conditions,

$$\begin{aligned}\frac{\partial \boldsymbol{u}}{\partial t} &= -\frac{1}{\rho}\nabla p + \nu\Delta \boldsymbol{u} + g\beta\theta \boldsymbol{e}_3,\\ \operatorname{div}\boldsymbol{u} &= 0,\\ \frac{\partial \theta}{\partial t} &= \alpha \boldsymbol{u}\cdot \boldsymbol{e}_3 + \chi\Delta\theta \text{ in } \Omega,\\ \boldsymbol{u} &= \boldsymbol{0},\ \theta = 0 \text{ on } S.\end{aligned} \tag{9.2}$$

for the velocity field $\boldsymbol{u}(t,x)$, the deviation $p(t,x)$ of the pressure $P(t,x)$ from its equilibrium pressure (9.1), and the deviation $\theta(t,x)$ of the temperature from the equilibrium temperature $T_m + T_0(x_3)$.

The linearized dynamic and kinematic conditions on Γ, which are similar for a unheated fluid, have the form

$$\begin{aligned}\rho\nu\left(\frac{\partial u_i}{\partial x_3} + \frac{\partial u_3}{\partial x_i}\right) &= 0, \qquad i = 1,2,\\ \frac{\partial \zeta}{\partial t} &= u_n,\\ p - 2\rho\nu\frac{\partial u_3}{\partial x_3} &= \rho g(1-\beta\alpha_0)\zeta \text{ on } \Gamma,\end{aligned} \tag{9.3}$$

where $\zeta = \zeta(t, x_1, x_2)$ is the deviation of the moving free surface.

Let us assume that the next general condition of heat exchange holds true on the perturbed moving boundary $\Gamma(t)$, that is,

$$\chi\frac{\partial}{\partial n}(T_m + T_0(x_3) + \theta(t,x)) + a(T_m + T_0(x_3) + \theta(t,x)) = \text{const},$$

where $a \geq 0$ is the interphase coefficient of heat exchange. For $a = 0$, we obtain from this condition a heat flow from fluid into gas. Let us assume that this general condition of heat exchange is satisfied also in the state of mechanical equilibrium. Then, after linearization and after taking into account (9.1) we obtain the following relation for the temperature field $\theta(t,x)$,

$$\chi\frac{\partial \theta}{\partial n} + a\theta = a\alpha\zeta \text{ on } \Gamma. \tag{9.4}$$

In order to complete the mathematical formulation of the initial boundary value problem (9.2)–(9.4) we need to take into account the initial conditions

$$\begin{aligned}\boldsymbol{u}(0,x) &= \boldsymbol{u}^0(x),\\ \theta(0,x) &= \theta^0(x),\\ \zeta(0,x_1,x_2) &= \zeta^0(x_1,x_2).\end{aligned} \tag{9.5}$$

Since usually $\beta\alpha_0 \ll 1$, then we can further assume that $\beta\alpha_0 = 0$.

8.9.2 Transition to a System of Operator Equations

As in Section 8.1, let us assume that $\boldsymbol{u}(t,x)$ is a function of variable t with vaules in $\boldsymbol{J}_{0,S}(\Omega)$. Applying the othoprojector $P_{0,S}$ (onto the subspace $\boldsymbol{J}_{0,S}(\Omega)$) to the first equation in (9.2) and then performing transitions similar to the ones in Sections 8.1.2–8.1.3 to equations (1.23), we obtain for the field $\boldsymbol{u}(t)$ the following equations from (9.2)–(9.3):

$$\nu A\boldsymbol{s} = -\frac{\mathrm{d}\boldsymbol{u}}{\mathrm{d}t} + g\beta P_{0.S}(\theta \boldsymbol{e}_3), \qquad \frac{\mathrm{d}\zeta}{\mathrm{d}t} = \gamma_n \boldsymbol{u},$$
$$\nu \boldsymbol{w} = -gT\zeta, \qquad \boldsymbol{u} = \boldsymbol{s} + \boldsymbol{w}, \tag{9.6}$$

where A, T are the operators in the boundary value problems I and II.

Let us perform for the temperature field $\theta(t,x)$ a similar transition in the last equation (9.2). With this purpose in mind, we consider the Hilbert space of scalar functions $E = L_2(\Omega)$ equipped with the ordinary scalar product and the space $F = H^1_{0,S}(\Omega)$ densely embedded into E with the following scalar product,

$$(\varphi,\psi)_{H^1_{0,S}(\Omega)} := \int_\Omega \nabla\varphi \cdot \overline{\nabla\psi}\mathrm{d}\Omega + \int_\Gamma a\chi^{-1}\varphi\overline{\psi}\mathrm{d}\Gamma, \qquad a \geq 0,\ \chi > 0.$$

The elements φ in $H^1_{0,S}(\Omega)$ satisfy the Dirichlet condition $\varphi = 0$ on S.

Since $F = H^1_{0,S}(\Omega)$ is compactly embedded into $E = L_2(\Omega)$, the Hilbert pair $(F;E)$ generates the unbounded positive definite operator A_1 that has a positive compact inverse, A_1^{-1}. A_1 is the resolving operator for the following boundary value problem III:

$$-\Delta\varphi = f \text{ in } \Omega,$$
$$\varphi = 0 \text{ on } S,$$
$$\frac{\partial\varphi}{\partial n} + a\chi^{-1}\varphi = 0 \text{ on } \Gamma. \tag{9.7}$$

For its eigenvalues, $\lambda_n = \lambda_n(A_1)$, we have the following asymptotic formula,

$$\lambda_n(A_1) = c_{A_1}^{-2/3} n^{2/3}[1 + \mathrm{o}(1)] \text{ as } n \to \infty, \qquad \text{where } c_{A_1} = \frac{\operatorname{mes}\Omega}{6\pi^2}. \tag{9.8}$$

From (9.8) it follows that $A_1^{-1} \in \mathfrak{S}_p$ for $p > 3/2$.

Let us introduce the operator T_1, which will make it possible to find a solution ψ of the following boundary value problem IV:

$$-\Delta\psi = 0 \text{ in } \Omega,$$
$$\psi = 0 \text{ on } S,$$
$$\frac{\partial\psi}{\partial n} + a\chi^{-1}\psi = \eta \text{ on } \Gamma, \tag{9.9}$$

using the known function η. From the abstract scheme in Section 1.8, it follows that the operator T_1 acts boundedly from $H^{-1/2}(\Gamma)$ to $H^1_{0.S}(\Omega)$. Here $\psi = T_1\eta$ is a generalized solution of problem (9.9).

The two operators A_1 and T_1 allow us to perform a transformation, along the same lines as the scheme in Section 8.1, from the equations and boundary conditions (9.2) and (9.4) for the temerature $\theta(t,x)$ to abstract equations of the form (9.6). Specifically, let us assume that $\theta(t,x)$ is a function of t with values in $L_2(\Omega)$ and it can be represented as $\theta(t,x) = \theta_1(t,x) + \theta_2(t,x)$, where θ_1 and θ_2 are the generalized solutions of the boundary value problems III and IV with the functions determined from (9.2) and (9.4). Then the last equation in (9.2) and the boundary condition (9.4) are equivalent to the system of operator equations

$$\begin{aligned} \chi A_1\theta_1 &= -\frac{\mathrm{d}\theta}{\mathrm{d}t} + \alpha \boldsymbol{u}\cdot \boldsymbol{e}_3, \qquad \theta = \theta_1 + \theta_2, \\ \chi\theta_2 &= \alpha a T_1\zeta. \end{aligned} \tag{9.10}$$

Hence, the problem (9.2)–(9.4) is equivalent to the system of equations (9.9) and (9.10). The initial conditions (9.5) should be added to this system.

8.9.3 Solvability of the Evolution Problem

If in (9.6) and (9.10) we perform the following substitutions

$$\boldsymbol{u} = A^{-1/2}\boldsymbol{\xi}, \quad \boldsymbol{s} = A^{-1/2}\boldsymbol{\eta}, \quad \boldsymbol{w} = A^{-1/2}\boldsymbol{\delta}, \quad \theta_i = A_1^{-1/2}v_i, \qquad i = 1,2, \tag{9.11}$$

then the problem (9.6), (9.10), and (9.5) takes the form

$$\begin{aligned} &\frac{\mathrm{d}\boldsymbol{\eta}}{\mathrm{d}t} + \nu A\boldsymbol{\eta} = g\nu^{-1}B(\boldsymbol{\eta}+\boldsymbol{\delta}) + g\beta A^{1/2}P_{0.S}\left(A_1^{-1/2}(v_1+v_2)\boldsymbol{e}_3\right), \\ &\nu\boldsymbol{\delta} = -gA^{1/2}T\zeta, \quad \frac{\mathrm{d}\zeta}{\mathrm{d}t} = Q(\boldsymbol{\eta}+\boldsymbol{\delta}), \quad \boldsymbol{\xi} = \boldsymbol{\eta}+\boldsymbol{\delta}, \\ &\frac{\mathrm{d}v_1}{\mathrm{d}t} + \chi A_1 v_1 = -\alpha a\chi^{-1}Q_1^*Q(\boldsymbol{\eta}+\boldsymbol{\delta}) + \alpha A_1^{1/2}\left(A^{-1/2}(\boldsymbol{\eta}+\boldsymbol{\delta})\cdot \boldsymbol{e}_3\right), \\ &v_2 = a\alpha\chi^{-1}Q_1^*\zeta, \qquad v = v_1 + v_2, \\ &\boldsymbol{\xi}(0) = A^{1/2}\boldsymbol{u}^0, \quad v(0) = v_1(0) + v_2(0) = A^{1/2}\theta^0, \quad \zeta(0) = \zeta^0, \end{aligned} \tag{9.12}$$

where we used the following notations

$$Q = \gamma_n A^{-1/2}, \quad Q^* = A^{1/2}T, \quad B = Q^*Q, \quad Q_1^* = A_1^{1/2}T_1, \tag{9.13}$$

that have been partially used in Section 8.1.

As in Section 8.4.3, the Cauchy problem (9.12) can be reduced to an equivalent system of integral Volterra equations. Indeed, from the third equation (9.12) and taking into consideration the initial condition, we have

$$\zeta(t) = \zeta^0 + \int_0^t Q(\boldsymbol{\eta} + \boldsymbol{\delta})(\tau)\mathrm{d}\tau,$$

and then the second equation in (9.12) takes the form

$$\boldsymbol{\delta}(t) = -g\nu^{-1}Q^*\zeta^0 - g\nu^{-1}\int_0^t B(\boldsymbol{\eta} + \boldsymbol{\delta})(\tau)\mathrm{d}\tau. \tag{9.14}$$

Further, in the first equation (9.12) we replace t by τ, apply the operator-valued function $\exp(-\nu(t-\tau)A)$ from the left, and integrate by τ between the limits 0 and t. Using the initial conditions derived from (9.12) and (9.14), we obtain the following equation

$$\begin{aligned}\boldsymbol{\eta}(t) &= \exp(-\nu tA)(A^{1/2}\boldsymbol{u}^0 + g\nu^{-1}Q^*\zeta^0) + g\nu^{-1}\int_0^t \exp(-\nu(t-\tau)A)B(\boldsymbol{\eta} + \boldsymbol{\delta})(\tau)\mathrm{d}\tau \\ &+ g\beta\int_0^t \exp(-\nu(t-\tau)A)A^{1/2}P_{0.S}\left(A_1^{-1/2}(v_1+v_2)(\tau)\boldsymbol{e}_3\right)\mathrm{d}\tau. \end{aligned} \tag{9.15}$$

A similar reasoning can be done for the second group of equations (9.12) and that leads to the following equations

$$v_2(t) = a\alpha\chi^{-1}Q_1^*\zeta^0 + a\alpha\chi^{-1}\int_0^t Q_1^*Q(\boldsymbol{\eta} + \boldsymbol{\delta})(\tau)\mathrm{d}\tau, \tag{9.16}$$

$$\begin{aligned} v_1(t) &= \exp(-\chi tA_1)\left(A_1^{1/2}\theta^0 - a\alpha\chi^{-1}Q_1^*\zeta^0\right) \\ &- a\alpha\chi^{-1}\int_0^t \exp(-\chi(t-\tau)A_1)Q_1^*Q(\boldsymbol{\eta} + \boldsymbol{\delta})(\tau)\mathrm{d}\tau \\ &+ \alpha\int_0^t \left[\exp(-\chi(t-\tau)A_1)A_1^{1/2}\right]\left(A^{-1/2}(\boldsymbol{\eta} + \boldsymbol{\delta})(\tau)\cdot\boldsymbol{e}_3\right)\mathrm{d}\tau. \end{aligned} \tag{9.17}$$

Since the operators Q, Q^*, $B = Q^*Q$, Q_1^*, $A^{-1/2}$, $A_1^{-1/2}$ are compact and the operator-valued functions $\exp(-\nu(t-\tau)A)A^{1/2}$ and $\exp(-\chi(t-\tau)A)A_1^{1/2}$ are continuous and have a weak singularity for $\tau \to t$, the equations (9.14)–(9.17) represent a system of integral Volterra equations with continuous operator kernels that have weak singularities. Therefore, if conditions

$$\zeta^0 \in H_\Gamma, \quad \boldsymbol{u}^0 \in \boldsymbol{J}_{0.S}^1(\Omega), \quad \theta^0 \in H_{0.S}^1(\Omega) \tag{9.18}$$

are satisfied, then the problem (9.14)–(9.17) is univalently solvable. In this case, $\boldsymbol{\delta}(t)$ and $\boldsymbol{\eta}(t)$ are continuous functions with values in $\boldsymbol{J}_{0.S}(\Omega)$ and $v_1(t)$, $v_2(t)$ are continuous functions with values in $L_2(\Omega)$.

Hence it appears that if conditions (9.18) are satisfied, then $\boldsymbol{\xi}(t) = \boldsymbol{\eta}(t)+\boldsymbol{\delta}(t)$ is a continuous function in $\boldsymbol{J}_{0.S}(\Omega)$ and $v(t) = v_1(t) + v_2(t)$ is a continuous function in $L_2(\Omega)$. Using now the substitutions in (9.11) we conclude that the initial boundary value problem (9.2)–(9.5) on convective movements of a fluid in a partially filled container is univalently solvable and has a generalized solution $\{\boldsymbol{u}(t,x);\ \theta(t,x);\ \zeta(t,x)\}$. For every t, the functions $\boldsymbol{u}(t,x)$, $\theta(t,x)$, $\zeta(t,x)|_{x\in\Gamma}$ belong to the spaces $\boldsymbol{J}^1_{0,S}(\Omega)$, $H^1_{0,S}(\Omega)$, H_Γ, respectively, and are continuous in t on these subspaces.

In conclusion, let us notice that the nonhomogeneous problem (9.2)–(9.5) can be treated similarly. For this problem, the small field of the external forces $\boldsymbol{f}(t,x)$ (in the equation for $\boldsymbol{u}$) and the density of the heat sources $\varphi(t,x)$ (in the equation for θ) should be defined in the right hand sides of the equations (9.2). In this case, if $\boldsymbol{f}(t,x)$ is a continuous function of t with values in $\boldsymbol{J}^1_{0.S}(\Omega)$, $\varphi(t,x)$ is a continuous function of t with values in $H^1_{0,S}(\Omega)$, and conditions (9.18) are satisfied, then the nonhomogeneous problem (9.2)–(9.5) is univalently solvable and its generalized solution possesses all the above mentioned properties.

8.9.4 Normal Convective Movements. Reduction to an Operator Pencil

Let us consider a solution of problem (9.12) that depends on t according to the law $\exp(-\lambda t)$. Then for the corresponding amplitude functions depending only on x, we obtain

$$\begin{aligned} -\lambda\zeta &= Q\boldsymbol{\xi}, \quad \boldsymbol{\xi} = \boldsymbol{\eta} + \boldsymbol{\delta}, \quad \nu\boldsymbol{\delta} = g\lambda^{-1}B\boldsymbol{\xi}, \\ \nu A\boldsymbol{\eta} &= \lambda(\boldsymbol{\eta}+\boldsymbol{\delta}) + g\beta A^{1/2}P_{0.S}\left(A_1^{-1/2}(v_1+v_2)\boldsymbol{e}_3\right), \\ \chi v_2 &= -\lambda^{-1}a\alpha Q_1^*Q(\boldsymbol{\eta}+\boldsymbol{\delta}), \qquad v = v_1 + v_2, \\ \chi A_1 v_1 &= \lambda(v_1+v_2) + \alpha A_1^{1/2}\left(A^{-1/2}(\boldsymbol{\eta}+\boldsymbol{\delta})\cdot\boldsymbol{e}_3\right). \end{aligned} \tag{9.19}$$

Hence, for $\boldsymbol{\xi} = \boldsymbol{\eta}+\boldsymbol{\delta}$ in $\boldsymbol{J}_{0.S}(\Omega)$ and $v = v_1+v_2$ in $L_2(\Omega)$ we obtain the system of equations

$$\begin{aligned} \boldsymbol{\xi} - g\beta\nu^{-1}Cv &= \lambda\nu^{-1}A^{-1}\boldsymbol{\xi} + \nu^{-1}g\lambda^{-1}B\boldsymbol{\xi}, \\ -\alpha\chi^{-1}C^*\boldsymbol{\xi} + v &= \lambda\chi^{-1}A_1^{-1}v - a\chi^{-1}\alpha\lambda^{-1}Q_1^*Q\boldsymbol{\xi}, \end{aligned} \tag{9.20}$$

with

$$Cv := A^{-1/2}P_{0,S}\left(A_1^{-1/2}v\boldsymbol{e}_3\right), \quad C^*\boldsymbol{\xi} := A_1^{-1/2}\left(A^{-1/2}\boldsymbol{\xi}\cdot\boldsymbol{e}_3\right). \tag{9.21}$$

For $\alpha = 0$, that is without heating, the system (9.20) splits into two independent problems. In the first problem, the solutions—nonperiodically fading heat waves—should be determined from the second equation in (9.20). Then we solve the first equation for the obtained λ and $v \neq 0$. Therefore, in the sequel, we are going to assume that $\alpha \neq 0$.

While we are still at it, let us note that in (9.20) the operators C and C^* defined by (9.21) are compact. It can be proved also that these operators are mutually adjoint. Indeed, for any $v \in L_2(\Omega)$ and any $\boldsymbol{\xi} \in \boldsymbol{J}_{0.S}(\Omega)$ we have

$$\begin{aligned}
(Cv, \boldsymbol{\xi})_{\boldsymbol{J}_{0.S}(\Omega)} &= \int_\Omega P_{0.S}\left(A_1^{-1/2} v \boldsymbol{e}_3\right) \cdot \overline{A^{-1/2}\boldsymbol{\xi}} \mathrm{d}\Omega \\
&= \int_\Omega \left(A_1^{-1/2} v \boldsymbol{e}_3\right) \cdot P_{0.S}(\overline{A^{-1/2}\boldsymbol{\xi}}) \mathrm{d}\Omega \\
&= \int_\Omega \left(A_1^{-1/2} v \boldsymbol{e}_3\right) \cdot (\overline{A^{-1/2}\boldsymbol{\xi}}) \mathrm{d}\Omega \\
&= \int_\Omega \left(A_1^{-1/2} v\right) \left(\overline{A^{-1/2}\boldsymbol{\xi}} \cdot \boldsymbol{e}_3\right) \mathrm{d}\Omega \\
&= \int_\Omega v \left(A_1^{-1/2}\left(\overline{A^{-1/2}\boldsymbol{\xi}} \cdot \boldsymbol{e}_3\right)\right) \mathrm{d}\Omega = (v, C^*\boldsymbol{\xi})_{L_2(\Omega)}.
\end{aligned}$$

To make the system (9.20) more symmetrical, we perform the substitution

$$v = \sqrt{\frac{|\alpha|\nu}{\chi g \beta}}\varphi. \tag{9.22}$$

The (9.20) becomes

$$\hat{I}_\varepsilon y = \lambda \hat{A} y + \lambda^{-1} \hat{B} y, \qquad y = (\xi; \varphi)^{\mathrm{t}}, \tag{9.23}$$

where the operator matrices are defined by

$$\begin{aligned}
\hat{I}_\varepsilon &:= \begin{pmatrix} I & -\varepsilon C \\ -\varepsilon \operatorname{sign} \alpha C^* & I_1 \end{pmatrix}, \qquad \varepsilon = \sqrt{\frac{|\alpha| g \beta}{\nu \chi}} > 0, \\
\hat{A} &:= \operatorname{diag}\left(\nu^{-1} A^{-1}; \chi^{-1} A_1^{-1}\right), \\
\hat{B} &:= \begin{pmatrix} g\nu^{-1} B & 0 \\ -a\varepsilon \operatorname{sign} \alpha B_1 & 0 \end{pmatrix},
\end{aligned} \tag{9.24}$$

where I_1 is the identity operator in $L_2(\Omega)$ and $B_1 := Q_1^* Q$.

Since A^{-1} and A_1^{-1} are compact positive operators from the class $\mathfrak{S}_p$ for $p > 3/2$, then these properties take place for the operator $\hat{A}$ in (9.24). Further, the operator $\hat{B}$ is compact because the operators B, Q_1^* and Q are compact. Obviously, the operator $\hat{I}_\varepsilon$ is equal to the difference between the identity operator $\hat{I} = \operatorname{diag}(I; I_1)$

and the compact operator

$$\varepsilon\hat{K} := \varepsilon\begin{pmatrix} 0 & C \\ \operatorname{sign}\alpha C^* & 0 \end{pmatrix}. \tag{9.25}$$

Further investigation of the normal oscillations is based on the properties of the operator pencil

$$L(\lambda) := \hat{I} - \varepsilon\hat{K} - \lambda\hat{A} - \lambda^{-1}\hat{B},$$

corresponding to problem (9.23). For $\varepsilon = 0$, that is, when $\hat{I}_\varepsilon = \hat{I}$, the pencil $L(\lambda)$ coincides with the operator pencil studied in Section 8.2.

8.9.5 Dissipatively Thermal and Surface Waves Under the General Law of Heat Transfer

If the system is heated from below, that is, when $\operatorname{sign}\alpha = 1$, then the operator $\hat{K}$ in (9.25) is self-adjoint. Since C is a compact operator, then $\hat{K}$ has a discrete spectrum with the limit point zero. Hence it appears that the kernel of the operator $\hat{I} - \varepsilon\hat{K}$ is no more than finite-dimensional for $\alpha > 0$. In the sequel, we are going to consider only the case of general assumption, that is, we will assume that, for $\alpha > 0$, the operator $\hat{I}_\varepsilon = \hat{I} - \varepsilon\hat{K}$ is invertible.

If the system is heated from above, that is, when $\alpha < 0$, we have

$$\hat{K} = -\mathrm{i}\hat{V}, \qquad \hat{V} = \begin{pmatrix} 0 & \mathrm{i}C \\ -\mathrm{i}C^* & 0 \end{pmatrix} = \hat{V}^*, \tag{9.26}$$

and the following representation takes place

$$\hat{I} - \varepsilon\hat{K} = \hat{I} + \mathrm{i}\varepsilon\hat{V} = \hat{I} - \varepsilon JV_1, \tag{9.27}$$

where

$$J = \begin{pmatrix} I & 0 \\ 0 & -I_1 \end{pmatrix}, \quad V_1 = \begin{pmatrix} 0 & C \\ C^* & 0 \end{pmatrix}.$$

Since $\hat{V} = \hat{V}^*$, then from the first formula in (9.26) it follows that $\hat{I} - \varepsilon\hat{K}$ is invertible. As in the cooresponding proof for the operator I_0 in Section 8.4.5, we can prove that when the system is heated from above, the inverse operator has the property

$$\left\|\hat{I}_\varepsilon^{-1}\right\| = \|(\hat{I} - \varepsilon\hat{K})^{-1}\| = 1, \qquad \alpha < 0. \tag{9.28}$$

Applying the operator $\hat{I}_\varepsilon^{-1}$ to (9.23) from the left, we obtain the problem

$$\hat{L}(\lambda)y := \left(\hat{I} - \hat{I}_\varepsilon^{-1}(\lambda\hat{A} + \lambda^{-1}\hat{B})\right)y = 0. \tag{9.29}$$

The pencil $\hat{L}(\lambda)$ in (9.29) differs from the pencil $L_{\nu,\omega_0}(\lambda)$ in (4.22), obtained by considering the normal oscillations of a rotating fluid, by notation only. The new feature in (9.29) is the following: operator $\hat{B}$ defined by the matrix (9.24) is not a nonnegative and self-adjoint operator, unlike in (4.22). That is why, the general reasoning stated in Sections 8.4 and 8.2 is not influenced by the new feature and can be applied to (9.29) as well.

Without presenting the detailed proofs, let us formulate the general properties of the pencil $\hat{L}(\lambda)$ and the properties of the solutions to problem (9.19).

1° The spectrum of problem (9.29) is no more than countable and can have $\lambda = 0$ and $\lambda = \infty$ as limit points. All the other points in the spectrum that are different from 0 and ∞ are eigenvalues with finite algebraic multiplicity.

2° If the following condition is satisfied,

$$4\left\|\hat{I}_\varepsilon^{-1}\right\| \cdot \|\hat{A}\| \cdot \|\hat{B}\| < 1, \tag{9.30}$$

then the pencil $\hat{L}(\lambda)$ admits the factorization

$$\hat{L}(\lambda) = \hat{I}_\varepsilon^{-1} X^{-1}(\hat{I} - \lambda^{-1} X\hat{B})(\hat{I} - \lambda X\hat{A}), \tag{9.31}$$

with

$$X = \hat{I}_\varepsilon^{-1}(\hat{I} + \hat{B}X\hat{A}X), \tag{9.32}$$

relatively to the circle $|\lambda| = r \in (r_-, r_+)$, where

$$r_\pm = \frac{1 \pm \sqrt{1 - 4\left\|\hat{I}_\varepsilon^{-1}\right\| \cdot \|\hat{A}\| \cdot \|\hat{B}\|}}{2\left\|\hat{I}_\varepsilon^{-1}\right\| \cdot \|\hat{A}\|}. \tag{9.33}$$

Here the operator X is invertible, $\hat{I} - \lambda^{-1}X\hat{B}$ is also invertible for $\lambda \geq r$, and the spectrum of the pencil $\hat{L}_1(\lambda) := \hat{I} - \lambda X\hat{A}$ is located outside the circle $|\lambda| < r_+$.

From (9.24) it follows that condition (9.30) is satisfied for sufficiently large values of viscousity $\nu > 0$ and the rest of the parameters of the system fixed.

3° Outside the circle $|\lambda| < r_+$, problem (9.29) has a discrete spectrum $\{\lambda_k^+\}_{k=1}^\infty$, with $\lim_{k\to\infty} \lambda_k^+ = \infty$. The eigen- and associated elements $\{y_{k,q}^+\}_{k=1}^\infty$, where $y_{k,q}^+ = (\boldsymbol{\xi}_{k,q}^+; \varphi_{k,q}^+)^t$, that correspond to this spectrum form a complete system in the space $\boldsymbol{J}_{0,S}(\Omega) \oplus L_2(\Omega)$.

Indeed, the problem on the spectrum of the pencil $\hat{L}_1(\lambda)$ is equivalent to the eigenvalue problem

$$X\hat{A}y = \lambda^{-1}y. \tag{9.34}$$

Therefore, Assertion 3° follows from the first Keldysh theorem because $\hat{A} > 0$, $\hat{A} \in \mathfrak{S}_p$ for $p > 3/2$, and $X = \hat{I} + X_1$, with $X_1 \in \mathfrak{S}_\infty$, is an invertible operator in virtue of (9.32) and the structure of $\hat{I}_\varepsilon^{-1} = \hat{I} + \hat{\Phi}$, with $\hat{\Phi} \in \mathfrak{S}_\infty$.

4° For any $\varepsilon > 0$, all the eigenvalues λ_k^+ except for maybe a finite number of them are located in the sector $|\arg\lambda| < \varepsilon$.

5° If condition (9.30) is not satisfied, then for any $\nu > 0$ and $r > 0$ the system of eigen- and associated elements $\{y_{k,q}\}_{k=1}^\infty$ corresponding to the eigenvalues λ_k^+ outside the circle $|\lambda| < r$ has no more than a finite defect in the space $\boldsymbol{J}_{0,S}(\Omega) \oplus L_2(\Omega)$.

6° The asymptotic behavior of the eigenvalues λ_k^+ is the following,

$$\lambda_k^+ = \lambda_k^{-1}(\hat{A})[1 + o(1)] = \lambda_k(\hat{A}^{-1})[1 + o(1)], \qquad k \to \infty, \tag{9.35}$$

where $\hat{A}^{-1} = \operatorname{diag}(\nu A; \chi A_1)$.

7° The above-mentioned class of normal movements corresponding to the spectrum $\{\lambda_k^+\}_{k=1}^\infty$ can be called naturally the *internal dissipatively thermal waves* of problem (9.19).

Indeed, for the normalized solutions $y = y_k^+$ of problem (9.23) we have

$$\|y\|^2 = \|\boldsymbol{\xi}\|^2_{\boldsymbol{J}_{0,S}(\Omega)} + \|\varphi\|^2_{L_2(\Omega)} = \|\boldsymbol{\eta} + \boldsymbol{\delta}\|^2_{\boldsymbol{J}_{0.S}(\Omega)} + \frac{\chi g\beta}{|\alpha|\nu}\|v_1 + v_2\|^2_{L_2(\Omega)} = 1.$$

For such solution from (9.19) we obtain

$$\begin{aligned}
&\|\boldsymbol{\delta}_k^+\| = g\nu^{-1}\left|\lambda_k^+\right|^{-1}\|B\boldsymbol{\xi}_k^+\| \to 0,\\
&\|v_{2k}\| = a|\alpha|\chi^{-1}\left|\lambda_k^+\right|^{-1}\|Q_1^*Q\boldsymbol{\xi}_k^+\| \to 0,\\
&\left\|\nu\left(\lambda_k^+\right)^{-1}A\boldsymbol{\eta}_k^+ - \boldsymbol{\eta}_k^+\right\| \to 0,\\
&\|\boldsymbol{\xi}_k^+\| = \left|\lambda_k^+\right|^{-1}\|Q\boldsymbol{\xi}_k^+\| \to 0,\\
&\left\|\chi\left(\lambda_k^+\right)^{-1}A_1 v_{1k}^+ - v_{1k}^+\right\| \to 0, \qquad k \to \infty.
\end{aligned} \tag{9.36}$$

Hence, the deviations $\boldsymbol{\xi}$ of the free surface of a fluid converge asymptotically to zero for $k \to \infty$ and problem (9.19) splits into two independent eigenvalue problems for the operators νA (the dissipative waves) and χA_1 (the thermal waves).

Let us consider now the existence of the surface waves under the general law of heat exchange ($a \geq 0$). Let $|\lambda| < r := \chi\|A_1^{-1}\|^{-1}$. Then from the second equation (9.23), written down in a vector-matrix form, we have

$$\varphi = \varepsilon \operatorname{sign}\alpha\left(I - \lambda\chi^{-1}A_1^{-1}\right)^{-1}\left(C^*\boldsymbol{\xi} - a\lambda^{-1}B_1\boldsymbol{\xi}\right).$$

Substituting this relation into the first equation we obtain the following spectral problem

$$\begin{aligned}&\left[g\nu^{-1}B - a\varepsilon^2 \operatorname{sign}\alpha C\left(I - \lambda\chi^{-1}A_1^{-1}\right)^{-1} B_1\right]\boldsymbol{\xi}\\ &\quad - \lambda\left[I - \varepsilon^2 \operatorname{sign}\alpha C\left(I - \lambda\chi^{-1}A_1^{-1}\right)^{-1} C^*\right]\boldsymbol{\xi} + \lambda^2\nu^{-1}A^{-1}\boldsymbol{\xi}\\ &=: \mathcal{L}(\lambda)\boldsymbol{\xi} = \mathbf{0}, \qquad \boldsymbol{\xi} \in \boldsymbol{J}_{0.S}(\Omega).\end{aligned}$$

Let us consider the shortened pencil

$$\mathcal{L}_0(\chi)\boldsymbol{\xi} := (g\nu^{-1}B - \lambda I)\boldsymbol{\xi} = \mathbf{0}, \qquad \boldsymbol{\xi} \in \boldsymbol{J}_{0,S}(\Omega).$$

As it follows from the previous facts, its nonzero eigenvalues $\lambda_k = g\nu^{-1}\lambda_k(B)$ form a nonincreasing sequence of positive numbers with the limit point zero and the eigenelements $\{\boldsymbol{\xi}_k(B)\}_{k=1}^{\infty}$ form an orthogonal basis in the subspace $\boldsymbol{M}_0(\Omega) \subset \boldsymbol{J}_{0,s}(\Omega)$.

Let us prove that $\mathcal{L}(\lambda)$ can be considered as an analytic perturbation of the pencil $\mathcal{L}_0(\lambda)$. Indeed, the expression in the first pair of brackets in the definition of $\mathcal{L}(\lambda)$ can be written down as

$$g\nu^{-1}\left[I - a\varepsilon^2\nu \operatorname{sign}\alpha g^{-1}CQ_1^*(Q^*)^{-1}\right]B\boldsymbol{\xi} + \lambda F_1(\lambda)\boldsymbol{\xi}, \qquad |\lambda| < r,$$

where $F_1(\lambda)$ is an analytic operator-valued function taking values in $\mathfrak{S}_\infty$. From the properties of the solutions of the auxiliary boundary value problems I–IV and from the abstract scheme in Section 1.8, it follows that $Q_1^*(Q^*)^{-1} = (A_1^{1/2}T_1)(T^{-1}A^{-1/2})$, with $T^{-1} = \partial$, acts boundedly from $\boldsymbol{M}_0(\Omega)$ to $L_2(\Omega)$. Since C is a compact operator from $L_2(\Omega)$ to $\boldsymbol{M}_0(\Omega)$, it follows that $K := CQ_1^*(Q^*)^{-1}$ is a compact operator acting on $\boldsymbol{M}_0(\Omega)$. The expression in the second pair of brackets in the definition of $\mathcal{L}(\lambda)$ has the form $I - \varepsilon^2 \operatorname{sign}\alpha C^*C + \lambda F_2(\lambda)$, where $C^*C \geq 0$ is a compact operator and $F_2(\lambda)$ is an analytic operator-valued function for $|\lambda| < r$ with values in $\mathfrak{S}_\infty$.

Hence, the pencil $\mathcal{L}(\lambda)$ can be written down as

$$g\nu^{-1}\left[I - a\varepsilon^2\nu g^{-1}\operatorname{sign}\alpha K\right]B - \lambda I + \lambda F(\lambda),$$

where $F(\lambda)$ is an analytic operator-valued function with compact values. Since $B \in \mathfrak{S}_p$ for $p > 2$ and $\operatorname{Ker} B = \boldsymbol{N}_0(\Omega)$, then the conclusions in Sections 1.6.4, 1.6.7, and 1.6.8 can be applied to the pencil $\mathcal{L}(\lambda)$ as well. Using those results, we can prove the following facts:

1° In problem (9.23) there is a branch of eigenvalues $\{\lambda_k^-\}_{k=1}^{\infty}$ with the limit point $\lambda = 0$.

2° For any $\varepsilon > 0$, all eigenvalues λ_k^-, except maybe for a finite number of them, are located inside the angle $|\arg\lambda| < \varepsilon$ and have the following asymptotic behavior,

$$\lambda_k^- = g\nu^{-1}\lambda_k(B)[1+\mathrm{o}(1)], \qquad k\to\infty.$$

3° If we project the eigen- and associated elements $\{\boldsymbol{\xi}_{k,q}^-\}$ corresponding to the branch $\{\lambda_k^-\}_{k=1}^\infty$ onto $\boldsymbol{M}_0(\Omega)$ we obtain a system that has no more than a finite defect in $\boldsymbol{M}_0(\Omega)$.

It may very well be that the thus obtained class of normal movements is connected with the surface waves, which are similar to the surface waves in a non-heated fluid.

8.9.6 Surface and Internal Waves for Heating From Above

Let us consider the properties of the solution of the problem on normal oscillations under the following conditions. On the free surface of the fluid, we have the condition of constant heat flow

$$\chi\frac{\partial}{\partial n}(T_m + T_0(x_3) + \theta(t,x)) = \text{const}$$

instead of the general heat exchange condition. Then, in the boundary condition (9.4) on Γ, we should assume in the sequel that the interphase coefficient of heat exchange a equals zero. In this case, all the conclusions in Sections 8.9.1–8.9.5 hold true and the operator $\hat{B}$ in (9.24) has the form

$$\hat{B} = \operatorname{diag}(g\nu^{-1}B; 0), \tag{9.37}$$

that is, $\hat{B}$ becomes a nonnegative compact operator with the following set as its kernel

$$\operatorname{Ker}\hat{B} = \boldsymbol{N}_0(\Omega)\oplus L_2(\Omega). \tag{9.38}$$

Let us use representation (9.27) for the operator $\hat{I}_\varepsilon$:

$$\hat{I}_\varepsilon = \hat{I} - \varepsilon J V_1 = J(J-\varepsilon V_1). \tag{9.39}$$

Further, let us notice that $J-\varepsilon V_1$, $J\hat{A} = \operatorname{diag}(\nu^{-1}A^{-1}; -\chi^{-1}A_1^{-1})$, and $J\hat{B} = g\nu^{-1}\operatorname{diag}(B;0) = \hat{B}$ are self-adjoint operators. Hence, it appears that the operator pencil

$$L_J(\lambda) := J - \varepsilon V_1 - \lambda J\hat{A} - \lambda^{-1}\hat{B} \tag{9.40}$$

that corresponds to problem (9.23) is self-adjoint, too. Therefore, the spectrum of the pencil $L_J(\lambda)$ is symmetrical relatively to the real axis.

If we consider the equation $(L(\lambda)y, y) = 0$, which holds true for the solutions of problem (9.23), we can prove that $\operatorname{Re}\lambda > 0$, that is, the normal movements of the system are stable.

As in Section 8.9.5, it is convenient to make the transition from the pencil $L(\lambda)$ to the problem (9.29) and the pencil $\hat{L}(\lambda)$ while assuming that $\hat{B} = \hat{B}^* = g\nu^{-1}\operatorname{diag}(B;0)$. This allows us to apply all the results in Sections 8.3 and 8.4 related to the problem on oscillations of a viscous fluid in both an immovable and a rotating container as well as the results in Section 8.2 to the new pencil $\hat{L}(\lambda)$.

Without listing in detail all the general properties of the solutions to the problem

$$\hat{L}(\lambda)y = 0, \qquad y = (\boldsymbol{\xi};\varphi)^{\mathrm{t}}, \tag{9.41}$$

let us point out the following main result. On one hand, the internal dissipatively thermal waves caused by the operator $\hat{A}$ in (9.29) are present in the system, and on the other hand, the system also has mechanical waves, for which (similar to the case of a heavy viscous nonheated fluid) the oscillations have a surface character and the themperature corresponding to the modes of normal oscillations converges asymptotically to the equilibrium value $T_m + T_0(x_3)$ for $k \to \infty$.

The reader is invited to formulate in detail the properties of the spectra of the two mentioned types of wave motion and the properties on completeness and basicity of the system of eigen- and associated elements coressponding to these waves for both large and arbitrary values of the fluid viscosity. Actually, it is sufficient to repeat the majority of the properties pertaining to the solutions of the problems presented in Sections 8.2–8.4.

8.9.7 Normal Oscillations for Heating from Below and for a Given Heat Flow on the Free Surface

Let us consider now the second main case when the system is heated from below, that is, when $\alpha > 0$ in (9.25). In this particular case, $\hat{K} = \hat{K}^* \in \mathfrak{S}_\infty$ and we can split it into two separate subcases.

We assume that the intensity of heating from below is

$$\varepsilon\lambda_{\max}(\hat{K}) < 1. \tag{9.42}$$

The definition of ε, (9.24), shows that (9.42) is possible only for sufficiently small values of the heat intensity α. If the intensity α is given, then (9.42) is possible either for sufficiently large values of the viscosity $\nu > 0$, or a large coefficient of heat conductivity $\chi > 0$.

If (9.42) is satisfied, operator $\hat{I}_\varepsilon$ in the pencil (9.23) has the property $\hat{I}_\varepsilon = \hat{I} - \varepsilon\hat{K} \gg 0$ and, therefore, it admits a bounded inverse operator $\hat{I}_\varepsilon^{-1} \gg 0$. If in (9.23) we perform the substitution

$$\left(\hat{I}_\varepsilon\right)^{1/2} y = z \qquad \left(y = \left(\hat{I}_\varepsilon\right)^{-1/2} z\right) \tag{9.43}$$

and then apply the operator $(\hat{I}_\varepsilon)^{-1/2}$ from the left, we obtain

$$z = \lambda\tilde{A}z + \lambda^{-1}\tilde{B}z, \tag{9.44}$$
$$\tilde{A} := (\hat{I}_\varepsilon)^{-1/2}\hat{A}(\hat{I}_\varepsilon)^{-1/2}, \quad \tilde{B} = (\hat{I}_\varepsilon)^{-1/2}\hat{B}(\hat{I}_\varepsilon)^{-1/2}.$$

Based on the properties of the operators $\hat{A}$, $\hat{B}$, and $(\hat{I}_\varepsilon)^{-1/2}$, it follows that $\tilde{A}$ is a compact positive operator and $\tilde{B}$ is a compact nonnegative operator with infinite-dimensional kernel and range. Hence, equation (9.44) coincides completely with the equations studied in detail in Section 8.2. Therefore, we can list without proof the main properties of the solutions of problem (9.44) and the initial problem (9.19) on normal oscillations.

1° If (9.42) is satisfied, then, in the case when the system is heated from below, the spectrum of problems (9.19), (9.23) is situated in the right complex half-plane, and consists of two branches of eigenvalues $\{\lambda_k^+\}_{k=1}^\infty$ and $\{\lambda_k^-\}_{k=1}^\infty$ situated on the real axis and of no more than a finite number of intermediate nonreal eigenvalues. For large values of viscosity there are no intermediate nonreal eigenvalues.

2° The eigenvalues $\{\lambda_k^+\}_{k=1}^\infty$ have the limit point $\lambda = \infty$ and correspond to the dissipatively thermal waves for which the asymptotic relations (9.36) are satisfied.

3° The eigenvalues $\{\lambda_k^-\}_{k=1}^\infty$ have the limit point $\lambda = 0$ and correspond to the surface mechanical waves with arbitrarily small fadding decrements.

4° For both types of waves movements mentioned above, the asymptotic formulas for the branches of eigenvalues, for which the main parts can be expressed by the eigenvalues of the operators $\hat{A}^{-1}$ and $\hat{B}$, hold true for $k \to \infty$.

5° The eigenelements $\{y_k^+\}_{k=1}^\infty$ and $\{y_k^-\}_{k=1}^\infty$ of problem (9.23) that correspond to these two branches of eigenvalues have either the properties of basicity or the properties of defect Riesz basicity that were described in details in Sections 8.2–8.3.

Let us assume now that the operator $\hat{I}_\varepsilon = \hat{I} - \varepsilon\hat{K}$ is invertible and the following condition is satisfied instead of (9.42),

$$\varepsilon\lambda_{\max}(\hat{K}) > 1, \tag{9.45}$$

Then the operator $\hat{I}_\varepsilon$ has a finite number of negative eigenvalues, denoted by $\kappa \geq 1$, and equation (9.23) is equivalent to equation (9.29)

$$\hat{L}(\lambda)y := \left(\hat{I} - \lambda\hat{I}_\varepsilon^{-1}\hat{A} - \lambda^{-1}\hat{I}_\varepsilon^{-1}\hat{B}\right) y = 0. \tag{9.46}$$

Since $\hat{I}_\varepsilon = \hat{I}_\varepsilon^*$, then there is a self-adjoint pencil corresponding to equation (9.23). Therefore, the spectrum of problem (9.23) and the one of the equivalent problem (9.46) are symmetrical relatively to the real axis. Moreover, the $\hat{I}_\varepsilon$-self-adjoint operators $\hat{I}_\varepsilon^{-1}\hat{A}$ and $\hat{I}_\varepsilon^{-1}\hat{B}$ are $\hat{I}_\varepsilon$-positive and $\hat{I}_\varepsilon$-nonnegative, respectively. Thus, for large values of ν and χ, that is, when the norm of $\hat{A}$ is sufficiently small and in (9.46) we can apply to the pencil the same two types of factorization as in Sections 8.2–8.4, we have two eigenvalue problems for operators that are the weak perturbations of the operators $\hat{I}_\varepsilon^{-1}\hat{A}$ and $\hat{I}_\varepsilon^{-1}\hat{B}$, respectively. From the latter it follows again that there are two branches of eigenvalues $\{\lambda_k^+\}_{k=1}^\infty$ (dissipatively thermal waves) and $\{\lambda_k^-\}_{k=1}^\infty$ (mechanical waves) and the corresponding systems of eigen- and associated elements have the completeness property. If the parameters of the system are arbitrary, then the completeness property can be replaced by the property of completeness with finite defect.

Under conditions (9.45), the new feature is as follows: The spectrum of the problem may not be situated in the right half-plane and in this case, no more than a finite number of eigenvalues can be situated in the left half-plane, which means that the normal oscillations are unstable.

8.10 Sufficient Conditions of Instability for Convective Movements of a Fluid

In this section we obtain a sufficient condition for the instability of the free convective movements of a fluid partially filling a container and heated from below. In this case, the critical value of the heating intensity depends only on the geometry of the region filled with fluid.

8.10.1 Transition to a Two-Parameter Pencil

We consider again the spectral problem in Section 8.9 corresponding to the case when the fluid is heated from below with a given heat flow on the free surface of the fluid. For $\alpha > 0$ and $a = 0$, from (9.23) and (9.25) we obtain the following problem,

$$L_\varepsilon(\lambda)y = 0, \qquad y = (\boldsymbol{\xi}; \varphi)^t \in \boldsymbol{J}_{0,S}(\Omega) \oplus L_2(\Omega) =: \hat{H},$$

$$L_\varepsilon(\lambda) = \hat{I} - \varepsilon\hat{K} - \lambda\hat{A} - \lambda^{-1}\hat{B}, \qquad \hat{K} = \begin{pmatrix} 0 & C \\ C^* & 0 \end{pmatrix} = \hat{K}^* \in \mathfrak{S}_\infty,$$
$$0 < \hat{A} = \operatorname{diag}\left(\nu^{-1}A^{-1}; \chi^{-1}A_1^{-1}\right) \in \mathfrak{S}_\infty,$$
$$0 \le \hat{B} = \operatorname{diag}(g\nu^{-1}B; 0) \in \mathfrak{S}_\infty, \tag{10.1}$$

where the operators C and C^* were defined in (9.21) and ε was given in (9.25).

Let us assume that the heating intensity ε satisfies the following condition

$$\varepsilon \ge \varepsilon_1 := (\lambda_{\max}(\hat{K}))^{-1} > 0, \tag{10.2}$$

which is necessary for the existence of eigenvalues λ of the pencil $L_\varepsilon(\lambda)$ in the left half-plane (see (9.42) and (9.45)). If such an eigenvalue λ_0 exists, then the initial boundary value problem (9.2)–(9.4) has a particular solution that increases exponentially with time when $t \to \infty$.

Together with the pencil $L_\varepsilon(\lambda)$, let us consider the pencil

$$\lambda L_\varepsilon(\lambda) = -\lambda^2\hat{A} - \hat{B} + \lambda(\hat{I} - \varepsilon\hat{K}), \tag{10.3}$$

and the two-parameter pencil

$$M_{\varepsilon,\eta}(\tau) := \tau\hat{A} - \hat{B} - \eta(\hat{I} - \varepsilon\hat{K}), \tag{10.4}$$

which is linear relatively to the spectral parameter τ. We notice that (10.4) can be obtain from (10.3) if

$$\tau = -\lambda^2, \qquad \eta = -\lambda. \tag{10.5}$$

Our goal is to obtain the conditions under which the pencil (10.4) has the eigenvalue $\tau = -\eta^2$ for some $\eta > 0$. Obviously, in this case, the spectral problem (10.1) for the pencil $L_\varepsilon(\lambda)$ will have the solution $\lambda = -\eta < 0$ that corresponds to an aperiodic solution of problem (9.2)–(9.4) which increases exponentially with t.

Let us point out some important properties for the pencil (10.4).

1° For any $\eta > 0$, the pencil $M_{\varepsilon,\eta}(\tau)$ has a real discrete spectrum $\{\tau_k\}_{k=1}^\infty$, that consists of finite-multiple eigenvalues τ_k with $\lim_{k\to\infty}\tau_k = +\infty$.

The proof of this property goes as follows:

a) We write down the eigenvalue problem for the pencil $M_{\varepsilon,\eta}(\tau)$ in the form

$$\tau\eta^{-1}\hat{A}y = \left(\hat{I} - \varepsilon\hat{K} + \eta^{-1}\hat{B}\right)y =: (\hat{I} - \hat{T})y, \tag{10.6}$$

where $\hat{T}$ is a compact self-adjoint operator.

If for some chosen ε and η the condition

$$\hat{I} - \hat{T} \gg 0 \tag{10.7}$$

is satisfied, then performing in (10.6) the substitution

$$(\hat{I} - \hat{T})^{1/2} y = z, \tag{10.8}$$

we obtain the problem

$$\tau\eta^{-1} R z := \tau\eta^{-1}(\hat{I} - \hat{T})^{-1/2}\hat{A}(\hat{I} - \hat{T})^{-1/2} z = z,$$

with a compact positive operator R. From the latter and the Hilbert–Schmidt theorem (see Section 1.1.11) Assertion 1° follows provided condition (10.7) is satisfied.

b) Let us consider now the opposite case by making the assumption that the kernel of the operator $\hat{I} - \hat{T}$ is not zero and this operator is indefinite. Suppose

$$\mathrm{Ker}(\hat{I} - \hat{T}) \neq \{0\}. \tag{10.9}$$

We show that in this case we can make a transition from problem (10.6) to a similar problem where the kernel of the operator $\hat{I} - \hat{T}$ is zero.

Indeed, if condition (10.9) is satisfied, then problem (10.6) has the trivial solution

$$\tau = \tau_0 = 0, \quad y = y_k(\hat{I} - \hat{T}), \qquad k = 1, 2, \ldots, \kappa_0, \ 0 < \kappa_0 < \infty, \tag{10.10}$$

where κ_0 is the multiplicity of the zero eigenvalue of the operator $\hat{I} - \hat{T}$ and $\{y_k\}_{k=1}^{\kappa_0}$ is the corresponding set of eigenelements that are orthogonal relatively to the form $(\hat{A}y, y)$. Any element in the kernel of the operator $\hat{I} - \hat{T}$ has the form $y_0 = \sum_{k=1}^{\kappa_0} c_k y_k(\hat{I} - \hat{T})$, where c_k are arbitrary coefficients. Let us represent now the solution y of problem (10.6) in the form of a vector-column $y = (y_0; \tilde{y})^{\mathrm{t}}$ where $y_0 \in \mathrm{Ker}(\hat{I} - \hat{T})$, $\tilde{y} \perp \mathrm{Ker}(\hat{I} - \hat{T})$, and write down (10.6) in the vector-matrix form

$$\tau\eta^{-1}\begin{pmatrix} A_{11} & A_{12} \\ A_{21} & A_{22} \end{pmatrix}\begin{pmatrix} y_0 \\ \tilde{y} \end{pmatrix} = \begin{pmatrix} 0 & 0 \\ 0 & \hat{I}_2 - \hat{T}_{22} \end{pmatrix}\begin{pmatrix} y_0 \\ \tilde{y} \end{pmatrix}, \tag{10.11}$$

which corresponds to the orthogonal decomposition

$$\boldsymbol{J}_{0,S}(\Omega) \oplus L_2(\Omega) = \hat{H} = H_0 \oplus \tilde{H}, \qquad H_0 = \mathrm{Ker}(\hat{I} - \hat{T}), \ \tilde{H} = \hat{H} \ominus H_0.$$

For $\tau \neq 0$, from the first equation in (10.11) we have

$$A_{11} y_0 + A_{12}\tilde{y} = 0. \tag{10.12}$$

Since $\dim H_0 = \kappa_0 < \infty$ and $\hat{A} > 0$, then the finite-dimensional operator (i.e., the κ_0-dimensional matrix) A_{11} is positive and, therefore, it admits a positive inverse bounded operator A_{11}^{-1}. Then, from (10.12) and the second relation in (10.11), we obtain

$$\tau\eta^{-1}\tilde{A}\tilde{y} = \left(\hat{I}_2 - \hat{T}_{22}\right)\tilde{y}, \qquad \tilde{A} := A_{22} - A_{21}A_{11}^{-1}A_{12}. \tag{10.13}$$

Since the operator $\hat{A}$ in (10.1) is compact and positive in $\hat{H}$, then, according to the statement in Section 1.4.3, $\tilde{A}$ is a compact positive operator in $\tilde{H}$. The operator $\hat{I}_2 - \hat{T}_{22}$ in (10.13) has a zero kernel and equation (10.13) is similar to (10.6). Therefore, without loss of generality, we can assume that the operator $\hat{I} - \hat{T}$ in equation (10.6) has a zero kernel too.

c) Taking into account this condition and assuming that $\hat{I} - \hat{T}$ is indefinite, let us represent this operator in the form

$$\hat{I} - \hat{T} = F^{1/2}J_{\kappa_-}F^{1/2}, \qquad F := |\hat{I} - \hat{T}| := [(\hat{I} - \hat{T})^2]^{1/2}, \tag{10.14}$$

where κ_- is the number of negative (with respect to multiplicity) eigenvalues of the operator $\hat{I} - \hat{T}$, J_{κ_-} is the signature operator with $J_{\kappa_-} = J_{\kappa_-}^{-1} = J_{\kappa_-}^*$, $J_{\kappa_-}^2 = \hat{I}$, and F is a positive definite bounded operator, the absolute value of the operator $\hat{I} - \hat{T}$.

After performing the substitution $F^{1/2}y = z$ in (10.6), we obtain the following problem

$$\tau\eta^{-1}R_1 z = J_{\kappa_-}z, \qquad R_1 := F^{-1/2}\hat{A}F^{-1/2},$$

where R_1 is a compact positive operator. Hence, we have the eigenvalue problem

$$J_{\kappa_-}R_1 z = \mu z, \mu = \eta\tau^{-1}, \tag{10.15}$$

for the J_{κ_-}-self-adjoint, J_{κ_-}-positive operator $J_{\kappa_-}R_1$.

Since in problem (10.15) the root subspace $L_0(J_{\kappa_-}R_1)$ corresponding to $\mu = 0$, is the zero subspace, then, according to the assertion in Section 1.4.7, problem (10.15) has exactly κ_- (with regard to multiplicities) negative eigenvalues and a countable number of positive eigenvalues $\{\mu_k^+\}_{k=1}^\infty$, with $\mu_k^+ = \eta(\tau_k^+)^{-1} \to 0$ as $k \to \infty$. The eigenelements of problem (10.15) form a J_{κ_-}-orthonormal basis that is a Riesz basis in $\hat{H}$.

The proof of Assertion 1° under the general assumption that the operator $\hat{I} - \hat{T}$ is indefinite and has a nonzero kernel follows from these facts and the previous reasoning.

2° For the eigenvalues $\{\tau_k\}_{k=1}^\infty$ of problem (10.6) counted in nondecreasing order with regard to their multiplicities, the following variational principle is satisfied:

$$\tau_k = \tau_k(\eta, \varepsilon) = \max_{L_{k-1}} \min_{y \in L_{k-1}} \frac{(\hat{B}y, y) + \eta((\hat{I} - \varepsilon\hat{K})y, y)}{(\hat{A}y, y)}, \qquad k \in \mathbb{N}, \tag{10.16}$$

where L_{k-1} is an arbitrary subspace in $\hat{H}$ of codimension $k-1$.

a) As in the proof of Property 1° we can assume that $\mathrm{Ker}(\hat{I}-\hat{T}) = \{0\}$. Then there exists the bounded inverse operator $(\hat{I}-\hat{T})^{-1} =: \hat{I} + T_1$, $T_1 = T_1^* \in \mathfrak{S}_\infty$, and (10.6) is equivalent to the problem of determining the characteristic numbers $\tau\eta^{-1}$ in the equation

$$\tau\eta^{-1}\hat{A}^{1/2}(\hat{I}+T_1)\hat{A}^{1/2}z = z, \qquad z = \hat{A}^{1/2}y,$$

with a compact self-adjoint operator that has a finite number of negative eigenvalues. Obviously, its inverse is an unbounded operator bounded from below with a discrete spectrum and a complete orthogonal system of eigenelements.

b) From these facts and from the corresponding max-min principle for eigenvalues of the above mentioned unbounded inverse operator, the assertions (10.16) easily follow after performing a transition to the eigenelements y in problem (10.6).

3° The eigenvalues $\tau_k(\eta, \varepsilon)$ of problem (10.6) are continuous functions of parameters η and ε.

This property is a corollary to the variational principle (10.16) because the numerator of the fraction in (10.16) depends continuously on η and ε, and the denominator does not depend on them at all.

8.10.2 On the Structure of the Kernels of Operator Coefficients

In this section we state some important properties of the operator coefficients $\hat{K}$ and $\hat{B}$ of the pencil $L_\varepsilon(\lambda)$ that are going to be used in the sequel. Specifically, we prove that

$$\mathrm{Ker}\,\hat{K} \subset \mathrm{Ker}\,\hat{B}, \quad \dim\left(\mathrm{Ker}\,\hat{B} \ominus \overline{\mathrm{Ker}\,\hat{K}}\right) = \infty. \tag{10.17}$$

The proof of properties (10.17) can be done in two steps.

a) Let $y = (\boldsymbol{\xi};\psi)^{\mathrm{t}} \in \mathrm{Ker}\,\hat{B}$, that is, $\hat{B}y = (g\nu^{-1}A^{1/2}T\gamma_n A^{-1/2}\boldsymbol{\xi};0)^{\mathrm{t}} = (\mathbf{0};0)^{\mathrm{t}}$. Here $\psi \in L_2(\Omega)$ is an arbitrary element, and from the condition $(A^{1/2}T)^* = \gamma_n A^{-1/2}$ in Section 8.1.3 and the equality of the first components we have $(A^{1/2}T\gamma_n A^{-1/2}\boldsymbol{\xi}, \boldsymbol{\xi})_{\boldsymbol{J}_{0,S}(\Omega)} = \|\gamma_n A^{-1/2}\boldsymbol{\xi}\|^2_{L_2(\Gamma)} = 0$. Hence, according to the orthogonal decomposition (3.24), $\boldsymbol{\xi} \in \boldsymbol{N}_0(\Omega)$.

Therefore,

$$\mathrm{Ker}\,\hat{B} = \left\{y = (\boldsymbol{\xi};\eta)^{\mathrm{t}} \in \hat{H} : \forall\, \boldsymbol{\xi} \in \boldsymbol{N}_0(\Omega),\ \forall\, \psi \in L_2(\Omega)\right\}. \tag{10.18}$$

b) Let $y = (\boldsymbol{\xi};\psi)^{\mathrm{t}} \in \mathrm{Ker}\,\hat{K}$. According to definition (10.1) of the operator matrix $\hat{K}$ and its coefficients (see (9.21)), we have

$$A^{-1/2}P_{0,S}\left(A_1^{-1/2}\psi \boldsymbol{e}_3\right) = \mathbf{0}, \quad A_1^{-1/2}\left(A^{-1/2}\boldsymbol{\xi}\cdot \boldsymbol{e}_3\right) = 0. \tag{10.19}$$

From the first equation we have $P_{0.S}(A_1^{-1/2}\psi \boldsymbol{e}_3) = \boldsymbol{0}$ and, according to the orthogonal decomposition (2.1.25), it follows that $A_1^{-1/2}\psi \boldsymbol{e}_3 = \nabla\varphi \in \boldsymbol{G}_{0,\Gamma}(\Omega)$. Then $\varphi = \varphi(x_3)$ and $A_1^{-1/2}\psi = \mathrm{d}\varphi/\mathrm{d}x_3$. Since for any $\psi \in L_2(\Omega)$ the element $A_1^{-1/2}\psi$ belongs to $H^1_{0,S}(\Omega)$, then $(A_1^{-1/2}\psi)_S = 0$ and, therefore, $\mathrm{d}\varphi/\mathrm{d}x_3 = 0$. That is why $A_1^{-1/2}\psi = 0$ and $\psi = 0$.

From the second equation in (10.19) we have $A^{-1/2}\boldsymbol{\xi} \cdot \boldsymbol{e}_3 = 0$. If $\boldsymbol{\xi} \in \boldsymbol{J}_{0,S}(\Omega)$, then $A^{-1/2}\boldsymbol{\xi} \in \boldsymbol{J}^1_{0,S}(\Omega)$ and then $\boldsymbol{v} := A^{-1/2}\boldsymbol{\xi} \in \boldsymbol{J}^1_{0,S}(\Omega)$ and $v_3|_\Omega \equiv 0$. All such vector fields form the subspace $\boldsymbol{N}_{1,0}(\Omega) \subset \boldsymbol{N}_1(\Omega)$ that corresponds to horizontal movements of the fluid. For example, this subspace can be descriebed by means of the stream function $\Psi(x_1, x_2, x_3)$ that depends on x_3 as if on the parameter. Hence, $\boldsymbol{N}_{1,0}(\Omega)$ is an infinite-dimensional subspace in $\boldsymbol{N}_1(\Omega)$ with an orthogonal complement in $\boldsymbol{N}_1(\Omega)$ that is obviously infinite dimensional too. Here we proved that $\boldsymbol{v} = A^{-1/2}\boldsymbol{x} \in \boldsymbol{N}_{1,0}(\Omega)$ and, therefore, $\boldsymbol{x} = A^{1/2}\boldsymbol{v} \in A^{1/2}\boldsymbol{N}_{1,0}(\Omega) =: \boldsymbol{N}_{0,0}(\Omega) \subset \boldsymbol{N}_0(\Omega)$. Finally,

$$\operatorname{Ker} \hat{K} = \left\{ y = (\boldsymbol{\xi}; \psi)^{\mathrm{t}} : \psi = 0,\ \boldsymbol{\xi} \in \boldsymbol{N}_{0,0}(\Omega) \subset \boldsymbol{N}_0(\Omega) \right\}. \tag{10.20}$$

c) The first expression in (10.17) follows now from (10.18) and (10.20) whereas the second one follows from the reasoning in b).

8.10.3 On the Existence of Eigenvalues in the Left Complex Half-Plane

Before we start finding sufficient conditions for the instablity of convective movements, let us prove that if the intensity of heating ε is sufficiently large, there exists an eigenvalue of problem (10.1) in the left complex half-plane.

We show that for an arbitrary negative number λ_0 there exists a countable set of such positive numbers $\varepsilon = \varepsilon_k^+(\lambda_0)$ for which the operator pencil $L_\varepsilon(\lambda_0)$ has an eigenelement $y = y_k^+(\varepsilon, \lambda_0)$.

To prove this fact, we write down equation (10.1) as

$$\hat{F}_0 y := \left(\hat{I} - \lambda_0\hat{A} - \lambda_0^{-1}\hat{B}\right) y = \varepsilon\hat{K}y, \tag{10.21}$$

and point out that for $\lambda_0 < 0$, the operator $\hat{F}_0$ is positive definite because $\hat{A} > 0$ and $\hat{B} \geq 0$. That is why (10.21) is equivalent to the equation

$$z = \varepsilon\hat{G}z, \qquad \hat{G} := \hat{F}_0^{-1/2}\hat{K}\hat{F}_0^{-1/2},\ z = \hat{F}_0^{-1/2}y, \tag{10.22}$$

where $\hat{G}$ is a compact self-adjoint operator that has an infinite-dimensional range (see the proof of properties (10.17) where this particular fact was proved for $\hat{K}$). Moreover, using the form of the operator $\hat{K}$ given in (10.1), we can show that its eigenvalues, $\varepsilon_n^\pm(\hat{K})$ have the property $\varepsilon_n^- = -\varepsilon_n^+$, $n \in \mathbb{N}$. Hence, according to the Hilbert-Schmidt theorem (see Section 1.1.11), we obtain that in problem (10.22) $\varepsilon = \varepsilon_k^\pm(\lambda_0)$, where $\{\varepsilon_k^\pm(\lambda_0)\}_{k=1}^\infty$ are the characteristic numbers of the operator $\hat{G}$ and $\varepsilon_k^-(\lambda_0) = -\varepsilon_k^+(\lambda_0)$, with $0 < \varepsilon_k^+(\lambda_0) \to +\infty$ as $k \to \infty$.

From the above proof we obtain as a corollary the following result: The value ε of the critical intensity of heating is located in the interval $(\varepsilon_1, \varepsilon_1^+(\lambda_0))$, where λ_0 is any negative number and ε_1 was defined in (10.2).

8.10.4 Double-Sided Estimates for Eigenvalues

While determining sufficient conditions for instability, we next obtain double-sided estimates for the eigenvalues $\tau_k(\eta, \varepsilon)$ of the linear operator pencil $M_{\varepsilon,\eta}(\tau)$ in (10.4) by using the variational principle (10.16) and the properties (10.16).

Specifically, let us prove that for any $\eta > 0$ and $\varepsilon > 0$, the following estimates take place,

$$\eta\lambda_k^{\mathrm{I}}(\varepsilon) \leq \tau_k(\eta, \varepsilon) \leq \eta\lambda_k^{\mathrm{II}}(\varepsilon), \qquad k \in \mathbb{N}, \tag{10.23}$$

where $\{\lambda_k^{\mathrm{I}}\}_{k=1}^{\infty}$ are eigenvalues of the problem

$$\lambda \hat{A} y = (\hat{I} - \varepsilon \hat{K}) y, \tag{10.24}$$

$\{\lambda_k^{\mathrm{II}}\}_{k=1}^{\infty}$ are eigenvalues of the problem

$$\lambda P_{\hat{B}} \hat{A} P_{\hat{B}} y = P_{\hat{B}} (\hat{I} - \varepsilon \hat{K}) P_{\hat{B}} y, \tag{10.25}$$

and $P_{\hat{B}}$ is the orthoprojector onto $\operatorname{Ker} \hat{B}$.

The proof has three steps.

a) Since $\hat{B} \geq 0$, from (10.16) we have

$$\tau_k(\eta, \varepsilon) \geq \eta \max_{L_{k-1}} \min_{y \in L_{k-1}} \frac{((\hat{I} - \varepsilon \hat{K}) y, y)}{(\hat{A} y, y)} =: \eta \lambda_k^{\mathrm{I}}(\varepsilon). \tag{10.26}$$

As in the proofs of Properties 1° and 2° in Section 8.10.1, we can prove that the max-min values of the variational relation (10.26) are the eigenvalues of problem (10.24), which has a real discrete spectrum with the accumulation point $+\infty$.

b) Similarly, by estimating the variational relation (10.16) on the kernel of the operator $\hat{B}$, we obtain

$$\tau_k(\eta, \varepsilon) \leq \eta \max_{\tilde{L}_{k-1}} \min_{y \in L_{k-1} \cap \operatorname{Ker} \hat{B} =: \tilde{L}_{k-1}} \frac{((\hat{I} - \varepsilon \hat{K}) y, y)}{(\hat{A} y, y)} =: \eta \lambda_k^{\mathrm{II}}(\varepsilon). \tag{10.27}$$

Since $y \in \operatorname{Ker} \hat{B}$, the variational relation in the right-hand side of (10.27) corresponds to the spectral problem (10.25), where $P_{\hat{B}} \hat{A} P_{\hat{B}}$ is compact and positive because $\hat{A} > 0$ and $\dim \overline{\operatorname{Ker} \hat{B}} = \infty$. According to (10.17), since $\dim(\operatorname{Ker} \hat{B} \ominus \overline{\operatorname{Ker} \hat{K}}) = \infty$, problem (10.25) has a discrete spectrum $\{\lambda_k^{\mathrm{II}}(\varepsilon)\}_{k=1}^{\infty}$ with the accumulation point $+\infty$ as well.

c) From inequalities (10.26) and (10.27), the estimates (10.23) and the conclusions regarding the properties of the solutions of problems (10.24) and (10.25) follow.

8.10.5 Derivation of a Sufficient Condition for Instability

Such a condition can be derived from the following statements.

1° If $\varepsilon > 0$ is such that for some $m \in \mathbb{N}$ the condition $\lambda_m^{\mathrm{II}}(\varepsilon) < 0$ holds true, then the spectral problem on convection (10.1) for the operator pencil $L_\varepsilon(\lambda)$ has at least m negative eigenvalues $\lambda_k(\varepsilon)$, $k = 1, 2, \ldots, m$.

The proof of this statement can be carried out graphically. First, let us note that if $\lambda_m^{\mathrm{II}}(\varepsilon) < 0$, then $\lambda_k^{\mathrm{I}}(\varepsilon) \le \lambda_k^{\mathrm{II}}(\varepsilon) < 0$, $k = 1, 2, \ldots, m$ and the reasoning can be carried out for any $k \le m$. In the plane $O\eta\tau$, for $\eta > 0$, let us draw the graphs of the functions $\tau = \lambda_k^{\mathrm{I}}(\varepsilon)\eta$, $\tau = \lambda_k^{\mathrm{II}}(\varepsilon)\eta$, and $\tau = -\eta^2$, $\tau = \tau_k(\eta, \varepsilon)$. Using the double-sided estimates (10.23), the fact that $\tau_k(\eta, \varepsilon)$ is continuous in η (see Property 3° in Section 8.10.2), and the inequalities $\lambda_k^{\mathrm{I}}(\varepsilon) \le \lambda_k^{\mathrm{II}}(\varepsilon) < 0$, we obtain that on the interval $[-\lambda_k^{\mathrm{II}}(\varepsilon), -\lambda_k^{\mathrm{I}}(\varepsilon)]$ the graphs of the functions $\tau_k(\eta, \varepsilon)$ and $\tau = -\eta^2$ have at least one intersection for any $k = 1, 2, \ldots, m$.

2° Problem (10.25) has m (with regard to multiplicities) negative eigenvalues $\lambda_k^{\mathrm{II}}(\varepsilon)$, $k = 1, 2, \ldots, m$, if and only if the operator $P_{\hat{B}}(\hat{I} - \varepsilon\hat{K})P_{\hat{B}}$ has the same properties.

The proof follows exactly the same reasoning as in Sections b) and c) of the proof of Property 1°.

3° The operator $P_{\hat{B}}(\hat{I} - \varepsilon\hat{K})P_{\hat{B}}$ has m negative eigenvalues if and only if the following conditions are satisfied,

$$1 - \varepsilon\mu_k^+ < 0, \quad 1 - \mu_{m+1}^+ \ge 0, \qquad k = 1, 2, \ldots, m, \tag{10.28}$$

where μ_k^+ are positive eigenvalues of the following auxiliary spectral problem

$$\begin{aligned} &-\Delta\boldsymbol{u} + \nabla p = \mu^{-1} v\boldsymbol{e}_3, \quad \operatorname{div}\boldsymbol{u} = 0 \text{ in } \Omega, \quad \boldsymbol{u} = \boldsymbol{0} \text{ on } S, \\ &u_n (= u_3) = 0, \quad \frac{\partial u_i}{\partial x_3} + \frac{\partial u_3}{\partial x_i} \text{ on } \Gamma, \qquad i = 1, 2, \\ &-\Delta v = \mu^{-1} u_3 \text{ in } \Omega, \quad v = 0 \text{ on } S, \quad \frac{\partial v}{\partial n} = 0 \text{ on } \Gamma. \end{aligned} \tag{10.29}$$

The proof has three steps.

a) Using Property 2°, let us consider the eigenvalue problem

$$P_{\hat{B}}(\hat{I} - \varepsilon\hat{K})P_{\hat{B}}y = \beta y, \qquad y \in \operatorname{Ker}\hat{B}, \tag{10.30}$$

or the problem

$$P_{\hat{B}}\hat{K}P_{\hat{B}}y = (1 - \beta)\varepsilon^{-1}y =: \alpha y. \tag{10.31}$$

Since, according to (10.17), the range of the operator $P_{\hat{B}}\hat{K}P_{\hat{B}}$ is infinite dimensional and this operator is self-adjoint, compact and indefinite, problem (10.31) has a countable set of positive and negative eigenvalues $\{\alpha_k^{\pm}\}_{k=1}^{\infty}$ with $\alpha_k^- = -\alpha_k^+ < 0$ (this property is a corollary of the structure of operator $\hat{K}$ from (10.1)) and $\alpha_k^{\pm} \to 0$ as $k \to \infty$.

b) Using the definition of operator $\hat{K}$ and the max-min principle in Section 1.4.1 for its positive eigenvalues, we obtain that the numbers α_k^+ are consecutive maxima of the variational ratio

$$\frac{\left(P_{\hat{B}}\hat{K}P_{\hat{B}}y, y\right)}{(y,y)} = \frac{(\hat{K}y,y)}{(y,y)} = \frac{(C\psi,\boldsymbol{\xi})_{\boldsymbol{L}_2(\Omega)} + (C^*\boldsymbol{\xi},\psi)_{L_2(\Omega)}}{\|\boldsymbol{\xi}\|^2_{\boldsymbol{L}_2(\Omega)} + \|\psi\|^2_{L_2(\Omega)}}, \qquad y \in (\boldsymbol{\xi},\psi)^{\mathrm{t}} \in \operatorname{Ker}\hat{B}. \tag{10.32}$$

Taking into account the structure (10.18) of the kernel of operator $\hat{B}$ and the relation between the spaces $\boldsymbol{N}_0(\Omega)$ and $\boldsymbol{N}_1(\Omega)$, after performing in (10.32) the substitutions $\boldsymbol{\xi} = A^{1/2}\boldsymbol{u}$, $\boldsymbol{u} \in \boldsymbol{J}^1_{0,S}(\Omega)$ and $\psi = A_1^{1/2}v$, $v \in H^1_{0,S}(\Omega)$, we obtain

$$\frac{(P_{0,S}(v\boldsymbol{e}_3),\boldsymbol{u})_{\boldsymbol{L}_2(\Omega)} + (\boldsymbol{u}\cdot\boldsymbol{e}_3, v)_{L_2(\Omega)}}{\|A^{1/2}\boldsymbol{u}\|^2_{\boldsymbol{L}_2(\Omega)} + \left\|A_1^{1/2}v\right\|^2_{L_2(\Omega)}} = \frac{2\operatorname{Re}\displaystyle\int_\Omega u_3\bar{v}\,\mathrm{d}\Omega}{E(\boldsymbol{u},\boldsymbol{u}) + \displaystyle\int_\Omega |\nabla v|^2 \mathrm{d}\Omega}. \tag{10.33}$$

The above-mentioned ratio should be considered on pairs of functions $(\boldsymbol{u}; v)^{\mathrm{t}}$ from $\boldsymbol{N}_1(\Omega) \oplus H^1_{0,S}(\Omega)$.

c) Using the Green formula (see (2.2.27))

$$-\int_\Omega \Delta\boldsymbol{u}\cdot\bar{\boldsymbol{w}}\,\mathrm{d}\Omega = E(\boldsymbol{u},\boldsymbol{w}) - \int_\Gamma \sum_{i=1}^{2}\left(\frac{\partial u_i}{\partial x_3} + \frac{\partial u_3}{\partial x_i}\right)\bar{w}_i\,\mathrm{d}\Gamma,$$

that holds true for (smooth) functions $\boldsymbol{u}$ and $\boldsymbol{w}$ in $\boldsymbol{N}_1(\Omega)$, the similar Green formula for the Laplace operator applied to (scalar) functions in $H^1_{0,S}(\Omega)$, and the orthogonal decomposition (2.1.25), by the reasoning in Section 2.2.7 and some ordinary methods of the variational calculus, from (10.33) we can infer that the spectral boundary value problem (10.29) for $\mu = \alpha$ corresponds to the variational relation (10.33). Therefore, considering the conection (10.31) between β and α, the requirements $\beta_m < 0$ and $\beta_{m+1} \geq 0$ in problem (10.30) are equivalent to conditions (10.28).

The following statement follows immediately from Properties 1°–3°.

If the intensity of heating, ε, satisfies the condition

$$\varepsilon > \varepsilon_2 := \left(\mu_1^+\right)^{-1}, \tag{10.34}$$

where μ_1^+ is the maximal eigenvalue of the auxiliary spectral problem (10.29), *then under the condition of constant heat flow on the free surface* Γ *and in the case of heating from below, the spectral problem of convection* (10.1) *has at least one negative eigenvalue and, therefore, it has an aperiodically increasing with time mode of normal convective movements.*

8.10.5 Remarks

In the conclusion of this section, we present some facts connected with the problem of instability of normal conective movements of a fluid in an open container.

1° First of all, let us point out that the number μ_1^+ and the critical value ε_2 of the intensity of heating depend only on the geometric characteristics of the region Ω that is filled with fluid in an immovable state and does not depend on the physical parameters of the considered problem.

2° Our investigation of the problem regarding instability was carried out using dimensional variables. The corresponding study in dimensionless variables leads to determining the boundary values of the dimensionless Rayleigh number $\mathrm{Ra} := g\beta a l^4/(\nu\chi)$, where l is a characteristic size of the problem. Ra is connected with ε by the equality $\mathrm{Ra} = \varepsilon^2 l^4$.

3° The obtained critical value ε_2 corresponds to the case when in the spectral problem of convection (10.1) the eigenvalue λ passes through zero for the first time as ε increases from zero.

Indeed, let us assume in (9.2)–(9.4) that $\boldsymbol{u}(t,x) \equiv \boldsymbol{u}(x)$, $p(t,x) \equiv p(x)$, $\theta(t,x) \equiv \theta(x)$, $\zeta(t,x_1,x_2) \equiv \zeta(x_1,x_2)$ that correspond to the dependence of t given by $\exp(0\cdot t)$. We perform the substitution $\theta = [a\nu/(\chi g\beta)]^{1/2}v$. For the functions $\boldsymbol{u}$ and v we obtain problem (10.29) with $\mu = \varepsilon^{-1}$. Since $\mu = \mu_1^+$ is the maximal positive eigenvalue of this spectral problem, $\varepsilon_2 = (\mu_1^+)^{-1}$ is the minimal value of the intensity of heating corresponding to the eigenvalue $\lambda = 0$. Other transitions of eigenvalues λ through zero take place for $\varepsilon = (\mu_k^+)^{-1}$, $k = 2,3,\ldots$.

4° A similar reasoning as in the proof of Property 3° in Section 8.10.5, assures that the number ε_1 in (10.2) equals the value $(\mu_1^+)^{-1}$ for the transformed problem (10.29) when for the field $\boldsymbol{u}$ the tangent and normal tensions equal zero on the boundary Γ. This new spectral problem is equivalent to the problem of the spectrum of the variational relation (10.33) considered on the pairs of functions $(\boldsymbol{u};v)^{\mathrm{t}}$ from $\boldsymbol{J}_{0,S}^1(\Omega) \oplus H_{0,S}^1(\Omega)$.

5° Let us introduce the number ε_3 that is the inverse to the maximal eigenvalue of the variational relation (10.33) which is being considered on the pairs of functions $(\boldsymbol{u};v)^{\mathrm{t}}$ from $\boldsymbol{J}_0^1(\Omega) \oplus H_{0,S}^1(\Omega)$. Here again we arrive at problem (10.29), but now with the condition $\boldsymbol{u} = \boldsymbol{0}$ on $\partial\Omega$.

It turns out that the numbers ε_k, $k = 1, 2, 3$ correspond to three different extensions of the main hydrodynamic Stokes operator $-\Delta \boldsymbol{u} + \nabla p$. Specifically, for $k = 1, 2$, and 3 we have three operators A, $\tilde{A}$, and A_0 corresponding to the three mentioned boundary conditions on Γ for the field $\boldsymbol{u}$. Here, considering the variational relation (10.33) on consequently reducing subspaces it follows that $\varepsilon_1 < \varepsilon_2 < \varepsilon_3$. Hence, it follows that the boundary of instability ε_3 of normal convective movements in a completely filled container is higher than the boundary for the same region Ω if there is a free surface and a constant heat flow on it. The same reasoning also shows that for $\varepsilon \in (\varepsilon_1, \varepsilon_2)$ the issue of instability of normal convective movements remains unsolved because the obtained boundaries of the intensity of heating that give the necessary (ε_1) and the suficient (ε_2) conditions of instability do nor coincide.

6° As mentioned in Section 7.7, for the problem of convection of a fluid in a completely filled container, the principle of stability change takes place, that is, whenever ε is increasing, the transition from the right half-plane to the left of the minimal eigenvalue of the problem takes place only through zero (since the problem is self-adjoint). In the problems of convection of fluids with free surfaces, the principle of stability change gets postulated. The reasoning previously carried out in this section shows that for $\varepsilon = \varepsilon_2$ the eigenvalue λ passes from the right half-plane to the left through zero, but it has not been yet proved whether there exist values ε in the interval $(\varepsilon_1, \varepsilon_2)$ such that the eigenvalue λ can be situated, for example, on the imaginary axis. Hence, the validity of the principle of stability change in the problem of convection with a free surface is an issue yet to be done.

Chapter 9

Oscillations of Capillary Viscous Fluids

In this chapter we consider a few problems on small movements and normal oscillations of a viscous fluid that partially fills a container under the conditions of low gravity, that is, when the forces of surface tension are to be taken into account. As it turns out, the capillary forces have a great influence on the structure of the spectrum of normal oscillations. The general scheme of approaching this type of problems will differ from the method developed for the problems in Chapter 8.

9.1 Statement of the Problem

In this section we will state the initial boundary value problem on small oscillations of a capillary fluid in an arbitrary container. Then we will give a qualitative description of the properties of the solutions of the problem on normal oscillations for two concrete examples, which admit separation of variables.

9.1.1 Basic Equations and Boundary Conditions

Let us assume that a capillary viscous fluid fills partially an arbitrary container and, in the equilibrium state, uniformly rotates around the axis Ox_3 with the angular velocity $\boldsymbol{\omega}_0 = \omega_0 \boldsymbol{e}_3$. The fluid occupies the region Ω that is bounded by the solid boundary S and the equilbrium free surface Γ. For simplicity, we assume that in the equilibrium state the system is influenced only by the homogeneous gravitation field with acceleration $\boldsymbol{g} = -g\boldsymbol{e}_3$.

The equilibrium state of the considered system was described in details in Section 6.3.1. In particular, the equilibrium pressure $P_0(x)$ is expressed by formula (6.3.1) and the equation of the free surface can be determined from (6.3.5), the additional condition (6.3.3) and the following requirement: The dihedral angle between Γ and S should be equal to a given constant δ defined by the characteristics of the three media in the neighborhood of the contact line, $\partial\Gamma$, of Γ and S.

Let us consider small movements of the fluid that are close to the uniform rotation of the system as a rigid body with angular velocity $\boldsymbol{\omega}_0 = \omega_0 \boldsymbol{e}_3$. Then, as in Section 8.4, for the field of relative velocities $\boldsymbol{u}(t,x)$ and the dynamic pressure $p(t,x)$ we have the following equations

$$\frac{\partial \boldsymbol{u}}{\partial t} - 2\omega_0 \boldsymbol{u} \times \boldsymbol{e}_3 = -\frac{1}{\rho}\nabla p + \nu\Delta\boldsymbol{u} + \boldsymbol{f}, \quad \text{div}\boldsymbol{u} = 0 \text{ in } \Omega \tag{1.1}$$

and the boundary stickiness conditions

$$\boldsymbol{u} = \boldsymbol{0} \text{ on } S. \tag{1.2}$$

As in the case of a heavy fluid, on the equilibrium surface Γ, the following kinematic and dynamic tangent conditions should be satisfied:

$$\frac{\partial \zeta}{\partial t} = u_n, \quad \rho\nu(u_{i,3} + u_{3,i}) = 0, \qquad i = 1,2; \text{ on } \Gamma. \tag{1.3}$$

The normal dynamic condition on Γ is different though from the corresponding condition (8.4.4) for a heavy fluid because in our case, much as in Section 6.3, we take into account the capillary leap of pressures:

$$-p + 2\rho\nu u_{3,3} = -\sigma\mathcal{L}\zeta := \sigma\Delta_\Gamma\zeta - a_\Gamma\zeta \text{ on } \Gamma, \quad \int_\Gamma \zeta d\Gamma = 0, \tag{1.4}$$

with

$$a_\Gamma := -\sigma\left(k_1^2 + k_2^2\right) + \rho g \cos(\widehat{\boldsymbol{n}, \boldsymbol{e}_3}) - \rho\omega_0^2 r \cos(\widehat{\boldsymbol{n}, \boldsymbol{e}_r}),$$

where $\sigma > 0$ is the coefficient of surface tension and Δ_Γ is the Laplace–Beltrami operator defined for functions precribed on Γ.

As in Sections 6.3 and 8.4, the boundary conditions (1.3) and (1.4) are written down in the curviliniar coordinate system $\tilde{O}\xi^1\xi^2\xi^3$ that was chosen such as the deviation of the moving free surface of the fluid $\Gamma(t)$ from the equilibrium surface Γ can be described by the equation $\xi^3 = \zeta(t, \hat{\xi})$, with $\hat{\xi} := (\xi^1, \xi^2) \in \Gamma$.

In the case of a viscous fluid, the formulation of the boundary condition on the wetting line $\partial\Gamma$ is rather difficult. It is obvious that if the velocity field is continuous near $\partial\Gamma$, then, in virtue of condition (1.2), this line remains unchanged during the process of small movements, that is, it does not depend on t. Therefore, the corresponding condition on $\partial\Gamma$ should have the following form:

$$\zeta = 0 \text{ on } \partial\Gamma. \tag{1.5'}$$

However, the hypothesis on $\partial\Gamma$ remaining unchanged does not hold true in the case of big movements of the fluid. For small movements and vanishing viscosity, $\nu \to 0$, condition (1.5) contradicts the property of preserving the wetting angle δ expressed by condition (4.1.13), namely,

$$\frac{\partial\zeta}{\partial e} + \chi\zeta = 0 \text{ on } \partial\Gamma, \tag{1.5''}$$

with

$$\chi := \frac{k_\Gamma \cos\delta - k_S}{\sin\delta},$$

where $\boldsymbol{e}$ is an external normal to $\partial\Gamma$ in the tangent plane to Γ.

Further we will use either condition (1.5′) or condition (1.5″), pointing out which one is employed in considering a specific problem. The issue of choosing a proper boundary condition on $\partial\Gamma$ for a viscous capillary fluid is still being discussed. Most probably, it can be solved somehow at the theoretical physics level by comparing expertimental data and computation results with regard to one of the conditions on $\partial\Gamma$.

To complete the statement of the initial boundary value problem on small movements of a viscous capillary rotating fluid, we should add to equations and boundary conditions (1.1)–(1.5) the initial conditions

$$\boldsymbol{u}(0,x) = \boldsymbol{u}^0(x), \quad \zeta(0,\hat{\xi}) = \zeta^0(\hat{\xi}). \tag{1.6}$$

9.1.2 Some Properties of Solutions to the Normal Oscillations Problem

Let us consider normal movements of the system, that is, such solutions of the problem (1.1)–(1.5) for $\boldsymbol{f}(t,x) \equiv \boldsymbol{0}$, that depend on t according to the law $\exp(-\lambda t)$. Excluding the deviation ζ from (1.1)–(1.5), we obtain the following problem:

$$\begin{aligned}
&-\nu\Delta\boldsymbol{u} + \frac{1}{\rho}\nabla p - 2\omega_0\boldsymbol{u}\times\boldsymbol{e}_3 = \lambda\boldsymbol{u}, \quad \operatorname{div}\boldsymbol{u} = 0 \text{ in } \Omega,\\
&\boldsymbol{u} = \boldsymbol{0} \text{ on } S,\\
&\rho\nu(u_{i,3} + u_{3,i}) = 0, \qquad i = 1,2; \text{ on } \Gamma,\\
&\lambda(-p + 2\rho\nu u_{3,3}) = \sigma\mathcal{L}u_n \text{ on } \Gamma, \quad \int_\Gamma u_n d\Gamma = 0,\\
&u_n = 0 \text{ on } \partial\Gamma \quad \text{or} \quad \frac{\partial u_n}{\partial e} + \chi u_n = 0, \text{ on } \partial\Gamma.
\end{aligned} \tag{1.7}$$

In this case, the modes of normal movements, that is, the functions $\boldsymbol{u}(x)$ and $p(x)$ are considered to be complex-valued functions. Let us note also that both the equation and the boundary condition on Γ contain the spectral parameter λ which makes the entire problem (1.7) nonself-adjoint.

If the solutions of the spectral problem (1.7) are classical, let us list their properties. Obviously, in this case, we should opt for the first boundary condition on $\partial\Gamma$, that is, the Dirichlet condition for u_n. Then, the expression $\sigma P_\Gamma \mathcal{L} u_n$ determined by the first part of (1.4) together with the condition $u_n = 0$ on $\partial\Gamma$ generates a unbounded self-adjoint operator of elliptical type denoted by σB_σ in the sequel.

Let us assume that the state of relative equilibrium for the rotating capillary fluid is statically stable in linear approximation, that is, the operator B_σ is positive definite in $L_{2,\Gamma}$, namely,

$$(B_\sigma\zeta,\zeta)_{L_{2,\Gamma}} \geq c_\sigma \|\zeta\|^2_{L_{2,\Gamma}}, \qquad \zeta \in \mathcal{D}(B_\sigma),\ c_\sigma > 0. \tag{1.8}$$

Then B_σ has an inverse operator B_σ^{-1}, which is positive and compact.

The following properties take place for problem (1.7).

1° The number $\lambda = 0$ is not an eigenvalue of problem (1.7). Indeed, if we assume $\lambda = 0$ in (1.7), then using the Green formula (2.2.10), the condition $u_n = 0$ on Γ, which is true in virtue of (1.7), and after scalar multiplication by $\boldsymbol{u}$ in $\boldsymbol{L}_2(\Omega)$, we obtain the following relation

$$\nu E(\boldsymbol{u},\boldsymbol{u}) - 4\mathrm{i}\,\omega_0(\operatorname{Im}\boldsymbol{u} \times \boldsymbol{e}_3, \operatorname{Re}\boldsymbol{u})_{\boldsymbol{L}_2(\Omega)} = 0$$

for the solutions

$$\boldsymbol{u}(x) = \operatorname{Re}\boldsymbol{u}(x) + \mathrm{i}\operatorname{Im}\boldsymbol{u}(x).$$

Extracting the real part and using the inequality

$$E(\boldsymbol{u},\boldsymbol{u}) = \|A^{1/2}\boldsymbol{u}\|^2_{\boldsymbol{L}_2(\Omega)} \geq c\|u\|^2_{\boldsymbol{L}_2(\Omega)}, \qquad c = \lambda_1(A) > 0, \tag{1.9}$$

we obtain that $\boldsymbol{u}(x) \equiv 0$.

2° All the eigenvalues of problem (1.7) are located in the right complex half-plane, that is, all the normal movements of the system are fading.

Indeed, multiplying equation (1.7) by $\boldsymbol{u}$ in $\boldsymbol{L}_2(\Omega)$ and applying again the Green formula (2.2.10) and the property (1.8), we have:

$$\lambda\rho\|\boldsymbol{u}\|^2_{\boldsymbol{L}_2(\Omega)} - \rho\nu E(\boldsymbol{u},\boldsymbol{u}) + 2\mathrm{i}\,\rho\omega_0\operatorname{Im}(\boldsymbol{u}\times\boldsymbol{e}_3,\boldsymbol{u})_{\boldsymbol{L}_2(\Omega)} + \sigma\lambda^{-1}\left\|B_\sigma^{1/2}u_n\right\|^2_{L_{2,\Gamma}} = 0, \tag{1.10}$$

whence

$$\operatorname{Re}\lambda = \frac{\rho\nu E(\boldsymbol{u},\boldsymbol{u})}{\rho\|\boldsymbol{u}\|^2_{\boldsymbol{L}_2(\Omega)} + \sigma|\lambda|^{-2}\left\|B_\sigma^{1/2}u_n\right\|^2_{L_{2,\Gamma}}} > 0.$$

For a nonrotating fluid, that is, for a fluid with $\omega_0 = 0$, we can obtain the next two results based on equation (1.10) that in this case can be reduced to a quadratic equation relatively to λ.

3° If the solution $\boldsymbol{u}(x)$ satisfies the following condition

$$\nu E(\boldsymbol{u},\boldsymbol{u}) \geq 2\left(\frac{\sigma}{\rho}\right)^{1/2}\|\boldsymbol{u}\|_{\boldsymbol{L}_2(\Omega)} \times \left\|B_\sigma^{1/2}u_n\right\|^2_{L_{2,\Gamma}},$$

then the corresponding eigenvalue λ is real.

4° If the opposite inequality holds true, that is,

$$\nu E(\boldsymbol{u},\boldsymbol{u}) < 2\left(\frac{\sigma}{\rho}\right)^{1/2}\|\boldsymbol{u}\|_{\boldsymbol{L}_2(\Omega)} \times \left\|B_\sigma^{1/2}u_n\right\|^2_{L_{2,\Gamma}},$$

then the pair of complex-conjugate roots λ and $\bar{\lambda}$ corresponds to the solution $\boldsymbol{u}(x)$ and

$$|\lambda| = \left(\frac{\sigma}{\rho}\right)^{1/2}\frac{\left\|B_\sigma^{1/2}u_n\right\|^2_{L_{2,\Gamma}}}{\|\boldsymbol{u}\|_{\boldsymbol{L}_2(\Omega)}} > \frac{\nu}{2}\frac{E(\boldsymbol{u},\boldsymbol{u})}{\|\boldsymbol{u}\|^2_{\boldsymbol{L}_2(\Omega)}} \geq \frac{\nu}{2}c > 0,$$

where $c = \lambda_1(A) > 0$ is the constant in inequality (1.9). Hence, the nonreal λ cannot be too small in absolute value.

Other properties pertaining to the spectrum of problem (1.7) will be illustrated by two examples.

9.1.3 On the Spectrum Structure of Normal Oscillations

Problem (1.7) has been previously investigated in details by using examples that admit the complete division of variables. Let us recall the main results of this investigation in the case of the normal oscillations for a viscous capillary self-gravitating drop and also in the case of the plane problem for a rotating fluid ring.

1° For arbitrary viscosity ν, the spectrum of the two above-mentioned problems is discrete and has a unique limit point $\lambda = \infty$.

2° In the absence of rotation, that is when $\omega_0 = 0$, and for sufficiently high viscosity ν, all the eigenvalues λ are real and positive.

3° For arbitrary viscosity ν and $\omega_0 = 0$, the nonreal eigenvalues are located in pairs symmetrically relatively to the real axis and there is no more than a finite number of such pairs that depend on viscosity.

4° By taking the rotation into account, that is $\omega_0 \neq 0$, the eigenvalues move away from the real axis and the pairs of nonreal eigenvalues are not symmetrical anymore, but this fact does not qualitatively change the general structure of the spectrum. Those eigenvalues that are large in absolute value move closer to the real axis.

It is natural to expect that these conclusions formulated for two concrete problems will hold true for problem (1.7) on the oscillations of a capillary viscous fluid that fills an arbitrary container. Some of these conclusions will be either proved or considered in details in Section 9.2.

9.2 Oscillations of Capillary Fluids in Arbitrary Containers

In this section we will consider the initial boundary value problem and the spectral problem on normal oscillations of a rotating viscous capillary fluid partially filling a container. Our investigation of these interesting problems will be different from the one considered by A. D. Myshkis, V. G. Babskii, N. D. Kopachevsky, L. A. Slobozhanin, and A. D. Tyuptsov in their monograph [1]. In [1], the authors considered only the case when the equilibrium free surface Γ of the fluid and its solid boundary S do not have any common points. This situation can occur under certain conditions that are close to low gravity, namely, the capillary and the self-gravitational forces keep the fluid in a stable state, and the fluid occupies a region of a spheric type with an internal solid boundary S and the external free surface Γ. Oceans are typical examples that illustrate this situation. They cover the Earth in layers of constant or variable thickness (because of the influence of the centrifugal forces). Since in this case $\partial\Gamma = \overline{\Gamma} \cap \overline{S} = \emptyset$, then, instead of conditions (1.5′) or (1.5″) for the operator B_σ we need to take into consideration the boundedness of the solutions of equation $B_\sigma u = v$ on the whole surface Γ as an additional condition.

9.2.1 Transition to a System of Operator Equations

Let us consider the problem (1.1)–(1.6) with the boundary condition (1.5′) and assume that the field of external forces $\boldsymbol{f}(t, x)$ is a function of the variable t with values in

$\boldsymbol{J}_{0,S}(\Omega)$. As in Section 8.1, let us assume that the velocity field $\boldsymbol{u}(t,x)$ is a function with values in $\boldsymbol{J}^1_{0,S}(\Omega)$.

Let us represent $\boldsymbol{u}(t,x)$ as a sum, that is, $\boldsymbol{u}(t,x) = \boldsymbol{s}(t,x) + \boldsymbol{w}(t,x)$, where $\boldsymbol{s}(t,x)$ is a solution of the boundary value problem I in Section 8.1.2 in which we substitute $\boldsymbol{f}$ for $\boldsymbol{f} + 2\omega_0\boldsymbol{u} \times \boldsymbol{e}_3 - \mathrm{d}\boldsymbol{u}/\mathrm{d}t$, and $\boldsymbol{w}(t,x)$ is a solution of the boundary value problem II for $\psi = -\sigma B_\sigma \zeta$. A similar reasoning as in Section 8.4.2 leads us to a system of operator equations of the form (8.4.6) in which we have substituted B_0 for σB_σ:

$$\nu A^{1/2}\boldsymbol{s} + A^{-1/2}\left(\frac{\mathrm{d}\boldsymbol{u}}{\mathrm{d}t} - 2\mathrm{i}\omega_0 S_0 \boldsymbol{u} - \boldsymbol{f}(t)\right) = \boldsymbol{0}, \tag{2.1}$$

$$\boldsymbol{u}(t) = \boldsymbol{s}(t) + \boldsymbol{w}(t), \qquad \frac{\mathrm{d}\zeta}{\mathrm{d}t} = \gamma_n \boldsymbol{u}, \qquad \nu\boldsymbol{w} + \sigma T B_\sigma \zeta = \boldsymbol{0}. \tag{2.2}$$

Let us recall that

$$S_0\boldsymbol{u} := -\mathrm{i}P_{0,S}(\boldsymbol{u} \times \boldsymbol{e}_3), \qquad S_0 = S_0^*, \qquad \|S_0\| = 1. \tag{2.3}$$

All the other notations in (2.1) and (2.2) are the same as in Chapter 8 and the operator B_σ was defined in Section 9.1.

In (2.1) and (2.2), let us omit the function $\zeta(t)$ and make the following substitutions:

$$\boldsymbol{s} = A^{-1/2}\boldsymbol{\xi}, \qquad \boldsymbol{w} = A^{-1/2}\boldsymbol{\eta}, \qquad \boldsymbol{u} = A^{-1/2}\boldsymbol{\delta}. \tag{2.4}$$

In addition, by formally applying the operator A to the left of the first equation (this fact will be proved in the sequel) and the operator $A^{1/2}$ to the second equation, we obtain the system of two ordinary differential equations of the first order

$$\begin{aligned} &\frac{\mathrm{d}\boldsymbol{\xi}}{\mathrm{d}t} + \nu A\boldsymbol{\xi} = \sigma\nu^{-1}B(\boldsymbol{\xi}+\boldsymbol{\eta}) + 2\mathrm{i}\omega_0 A^{1/2}S_0A^{-1/2}(\boldsymbol{\xi}+\boldsymbol{\eta}) + A^{1/2}\boldsymbol{f}(t),\\ &\frac{\mathrm{d}\boldsymbol{\eta}}{\mathrm{d}t} = -\sigma\nu^{-1}B(\boldsymbol{\xi}+\boldsymbol{\eta}), \end{aligned} \tag{2.5}$$

$$B := Q^*B_\sigma Q, \qquad Q^* = A^{1/2}T, \qquad Q = \gamma_n A^{-1/2}. \tag{2.6}$$

The following initial conditions

$$\boldsymbol{\eta}(0) = -\sigma\nu^{-1}A^{1/2}TB_\sigma\zeta^0, \qquad \boldsymbol{\xi}(0) = A^{1/2}\boldsymbol{u}^0 - \boldsymbol{\eta}(0) \tag{2.7}$$

that follow from conditions (1.6) and from (2.2) and (2.4) should be added to equations (2.5).

Thus, the initial boundary value problem (1.1)–(1.6) on small movements of a capillary viscous fluid in an arbitrary container is reduced to the Cauchy problem (2.5)–(2.7). Here, $\boldsymbol{\xi}(t)$ is a function with values in the space $\boldsymbol{J}_{0,S}(\Omega)$ and $\boldsymbol{\eta}(t)$ is a function with values in the subspace $\boldsymbol{M}_0(\Omega) \subset \boldsymbol{J}_{0,S}(\Omega)$ defined in Chapter 8 (see the orthogonal decompositions (8.3.23) and (8.3.24)).

To conclude this section, let us point out the following remark which will be of importance in the sequel. As it was mentioned in Section 9.1, a capillary viscous fluid situated in a container in a state of relative equilibrium is influenced both by the gravitation and the centrifugal forces that are caused by the rotation of the system. Since these forces are infinitely differentiable functions, then the function that defines the free surface Γ of a fluid in a nonperturbed state, is infinitely differentiable, too. Further, we will assume that the wetting line $\partial\Gamma$ is also a sufficiently smooth curve.

9.2.2 Normal Oscillations. Properties of the Operators of the Oscillations Problem

In (2.5), let us assume $\boldsymbol{f}(t,x) \equiv \boldsymbol{0}$ and consider those solutions that depend on time according to the law $\exp(-\lambda t)$. We have

$$\nu A\boldsymbol{\xi} - \sigma\nu^{-1}B(\boldsymbol{\xi}+\boldsymbol{\eta}) - 2\mathrm{i}\omega_0 A^{1/2}S_0A^{-1/2}(\boldsymbol{\xi}+\boldsymbol{\eta}) = \lambda\boldsymbol{\xi},$$
$$\sigma\nu^{-1}(\boldsymbol{\xi}+\boldsymbol{\eta}) = \lambda\boldsymbol{\eta}. \tag{2.8}$$

Here, the operator B is defined by (2.6) and acts on the space $\boldsymbol{M}_0(\Omega) \subset \boldsymbol{J}_{0,S}(\Omega)$ because the elements $\boldsymbol{v}$ from the orthogonal complement $\boldsymbol{N}_0(\Omega)$ that can be represented as $\boldsymbol{v} = A^{1/2}\boldsymbol{u}$, with $\boldsymbol{u} \in \boldsymbol{N}_1(\Omega)$ and $\gamma_n\boldsymbol{u} = 0$, are turned to zero by B. Therefore, we have

$$B = I_0B = I_0BI_0, \tag{2.9}$$

where $I_0 := Q_{\boldsymbol{M}_0}$ is the orthoprojector onto $\boldsymbol{M}_0(\Omega)$.

To investigate the properties of the operator B, let us consider the eigenvalue problem

$$\sigma B\boldsymbol{\eta} := \sigma A^{1/2}TB_\sigma\gamma_nA^{-1/2}\boldsymbol{\eta} = \lambda\boldsymbol{\eta}, \qquad \boldsymbol{\eta} \in \boldsymbol{M}_0(\Omega). \tag{2.10}$$

Taking into account the definitions of the operators in (2.10) and the equations of the boundary value problem II, we conclude that the function $\boldsymbol{w} = A^{-1/2}\boldsymbol{\eta} \in \boldsymbol{M}_1(\Omega) \subset \boldsymbol{J}^1_{0,S}(\Omega)$ should be a solution of the following spectral boundary problem,

$$-\Delta\boldsymbol{w} + \nabla p = \boldsymbol{0}, \quad \mathrm{div}\boldsymbol{w} = 0 \text{ in } \Omega,$$
$$\boldsymbol{w} = \boldsymbol{0} \text{ on } S, \quad w_{j,3} + w_{3,j} = 0 \text{ on } \Gamma, \qquad j = 1,2,$$
$$\lambda(-p + 2w_{3,3}) = \sigma B_\sigma w_n \text{ on } \Gamma. \tag{2.11}$$

Since by (1.8) the operator B_σ has an inverse B_σ^{-1}, which is a compact positive operator, we can apply B_σ^{-1} to both the left and right side of the last condition in (2.11). Thus we obtain the equivalent spectral problem

$$\begin{aligned} &-\Delta \boldsymbol{w} + \nabla p = \boldsymbol{0}, \quad \operatorname{div} \boldsymbol{w} = 0 \text{ in } \Omega, \\ &\boldsymbol{w} = \boldsymbol{0} \text{ on } S, \quad w_{j.3} + w_{3.j} = 0 \text{ on } \Gamma, \qquad j = 1, 2, \\ &\lambda \sigma^{-1} B_\sigma^{-1}(-p + 2w_{3.3}) = w_n \text{ on } \Gamma. \end{aligned} \tag{2.12}$$

It is easy to prove using the Green formula (2.2.10) and condition (1.8) that the eigenvalues λ of problem (2.11) are positive and can be obtain from the solution $\boldsymbol{w}$ as ratios of the form

$$\lambda = \frac{\sigma \left\| B_\sigma^{1/2} w_n \right\|^2_{L_{2.\Gamma}}}{\|A^{1/2}\boldsymbol{w}\|^2_{\boldsymbol{M}_0(\Omega)}} = \frac{\int_\Gamma \left[\sigma \nabla_\Gamma(w_n, w_n) + a|w_n|^2\right] d\Gamma}{E(\boldsymbol{w}, \boldsymbol{w})}. \tag{2.13}$$

For problem (2.12) we have, respectively,

$$\frac{1}{\lambda} = \frac{\sigma^{-1} \left\| B_\sigma^{-1/2} \partial \boldsymbol{w} \right\|^2_{L_{2.\Gamma}}}{E(\boldsymbol{w}, \boldsymbol{w})}, \qquad \partial \boldsymbol{w} := -p + 2w_{3.3}. \tag{2.14}$$

Moreover, let us note that since $T^{-1} = \partial$ (for elements in $\boldsymbol{M}_1(\Omega)$, see the abstract scheme in Section 1.8), problem (2.12) can be rewritten as

$$\sigma^{-1} B_\sigma^{-1} \boldsymbol{w} = \tilde{\lambda} \gamma_n \boldsymbol{w}, \qquad \boldsymbol{w} \in \boldsymbol{M}_1(\Omega),\ \tilde{\lambda} := \lambda^{-1}, \tag{2.15}$$

which is equivalent to the eigenvalue problem for the operator $\sigma^{-1}B^{-1}$.

It can be proved that the variational relation (2.14) generates a compact positive operator that is obviously equal to $\sigma^{-1}B^{-1}$. It is also easy to prove the compactness of B^{-1} if $\overline{\Gamma} \cap \overline{S} = \emptyset$ and Γ and S are sufficiently smooth. Indeed, if Γ and S are sufficiently smooth, then the solutions $\boldsymbol{w}$ of problem (2.11) that belong to the class of the so-called elliptic Duglis-Nirenberg equations (see the articles by V. A. Solonnikov [2]) are infinitely differentiable in any subregion Ω' of Ω that is internal for Ω and in subregions Ω_S and Ω_Γ adjacent to either S or Γ, respectively. These facts follow from the properties of the solutions of equation (2.11) that were investigated in details in O. A. Ladyzhenskaja's book [1] and in the article by V. A. Solonnikov and V. E. Schadilov [1].

Using these properties and the formula

$$\sigma^{-1}B^{-1} = A^{1/2}(\gamma_n)^{-1}\left(\sigma^{-1}B_\sigma^{-1}\right)\partial \cdot A^{-1/2}, \qquad \partial = T^{-1},$$

we obtain that for any $\boldsymbol{\eta} \in \boldsymbol{M}_0(\Omega)$

$$A^{-1/2}\boldsymbol{\eta} \in \boldsymbol{M}_1(\Omega), \qquad \partial(A^{-1/2}\boldsymbol{\eta}) \in H_\Gamma^{-1/2}, \qquad \sigma^{-1}B_\sigma^{-1}\partial(A^{-1/2}\boldsymbol{\eta}) \in H_\Gamma^{3/2},$$

and

$$\boldsymbol{w} := (\gamma_n)^{-1}\left(\sigma^{-1}B_\sigma^{-1}\right)\partial(A^{-1/2}\boldsymbol{\eta}) \in \boldsymbol{H}^2(\Omega).$$

Finally,

$$\sigma^{-1}B^{-1}\boldsymbol{\eta} = A^{1/2}\boldsymbol{w} \in \boldsymbol{M}_1(\Omega) \subset \mathcal{D}(A^{1/2}) = \boldsymbol{J}_{0,S}^1(\Omega).$$

Since $\boldsymbol{M}_0(\Omega) = A^{1/2}\boldsymbol{M}_1(\Omega)$ and $A^{-1/2}$ is a compact operator in $\boldsymbol{J}_{0,S}(\Omega)$, then $\sigma^{-1}B^{-1}$ is a compact operator in $\boldsymbol{M}_0(\Omega)$.

In the more interesting case when $\overline{\Gamma} \cap \overline{S} = \partial\Gamma \neq \emptyset$, Γ and S are smooth, and the boundary condition (1.5′) is satisfied on $\partial\Gamma$, a similar relation as the one in (2.14) generates a compact positive operator $\sigma^{-1}B^{-1}$ as well. This fact was proved by T. A. Suslina in [3], [6] (see also [4], [5], and [7]) using the theory of pseudodifferential operators. For the eigenvalues $\tilde{\lambda}_k$, Suslina obtained an asymptotic formula that for $\lambda_k = \lambda_k(B)$ leads to

$$\lambda_k(B) = \left(\frac{1}{\pi}\operatorname{mes}\Gamma\right)^{-1/2} k^{1/2}[1 + o(1)], \qquad k \to \infty. \tag{2.16}$$

Hence, it follows that the operator B^{-1} belongs to the class of compact operators $\mathfrak{S}_p$ for $p > 2$.

The same reasoning shows that the eigenelements $\{\eta_k(B)\}_{k=1}^\infty$ of the operator B form an orthogonal basis in $\boldsymbol{M}_0(\Omega)$ and, therefore, the solutions $\{\boldsymbol{w}_k\}_{k=1}^\infty$ of the spectral problems (2.11) and (2.12) form an orthogonal basis in the subspace $\boldsymbol{M}_1(\Omega) \subset \boldsymbol{J}_{0,S}^1(\Omega)$.

Let us recall that A, the operator in problem (2.8), has a compact positive inverse A^{-1} in the space $\boldsymbol{J}_{0,S}(\Omega)$, whose eigenvalues behave according to (8.1.12), that is,

$$\lambda_k(A^{-1}) = \left(\frac{1}{3\pi^2}\operatorname{mes}\Omega\right)^{2/3} k^{-2/3}[1 + o(1)], \qquad k \to \infty, \tag{2.17}$$

and, therefore, $A^{-1} \in \mathfrak{S}_p$ for $p > 3/2$. The eigenelements $\{\boldsymbol{s}_k(A)\}_{k=1}^\infty$ of operator A form an orthogonal basis in $\boldsymbol{J}_{0,S}^1(\Omega)$.

Let us rewrite (2.8) in some other equivalent forms. Since $\boldsymbol{\xi} \in \boldsymbol{J}_{0,S}(\Omega)$ and $\boldsymbol{\eta} \in \boldsymbol{M}_0(\Omega)$, let us assume that

$$y := (\boldsymbol{\xi}, \boldsymbol{\eta})^t \in E := \boldsymbol{J}_{0,S}(\Omega) \oplus \boldsymbol{M}_0(\Omega).$$

From (2.9), we infer that the left side of (2.8) is the result of applying the unbounded matrix operator

$$\mathcal{A} := \mathcal{A}_0 - 2\mathrm{i}\omega_0\mathcal{F}, \tag{2.18}$$

with

$$\mathcal{A}_0 := \begin{pmatrix} \nu A - \sigma\nu^{-1}B & -\sigma\nu^{-1}B \\ \sigma\nu^{-1}B & \sigma\nu^{-1}B \end{pmatrix},$$
$$\mathcal{F} := \begin{pmatrix} A^{1/2}S_0A^{-1/2} & A^{1/2}S_0A^{-1/2} \\ 0 & 0 \end{pmatrix}$$

that acts in $\boldsymbol{E}$, to the elements $y \in \mathcal{D}(\mathcal{A}) \subset \boldsymbol{E}$. Here, the orthoprojectors $I_0 = Q_{\boldsymbol{M}_0}$ that characterize the properties (2.9) are not written down in order to simplify the form of (2.18).

In (2.8), let us make the following substitutions:

$$\nu^{1/2}A^{1/2}\boldsymbol{\xi} = \boldsymbol{\varphi}, \qquad \left(\frac{\sigma}{\nu}\right)^{1/2} B^{1/2}\boldsymbol{\eta} = \boldsymbol{\psi}. \tag{2.19}$$

Therefore,

$$\begin{aligned}
&\left(\nu^{1/2}A^{1/2} - \left(\frac{\sigma}{\nu}\right)\nu^{-1/2}BA^{-1/2}\right)\boldsymbol{\varphi} - \left(\frac{\sigma}{\nu}\right)^{1/2} B^{1/2}\boldsymbol{\psi} \\
&\qquad - 2\mathrm{i}\omega_0 A^{1/2}S_0A^{-1/2}\left(\nu^{-1/2}A^{-1/2}\boldsymbol{\varphi} + \left(\frac{\sigma}{\nu}\right)^{-1/2} B^{-1/2}\boldsymbol{\psi}\right) \\
&\quad = \lambda\nu^{-1/2}A^{-1/2}\boldsymbol{\varphi}, \\
&\left(\frac{\sigma}{\nu}\right) B(\nu^{-1/2}A^{-1/2}\boldsymbol{\varphi}) + \left(\frac{\sigma}{\nu}\right)^{1/2} B^{1/2}\boldsymbol{\psi} \\
&\quad = \lambda\left(\frac{\sigma}{\nu}\right)^{-1/2} B^{-1/2}\boldsymbol{\psi}.
\end{aligned} \tag{2.20}$$

Applying now the operator $\nu^{-1/2}A^{-1/2}$ to the first equation and the operator $(\sigma/\nu)^{-1/2}B^{-1/2}$ to the second equation in (2.20) we obtain instead a system of two equations, or a vector-matrix equation of the form

$$(\mathcal{I} - \mathcal{R} - 2\mathrm{i}\omega_0\mathcal{T}_0)z = \lambda\mathcal{D}z, \qquad z = (\boldsymbol{\varphi}; \boldsymbol{\psi})^{\mathrm{t}} \in \boldsymbol{E}, \tag{2.21}$$

with

$$\begin{aligned}
&\mathcal{I} := \operatorname{diag}(I; I_0), \\
&\mathcal{D} := \operatorname{diag}\left(\nu^{-1}A^{-1}; \left(\frac{\sigma}{\nu}\right)^{-1} B^{-1}\right), \\
&\mathcal{R} := \begin{pmatrix} \dfrac{\sigma}{\nu^2}R^*R & \dfrac{\sigma^{1/2}}{\nu}R^* \\ -\dfrac{\sigma^{1/2}}{\nu}R & 0 \end{pmatrix}, \\
&\mathcal{T}_0 := \begin{pmatrix} \nu^{-1}S_0A^{-1} & \sigma^{-1/2}S_0A^{-1/2}B^{-1/2} \\ 0 & 0 \end{pmatrix},
\end{aligned} \tag{2.22}$$

where

$$R := B^{1/2}A^{-1/2}, \qquad R^* := A^{-1/2}B^{1/2}, \qquad I_0 = Q_{M_0}. \tag{2.23}$$

Let us find out the general properties of the operators involved in problem (2.21). Since A^{-1} is a compact positive operator from the class $\mathfrak{S}_p$ for $p > 3/2$ and B^{-1} is an operator with the same properties from the class $\mathfrak{S}_p$ for $p > 2$, the operator $\mathcal{D} = \operatorname{diag}(\nu^{-1}A^{-1}; (\sigma/\nu)^{-1}B^{-1})$ is compact and positive and belongs to the class $\mathfrak{S}_p$ for $p > 2$. Further, since $S_0 \in \mathcal{L}(\boldsymbol{J}_{0.S}(\Omega))$, $A^{-1} \in \mathfrak{S}_\infty$, and $B^{-1} \in \mathfrak{S}_\infty$, the operator matrix $\mathcal{T}_0$ in (2.23) is a compact operator acting in $\boldsymbol{E} = \boldsymbol{J}_{0.S}(\Omega) \oplus \boldsymbol{M}_0(\Omega)$.

A remarkable property is the following:

The mutually adjoint operators R^* and R acting from $\boldsymbol{M}_0(\Omega)$ to $\boldsymbol{J}_{0,S}(\Omega)$ and from $\boldsymbol{J}_{0,S}(\Omega)$ to $\boldsymbol{M}_0(\Omega)$, respectively, are compact.

To prove this statement, let us consider the operator R^*R, which, in virtue of (2.6), equals

$$R^*R = A^{-1/2}BA^{-1/2} = A^{-1/2}\left(A^{1/2}TB_\sigma\gamma_n A^{-1}\right). \tag{2.24}$$

Let us consider first the case when $\overline{\Gamma} \cap \overline{S} = \emptyset$ and both the surfaces Γ and S are sufficiently smooth. Then, for any $\boldsymbol{\xi} \in \boldsymbol{J}_{0.S}(\Omega)$, we have

$$A^{-1}\boldsymbol{\xi} \in \mathcal{D}(A) \subset \boldsymbol{H}^2(\Omega), \qquad \gamma_n A^{-1}\boldsymbol{\xi} \in H_\Gamma^{3/2}, \qquad B_\sigma\gamma_n A^{-1}\boldsymbol{\xi} \in H_\Gamma^{-1/2},$$

and, therefore,

$$T\left(B_\sigma\gamma_n A^{-1}\boldsymbol{\xi}\right) \in \boldsymbol{M}_1(\Omega), \qquad A^{1/2}T\left(B_\sigma\gamma_n A^{-1}\right)\boldsymbol{\xi} \in \boldsymbol{M}_0(\Omega) \subset \boldsymbol{J}_{0,S}(\Omega).$$

This proves that the operator $A^{1/2}TB_\sigma\gamma_n A^{-1}$ acts boundedly from $\boldsymbol{J}_{0,S}(\Omega)$ to $\boldsymbol{M}_0(\Omega)$ and since $A^{-1/2} \in \mathfrak{S}_\infty$, then R^*R is a compact, self-adjoint nonnegative operator. Hence, R^* and R are compact operators.

If $\overline{\Gamma} \cap \overline{S} = \partial\Gamma \neq \emptyset$ and condition (1.5′) is satisfied, then we have the same properties of compactness for the operators R^* and R for $0 < \delta < \delta_* \approx 0.354\pi$. Here, the eigenvalues $\lambda_k(R^*R)$ of the operator R^*R are the successive minima of the variational ratio

$$\frac{\|R\boldsymbol{\xi}\|^2_{\boldsymbol{J}_{0.S}(\Omega)}}{\|\boldsymbol{\xi}\|^2_{\boldsymbol{J}_{0.S}(\Omega)}} = \frac{\left\|B_\sigma^{1/2}\gamma_n \boldsymbol{v}\right\|^2_{L_{2.\Gamma}}}{\|A\boldsymbol{v}\|^2_{\boldsymbol{J}_{0.S}(\Omega)}}, \qquad \boldsymbol{v} \in A^{-1}\boldsymbol{\xi}. \tag{2.25}$$

In classical terms, we are talking about the following variational ratio

$$\frac{\sigma^{-1}\displaystyle\int_{\Gamma}\left[\sigma\nabla_{\Gamma}(v_n,v_n)+a|v_n|^2\right]\mathrm{d}\Gamma}{\displaystyle\int_{\Omega}|\boldsymbol{\xi}|^2\mathrm{d}\Omega}, \tag{2.26}$$

where $\boldsymbol{v}$ can be found using $\boldsymbol{\xi}\in\boldsymbol{J}_{0.S}(\Omega)$ as a solution of the boundary value problem I,

$$\begin{aligned}
&-\Delta\boldsymbol{v}+\nabla p=\boldsymbol{\xi},\quad \operatorname{div}\boldsymbol{v}=0\text{ in }\Omega,\\
&\boldsymbol{v}=\boldsymbol{0}\text{ on }S,\\
&v_{j.3}+v_{3.j}=0,\quad j=1,2,\qquad -p+2v_{3.3}=0\text{ on }\Gamma.
\end{aligned} \tag{2.27}$$

For $0<\delta<\delta_*$, the eigenvalues $\lambda_k=\lambda_k(R^*R)$ of the variational problem (2.26) and (2.27) have the following asymptotic behavior

$$\lambda_k(R^*R)=\left(\frac{c}{k}\right)^{1/2}[1+\mathrm{o}(1)]\text{ as }k\to\infty,\qquad c=\sigma^2\frac{9}{256\pi}\operatorname{mes}\Gamma>0. \tag{2.28}$$

From (2.28) it follows that $R^*R\in\mathfrak{S}_p$ for $p>2$.

Using the previously proved properties we can conclude that in the two main versions of the problem that we have considered (i.e., $\bar\Gamma\cap\bar S=\emptyset$ and $\bar\Gamma\cap\bar S=\partial\Gamma\neq\emptyset$), the operators R^* and R are compact and, therefore, $\mathcal{R}$ in (2.21) and (2.23) is compact as well.

Let us now prove that the operator $\mathcal{I}-2\mathrm{i}\omega_0\mathcal{T}_0-\mathcal{R}$ in (2.21) is invertible. Since $\mathcal{R}$ and $\mathcal{T}_0$ belong to $\mathfrak{S}_\infty$, it is sufficient to prove that the equation

$$(\mathcal{I}-\mathcal{R}-2\mathrm{i}\omega_0\mathcal{T}_0)z=0,\qquad z\in E, \tag{2.29}$$

has only the trivial solution. Writing down (2.29) by components we obtain

$$\begin{aligned}
\varphi-\frac{\sigma}{\nu^2}R^*R\varphi-\frac{\sigma^{1/2}}{\nu}R^*\psi-2\mathrm{i}\omega_0\left(\nu^{-1}S_0A^{-1}\varphi+\sigma^{1/2}S_0A^{-1/2}B^{-1/2}\psi\right)&=\boldsymbol{0},\\
\frac{\sigma^{1/2}}{\nu}R\varphi+\psi&=\boldsymbol{0}.
\end{aligned}$$

From the second equation we obtain that $\psi=-\sigma^{1/2}\nu^{-1}R\varphi$ and from the first equation we get

$$\begin{aligned}
&\varphi-\frac{\sigma}{\nu^2}R^*R\varphi-\frac{\sigma^{1/2}}{\nu}R^*\left(-\frac{\sigma^{1/2}}{\nu}R\varphi\right)\\
&\quad-2\mathrm{i}\omega_0\left(\nu^{-1}S_0A^{-1}\varphi+\frac{1}{\sigma^{1/2}}S_0A^{-1/2}B^{-1/2}\left(-\frac{\sigma^{1/2}}{\nu}B^{1/2}A^{-1/2}\varphi\right)\right)=\varphi=\boldsymbol{0}.
\end{aligned}$$

Finally, we notice that (2.29) has only the trivial solution and, therefore, there exists an inverse operator

$$(\mathcal{I} - \mathcal{R} - 2\mathrm{i}\omega_0 \mathcal{T}_0)^{-1} = \mathcal{I} + \Phi, \qquad \Phi \in \mathfrak{S}_\infty. \tag{2.30}$$

Applying the operator $\mathcal{I} + \Phi$ to (2.21) we obtain

$$(\mathcal{I} - \lambda(\mathcal{I} + \Phi)\mathcal{D})z = 0. \tag{2.31}$$

Let us derive one more form of the equation of the problem on normal oscillations. A direct proof shows (see Section 9.3.4) that the unbounded operator $\mathcal{A}$ in (2.18) has an inverse operator $\mathcal{A}^{-1}$ that can be calculated using the following formula,

$$\begin{aligned}
\mathcal{A}^{-1} &= \mathcal{A}_0^{-1} + 2\mathrm{i}\omega_0 \mathcal{F}_{-1}, \\
\mathcal{A}_0^{-1} &= \begin{pmatrix} \nu^{-1}A^{-1} & \nu^{-1}A^{-1}I_0 \\ -\nu^{-1}I_0A^{-1} & \sigma\nu^{-1}B^{-1} - \nu^{-1}I_0A^{-1}I_0 \end{pmatrix}, \\
\mathcal{F}_{-1} &= \mathcal{F}_{-1}\left(\omega_0; A^{-1/2}, B^{-1}, S_0\right), \qquad \mathcal{F}_{-1}\Big|_{\omega_0=0} = 0.
\end{aligned} \tag{2.32}$$

Since B^{-1} and A^{-1} are compact and S_0 is a bounded operator ($\|S_0\| = 1$), then $\mathcal{A}^{-1}$ is a compact operator as well. In this case, the eigenvalue problem (2.8) is equivalent to the problem on the spectrum of a Fredholm linear pencil:

$$(\mathcal{I} - \lambda\mathcal{A}^{-1})y = 0, \qquad y = (\boldsymbol{\xi}, \boldsymbol{\eta})^{\mathrm{t}} \in \boldsymbol{J}_{0.S}(\Omega) \oplus \boldsymbol{M}_0(\Omega) = \boldsymbol{E}. \tag{2.33}$$

9.2.3 Normal Oscillations of a Rotating Fluid

Using equations (2.8), (2.21), and (2.33), let us consider now the properties of normal oscillations of a fluid in a rotating container.

1° Under the boundary condition $u_n = 0$ (on $\partial\Gamma$) problem (1.7) has a discrete spectrum $\{\lambda_j\}_{j=1}^{\infty}$ that is located in the right half-plane and has a unique limit point $\lambda = \infty$. This fact is caused by the fundamental influence of the capillary forces on the spectrum structure. These forces exclude the limit point $\lambda = 0$ which appeared in the problems studied in Chapter 8 for $\sigma = 0$.

Indeed, since the operator pencil (2.33) is invertible for $\lambda = 0$, does not have any singularities at finite points of the complex plane, and the operator $\mathcal{A}^{-1} \in \mathfrak{S}_\infty$ is infinite-dimensional, Assertion 1° follows from the same type of consideration as in Section 1.6.4.

2° For any $\varepsilon > 0$, all the eigenvalues λ_j, with the exception of a finite number of them, are located in the sector $|\arg\lambda| < \varepsilon$, i.e., they are close to the positive half-axis.

Indeed, (2.31) is an eigenvalue problem for $\tilde{\lambda} = 1/\lambda$ for the weakly perturbed self-adjoint operator $(\mathcal{I} + \Phi)\mathcal{D}$, where $0 < \mathcal{D} \in \mathfrak{S}_p$ for $p > 2$, and $\mathcal{I} + \Phi$ is invertible, $\Phi \in \mathfrak{S}_\infty$. That is why Assertion 2° follows from the first Keldysh theorem (see Section 1.6.5).

At the same time, the next third property can be obtained.

3° The set $\{z_{j,q}\}_{j=1}^{\infty}$ containing the eigen- and associated elements $z_{j,q} = (\boldsymbol{\varphi}_{j,q}; \boldsymbol{\psi}_{j,q})^{\mathrm{t}} \in \boldsymbol{E} = \boldsymbol{J}_{0,S}(\Omega) \oplus \boldsymbol{M}_0(\Omega)$ forms a complete system in the space $\boldsymbol{E}$.

The fourth property is a corollary of Assertion 3°.

4° The set $\{y_{j,q}\}_{j=1}^{\infty}$, $y_{j,q} = (\boldsymbol{\xi}_{j,q}; \boldsymbol{\eta}_{j,q})^{\mathrm{t}}$, with $\boldsymbol{\xi}_{j,q} = \nu^{-1/2}A^{-1/2}\boldsymbol{\varphi}_{j,q}$ and $\boldsymbol{\eta}_{j,q} = (\sigma/\nu)^{-1/2}B^{1/2}\psi_{j,q}$, forms a complete system of elements both in the space $\mathcal{D}(A^{1/2}) \oplus \mathcal{D}(B^{1/2})$ and in the space $\boldsymbol{J}_{0,S}^1(\Omega) \oplus \boldsymbol{M}_0(\Omega)$. Thus, there is a set $\{\boldsymbol{u}_{j,q}\}_{j=1}^{\infty}$ of solutions to problem (1.7) that can be represented as the sum $\boldsymbol{u}_{j,q} = \boldsymbol{s}_{j,q} + \boldsymbol{w}_{j,q}$, with $\{(\boldsymbol{s}_{j,q}; \boldsymbol{w}_{j,q})^{\mathrm{t}}\}_{j=1}^{\infty}$ a complete system in $\mathcal{D}(A) \oplus \boldsymbol{M}_1(\Omega)$ and $\{(\boldsymbol{s}_{j,q}; \gamma_n\boldsymbol{w}_{j,q})^{\mathrm{t}}\}_{j=1}^{\infty}$ a complete system in the space $\mathcal{D}(A) \oplus \mathcal{D}(B_\sigma^{1/2})$, where $\mathcal{D}(A) \subset \boldsymbol{J}_{0,S}^1(\Omega)$, and $\mathcal{D}(B_\sigma^{1/2}) = H_{0,\Gamma}^1(\Omega)$ (for $\bar{\Gamma} \cap \bar{S} = \partial\Gamma \neq \emptyset$) and $\mathcal{D}(B_\sigma^{1/2}) = H_\Gamma^1(\Omega)$ (for $\bar{\Gamma} \cap \bar{S} = \emptyset$).

The proof of Property 4° follows from Property 3° by using the substitutions (2.19) and (2.4). Indeed, the completeness of the set of elements $\{y_{j,q}\}_{j=1}^{\infty}$ in $\mathcal{D}(A^{1/2}) \oplus \mathcal{D}(B^{1/2})$ is a direct corollary of formulas (2.19). Further, according to the previous considerations, the set $\mathcal{D}(B^{1/2}) \subset \boldsymbol{M}_0(\Omega)$ is dense in $\boldsymbol{M}_0(\Omega)$. This fact proves that the set $\{y_{j,q}\}_{j=1}^{\infty}$ is complete in $\boldsymbol{J}_{0,S}^1(\Omega) \oplus \boldsymbol{M}(\Omega)$ (because $\mathcal{D}(A^{1/2}) = \boldsymbol{J}_{0,S}^1(\Omega)$). Now using the substitutions (2.4) and the equality $A^{-1/2}\boldsymbol{M}_0(\Omega) = \boldsymbol{M}_1(\Omega)$ we prove the completeness of the set of elements $\{(\boldsymbol{s}_{j,q}; \boldsymbol{w}_{j,q})^{\mathrm{t}}\}_{j=1}^{\infty}$ in $\mathcal{D}(A) \oplus \boldsymbol{M}_1(\Omega)$.

Let us note next that the elements $\boldsymbol{\eta}_{j,q} = A^{1/2}\boldsymbol{w}_{j,q}$ have the property $\boldsymbol{\eta}_{j,q} \in \mathcal{D}(B^{1/2})$ and, therefore,

$$\left\|B^{1/2}\boldsymbol{\eta}_{j,q}\right\|_{\boldsymbol{J}_{0,S}(\Omega)}^2 = \left(B^{1/2}\boldsymbol{\eta}_{j,q}, B^{1/2}\boldsymbol{\eta}_{j,q}\right)_{\boldsymbol{J}_{0,S}(\Omega)} < \infty.$$

Let us recall that $\mathcal{D}(B^{1/2})$ can be obtained by closing the set $\mathcal{D}(B)$ in the norm $\|B^{1/2}\boldsymbol{\eta}\|_{\boldsymbol{J}_{0,S}(\Omega)}$, and $B := A^{1/2}TB_\sigma\gamma_n A^{-1/2}$, $(A^{1/2}T)^* = \gamma_n A^{-1/2}$. Therefore, for elements $\boldsymbol{\eta}$ in $\mathcal{D}(B)$, we have

$$(B^{1/2}\boldsymbol{\eta}, B^{1/2}\boldsymbol{\eta})_{\boldsymbol{J}_{0,S}(\Omega)} = (B\boldsymbol{\eta}, \boldsymbol{\eta})_{\boldsymbol{J}_{0,S}(\Omega)} = \left\|B_\sigma^{1/2}\gamma_n A^{-1/2}\boldsymbol{\eta}\right\|_{L_{2,\Gamma}}^2 < \infty.$$

Hence, after closing we obtain

$$\|B^{1/2}\boldsymbol{\eta}\|^2_{\boldsymbol{J}_{0,S}(\Omega)} = \left\|B_\sigma^{1/2}\gamma_n A^{-1/2}\boldsymbol{\eta}\right\|^2_{L_{2,\Gamma}} < \infty$$

for all elements from $\mathcal{D}(B^{1/2})$.

Hence, the elements $\boldsymbol{\eta}_{j,q} = A^{1/2}\boldsymbol{w}_{j,q}$ have the property $B_\sigma^{1/2}\gamma_n A^{-1/2}\boldsymbol{\eta}_{j,q} = B_\sigma^{1/2}\gamma_n \boldsymbol{w}_{j,q} \in L_{2,\Gamma}$ and, therefore, $\gamma_n w_{j,q} \in \mathcal{D}(B_\sigma^{1/2})$. Assertion 4° follows now from the latter if we take into account that $\mathcal{D}(B_\sigma^{1/2}) = H^1_{0,\Gamma}(\Omega)$ for $\bar\Gamma \cap \bar S = \partial\Gamma \neq \emptyset$ and $\mathcal{D}(B_\sigma^{1/2}) = H^1_\Gamma(\Omega)$ for $\bar\Gamma \cap \bar S = \emptyset$.

5° For $j \to \infty$, the eigenvalues λ_j of problem (1.7) have the following asymptotic behavior,

$$\lambda_j = \frac{\sigma}{\nu}\lambda_j(B)[1 + \mathrm{o}(1)], \qquad j \to \infty. \tag{2.34}$$

This fact follows from equation (2.21) by taking into account the result in Section 1.6.11 and the equality

$$\lambda_j(\mathcal{D}^{-1}) = \frac{\sigma}{\nu}\lambda_j(B)[1 + \mathrm{o}(1)], \qquad j \to \infty,$$

with

$$\mathcal{D}^{-1} = \mathrm{diag}(\nu A; \sigma\nu^{-1}B),$$

which is a corollary of formulas (2.16) and (2.17) and the diagonal structure of operator $\mathcal{D}^{-1}$.

9.2.4 Normal Oscillations of a Nonrotating Fluid

The results in Section 9.2.3 for a rotating system can also take place for $\omega_0 = 0$, but in this case there are new properties for the solutions, which will be stated in the sequel. Here we are going to make extensive use of some results from the theory of J-self-adjoint operators (see Section 1.3).

1° The spectrum of problems (2.33) and (2.32) for $\omega_0 = 0$, that is, of the problem

$$\begin{pmatrix}\boldsymbol{\xi}\\ \boldsymbol{\eta}\end{pmatrix} = \lambda \begin{pmatrix} \nu^{-1}A^{-1} & \nu^{-1}A^{-1}I_0 \\ -\nu^{-1}I_0A^{-1} & (\sigma/\nu)^{-1}B^{-1} - \nu^{-1}I_0A^{-1}I_0 \end{pmatrix}\begin{pmatrix}\boldsymbol{\xi}\\ \boldsymbol{\eta}\end{pmatrix}, \tag{2.35}$$

is symmetrical relatively to the real axis.

Indeed, the linear operator pencil corresponding to problem (2.35) is given by

$$\mathcal{L}(\lambda) := \mathcal{I} - \lambda\mathcal{A}_0^{-1}, \tag{2.36}$$

where $\mathcal{I} = \operatorname{diag}(I; I_0)$ is the identical operator in $\boldsymbol{E} = \boldsymbol{J}_{0,S}(\Omega) \oplus \boldsymbol{M}_0(\Omega)$ and $\mathcal{A}_0^{-1} \in \mathfrak{S}_\infty$ is the operator defined in (2.32). From (2.35) it follows that $\mathcal{A}_0^{-1}$ is a $\mathcal{J}$-self-adjoint operator,

$$\left(\mathcal{J}\mathcal{A}_0^{-1}\right)^* = \mathcal{J}\mathcal{A}_0^{-1}, \qquad \mathcal{J} := \operatorname{diag}(I; -I_0). \tag{2.37}$$

Therefore, the self-adjoint linear operator pencil corresponding to problem (2.35),

$$\mathcal{J}\mathcal{L}(\lambda) = \mathcal{J} - \lambda \mathcal{J}\mathcal{A}_0^{-1}, \tag{2.38}$$

has a spectrum symmetrical relatively to the real axis.

2° There exist only neutral eigenelements corresponding to the nonreal eigenvalues $\lambda = \lambda_0$, that is, such elements $y = y_0 = (\boldsymbol{\xi}_0; \boldsymbol{\eta}_0)^{\mathrm{t}} \in \boldsymbol{E}$, for which

$$[y_0, y_0] := (\mathcal{J} y_0, y_0)_{\boldsymbol{E}} = \|\boldsymbol{\xi}_0\|^2_{\boldsymbol{J}_{0,S}(\Omega)} - \|\boldsymbol{\eta}_0\|^2_{\boldsymbol{M}_0(\Omega)} = 0. \tag{2.39}$$

Indeed, applying the pencil (2.38) for $\lambda = \lambda_0$ to the element $y = y_0$, we have

$$[y_0, y_0] - \operatorname{Re}\lambda_0[\mathcal{A}_0^{-1} y_0, y_0] - \mathrm{i}\operatorname{Im}\lambda_0[\mathcal{A}_0^{-1} y_0, y_0] = 0;$$

since $[y_0, y_0] \in \mathbb{R}$, $[\mathcal{A}_0^{-1} y_0, y_0] \in \mathbb{R}$, and $\operatorname{Im}\lambda_0 \neq 0$, then $[\mathcal{A}_0^{-1} y_0, y_0] = 0$ and, therefore, $[y_0, y_0] = 0$.

For a real (and positive, in virtue of previous considerations) eigenvalue λ_0, the pair (λ_0, y_0) is called a *positive pair for the operator* $\mathcal{A}_0^{-1}$ if $[y_0, y_0] > 0$. Similarly, (λ_0, y_0) is called *negative* (*neutral*) if $[y_0, y_0] < 0$ ($[y_0, y_0] = 0$).

3° Definite (i.e., positive or negative) pairs of solutions do not have associated elements.

Indeed, let $\lambda_0 \in \mathbb{R}$ and assume that the eigenelement y_0 has an associated element y_1. Then the following relations take place,

$$\mathcal{J} y_0 - \lambda_0 \mathcal{J}\mathcal{A}_0^{-1} y_0 = 0,$$
$$\mathcal{J} y_1 - \lambda_0 \mathcal{J}\mathcal{A}_0^{-1} y_1 - \mathcal{J}\mathcal{A}_0^{-1} y_0 = 0.$$

Multiplying scalarly (in $\boldsymbol{E}$) the second equation by y_0 and using the first equation, we have

$$\begin{aligned}(\mathcal{J} y_1, y_0)_{\boldsymbol{E}} &- \lambda_0(\mathcal{J}\mathcal{A}_0^{-1} y_1, y_0)_{\boldsymbol{E}} \\ &= (y_1, \mathcal{J} y_0 - \lambda_0 \mathcal{J}\mathcal{A}_0^{-1} y_0)_{\boldsymbol{E}} = 0 = (\mathcal{J}\mathcal{A}_0^{-1} y_0, y_0)_{\boldsymbol{E}} = \lambda_0^{-1}(\mathcal{J} y_0, y_0)_{\boldsymbol{E}}.\end{aligned}$$

This is a contradiction because $\lambda_0^{-1} > 0$, $(\mathcal{J} y_0, y_0)_{\boldsymbol{E}} \neq 0$.

Let us represent the nonreal spectrum σ_{non} of problem (2.35) as a union,

$\sigma_{\text{non}} = \Lambda \cup \bar{\Lambda}$, where Λ is the set of nonreal eigenvalues located in the upper half-plane and $\bar{\Lambda}$ is the complex-conjugate set.

4° The operator $\mathcal{A}_0^{-1}$ in (2.35) and (2.36) has a $\mathcal{J}$-nonnegative maximal invariant subspace $L_+ \subset \boldsymbol{E}$ and a $\mathcal{J}$-nonpositive maximal invariant subspace $L_- \subset \boldsymbol{E}$. Moreover, the spectra of the restrictions have the following forms:

$$\sigma_{\text{non}}\left(\mathcal{A}_0^{-1} \mid L_+\right) = \Lambda, \quad \sigma_{\text{non}}\left(\mathcal{A}_0^{-1} \mid L_-\right) = \bar{\Lambda}. \tag{2.40}$$

This property is a consequence of the facts that the operator $\mathcal{A}_0^{-1}$ is compact and $\mathcal{J}$-self-adjoint (see Section 1.3.5).

The subspace $L_+ \subset \boldsymbol{E}$ is the set of elements of the form

$$L_+ = \{y \in \boldsymbol{E} : y = (\boldsymbol{\xi}; K_+\boldsymbol{\xi})^{\mathrm{t}},\ \boldsymbol{\xi} \in \boldsymbol{J}_{0,S}(\Omega)\}, \tag{2.41}$$

where $K_+ : \boldsymbol{J}_{0,S}(\Omega) \to \boldsymbol{M}_0(\Omega)$ is the angular operator of the subspace L_+. Similarly, L_- is a set of elements of the form

$$L_- = \{y \in \boldsymbol{E} : y = (K_-\boldsymbol{\eta}; \boldsymbol{\eta})^{\mathrm{t}},\ \boldsymbol{\eta} \in \boldsymbol{M}_{0,S}(\Omega)\}, \tag{2.42}$$

where $K_- : \boldsymbol{M}_0(\Omega) \to \boldsymbol{J}_{0,S}(\Omega)$ is the angular operator of the subspace L_-. Here $\|K_+\| \le 1$, $\|K_-\| \le 1$.

These properties of the angular operators $K_\pm$ corresponding to the maximal definite subspaces $L_\pm$ have been mentioned in Section 1.3.1.

6° The angular operator K_+ is a solution of the operator equation

$$\begin{aligned} B^{-1}K_+ &= \alpha\left(K_+A^{-1} + I_0A^{-1}I_0K_+ + K_+A^{-1}I_0K_+ + I_0A^{-1}\right), \\ \alpha &= \sigma/\nu^2, \end{aligned} \tag{2.43}$$

in the operator sphere $\|K\| \le 1$. Similarly, the operator K_- is a solution of the operator equation

$$K_-B^{-1} = \alpha\left(A^{-1}K_- + K_-I_0A^{-1}I_0 + K_-I_0A^{-1}K_- + A^{-1}I_0\right), \tag{2.44}$$

in the same sphere.

To obtain equation (2.43), let us write down the operator in (2.35) in a form that explicitly takes into account the effect of the orthoprojector $I_0 = Q_{\boldsymbol{M}_0}$ on the subspace $\boldsymbol{M}_0(\Omega)$ (see (2.9)). We have

$$\mathcal{A}_0^{-1} = \begin{pmatrix} \nu^{-1}A^{-1} & \nu^{-1}A^{-1}I_0 \\ -\nu^{-1}I_0A^{-1} & (\sigma/\nu)^{-1}B^{-1} - \nu^{-1}I_0A^{-1}I_0 \end{pmatrix}. \tag{2.45}$$

Let $y = (\boldsymbol{\xi}; K_+\boldsymbol{\xi})^{\mathrm{t}} \in L_+$. Then, since L_+ is invariant, $\mathcal{A}_0^{-1}y \in L_+$, that is, $\mathcal{A}_0^{-1}y = y_1 = (\boldsymbol{\xi}_1; K_+\boldsymbol{\xi}_1)^{\mathrm{t}}$. Hence we obtain the following, in a vector-matrix form,

$$\nu^{-1}A^{-1}\boldsymbol{\xi} + \nu^{-1}A^{-1}I_0K_+\boldsymbol{\xi} = \boldsymbol{\xi}_1,$$
$$-\nu^{-1}I_0A^{-1}\boldsymbol{\xi} + (\sigma/\nu)^{-1}B^{-1}K_+\boldsymbol{\xi} - \nu^{-1}I_0A^{-1}I_0K_+\boldsymbol{\xi} = K_+\boldsymbol{\xi}_1. \tag{2.46}$$

By substituting the element $\boldsymbol{\xi}_1$ from the first equation into the second one and using the fact that the element $\boldsymbol{\xi} \in \boldsymbol{J}_{0,S}(\Omega)$ is arbitrary, we obtain the following equation

$$\nu^{-1}K_+A^{-1} + \nu^{-1}A^{-1}I_0K_+ = -\nu^{-1}I_0A^{-1} + (\sigma/\nu)^{-1}B^{-1}K_+ - \nu^{-1}I_0A^{-1}I_0K_+,$$

which, in turn, leads to (2.43).

A similar reasoning for elements of the form $y = (K_-\boldsymbol{\eta}; \boldsymbol{\eta})^{\mathrm{t}}$ with arbitrary $\boldsymbol{\eta} \in \boldsymbol{M}_0(\Omega)$ leads to equation (2.44). From (2.43) and (2.44) it follows that if the operator K_+ is a solution of equation (2.43), then the operator $(K_+)^*$ satisfies equation (2.44). The converse is also true.

9.2.5 The Matrix Structure of the Main Operator

Using the properties of the solutions of problem (2.35), let us find out the matrix structure of the operator $\mathcal{A}_0^{-1}$. Let us first extract from the maximal nonnegative subspace L_+ which is invariant for $\mathcal{A}_0^{-1}$ its isotropic part $L_+^0 = L_+ \cap L_+^{[\perp]}$. Then $L_+ = L_+^0 \oplus L_+^1$, where L_+^1 is a positive subspace. Therefore, the entire space $\boldsymbol{E} = \boldsymbol{J}_{0,S}(\Omega) \oplus \boldsymbol{M}_0(\Omega)$ has an orthogonal decomposition

$$\boldsymbol{E} = L_+^0 \oplus L_+^1 \oplus L_+^{\perp}, \tag{2.47}$$

where $L_+^{\perp}$ is an orthogonal complement to L_+ in $\boldsymbol{E}$.

In this decomposition not only is $L_+ = L_+^0 \oplus L_+^1$ invariant for $\mathcal{A}_0^{-1}$, but L_+^0 is invariant for $\mathcal{A}_0^{-1}$ as well. To prove that, let us choose some $y_0 \in L_+^0$ and an arbitrary element $y \in L_+$. Then

$$[\mathcal{A}_0^{-1}y_0, y] = [y_0, \mathcal{A}_0^{-1}y] = 0,$$

because $\mathcal{A}_0^{-1}y \in L_+$ and $[y_0, z] = 0$ for all $y_0 \in L_+^0$ and $z \in L_+$.

This last equality can be a result of the following reasoning. Since L_+ is a nonnegative subspace, then $[z, z] \geq 0$ for any z in $L_+ = L_+^0 \oplus L_+^1$. Therefore, from the Bunyakovsky–Schwartz inequality apllied to nonnegative elements in indefinite scalar product, we obtain

$$|[y_0, z]|^2 \leq [y_0, y_0] \cdot [z, z] = 0$$

for any $y_0 \in L_+^0$, because $[y_0, y_0] = 0$.

The decomposition (2.47) shows that the matrix decomposition of the operator $\mathcal{A}_0^{-1}$ should have the form

$$\mathcal{A}_0^{-1} = \begin{pmatrix} A_1 & A'_{12} & A'_{13} \\ 0 & A_2 & A'_{23} \\ 0 & 0 & A'_3 \end{pmatrix}. \tag{2.48}$$

Let us use now the fact that the subspace $L_+^\perp$ is invariant relatively to $(\mathcal{A}_0^{-1})^*$, that is, $(\mathcal{A}_0^{-1})^* L_+^\perp \subset L_+^\perp$. Let us extract from $L_+^\perp$ its isotropic part: $L_+^\perp = (L_+^\perp)^0 \oplus (L_+^\perp)_0^1$. Here it turns out that $(L_+^\perp)^0 = \mathcal{J} L_+^0$. Indeed, L_+^0 is the isotropic part of $L_+^{[\perp]}$, and $L_+^\perp = \mathcal{J} L_+^{[\perp]}$.

From these conclusions it follows that the subspaces L_+^0 and $(L_+^\perp)^0 =: \hat{L}_+^0$ have the same dimension. Further, similarly to the case of operator $\mathcal{A}_0^{-1}$ it can be proved that $(\mathcal{A}_0^{-1})^* \hat{L}_+^0 \subset \hat{L}_+^0$ and, therefore, for $\mathcal{A}_0^{-1} = ((\mathcal{A}_0^{-1})^*)^*$ the subspace $(\hat{L}_+^0)^\perp$ is invariant: $\mathcal{A}_0^{-1}(\hat{L}_+^0)^\perp \subset (\hat{L}_+^0)^\perp$. On the other hand, from the decomposition (2.47), and the fact that $L_+^\perp = \hat{L}_+^0 \oplus \hat{L}_+^1$, $\hat{L}_+^1 := (L_+^\perp)^1$, it follows that $(\hat{L}_+^0)^\perp = L_+^0 \oplus L_+^1 \oplus \hat{L}_+^1$, and that is why

$$\boldsymbol{E} = L_+^0 \oplus L_+^1 \oplus \hat{L}_+^1 \oplus \hat{L}_+^0. \tag{2.49}$$

Since the subspaces L_+^0, $L_0 \oplus L_+^1$, and $L_+^0 \oplus L_+^1 \oplus \hat{L}_+^1$ are invariant for $\mathcal{A}_0^{-1}$, then instead of (2.48) we have

$$\mathcal{A}_0^{-1} = \begin{pmatrix} A_1 & A_{12} & A_{13} & A_{14} \\ 0 & A_2 & A_{23} & A_{24} \\ 0 & 0 & A_3 & A_{34} \\ 0 & 0 & 0 & A_4 \end{pmatrix}. \tag{2.50}$$

Let us point out again that the subspaces L_+^1, $\hat{L}_+^1$ are definite, and L_+^0, $\hat{L}_+^0$ are isotropic.

9.2.6 On the Finitness of the Number of Nonreal Eigenvalues

As mentioned in Section 9.1.3, in some instances under arbitrary viscousity ν and $\omega_0 = 0$, problem (1.7) can have only a finite amount of nonreal eigenvalues and if the viscousity is sufficiently big, then there are no nonreal eigenvalues at all. Though this problem was not completely solved in the case of an arbitrary container, there are some interesting facts that will be discussed in the sequel. They will be considered as additional properties to Properties 1°–6° in Section 9.2.4.

7° If the angular operator K_+ of the nonnegative maximal subspace $L_+ \subset \boldsymbol{E}$, which is invariant for $\mathcal{A}_0^{-1}$, is compact, then, for any $\nu > 0$, the problem (2.35) can have only a finite amount of nonreal eigenvalues.

To prove this assertion, let us note that on the isotropic invariant subspace L_+^0 operator K_+ acts as an isometry because if $y_0 = (\boldsymbol{\xi}_0; \boldsymbol{\eta}_0)^{\mathrm{t}} \in L_+^0$,

$$[y_0, y_0] = \|\boldsymbol{\xi}_0\|^2_{\boldsymbol{J}_{0,S}(\Omega)} - \|\boldsymbol{\eta}_0\|^2_{\boldsymbol{M}_0(\Omega)} = \|\boldsymbol{\xi}_0\|^2_{\boldsymbol{J}_{0,S}(\Omega)} - \|K_+\boldsymbol{\xi}_0\|^2_{\boldsymbol{M}_0(\Omega)} = 0. \tag{2.51}$$

However, since the operator K_+ is compact, then the dimension of the space L_+^0 should be finite and, therefore,

$$\dim L_+^0 = \dim \hat{L}_+^0 =: \kappa < \infty. \tag{2.52}$$

Hence it follows that the operators A_1 and A_4 in representation (2.50) are precisely κ-dimensional.

Let $\lambda = \lambda_0$ be an eigenvalue of the operator $\mathcal{A}_0^{-1}$. Then, from (2.50) it follows that λ_0 is also an eigenvalue of one of the operators A_k, $k = 1, \dots, 4$. This fact can be easily proved if we assume that the opposite is true. Let us write down the equation $\mathcal{A}_0^{-1} y = \lambda y$ in the decomposition (2.50) and consequently solve the triangular system under the condition that λ_0 is not an eigenvalue of any of the operators A_k, $k = 1, \dots, 4$. Finally, we obtain the trivial solution $y = 0$.

Since L_+^1 and $\hat{L}_+^1$ are definite subspaces, then, if $\lambda = \lambda_0 \neq \bar{\lambda}_0$ is a nonreal eigenvalue of the operator $\mathcal{A}_0^{-1}$, then λ_0 cannot be an eigenvalue of either of the operators A_2 or A_3. Hence it appears that λ_0 is an eigenvalue of either operator A_1 or A_4. Therefore, the amount of nonreal eigenvalues of problem (2.35) is less or equal to $\dim L_+^0 + \dim \hat{L}_+^0 = 2\kappa < \infty$, that is, there are no more than κ pairs of complex-conjugate solutions.

In order to conclude the proof of 7°, let us check that a nonreal eigenvalue λ_0 of problem (2.35) cannot be an eigenvalue of either operator A_2 or A_3. For example, let $A_2 y_2 = \lambda_0 y_2$. It is easy to see that in this case the element $y = (-(A_1 - \lambda_0 I_1)^{-1} A_{12} y_2; y_2; 0; 0)^{\mathrm{t}} \in L_+$ is an eigenelement of operator $\mathcal{A}_0^{-1}$. Indeed,

$$\begin{aligned}(\mathcal{A}_0^{-1} - \lambda_0 \mathcal{I})y &= \left(-(A_1 - \lambda_0 I_1)(A_1 - \lambda_0 I_1)^{-1} A_{12} y_2 + A_{12} y_2; (A_2 - \lambda_0 I_2) y_2; 0; 0\right)^{\mathrm{t}} \\ &= (0;0;0;0)^{\mathrm{t}} = 0.\end{aligned}$$

Since $\lambda_0 \neq \bar{\lambda}_0$, then $[y, y] = 0$. Then, from the fact that $y \in L_+$, it follows that $y \in L_+^0$. Thus, $y_2 = 0$ and, therefore, $y = 0$. This is a contradiction that shows that λ_0 cannot be an eigenvalue of operator A_2.

Similarly we can prove the same property for operator A_3. In this case, we should consider the operator $(\mathcal{A}_0^{-1})^*$ instead of $\mathcal{A}_0^{-1}$ and a matrix representation for $(\mathcal{A}_0^{-1})^*$ similar to (2.50) should be taken into account. This proves Property 7° completely.

As a remark, let us point out that Property 7° holds true even in the more general case that was used for its proof, that is, whenever for the operator $\mathcal{A}_0^{-1}$ there exists a nonnegative maximal invariant subspace $L_+ = L_+^0 \oplus L_+^1$ for which $\kappa := \dim L_+^0 < \infty$. Here the property of compactness of the corresponding angular operator K_+ for L_+ is not required.

8° If the condition $\|K_+\| < 1$ holds true, problem (2.35) does not have any nonreal eigenvalues.

Indeed, for the solutions $y_0 = (\boldsymbol{\xi}_0; K_+\boldsymbol{\xi}_0)^t$ corresponding to a nonreal eigenvalue λ_0, condition (2.51)should be valid. This condition contradicts $\|K_+\| < 1$, therefore, $\dim L_+^0 = \dim \hat{L}_+^0 = \kappa = 0$.

9° For $\|K_+\| < 1$, the maximal invariant subspace is uniformly positive and projectively complete. Here the $\mathcal{J}$-orthogonal complement $L_+^{[\perp]} = L_-$ is a maximal uniformly negative invariant subspace for $\mathcal{A}_0^{-1}$, $K_- = (K_+)^*$ and

$$\boldsymbol{E} = L_+ \, [+] \, L_-. \tag{2.53}$$

These facts follow from the statements in Section 1.3.5.

10° For $\|K_+\| < 1$, the spectrum $\{\lambda_j\}_{j=1}^{\infty}$ of problem (2.35) is a union of the spectra $\{\lambda_k^{\pm}\}_{k=1}^{\infty}$ of the operators $\mathcal{A}_0^{-1} \mid L_{\pm}$, the restrictions of $\mathcal{A}_0^{-1}$ to the invariant subspaces $L_{\pm}$. Here, the corresponding pairs $\{(\lambda_k^{\pm}; y_k^{\pm})\}_{k=1}^{\infty}$ are positive in L_+ and negative in L_-.

The first statement follows from the fact that—in virtue of the previously proved facts—the operator $\mathcal{A}_0^{-1}$ has the matrix representation

$$\mathcal{A}_0^{-1} = \operatorname{diag}\left(\mathcal{A}_0^{-1} \mid L_+; \mathcal{A}_0^{-1} \mid L_-\right) \tag{2.54}$$

in the $\mathcal{J}$-orthogonal decomposition (2.53). The other statement in 10° is obvious.

11° For $\|K_+\| < 1$, the eigenelements $\{y_k^+\}_{k=1}^{\infty}$ are $\mathcal{J}$-orthonormal and form a Riesz basis in the subspace L_+. The eigenelements $\{y_k^-\}_{k=1}^{\infty}$ are $\mathcal{J}$-orthonormal and form a Riesz basis in the subspace L_-.

Indeed, in virtue of (2.54), problem (2.35) splits into two independent problems

$$\lambda^{\pm}\left(\mathcal{A}_0^{-1} \mid L_{\pm}\right) y^{\pm} = y^{\pm}, \qquad y^{\pm} \in L_{\pm}. \tag{2.55}$$

In $L_{\pm}$, the scalar square $\pm[y^{\pm}, y^{\pm}]$ generates a squared norm that is equivalent to the ordinary norm of the space $\boldsymbol{E}$ and the corresponding operators $\mathcal{A}_0^{-1} \mid L_{\pm}$ in $L_{\pm}$ are self-adjoint compact operators. Therefore, Property 11° is a corollary of the Hilbert–Schmidt theorem and the property of norm equivalence mentioned previously.

12° The following fact is a corollary of Property 11° and decomposition (2.53). For $\|K_+\| < 1$, the eigenelements $\{y_j\}_{j=1}^{\infty}$, with $y_j = (\boldsymbol{\xi}_j; \boldsymbol{\eta}_j)^{\mathrm{t}}$, of the problem (2.53) form a $\mathcal{J}$-orthonormal basis and a Riesz basis in the space $\boldsymbol{E} = \boldsymbol{J}_{0,S}(\Omega) \oplus \boldsymbol{M}_0(\Omega)$.

Let us note one more interesting property of separate completeness that takes place for solutions of problem (2.35).

13° If the conditions

$$K_+ \in \mathfrak{S}_\infty, \qquad \|K_+\| < 1, \tag{2.56}$$

are true, then the set $\{\boldsymbol{\xi}_k^+\}_{k=1}^{\infty}$ of the first components of the eigenelements $y_k^+ = (\boldsymbol{\xi}_k^+; \boldsymbol{\eta}_k^+)^{\mathrm{t}}$ (in L_+) form a system of elements that is complete in $\boldsymbol{J}_{0,S}(\Omega)$. Similarly, the set $\{\boldsymbol{\eta}_k^-\}_{k=1}^{\infty}$ of the second components of the negative eigenelements $y_k^- = (\boldsymbol{\xi}_k^-; \boldsymbol{\eta}_k^-)^{\mathrm{t}}$ (in L_-) form a complete system of elements in the subspace $\boldsymbol{M}_0(\Omega) \subset \boldsymbol{J}_{0,S}(\Omega)$.

Indeed, for $y^+ = (\boldsymbol{\xi}; K_+\boldsymbol{\xi})^{\mathrm{t}}$ in L_+, the first equation in (2.35) gives

$$\boldsymbol{\xi} = \lambda\nu^{-1}A^{-1}(I + K_+)\boldsymbol{\xi}, \qquad \boldsymbol{\xi} \in \boldsymbol{J}_{0,S}(\Omega), \tag{2.57}$$

and the second equation in (2.35) is a corollary of (2.57) in virtue of (2.43). Since conditions (2.56) hold true, then there exists an inverse operator $(I+K_+)^{-1} = I+K_+^1$, with $K_+^1 \in \mathfrak{S}_\infty$. Therefore, from (2.57), we obtain that

$$(I + K_+^1 - \lambda\nu^{-1}A^{-1})\boldsymbol{\xi}_1 = \boldsymbol{0}, \qquad \boldsymbol{\xi}_1 = (I + K_+)\boldsymbol{\xi}, \tag{2.58}$$

where $K_+^1 \in \mathfrak{S}_\infty$, $0 < A^{-1} \in \mathfrak{S}_p$. According to Keldysh theorem, from (2.58) it follows that the system of elements $\{\boldsymbol{\xi}_{1k}\}_{k=1}^{\infty}$ and, therefore, the system $\{\boldsymbol{\xi}_k\}_{k=1}^{\infty}$ is complete in $\boldsymbol{J}_{0,S}(\Omega)$.

The second statement in Property 13° can be proved in a similar way. For this we should use equation (2.8) for $\omega_0 = 0$:

$$\begin{aligned} \nu A\boldsymbol{\xi} - \sigma\nu^{-1}B(\boldsymbol{\xi} + \boldsymbol{\eta}) &= \lambda\boldsymbol{\xi}, \\ \sigma\nu^{-1}B(\boldsymbol{\xi} + \boldsymbol{\eta}) &= \lambda\boldsymbol{\eta}. \end{aligned} \tag{2.59}$$

Assuming $\boldsymbol{\xi} = K_-\boldsymbol{\eta}$, from the second relation we obtain

$$\sigma\nu^{-1}B(I + K_-)\boldsymbol{\eta} = \lambda\boldsymbol{\eta};$$

from the latter the linear spectral problem follows

$$\left(I + K_- - \lambda\nu\sigma^{-1}B^{-1}\right)\boldsymbol{\eta} = \boldsymbol{0}. \tag{2.60}$$

14° If conditions (2.56) hold true, then the eigenvalues λ_k^+ have the following asymptotic behavior,

$$\lambda_k^+ = \nu\lambda_k(A)[1 + o(1)], \qquad k \to \infty, \tag{2.61}$$

and the eigenvalues λ_k^- have the following behavior,

$$\lambda_k^- = \sigma\nu^{-1}\lambda_k(B)[1 + o(1)] \to \infty, \qquad k \to \infty, \tag{2.62}$$

where the asymptotics of the numbers $\lambda_k(A)$ and $\lambda_k(B)$ are determined by formulas (2.16) and (2.17), respectively.

Indeed, formula (2.61) follows from the assertion in Section 1.6.11, (2.17), and (2.58), and formula (2.62) follows from Section 1.6.11, (2.16), and (2.60).

Let us state now without proof one of the most important property of the eigenvalues of problem (2.35).

15° For decreasing values of the viscousity ν , the eigenvalues λ_k^+ move to the left along the positive half-axis, whileas the eigenvalues λ_k^- move to the right along the same half-axis.

9.2.7 Heuristic Considerations. The Abstract Spectral Problem

Using formula (2.45), let us consider now the matrix compact operator

$$\mathcal{A}_\alpha = \begin{pmatrix} \alpha A & \alpha A \\ -\alpha A & B - \alpha A \end{pmatrix} \tag{2.63}$$

in the infinite-dimensional Hilbert space $\tilde{E} = E \oplus E$, where $\alpha \geq 0$ is a nonnegative parameter, and the self-adjoint operators A and B have the properties

$$0 < A \in \mathfrak{S}_\infty, \qquad 0 < B. \tag{2.64}$$

Obviously, the operator $\mathcal{A}_\alpha$ is a $\mathcal{J}$-self-adjoint operator for $\mathcal{J} = \operatorname{diag}(I; -I)$. Since $A \in \mathfrak{S}_\infty$, then $\mathcal{A}_\alpha$ has a nonnegative maximal invariant subspace L_+. The angular operator $K = K(\alpha) = K_+(\alpha)$ of the subspace L_+ satisfies the equation

$$BK = \alpha(I + K)A(I + K), \tag{2.65}$$

which has a solution in the sphere $\|K\| \leq 1$ for any $\alpha \geq 0$. For $\alpha = 0$, we have $BK(0) = 0$, and since $B > 0$, then there exists an inverse operator B^{-1}; therefore $K(0) = 0$. Assuming that the solution $K(\alpha)$ is analytical for small $\alpha > 0$, we obtain the first several terms of the decomposition,

$$K(\alpha) = \alpha B^{-1}A + \alpha^2\left((B^{-1}A)^2 + B^{-2}A^{-2}\right) + O(\alpha^3). \tag{2.66}$$

In addition to (2.64), let us assume that all the coefficients of the different powers of α in (2.66) are compact operators. In particular, such a situation occurs if $B^{-1} \in \mathcal{L}(E)$ or $B = A^{-\gamma}$, with $0 \le \gamma < 1$. It is easy to see that the amount c_k of such terms for the power α^k is equal to the coefficients in the series expansion of the solution $x(\alpha)$ to the scalar quadratic equation

$$x = \alpha b^{-1} a(1+x)^2, \qquad a > 0,\ b > 0, \tag{2.67}$$

which is associated with (2.66), namely

$$x(\alpha) = \sum_{k=1}^{\infty} c_k (b^{-1}a)^k \alpha^k, \qquad c_k = \frac{1}{2} \cdot \frac{(2k+2)!}{(2k+1)[(k+1)!]^2}. \tag{2.68}$$

Let us denote the operator coefficients in the expansion of $K = K(\alpha)$ for small α by $\hat{c}_k = \hat{c}_k(B^{-1}; A)$:

$$K(\alpha) = \sum_{k=1}^{\infty} \hat{c}_k(B^{-1}; A)\alpha^k. \tag{2.69}$$

Easy calculations show that the coefficients $\hat{c}_k$ satisfy the following recurrent relations:

$$\hat{c}_k = \sum_{j=0}^{k-1} B^{-1}\hat{c}_j A \hat{c}_{k-j-1}, \qquad \hat{c}_0 = I. \tag{2.70}$$

Hence it appears that each operator $\hat{c}_k$ is the sum of c_k terms; each of these terms is the product of k operators B^{-1} and k operators A, taken in such order as to each operator A there coresponds exactly one operator B^{-1} on the left side (see (2.66)).

Further considerations of the properties of operator $\mathcal{A}_\alpha$ in (2.63) are based on an additional condition called the *condition of uniform boundedness*, namely, there exists a constant $q > 0$ such that for all $k = 1, 2, \ldots$ the following estimations take place

$$\left\|\hat{c}_k(B^{-1}; A)\right\| \le c_k q^k. \tag{2.71}$$

Obviously, condition (2.71) holds true if $B^{-1} \in \mathcal{L}(E)$, when $\|B^{-1}\| \cdot \|A\|$ could be taken as q. If $B = A^{-\gamma}$, with $0 \le \gamma < 1$, then there can be assumed that $q = \|A\|^{1-\gamma}$.

If conditions (2.71) hold true, then the operator series (2.69) for $K(\alpha)$ can be estimated from above by the scalar number series:

$$\|K(\alpha)\| \le \sum_{k=1}^{\infty} c_k q^k \alpha^k = \frac{1 - 2\alpha q - \sqrt{1 - 4\alpha q}}{2\alpha q}. \tag{2.72}$$

Hence, we obtain that, for $\alpha < 1/(4q)$, the series (2.69) converges and $\|K(\alpha)\| < 1$.

Since, according to our assumption, each term in (2.69) is a compact operator and the series (2.69) converges uniformly by α for $0 \le \alpha < 1/(4q)$, then $K(\alpha)$ is a compact operator. Therefore, in the eigenvalue problem, the properties of the solutions for the pencil $\mathcal{I} - \lambda\mathcal{A}_\alpha$, for sufficiently small $\alpha > 0$ take place. These properties were listed in Sections 9.2.4–9.2.6 as they were done for problem (2.35). In particular, if the conditions $0 < \alpha < 1/(4q)$ hold true, then the operator (2.63) does not have nonreal values.

9.2.8 Heuristic Considerations. Physical Conclusions

The conclusions in Sections 9.2.6–9.2.7 and the analysis of the examples considered in Section 9.1 make it possible to point out some additional assumptions on the properties of the solutions of problem (2.35) to Properties 1°–6° in Section 9.2.4. These new assumptions have a heuristic character and will not be completely proved here. Let us state these properties and explain their physical meaning.

1° For sufficiently large values of the viscousity ν, the spectrum of the problem on the oscillations of a capillary viscous fluid is located on the positive half-axis and can be naturally split into two sets $\{\lambda_k^+\}_{k=1}^\infty$ and $\{\lambda_k^-\}_{k=1}^\infty$. Each of these sets has the limit point $\lambda = +\infty$. There are positive pairs of solutions $(\lambda_k^+; y_k^+)$ corresponding to λ_k^+ and negative pairs of solutions $(\lambda_k^-; y_k^-)$ corresponding to λ_k^-. The following properties hold true in this case: Properties 8°–14° in Section 9.2.6 on separate Riesz basicity in the definite invariant subspaces L_+ and L_-, on Riesz basicity of the union system $\{y_k^+\}_{k=1}^\infty \cup \{y_k^-\}_{k=1}^\infty$ in the entire space $\boldsymbol{E} = \boldsymbol{J}_{0,S}(\Omega) \oplus \boldsymbol{M}_0(\Omega) = L_+ [+] L_-$, and the asymptotic formulas (2.61) and (2.62) for the eigenvalues λ_k^+ and λ_k^- as $k \to \infty$.

2° To the positive solutions $(\lambda_k^+; y_k^+)$, with $y_k^+ = (\boldsymbol{\xi}_k^+; \boldsymbol{\eta}_k^+)^{\mathrm{t}} \in \boldsymbol{E}$, there correspond dissipative internal waves that are allied with similar waves occuring in a heavy viscous fluid partially filling a container and with the dissipative waves in an entirely filled container (see Chapters 8 and 7).

To the negative solutions $(\lambda_k^-; y_k^-)$ there correspond surface capillary gravitational waves that have large fading decrements (unlike the ones in the case of heavy fluids).

3° For $\nu \to \infty$, the asymptotic behavior of the positive eigenvalues $\lambda_k^+ = \lambda_k^+(\nu)$—assuming that they are one-multiple—is the following

$$\lambda_k^+(\nu) = \nu\lambda_k^+(A) - \frac{\sigma}{\nu}\lambda_k^{1/2}(A)\left\|B^{1/2}A^{-1/2}\boldsymbol{\xi}_k(A)\right\|^2_{\boldsymbol{J}_{0,S}(\Omega)} + O(\nu^{-3}), \tag{2.73}$$

where $\lambda_k^+(A)$ are eigenvalues and $\boldsymbol{\xi}_k(A)$ are the normalized eigenelements of operator A. A similar formula for $\lambda_k^-(\nu)$ has the form

$$\lambda_k^-(\nu) = \sigma\nu^{-1}\lambda_k(B) + \sigma^2\nu^{-3}\lambda_k^{1/2}(B)\left\|A^{-1/2}B^{1/2}\boldsymbol{\eta}_k(B)\right\|^2_{\boldsymbol{M}_0(\Omega)} + O\left(\frac{\sigma^3}{\nu^5}\right), \tag{2.74}$$

where $\lambda_k(B)$ are eigenvalues and $\eta_k(B)$ are the normalized eigenelements of operator B.

Formulas (2.73) show that, for $\nu \to \infty$, the positive eigenvalues $\lambda_k^+(\nu)$ are determined mainly by the eigenvalues of operator A, and the capillary forces make the fading decrements decrease insignificantly. Here $\lambda_k^+(\nu) \to \infty$ for any $k = 1, 2, \ldots$. From (2.74) it follows that the fading decrements $\lambda_k^-(\nu)$ are determined mainly by the eigenvalues of operator B and the dissipative internal waves make these magnitudes increase insignificantly. Here, $\lambda_k^-(\nu) \to 0$ as $\nu \to \infty$, for any $k = 1, 2, \ldots$.

Let us next state some facts that became clear while we deduced formulas (2.73) and (2.74). For big ν, there are normal movements of the fluid corresponding to the positive solutions. These movements take place mainly inside region Ω occupied by the fluid and the deviation of the free moving surface is sufficiently small for such solutions. On the other hand, for big ν, there are normal movements corresponding to the negative solutions, for which the free surface essentially deviates from its equilibrium position, and the movements that take place inside the region Ω are sufficiently small.

Let us state now without proof a similar result for the case of a rotating fluid and compare it with (2.73) and (2.74). For $\nu \to \infty$, the asymptotic behavior of the eigenvalues $\lambda = \lambda_k^+(\nu)$ are the following

$$\begin{aligned}
&\lambda_k^+(\nu) = \nu\lambda_k(A) - 2i\omega_0(S_0\boldsymbol{\xi}_k(A), \boldsymbol{\xi}_k(A))_{J_{0.S}(\Omega)} + O(\nu^{-1}), \qquad k = 1, 2, \ldots \\
&\boldsymbol{\xi}_k^+(\nu) = \boldsymbol{\xi}_k(A) + O(\nu^{-1}), \\
&\boldsymbol{\eta}_k^+(\nu) = O(\nu^{-2}), \qquad \nu \to \infty.
\end{aligned} \tag{2.75}$$

For the branch $\lambda = \lambda_k^-(\nu)$ we have

$$\begin{aligned}
\lambda_k^-(\nu) &= \sigma\nu^{-1}\lambda_k(B) + \sigma\nu^{-2}2i\omega_0\lambda_k(B)(S_0A^{-1/2}\boldsymbol{\eta}_k(B), A^{-1/2}\boldsymbol{\eta}_k(B))_{J_{0,S}(\Omega)} \\
&\quad + O(\nu^{-3}), \qquad k = 1, 2, \ldots \\
\boldsymbol{\xi}_k^-(\nu) &= O(\nu^{-1}), \\
\boldsymbol{\eta}_k^-(\nu) &= \boldsymbol{\eta}_k(B) + O(\nu^{-1}), \qquad \nu \to \infty.
\end{aligned} \tag{2.76}$$

Formulas (2.75) and (2.76) show that taking into account the Coriolis forces changes the character of asymptotic behavior of the branches of eigenvalues. Thus, for the branch $\lambda_k^+(\nu)$ we have

$$|\operatorname{Im}\lambda_k^+(\nu)| \le 2\omega_0[1 + O(\nu^{-1})], \qquad \nu \to \infty$$

and for $\lambda_k^-(\nu)$,

$$|\operatorname{Im}\lambda_k^-(\nu)| = O(\nu^{-2}).$$

4° For sufficiently large values of the viscousity ν, all the pairs of solutions $y_k^{\pm}(\nu)$ are definite, that is, each pair is either positive or negative. As the viscousity ν decreases, the eigenvalues $\lambda_k^+(\nu)$ move to the left and the eigenvalues $\lambda_k^-(\nu)$ move to the right.

When the parameter ν reaches a certain critical value $\nu = \nu_*$, then a finite amount of neutral pairs of solutions and a finite amount of eigenvalues that have associated elements, begin to show up. If in this case multiple eigenvalues appear, that is, for $\nu = \nu_*$, collisions of multiple eigenvalues of different types take place, then for further decrease of ν they move out of the real axis and enter the complex plane as complex-adjoint pairs. If ν continues to decrease untill it reaches the given value, ν_0 then these pairs of eigenvalues do not return on the real positive half-axis anymore, and the amount of other collisions is finite.

Thus, we came to the following conclusion: for any viscousity ν the amount of nonreal eigenvalues of problem (2.35) is no more than finite.

5° For arbitrary ν, the ($\mathcal{J}$-orthonormal) system of eigenelements $\{y_k^+\}_{k=1}^{\infty}$ corresponding to the positive pairs of solutions $\{(\lambda_k^+; y_k^+)\}$ forms a Riesz basis with accuracy up to a finite defect in the nonnegative subspace L_+. The negative pairs of solutions $\{(\lambda_k^-; y_k^-)\}$ have the similar property in the subspace L_-; moreover, the reunion of all such pairs also has this property in the entire space $\boldsymbol{E}$.

6° There exists a rule of selection. According to this rule, the above mentioned $\mathcal{J}$-orthonormal system of elements can be extended to a complete system of eigen- and associated elements that correspond to nonreal eigenvalues and neutral pairs. In this case, for the subspace L_+, we can select eigen- and associated elements corresponding to eigenvalues in the upper half-plane, and, for L_-, we can choose those eigen- and associated elements that correspond to the eigenvalues in the lower half-plane.

7° For any viscousity $\nu > 0$, the asymptotic behavior of the eigenvalues $\lambda_k^{\pm}$, for $k \to \infty$, corresponding to the subspaces $L_{\pm}$ is defined by the formulas (2.61) and (2.62).

8° If the viscousity decreases to zero, then the amount of pairs of nonreal eigenvalues increases infinitely and becomes infinite for $\nu = 0$. In this case, after transition to the limit, all the nonreal eigenvalues occur on the imaginary axis and become equal with the magnitudes $\pm i\omega_k$, where ω_k are oscillation frequencies of an ideal capillary fluid in a container.

At the conclusion of this section, let us point out some considerations of a mathematical nature. First let us notice that if the surfaces Γ and S are infinitely smooth and $\bar{\Gamma} \cap \bar{S} = \emptyset$, then it can be proved using the theory of scales of Hilbert spaces that are build according to operators A and B, that all the operators occurring in the representation (2.66) of the angular operator $K_+(\alpha)$—the solution of equation (2.43)—are compact. However, this condition, which is is equivalent to the condition of uniform boundedness, (2.71), has not been either proved or refuted in relation to equation (2.43).

In problem (2.35), for arbitrary $\nu > 0$, the angular operator $K_+ = K_+(\alpha)$ has probably the following property: it can be represented as a sum of a finite-dimensional isometric operator and an operator whose norm is less than 1. In this case, the dimension $\kappa = \kappa(\nu)$ of the subspaces L_0^+ and $\hat{L}_0^+$ is finite for any $\nu > 0$ (see representation (2.49)), and probably $\kappa(\nu) \to +\infty$ as $\nu \to +0$.

9.2.9 The Solvability of the Evolution Problem

Let us now consider again the Cauchy problem (2.5)–(2.7) and investigate the existence and uniqueness of its generalized solution. Let us write down (2.5)–(2.7) in the form

$$\frac{\mathrm{d}y}{\mathrm{d}t} + \mathcal{A}y = f(t), \qquad y(0) = y^0, \tag{2.77}$$

where $y(t) = (\boldsymbol{\xi}(t); \boldsymbol{\eta}(t))^{\mathrm{t}} \in \boldsymbol{E} = \boldsymbol{J}_{0,S}(\Omega) \oplus \boldsymbol{M}_0(\Omega)$, $y^0 = (\boldsymbol{\xi}(0); \boldsymbol{\eta}(0))^{\mathrm{t}}$, $f(t) = (A^{1/2}\boldsymbol{f}(t); \mathbf{0})^{\mathrm{t}}$ and operator matrix $\mathcal{A}$ is defined by the formulas (2.18), (2.19).

First of all let us show that for sufficiently big ν, the operator $\mathcal{A}$ admits the estimate

$$\mathrm{Re}(\mathcal{A}y, y)_{\boldsymbol{E}} \geq c\|y\|^2_{\boldsymbol{E}}, \qquad y \in \mathcal{D}(\mathcal{A}), \tag{2.78}$$

with some constant c, which depends on the parameters ν, ω_0, and the characteristics of the region Ω. Taking into account that $\|S_0\| = 1$, we have

$$\begin{aligned}\mathrm{Re}(\mathcal{A}y, y)_{\boldsymbol{E}} = \mathrm{Re}\Big\{&\nu(A\boldsymbol{\xi}, \boldsymbol{\xi}) - \sigma\nu^{-1}(B\boldsymbol{\xi}, \boldsymbol{\xi}) - \sigma\nu^{-1}(B\boldsymbol{\eta}, \boldsymbol{\xi}) \\ &- 2\mathrm{i}\omega_0\left(S_0 A^{-1/2}(\boldsymbol{\xi} + \boldsymbol{\eta}), A^{-1/2}\boldsymbol{\xi}\right) + \sigma\nu^{-1}(B\boldsymbol{\xi}, \boldsymbol{\eta}) + \sigma\nu^{-1}(B\boldsymbol{\eta}, \boldsymbol{\eta})\Big\} \\ \geq\ &\nu\|A^{1/2}\boldsymbol{\xi}\| - \sigma\nu^{-1}\|B^{-1/2}\boldsymbol{\xi}\|^2 - 2\omega_0\|A^{1/2}\boldsymbol{\xi}\|\left(\|A^{-1/2}\boldsymbol{\xi}\| + \|A^{-1/2}\boldsymbol{\eta}\|\right) \\ &+ \sigma\nu^{-1}\|B^{1/2}\boldsymbol{\eta}\|^2.\end{aligned} \tag{2.79}$$

In the second term in the right hand side let us represent the operator $B^{1/2}$ in the form $B^{1/2} = RA^{1/2}$, where $R = B^{1/2}A^{-1/2}$ is a compact operator whose properties have been considered in Section 9.2.2 (see (2.23), (2.25), (2.28)). Further, let us use the elementary inequality $2ab \leq \varepsilon a^2 + \varepsilon^{-1}b^2$, $\varepsilon > 0$ for the forth term in the right part of (2.79). Then the right part of (2.79) can be estimated from below by the following quantity,

$$(\nu - \sigma\nu^{-1}\|R\|^2)\|A^{1/2}\boldsymbol{\xi}\|^2 - 2\omega_0\varepsilon\|A^{1/2}\boldsymbol{\xi}\|^2 - \omega_0\|A^{-1}\|\varepsilon^{-1}(\|\boldsymbol{\xi}\|^2 + \|\boldsymbol{\eta}\|^2)$$

$$+\sigma\nu^{-1}\lambda_1(B)\|\boldsymbol{\eta}\|^2.$$

If the condition

$$\nu - \sigma\nu^{-1}\|R\|^2 - 2\omega_0\varepsilon > 0 \tag{2.80}$$

holds true, then the previously mentioned expression can be estimated from above by the follwing quantity,

$$(\nu - \sigma\nu^{-1}\|R\|^2) - 2\omega_0\varepsilon)\lambda_1(A)\|\boldsymbol{\xi}\|^2 + \sigma\nu^{-1}\lambda_1(B)\|\boldsymbol{\eta}\|^2 - \frac{\omega_0}{\varepsilon\lambda_1(A)}(\|\boldsymbol{\xi}\|^2 + \|\boldsymbol{\eta}\|^2).$$

Let us choose the number $\varepsilon > 0$ from the condition

$$(\nu - \sigma\nu^{-1}\|R\|^2) - 2\omega_0\varepsilon)\lambda_1(A) = \sigma\nu^{-1}\lambda_1(B)\|\boldsymbol{\eta}\|^2 > 0.$$

This can be done for sufficiently large values of the viscousity ν, that is, whenever

$$\beta := 1 - \frac{\sigma}{\nu^2}\left(\|R\|^2 + \frac{\lambda_1(B)}{\lambda_1(A)}\right) > 0, \tag{2.81}$$

a condition that is assumed valid in the sequel. Then, $\varepsilon = \nu\beta/(2\omega_0) > 0$ and condition (2.80) holds true automatically. The previously mentioned reasoning leads to the estimate (2.78) with the constant

$$c = \sigma\nu^{-1}\lambda_1(B) - \frac{\omega_0}{\varepsilon\lambda_1(A)} = \frac{\sigma\lambda_1(A)\lambda_1(B)\beta - 2\omega_0^2}{\nu\beta\lambda_1(A)}. \tag{2.82}$$

A similar reasoning for operator $\mathcal{A}^*$ gives the estimate

$$\operatorname{Re}(\mathcal{A}^* z, z)_{\boldsymbol{E}} \geq c\|z\|^2_{\boldsymbol{E}}, \qquad z \in \mathcal{D}(\mathcal{A}^*), \tag{2.83}$$

with the same constant c as in (2.82).

Hence, if the value of the viscousity ν is so large that condition (2.81) holds true, then the inequalities (2.78) and (2.82) with the constant c as in (2.82) are valid as well. Thus, it follows that $-\mathcal{A} + c\mathcal{I}$ is a maximal dissipative operator and the following estimate

$$\|\mathcal{U}(t)\| \leq \exp(-ct) \tag{2.84}$$

is true for the semigroup $\mathcal{U}(t)$ that corresponds to the operator $-\mathcal{A}$.

Using these results and the facts stated in Section 1.5 we obtain the following statements:

1° If condition (2.81) is satisfied, then the homogeneous problem (2.77) is uniformly correct and its solution is given by the formula

$$y(t) = \mathcal{U}(t)y^0, \tag{2.85}$$

where $\mathcal{U}(t)$ is a semigroup that satisfies (2.84).

2° If $f(t)$ is a continuous function with values in $\boldsymbol{E}$ and $y^0 \in \boldsymbol{E}$, then the nonhomogeneous problem (2.77) has a generalized solution $y(t)$ that can be determined by the formula:

$$y(t) = \mathcal{U}(t)y^0 + \int_0^t \mathcal{U}(t-\tau)f(\tau)d\tau. \tag{2.86}$$

3° If the system is not rotating, (i.e., $\omega_0 = 0$), then the constant c in (2.82) is positive and the property of asymptotic stability of solutions of the homogeneous problem (2.77) is satisfied. If the angular velocity of the system is small, that is, when it satisfies the inequality

$$\omega_0^2 < \sigma\lambda_1(B)\lambda_1(A)\frac{\beta}{2}, \tag{2.87}$$

this fact also takes place.

4° If for the initial boundary value problem (1.1)–(1.6) (or for the problem (2.5)–(2.7)) the conditions

$$\boldsymbol{u}^0 \in \boldsymbol{J}^1_{0,S}(\Omega), \qquad \zeta^0 \in H^{3/2}_\Gamma(\Gamma) \tag{2.88}$$

hold true and $\boldsymbol{f}(t,x)$ is a continuous function of t with values in $\boldsymbol{J}^1_{0,S}(\Omega)$, then this problem has the only generalized solution $\boldsymbol{u}(t,x) = A^{-1/2}(\boldsymbol{\xi}(t,x) + \boldsymbol{\eta}(t,x))$, which is a continuous function of t with values in $\boldsymbol{J}^1_{0.S}(\Omega)$.

Indeed, if $\boldsymbol{f}(t,x)$ satisfies the previously formulated conditions, then $f(t) = (A^{1/2}\boldsymbol{f}(t); 0)^{\mathrm{t}}$ satisfies condition 2°. Further, for $\boldsymbol{u}^0 \in \boldsymbol{J}^1_{0.S}(\Omega) = \mathcal{D}(A^{1/2})$ we have $A^{1/2}\boldsymbol{u}^0 \in \boldsymbol{J}_{0,S}(\Omega)$. Moreover, if $\zeta^0 \in H^{3/2}_\Gamma(\Gamma)$, then $B_\sigma\zeta^0 \in H^{-1/2}_\Gamma$, $TB_\sigma\zeta^0 \in \boldsymbol{M}_1(\Omega) = A^{-1/2}\boldsymbol{M}_0(\Omega)$. Therefore, according to (2.7), we finally obtain

$$\boldsymbol{\eta}(0) = -\sigma\nu^{-1}A^{1/2}TB_\sigma\zeta^0 \in \boldsymbol{M}_0(\Omega),$$
$$\boldsymbol{\xi}(0) = A^{1/2}\boldsymbol{u}^0 - \boldsymbol{\eta}(0) \in \boldsymbol{J}_{0.S}(\Omega)$$

and all the conditions 2° hold true.

9.3 The Inverse of the Lagrange Theorem on Stability

In this section, we consider the problem (9.2.8) on normal oscillations of a capillary viscous fluid rotating in an arbitrary container under the assumption that the operator of potential energy of the system has a negative minimum eigenvalue. In this case, we will prove that problem (9.2.8) has at least one eigenvalue in the left complex half-plane, which accounts for the instability of the normal oscillation modes.

9.3.1 Formulating the Theorem

Let us return to the problem (9.2.8) on normal oscillations, namely, to a spectral problem of the following form

$$\nu A\boldsymbol{\xi} - \sigma\nu^{-1}B(\boldsymbol{\xi}+\boldsymbol{\eta}) - 2\mathrm{i}\omega_0 A^{1/2}S_0A^{-1/2}(\boldsymbol{\xi}+\boldsymbol{\eta}) = \lambda\boldsymbol{\xi},$$
$$\sigma\nu^{-1}B(\boldsymbol{\xi}+\boldsymbol{\eta}) = \lambda\boldsymbol{\eta}, \tag{3.1}$$

where $\boldsymbol{\xi} \in \mathcal{D}(A) \subset \boldsymbol{J}^1_{0,S}(\Omega) \subset \boldsymbol{J}_{0,S}(\Omega)$, $\boldsymbol{\eta} \in \boldsymbol{M}_0(\Omega) := I_0\boldsymbol{J}_{0,S}(\Omega)$, $B = I_0B = BI_0 = I_0BI_0 = Q^*B_\sigma Q$, $Q^* = A^{1/2}T$, $Q = \gamma_n A^{-1/2}$, and B_σ is the operator of potential energy of the considered hydrodynamical system. We need to recall that in (3.1) the operators A and B are unbounded, $A \gg 0$, $A^{-1} \in \mathfrak{S}_\infty$, and $B = B^*$.

In the previous sections of the book, it was assumed that the operator of potential energy B_σ is positive definite and, therefore, the system is statically stable in linear approximation. As it was proved in Section 9.2, in this case all the normal oscillation modes in the evolution problem (9.2.5) are stable because the spectrum of the problem (3.1) consists of finitely multiple eigenvalues located in the right complex half-plane with a unique limit point $\lambda = \infty$.

If the condition of static stability in linear approximation is nor satisfied, the the spectrum of problem (3.1) may become unstable, that is, some eigenvelues λ may occur in the left complex half-plane.

Let us formulate the main result to be proved in the sequel in this case.

Suppose the physical parameters of problem (3.1) are such that the eigenvalues $\lambda_k(B_\sigma)$ of the operator of potential energy B_σ satisfy the conditions

$$\lambda_{\min}(B_\sigma) =: \lambda_1(B_\sigma) \le \cdots \le \lambda_\kappa(B_\sigma) < 0$$
$$= \lambda_{\kappa+1}(B_\sigma) = \cdots = \lambda_{\kappa+q}(B_\sigma) < \lambda_{\kappa+q+1}(B_\sigma) \le \cdots . \tag{3.2}$$

Then problem (3.1) has exactly κ (with due account of multiplicity) eigenvalues in the left complex half-plane.

Note that conditions (3.2) are quite natural from a physical point of view if we consider the stability or instability of the system. Because the potential energy of the system is equal to $(B_\sigma\zeta,\zeta)_{L_2,\Gamma}/2$ and the self-adjoint operator B_σ has a discrete spectrum with a limit point $+\infty$, then its eigenvalues are finitely multiple and the minimum one may be both positive (the potential energy has a minimum) and negative (the potential energy has no minimum in the given state of equilibrium). If, for example, a capillary fluid is kept by the surface forces at the top of a container shaped as a glass, then it means that $\lambda_1(B_\sigma) > 0$. If we increase the intensity of the gravitational field acting from above downwards, the potential energy of the system may not have a minimum in the considered equilibrium state and the fluid will spill on the glass bottom. Mathematically this situation reduces to the fact that $\lambda_{\min}(B_\sigma) < 0$.

9.3.2 Auxiliary Propositions

To prove the theorem stated in the previous section, we formulate first some auxiliary propositions.

1° The eigenvalues $\lambda_k(B)$ of the operator B satisfy inequalities (3.2) as well.

To prove this property, let us consider the following spectral problem

$$B\boldsymbol{\eta} := Q^* B_\sigma Q\boldsymbol{\eta} = \lambda\boldsymbol{\eta}, \qquad \boldsymbol{\eta} \in \mathcal{D}(B) \subset \boldsymbol{M}_0(\Omega). \tag{3.3}$$

If we substitute $Q\boldsymbol{\eta}$ by ζ and use operator Q in the left, we obtain the equation

$$CB_\sigma\zeta = \lambda\zeta, \qquad \zeta \in H_\Gamma = L_{2.\Gamma}, \; C = \gamma_n T,$$

where $0 < C \in \mathfrak{S}_\infty$ is the operator occurring in the abstract scheme of Section 1.8. Thus it follows that $\zeta \in \mathcal{D}(C^{-1})$ and, therefore,

$$B_\sigma\zeta = \lambda C^{-1}\zeta, \qquad \zeta \in \mathcal{D}(B_\sigma) \subset \mathcal{D}(C^{-1}) \subset L_{2.\Gamma}, \tag{3.4}$$

where $C^{-1} \gg 0$ is an unbounded operator with a discrete spectrum.

We consider next the problem (3.4) whose spectrum coincides with the spectrum of problem (3.3). Since, for the operator B_σ, (3.2) are satisfied, then it is obvious that problem (3.4) has the eigenvalue $\lambda_0 = 0$. The eigenelements $\{\zeta_k^0\}_{k=\kappa+1}^{\kappa+q}$ of B_σ corresponding to that eigenvalue form an orthogonal basis in the subspace H_0, with $\dim H_0 = q$.

Let

$$L_{2.\Gamma} = \tilde{H} \oplus H_0, \tag{3.5}$$

where H_0 is the just introduced subspace. In the resolution (3.5), the operator B_σ has the form $\operatorname{diag}(\tilde{B}_\sigma; 0)$ and the operator $C^{-1} = (C_{ik})_{i.k=1}^2$ has a general form. Therefore, representing ζ as $\tilde{\zeta} + \zeta_0$, with $\tilde{\zeta} \in \tilde{H}$ and $\zeta_0 \in H_0$, from (3.4) we deduce the following system of equations

$$\begin{aligned} \tilde{B}_\sigma\tilde{\zeta} &= \lambda(C_{11}\tilde{\zeta} + C_{12}\zeta_0) \\ 0 &= \lambda(C_{21}\tilde{\zeta} + C_{22}\zeta_0). \end{aligned} \tag{3.6}$$

If $\lambda = 0$, then, by virtue of the property $\operatorname{Ker}\tilde{B}_\sigma = \{0\}$, we get that the solution of problem (3.6)—as expected—has the form $\tilde{\zeta} = 0$, for all $\zeta_0 \in H_0$. For $\lambda \neq 0$, from the second equation in (3.6) it follows that $C_{22}\zeta_0 = -C_{21}\tilde{\zeta}$. Since the operator C^{-1} is positive definite, then the operator C_{22}—a q-dimensional matrix acting in H_0—is positive definite too and its inverse operator C_{22}^{-1} is bounded. Using the substitution $\zeta_0 = -C_{22}^{-1}C_{21}\tilde{\zeta}$ in the first equation in (3.6) we get the problem

$$\tilde{B}_\sigma\tilde{\zeta} = \lambda\tilde{C}\tilde{\zeta}, \qquad \tilde{C} := C_{11} - C_{12}C_{22}^{-1}C_{21}, \; \tilde{\zeta} \in \mathcal{D}(\tilde{B}_\sigma) \subset \mathcal{D}(\tilde{C}) \subset \tilde{H}, \tag{3.7}$$

where $\tilde{C}$—as stated in Section 1.4.3—is a positive operator. Since $\tilde{C}$ is equal to the sum of the positive definite unbounded operator C_{11} and the finite-dimensional operator $-C_{12}C_{22}^{-1}C_{21}$, then it is an unbounded positive definite operator.

Thus, we are making the transition from equation (3.4) to equation (3.7), where $\tilde{C}$ has the same general properties as the operator C^{-1}, and the operator $\tilde{B}_\sigma$ unlike B_σ has a zero kernel. Its eigenvalues are obviously the following:

$$\lambda_k(\tilde{B}_\sigma) = \begin{cases} \lambda_k(B_\sigma), & \text{for } k = 1, \dots, \kappa \\ \lambda_{k+q}(B_\sigma), & \text{for } k = \kappa + 1, \dots . \end{cases}$$

We can represent $\tilde{B}_\sigma$ as $\tilde{B}_\sigma = |\tilde{B}_\sigma|^{1/2} J_\kappa |\tilde{B}_\sigma|^{1/2}$, where $0 \ll |\tilde{B}_\sigma| = (\tilde{B}_\sigma^2)^{1/2}$, and J_κ is an operator of signature: $J_\kappa = J_\kappa^{-1} = J_\kappa^*$, $J_\kappa^2 = I$. Making then the substituition $|\tilde{B}_\sigma|^{1/2}\zeta = w$ in (3.7), we get the problem

$$w = \lambda J_\kappa |\tilde{B}_\sigma|^{-1/2} \tilde{C} |\tilde{B}_\sigma|^{-1/2} w =: \lambda J_\kappa R w \tag{3.8}$$

on the definition of characteristic numbers of J_κ-positive operator $J_\kappa R$ acting in a Pontryagin space Π_κ.

As it follows from Section 1.3.6, in this case problem (3.8) has a real spectrum and the J_κ-orthogonal resolution $\tilde{H} = \tilde{H}_1 [+] \tilde{H}_2$ takes place, where $\tilde{H}_1$ is a κ-dimensional nonpositive invariant subspace with respect to the operator $J_\kappa R$, and $\tilde{H}_2$ is an infinite-dimensional nonnegative invariant subspace. The J_κ-orthogonal resolution mentioned previously leads to a spliting of problem (3.8) into two problems,

$$w_1 = \lambda (J_\kappa R)_1 w_1, \quad w_2 = \lambda (J_\kappa R)_2 w_2, \qquad w_i \in \tilde{H}_i,\ i = 1, 2. \tag{3.9}$$

In subspace $\tilde{H}_1$ the operator $(J_\kappa R)_1$ is negative and, therefore, the first problem in (3.9) has exactly κ eigenvalues (with due account of multiplicities): $\lambda_1^- \le \lambda_2^- \le \cdots \le \lambda_\kappa^- < 0$. In subspace $\tilde{H}_2$, the operator $(J_\kappa R)_2$ is positive. Making the inverse changes we come to the conclusion that the second problem in (3.9) is equivalent to the problem on the spectrum of the variational ratio

$$\frac{(B_\sigma \gamma_n \boldsymbol{u}, \gamma_n \boldsymbol{u})_{L_{2.\Gamma}}}{E(\boldsymbol{u}, \boldsymbol{u})}, \qquad \boldsymbol{u} \in \boldsymbol{M}_1(\Omega), \tag{3.10}$$

considered in the subspace of codimension $\kappa + q$ of those elements $\boldsymbol{u} \neq \boldsymbol{0}$ from the space $\boldsymbol{M}_1(\Omega) \subset \boldsymbol{J}_{0,S}^1(\Omega)$, for which the conditions $(B_\sigma \gamma_n \boldsymbol{u}, \gamma_n \boldsymbol{u})_{L_{2,\Gamma}} > 0$ are fulfilled. Therefore, the form (3.10) takes positive values and—as it was proved by T. A. Suslina [3] and [6]—the discrete positive spectrum $\{\lambda_k^+\}$, with a limit point $\lambda = +\infty$, corresponds to it.

Therefore, problem (3.9) has the eigenvalues (with due account of multiplicities) $\{\lambda_k^-\}_{k=1}^{\kappa}$, with $\lambda_k^- = \lambda_k^-(B)$, and also $\{\lambda_k^+\}_{k=1}^{\kappa}$, with $\lambda_k^+ = \lambda_k^+(B)$, where it is convenient to think that $k = \kappa+q+1, \kappa+q+2, \dots$. Then, as we previously mentioned, the problem (3.4) and, therefore, problem (3.3) has the eigenvalues

$$\lambda_1^-(B) \le \cdots \le \lambda_\kappa^-(B) < 0 = \lambda_{\kappa+1}^0(B) = \cdots = \lambda_{\kappa+q}^0(B) < \lambda_{\kappa+q+1}^+(B) \le \cdots \tag{3.11}$$

Inequalities (3.11) for the operator B have quite the same form as the inequalities (3.2) for the operator B_σ and, therefore, Statement 1° has been proved.

2° If $\operatorname{Ker} B \neq \{0\}$, that is, in conditions (3.2) and, therefore, in (3.11) $q > 0$, then problem (3.1) has the solution

$$\begin{aligned}
&\lambda = \lambda_0 = 0,\\
&\boldsymbol{\eta} = \boldsymbol{\eta}_0 := \left(I_0 \left(I - 2\mathrm{i}\omega_0\nu^{-1}S\right)^{-1} I_0\right)^{-1} \boldsymbol{\psi}, \quad \text{for all } \boldsymbol{\psi} \in \operatorname{Ker} B,\\
&\boldsymbol{\xi} = \boldsymbol{\xi}_0 := 2\mathrm{i}\omega_0\nu^{-1}S\left(I - 2\mathrm{i}\omega_0\nu^{-1}S\right)^{-1} \boldsymbol{\eta}_0, \qquad S := A^{-1/2}S_0A^{-1/2},
\end{aligned}$$

which we call a *transformation solution*. Here, I_0 is the orthoprojector in $\boldsymbol{J}_{0,S}(\Omega)$ onto $\boldsymbol{M}_0(\Omega)$.

To prove this we let $\lambda = 0$ into (3.1) and get

$$A\left(\boldsymbol{\xi} - 2\mathrm{i}\omega_0\nu^{-1}S(\boldsymbol{\xi} + \boldsymbol{\eta})\right) = \mathbf{0}, \quad B(\boldsymbol{\xi} + \boldsymbol{\eta}) = \mathbf{0}. \tag{3.13}$$

From the first equation we obtain

$$\left(I - 2\mathrm{i}\omega_0\nu^{-1}S\right)\boldsymbol{\xi} = 2\mathrm{i}\omega_0\nu^{-1}S\boldsymbol{\eta}.$$

Since $S = S^* \in \mathfrak{S}_\infty$, then there exists an inverse operator $(I - 2\mathrm{i}\omega_0\nu^{-1}S)^{-1}$ and, therefore,

$$\boldsymbol{\xi} = 2\mathrm{i}\omega_0\nu^{-1}S\left(I - 2\mathrm{i}\omega_0\nu^{-1}S\right)\boldsymbol{\eta}. \tag{3.14}$$

Taking into account that $B = BI_0 = I_0BI_0$, from the second equation (3.11), we have $I_0\boldsymbol{\xi} + \boldsymbol{\eta} = \boldsymbol{\psi}$, for all $\boldsymbol{\psi} \in \operatorname{Ker} B$. Substituting here (3.14) and considering that $\boldsymbol{\eta} = I_0\boldsymbol{\eta} = I_0^2\boldsymbol{\eta}$, $I + K(I-K)^{-1} = (I-K)^{-1}$, and $K = 2\mathrm{i}\omega_0\nu^{-1}S$, we come to the following relation

$$T\boldsymbol{\eta} := I_0\left(I - 2\mathrm{i}\omega_0\nu^{-1}S\right)I_0\boldsymbol{\eta} = \boldsymbol{\psi}. \tag{3.15}$$

Note now that the operator T is invertible in $\boldsymbol{M}_0(\Omega)$. Indeed, if $T\boldsymbol{\eta} = \mathbf{0}$, with $\boldsymbol{\eta} \in \boldsymbol{M}_0(\Omega)$, then

$$\begin{aligned}
(T\boldsymbol{\eta}, \boldsymbol{\eta})_{\boldsymbol{L}_2(\Omega)} &= \left(\left(I - 2\mathrm{i}\omega_0\nu^{-1}S\right)^{-1} I_0\boldsymbol{\eta}, I_0\boldsymbol{\eta}\right)_{\boldsymbol{L}_2(\Omega)}\\
&= \left(\left(I - 2\mathrm{i}\omega_0\nu^{-1}S\right)^{-1} \boldsymbol{\eta}, \boldsymbol{\eta}\right)_{\boldsymbol{L}_2(\Omega)}\\
&= 0.
\end{aligned}$$

After changes $(I - 2\mathrm{i}\omega_0\nu^{-1}S)^{-1}\boldsymbol{\eta} = \boldsymbol{\delta}$ and hence we get

$$\|\boldsymbol{\delta}\|^2_{\boldsymbol{L}_2(\Omega)} - 2\mathrm{i}\omega_0\nu^{-1}(S\boldsymbol{\delta}, \boldsymbol{\delta})_{\boldsymbol{L}_2(\Omega)} = 0,$$

and since $S = S^*$, then we have $\|\boldsymbol{\delta}\|^2_{\boldsymbol{L}_2(\Omega)} = 0$ and, therefore, $\boldsymbol{\delta} = \mathbf{0}$ and $\boldsymbol{\eta} = \mathbf{0}$.

Using T^{-1} for (3.15) we come to the formula for $\boldsymbol{\eta} = \boldsymbol{\eta}_0$ from (3.12) and then (3.14) gives the formula for $\boldsymbol{\xi} = \boldsymbol{\xi}_0$. Statement 2° is now proved.

9.3.3 The Principle of Changing Stability

As it can be seen from the second equation in (3.1), for $\lambda \neq 0$ the element $\boldsymbol{\eta}$ belongs to the range of values of operator B, that is, $\boldsymbol{\eta} \in \boldsymbol{M}_0(\Omega) \ominus \operatorname{Ker} B =: \boldsymbol{H}_0$. Then in (3.1) instead of B we can put the operator $B_0 := P_0 B|_{\boldsymbol{H}_0}$, where P_0 is the orthoprojector of the space $\boldsymbol{M}_0(\Omega)$ onto $\boldsymbol{H}_0$. Thus, for $\lambda \neq 0$ we come to the problem

$$\nu A\boldsymbol{\xi} - \sigma\nu^{-1}B_0(\boldsymbol{\xi}+\boldsymbol{\eta}) - 2\mathrm{i}\omega_0 A^{1/2} S_0 A^{-1/2}(\boldsymbol{\xi}+\boldsymbol{\eta}) = \lambda\boldsymbol{\xi},$$
$$\sigma\nu^{-1}B_0(\boldsymbol{\xi}+\boldsymbol{\eta}) = \lambda\boldsymbol{\eta}, \tag{3.16}$$

where $\operatorname{Ker} B_0 = \{0\}$, $\boldsymbol{\eta} \in \boldsymbol{H}_0$, $B_0 = P_0 B P_0 = P_0 B_0 = B_0 P_0$.

From Property 1° in Section 9.3.2 and the previous remarks it follows that operator B_0 satisfies the following inequalities instead of (3.11),

$$\lambda_1(B_0) \leq \cdots \leq \lambda_\kappa(B_0) < 0 < \lambda_{\kappa+1}(B_0) \leq \cdots, \tag{3.17}$$

because $\lambda_k(B_0) = \lambda_k(B)$, for $k = 1, \ldots, \kappa$, and $\lambda_k(B_0) = \lambda_{\kappa+q}(B)$, for $k > \kappa$.

For further considerations it is important to mention the following statement, which may be called the *principle of changing stability* as in the problem on convective motions of a viscous fluid in an open container.

3° The eigenvalues λ of problem (3.16) may transform from the right complex half-plane to the left one only by passing through zero.

To prove this property we apply the operator A^{-1} to both parts of the first equation (3.16) and introduce a new unknown element $\boldsymbol{\delta} := \boldsymbol{\xi} + \boldsymbol{\eta}$. We obtain

$$\boldsymbol{\delta} = \lambda\nu^{-1}A^{-1}\boldsymbol{\delta} + (\lambda\nu)^{-1}\sigma B_0\boldsymbol{\delta} + 2\mathrm{i}\omega_0\nu^{-1}S\boldsymbol{\delta}. \tag{3.18}$$

If we put $\lambda = \mathrm{i}\gamma$, $0 \neq \gamma \in \mathbb{R}$, then, after scalar multiplication by $\boldsymbol{\delta}$ (in $\boldsymbol{J}_{0,S}(\Omega)$), the left part turns out to be real, whileas the right side is purely imaginary since A^{-1}, B_0, and S are self-adjoint operators. It means $\boldsymbol{\delta} = \mathbf{0}$, that is, problem (3.16) has no solutions on the imaginary axis besides zero.

This fact is a consequence of Properties 1°–3°.

4° In loosing the stability because of changes in physical parameters of the considered hydrodynamic system, the transition of the eigenvalues λ from the right complex half-plane to the left one passes through the zero (origin) of the complex plane and only under the condition $\operatorname{Ker} B_\sigma \neq \{0\}$.

If we consider a nonrotating capillary viscous fluid ($\omega_0 = 0$), then the character of the eigenvalue transition may be specified in more detail. In this case, the following statement takes place.

5° The nonreal eigenvalues of problem (3.16) for $\omega_0 = 0$, as well as the real eigenvalues to which associated elements correspond, are situated in the half-plane

$$\operatorname{Re}\lambda \geq \nu \frac{\lambda_{\min}(A)}{2}. \tag{3.19}$$

To prove this, let us set $\omega_0 = 0$ in (3.18) and note that the problem we obtain this way, coincides—except the notations—with problem (8.2.1), where the basic operator pencil in the problem on oscillations of a heavy (non-capillary) fluid is consider in an open container. Following the results in Section 8.2.1, we obtain, in particular, that for a nonreal λ and according to (8.2.8)

$$\operatorname{Re}\lambda = \frac{(\boldsymbol{\delta}, \boldsymbol{\delta})_{\boldsymbol{L}_2(\Omega)}}{2\nu^{-1}(A^{-1}\boldsymbol{\delta}, \boldsymbol{\delta})_{\boldsymbol{L}_2(\Omega)}} \geq \nu \frac{\lambda_{\min}(A)}{2}.$$

For those real eigenvalues λ that have associated elements, in Section 8.2.1 we obtained the formula

$$\lambda = \frac{(\boldsymbol{\delta}, \boldsymbol{\delta})_{\boldsymbol{L}_2(\Omega)}}{2\nu^{-1}(A^{-1}\boldsymbol{\delta}, \boldsymbol{\delta})_{\boldsymbol{L}_2(\Omega)}} \geq \nu \frac{\lambda_{\min}(A)}{2}.$$

whence (3.19) follows again because, in this case, $\lambda = \operatorname{Re}\lambda$.

Thus, according to Property 5°, for a nonrotating capillary fluid, the transition of the eigenvalues from the right complex half-plane to the left one occurs along the real axis, where the eigenelements corresponding to such λ's have no associated elements.

9.3.4 Transition to an Equation with a Compact Operator

Let $U := U(\omega_0\nu^{-1}) := (-2i\omega_0\nu^{-1}S)^{-1}$ and P_0 be the orthoprojector on $\boldsymbol{H}_0 = \boldsymbol{M}_0(\Omega) \ominus \operatorname{Ker} B$. We introduce an auxiliary operator

$$R = R\left(\omega_0\nu^{-1}\right) := I + 2i\omega_0\nu^{-1}U\left(\omega_0\nu^{-1}\right)SP_0, \tag{3.21}$$

acting on the space $\boldsymbol{J}_{0,S}(\Omega)$. For $\omega_0 = 0$, the operator $R(\omega_0\nu^{-1})$ turns into the identity operator $R(0) = I$.

6° For any $\omega_0 \in \mathbb{R}$, the operator $R(\omega_0\nu^{-1})$ is invertible and $R^{-1}(\omega_0\nu^{-1})$ has the structure

$$R^{-1}\left(\omega_0\nu^{-1}\right) = I + T_1\left(\omega_0\nu^{-1}\right), \qquad T_1\left(\omega_0\nu^{-1}\right) \in \mathfrak{S}_\infty,\ T_1(0) = 0. \tag{3.22}$$

To prove Property 6°, we represent $R(\omega_0\nu^{-1})$ as $P_0 + Q_0 + 2i\omega_0 U(\omega_0\nu^{-1})SP_0$, where $Q_0 = I - P_0$. Since $I + 2i\omega_0\nu^{-1}U(\omega_0\nu^{-1})S = U(\omega_0\nu^{-1})$, then $R(\omega_0\nu^{-1}) = Q_0 + U(\omega_0\nu^{-1})P_0$.

We consider the equation $R(\omega_0\nu^{-1})\boldsymbol{\xi} = \mathbf{0}$ and represent the element $\boldsymbol{\xi} \in \boldsymbol{J}_{0,S}(\Omega)$ as $\boldsymbol{\xi} = \tilde{\boldsymbol{\xi}} + \boldsymbol{\xi}_0$, $\boldsymbol{\xi}_0 = P_0\boldsymbol{\xi}$. Thus, we have

$$Q_0\tilde{\boldsymbol{\xi}} + U(\omega_0\nu^{-1})P_0\boldsymbol{\xi}_0 = \tilde{\boldsymbol{\xi}} + Q_0U(\omega_0\nu^{-1})P_0\boldsymbol{\xi}_0 + P_0U(\omega_0\nu^{-1})P_0\boldsymbol{\xi}_0 = \mathbf{0}.$$

Since Q_0 and P_0 are mutually orthogonal orthoprojectors, we get

$$P_0U(\omega_0\nu^{-1})P_0\boldsymbol{\xi}_0 = \mathbf{0}, \qquad \tilde{\boldsymbol{\xi}} = -Q_0U(\omega_0\nu^{-1})P_0\boldsymbol{\xi}_0. \tag{3.23}$$

However, the operator $P_0U(\omega_0\nu^{-1})P_0 = P_0(I - 2\mathrm{i}\omega_0\nu^{-1}S)^{-1}P_0$ is invertible in $\boldsymbol{H}_0$. The proof of such a fact has been already encountered previously in (3.15), where we applies it to the orthoprojector I_0 and the invertibility in $\boldsymbol{M}_0(\Omega) \supset \boldsymbol{H}_0$. Then from the first equation in (3.23) it follows that $\boldsymbol{\xi}_0 = \mathbf{0}$ and from the second we get $\tilde{\boldsymbol{\xi}} = \mathbf{0}$, that is, $\boldsymbol{\xi} = \tilde{\boldsymbol{\xi}} + \boldsymbol{\xi}_0 = \mathbf{0}$.

Thus, the operator $R(\omega_0\nu^{-1})$ is invertible and since $U(\omega_0\nu^{-1})SP_0$ is a compact operator by virtue of the fact that S is compact, (3.22) is satisfied and Property 6° is proved.

Based on Property 6° we come from the system of equations (3.16) with an unbounded matrix operaor in the left side to the problem on eigenvalues for its inverse operator, which as it will be seen next, is compact.

7° Problem (3.16) is equivalent to a spectral problem of the form

$$\begin{aligned}
&\sigma\nu^{-2}R^{-1}(\omega_0\nu^{-1})U(\omega_0\nu^{-1})A^{-1}\boldsymbol{\xi}\\
&\qquad + R^{-1}(\omega_0\nu^{-1})\left[2\mathrm{i}\omega_0\nu^{-1}U(\omega_0\nu^{-1})SB_0^{-1} + \sigma\nu^{-2}U(\omega_0\nu^{-1})A^{-1}P_0\right]\boldsymbol{\eta}\\
&\quad = \tilde{\lambda}\boldsymbol{\xi}, \qquad \tilde{\lambda} = \frac{\sigma}{\nu\lambda},\\
&B_0^{-1}\boldsymbol{\eta} - \sigma\nu^{-2}P_0R^{-1}(\omega_0\nu^{-1})U(\omega_0\nu^{-1})A^{-1}\boldsymbol{\xi}\\
&\qquad - P_0R^{-1}(\omega_0\nu^{-1})\left[\sigma\nu^{-2}U(\omega_0\nu^{-1})A^{-1}P_0 + 2\mathrm{i}\omega_0\nu^{-1}U(\omega_0\nu^{-1})SB_0^{-1}\right]\boldsymbol{\eta}\\
&\quad = \tilde{\lambda}\boldsymbol{\eta}.
\end{aligned} \tag{3.24}$$

To get the system of equations (3.24) we apply the operator $\nu^{-1}U(\omega_0\nu^{-1})A^{-1}$ to both sides of the first equation and the operator B_0^{-1}, which exists because $\operatorname{Ker} B_0 = \{0\}$ and is compact, to the second equation. By using the definition of S, we obtain

$$\begin{aligned}
\boldsymbol{\xi} &= \lambda\nu^{-1}U(\omega_0\nu^{-1})A^{-1}(\boldsymbol{\xi} + \boldsymbol{\eta}) + 2\mathrm{i}\omega_0\nu^{-1}U(\omega_0\nu^{-1})S\boldsymbol{\eta},\\
\boldsymbol{\eta} &= -P_0\boldsymbol{\xi} + \lambda\nu\sigma^{-1}B_0^{-1}\boldsymbol{\eta}.
\end{aligned} \tag{3.25}$$

Putting the expression of $\boldsymbol{\eta}$ from the second equation into the first one we get

$$R(\omega_0\nu^{-1})\boldsymbol{\xi} = 2\mathrm{i}\omega_0\lambda\sigma^{-1}U(\omega_0\nu^{-1})SB_0^{-1}\boldsymbol{\eta} + \lambda\nu^{-1}U(\omega_0\nu^{-1})A^{-1}(P_0\boldsymbol{\eta} + \boldsymbol{\xi}). \tag{3.26}$$

Since, according to Property 6° the operator $R(\omega_0\nu^{-1})$ is invertible, from (3.26) after introducing a new spectral parameter $\tilde{\lambda} = \sigma/(\nu\lambda)$ we get the first equation (3.24) and after substituting the expression for $\boldsymbol{\xi}$ from (3.26) in the second equation (3.25) we get the second equation (3.24).

It is easy to see that all the coefficients of the operator-matrix standing in the left part of (3.24) are compact, since each of them consists of a product of bounded operators and one of the factors in the form A^{-1} or B_0^{-1} , which are compact operators.

9.3.5 Application of Perturbation Theory

Considering problem (3.24) from the perturbation theory of bounded operators leads to some preliminary remarks. The problem (3.24) contains several physical parameters: σ, ν and ω_0. If σ or ω_0 are changed, the configuration of the domain Ω filled with fluid is essentially changed. These issues are discussed in great detail in the book by A. D. Myshkis, V. G. Babsky, N. D. Kopachevsky, L. A. Slobozhanin, and A. D. Tyuptzov [1] and also partially in Sections 4.1 and 6.3 of the present book. Hence it follows that the operators presented in (3.24) are implicit functions of the parameters σ and ω_0. However, the configuration of the domain Ω and all the operators in (3.24) do not depend on the fluid's viscousity, ν, since the state of relative equilibrium of a uniformly rotating fluid in a container or that of a fluid at rest is determined by the interaction of the gravitational, capillary and centrifugal forces. The viscous forces acting in a fluid are displayed in dynamics only.

These considerations show that as a basic physical parameter, which may be taken into account in applying the perturbation theory in problem (3.24) it is necessary to take the kinetic viscousity ν, where we consider it as a large parameter, that is ν^{-1} is small. Moreover, since for large values of viscousity, ν, the operators $U(\omega_0\nu^{-1})$ and $R(\omega_0\nu^{-1})$ are close to the identity operators, then we may choose the following problem

$$\sigma\nu^{-2}A^{-1}\boldsymbol{\xi} = \tilde{\lambda}\boldsymbol{\xi}, \qquad B_0^{-1}\boldsymbol{\eta} = \tilde{\lambda}\boldsymbol{\eta}, \tag{3.27}$$

as the nonperturbed problem corresponding to (3.24), which was already studied previously.

8° If the viscousity ν is large enough that condition

$$\psi(\nu) := \frac{2\sigma\nu^{-2}\|A^{-1}\|(1+\omega_0\nu^{-1}\|A^{-1}\|) + (\sigma\nu^{-2} + 4\omega_0\nu^{-1}\|B_0^{-1}\|)\|A^{-1}\|}{1 - 2\omega_0\nu^{-1}\|A^{-1}\|)}$$
$$< (2|\lambda_{\min}(B_0)|)^{-1} \tag{3.28}$$

is satisfied, then problem (3.24) has at least κ eigenvalues in the left complex half-plane.

To prove this statement, let us note first that the difference between the operator matrix in problem (3.24) and the diagonal matrix $\operatorname{diag}(\sigma\nu^{-2}A^{-1}; B_0^{-1})$ in problem (3.27), that is, the operator matrix of perturbation $F = F(\omega_0, \nu)$, has the form $F = (F_{ik})_{i,k=1}^2$, where

$$\begin{aligned} F_{11} &= \sigma\nu^{-2}(R^{-1}(\omega_0\nu^{-1})U(\omega_0\nu^{-1}) - I)A^{-1}, \\ F_{12} &= R^{-1}(\omega_0\nu^{-1})U(\omega_0\nu^{-1})(2\mathrm{i}\omega_0\nu^{-1}SB_0^{-1} + \sigma\nu^{-2}A^{-1}P_0), \\ F_{21} &= -\sigma\nu^{-2}P_0R^{-1}(\omega_0\nu^{-1})U(\omega_0\nu^{-1})A^{-1}, \\ F_{22} &= -P_0R^{-1}(\omega_0\nu^{-1})U(\omega_0\nu^{-1})(\sigma\nu^{-2}A^{-1}P_0 + 2\mathrm{i}\omega_0\nu^{-1}SB_0^{-1}). \end{aligned} \tag{3.29}$$

To estimate the norm for small ν^{-1}, we need to transform the expressions F_{ik}. Let us represent F_{11} as

$$F_{11} = \sigma\nu^{-2}D_{11}A^{-1}, \qquad D_{11} = R^{-1}(\omega_0\nu^{-1})U(\omega_0\nu^{-1}) - I. \tag{3.30}$$

Then

$$\begin{aligned} R(\omega_0\nu^{-1})D_{11} &= U(\omega_0\nu^{-1}) - R(\omega_0\nu^{-1}) \\ &= (I - 2\mathrm{i}\omega_0\nu^{-1}S)^{-1} - (I + 2\mathrm{i}\omega_0\nu^{-1}(I - 2\mathrm{i}\omega_0\nu^{-1}S)^{-1}SP_0) \\ &= (I - 2\mathrm{i}\omega_0\nu^{-1}S)^{-1}(I - (I - 2\mathrm{i}\omega_0\nu^{-1}S + 2\mathrm{i}\omega_0\nu^{-1}SP_0)) \\ &= (I - 2\mathrm{i}\omega_0\nu^{-1}S)^{-1}2\mathrm{i}\omega_0\nu^{-1}S(I - P_0). \end{aligned} \tag{3.31}$$

Since

$$\begin{aligned} R(\omega_0\nu^{-1}) &= (I - 2\mathrm{i}\omega_0\nu^{-1}S)^{-1}(I - 2\mathrm{i}\omega_0\nu^{-1}S + 2\mathrm{i}\omega_0\nu^{-1}SP_0) \\ &= (I - 2\mathrm{i}\omega_0\nu^{-1}S)^{-1}(I - 2\mathrm{i}\omega_0\nu^{-1}S(I - P_0)), \end{aligned}$$

then from (3.31) we have

$$(I - 2\mathrm{i}\omega_0\nu^{-1}SQ_0)D_{11} = 2\mathrm{i}\omega_0\nu^{-1}SQ_0, \qquad Q_0 = I - P_0. \tag{3.32}$$

Therefore, (3.30) and (3.32) lead to the following relation,

$$F_{11} = F_{11}(\omega_0, \nu) = \sigma\nu^{-2}(I - 2\mathrm{i}\omega_0\nu^{-1}SQ_0)^{-1}2\mathrm{i}\omega_0\nu^{-1}SQ_0A^{-1}. \tag{3.33}$$

Using also the connection $S = A^{-1/2}S_0A^{-1/2}$, with $\|S_0\| = 1$, from (3.33) we have

$$\|D_{11}\| = \|(I - 2\mathrm{i}\omega_0\nu^{-1}SQ_0)^{-1}2\mathrm{i}\omega_0\nu^{-1}SQ_0\| \le \frac{2\omega_0\nu^{-1}\|A^{-1}\|}{1 - 2\omega_0\nu^{-1}\|A^{-1}\|}, \tag{3.34}$$

hence

$$\|F_{11}(\omega_0, \nu)\| \le \frac{2\sigma\omega_0\nu^{-3}\|A^{-1}\|^2}{1 - 2\omega_0\nu^{-1}\|A^{-1}\|} = \mathrm{O}(\nu^{-3}), \qquad \nu \to \infty. \tag{3.35}$$

Let us now represent $F_{21}(\omega, \nu)$ in (3.29) as

$$F_{21} = -\sigma\nu^{-2}P_0(D_{11} + I)A^{-1}.$$

Then using the estimate (3.34) we have

$$\|F_{21}\| \le \sigma\nu^{-2}\|A^{-1}\| \frac{1 + 2\omega\nu^{-1}\|A^{-1}\|}{1 - 2\omega_0\nu^{-1}\|A^{-1}\|} = \frac{\sigma\nu^{-2}\|A^{-1}\|}{1 - 2\omega_0\nu^{-1}\|A^{-1}\|} = \mathrm{O}(\nu^{-2}), \qquad \nu \to \infty. \tag{3.36}$$

Similar considerations for the operator $F_{12}(\omega_0, \nu)$ in (3.29) lead to the following formula

$$F_{12}(\omega_0, \nu) = (I - 2\mathrm{i}\omega_0\nu^{-1}SQ_0)^{-1}(2\mathrm{i}\omega_0\nu^{-1}SB_0^{-1} + \sigma\nu^2A^{-1}P_0),$$

and, therefore,

$$\|F_{12}\| \le \frac{(2\omega_0\nu^{-1}\|B_0^{-1}\| + \sigma\nu^{-2})\|A^{-1}\|}{1 - 2\omega_0\nu^{-1}\|A^{-1}\|} = \mathrm{O}(\nu^{-1}), \qquad \nu \to \infty. \tag{3.37}$$

Finally, since $F_{22} = -P_0F_{12}$ and using the estimate (3.37), we have

$$\|F_{22}\| \le \|F_{12}\| = \mathrm{O}(\nu^{-1}), \qquad \nu \to \infty. \tag{3.38}$$

From the estimates (3.35)–(3.38), we can deduce the following rough estimate of the norm of the matrix operator $F = (F_{ik})_{i.k=1}^2$,

$$\|F\| \le \sum_{i,k=1}^{2} \|F_{i,k}\| \le \psi(\nu) = \mathrm{O}(\nu^{-1}), \qquad \nu \to \infty, \tag{3.39}$$

where the function $\psi(\nu)$ is defined by the left side of inequality (3.28).

From the estimate (3.39) of the norm of the operator matrix of perturbation F in the spectral problem (3.27), we can immediately derive an application of perturbation theory. The spectrum of problem (3.27) consists obviously of a set of positive eigenvalues that is the union of the set of eigenvalues $\{\sigma\nu^{-2}\lambda_k(A^{-1})\}_{k=1}^{\infty}$ with a limit point at zero and the set of eigenvalues $\{\lambda_{k+q}^{-1}(B_0)\}_{k=1}^{\infty}$ of operator B_0^{-1} with a limit point at zero, too. Aditional eigenvalues of problem (3.27) are situated on the negative half-axis: $\lambda_\kappa^{-1}(B_0) \le \cdots \le \lambda_1^{-1}(B_0) < 0$. In this case, the distance between the set of negative eigenvalues to the spectrum in the right half-plane is equal to $|\lambda_1^{-1}(B_0)| = 1/|\lambda_{\min}(B_0)|$. Therefore, for satisfying the condition $\psi(\nu) \le r < 1/(2|\lambda_{\min}(B_0)|)$, that is, condition (3.28), the spectrum of the perturbed problem (3.24) contains at least κ eigenvalues (with due account of their multiplicities) situated in the left half-plane.

Statement 8° is now proved because $\psi(\nu) \to 0$ as $\nu \to \infty$ and, therefore, condition (3.28) is satisfied for large enough values of the viscousity ν.

9.3.6 The Existence of Eigenvalues in the Left Complex Half-Plane for an Arbitrary Viscosity Value

Let us state as a preliminary yet another property of the spectrum of problem (3.24) for large values of ν.

9° If condition (3.28) is satisfied, then problem (3.24) has exactly κ eigenvalues (with the account of their multiplicities) in the left half-plane.

Using Property 8°, it is enough to state that there could be no more than κ such eigenvalues. To this point, let us go back to equation (3.18), multiply scalarly its both sides by $\boldsymbol{\delta}$ and compute $\operatorname{Re}\lambda$; we will get

$$\operatorname{Re}\lambda = \frac{\|\boldsymbol{\delta}\|^2_{\boldsymbol{L}_2(\Omega)}\nu^{-1}}{\|A^{-1/2}\boldsymbol{\delta}\|^2_{\boldsymbol{L}_2(\Omega)} + |\lambda|^{-2}\sigma(B_0\boldsymbol{\delta}, \boldsymbol{\delta})_{\boldsymbol{L}_2(\Omega)}}. \tag{3.40}$$

Since in this expression the numerator is positive and the denominator— in virtue of condition (3.17)—can take negative values on a subspace whose dimension is no more than κ, then in the left half-plane there cannot be more than κ eigenvalues.

10° For any value of ν, problem (3.24) has exactly κ eigenvalues in the left half-plane.

For large enough values of the viscousity, $\nu = \nu_0$, this property has been already stated and proved provided condition (3.28) is satisfied. Let us now decrease the value of ν from ν_0 to any positive value $\nu = \nu_1$. Then, on one hand, the κ eigenvalues situated in the left half-plane are continuous functions of the parameter ν and, on the other hand, taking into account Properties 2° and 3° applied to problem (3.16), and thus to problem (3.24), the transition of the eigenvalues of problem (3.24) from one half-plane to the other passes through zero if $\operatorname{Ker} B_0 \neq \{0\}$, a condition that does not take place for the operator B_0. Hence, it follows that for problem (3.24), similar to problem (3.16), such a transition is impossible, that is, in the left complex half-plane there are exactly κ eigenvalues for an arbitrary value of $\nu > 0$.

Thus, the main theorem formulated in Section 9.3.1 is proved completely. It is enough to use consequently the properties 1°–10° for the solution of problem (3.1). In this case, according to Property 5°, for a nonrotating capillary fluid, the eigenvalues in the left half-plane are situated on the real axis.

A corollary of this main theorem—called in mechanics the *inverse of the Lagrange theorem on stability*—is the following statement:

If the potential energy of a uniformly rotating capillary viscous fluid in a container has no minimum (therefore, the quadratic form of the operator B_σ takes negative values), then problem (3.1) has at least one solution, for which $\operatorname{Re}\lambda < 0$. *An unstable mode of normal motion corresponds to this solution.*

9.4 Motions of a Rigid Body Containing a Cavity Filled with a Capillary Fluid under Conditions of Complete Low Gravity

The same method used to study the case of an immovable container or a container rotating relatively to a fixed axis can be also used to study the problem on dynamics of a body filled with a capillary viscous fluid.

9.4.1 Statement of the Problem

Let us assume that in the nonperturbed state the body with a cavity partially filled with a capillary fluid does not move, is fixed at the pole O, and there are no gravitational or other mass forces ($g = 0$). We will also assume that the free surface Γ of the fluid is horizontal and has the equation $x_3 = \text{const} = b$. The pressure in the fluid is constant and equals the external atmospheric pressure p_{a}. Such an equilibrium state requires also a special choice of the wetting angle (boundary angle) on the boundary $\partial\Gamma$ between Γ and the solid boundary S; this condition is supposed to take place in the sequel.

Let us consider small movements of the just introduced system. The statement of this problem can be derived from the initial boundary value problem (8.6.1)–(8.6.4), in which we formally assume $g = 0$ and change in the last condition (8.6.3)

$$\rho g(\zeta - \delta_2 x_1 + \delta_1 x_2)$$

with

$$\sigma B_\sigma(\zeta - \delta_2 x_1 + \delta_1 x_2),$$

where

$$\sigma B_\sigma u := -\sigma\Delta_\Gamma u \quad \text{on } \Gamma,$$
$$u = 0 \quad \text{on } \partial\Gamma.$$

Since $\Delta_\Gamma = \partial^2/\partial x_1^2 + \partial^2/\partial x_2^2$ for a horizontal Γ, then $\Delta_\Gamma(-\delta_2 x_1 + \delta_1 x_2) \equiv 0$ and the final formulation of the problem on small movements of the considered system is the following:

$$\boldsymbol{J}\frac{\mathrm{d}\boldsymbol{\omega}}{\mathrm{d}t} + \rho\int_\Omega \left(\boldsymbol{r} \times \frac{\partial \boldsymbol{u}}{\partial t}\right) \mathrm{d}\Omega = \boldsymbol{M}(t), \tag{4.1}$$

$$\frac{\partial \boldsymbol{u}}{\partial t} + \frac{\mathrm{d}\boldsymbol{\omega}}{\mathrm{d}t} \times \boldsymbol{r} = -\frac{1}{\rho}\nabla p + \nu\Delta\boldsymbol{u} + \boldsymbol{f}(t,x), \quad \mathrm{div}\boldsymbol{u} = 0 \quad \text{in } \Omega,$$

$$\boldsymbol{u} = 0 \quad \text{on } S, \qquad \frac{\partial \zeta}{\partial t} = u_n = u_3 \quad \text{on } \Gamma,$$

$$\rho\nu(u_{i,3} + u_{3,i}) = 0, \quad i = 1,2; \quad \text{on } \Gamma,$$

$$p - 2\rho\nu u_{3,3} = \sigma B_\sigma \zeta \quad \text{on } \Gamma, \qquad \zeta = 0 \quad \text{on } \partial\Gamma, \tag{4.2}$$

$$\boldsymbol{u}(0,x) = \boldsymbol{u}^0(x), \quad \zeta(0,x_1,x_2) = \zeta^0(x_1,x_2), \quad \boldsymbol{\omega}(0) = \boldsymbol{\omega}^0. \tag{4.3}$$

Here $\boldsymbol{\omega} = \boldsymbol{\omega}(t)$ is the angular velocity of rotation, $\boldsymbol{J}$ is the tensor of inertia of the system in the nonperturbed state, $\boldsymbol{M}(t)$ is the given moment of external forces relatively to the pole O, and $\boldsymbol{f}(t,x)$ is a given small field of volume forces influencing the fluid. All the other notations are the same as in Section 9.2.

If we assume $g = 0$ and change the term $\rho g \int_\Gamma |\zeta|^2 \mathrm{d}\Gamma$ for the expression of doubled potential energy of the system with $\sigma \int_\Gamma \nabla_\Gamma(\zeta,\zeta)\mathrm{d}\Gamma$, we realize that the law of full energy balance as expressed in (8.6.5) holds true for the classical solutions of problem (4.1)–(4.3).

9.4.2 Transition to a System of Operator Equations

As in Section 9.2, let us assume that the velocity field $\boldsymbol{u}(t,x)$ is a function of t with values in the space $\boldsymbol{J}^1_{0,S}(\Omega)$. Let us represent this field as a sum $\boldsymbol{u}(t,x) = \boldsymbol{s}(t,x) + \boldsymbol{w}(t,x)$, where $\boldsymbol{s}(t,x)$ is a solution of the boundary value problem I (see Section 8.1.2), in which we substituted $\boldsymbol{f}$ for $\boldsymbol{f} - \mathrm{d}\boldsymbol{u}/\mathrm{d}t - \mathrm{d}\boldsymbol{\omega}/\mathrm{d}t \times \boldsymbol{r}$, and $\boldsymbol{w}(t,x)$ is a solution of the boundary value problem II for $\psi = -\sigma B_\sigma \zeta$. Then the equations and the boundary conditins (4.2) give the following operator relations,

$$\nu A\boldsymbol{s} = -\frac{\mathrm{d}\boldsymbol{u}}{\mathrm{d}t} - P_{0,S}\left(\frac{\mathrm{d}\boldsymbol{\omega}}{\mathrm{d}t} \times \boldsymbol{r}\right) + P_{0,S}\boldsymbol{f}, \qquad \boldsymbol{u} = \boldsymbol{s} + \boldsymbol{w},$$

$$\nu A^{1/2}\boldsymbol{w} + \sigma A^{1/2} T B_\sigma \zeta = 0, \qquad \frac{\mathrm{d}\zeta}{\mathrm{d}t} = \gamma_n \boldsymbol{u} (= u_3), \tag{4.4}$$

where A and T are operators of the boundary value problems I and II and $P_{0,S}$ is the orthoprojector onto $\boldsymbol{J}_{0,S}(\Omega)$.

Hence, problem (4.1)–(4.3) is reduced to the evolution problem (4.1) and (4.4) with the initial conditions following from (4.3)

$$\boldsymbol{w}(0) = -\sigma\nu^{-1} T B_\sigma \zeta^0,$$
$$\boldsymbol{s}(0) = \boldsymbol{u}^0 - \boldsymbol{w}(0),$$
$$\zeta(0) = \zeta^0,$$
$$\boldsymbol{\omega}(0) = \boldsymbol{\omega}^0. \tag{4.5}$$

9.4.3 Normal Oscillations. Transition to an Equation with a Dissipative Operator

Let us consider the solutions of problem (4.1) and (4.4) for $\boldsymbol{M}(t) \equiv 0$ and $\boldsymbol{f}(t, x) \equiv \boldsymbol{0}$, that depend on t according to the law $\exp(-\lambda t)$. Then for the aplitude functions we obtain

$$\begin{aligned} &\nu A\boldsymbol{s} = \lambda(\boldsymbol{u} + P_{0.S}(\boldsymbol{\omega} \times \boldsymbol{r})), \qquad \gamma_n \boldsymbol{u} = -\lambda\zeta, \\ &\nu A^{1/2}\boldsymbol{w} + \sigma A^{1/2} T B_\sigma \zeta = 0, \qquad \boldsymbol{u} = \boldsymbol{s} + \boldsymbol{w}, \\ &\lambda\left(\boldsymbol{J}\boldsymbol{\omega} + \rho \int_\Omega (\boldsymbol{r} \times \boldsymbol{u}) d\Omega\right) = \boldsymbol{0}. \end{aligned} \tag{4.6}$$

Let us find the stationary solutions of problem (4.6), that is, solutions corresponding to the value $\lambda = 0$. We have

$$\begin{aligned} &\nu A\boldsymbol{s} = \boldsymbol{0}, \qquad \gamma_n \boldsymbol{u} = 0, \qquad \boldsymbol{u} = \boldsymbol{s} + \boldsymbol{w}, \\ &\nu A^{1/2}\boldsymbol{w} + \sigma A^{1/2} T B_\sigma \zeta = \boldsymbol{0}. \end{aligned} \tag{4.7}$$

Since $A \gg 0$, then from the first relation we obtain $\boldsymbol{s} = \boldsymbol{0}$ and, thus, $\boldsymbol{u} = \boldsymbol{w}$. However, in this case, from the second relation it follows that $\gamma_n \boldsymbol{u} = \gamma_n \boldsymbol{w} = 0$ and since the operator γ_n is invertible in the subspace $\boldsymbol{M}_1(\Omega)$ (see (2.2.36)), then $\boldsymbol{w} = \boldsymbol{0}$. By the invertibility of the operators $A^{1/2}$, T, and B_σ, from the second equality (4.7) we obtain that $\zeta = 0$.

Hence, the number $\lambda = 0$ is an eigenvalue of the problem (4.6); for this eigenvalue, $\boldsymbol{u} = \boldsymbol{0}$, $\zeta = 0$ and $\boldsymbol{\omega} = \boldsymbol{\omega}_0$ is an arbitrary vector. A slow rotation with angular velocity $\boldsymbol{\omega}_0$ around an arbitrary axis corresponds to this solution. Such an effect can take place in this case because the lack of gravitational forces and the system exists under the conditions of complete low-gravity.

Let us consider now nonstationary normal movements of the system, that is, the case when $\lambda \neq 0$. Then from the last condition (4.6) we have

$$\boldsymbol{\omega} = -\rho \boldsymbol{J}^{-1} \int_\Omega (\boldsymbol{r} \times \boldsymbol{u}) d\Omega.$$

Excluding the variable ζ, we obtain the following problem,

$$\begin{aligned} &\nu A\boldsymbol{s} = \lambda(I - \Pi)\boldsymbol{u}, \qquad \boldsymbol{u} = \boldsymbol{s} + \boldsymbol{w}, \\ &\sigma\nu^{-1} A^{1/2} T B_\sigma \gamma_n \boldsymbol{u} = \lambda A^{1/2}\boldsymbol{w}, \end{aligned} \tag{4.8}$$

where Π is the translation operator defined by formula (8.6.11). As an operator acting in $\boldsymbol{J}_{0,S}(\Omega)$, Π has the property $0 \leq \Pi < I$, and is a finite dimensional (three-dimensional) operator.

Let us make the following substitutions in (4.8),

$$A^{1/2}\boldsymbol{s} = \boldsymbol{\xi} \in \boldsymbol{J}_{0,S}(\Omega), \quad A^{1/2}\boldsymbol{w} = \boldsymbol{\eta} \in \boldsymbol{M}_0(\Omega), \tag{4.9}$$

perform the transition to a system with the unknown elements $\boldsymbol{\xi}$, $\boldsymbol{\eta}$ and apply the operator $A^{1/2}(I-\Pi)^{-1}$ to the first equation. Thus, we obtain the following eigenvalue problem,

$$\begin{aligned} \nu\tilde{A}\boldsymbol{\xi} - \sigma\nu^{-1}B(\boldsymbol{\xi}+\boldsymbol{\eta}) &= \lambda\boldsymbol{\xi}, \\ \sigma\nu^{-1}B(\boldsymbol{\xi}+\boldsymbol{\eta}) &= \lambda\boldsymbol{\eta}, \end{aligned} \tag{4.10}$$

where $B := A^{1/2}TB_\sigma\gamma_n A^{-1/2}$ is the operator that was investigated in details in Section 9.2 and the operator $\tilde{A}$ is defined by (see also (8.7.17))

$$\tilde{A} := A^{1/2}(I-\Pi)^{-1}A^{1/2}. \tag{4.11}$$

Let us consider the properties of $\tilde{A}$. It is defined on the dense set $\mathcal{D}(\tilde{A}) \subset \boldsymbol{J}_{0,S}(\Omega)$, which is the range $\mathcal{R}(\tilde{A}^{-1})$ of the compact positive operator $\tilde{A}^{-1} = A^{-1/2}(I-\Pi)A^{-1/2}$. The positivity of operator $\tilde{A}^{-1}$ follows from the fact that $A^{-1/2} > 0$ and $I-\Pi > 0$. Here, operator Π is three-dimensional. Furthermore, since $I-\Pi \leq I$, then the eigenvalues of $\tilde{A}^{-1}$ and A^{-1} satisfy $\lambda_k(\tilde{A}^{-1}) \leq \lambda_k(A^{-1})$, with $k = 1, 2, \ldots$, and, therefore,

$$\lambda_k(A) \leq \lambda_k(\tilde{A}), \qquad k = 1, 2, \ldots$$

The eigenvalues $\lambda_k(\tilde{A}^{-1})$ can be found as consecutive maxima of the variational ratio

$$\frac{((I-\Pi)\boldsymbol{v}, \boldsymbol{v})_{\boldsymbol{J}_{0,S}(\Omega)}}{E(\boldsymbol{v}, \boldsymbol{v})}, \qquad \boldsymbol{v} \in \boldsymbol{J}^1_{0,S}(\Omega),$$

and because Π is finite dimensional these eigenvalues have the same asymptotic behavior as $\lambda_k(A^{-1})$. Therefore, we have the following relation,

$$\lambda_k(\tilde{A}^{-1}) = \lambda_k(A^{-1})[1 + o(1)], \qquad k \to \infty, \tag{4.12}$$

where the asymptotics of the numbers $\lambda_k(A^{-1})$ are defined by (2.17). Hence, $\tilde{A}^{-1} \in \mathfrak{S}_p$ for $p > 3/2$.

Let us note now that the system of equations (4.10) has the same form as the system (2.8) for $\omega_0 = 0$. Here the difference is the following: A, the operator in (2.8), is replaced by $\tilde{A}$, the operator in (4.11), whose properties are closed to those of operator A. Hence we get that (4.10) is an eigenvalue problem for the operator

$$\tilde{\mathcal{A}}_0 := \begin{pmatrix} \nu\tilde{A} - \sigma\nu^{-1}B & -\sigma\nu^{-1}B \\ \sigma\nu^{-1}B & \sigma\nu^{-1}B \end{pmatrix}, \tag{4.13}$$

which—similarly to the operator $\mathcal{A}_0$ in (2.19)—is a dissipative operator for large values of viscousity.

9.4.4 Properties of Normal Movements

Using the conclusions of Section 9.2 on the properties of the solutions to equation (2.8) for $\omega_0 = 0$, or, similarly, to equation (2.59), let us formulate the corresponding properties of the solutions of problem (4.10).

1° The spectrum of problem (4.10) is discrete, it consists of the three-multiple eigenvalue $\lambda_0 = 0$ and the set $\{\lambda_j\}_{j=1}^{\infty}$ located in the right half-plane symmetrically relatively to the real axis, and $\lim_{j\to\infty}\lambda_j = \infty$. A slow uniform rotation of the system considered as a rigid body with an arbitrary angular velocity, $\boldsymbol{\omega}_0$, coresponds to the number $\lambda_0 = 0$. Here, the free surface of the fluid does not change and the field of relative velocity equals zero inside the region Ω.

Similarly to the problem in Section 9.2, in our problem, fading decrements $\lambda = \lambda_j$, with the unique limit point $\lambda = \infty$, correspond to the normal nonstationary movements.

2° For any $\varepsilon > 0$, all the eigenvalues λ_j except for maybe a finite number of them are located in the sector $|\arg\lambda| < \varepsilon$.

3° For the set $\{y_{j,q}\}_{j=1}^{\infty}$ of all the eigen- and associated elements $y_{j,q} = \{\boldsymbol{\xi}_{j,q};\boldsymbol{\eta}_{j,q}\}^{\mathrm{t}} \in \boldsymbol{E} = \boldsymbol{J}_{0,S}(\Omega) \oplus \boldsymbol{M}_0(\Omega)$, the assertions 4°–5° in Section 9.2.3 on completeness in the corresponding spaces hold true and the asymptotic formula (2.3.4) is valid for the numbers λ_j.

4° For the solutions of problem (4.10), all the mathematical and physical conclusions in Sections 9.2.4.–9.2.8, as well as the heuristic considerations on the spectrum for arbitrary values of the viscousity of a fluid hold true. Here, the operator A should be replaced by $\tilde{A}$ in all formulas, while considering that (4.12) takes place.

9.4.5 The Existence of a Generalized Solution to the Nonstationary Problem

In the evolution problem (4.1), (4.4) and (4.5), let us perform the transformations that have been already made while performing the transition from equations (4.6) to (4.10). We are obtaining thus the following Cauchy problem,

$$\frac{\mathrm{d}y}{\mathrm{d}t} + \tilde{A}_0 y = f(t), \qquad y(0) = y^0 \tag{4.14}$$

$$y(t) = (\boldsymbol{\xi}(t);\boldsymbol{\eta}(t))^{\mathrm{t}} \in \boldsymbol{E}, \qquad y(0) = (A^{1/2}\boldsymbol{s}(0); A^{1/2}\boldsymbol{w}(0))^{\mathrm{t}},$$

$$f(t) = \left(A^{1/2}(I-\Pi)^{-1}P_{0,S}(\boldsymbol{f}(t) - \boldsymbol{J}^{-1}\boldsymbol{M}(t)\times\boldsymbol{r});\boldsymbol{0}\right)^{\mathrm{t}}, \tag{4.15}$$

$$y^0 = (\boldsymbol{\xi}(0); \boldsymbol{\eta}(0))^{\mathrm{t}}, \qquad \boldsymbol{\eta}(0) = -\sigma\nu^{-1}A^{1/2}TB_\sigma\zeta^0,$$
$$\boldsymbol{\xi}(0) = A^{1/2}\boldsymbol{u}^0 - \boldsymbol{\eta}(0), \tag{4.16}$$

where the operator $\tilde{\mathcal{A}}_0$ is defined by the formula (4.13).

Using the conclusions of Section 9.2.9 for $\omega_0 = 0$, let us formulate the assertions on the solvability of the problem (4.14)–(4.16). Suppose the viscousity of the fluid is large enough so that

$$1 - \frac{\sigma}{\nu^2}\left(\|\tilde{R}\|^2 + \frac{\lambda_1(B)}{\lambda_1(\tilde{A})}\right) > 0, \qquad \tilde{R} := B^{1/2}\tilde{A}^{-1/2} \in \mathfrak{S}_\infty. \tag{4.17}$$

Then the operators $\tilde{\mathcal{A}}_0$ and $\tilde{\mathcal{A}}_0^*$ admit the estimates

$$\begin{aligned} \operatorname{Re}(\tilde{\mathcal{A}}_0 y, y)_{\boldsymbol{E}} &\geq c\|y\|^2_{\boldsymbol{E}}, \qquad y \in \mathcal{D}(\tilde{\mathcal{A}}_0), \\ \operatorname{Re}(\tilde{\mathcal{A}}_0^* z, z)_{\boldsymbol{E}} &\geq c\|z\|^2_{\boldsymbol{E}}, \qquad z \in \mathcal{D}(\tilde{\mathcal{A}}_0^*), \\ c &= \sigma\nu^{-1}\lambda_1(B) > 0. \end{aligned} \tag{4.18}$$

Further, the solution of the homogeneous problem (4.14) is given by $y(t) = \mathcal{U}(t)y^0$, where $\mathcal{U}(t)$ is a contractive semigroup that admits the estimate

$$\|\mathcal{U}(t)\| \leq \exp\left(-\frac{\sigma}{\nu}\lambda_1(B)t\right). \tag{4.19}$$

The property of exponential stability of the solutions of the homogeneous Cauchy problem (4.14) for $\omega^0 = 0$ follows from (4.19).

If the function $f(t)$ has values in $\boldsymbol{E}$, is continuous, and $y^0 \in \boldsymbol{E}$, then the problem (4.14)–(4.16) has a unique generalized solution $\boldsymbol{y}(t)$ that is also continuous in $\boldsymbol{E}$. Hence we obtain the following assertion on univalent solvability of the initial boundary value problem (4.1)–(4.3).

If the following conditions are satisfied

(a) $\boldsymbol{u}^0 \in \boldsymbol{J}^1_{0,S}(\Omega)$, $\zeta^0 \in H^{3/2}_\Gamma(\Gamma)$;

(b) $\boldsymbol{M}(t)$ *is a continuous function with values in* $\mathbb{R}^3$;

(c) $\boldsymbol{f}(t,x)$ *is a continuous function of* t *with values in* $\boldsymbol{J}^1_{0,S}(\Omega)$,

then the problem (4.1)–(4.3) has a generalized solution for which the function $\boldsymbol{u}(t,x) = \boldsymbol{s}(t,x) + \boldsymbol{w}(t,x) \in \boldsymbol{J}^1_{0,S}(\Omega)$ *is continuous in* t, *and the angular velocity* $\boldsymbol{\omega}(t)$ *of the system's rotation is given by the formula*

$$\boldsymbol{\omega}(t) = \boldsymbol{\omega}^0 + \int_0^t \boldsymbol{J}^{-1}\boldsymbol{M}(t)\mathrm{d}t - \rho\boldsymbol{J}^{-1}\int_\Omega (\boldsymbol{r} \times (\boldsymbol{u}(t) - \boldsymbol{u}^0))\mathrm{d}\Omega$$

and is a continuous function from $\mathbb{R}^3$. *Here, the deviation* $\zeta(t,x_1,x_2)$ *of the free surface of the fluid from the equilibrium surface* Γ *during the process of oscillations equals*

$$\zeta(t,x_1,x_2) = \zeta^0(x_1,x_2) + \int_0^t u_3(t,x_1,x_2,b)\mathrm{d}t.$$

In conclusion, let us point out that the previously mentioned reasoning and investigations of the evolution and the spectral problems admit natural generalizations for the following cases: (a) oscillations of the system under conditions of complete low gravity and nonplane surface Γ (when Γ is a part of a sphere); (b) motion under the influence of a homogeneous gravitation field along the axis Ox_3 ($\boldsymbol{g} = -g\boldsymbol{e}_3$, $g \neq 0$); (c) oscillations of a rotating system ($\boldsymbol{\omega}_0 \neq \mathbf{0}$).

Appendix C

Remarks and Reference Comments to Part III

C.1 Chapter 7

7.1. The equation of fluid motion in the form (7.1.4) was considered for the first time by S. G. Krein [1]. This work also studies the corresponding nonlinear equation, for which the existence theorem of a local in time solution was proved.

The spectrum asymptotics of the Stokes operator in the form (7.1.11) follows from the G. Metivier's results in [1]. Further research leading to formula (7.1.12) was done by A. N. Kozhevnikov [1]. The formula for the inverse operator of the Stokes operator A_0 that he obtained is interesting,

$$A_0^{-1} = G_1(I - 2\nabla(I + G_2K_{-1}\gamma)\mathrm{div}G_1).$$

Here, G_1 and G_2 are operators that solve the classical Dirichlet problems for the Poisson and Laplace equations, respectively; γ is the trace operator, that is, function restriction to the boundary $\partial\Omega$, which is supposed to be infinitely smooth; and K_{-1} is a pseudodifferential operator on $\partial\Omega$ of order -1, that is, an integral operator with a weak singularity. This description allows us to obtain the solution estimates for the Stokes problem in Sobolev spaces.

The disturbance of the stationary motion of a fluid (see Section 7.1.4) was studied by S. G. Krein in [2]. He used there for the first time the Keldysh theorems to prove that the system of eigen- and associated functions of a hydrodynamic problem is complete. V. I. Yudovich studied this problem further in more detail. He proved rigorously that it is possible to decide on the stability or the instability of a stationary flow of a viscous incompressible fluid by analyzing the location of the spectrum of the operator $-\nu A_0 - R$. Here, the stability is studied in terms of the $L_p(\Omega)$ spaces, for various values of p.

In particular, he studied the action of the operator P_0 in these spaces. A more detalied presentation of these problems can be found in V. I. Yudovich's book [1].

Spectrum properties for the rotating fluid were studied by N. D. Kopachevsky in [4].

7.2. The translation operator B was independently introduced by A. I. Kobrin [1] and by Ngo Zui Kan [1]. Its most important property, $I - B \gg 0$, was proved by A. I. Kobrin (Ngo Zui Kan introduced some redundant additional restrictions while proving that property).

The existence theorem was proved by Ngo Zui Kan [1]. A similar theorem was later on proved by I. G. Zagnibeda [1], under more rigid restrictions. The spectral problem (7.2.18) on normal oscillations was studied by N. D. Kopachevsky.

7.3. The problem (7.3.1)–(7.3.3) was first studied by Ngo Zui Kan [3], who proved the existence theorem of solution. In our present monograph, we considered this problem under a rather different form than the one given by Ngo Zui Kan in [3]. We added the spectral problem introduced by N. D. Kopachevsky.

Among the first problems that dealt with the dynamics of a body with a cavity filled with a viscous fluid we note the problems on oscillations of a pendulum with a sherical cavity that were studied by P. S. Krasnoschokov [1], [2] (in the case of small values of viscosity) and by O. B. Ievleva [1], [2] (for arbitrary values of viscosity).

7.4. The study of the asymptotic solutions of the problems presented in this chapter in case of large values of viscosity (small Reinolds numbers) was originated by F. L. Chernousko [1]. For $\omega_0 = 0$, he obtained the main term of the asymtotics, introduced the tensor $P^{(1)}$, proved its positive definiteness, and carried out the calculations in the case of an elliptic cavity.

The more general case presented in Section 7.4 was considered by S. G. Krein and Ngo Zui Kan [3]. A. I. Kobrin [1] proved that in the case of large values of viscosity, one can apply the method of studying the systems with small parameters for higher order derivatives and he gave an algorithm for constructing the asymptotic expansion according to the scheme presented by A. B. Vasilieva and V. F. Butuzov [1].

It is necessary to remark that F. L. Chernousko [3] considered solution asymptotics for large Reinolds numbers (small values of viscosity) as well. Those results were not included in our monograph because, as far as we know, there is no adequate operator scheme for the method of a boundary layer on space variables that Chernousko used in [3].

7.5. The study of the problem on oscillations of a pendulum with a cavity completely filled with a viscous fluid by methods of functional analysis was considered by Ngo Zui Kan [1] (see also [2]–[7]) and more recently by N. D. Kopachevsky and E. D. Volodkovich [1]. Our Section 7.5 was based on their studies. The study by numerical methods of the same problem was carried out by M. Ya. Barnyak [2] and by M. Ya. Barnyak and R. I. Tzebrii [1], [2].

It is necessary to remark that in the case considered in Section 7.5, the mass center C of the system does not coincide with the suspension center O. Moreover, the problem is studied without using the translation operator Π that was mentioned in Section 7.2, but rather with the help of the joint study of the equation of the fluid motion and the equation of the body motion.

We note that the three-dimensional problem, that is, the problem of a spherical pendulum with a cavity completely filled with a visous fluid can be carried out along the same lines with insignificant complications.

7.6. The study of the problem on a viscous fluid flowing through a given region was based on the work of N. D. Kopachevsky and S. G. Krein [1]. Chan Thu Ha [2] studied the same problem in the case of a heated fluid.

7.7. The convective motions of a fluid in a closed container was studied by many authors. We mention here the well-known books by G. Z. Gershuny and E. M. Zhukhovitskii [1] and by G. Z. Gershuny, E. M. Zhukhovitskii, and A. A. Nepomnyatshii [1].

Section 7.7 was written based on the work of Ngo Zui Kan [6]. However, the results stated there were presented following the scheme developed by N. D. Kopachevsky, which enabled us to go beyond Ngo Zui Kan's work. Formula (7.7.11) is a classical result of H. Weyl. The asymptotic formula (7.7.27) is a consequence of formulas (7.1.11), and also the form of the operator in (7.7.24).

In connection with this problem we have to mention also the works of A. G. Zarubin [1], A. G. Zarubin and Ngo Zui Kan [1], [2], and E. L. Tarunin [1].

The problem on the transition of the eigenvalues to the left half-plane in the case of a fluid heated from below was studied by N. D. Kopachevsky and is presented here for the first time.

C.2 Chapter 8

8.1. The linearized problem on the motion of a viscous incompressible fluid in an open container was studied for the first time in its general case by S. G. Krein [1], [2]. These two works also propose a method to reduce the investigation of this problem to solving an evolution equation in the Hilbert space $\boldsymbol{J}_{0,S}(\Omega) \oplus \boldsymbol{J}_{0,S}(\Omega)$, a method that can be used in other hydrodynamics problems and in the theory of elasticity. These results were presented in detail in S. G. Krein's and G. I. Laptev's article [1]. The approach in our monograph is based on the application of the abstract scheme developed in Section 1.8 and the results in S. G. Krein's monograph [K].

The asymptoptic formula (8.1.12) for the eigenvalues of the operator A follows from G. Metivier's results in [1]. The asymptotic formula (8.1.33) for the eigenvalues of the operator B was obtained using a heuristic method by N. D. Kopachevsky and thoroughly proved by N. A. Karazeeva and M. Z. Solomyak [1].

8.2. The main operator pencil (8.2.1) was obtained by S. G. Krein [3], [4] who also investigated properties of its spectrum. The pencil linearization method (see Section 8.2.2) was suggested by G. I. Laptev. In the joint work by N. K. Askerov, S. G. Krein, and G. I. Laptev [1] (see also [2]) the authors studied the system of eigen- and associated elements of a pencil, and using M. V. Keldysh theorem (see Section 1.6.5), they proved the two multiple completeness of this system in an original way. Some corrections to those properties were given by V. M. Greenlee [1], [2], and they are reflected in our monograph. Specifically, the pencil linearization presented in V. M. Greenlee [1] is somewhat changed and it is used in Section 8.2.

It is necessary to point out that by the time pencil (8.2.1) was used in the problem on hydrodynamics, the theory on quadratic operator pencils has been already created, essentially by M. G. Krein and H. Langer [1], [2]. However, the pencil (8.2.1) does not fall within the range of that theory. Additional investigations were still needed.

The first to do that were I. Gohberg and M. G. Krein, who in their work [GK] dealt with the case of a strongly damping pencil for which inequality (8.2.7) is satisfied. They obtained theorems on the existence of the two Riesz bases of eigen-elements corresponding to eigenvalues of the first and second order and studied the asymptotics of these eigenvalues.

E. Z. Mogulskii [1] obtained theorems on n-multiple completeness of eigen- and associated elements for a broad class of polynomial pencils, among which the pencils of type (8.2.1) were a particular case.

A further step was done by E. A. Larionov [1]. He proved in a general setting the existence of two Riesz bases for the pencil (8.2.1) distinguishing between the eigenelements of the first order, neutral, and the eigenelements of the second order. Those Riesz bases contained eigenelements of the first (second) order and a part of the neutral root elements. He also obtained, among other results, a theorem on the length of chains of eigen- and associated elements corresponding to eigenvalues of the pencil. Larionov's results were further developed by V. M. Greenlee [1].

The proof of property (8.2.26) on the separation zones Δ_- and Δ_+ under the condition of strong damping (8.2.7) was done by A. S. Markus, V. I. Matsaev, and G. I. Russu [1]. There they obtained also the theorem on the factorization of pencil (8.2.1).

The two-sided estimates for the two branches of eigenvalues of the pencil (8.2.1) were obtained by V. A. Grinstein and N. D. Kopachevsky [1], [2].

Additional fundamental results in the study of the pencil (8.2.1), and of other pencils of the same type, were obtained by A. S. Marcus and V. I. Matsaev [1]–[5], A. A. Shkalikov and V. T. Pliev [1], A. A. Shkalikov [1]–[7], A. G. Kostuchenko and M. B. Orazov [1], [2], A. G. Kostuchenko and A. A. Shkalikov [1], [2], T. Ya. Azizov and Iohvidov [AI], G. V. Radzievskii [1]–[4], T. Ya. Azizov and N. D. Kopachevsky [1], T. Ya. Azizov and L. I. Sukhocheva [1], Chan Thu Ha [1], A. G. Garadzhaev [1], V. A. Grinstein [1], [2], N. D. Kopachevsky [4]–[7], [11], L. A. Kotko and S. G. Krein [1], A. N. Kozhevnikov [2], M. B. Orazov [1]–[3], M. B. Orazov and K. A. Shukurov [1], S. Ya. Yakubov [1], V. I. Yudovich [2], and by many others.

8.3. The problem on normal oscillations of a viscous fluid in an open container was studied by S. G. Krein [3], and N. K. Askerov, S. G. Krein, and G. I. Laptev [1], and also by N. D. Kopachevsky [4]–[7], A. G. Garadzhaev [1], [2], M. Ya. Barnyak [1], [3], [4], and F. L. Chernousko [2], [4]–[6].

Thus, in his work [1], A. G. Garadzhaev studied the behavior of the eigenvalues of the spectral problem in the case when the viscosity of the fluid is decreasing. He stated that if the viscosity ν is decreasing, the eigenvalues $\lambda_k^-(\nu)$ move to the right, and the eigenvalues $\lambda_k^+(\nu)$ move to the left. In their collision, some nonreal or multiple real eigenvalues may appear and there are specific examples when these situations take place. In order to establish these facts, A. G. Garadzhaev used the methods developed by A. G. Kostuchenko and M. B. Orazov [1], [2] in the problem on the theory of elasticity, as well as general results obtained by A. G. Kostuchenko and A. A. Shkalikov [1], [2] on quadratic pencils.

The case of a fluid with low viscosity was considered by F. L. Chernousko [3]. Formula (8.3.12) in our monograph was obtained by him using the methods of a boundary layer.

The properties of surface and internal waves were studied by N. D. Kopachevsky [4], [5].

We acknowledge also the interesting results obtained by M. Ya. Barnyak [1], [3], [4] on the development of numerical methods in solving the problem on normal oscillations of a heavy viscous fluid in a container. In particular, he developed an approach for calculating the multiple real eigenvalues that appear before their yield in the nonreal region of the complex plane by using the Galerkin method.

8.4. The statement, the investigation and the basic results in the problem on the oscillations of a rotating heavy fluid were obtained by N. D. Kopachvsky [4], [5].

8.5. The asymptotic solutions of the evolution and spectral problems on oscillations of a viscous fluid in an open container are based on the general theory developed by S. G. Krein [K] that was applied to a hydrodynamic problem by S. G. Krein and Ngo Zui Kan [1], [3]. Let us notice that the spectrum asymptotics of a rotating fluid with low viscosity was studied by N. K. Radyakin [1], [2].

8.6. The problem on the oscillations of a system of nonmixing viscous fluids with high viscosity was studied by N. D. Kopachevsky [4], [5]. The asymptotic formulas (8.6.18), (8.6.21), and (8.6.23) were obtained by T. A. Suslina [3]–[5], [9] (see also [1], [2], [6]–[8], [10], and [11]).

8.7. The problem on small oscillations arround a fix point of a body with a cavity partially filled with a viscous fluid was originated by S. G. Krein and Ngo Zui Kan [2].

8.8. The small motions of a plane pendulum with a cavity partially filled with fluid were studied without using the translation operator by N. D. Kopachevsky and E. D. Volodkovich [2]. Their study was based on the imediate analysis of the equations for motion and kinetic moment. A similar problem was investigated by D. N. Kopachevsky and Vadiaa Ali [1]–[4] for a system of nonmixing fluids that fill a cavity of a plane pendulum.

8.9. The small convective motions of a fluid in a partially filled bounded container were studied by many authors. We mention here, among others, the monographs by G. Z. Gershuny and E. M. Zhukhovitskii [1] and G. Z. Gershuny, E. M. Zhukhovitskii and A. A. Nepomnyaschii [1]. Section 8.9 was written based on the work of N. D. Kopachevsky and Ngo Zui Kan [1], where the evolution problem had been studied. The properties of the solutions of the spectral problem in the cases of a fluid heated from above or from below were investigated by N. D. Kopachevsky and they are reflected in this section. The same problem was studied further by Chan Thu Ha [3], [4].

8.10. The problem of obtaining simple sufficient conditions for the instability of the convective motions of a fluid in an arbitrary container was stated in the monograph by N. D. Kopachevsky, S. G. Krein, and Ngo Zui Kan [KKN]. The solution of this problem was obtained by N. D. Kopachevsky and V. N. Pivovarchik [1], [2]. It involved the use of a transition method to a two-parameter operator pencil [see (8.10.4)]. V. N. Pivovarchik also used it in his other works on the theory of quadratic operator pencils, [1] and [2].

Let us notice that the spectral asymptotics of the additional problem (8.10.29) were studied by T. A. Suslina and A. B. Alexeev. In particular, they obtained the asymptotic formula

$$\lim_{k\to\infty} k\left|\mu_k^{\pm}\right|^{3/2} = \frac{\sqrt{2}\mathrm{mes}\,\Omega}{5\sqrt{\pi}\,\Gamma^2\left(\dfrac{1}{4}\right)},$$

which was further justified by T. A. Suslina [10].

C.3 Chapter 9

During the period of time between the All-Union Congress on Mechanics (USSR, Moscow, 1964) and the International Mathematical Congress (USSR, Moscow, 1966), when an intensive study of the behavior of an incompressible fluid under conditions close to weightlessness began, researchers raised the question on how do the capillary forces influence the spectral structure of the normal oscillations of a viscous fluid in a partially filled container.

In his report at the Congress on Mechanics, S. G. Krein stated the hypothesis that the surface forces remove the limit point of the spectrum at zero in the complex plane. Namely, the spectrum of the considered problem, unlike the results in Chapter 8 for a heavy fluid, is discrete in the case of a capillary fluid, and its only limit point is at infinity. In further discussions centered around the same problem, S. G. Krein, N. D. Kopachevsky, and A. D. Myshkis suggested also that with regard to nonreal eigenvalues associated with arbitrary values of viscosity, for a capillary fluid with low viscosity there are no more than a finite number of nonreal eigenvalues, whileas in the case of a capillary fluid with high viscosity, those eigenvalues are absent. These statements were proved for the simplest model problem by N. D. Kopachevsky and A. D. Myshkis [1], [2]. Later on, N. D. Kopachevsky [1] studied also a general scalar problem that reflected all the peculiarities of the problem and addressed the case of sufficiently smooth uncrossed free surface Γ and solid wall S of the container Ω.

9.1. The statement of the problem on small oscillations of a capillary viscous fluid in an arbitrary container was advanced in the work of N. D. Kopachevsky and A. D Myshkis [1]. Because there is still a dispute going on about the way of selecting boundary conditions on the contour of moistening $\partial\Gamma$, we presented in Section 9.1 two basic possible variants (9.1.5′) and (9.1.5″), but in the sequel we dealt with the first one only.

The main reason that gave us confidence in the fact that the previously mentioned hypotheses are valid, were two specific problems studied by N. D. Kopachevsky and A. D Myshkis [1], [2]. They are described in complete detail in Sections 7.3 and 7.4 of the monograph by A. D Myshkis, V. G. Babskii, N. D. Kopachevsky, L. A. Slobozhanin, and A. D. Tyuptsov [1] (see also a previous monopraph by V. G. Babskii, N. D. Kopachevsky, A. D Myshkis, L. A. Slobozhanin, and A. D. Tyuptsov [1] and the following book by A. D Myshkis, V. G. Babskii, M. Y. Zhukov, N. D. Kopachevsky, L. A. Slobozhanin, and A. D. Tyuptsov [1]). In that monograph, in the problem on oscillations of a capillary viscous drop, the authors managed to separate the variables of generalized spherical functions. The same procedure for a Helmholtz vector equation for a solenoidal field is discussed in the book by I. M. Gelfand, P. A Milnos, and Z. Ya. Shapiro [1]. In our Section 9.1.3, we presented only the basic conclusions from the previously mentioned work of N. D. Kopachevsky and A. D. Myshkis [1], [2]. We notice that the oscillations of a self-gravitating noncapilary viscous drop were studied by S. Chandrasekhar [1].

9.2. Section 9.2 is the most important one in Chapter 9. Earlier on, N. D. Kopachevsky [1], [2] studied the initial boundary value problem (9.1.1)–(9.1.6) as well as the spectral problem (9.1.17) using operator methods (see also the presentation of the problem in Section 7.9 of the monograph by A. D Myshkis, V. G. Babskii, N. D. Kopachevsky, L. A. Slobozhanin, and A. D. Tyuptsov [1]). All those early works were still assuming that there was no intersection between the free equilibrium surface Γ of the fluid, and the solid wall S of the container Ω, and the equilibrium state of the fluid is statically stable in linear approximation.

In Section 9.2 we present brand new results obtained by N. D. Kopachevsky. Here, the same problem is considered simultaneously in two different situations: For smooth and nonintersecting Γ and S and for smooth intersecting Γ and S, with the boundary condition (9.1.5′). The second part of the research scheme is different from the scheme presented earlier by N. D. Kopachevsky [1], [2] (see also the previously mentioned monograph by A. D Myshkis, V. G. Babskii, N. D. Kopachevsky, L. A.

Slobozhanin, and A. D. Tyuptsov [1]). When the system is not rotating, that is, for $\omega_0 = 0$, the results of the theory of linear operators in a space with indefinite metrics are essentially used. Thus, for sufficiently large values of the viscosity of a fluid, it is possible to get a theorem of existence of not only weak, but also of generalized solutions. Though the results obtained in Sections 9.2.6 and 9.2.7 are of a limited character, we think that the previously mentioned hypothesis on the finiteness of the number of nonreal eigenvalues is still true.

The asymptotic formula (9.2.16) was obtained by N. D. Kopachevsky by a heuristic method and proved by T. A. Suslina [3]–[6], [9]. At N. D. Kopachevsky's request, formula (9.2.28) was found and proved by T. A. Suslina for $0 < \delta < \delta^*$.

The results presented in Section 9.2.5 as well as Property 7° in Section 9.2.6 were obtained at the authors' request by T. Ya. Azizov (see also the monograph by T. Ya. Azizov and I. S. Iohvidov [AI]).

Property 15° in Section 9.2.6 was proved by M. B. Orazov [2]. In this work, M. B. Orazov studied in detail the behavior of the two branches of eigenvalues λ_k^+ and λ_k^- as functions of the parameter $\lambda = \sigma\nu^{-2} > 0$. The transition mechanism of pairs of real eigenvalues (after their collision) in the complex space was considered there as well (see Property 4° in Section 9.2.8).

A similar condition to the one of uniform boundedness for nonintersecting Γ and S was used also by M. B. Orazov [2]. The selection law mentioned in Property 6° (Section 9.2.8) was pointed out by M. B. Orazov [2] as well (see also A. G. Kostuchenko and M. B. Orazov [1], [2], A. G. Kostuchenko and A. A. Shkalikov [1], [2], and A. A. Shkalikov [2]–[5]).

As for the problem on the finiteness of the number of nonreal eigenvalues, we have to mention a result due to S. A. Stepin [1], [2]. For the scalar function $u = u(r, \varphi)$ in the unit circle Ω he considered the system

$$-\Delta u = \lambda u \quad \text{in } \Omega,$$
$$\lambda \left(\frac{\partial u}{\partial n}\right)_{r=1} = \alpha \left(-\mathrm{i}\frac{\partial}{\partial \varphi}\right)^{\chi} u\bigg|_{r=1} \qquad \text{on } \Gamma = \partial\Omega,$$

that models the problem on normal oscillations of a capillary viscous fluid. The main result related to this problem in S. A. Stepin [2] is the following.

The set of nonreal eigenvalues is (a) finite if $\chi < 3$; (b) infinite if $\chi > 3$ or $\chi = 3$, $\alpha > 1$ and (c) the spectrum of the problem is real if $\chi \le 3$; and $\alpha \in [0, 1/4]$.

The proof of the existence of a generalized solution of the initial boundary value problem (Section 9.2.9) based on the property of maximal dissipativity of the operator coefficient $\mathcal{A}$ (see (9.2.78), (9.2.83)) was done by N. D. Kopachevsky. We also note that A. M. Gomilko (private communication) studied problem (9.1.1)–(9.1.6) in negative spaces and with a boundary condition (9.1.5″). The theorem on the existence of a weak solution was proved by A. M. Gomilko using the property of dissipativity of the main operator acting on a space with a negative norm.

Finally, let us acknowledge the work of A. V. Trubachev [1], where it is stated that the spectrum of the problem studied in Section 9.2 lies in the region

$$|\operatorname{Im}\lambda| \leq c/|\lambda|, \qquad \operatorname{Re}\lambda > 0.$$

9.3. The inverse of Lagrange theorem on stability for a capillary viscous fluid that partially fills an arbitrary stationary container was proved by N. D. Kopachevsky [3]. In the case when the container is rotating uniformly around a fixed axis, a similar theorem was obtained by N. D. Kopachevsky and E. D. Volodkovich [3].

9.4. The problem on the motion of a rigid body, with a cavity partially filled with a capillary viscous fluid, and in a state of complete weightlessness was considered by Ngo Zui Kan [4]. Here, it was suggested that $\Gamma \cap S = \emptyset$. The presentation of the results in Section 9.4 goes along the same lines as the one in Section 9.2 and was initiated by N. D. Kopachevsky.

In connection with the problems discussed in Chapter 9, we would like to mention also the work of V. V. Rumyantzev [1], V. A. Solonikov [1]–[3], and A. G. Shmidt [1]. We notice in addition that in the work of N. D. Kopachevsky [12] the proof of the existence of a generalized solution to problem (9.1.1)–(9.1.6) is based on Galerkin's method.

Finally, let us look briefly at some other research dealing with the issues presented in Chapters 7 through 9. The initial boundary-value problem for the Navier-Stokes linearized equations was studied by A. A. Kiselev [1], and the case of nonlinear equations was considered by O. A. Ladyzhenskaya [1] and by R. Temam [1]. The problem on the oscillations of an weakly viscous fluid was studied by N. N. Moiseev [1], F. L. Chernousko [3], [6], and by N. D. Kopachevsky and N. K. Radyakin [1]. The dynamics of a body with a cavity filled with one or several viscous fluids was considered by F. L. Chernousko [1]–[6], S. I. Krushinskaya [1], Ngo Zui Kan [1], [3]–[6], and E. P. Smirnova [1]. The convection problems in a fluid were dealt with by N. K. Korenev [1]. The stability of hydrodynamic systems and the operator generalizations of these problems were studied by V. N. Pivovarchik [1], [2], S. M. Barkar [1]–[3], and

[7]. Finally, let us notice that as it was shown by N. D. Kopachevsky and A. N. Temnov [1], [2], as well as by A. N. Temnov [1], [2], the main operator pencil (8.2.1) also arises in the problem of small oscillations of stratified or nonhomogeneous fluids in a container.

PART IV

SMALL OSCILLATIONS OF COMPLEX HYDRODYNAMIC SYSTEMS

This part of the second volume is a collection of recently published results on small motions and normal oscillations of hydrodynamic systems that are even more complicated than the ones presented in Part III. These hydrodynamic systems are partially conservative and thus similar to the ones considered in Part II of the first volume and partially dissipative (see Part III). The union of such subsystems forms a partially dissipative hydrosystem that inherits both their particularities. Thus, the common properties of the two subsystems having a discrete spectrum with various physical meanings hold true. However, the spectrum of the whole system is compound and partially deformed. These and other effects related to the properties of completeness and basicity of the system of eigen- and associated elements of the common hydrodynamic system are discussed in Chapter 10, where we study the problem on the oscillations of a system formed by a viscous fluid plus an ideal fluid in an arbitrary container. It is necessary to point out that if the ideal fluid does not have a free surface (e.g., if the ideal fluid is underneath a viscous one, or if the given hydrosystem completely fills the container), then the problems become similar to the ones discussed in Part III. Therefore, in Chapter 10, we study the more complex and more interesting problems when every subsystem of the common hydrosystem makes its own contribution to the properties of the whole system.

In Chapter 11, we study problems on the oscillations of a fluid that is not yet a homogeneous viscous one, but has a more complicated physical nature. We first discuss the problems on motions of a visco-elastic fluid. In this case, contrary to the problems considered in Chapters 7–8, the physical properties of the fluid generate additional mathematical and physical effects. Thus, in the spectral problems the property of the spectrum discreteness is preserved, however along with the spectrum limit points, $\lambda = 0$ and $\lambda = \infty$, some new limit points and branches of the eigenvalues can show up. The number of such limit points depends on the choice of the visco-elastic fluid. In the last section of Chapter 11 we consider the problem on small motions of an ideal relaxing fluid. In this problem, the study of normal oscillations involves a polynomial operator pencil of the third degree.

Chapter 10

Oscillations of Partially Dissipative Hydrosystems

In this chapter we consider the problem on small motions and normal oscillations of a hydrosystem consisting of two fluid layers each one of a different physical nature: The lower layer is a homogeneous, viscous fluid and the upper one is an ideal homogeneous fluid with a free surface.

In Section 10.1, we give the complete statement of the initial boundary value problem on the motion of such a hydrosystem. We infer the law of balance of full energy and formulate the problem on normal oscillations.

In Section 10.2, we introduce several auxiliary problems and their corresponding operators that help us switch from the initial boundary value problem to a differential-operator equation of the first order in a Hilbert space. Then, based on the property of dissipativeness of the operator coefficient of the equation, we prove the theorem on correct solvability of the initial boundary value problem.

To study the possible characteristics of the spectrum of the problem on normal oscillations of a general hydrosystem, in Section 10.3 we first consider a model spectrum problem containing the general peculiarities of the original problem and allowing the separation of variables in a rectangular region. The properties of the solutions of this model problem allow us to state a hypothesis on the spectrum structure in the problem on normal oscilations.

Based on these preliminary considerations, in Section 10.4, we study the properties of the spectrum of the investigated problem on the oscillations of a partially dissipative hydrosystem in an arbitrary container. We deal here with an operator pencil that generalizes the basic operator pencil in Chapter 8. The study of this operator pencil leads to the proof of a theorem on the spectrum localization in the complex plane, its separation in four branches and some properties of eigenvalues.

In Section 10.5 we consider the issues of multiple system completeness of eigen- and associated elements of the problem on normal oscillations of a hydrosystem. We follow Keldysh's scheme to prove our theorem and we discuss the existence of three branches of eigenvalues with the limit point $\lambda = \infty$ and their physical nature.

10.1 Statement of the Problem

10.1.1 The Classical Statement of the Problem

Let us assume that a certain container $\Omega \subset \mathbb{R}^3$ (with a Lipschitz boundary $\partial\Omega$) is partially filled with two nonmixing homogeneous heavy fluids with densities ρ_1 and ρ_2 for the lower and upper fluid, respectively, with $\rho_2 < \rho_1$. The lower fluid is viscous with the coefficient of dynamical viscosity $\mu = \rho_1\nu > 0$ and the upper fluid is ideal.

At rest the lower fluid takes up the region $\Omega_1 \subset \Omega$ limited by the solid wall of the container $S_1 \subset \partial\Omega$ and the horizontal boundary surface separating the two fluids Γ_1. The upper fluid takes up the region $\Omega_2 \subset \Omega$ and is bounded by the solid wall $S_2 \subset \partial\Omega$ and by Γ_1 and the free horizontal surface Γ_2.

We introduce a coordinate system $Ox_1x_2x_3$, where the direction of the axis Ox_3 is opposite to that of the vector $\boldsymbol{g}$, the acceleration of the gravity force $\boldsymbol{g} = -g\boldsymbol{e}_3$, $g > 0$, and the origin O is placed on the free surface Γ_2. The equilibrium pressures $P_i(x_3)$ in each fluid are obtained by

$$\begin{aligned} P_2(x_3) &= p_{\mathrm{a}} - \rho_2 g x_3, \\ P_1(x_3) &= p_{\mathrm{a}} - \rho_1 g x_3 - (\rho_1 - \rho_2)h, \end{aligned} \tag{1.1}$$

where h is the layer thickness of the second fluid, and p_{a} is the external constant atmospheric pressure.

Let us consider small (linear) motions of the given hydrosystem with respect to the state of rest. Denoting the velocity fields of each fluid by $\boldsymbol{u}_i(t,x)$, $x = (x_1, x_2, x_3) \in \Omega_i$, and the deviation of the pressure fields from of their equilibrium values (1.1) by $p_i(t,x)$, we obtain a system of equations consisting of a linear Navier-Stokes equation in the region Ω_1, and an Euler equation in the region Ω_2,

$$\rho_1 \frac{\partial \boldsymbol{u}_1}{\partial t} = -\nabla p_1 + \mu \Delta \boldsymbol{u}_1 + \rho_1 \boldsymbol{f}(t,x), \quad \operatorname{div} \boldsymbol{u}_1 = 0 \qquad \text{in } \Omega_1, \tag{1.2}$$

$$\rho_2 \frac{\partial \boldsymbol{u}_2}{\partial t} = -\nabla p_2 + \rho_2 \boldsymbol{f}(t,x), \quad \operatorname{div} \boldsymbol{u}_2 = 0 \qquad \text{in } \Omega_2, \tag{1.3}$$

where $\boldsymbol{f}(t,x)$ is the small field of external forces.

Obviously, on the rigid wall of the container, the condition of stickiness (for the viscous fluid) and the nonleakage condition (for the ideal fluid) are satisfied, that is,

$$\boldsymbol{u} = \boldsymbol{0} \qquad \text{on } S_1,$$
$$u_{2n} := \boldsymbol{u}_2 \cdot \boldsymbol{n} = 0 \qquad \text{on } S_2, \tag{1.4}$$

where $\boldsymbol{n}$ is the external normal to $\partial\Omega$.

The normals $\boldsymbol{n}_i$ to the surfaces Γ_i are aligned along the axis Ox_3. The kinematic conditions on Γ_i are the following,

$$\frac{\partial \zeta_1}{\partial t} = \gamma_1 \boldsymbol{u}_1 := \boldsymbol{u}_1 \cdot \boldsymbol{n}_1 = \gamma_1 \boldsymbol{u}_2 \, (:= \boldsymbol{u}_2 \cdot \boldsymbol{n}_1 = (u_1)_3 = (u_2)_3) \qquad \text{on } \Gamma_1, \tag{1.5}$$
$$\frac{\partial \zeta_2}{\partial t} = \gamma_2 \boldsymbol{u}_2 := \boldsymbol{u}_2 \cdot \boldsymbol{n}_2 \, (= (u_2)_3) \qquad \text{on } \Gamma_2. \tag{1.6}$$

Here the functions $\zeta_i(t, x_1, x_2)$ describe the deviations along the vertical $\boldsymbol{e}_3$ of the free moving surfaces $\Gamma_i(t)$ from their horizontal equilibrium positions Γ_i, $i = 1, 2$.

We denote the elements of the stress tensor corresponding to the flow $\{\boldsymbol{u}; p\}$ with $\tau_{ik}(\boldsymbol{u}) := -p\delta_{ik} + \mu\tilde{\tau}_{ik}(\boldsymbol{u})$, $\tilde{\tau}_{ik} := \partial u_i/\partial x_k + \partial u_k/\partial x_i$ and write down the dynamic boundary value conditions on Γ_1 and Γ_2. Since the layers of an ideal fluid may slip freely along the boundary Γ_1 with the viscous fluid, then the tangent stresses in the viscous fluid are zero along the boundary and the normal stresses are compensated by a gravitational jump of pressures,

$$\tau_{i3}(\boldsymbol{u}_1) = 0 \qquad i = 1, 2,$$
$$\tau_{33}(\boldsymbol{u}_1) + p_2 + g(\Delta\rho)\zeta_1 = 0 \qquad \text{on } \Gamma_1, \tag{1.7}$$

where $\Delta\rho := \rho_1 - \rho_2 > 0$. The dynamic boundary value condition for the ideal fluid on the free surface Γ_2 has a trivial form (see Section 3.3),

$$p_2 = \rho_2 g \zeta_2 \qquad \text{on } \Gamma_2. \tag{1.8}$$

Let us note also that the conditions

$$\int_{\Gamma_1} \zeta_1 \mathrm{d}\Gamma_1 = 0, \quad \int_{\Gamma_2} \zeta_2 \mathrm{d}\Gamma_2 = 0 \tag{1.9}$$

are due to the preservation of volumes of each fluid in the process of oscillations.

Moreover, from (1.7)–(1.9) and the formula

$$\int_{\Gamma_1} \tau_{33}(u_1)\mathrm{d}\Gamma_1 = 2\int_{\Gamma_1} \frac{\partial(u_1)_3}{\partial x_3}\mathrm{d}\Gamma_1 = 0,$$

which is an analog of formula (2.2.25), it follows that the normalization conditions

$$\int_{\Gamma_2} p_2 \mathrm{d}\Gamma_2 = 0, \quad \int_{\Gamma_1} (p_1 - p_2) \mathrm{d}\Gamma_1 = 0. \tag{1.10}$$

are satisfied for the fields p_1 and p_2.

At the initial moment of time it is necessary to specify the velocity fields in the regions Ω_1 and Ω_2 as well as the vertical surface deviatiations of Γ_1 and Γ_2,

$$\begin{aligned} \boldsymbol{u}_1(0,x) &= \boldsymbol{u}_1^0(x), \\ \boldsymbol{u}_2(0,x) &= \boldsymbol{u}_2^0(x), \\ \zeta_1(0,x_1,x_2) &= \zeta_1^0(x_1,x_2), \\ \zeta_2(0,x_1,x_2) &= \zeta_2^0(x_1,x_2), \end{aligned} \tag{1.11}$$

This way, the considered initial boundary value problem on the induced oscillations of a partially dissipative hydrosystem consists in finding the functions $\boldsymbol{u}_i(t,x)$, $p_i(t,x)$, and $\zeta_i(t,x_1,x_2)$, for $i = 1,2$, in equations (1.2) and (1.3), the boundary value conditions (1.4)–(1.10), and the initial conditions (1.11). If $\boldsymbol{f}(t,x) \equiv \boldsymbol{0}$, then we are obtaining the problem on free system oscilations in a homogeneous gravitational field.

10.1.2 The Law of Full Energy Balance. Definition of a Generalized Solution

We call a *classical solution* of the problem (1.2)–(1.11) the functions $\boldsymbol{u}_i(t,x)$, $p_i(t,x)$, and $\zeta_i(t,x_1,x_2)$, for $i = 1,2$, that are continuous in their variables with all derivatives in (1.2)–(1.11) and satisfy the equations and the boundary value and initial conditions of the problem.

We will next show that a classical solution of the problem (1.2)–(1.11) meets the balance law of full energy of the system, and also give the definition of a generalized solution that can be obtained by using some integral identities the same way we did in Section 8.3.3.

Let us multiply scalarly in $\mathbb{R}^3$ both parts of the first equation in (1.2) by a function $\boldsymbol{v}_1(x)$ from the subspace $\boldsymbol{J}^1_{0,S_1}(\Omega_1)$, for which the following conditions are obviously satisfied,

$$\begin{aligned} \operatorname{div} \boldsymbol{v}_1 &= 0 \quad && \text{in } \Omega_1, \\ \boldsymbol{v}_1 &= \boldsymbol{0} \quad && \text{on } S_1. \end{aligned} \tag{1.12}$$

Integrating then on Ω_1 and using the Green formula (2.2.10), that is,

$$\int_{\Omega_1} (\nabla p_1 - \mu \Delta \boldsymbol{u}_1) \cdot \boldsymbol{v}_1 \mathrm{d}\Omega_1 = \mu E(\boldsymbol{u}_1, \boldsymbol{v}_1) - \int_{\Gamma_1} \left(\sum_{k=1}^{3} \tau_{k3}(\boldsymbol{u}_1) v_1^k \right) \mathrm{d}\Gamma_1, \tag{1.13}$$

where

$$E(\boldsymbol{u}_1, \boldsymbol{v}_1) := \int_{\Omega_1} \left(\sum_{i,k=1}^{3} \tilde{\tau}_{ik}(\boldsymbol{u}_1) \tilde{\tau}_{ik}(\boldsymbol{v}_1) \right) \mathrm{d}\Omega_1,$$

for $\boldsymbol{v}_1$ satisfying conditions (1.12), we obtain the following integral identity

$$\rho_1 \int_{\Omega_1} \frac{\partial \boldsymbol{u}_1}{\partial t} \cdot \boldsymbol{v}_1 \mathrm{d}\Omega_1 + \mu E(\boldsymbol{u}_1, \boldsymbol{v}_1) - \int_{\Gamma_1} \left(\sum_{k=1}^{3} \tau_{k3}(\boldsymbol{u}_1) v_1^k \right) \mathrm{d}\Gamma_1 = \rho_1 \int_{\Omega_1} \boldsymbol{f} \cdot \boldsymbol{v}_1 \mathrm{d}\Omega_1. \tag{1.14}$$

A similar procedure applied to equation (1.3) with a function $\boldsymbol{v}_2(x) \in \boldsymbol{J}_{0,S_2}(\Omega_2)$ satisfying obviously the conditions

$$\begin{aligned} \operatorname{div} \boldsymbol{v}_2 &= 0 \qquad \text{in } \Omega_2, \\ v_{2n} &= 0 \qquad \text{on } S_2, \end{aligned} \tag{1.15}$$

leads to the relation

$$\rho_2 \int_{\Omega_2} \frac{\partial \boldsymbol{u}_2}{\partial t} \cdot \boldsymbol{v}_2 \mathrm{d}\Omega_2 - \int_{\Gamma_1} p_2(\gamma_1 \boldsymbol{v}_2) \mathrm{d}\Gamma_1 + \int_{\Gamma_2} p_2(\gamma_2 \boldsymbol{v}_2) \mathrm{d}\Gamma_2 = \rho_2 \int_{\Omega_2} \boldsymbol{f} \cdot \boldsymbol{v}_2 \mathrm{d}\Omega_2. \tag{1.16}$$

In addition, we require that the fields $\boldsymbol{v}_1$ and $\boldsymbol{v}_2$ in (1.14) and (1.16) respectively, should be connected by the following relation,

$$\gamma_1 \boldsymbol{v}_1 = \gamma_1 \boldsymbol{v}_2 \qquad \text{on } \Gamma_1, \tag{1.17}$$

that is, the pair $\{\boldsymbol{v}_1; \boldsymbol{v}_2\} =: \hat{v}$ is an element of the space $\hat{\boldsymbol{J}}_{0,S}(\Omega)$ (see Section 2.2.8), where $\boldsymbol{v}_1 \in \boldsymbol{J}^1_{0,S_1}(\Omega_1)$. Then, adding up the left and the right sides in (1.14) and (1.16) we have

$$\begin{aligned} &\rho_1 \int_{\Omega_1} \frac{\partial \boldsymbol{u}_1}{\partial t} \cdot \boldsymbol{v}_1 \mathrm{d}\Omega_1 + \rho_2 \int_{\Omega_2} \frac{\partial \boldsymbol{u}_2}{\partial t} \cdot \boldsymbol{v}_2 \mathrm{d}\Omega_2 + \mu E(\boldsymbol{u}_1, \boldsymbol{v}_1) \\ &\quad - \int_{\Gamma_1} \left(\tau_{13}(\boldsymbol{u}_1) v_1^1 + \tau_{23}(\boldsymbol{u}_1) v_1^2 + (\tau_{33}(\boldsymbol{u}_1) + p_2) v_1^3 \right) \mathrm{d}\Gamma_1 + \int_{\Gamma_2} p_2(\gamma_2 \boldsymbol{v}_2) \mathrm{d}\Gamma_2 \\ &= \rho_1 \int_{\Omega_1} \boldsymbol{f} \cdot \boldsymbol{v}_1 \mathrm{d}\Omega_1 + \rho_2 \int_{\Omega_2} \boldsymbol{f} \cdot \boldsymbol{v}_2 \mathrm{d}\Omega_2. \end{aligned} \tag{1.18}$$

Using the boundary value conditions (1.7) and (1.8) we get

$$\sum_{k=1}^{2}\rho_k\int_{\Omega_k}\frac{\partial \boldsymbol{u}_k}{\partial t}\cdot\boldsymbol{v}_k\mathrm{d}\Omega_k+\mu E(\boldsymbol{u}_1,\boldsymbol{u}_1)+g\sum_{k=1}^{2}(\Delta\rho)_k\int_{\Gamma_k}\zeta_k(\gamma_k\boldsymbol{v}_k)\mathrm{d}\Gamma_k$$
$$=\sum_{k=1}^{2}\rho_k\int_{\Omega_k}\boldsymbol{f}\cdot\boldsymbol{v}_k\mathrm{d}\Omega_k, \tag{1.19}$$

where $(\Delta\rho)_1=\rho_1-\rho_2$ and $(\Delta\rho)_2=\rho_2-0>0$.

We note that, based on the kinematic relationships (1.5) and (1.6), and the initial conditions (1.11), we have

$$\zeta_k(t,x_1,x_2)=\zeta_k^0(x_1,x_2)+\int_0^{\mathrm{t}}(\gamma_k\boldsymbol{u}_k)(\tau,x_1,x_2)\mathrm{d}\tau,\qquad k=1,2. \tag{1.20}$$

Taking into account identity (1.19), we derive the law of balance of full energy in problem (1.2)–(1.11). We set in (1.19) $\boldsymbol{v}_k(t,x)=\boldsymbol{u}_k(t,x)$, $k=1,2$, which is possible to do because the functions $\boldsymbol{u}_k(t,x)$, $k=1,2$, satisfy conditions (1.12), (1.15), and (1.17). Then by (1.5) and (1.6) we get

$$\frac{1}{2}\frac{\mathrm{d}}{\mathrm{d}t}\left(\sum_{k=1}^{2}\rho_k\int_{\Omega_k}|\boldsymbol{u}_k|^2\mathrm{d}\Omega_k+g\sum_{k=1}^{2}(\Delta\rho)_k\int_{\Gamma_k}|\zeta_k|^2\mathrm{d}\Gamma_k\right)$$
$$=-\mu E(\boldsymbol{u}_1,\boldsymbol{u}_1)+\sum_{k=1}^{2}\rho_k\int_{\Omega_k}\boldsymbol{f}\cdot\boldsymbol{u}_k\mathrm{d}\Omega_k.$$

Hence, based on the initial conditions (1.11), we get the law of balance of full (kinetic and potential) energy of the hydrosystem,

$$\frac{1}{2}\left(\sum_{k=1}^{2}\rho_k\int_{\Omega_k}|\boldsymbol{u}_k(t,x)|^2\mathrm{d}\Omega_k+g\sum_{k=1}^{2}(\Delta\rho)_k\int_{\Gamma_k}|\zeta_k(t,x_1,x_2)|^2\mathrm{d}\Gamma_k\right)$$
$$=\frac{1}{2}\left(\sum_{k=1}^{2}\rho_k\int_{\Omega_k}|\boldsymbol{u}_k^0(x)|^2\mathrm{d}\Omega_k+g\sum_{k=1}^{2}(\Delta\rho)_k\int_{\Gamma_k}|\zeta_k^0(x_1,x_2)|^2\mathrm{d}\Gamma_k\right)$$
$$-\mu\int_0^{\mathrm{t}}E(\boldsymbol{u}_1(s,x),\boldsymbol{u}_1(s,x))\mathrm{d}s+\int_0^{\mathrm{t}}\left(\sum_{k=1}^{2}\rho_k\int_{\Omega_k}\boldsymbol{f}(s,x)\cdot\boldsymbol{u}_k(s,x)\mathrm{d}\Omega_k\right)\mathrm{d}s. \tag{1.21}$$

This formula shows that the change in the full energy of the system is due to the work of dissipation forces (in the first fluid) and the work (in the entire system) of external forces with mass density $\boldsymbol{f}(t,x)$.

The previous statement is a consequence of (1.21). If there exists a classical solution to problem (1.2)–(1.11), then that solution is unique.

Indeed, if problem (1.2)–(1.11) has two solutions, then their difference leads to a solution of the homogeneous problem (1.2)–(1.11) with zero initial conditions. Then from (1.21) for $\boldsymbol{f}(t,x) \equiv \mathbf{0}$, $\boldsymbol{u}_k^0(x) \equiv \mathbf{0}$, $\zeta_k^0(x_1,x_2) \equiv 0$, we get that the previously mentioned differences are identically zero functions: $\boldsymbol{u}_k(t,x) \equiv 0$, $\zeta_k(t,x_1,x_2) \equiv 0$. Whence it also follows that $p_k(t,x) \equiv 0$, $k = 1,2$.

We are now in a position to introduce a new definition. We call a *generalized solution* of problem (1.2)–(1.11) any pairs of functions $\boldsymbol{u}_i(t,x), \zeta_i(t,x_1,x_2)$ such that the integrals

$$\int_{\Omega_k} \left|\frac{\partial \boldsymbol{u}_k}{\partial t}\right|^2 \mathrm{d}\Omega_k, \quad E(\boldsymbol{u}_1,\boldsymbol{u}_1), \quad \int_{\Gamma_k} |\zeta_k|^2 \mathrm{d}\Gamma_k, \qquad k = 1,2,$$

are continuous in t, and, for which the initial conditions $\boldsymbol{u}_1(0,x) = \boldsymbol{u}_1^0(x)$, $\boldsymbol{u}_2(0,x) = \boldsymbol{u}_2^0(x)$, equation (1.20) and identity (1.19) are satisfied for all $\boldsymbol{v}_i(x)$, $i = 1,2$, subject to the constrains (1.12), (1.15), (1.17) and the conditions

$$\int_{\Omega_k} |\boldsymbol{v}_k|^2 \mathrm{d}\Omega_k < \infty, \quad E(\boldsymbol{v}_1,\boldsymbol{v}_1) < \infty.$$

From the last conditions it follows that a classical solution of problem (1.2)–(1.11) is generalized in the sense of the just introduced definition. Inversely, a generalized solution of problem (1.2)–(1.11), for which all the functions in the equations, boundary value and intial conditions of the problem are continuous, is classical. The proof of this statement is left to the readers.

Let us note that for a generalized solution of problem (1.2)–(1.11) the law of balance of full energy (1.21) is fulfilled. This ensures that if a generalized solution to problem (1.2)–(1.11) exists, then it is unique.

It will be stated in the sequel (see Section 10.2) that for certain conditions on the initial data and on the function $\boldsymbol{f}(t,x)$, the problem (1.2)–(1.11) has a (unique) generalized solution.

10.1.3 Normal Oscillations. Statement of the Problem

Next, associated with problem (1.2)–(1.11) we will consider the free normal oscillations of the system, that is, solutions for $\boldsymbol{f}(t,x) \equiv 0$ that depend on t by the law $\exp(-\lambda t)$:

$$\begin{aligned} \boldsymbol{u}_i(t,x) &= \exp(-\lambda t)\boldsymbol{u}_i(x), \\ p_i(t,x) &= \exp(-\lambda t)p_i(x), \\ \zeta_i(t,x_1,x_2) &= \exp(-\lambda t)\zeta_i(x_1,x_2), \qquad i = 1,2. \end{aligned} \tag{1.22}$$

Here $\boldsymbol{u}_i(x)$, $p_i(x)$, and $\zeta_i(x_1, x_2)$ are the coomplex amplitude functions and λ is the complex oscillation frequency.

For the amplitude functions and the spectral parameter λ from the equations and boundary value conditions (1.2)–(1.10) and using (1.22) we deduce the homogeneous boundary value problem

$$
\begin{aligned}
& -\mu\Delta\boldsymbol{u}_1 + \nabla p_1 = \lambda\rho_1\boldsymbol{u}_1, \quad \operatorname{div}\boldsymbol{u}_1 = 0 \qquad \text{in } \Omega_1, \\
& \nabla p_2 = \lambda\rho_2\boldsymbol{u}_2, \quad \operatorname{div}\boldsymbol{u}_2 = 0 \qquad \text{in } \Omega_2, \\
& \boldsymbol{u}_1 = 0 \quad \text{on } S_1, \qquad \boldsymbol{u}_2\cdot\boldsymbol{n} = 0 \quad \text{on } S_2, \\
& \tau_{i3}(\boldsymbol{u}_1) = 0, \qquad i = 1,2, \quad \tau_{33}(\boldsymbol{u}_1) + p_2 + g(\Delta\rho)\zeta_1 = 0, \\
& \gamma_1\boldsymbol{u}_1 = \gamma_1\boldsymbol{u}_2 = -\lambda\zeta_1, \quad \text{on } \Gamma_1, \\
& p_2 = g\rho_2\zeta_2, \quad \gamma_2\boldsymbol{u}_2 = -\lambda\zeta_2 \qquad \text{on } \Gamma_2, \\
& \int_{\Gamma_1}\zeta_1 d\Gamma_1 = 0, \quad \int_{\Gamma_2}\zeta_2 d\Gamma_2 = 0 \qquad \left(\int_{\Gamma_2} p_2 d\Gamma_2 = 0, \quad \int_{\Gamma_1}(p_1 - p_2)d\Gamma_1 = 0\right).
\end{aligned}
\tag{1.23}
$$

Note that in problem (1.23) the spectral parameter λ is involved in the equations that take place on Ω_1 and Ω_2, and also in the boundary value conditions on the surfaces Γ_1 and Γ_2.

The study of problem (1.23) will make the object of some of the sections in this chapter. Here we will mention only two important properties of its solution.

1° The number $\lambda = \lambda_0 = 0$ is an eigenvalue of the problem (1.23) with infinite multiplicity. The trivial solution of the form

$$
\begin{aligned}
& \boldsymbol{u}_1(x) \equiv 0, \quad \boldsymbol{u}_2(x) \in \boldsymbol{J}_0(\Omega_2) \quad \text{arbitrary}, \\
& p_1(x) \equiv 0, \quad p_2(x) \equiv 0, \\
& \zeta_1(x_1, x_2) \equiv 0, \quad \zeta_2(x_1, x_2) \equiv 0,
\end{aligned}
\tag{1.24}
$$

corresponds to it. From the physical point of view, the trivial solution accounts for system motions for which the lower fluid and the boundaries Γ_1 and Γ_2 are in an immovable state, each fluid is under an equilibrium pressure, and in the ideal fluid—the upper one—there is an arbitrary vortex motion that does not perturb the boundaries Γ_1 and Γ_2. Because these motions are described by functions from the infinite dimensional subspace $\boldsymbol{J}_0(\Omega_2)$, the eigenvalue $\lambda_0 = 0$ has infinite multiplicity.

To prove Propery 1° we set $\lambda = 0$ in (1.23)

$$\begin{aligned}
&-\mu\Delta \boldsymbol{u}_1 + \nabla p_1 = 0, \quad \operatorname{div} \boldsymbol{u}_1 = 0 \qquad \text{in } \Omega_1,\\
&\boldsymbol{u}_1 = 0 \qquad \text{on } S_1,\\
&\tau_{i3}(\boldsymbol{u}_1) = 0, \qquad i = 1,2,\\
&\tau_{33}(\boldsymbol{u}_1) + p_2 + g(\Delta\rho)\zeta_1 = 0, \quad \text{on } \Gamma_1,\\
&\gamma_1 \boldsymbol{u}_1 = 0 = \gamma_1 \boldsymbol{u}_2, \quad \text{on } \Gamma_1,\\
&\nabla p_2 \equiv \mathbf{0}, \quad \operatorname{div} \boldsymbol{u}_2 = 0 \qquad \text{in } \Omega_2,\\
&p_2 = \rho_2 g \zeta_2, \quad \gamma_2 \boldsymbol{u}_2 = 0 \qquad \text{on } \Gamma_2,\\
&\int_{\Gamma_1} (p_1 - p_2)\mathrm{d}\Gamma_1 = 0, \quad \int_{\Gamma_2} \zeta_2 \mathrm{d}\Gamma_2 = 0, \quad \int_{\Gamma_2} p_2 \mathrm{d}\Gamma_2 = 0.
\end{aligned} \tag{1.25}$$

Using the Green formula (1.12) we obtain

$$\begin{aligned}
0 &= \int_{\Omega_1} (-\mu\Delta \boldsymbol{u}_1 + \nabla p_1) \cdot \overline{\boldsymbol{u}_1} \mathrm{d}\Omega_1\\
&= \mu E(\boldsymbol{u}_1, \boldsymbol{u}_1) - \int_{\Gamma_1} \left(\sum_{k=1}^{3} \tau_{k3}(\boldsymbol{u}_1) \overline{u_1^k} \right) \mathrm{d}\Gamma_1\\
&= \mu E(\boldsymbol{u}_1, \boldsymbol{u}_1),
\end{aligned}$$

where we took into account that $\gamma_1 \boldsymbol{u}_1 = u_1^3 = 0$. Whence, it follows that $\boldsymbol{u}_1 \equiv \mathbf{0}$ by the positive definiteness of the quadratic form $E(\boldsymbol{u}_1, \boldsymbol{u}_1)$. Then $\nabla p_1 \equiv \mathbf{0}$ and $p_1(x) = \text{const} = C_1$. Similarly $\nabla p_2 \equiv \mathbf{0}$ and $p_2(x) = \text{const} = C_2$. Since $\int_{\Gamma_2} p_2 \mathrm{d}\Gamma_2 = 0$, then $C_2 = 0$ and from condition

$$\int_{\Gamma_1} (p_1 - p_2)\mathrm{d}\Gamma_1 = \int_{\Gamma_1} p_1 \mathrm{d}\Gamma_1 = 0$$

we get that $C_1 = 0$ and, therefore, $p_1(x) \equiv 0$.

Going back to the equalities $p_2 = g\rho_2\zeta_2$ on Γ_2 and $\tau_{33}(\boldsymbol{u}_1) + p_2 + g(\Delta\rho)\zeta_1 = 0$ on Γ_1, we get that $\zeta_2(x_1, x_2) \equiv 0$, $\zeta_1(x_1, x_2) \equiv 0$. The following conditions

$$\begin{aligned}
&\operatorname{div} \boldsymbol{u}_2 = 0 \qquad \text{in } \Omega_2,\\
&(u_2)_n = 0 \qquad \text{on } \partial\Omega_2 = S_2 \cup \Gamma_1 \cup \Gamma_2,
\end{aligned}$$

are satisfied for the field $\boldsymbol{u}_2(x)$, meaning that $\boldsymbol{u}_2(x)$ can be an arbitrary function on $\boldsymbol{J}_0(\Omega_2)$. This completely proves Property 1°.

2° The nonzero eigenvalues λ of problem (1.23) are situated in the open right half-plane symmetrically with respect to the real axis.

To prove it, let us set $\boldsymbol{f}(t,x) \equiv \mathbf{0}$ in (1.19) and do the transition to the amplitude functions by the formulas (1.22). Using $\zeta_i = -\lambda^{-1}\gamma_i \boldsymbol{u}_i$, $i = 1,2$, and assuming $\boldsymbol{v}_i = \boldsymbol{u}_i$ we get

$$\lambda\left(\sum_{k=1}^{2}\rho_k \int_{\Omega_k} |\boldsymbol{u}_k|^2 d\Omega_k\right) - \mu E(\boldsymbol{u}_1,\boldsymbol{u}_1) + g\lambda^{-1}\left(\sum_{k=1}^{2}(\Delta\rho)_k \int_{\Gamma_k} |\gamma_k \boldsymbol{u}_k|^2 d\Gamma_k\right) = 0. \tag{1.26}$$

Hence, solving a quadratic equation relative to λ, we check to see, first that the nonreal eigenvalues are situated symmetrically to the real axis and, then, that

$$\operatorname{Re}\lambda = \frac{\mu E(\boldsymbol{u}_1,\boldsymbol{u}_1)}{\sum_{k=1}^{2}\rho_k \int_{\Omega_k} |\boldsymbol{u}_k|^2 d\Omega_k + g|\lambda|^{-2}\sum_{k=1}^{2}(\Delta\rho)_k \int_{\Gamma_k} |\gamma_k \boldsymbol{u}_k|^2 d\Gamma_k}, \tag{1.27}$$

whence it follows that $\operatorname{Re}\lambda \geq 0$. If $\operatorname{Re}\lambda = 0$, then $E(\boldsymbol{u}_1,\boldsymbol{u}_1) = 0$ and by (1.26) we get $\operatorname{Im}\lambda = 0$, which contradicts the fact that $\lambda \neq 0$.

The proof of Property 2° states simultaneously that in problem (1.23) the nonzero eigenvalues λ correspond to solutions for which $E(\boldsymbol{u}_1,\boldsymbol{u}_1) > 0$ and, therefore, the velocity field in the viscous fluid cannot be zero.

10.2 Studying an Initial Boundary Value Problem

In this section we state a theorem on the correct solvability of the initial boundary value problem (1.2)–(1.11). The approach we will use is somewhat different from the common scheme developed in Section 1.8 and from the considerations in Section 8.1.

10.2.1 Projections of Euler and Navier-Stokes Equations on Orthogonal Subspaces

We now return to the study of the initial boundary value problem (1.2)–(1.11) and address the issue of the existence of its solutions. From the equations (1.2) and (1.3) and the boundary conditions (1.4) it follows that the function $\boldsymbol{u}_2(t,x)$ may be obtained, for every t, as an element of the subspace $\boldsymbol{J}_{0,S_2}(\Omega_2)$, and the function $\boldsymbol{u}_1(t,x)$ (with due account of the dissipative term $\mu\Delta\boldsymbol{u}_1$ in (1.2)) is naturally, for any t, an element of the space $\boldsymbol{J}^1_{0,S_1}(\Omega_1)$ (the energy dissipative velocity is finite for it). Furthermore, by virtue of the kinematic relationship $\gamma_1\boldsymbol{u}_1 = \gamma_1\boldsymbol{u}_2$ imposed on Γ_1, we get that the pair of functions $\{\boldsymbol{u}_1;\boldsymbol{u}_2\}$ belongs to the space $\hat{\boldsymbol{J}}_{0,S}(\Omega)$ (see Sections 2.1.11 and 2.1.12).

In problem (1.2)–(1.11) just as in the spectral problem (1.23), a trivial solution can be derived similarly to Property 1° in Section 10.1.3. We next apply the orthoprojector $P_{0,2}$ on to the subspace $\boldsymbol{J}_0(\Omega_2)$ to both sides of equation (1.3). Then for the function $\boldsymbol{w}_2(t,x) = P_{0,2}\boldsymbol{u}_2(t,x)$ we obtain the problem

$$\frac{\partial \boldsymbol{w}_2}{\partial t} = P_{0,2}\boldsymbol{f}(t,x), \qquad \boldsymbol{w}_2(0,x) = P_{0,2}\boldsymbol{u}_2^0(x), \tag{2.1}$$

with the obvious solution

$$\boldsymbol{w}_2(t,x) = P_{0,2}\boldsymbol{u}_2^0 + \int_0^t P_{0.2}\boldsymbol{f}(s,x)\mathrm{d}s. \tag{2.2}$$

Let us recall that the space $\boldsymbol{L}_2(\Omega_2)$ allows the orthogonal decomposition

$$\boldsymbol{L}_2(\Omega_2) = \boldsymbol{J}_0(\Omega_2) \oplus \boldsymbol{G}_{\mathrm{h}.S_2}(\Omega_2) \oplus \boldsymbol{G}_{0.\Gamma}(\Omega_2), \tag{2.3}$$

with

$$\boldsymbol{G}_{\mathrm{h},S_2}(\Omega_2) = \left\{\boldsymbol{v} = \nabla\Phi : \Delta\Phi = 0 \text{ in } \Omega_2, \frac{\partial\Phi}{\partial n} = 0 \text{ on } S_2, \int_{\Gamma_2} \Phi \mathrm{d}\Gamma_2 = 0\right\},$$
$$\boldsymbol{G}_{0,\Gamma}(\Omega_2) = \{\boldsymbol{u} = \nabla\kappa : \kappa = 0 \text{ on } \Gamma := \Gamma_1 \cup \Gamma_2\}.$$

We denote the orthoprojectors onto $\boldsymbol{G}_{\mathrm{h}.S_2}(\Omega_2)$ and $\boldsymbol{G}_{0,\Gamma}(\Omega_2)$ by $P_{h.S_2}$ and $P_{0,\Gamma}$, respectively. Projecting (1.3) on these subspaces we obtain

$$\begin{aligned}
\rho_2\frac{\partial}{\partial t}\nabla\Phi &= -\nabla\tilde{p}_2 + \rho_2 P_{h.S_2}\boldsymbol{f}, \\
\boldsymbol{0} &= -\nabla\kappa + \rho_2 P_{0.\Gamma}\boldsymbol{f}, \\
\nabla\Phi &= P_{h.S_2}\boldsymbol{u}_2, \\
\nabla\kappa &= P_{0.\Gamma}\nabla p_2, \\
\nabla\tilde{p}_2 &= P_{h.S_2}\nabla p.
\end{aligned} \tag{2.4}$$

From the second equation (2.4) it follows that $\nabla\kappa$ is uniquely defined by the field of external forces $\boldsymbol{f}(t,x)$. Therefore, taking into account the already deduced solution (2.1), (2.2) we can consider in the sequel that in problem (1.2)–(1.11) we have the field $\nabla\Phi(t,x) \in \boldsymbol{G}_{\mathrm{h}.S_2}(\Omega)$ instead of $\boldsymbol{u}_2(t,x)$, that is, the motion in the ideal fluid is potential. Further, since $\kappa = 0$ on $\Gamma = \Gamma_1 \cup \Gamma_2$, in problem (1.2)–(1.11) the function $p_2(t,x)$ may be replaced by the function $\tilde{p}_2(t,x)$. The initial condition for $\boldsymbol{u}_2(0,x)$ is replaced correspondingly, that is, the function $\boldsymbol{u}_2^0(x)$ is substituted by $P_{h,S_2}\boldsymbol{u}_2^0$. It is also obvious that now the field $\boldsymbol{f}(t,x)$ in the region Ω_2 is substituted by $P_{h,S_2}\boldsymbol{f}$.

Similar transformations can be done with the fields $\boldsymbol{u}_1(t,x)$, $\nabla p_1(t,x)$ based on the orthogonal decomposition

$$\boldsymbol{L}_2(\Omega_1) = \boldsymbol{J}_{0.S_1}(\Omega_1) \oplus \boldsymbol{G}_{0.\Gamma_1}(\Omega_1), \tag{2.5}$$

with

$$\boldsymbol{J}_{0,S_1}(\Omega_1) = \{\boldsymbol{v} : \operatorname{div} \boldsymbol{v} = 0 \text{ in } \Omega_1,\ v_n = 0 \text{ on } S_1\},$$
$$\boldsymbol{G}_{0,\Gamma_1}(\Omega_1) = \{\nabla p : p = 0 \text{ on } \Gamma_1\}.$$

Let us denote by P_{0,S_1} the orthoprojector onto $\boldsymbol{J}_{0.S_1}(\Omega_1)$. If all the terms in (1.2) are elements in $\boldsymbol{L}_2(\Omega_1)$, then projecting on $\boldsymbol{J}_{0.S_1}(\Omega)$ we get

$$\rho_1 \frac{\partial \boldsymbol{u}_1}{\partial t} = -\nabla \tilde{p}_1 + \mu P_{0.S_1} \Delta \boldsymbol{u}_1 + \rho_1 P_{0.S_1} \boldsymbol{f}(t,x), \tag{2.6}$$

where $\nabla \tilde{p}_1 = P_{0,S_1} \nabla p_1$. The projection on $\boldsymbol{G}_{0.\Gamma_1}(\Omega_1)$ gives the relationship

$$0 = -\nabla(p_1 - \tilde{p}_1) + \mu(I - P_{0.S_1})\Delta \boldsymbol{u}_1 + \rho_1(I - P_{0.S_1})\boldsymbol{f}(t,s) \tag{2.7}$$

from which it is clear that the field $\nabla(p_1, -\tilde{p}_1)$ is obtained from $\boldsymbol{u}_1(t,x)$ and $\boldsymbol{f}(t,x)$. Thus, only equation (2.6) needs to be considered further.

Here is the final formulation of the initial-boundary value problem (1.2)–(1.11) after projecting on the derived orthogonal subspaces:

$$\begin{aligned}
&\rho_1 \frac{\partial \boldsymbol{u}_1}{\partial t} = -\nabla \tilde{p}_1 + \mu P_{0.S_1} \Delta \boldsymbol{u}_1 + \rho_1 P_{0.S_1} \boldsymbol{f}(t,x) \quad \text{in } \Omega_1,\\
&\rho_2 \frac{\partial}{\partial t} \nabla \Phi = -\nabla \tilde{p}_2 + \rho_2 P_{h.S_2} \boldsymbol{f}(t,x), \qquad \Delta \Phi = 0 \quad \text{in } \Omega_2,\\
&\boldsymbol{u}_1 = 0 \quad \text{on } S_1, \qquad \frac{\partial \Phi}{\partial n} = 0 \quad \text{on } S_2,\\
&\frac{\partial \zeta_1}{\partial t} = \gamma_1 \boldsymbol{u}_1 = \frac{\partial \Phi}{\partial n_1} \quad \text{on } \Gamma_1, \qquad \frac{\partial \zeta_2}{\partial t} = \gamma_2 \boldsymbol{u}_2 = \frac{\partial \Phi}{\partial n_2} \quad \text{on } \Gamma_2,\\
&\tau_{i3}(\boldsymbol{u}_1) = 0, \qquad i = 1,2,\\
&\mu \tilde{\tau}_{33}(\boldsymbol{u}_1) - \tilde{p}_1 + \tilde{p}_2 + g(\Delta\rho)\zeta_1 = 0 \quad \text{on } \Gamma_1,\\
&\tilde{p}_2 = \rho_2 g \zeta_2 \quad \text{on } \Gamma_2,\\
&\int_{\Gamma_1} \zeta_1 d\Gamma_1 = 0, \qquad \int_{\Gamma_2} \zeta_2 d\Gamma_2 = 0, \qquad \int_{\Gamma_2} \tilde{p}_2 d\Gamma_2 = 0, \qquad \int_{\Gamma_1} (\tilde{p}_1 - \tilde{p}_2) d\Gamma_1 = 0,\\
&\boldsymbol{u}_1(0,x) = \boldsymbol{u}_1^0(x), \qquad \boldsymbol{u}_2(0,x) = \nabla\Phi(0,x) = (P_{h.S_2} \boldsymbol{u}_2^0)(x),\\
&\zeta_1(0,x_1,x_2) = \zeta_1^0(x_1,x_2), \qquad \zeta_2(0,x_1,x_2) = \zeta_2^0(x_1,x_2).
\end{aligned} \tag{2.8}$$

10.2.2 Auxiliary Boundary Value Problems

Let us represent the pressure field $\nabla\tilde{p}_1$ in the region Ω_1 as a sum of two fields from $\boldsymbol{G}_{\mathrm{h},S_1}(\Omega_1)$, $\nabla\tilde{p}_1 = \nabla\varphi_1 + \nabla\varphi_2$, and choose $\nabla\varphi_1$ in such a way that $\boldsymbol{u}_1$ and $\nabla\varphi_1$ are a solution of the following auxiliary problem of the form (2.2.30).

Problem I

$$\begin{aligned}
&-\mu P_{0,S_1}\Delta\boldsymbol{u}_1 + \nabla\varphi_1 =: \mu A\boldsymbol{u}_1 = \boldsymbol{\eta} := \left(-\rho_1\frac{\partial \boldsymbol{u}_1}{\partial t} - \nabla\varphi_2 + \rho_1 P_{0,S_1}\boldsymbol{f}\right),\\
&\operatorname{div}\boldsymbol{u}_1 = 0 \quad \text{in } \Omega_1, \qquad \boldsymbol{u}_1 = \boldsymbol{0} \quad \text{on } S_1,\\
&\tau_{i3}(\boldsymbol{u}_1) = 0, \qquad i = 1,2,\\
&\tau_{33}(\boldsymbol{u}_1) \equiv -\varphi_1 + 2\mu\frac{\partial u_{1,3}}{\partial x_3} = 0 \quad \text{on } \Gamma_1.
\end{aligned} \tag{2.9}$$

As it was previously discussed in the book (see Sections 2.2, 8.1, and others), for any $\boldsymbol{\eta} \in \boldsymbol{J}_{0,S_1}(\Omega_1)$ the problem (2.9) has a unique generalized solution

$$\mu\boldsymbol{u}_1 = A^{-1}\boldsymbol{\eta} = A^{-1}\left(-\rho_1\frac{\partial \boldsymbol{u}_1}{\partial t} - \nabla\varphi_2 + \rho_1 P_{0,S_1}\boldsymbol{f}\right), \tag{2.10}$$

belonging to $\mathcal{D}(A) \subset \boldsymbol{J}^1_{0,S_1}(\Omega_1) = \mathcal{D}(A^{1/2})$.

Now we formulate the second auxiliary problem for a definition of the field $\nabla\varphi_2 \in \boldsymbol{G}_{\mathrm{h},S_1}(\Omega_1)$.

Problem II

$$\begin{aligned}
&\Delta\varphi_2 = 0 \quad \text{in } \Omega_1, \qquad \frac{\partial\varphi_2}{\partial n} = 0 \quad \text{on } S_1,\\
&\varphi_2 = \psi := \tilde{p}_2 + g(\Delta\rho)\zeta_1 \quad \text{on } \Gamma_1.
\end{aligned} \tag{2.11}$$

As stated in Section 1.8.6 when we considered Example 3°, problem (2.11), also called a *Zaremba problem* [see (1.8.37)], has a unique solution $\varphi_2(x) \in H_1(\Omega_1)$ and it may be assumed that

$$\nabla\varphi_2 := G_1\psi = G_1(\tilde{p}_2 + g(\Delta\rho)\zeta_1). \tag{2.12}$$

Here the operator G_1 boundedly acts from $H^{1/2}(\Gamma_1)$ into $\boldsymbol{G}_{\mathrm{h}.S_1}(\Omega_1)$; for a special choice of the norms in the spaces mentioned [see Section 1.8.6 and the formula (1.8.38)] it may be considered to be isometric.

By means of the operators A and G_1 of Problems I and II, the equations and boundary value conditions from (2.8) containing the functions $\boldsymbol{u}_1$ and $\tilde{p}_1 = \varphi_1 + \varphi_2$, may be written in the form

$$\mu\boldsymbol{u}_1 = A^{-1}\left(-\rho_1\frac{\partial \boldsymbol{u}_1}{\partial t} + \rho_1 P_{0.S_1}\boldsymbol{f} - G_1(\tilde{p}_2 + g(\Delta\rho)\zeta_1)\right). \tag{2.13}$$

Now we come to consider auxiliary problems in the region Ω_2 where the velocity fields and pressures of the ideal fluid are looked for. For this purpose, as in Section 4.2, we formulate the two following problems.

Problem III

$$\Delta\Phi_1 = 0 \quad \text{in } \Omega_2, \qquad \frac{\partial \Phi_1}{\partial n} = 0 \quad \text{on } S_2,$$
$$\frac{\partial \Phi_1}{\partial n_2} = 0 \quad \text{on } \Gamma_2,$$
$$\frac{\partial \Phi_1}{\partial n_1} = \eta_1 := \gamma_1 \boldsymbol{u}_1 \quad \text{on } \Gamma_1, \qquad \int_{\Gamma_2} \Phi_1 d\Gamma_2 = 0. \tag{2.14}$$

Problem IV

$$\Delta\Phi_2 = 0 \quad \text{in } \Omega_2, \qquad \frac{\partial \Phi_2}{\partial n} = 0 \quad \text{on } S_2,$$
$$\frac{\partial \Phi_2}{\partial n_1} = 0 \quad \text{on } \Gamma_1,$$
$$\frac{\partial \Phi_2}{\partial n_2} = \eta_2 := \gamma_2 \boldsymbol{u}_2 \quad \text{on } \Gamma_2, \qquad \int_{\Gamma_2} \Phi_2 d\Gamma_2 = 0. \tag{2.15}$$

If $\eta_1 \in H_{\Gamma_1}^{-1/2}(\Omega_2)$, then Problem III has a unique generalized solution $\Phi_1 \in H_{\Gamma_1}^{-1/2}(\Omega_2)$; similarly, for $\eta_2 \in H_{\Gamma_2}^{-1/2}(\Omega_2)$, Problem IV has a unique generalized solution $\Phi_2 \in H_{\Gamma_2}^{-1/2}(\Omega_2)$. On these solutions we introduce the operators C_{ik} according to the following formulas:

$$\Phi_1|_{\Gamma_1} =: -C_{11}\eta_1, \qquad \Phi_1|_{\Gamma_2} =: C_{21}\eta_1,$$
$$\Phi_2|_{\Gamma_1} =: -C_{12}\eta_2, \qquad \Phi_2|_{\Gamma_2} =: C_{22}\eta_2. \tag{2.16}$$

In problem (2.8), from the Euler equation in the region Ω_2 it follows that the next Cauchy-Lagrange integral is valid,

$$\tilde{p}_2 + \rho_2 \frac{\partial \Phi}{\partial t} - \rho_2 \psi_{\boldsymbol{f}} = c(t), \qquad x \in \Omega_2, \tag{2.17}$$

where $\psi_{\boldsymbol{f}}$ is the field potential $P_{0.S_2}\boldsymbol{f}$ and $c(t)$ is an arbitrary function of t. Further, from (2.8), (2.14), (2.15) it follows that the velocity potential $\Phi(t,x)$ in the region Ω_2 may be represented in the form of a sum

$$\Phi(t,x) = \Phi_1(t,x) + \Phi_2(t,x), \tag{2.18}$$

if $\eta_i = \gamma_i \boldsymbol{u}_i (= \boldsymbol{u}_i \cdot \boldsymbol{n}_i)$. Then from (2.17), (2.18) and (2.16) we get that the boundary value condition $\tilde{p}_2 = \rho_2 g \zeta_2$ from (2.8) may be rewritten in the form

$$-\rho_2 \frac{\partial}{\partial t}(C_{21}\gamma_1 \boldsymbol{u}_1 + C_{22}\gamma_2 \boldsymbol{u}_2) + \rho_2 \psi_{\boldsymbol{f}} + c(t) = \rho_2 g \zeta_2 \quad \text{on } \Gamma_2. \tag{2.19}$$

We will consider one more auxiliary Zaremba problem analogous to Problem II.

Problem V

$$\Delta \Psi = 0 \quad \text{in } \Omega_2, \qquad \frac{\partial \Psi}{\partial n} = 0 \quad \text{on } S_2, \qquad \frac{\partial \Psi}{\partial n_1} = 0 \quad \text{on } \Gamma_1,$$
$$\Psi = \psi := \rho_2 g \zeta_2 \quad \text{on } \Gamma_2. \tag{2.20}$$

If $\psi \in H^{1/2}(\Gamma_2)$, then problem (2.20) has a unique generalized solution $\Psi \in H^1(\Omega_2)$ and by virtue of the relationship (2.19) we may consider that

$$\nabla \Psi =: G_2 \psi = G_2(\rho_2 g \zeta_2) = G_2\left(-\rho_2 \frac{\partial}{\partial t}(C_{21}\gamma_1 \boldsymbol{u}_1 + C_{22}\gamma_2 \boldsymbol{u}_2) + \rho_2 \psi_{\boldsymbol{f}}\right), \tag{2.21}$$

since $G_2 c(t) \equiv \boldsymbol{0}$. For this all terms in (2.21) are elements of the subspace $\boldsymbol{G}_{\mathrm{h},S_2}(\Omega_2)$.

Based on the relationships (2.13), (2.21), the previously introduced auxiliary Problems I–V and the operators corresponding to them, we come to the conclusion that the classical solution of problem (2.8) satisfies the following system of equations,

$$\begin{aligned}
&\frac{\mathrm{d}}{\mathrm{d}t}(\rho_1 \boldsymbol{u}_1 + \rho_2(G_1 C_{11}\gamma \boldsymbol{u}_1 + G_1 C_{12}\gamma_2 \boldsymbol{u}_2)) + \mu A \boldsymbol{u}_1 + g(\Delta\rho) G_1 \zeta_1 \\
&\qquad = \rho_2 G_1 \psi_{\boldsymbol{f}} + \rho_1 P_{0,S_1} \boldsymbol{f}, \\
&\frac{\mathrm{d}}{\mathrm{d}t}(\rho_2(G_2 C_{21}\gamma_1 \boldsymbol{u}_1 + G_2 C_{22}\gamma_2 \boldsymbol{u}_2)) + g\rho_2 G_2 \zeta_2 = \rho_2 G_2 \psi_{\boldsymbol{f}}, \\
&\frac{\mathrm{d}}{\mathrm{d}t}(g(\Delta\rho)\zeta_1) - g(\Delta\rho)\gamma_1 \boldsymbol{u}_1 = 0, \\
&\frac{\mathrm{d}}{\mathrm{d}t}(g\rho_2 \zeta_2) - g\rho_2 \gamma_2 \boldsymbol{u}_2 = 0, \\
&\boldsymbol{u}_1(0,x) = \boldsymbol{u}_1^0(x), \\
&\boldsymbol{u}_2(0,x) = \nabla\Phi(0,x) = (P_{h.S_2} \boldsymbol{u}_2^0)(x), \\
&\zeta_1(0,x_1,x_2) = \zeta_1^0(x_1,x_2), \\
&\zeta_2(0,x_1,x_2) = \zeta_2^0(x_1,x_2).
\end{aligned} \tag{2.22}$$

Here it may be assumed that $\boldsymbol{u}_1 = \boldsymbol{u}_1(t)$ is a function of the variable t with values in the space $\boldsymbol{J}_{0,S_1}(\Omega_1), \boldsymbol{u}_2 = \boldsymbol{u}_2(t) = (\nabla\Phi)(t)$ is a function with values in $\boldsymbol{G}_{\mathrm{h},S_2}(\Omega_2)$, and $\zeta_i(t)$ is a function with values in $H_i := L_2(\Gamma_i) \ominus \{1_i\}$. The last two equations (2.22) are the kinematic relationships on Γ_1 and Γ_2, respectively, where in the former we multiplied all terms by $g(\Delta\rho)$, and in the latter we multiplied all terms by $g\rho_2$.

Problem (2.22) may be written in a vector and block-matrix form in the following way:

$$\begin{pmatrix} C & 0 \\ 0 & gB \end{pmatrix} \frac{\mathrm{d}}{\mathrm{d}t} \begin{pmatrix} \boldsymbol{u} \\ \zeta \end{pmatrix} + \begin{pmatrix} \mu\tilde{A} & gF_1 \\ gF_2 & 0 \end{pmatrix} \begin{pmatrix} \boldsymbol{u} \\ \zeta \end{pmatrix} = \begin{pmatrix} \boldsymbol{f}_1 \\ 0 \end{pmatrix},$$

$$\begin{pmatrix} \boldsymbol{u}(0) \\ \zeta(0) \end{pmatrix} = \begin{pmatrix} \boldsymbol{u}^0 \\ \zeta^0 \end{pmatrix},$$

$$C = \begin{pmatrix} \rho_1 I_1 + \rho_2 G_1 C_{11}\gamma_1 & \rho_2 G_1 C_{12}\gamma_2 \\ \rho_2 G_2 C_{21}\gamma_1 & \rho_2 G_2 C_{22}\gamma_2 \end{pmatrix},$$

$$\boldsymbol{u} = \begin{pmatrix} \boldsymbol{u}_1 \\ \boldsymbol{u}_2 = \nabla\Phi \end{pmatrix},$$

$$\zeta = \begin{pmatrix} \zeta_1 \\ \zeta_2 \end{pmatrix},$$

$$f_1 = \begin{pmatrix} \rho_2 G_1 \psi_{\boldsymbol{f}} + \rho_1 P_{0,S_1}\boldsymbol{f} \\ \rho_2 G_2 \psi_{\boldsymbol{f}} \end{pmatrix},$$

where

$$\begin{aligned} B &:= \operatorname{diag}((\Delta\rho) I_{\Gamma_1}; \rho_2 I_{\Gamma_2}), \\ \tilde{A} &:= \operatorname{diag}(A; 0), \\ F_1 &:= \operatorname{diag}((\Delta\rho) G_1; \rho_2 G_2), \\ F_2 &:= -\operatorname{diag}((\Delta\rho)\gamma_1; \rho_2\gamma_2), \\ \boldsymbol{u}(0) &:= \boldsymbol{u}^0 = (\boldsymbol{u}_1^0; P_{h,S_2}\boldsymbol{u}_2^0)^{\mathrm{t}}, \\ \zeta(0) &:= (\zeta_1^0; \zeta_2^0)^{\mathrm{t}}. \end{aligned} \tag{2.23}$$

10.2.3 Properties of Matrix Blocks and Their Physical Meanings

We will next describe properties of the operator coefficients of the differential equation (2.23) with the unknown function $(\boldsymbol{u}; \zeta)^{\mathrm{t}} =: y = (\boldsymbol{u}_1; \boldsymbol{u}_2; \zeta_1; \zeta_2)^{\mathrm{t}}$ taking values in the Hilbert space

$$\tilde{H} := \boldsymbol{J}_{0,S_1}(\Omega_1) \oplus \boldsymbol{G}_{h,S_2}(\Omega_2) \oplus H_1 \oplus H_2 \tag{2.24}$$

with the scalar product

$$(y, z)_{\tilde{H}} := (\boldsymbol{u}_1, \boldsymbol{v}_1)_{\boldsymbol{J}_{0,S_1}(\Omega_1)} + (\boldsymbol{u}_2, \boldsymbol{v}_2)_{\boldsymbol{G}_{h,S_2}(\Omega_2)} + (\zeta_1, \eta_1)_{H_1} + (\zeta_2, \eta_2)_{H_2}, \tag{2.25}$$

where $z = (v_1; v_2; \eta_1; \eta_2)^{\mathrm{t}}$. Here, as in the previous problems, it is assumed that we already carried out a transition to dimensionless variables and the new variables are denoted by the same symbols.

We note first of all that the operator C is a bounded operator acting in the space $\boldsymbol{J}_{0,S_1}(\Omega_1) \oplus \boldsymbol{G}_{\mathrm{h},S_2}(\Omega_2)$, since its matrix coefficients are bounded. Let us check, for example, that the operator $G_1 C_{12}\gamma_2$ is bounded. In fact, for any element $\boldsymbol{u}_2 = \nabla\Phi \in \boldsymbol{G}_{\mathrm{h},S_2}(\Omega_2)$ we have $\eta_2 := \gamma_2 \boldsymbol{u}_2 = \partial\Phi/\partial n_2 \in H^{-1/2}(\Gamma_2)$. Then, by means of the auxiliary Problem I, we get that its solution $\Phi_2 \in H^1_{\Gamma_2}(\Omega_2)$ and, therefore, $C_{12}\eta_2 = C_{12}\gamma_2 u_2 = -\Phi_2|_{\Gamma_1}$ is in $H^{1/2}(\Omega_1)$. Considering now Problem II for $\psi = C_{12}\eta_2$ we proved that its solution $G_1\psi = G_1 C_{12}\gamma_2 \boldsymbol{u}_2$ is in $\boldsymbol{G}_{\mathrm{h},S_1}(\Omega_1)$. Thus, the operator $G_1 C_{12}\gamma_2$ acts boundedly from $\boldsymbol{G}_{\mathrm{h},S_2}(\Omega_2)$ into $\boldsymbol{G}_{\mathrm{h},S_1}(\Omega_1) \subset \boldsymbol{J}_{0,S_1}(\Omega_1)$.

Similarly, we may check that operators $G_1 C_{11}\gamma_1, G_2 C_{21}\gamma_1$ and $G_2 C_{22}\gamma_2$ are also bounded.

To verify the physical meaning of the operator C we calculate its quadratic form in the (complex) Hilbert space $\boldsymbol{J}_{0,S_1}(\Omega_1) \oplus \boldsymbol{G}_{\mathrm{h},S_2}(\Omega_2)$:

$$
\begin{aligned}
(C\boldsymbol{u}, \boldsymbol{u}) = \rho_1 \int_{\Omega_1} & |\boldsymbol{u}_1|^2 \mathrm{d}\Omega_1 \\
& + \rho_2 \left\{ \int_{\Omega_1} (G_1 C_{11}\gamma_1 \boldsymbol{u}_1) \cdot \bar{\boldsymbol{u}}_1 \mathrm{d}\Omega_1 + \int_{\Omega_1} (G_1 C_{12}\gamma_2 \boldsymbol{u}_2) \cdot \bar{\boldsymbol{u}}_1 \mathrm{d}\Omega_1 \right. \\
& \qquad \left. + \int_{\Omega_2} (G_2 C_{21}\gamma_1 \boldsymbol{u}_1) \cdot \bar{\boldsymbol{u}}_2 \mathrm{d}\Omega_2 + \int_{\Omega_2} (G_2 C_{22}\gamma_2 \boldsymbol{u}_2) \cdot \bar{\boldsymbol{u}}_2 \mathrm{d}\Omega_2 \right\}. \qquad (2.26)
\end{aligned}
$$

For our further discussion it is important to notice that the operators $G_1 : H^{1/2}(\Gamma_1) \to \boldsymbol{G}_{\mathrm{h},S_1}(\Omega_1)$ and $\gamma_1 : \boldsymbol{G}_{\mathrm{h},S_1}(\Omega_1) \to H^{-1/2}(\Gamma_1)$ are mutually adjoint. Indeed, let $\boldsymbol{v} = \nabla w$ be an arbitrary element from $\boldsymbol{G}_{\mathrm{h},S_1}(\Omega_1)$, and suppose φ_2 is the solution of Problem II for $\psi \in H^{1/2}(\Omega_1)$. Since $\Delta w = 0$ (in Ω_1), and $\partial w/\partial n = 0$ (on S_1), then by the first Green formula for the Laplace operator we get the identity

$$
(\nabla\varphi_2, \boldsymbol{v})_{\boldsymbol{G}_{\mathrm{h},S_1}(\Omega_1)} = (G_1\psi, \boldsymbol{v})_{\boldsymbol{G}_{\mathrm{h},S_1}(\Omega_1)} = (\psi, \gamma_1 \boldsymbol{v})_{L_2(\Gamma_1)} \qquad (2.27)
$$

for any $\psi \in H^{1/2}(\Gamma_1)$, and $\boldsymbol{v} \in \boldsymbol{G}_{\mathrm{h},S_1}(\Omega_1)$, whence the proof of the statement follows. Further, since for the elements $\boldsymbol{u} \in \boldsymbol{J}_0(\Omega_1)$ we have $\gamma_1 \boldsymbol{u} := \boldsymbol{u} \cdot \boldsymbol{n}_1 = 0$ (on Γ_1) and the subspaces $\boldsymbol{G}_{\mathrm{h},S_1}(\Omega_1)$ and $\boldsymbol{J}_0(\Omega_1)$ are orthogonal, then identity (2.27) can be extended to any $\boldsymbol{v} \in \boldsymbol{J}_{0,S_1}(\Omega_1) = \boldsymbol{G}_{\mathrm{h},S_1}(\Omega_1) \oplus \boldsymbol{J}_0(\Omega_1)$. Thus, it may be considered in view of this observation, that $\gamma_1 : \boldsymbol{J}_{0,S_1}(\Omega_1) \to H^{-1/2}(\Gamma_1)$ and $G_1 : H^{1/2}(\Gamma_1) \to \boldsymbol{G}_{\mathrm{h},S_1}(\Omega_1) \subset \boldsymbol{J}_{0,S_1}(\Omega_1)$ are mutually adjoint operators.

Based of the auxiliary Problem V it can be checked similarly that the operators $\gamma_2 : \boldsymbol{G}_{\mathrm{h},S_2}(\Omega_2) \to H^{-1/2}(\Gamma_2)$ and $G_2 : H^{1/2}(\Gamma_2) \to \boldsymbol{G}_{\mathrm{h},S_2}(\Omega_2)$ are also mutually adjoint, that is,

$$
(G_2\psi, \boldsymbol{v})_{\boldsymbol{G}_{\mathrm{h},S_2}(\Omega_2)} = (\psi, \gamma_2 \boldsymbol{v})_{L_2(\Gamma_2)}, \qquad \psi \in H^{1/2}(\Gamma_2),\ \boldsymbol{v} \in \boldsymbol{G}_{\mathrm{h},S_2}(\Omega_2). \qquad (2.28)
$$

Taking into account the identities (2.27), (2.28) and definitions (2.16) of the operators C_{ik} we transform the right hand side of (2.26) and have the following chain of equations

$$
\begin{aligned}
(C\boldsymbol{u},\boldsymbol{u}) &= \rho_1\int_{\Omega_1}|\boldsymbol{u}_1|^2\mathrm{d}\Omega_1 \\
&\quad+\rho_2\left\{\int_{\Gamma_1}(C_{11}\gamma_1\boldsymbol{u}_1)\overline{\gamma_1\boldsymbol{u}_1}\mathrm{d}\Gamma_1+\int_{\Gamma_1}(C_{12}\gamma_2\boldsymbol{u}_2)\overline{\gamma_1\boldsymbol{u}_1}\mathrm{d}\Gamma_1\right.\\
&\qquad\left.+\int_{\Gamma_2}(C_{21}\gamma_1\boldsymbol{u}_1)\overline{\gamma_2\boldsymbol{u}_2}\mathrm{d}\Gamma_2+\int_{\Gamma_2}(C_{22}\gamma_2\boldsymbol{u}_2)\overline{\gamma_2\boldsymbol{u}_2}\mathrm{d}\Gamma_2\right\}\\
&=\rho_1\int_{\Omega_1}|\boldsymbol{u}_1|^2\mathrm{d}\Omega_1\\
&\quad+\rho_2\left\{-\int_{\Gamma_1}\Phi_1\overline{\frac{\partial\Phi_1}{\partial n_1}}\mathrm{d}\Gamma_1-\int_{\Gamma_1}\Phi_2\overline{\frac{\partial\Phi_1}{\partial n_1}}\mathrm{d}\Gamma_1+\int_{\Gamma_2}\Phi_1\overline{\frac{\partial\Phi_2}{\partial n_2}}\mathrm{d}\Gamma_2+\int_{\Gamma_2}\overline{\frac{\partial\Phi_2}{\partial n_2}}\mathrm{d}\Gamma_2\right\}\\
&=\rho_1\int_{\Omega_1}|\boldsymbol{u}_1|^2\,d\Omega_1+\rho_2\int_{\partial\Omega_2}(\Phi_1+\Phi_2)\overline{\frac{\partial}{\partial n}(\Phi_1+\Phi_2)}\mathrm{d}S\\
&=\rho_1\int_{\Omega_1}|\boldsymbol{u}_1|^2\mathrm{d}\Omega_1+\rho_2\int_{\Omega_2}|\nabla\Phi|^2\mathrm{d}\Omega_2\\
&=\rho_1\int_{\Omega_1}|\boldsymbol{u}_1|^2\mathrm{d}\Omega_1+\rho_2\int_{\Omega_2}|\boldsymbol{u}_2|^2\mathrm{d}\Omega_2,\qquad \boldsymbol{u}_2=\nabla\Phi.
\end{aligned}
\tag{2.29}
$$

In deducing (2.29), we used the properties of the solutions of the auxiliary Problems III and IV for the functions Φ_1 and Φ_2, and also relationship (2.18). The right hand side of (2.29) obviously equals twice the kinetic energy of the connected part of the partially dissipative hydrosystem, that is, of the part which was obtained after separating the trivial solution (2.2). Therefore, it is natural to call operator C the *operator of kinetic energy* of the system. As it follows from (2.29), the operator C is positive definite in the space $\hat{J}_{0,S}(\Omega)$ and consequently it has a bounded inverse operator which is also positive definite since C is bounded.

Now we formulate the properties of the remaining matrix blocks in equation (2.23). Above all, the operator B is obviously a bounded and positive definite operator acting in $H=H_1\oplus H_2, H_i=L_2\ominus\{1_i\}$. Since the quadratic form

$$
(gB\zeta,\zeta)_H=g\left(\Delta\rho\int_{\Gamma_1}|\zeta_1|^2\mathrm{d}\Gamma_1+\rho_2\int_{\Gamma_2}|\zeta_2|^2\mathrm{d}\Gamma_2\right)
\tag{2.30}
$$

is equal to twice the potential energy of the system, then gB may be called the *operator of potential energy*. As a conclusion, the operator $\mathrm{diag}(C;gB)$ acting in $\hat{J}_{0,S}(\Omega)\oplus H$ is the operator of the full (kinetic plus potential) energy of the system.

Furthermore, the unbounded operator $\tilde{A} := \operatorname{diag}(A; 0)$ is the operator of dissipation, since the form $\mu(\tilde{A}\boldsymbol{u}, \boldsymbol{u})_{\hat{J}_{0,S}(\Omega)} \geq 0$ is equal to the energy dissipation velocity.

We also note that by virtue of the proved properties $(\gamma_1)^* = G_1, \gamma_2^* = G_2$ [see (2.27), (2.28)] and formulas (2.23) the connection $F_2 = -F_1^*$ is valid, where $F_1 : H^{1/2}(\Gamma_1) \oplus H^{1/2}(\Gamma_2) \to \hat{G}_{\mathrm{h},S}(\Omega) = \{\hat{\boldsymbol{u}} := (\boldsymbol{u}_1; \boldsymbol{u}_2)^{\mathrm{t}} : \boldsymbol{u}_1 \in \boldsymbol{G}_{\mathrm{h},S_1}(\Omega_1), \boldsymbol{u}_2 \in \boldsymbol{G}_{\mathrm{h},S_2}(\Omega_2),\ \gamma_1 \boldsymbol{u}_1 = \gamma_1 \boldsymbol{u}_2\}$, $F_2 : \hat{\boldsymbol{G}}_{\mathrm{h},S}(\Omega) \to H^{-1/2}(\Gamma_1) \oplus H^{-1/2}(\Gamma_2)$. If we consider that F_1 acts from H in $\hat{G}_{\mathrm{h},S}(\Omega)$, then it is defined on the set $\mathcal{D}(F_1) = H^{1/2}_{\Gamma_1} \oplus H^{1/2}_{\Gamma_2}$ dense in H and is an unbounded operator. For the operator $F_2 = -F_1^*$ the following statement is valid: as an operator from $\hat{\boldsymbol{G}}_{\mathrm{h},S}(\Omega)$ in H it is unbounded.

In the conclusion of this section we note that whenever $F_1 = -F_2^* = 0$, the problem (2.23) on finding the velocity fields $\boldsymbol{u} = (\boldsymbol{u}_1; \boldsymbol{u}_2)^{\mathrm{t}}$ and deviation fields $\zeta = (\zeta_1; \zeta_2)^{\mathrm{t}}$ splits into two independent problems.

10.2.4 Theorem on Correct Solvability of the Initial Boundary Value Problem

Based on the ascertained properties of the operator coefficients in problem (2.23), we next point out several important properties of the operator matrices of this differential equation. Above all, the operator $\mathcal{T} := \operatorname{diag}(C; gB)$ of the full energy of the system is a bounded positive definite operator acting in $\tilde{H} = \hat{\boldsymbol{J}}_{0,S}(\Omega) \oplus H$.

Furthermore, let us introduce the operator matrix

$$\mathcal{A} := \begin{pmatrix} \mu \tilde{A} & gF_1 \\ -gF_1^* & 0 \end{pmatrix}, \tag{2.31}$$

acting in $\tilde{H}$. Its domain of definition, $\mathcal{D}(\mathcal{A})$, obviously consists of those elements $y = (\boldsymbol{u}; \zeta)^{\mathrm{t}}$ from $\tilde{H}$ for which $\mathcal{A}y \in \tilde{H}$, that is,

$$\begin{aligned} \tilde{A}\boldsymbol{u} &= (A\boldsymbol{u}_1; 0)^{\mathrm{t}} \in \hat{\boldsymbol{J}}_{0,S}(\Omega), \\ F_1\zeta &= ((\Delta\rho)G_1\zeta_1; \rho_2 G_2 \zeta_2)^{\mathrm{t}} \in \hat{\boldsymbol{J}}_{0,S}(\Omega), \\ -gF_1^*\boldsymbol{u} &= ((\Delta\rho)\gamma_1\boldsymbol{u}_1; \rho_2\gamma_2\boldsymbol{u}_2)^{\mathrm{t}} \in H = H_1 \oplus H_2. \end{aligned}$$

Whence it follows that the operator $\mathcal{A}$ is defined on set a dense in $\tilde{H}$ given by

$$\mathcal{D}(\mathcal{A}) = \mathcal{D}(A) \oplus W_2 \oplus H^{1/2}_{\Gamma_1} \oplus H^{1/2}_{\Gamma_2}, \tag{2.32}$$

where

$$\begin{aligned} W_2 = \{\boldsymbol{u}_2 \in \boldsymbol{G}_{\mathrm{h},S_2}(\Omega_2) : (\boldsymbol{u}_1; \boldsymbol{u}_2)^{\mathrm{t}} \in \hat{\boldsymbol{J}}_{0,S}(\Omega), \boldsymbol{u}_1 \in \mathcal{D}(A), \\ \gamma_1\boldsymbol{u}_1 = \gamma_1\boldsymbol{u}_2, \gamma_2\boldsymbol{u}_2 \in H^{1/2}_{\Gamma_2}\}. \end{aligned} \tag{2.33}$$

Taking into account the previously introduced notations, problem (2.23) may be written in the form of a Cauchy problem for the differential operator equation

$$\mathcal{T}\frac{\mathrm{d}y}{\mathrm{d}t} + \mathcal{A}y = f(t), \qquad y(0) = y^0, \qquad f(t) = (\boldsymbol{f}_1(t); 0)^{\mathrm{t}}, \tag{2.34}$$

relative to the unknown function $y(t) = (\boldsymbol{u}; \zeta)^{\mathrm{t}} = (\boldsymbol{u}_1; \boldsymbol{u}_2; \zeta_1; \zeta_2)^{\mathrm{t}}$ with values in $\tilde{H}$. It is natural to call this function the *function of the state of the system.*

As it was mentioned above, the operator of full energy $\mathcal{T}$ in (2.34) is bounded and positive definite. Since $\tilde{A} = \operatorname{diag}(A; 0)$, then operator $\mathcal{A}$ from (2.31) possesses the property

$$\operatorname{Re}(\mathcal{A}y, y)_{\tilde{H}} \geq 0, \qquad y \in \mathcal{D}(\mathcal{A}). \tag{2.35}$$

In fact, it can be represented in the form

$$\begin{aligned} &\mathcal{A} = \mathcal{A}_0 + \mathrm{i}\mathcal{F}, \qquad \mathcal{A}_0 = \operatorname{diag}(\mu\tilde{A}; 0) \geq 0, \\ &\mathcal{F} = \begin{pmatrix} 0 & -\mathrm{i}F_1 \\ \mathrm{i}F_1^* & 0 \end{pmatrix} = \mathcal{F}^*. \end{aligned} \tag{2.36}$$

Similarly, it is easily checked that

$$\operatorname{Re}(\mathcal{A}^* z, z)_{\tilde{H}} \geq 0, \qquad z \in \mathcal{D}(\mathcal{A}^*), \tag{2.37}$$

whence it follows that the operator $-\mathcal{A}$ is in essence a maximal dissipative operator defined on the set (2.32).

We introduce on $\tilde{H}$ a norm equivalent to the initial one by defining the scalar product

$$\langle y_1, y_2 \rangle := (\mathcal{T}y_1, y_2)_{\tilde{H}} = (\mathcal{T}^{1/2}y_1, \mathcal{T}^{1/2}y_2)_{\tilde{H}}, \tag{2.38}$$

and rewrite equation (2.34) in the equivalent form

$$\frac{\mathrm{d}y}{\mathrm{d}t} = -\mathcal{T}^{-1}\mathcal{A}y + \mathcal{T}^{-1}f(t), \qquad y(0) = y^0. \tag{2.39}$$

It is obvious that the operator $-\mathcal{T}^{-1}\mathcal{A}$ is a maximal dissipative operator in the new scalar product. Therefore, according to the conclusions of Section 1.5.4, the homogeneous Cauchy problem (2.39) is uniformly correct and its corresponding semigroup $\mathcal{U}(t)$ is contractive. If $y^0 \in \mathcal{D}(\mathcal{A})$ and $f(t)$ is a continuously differentiable function with values in $\tilde{H}$, then problem (2.39) has a solution $y(t)$ on any interval $[0, T]$, expressed by the formula

$$y(t) = \mathcal{U}(t)y^0 + \int_0^t \mathcal{U}(t-\tau)\mathcal{T}^{-1}f(\tau)\mathrm{d}\tau. \tag{2.40}$$

If $y^0 \in \tilde{H}$, and $f(t)$ is a continuous function with values in $\tilde{H}$, then problem (2.39) has a generalized solution in the sense of the definition presented in Section 1.5.7. This generalized solution is found also by the formula (2.40).

10.3 Model Problem on Normal Oscilations of Partially Dissipative Hydrosystems

Here a model spectral problem preserving all the peculiarities of problem (1.23) on normal oscillations of a partially dissipative hydrosystem is considered, where capillary forces acting on the two separation boundaries are additionally taken into account. A qualitative and asymptotic investigation of the spectrum of the problem is carried out on the base of a study of the transcendent characteristic equation for the complex fading decrement of normal oscillations.

10.3.1 Statement of the Model Problem

The spectral problem (1.23) containing the spectral parameter both in equations and boundary conditions is rather complicated to research. To understand the spectrum structure and other properties of its solutions for an arbitrary region $\Omega = \Omega_1 \cup \Omega_2$ in a three-dimensional case, it is useful to consider a preliminary two-dimensional (plane) problem with some simplified assumptions, which presumably do not change substantially the general spectrum structure.

For this purpose, let us consider the plane (two-dimensional) problem for two fluids situated in a rectangular container of a width l. We suppose that the lower viscous fluid takes the region $\Omega_1 := \{(x; y) : 0 < x < l, -h_1 < y < 0\}$ and the upper ideal fluid takes the region $\Omega_2 := \{(x; y) : 0 < x < l, 0 < y < h_2\}$. The boundary Γ_1 has the equation $y = 0$, and the free surface Γ_2 of the ideal fluid has the equation $y = h_2$.

Suppose that the homogeneous gravitational field with the acceleration $\boldsymbol{g} = -\boldsymbol{g}e_2$ acts on the fluid system opposite to the direction of the axis Oy and capillary forces. Further, two cases will be considered: a) fluids are considered to be "heavy" and the capillary forces are not taken into account; b) fluids are considered to be "capillary", that is, being in a state close to weightlessness. In the second case the coefficients of surface tension $\sigma_i > 0$ on the fluid boundaries Γ_i are known physical constants, and the wetting angles (contact angles) between surfaces Γ_i and the rigid wall S of the vessel are right angles.

Using the equations and boundary value conditions of problem (1.23) we will model the velocity speed of normal oscillations of the viscous fluid in the region Ω_1 by a scalar field $u(x, y)$ for which the following equation is fulfilled,

$$-\mu\Delta u = \lambda\rho_1 u \qquad \text{in } \Omega_1, \tag{3.1}$$

where λ is the unknown complex frequency of oscillations of the system ($\operatorname{Re}\lambda$ is the fading decrement, and $\operatorname{Im}\lambda$ is the frequency), $\Delta := \partial^2/\partial x^2 + \partial^2/\partial y^2$ is the two-dimensional laplacean, and μ and ρ_1 are the viscosity and density of the first (lower) fluid. The equation (3.1) is a scalar analoque of the linearized Navier-Stokes equation from (1.23).

The motion of the ideal fluid in the region Ω_2 will be characterized by a velocity potential $\Phi = \Phi(x, y)$. By virtue of the continuity equation the function Φ should satisfy the Laplace equation

$$\Delta\Phi = 0 \qquad \text{in } \Omega_2. \tag{3.2}$$

At the boundary Γ_i the kinematic and dynamic conditions of (1.23) should be fulfilled. Here the kinematic conditions on Γ_i take the form

$$\begin{aligned} u = \frac{\partial\Phi}{\partial y} &= -\lambda\zeta_1 \qquad \text{on } \Gamma_1, \\ \frac{\partial\Phi}{\partial y} &= -\lambda\zeta_2 \qquad \text{on } \Gamma_2, \end{aligned} \tag{3.3}$$

where $\zeta_i(x)$ is the vertical boundary deviation in oscillating. Modelling the normal stress in the viscous fluid with the value $\mu\partial u/\partial y$, the pressure in the ideal fluid as a function $p_2 = \lambda\rho_2\Phi$, excluding the functions $\zeta_i(x)$ from the kinematic and dynamic conditions and taking into account the capillary forces parallel with gravitational forces, instead of the corresponding dynamic conditions (1.23) on Γ_1 and Γ_2 we get the conditions

$$\lambda\left(\mu\frac{\partial u}{\partial y} + \lambda\rho_2\Phi\right) = \left(g(\Delta\rho) - \sigma_1\frac{\partial^2}{\partial x^2}\right)u \qquad \text{on } \Gamma_1, \tag{3.4}$$

$$\lambda^2\rho_2\Phi = \left(\rho_2 g - \sigma_2\frac{\partial^2}{\partial x^2}\right)\frac{\partial\Phi}{\partial y} \qquad \text{on } \Gamma_2, \tag{3.5}$$

where $\Delta\rho := \rho_1 - \rho_2 > 0$.

We choose as boundary conditions on the rigid wall S of the vessel, instead of the stickness conditions for the viscous fluid and nonleaking conditions for the ideal fluid such analogues of these conditions which allow to separate variables in the given model problem:

$$u(0, y) = u(l, y) = u(x, -h_1) = 0 \tag{3.6}$$

$$\Phi(0, y) = \Phi(l, y) = 0. \tag{3.7}$$

Thus, we consider further the model boundary value problem (3.1)–(3.7) on finding the eigenfunctions $u(x, y), \Phi(x, y)$ and the spectral parameter λ. Its study will allow us to predict, at least approximately, the spectrum structure and the properties of the solutions three-dimensional spectral problem (1.23), and its analogues when one takes into account the actions of capillary forces on the boundary.

In the equations and boundary conditions (3.1)–(3.7) we may switch to dimensionless variables. In the sequel we will assume that $h_2 = h_1 =: h$, and the characteristic length of the problem is equal to $l_0 = l/\pi$. Then the dimensionless container width is equal to π. We choose again, for the sake of simplicity, $h = l_0$, and then the dimensionless fluid heights are equal to 1. If we take the value ρ_1 to be the characteristic density, and the corresponding dimensional combinations from the above-mentioned characteristic ones for the remaining values, then after a transition to dimensionless variables it can be formally considered that the conditions of the problem (3.1)–(3.7) are kept, where $\mu = 1, \ \rho_1 = 1$. Finally, we come to the following problem,

$$-\Delta u = \lambda u, \qquad 0 < x < \pi, \ -1 < y < 0, \tag{3.8}$$

$$\Delta \Phi = 0, \qquad 0 < x < \pi, \ 0 < y < 1, \tag{3.9}$$

$$u(0, y) = u(\pi, y) = u(x, -1) = 0, \tag{3.10}$$

$$\Phi(0, y) = \Phi(\pi, y) = 0, \quad u = \frac{\partial \Phi}{\partial y}, \qquad y = 0, \tag{3.11}$$

$$\left(\rho_2 g - \sigma_2 \frac{\partial^2}{\partial x^2}\right) \frac{\partial \Phi}{\partial y} + \lambda^2 \rho_2 \Phi = 0, \qquad y = 1, \tag{3.12}$$

$$\lambda\left(\frac{\partial u}{\partial y} + \lambda \rho_2 \Phi\right) = \left(g(\Delta\rho) - \sigma_1 \frac{\partial^2}{\partial x^2}\right) u, \qquad y = 0, \tag{3.13}$$

10.3.2 Obtaining the Characteristic Equation

We will find the solutions of the problem (3.8)–(3.13) in the form of Fourier series in the orthogonal function system $\{\sin nx\}_{n=1}^{\infty}$ on the segment $[0, \pi]$, satisfying the conditions (3.10) and (3.11) for $x = 0$ and $x = \pi$:

$$u = u_n(x, y) = u_n(y) \sin nx, \quad \Phi = \Phi_n(x, y) = \Phi_n(y) \sin nx, \qquad n \in \mathbb{N}. \tag{3.14}$$

Then, equation (3.9) gives the solution

$$\Phi_n(y) = C_{1n} \sinh(ny) + C_{2n} \cosh(ny), \tag{3.15}$$

with arbitrary constants C_{1n} and C_{2n}.

For the function $u_n(y)$ from (3.8) we get $u_n'' + (\lambda - n^2)u_n = 0$, and using the boundary condition (3.10) for $y = -1$ we have

$$u_n(y) = C_{3n} \sin\left(\sqrt{\lambda - n^2}(y+1)\right), \tag{3.16}$$

where it is assumed that $\operatorname{Re}\sqrt{z} \geq 0$, $-\pi < \arg z \leq \pi$.

From the kinetic condition (3.11) we get

$$C_{3n} \sin\sqrt{\lambda - n^2} = C_{2n} n, \qquad n \in \mathbb{N}. \tag{3.17}$$

Substitutions of the solutions (3.15) and (3.16) into the conditions (3.12) and (3.13) lead to the relationships

$$n(\rho_2 g + \sigma_2 n^2)(C_{1n}\cosh n + C_{2n}\sinh n) + \rho_2\lambda^2(C_{1n}\sinh n + C_{2n}\cosh n) = 0, \tag{3.18}$$

$$\lambda\left(\sqrt{\lambda - n^2}C_{3n}\cos\sqrt{\lambda - n^2} + \lambda\rho_2 C_{2n}\right) = (g(\Delta\rho) + \sigma_1 n^2)C_{3n}\sin\sqrt{\lambda - n^2}. \tag{3.19}$$

The system of equations (3.17)–(3.19) is a homogeneous system of linear algebraic equations with respect to the unknowns C_{in}, $i = 1, 2, 3$. For the existence of nontrivial solutions of the system it is necessary that its determinant is equal to zero. This requirement leads to the characteristic equation of the problem (3.18)–(3.13):

$$\begin{aligned} n &\left[n(\rho_2 g + \sigma_2 n^2)\sinh n + \rho_2\lambda^2\cosh n\right] \\ &\times \left[\lambda(\sqrt{\lambda - n^2}\cos\sqrt{\lambda - n^2} - (g(\Delta\rho) + \sigma_1 n^2)\sin\sqrt{\lambda - n^2}\right] \\ &- \sin\sqrt{\lambda - n^2}\rho_2\lambda^2[n(\rho_2 g + \sigma_2 n^2)\cosh n + \rho_2\lambda^2\sinh n] = 0, \qquad n \in \mathbb{N}. \end{aligned} \tag{3.20}$$

Now we notice that for the solution of problem (3.8)–(3.13) the conditions $\operatorname{Re}\lambda > 0$, $\cos\sqrt{\lambda - n^2} \neq 0$ are fulfilled. Without checking the first property let us suppose that $\cos\sqrt{\lambda - n^2} = 0$. Then from (3.20) we get the biquadratic equation

$$d_n(\lambda) := \rho_2^2\lambda^4\tanh n + \lambda^2\rho_2(d_n\coth n + \beta_n) + \alpha_n\beta_n = 0, \tag{3.21}$$

where

$$\begin{aligned} \alpha_n &:= n(\rho_2 g + \sigma_2 n^2)\tanh n, \\ \beta_n &:= n(g(\Delta\rho) + \sigma_1 n^2), \end{aligned} \tag{3.22}$$

which has purely imaginary solutions λ, a conclusion that contradicts the property $\operatorname{Re}\lambda = 0$.

It should be noticed also that equation (3.20) has the obvious solution $\lambda = n^2$, for which $C_{1n} = C_{2n} = 0$, $C_{3n} \neq 0$; for such solutions by virtue of (3.14)–(3.16) we get the trivial solution of the initial problem, and therefore we will consider further that $\lambda \neq n^2$.

Dividing both sides of equation (3.20) by $\sqrt{\lambda - n^2}\cos\sqrt{\lambda - n^2} \neq 0$, we obtain, after some simple transformations, the desired characteristic equation

$$\frac{\tan\sqrt{\lambda - n^2}}{\sqrt{\lambda - n^2}} = \frac{n\lambda(\rho_2\lambda^2 + \alpha_n)}{d_n(\lambda)}, \qquad n \in \mathbb{N}, \tag{3.23}$$

where $d_n(\lambda)$ is defined in (3.21) and (3.22).

10.3.3 Studying the Characteristic Equation

It is convenient to study equation (3.23) graphically by setting

$$\begin{aligned}\frac{\tan\sqrt{\lambda - n^2}}{\sqrt{\lambda - n^2}} &=: F_n(\lambda) = \frac{\tanh\sqrt{n^2 - \lambda}}{\sqrt{n^2 - \lambda}} \\ &= \sum_{k=1}^{\infty} \frac{2}{\left(\left(k - \frac{1}{2}\right)\pi\right)^2 + n^2 - \lambda},\end{aligned} \tag{3.24}$$

$$\frac{n\lambda\left(\rho_2\lambda^2 + \alpha_n\right)}{d_n(\lambda)} =: \Phi_n(\lambda), \tag{3.25}$$

$$F_n(\lambda) = \Phi_n(\lambda), \qquad n \in \mathbb{N}. \tag{3.26}$$

In the representation of $F_n(\lambda)$, (3.24), we used the decomposition of the function $\tan z$ into simple fractions.

The graphs of the two functions $F_n(\lambda)$ and $\Phi_n(\lambda)$ in two typical cases are qualitatively presented in Figures 10.3.1 and 10.3.2.

We first notice some obvious properties of these functions.

(a) The graph of $F_n(\lambda)$ has vertical asymptotes at the points

$$\lambda_{nk}^0 = \left(\left(k - \frac{1}{2}\right)\pi\right)^2 + n^2, \qquad k, n \in \mathbb{N}.$$

(b) $F_n(0) = (\tanh n)/n = O(1/n)$ as $n \to \infty$, $F_n(\lambda_{nk}^0 \pm 0) = \mp\infty$, $F_n(-\infty) = 0$.
(c) Function $\Phi_n(\lambda)$ is odd, positive for $\lambda \in (0, \infty)$, and $\Phi_n(+\infty) = 0$.

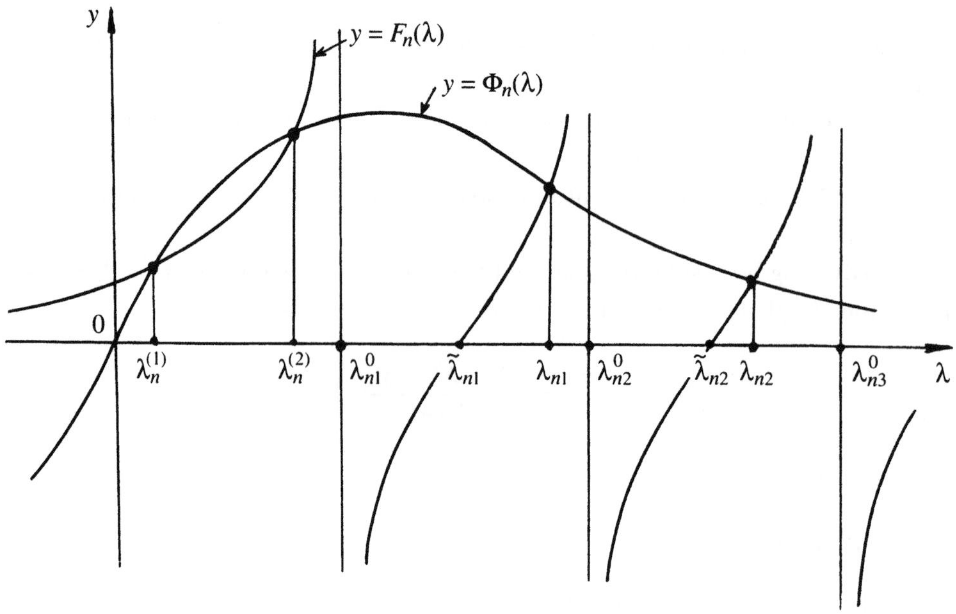

Figure 10.3.1

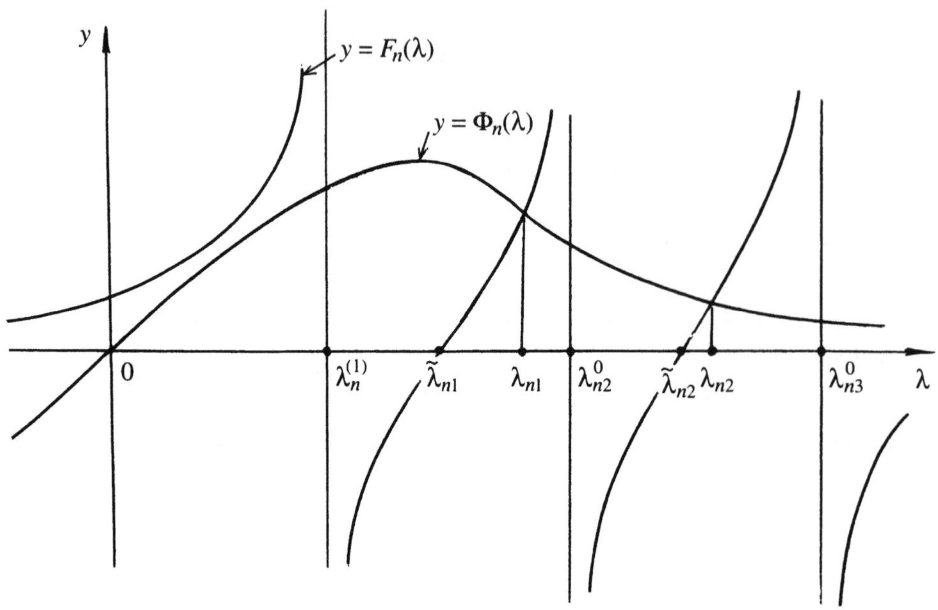

Figure 10.3.2

(d) Function $F_n(\lambda)$ monotonically increases at any point on the real axis that is not a point of discontinuity because

$$F_n'(\lambda) = \sum_{k=1}^{\infty} \frac{2}{\left[\left(\left(k-\frac{1}{2}\right)\pi\right)^2 + n^2 - \lambda\right]^2} > 0.$$

(e) Function $\Phi_n(\lambda)$ has no discontinuity points on the real axis.

The following are consequences of properties (a)–(e).

1° For any $n \in \mathbb{N}$, equation (3.23) has a countable set of real roots $\{\lambda_{nk}\}_{k=1}^{\infty}$, $\lambda_{nk} \to \infty$ as $n \to \infty$. In this set, the root λ_{nk} is situated in the interval $(\tilde{\lambda}_{nk}, \lambda_{n,k+1}^0)$, where $\tilde{\lambda}_{nk} = \pi^2 k^2 + n^2$ are zeros of the function $F_n(\lambda)$, that is, those values of λ, for which $\sin\sqrt{\lambda - n^2} = 0$. For $k \to \infty$, the asymptotic formula $\lambda_{nk} = \tilde{\lambda}_{nk} + o(1)$ is valid.

2° The two variants of the graphic solution of equation (3.23) depicted in Figures 10.3.1 and 10.3.2 show that if the graphs of the functions $F_n(\lambda)$ and $\Phi_n(\lambda)$ intersect in the interval $(0, \lambda_{n1}^0)$, then two more real roots, $\lambda_n^{(1)}$ and $\lambda_n^{(2)}$, can be found in that very same interval. If the graphs do not intersect in $(0, \lambda_{n1}^0)$, then instead of $\lambda_n^{(1)}$ and $\lambda_n^{(2)}$ there is a pair of nonreal complex conjugate roots of equation (3.23).

Switching now from the graphic to the qualitative study of equation (3.23), we can state more new properties of its roots.

3° For any $n \in \mathbb{N}$, equation (3.23) has at least one but no more than two pairs of nonreal complex conjugate roots.

To prove this property, we introduce the notation $\zeta := \sqrt{\lambda - n^2} = p + iq$ and consider in the plane ζ a contour Γ_ζ composed by the segments $0 \le p \le N$, $q = N$; $-N \le q \le N$, $p = N$; $0 \le p \le N$, $q = -N$, respectively, where N is a sufficiently large chosen number. It follows imediately that the value of $|\tan\zeta|$ on Γ_ζ is of order 1 for $N \to \infty$.

In mapping $\lambda = n^2 + \zeta^2$ the contour Γ_ζ transforms into a closed contour Γ_λ composed of two parabolas and having the property that the distance from points situated on Γ_λ to zero in the complex λ-plane has a value of order N^2. Since here $\Phi_n(\lambda) = o(\lambda^{-2}) = o(N^{-2})$, then by the Rouchet theorem we get that the differences between the number of zeroes and the number of poles for the functions $F_n(\lambda)$ and $F_n(\lambda) - \Phi_n(\lambda)$ that are contained inside Γ_λ are equal.

However, $F_n(\lambda) - \Phi_n(\lambda)$ has four more poles than $F_n(\lambda)$ and these new poles are situated on the imaginary axis at the points where $d_n(\lambda)$ turns into zero. It means that if the graphs of $F_n(\lambda)$ and $\Phi_n(\lambda)$ do not intersect in the interval $(0, \lambda_{n1}^0)$, then $F_n(\lambda) - \Phi_n(\lambda)$ has two more pairs of nonreal complex conjugate roots beside the obvious real ones: It is enough to choose $N = N_k = \pi(k+\delta)$, $0 < \delta < \pi/4$, where k is large. If the graphs of $F_n(\lambda)$ and $\Phi_n(\lambda)$ intersect in the interval $(0, \lambda_{n1}^0)$, then there is an additional pair of real roots, which means that equation (3.23) has only one pair of nonreal complex conjugate roots.

4° For sufficiently large n, equation (3.23) has only one pair of nonreal roots.

Indeed, employing the considerations needed to prove Property 3°, it is enough to prove that the graphs of the functions $F_n(\lambda)$ and $\Phi_n(\lambda)$ intersect in the interval $(0, \lambda_{n1}^0)$, with $\lambda_{n1}^0 = (\pi/2)^2 + n^2$. We are going to show that for sufficiently large n.

$$F_n(n^{3/2}) < \Phi_n(n^{3/2}), \tag{3.27}$$

and, therefore, the stated fact takes place because

$$F_n(0) = \frac{1}{n}\tanh n > \Phi_n(0) = 0.$$

In fact,

$$F_n(n^{3/2}) := \frac{\tanh\sqrt{n^2 - n^{3/2}}}{\sqrt{n^2 - n^{3/2}}} = \mathrm{O}\left(\frac{1}{n}\right), \qquad n \to \infty,$$

$$\Phi_n(n^{3/2}) := \frac{n n^{3/2}(\rho_2 n^3 + \alpha_n)}{\rho_2^2 n^6 \tanh n + n^3\rho_2(\alpha_n \coth n + \beta_n) + \alpha_n\beta_n}.$$

If $\sigma_1 > 0$ and $\sigma_2 > 0$, then using (3.22) we get that $\alpha_n = \mathrm{O}(n^3)$ and $\beta_n = \mathrm{O}(n^3)$. For $\sigma_1 = 0$ and $\sigma_2 = 0$ we have $\alpha_n = \mathrm{O}(n)$ and $\beta_n = \mathrm{O}(n)$. In both cases we obtain that

$$\Phi_n(n^{3/2}) \sim cn^{-1/2}, \qquad c > 0,\ n \to \infty,$$

and inequality (3.27) is satisfied for sufficiently large n.

5° For the roots $\lambda_n^{(1)}$ and $\lambda_n^{(2)}$ of equation (3.23) situated in the interval $(0, \lambda_{n1}^0)$ the following asymptotic formulas take place:

$$\lambda_n^{(1)} = \sigma_1 n + \sigma_1^2\left(\rho_2 + \frac{1}{2}\right) + \left[g(\Delta\rho) + \sigma_1^3\left(2\left(\rho_2 + \frac{1}{2}\right)^2 + \frac{1}{8}\right)\right] n^{-1} + \mathrm{O}(n^{-2}), \tag{3.28}$$

$$\lambda_n^{(2)} = \frac{\sqrt{1+4\rho_2^2} - 1}{2\rho_2^2} n^2[1 + \mathrm{o}(1)], \qquad n \to \infty. \tag{3.29}$$

Let us show the inference of the formulas (3.28) and (3.29). Since the root $\lambda_n^{(1)}$ lies to the left of the point $n^{3/2}$ for $n \to \infty$, then it is natual to find it in the form

$$\lambda = \lambda_n^{(1)} = c_1 n + c_0 + c_{-1} n^{-1} + \mathrm{O}(n^{-2}), \qquad n \to \infty. \tag{3.30}$$

Let us note that $\tanh n \sim 1 \sim \coth n$, $n \to \infty$ and, therefore, employing (3.30), the right-hand side of (3.23) is equivalent to the function $n\lambda_n^{(1)}/[\rho_2(\lambda_n^{(1)})^2 + \beta_n]$. Similarly we get that $\tanh \sqrt{n^2 - \lambda_n^{(1)}} \sim 1$. Whence it follows that the unknown root can be found from the equation

$$n\lambda_n^{(1)}\sqrt{n^2 - \lambda_n^{(1)}} = \rho_2(\lambda_n^{(1)})^2 + \beta_n, \qquad \beta_n = \sigma_1 n^3 + g(\Delta\rho)n.$$

within the desired accuracy of omitted exponential terms.

Comparing the coefficients of (3.30) with the ones in the preceding equation, we obtain that $c_1 = \sigma_1$, $c_0 = \sigma_1^2(\rho_2 + 1/2)$, and $c_{-1} = g(\Delta\rho) + \sigma_1^3[2(\rho_2 + 1/2)^2 + 1/8]$, that is, formula (3.28).

It is natural to find the second root $\lambda_n^{(2)}$ in the form $\lambda_n^{(2)} = \delta n^2[1 + \mathrm{o}(1)]$, $n \to \infty$. The comparison of the main terms in the left- and right-hand sides of equation (3.23) shows that δ should be chosen equal to $(\sqrt{1 + 4\rho_2^2} - 1)/(2\rho_2^2) < 1$, whence formula (3.23) follows.

We will study now properties of the nonreal roots of equation (3.23).

6° The nonreal root λ_n^+ of equation (3.23), whose existence and uniqueness in the upper half-plane for sufficiently large n was stated in 3°, is situated in the region

$$\left|\lambda - \mathrm{i}\sqrt{\frac{\alpha_n}{\rho_2}}\right| < r_n, \tag{3.31}$$

with

$$r_n = \mathrm{O}\left(\alpha_n \frac{1 - \tanh n}{n^{2-\varepsilon}}\right), \qquad \text{for all } \varepsilon > 0. \tag{3.32}$$

To prove this statement, let us rewrite equation (3.23) as

$$\begin{aligned} n\lambda(\rho_2\lambda^2 + \alpha_n) \\ = \frac{\tanh\sqrt{n^2 - \lambda}}{\sqrt{n^2 - \lambda}} \\ \times \left[(\rho_2\lambda^2 + \alpha_n)(\rho_2\lambda^2 + \beta_n) + \lambda^4\rho_2^2(\tanh n - 1) + \lambda^2\rho_2\alpha_n(\coth n - 1)\right]. \end{aligned} \tag{3.33}$$

For $\lambda = \mathrm{i}\sqrt{\alpha_n/\rho_2} + z$, $|z| = r_n$, the left side of (3.33) is estimated in the following way,

$$\left|n\lambda\left(\rho_2\lambda^2+\alpha_n\right)\right| = n\left|\mathrm{i}\sqrt{\frac{\alpha_n}{\rho_2}}+z\right|\cdot\rho_2\left|\lambda-\mathrm{i}\sqrt{\frac{\alpha_n}{\rho_2}}\right|\cdot\left|\lambda+\mathrm{i}\sqrt{\frac{\alpha_n}{\rho_2}}\right|$$
$$\geq \mathrm{const}\, n\sqrt{\alpha_n}r_n\sqrt{\alpha_n} = \mathrm{const}\, n\alpha_n r_n, \qquad (3.34)$$

where we took into account that $r_n = \mathrm{o}(n^{3/2})$ as $n \to \infty$.

Let us now estimate from above on the same circle $|z| = r_n$ the terms in the right-hand side of (3.33). We have

$$\left|\frac{\tanh\sqrt{n^2-\lambda}}{\sqrt{n^2-\lambda}}\right| \leq c_1 n^{-1}, \qquad n\to\infty, \qquad (3.35)$$

$$\left|\left(\rho_2\lambda^2+\alpha_n\right)\left(\rho_2\lambda^2+\beta_n\right)\right| \leq \rho_2^2\left|\left(\lambda-\mathrm{i}\sqrt{\frac{\alpha_n}{\rho_2}}\right)\left(\lambda+\mathrm{i}\sqrt{\frac{\alpha_n}{\rho_2}}\right)\left(\lambda^2+\frac{\beta_n}{\rho_2}\right)\right|$$
$$\leq c_2 r_n\sqrt{\alpha_n}\beta_n = \mathrm{O}\left(\alpha_n^{3/2}r_n\right), \qquad n\to\infty. \qquad (3.36)$$

In the deduction of (3.35) and (3.36) we took formulas (3.22) into account.

Similarly we get that

$$\left|\lambda^4\rho_2^2(\tanh n - 1)\right| \leq c_3\alpha_n^2(1-\tanh n),$$
$$\left|\lambda^2\rho_2\alpha_n(\coth n - 1)\right| \leq c_4\alpha_n^2(1-\tanh n), \qquad (3.37)$$

for which we took into account that $1-\tanh n \sim \coth n - 1 \sim 2\mathrm{e}^{-2n}$ as $n\to\infty$.

From (3.35)–(3.37) it follows that the orders of the terms in the right-hand side of (3.33) are

$$\alpha_n^{3/2}r_n n^{-1}, \quad \alpha_n^2 n^{-1}(1-\tanh n), \quad \alpha_n^2 n^{-1}(1-\tanh n).$$

Whence and from (3.34) it follows that for $n\to\infty$ the first term on the right side of (3.33) is of order $\mathrm{o}(nr_n\alpha_n)$ on any circle $|z| = |\lambda - \mathrm{i}\sqrt{\alpha_n/\rho_2}| = r_n$. Now, if we choose r_n such that $\alpha_n^2 n^{-1}(1-\tanh n)/(n\alpha_n r_n) = \mathrm{o}(1)$, $n\to\infty$, or, in particular, if r_n is as in (3.32), then using the Rouchet theorem we get that equation (3.33) has as many zeroes as its shorter version, $n\lambda(\rho_2\lambda^2+\alpha_n) = 0$, namely one, inside the circle (3.31).

Let us deduce now the first term of the asymptotic resolution for the root λ_n^+ as $n\to\infty$.

7° The next asymptotic formula takes place

$$\lambda_n^+ = \mathrm{i}\sqrt{\frac{\alpha_n}{\rho_2}} + \gamma_n[1+\mathrm{o}(1)], \qquad n\to\infty, \qquad (3.38)$$

with

$$\gamma_n = 2\alpha_n n^{-2} e^{-2n} = 2n^{-2} e^{-2n} \begin{cases} \sigma_2 n^3 & \text{for } \sigma_1 > 0,\ \sigma_2 > 0, \\ \rho_2 g n & \text{for } \sigma_1 = \sigma_2 = 0. \end{cases} \tag{3.39}$$

To prove formula (3.38) we represent $\lambda = \lambda_n^+ = \mathrm{i}\sqrt{\alpha_n/\rho_2} + w_n$ and then, applying 6° to w_n we have the estimate

$$|w_n| \le c\left[\frac{\alpha_n(1-\tanh n)}{n^{2-\varepsilon}}\right], \qquad \text{for each } \varepsilon > 0. \tag{3.40}$$

Substituting the representation for λ in (3.33) and taking into account that $\rho_2\lambda^4 = \alpha_n^2[1+\mathrm{o}(1)]$ as $n \to \infty$ and $(\tanh\sqrt{n^2-\lambda})/\sqrt{n^2-\lambda} = n^{-1}[1+\mathrm{o}(1)]$, we get

$$\begin{aligned} n&\left(\mathrm{i}\sqrt{\frac{\alpha_n}{\rho_2}} + w_n\right)\left(2\sqrt{\rho_2\alpha_n}w_n\mathrm{i} + \rho_2 w_n^2\right) \\ &\quad - n^{-1}[1+\mathrm{o}(1)]\left(2\sqrt{\rho_2\alpha_n}\mathrm{i} + \rho_2 w_n\right) w_n \left(\beta_n - \alpha_n + 2\sqrt{\rho_2\alpha_n}w_n\mathrm{i} + \rho_2 w_n^2\right) \\ &= n^{-1}[1+\mathrm{o}(1)]\left(-\alpha_n^2[1+\mathrm{o}(1)](1-\tanh n) - \alpha_n^2[1+\mathrm{o}(1)]\alpha_n^2(1-\tanh n)\right) \\ &= -2n^{-1}\alpha_n^2(1-\tanh n)[1+\mathrm{o}(1)]. \end{aligned} \tag{3.41}$$

From (3.22) and (3.40) it follows that for $n \to \infty$ the second term in the left-hand side of (3.41) has a higher infinitesimal order than the first one. Therefore, deriving the main terms we have

$$w_n = -n^{-1} 2\alpha_n \frac{1-\tanh n}{\left[n\sqrt{\dfrac{\alpha_n}{\rho_2}}\mathrm{i}\ 2\sqrt{\dfrac{\alpha_n}{\rho_2}}\mathrm{i}\right]}[1+\mathrm{o}(1)],$$

whence we get formula (3.38) with the coefficient (3.39).

10.3.4 General Conclusions and Hypotheses on the Structure of the Spectrum of the Hydrodynamics Problem

The previous considerations allow us to draw the following conclusions on the properties of the solutions to the model problem (3.8)–(3.13):

1° This problem has a discrete spectrum situated in the right half-plane.

2° All eigenvalues λ can be separated into several countable sets (branches) corresponding to different—from the physical point of view—forms of normal oscillations.

3° The branch $\{\lambda_{nk}\}_{n,k=1}^{\infty}$, $\lambda_{nk} > 0$ corresponds to those solutions of the evolution problem that are periodically fading in time, with fading decrements as large as possible ($\lambda_{nk} \to \infty$ for $n, k \to \infty$) that depend weakly on ρ_2, that is, on the presence or absence of an upper layer of an ideal fluid.

4° There are also two different branches of eigenvalues, $\{\lambda_n^{(1)}\}_{n=1}^{\infty}$ and $\{\lambda_n^{(2)}\}_{n=1}^{\infty}$, that correspond to the boundary waves on the boundary of the two media. The branch $\{\lambda_n^{(2)}\}_{n=1}^{\infty}$ corresponds to aperiodic solutions for which the fading decrements are substantially dependent on the presence of an upper fluid layer with density ρ_2 [see (3.29)].

5° The branch $\{\lambda_n^{(1)}\}_{n=1}^{\infty}$ corresponds to those aperiodic solutions for which the fading decrements depend substantially on whether or not the capillary forces are taken into account. In the first case ($\sigma_1 > 0$ and $\sigma_2 > 0$) we have $\lim_{n\to\infty} \lambda_n^{(1)} = +\infty$ [see (3.28)], that is, the solutions correspond to oscillations that are fading as quickly as possible. In the second case ($\sigma_1 = 0$ and $\sigma_2 = 0$) we have $\lim_{n\to\infty} \lambda_n^{(1)} = 0$ [see (3.28)], that is, the corresponding solutions of the evolution problem can fade in time as slowly as possible.

6° The two branches $\{\lambda_n^{\pm}\}_{n=1}^{\infty}$ of nonreal eigenvalues correspond to those oscillating modes for which the frequencies equal asymptotically $\sqrt{\alpha_n/\rho_2}$ and the fading decrements $\gamma_n > 0$ [see (3.38) and (3.39)]. Thus, it seems possible that the surface layer oscillations of the upper fluid correspond to the $\lambda_n^{\pm}$ eigenvalues. Since $\lim_{n\to\infty} \operatorname{Re} \lambda_n^{\pm} = \lim_{n\to\infty} \gamma_n = 0$, then the influence of the lower layer fluid on the fading decrements of these solutions is negligibly small for sufficiently large n.

These conclusions obtained in the case of the special model problem (3.8)–(3.13) lead to the following hypotheses on the spectrum structure of a real hydrodynamic problem (1.23), in which we study the normal oscillations of the two fluid layers situated in an arbitrary container (the lower layer consists of a viscous fluid and the upper layer of an ideal one).

Hypothesis 1. The spectrum of problem (1.23) is discrete, is situated in the right half-plane, and can have $\lambda = 0$ and $\lambda = \infty$ as limit points. If the fluids are capillary and we take into account the capillary forces, then the spectrum has only one limit point, $\lambda = \infty$. If the fluids are heavy, that is, the capillary forces are not taken into account, the spectrum has one more limit point, $\lambda = 0$.

Hypothesis 2. For capillary fluids, there are three branches of eigenvalues corresponding to the dissipative waves in the lower fluid layer, to the boundary waves on the boundary of the two fluid layers, and to the surface waves on the free surface of the upper fluid layer, respectively. The first two branches are situated on the positive semi-axis, and the third one is situated in the neighborhood of the imaginary axis.

Hypothesis 3. For heavy fluids, there exists a branch of eigenvalues corresponding to the waves on the media boundary that has zero as a limit point.

In the next section, we will prove that some of the statements presented in Hypotheses 1 through 3 actually take place. Some other statements on the completeness of the system of eigen- and associated functions of problem (1.23) will be obtained as well.

10.4 Normal Oscillations of a Partially Dissipative Hydrosystem in an Arbitrary Domain

In this section the solution properties of the spectral problem (1.23) for a hydrosystem situated in an arbitrary domain Ω are considered. The approach employed here is similar to the one in Section 10.2 for the initial-boundary value problem (1.2)–(1.11).

10.4.1 Transition to an Operator Pencil with Bounded Operator Coefficients

Let us go back to problem (1.23) for an arbitrary domain Ω and, after deriving the trivial solution (1.24), assume that the spectral parameter is $\lambda \neq 0$ and $\boldsymbol{u}_2(x) \in \boldsymbol{G}_{\mathrm{h},S_2}(\Omega_2)$. The latter condition follows from

$$\boldsymbol{u}_2 = (\lambda\rho_2)^{-1}\nabla p_2, \quad \operatorname{div}\boldsymbol{u}_2 = 0 \quad \text{in } \Omega_2 \qquad \boldsymbol{u}_2\cdot\boldsymbol{n} = 0 \quad \text{on } S_2.$$

Problem (1.23) leads to a system of operator equations obtained by repeating the same transformations that were done in Section 10.2 for the initial boundary value problem (1.2)–(1.11). It is obvious that the derivatives with respect to t are now substituted with multiplication by $(-\lambda)$ and the field of external forces $\boldsymbol{f}(t,x)$ should be set equal to zero. Thus, instead of the evolution problem (2.22) we come to the following system of equations,

$$\begin{aligned}
\mu A\boldsymbol{u}_1 + g(\Delta\rho)G_1\zeta_1 &= \lambda(\rho_1\boldsymbol{u}_1 + \rho_2(G_1C_{11}\gamma_1\boldsymbol{u}_1 + G_1C_{12}\gamma_2\boldsymbol{u}_2)),\\
g\rho_2\zeta_2 &= \lambda(\rho_2C_{21}\gamma_1\boldsymbol{u}_1 + \rho_2C_{22}\gamma_2\boldsymbol{u}_2),\\
\gamma_1\boldsymbol{u}_1 &= -\lambda\zeta_1,\\
\gamma_2\boldsymbol{u}_2 &= -\lambda\zeta_2.
\end{aligned} \tag{4.1}$$

[Here, instead of the second equation in (2.22), we are using the initial relation (2.19) from which the second equation in (2.22) is obtained by applying the operator G_2.]

Since $\lambda \neq 0$, then we can exclude from (4.1) both ζ_1 and ζ_2 and obtain a system of two equations with two unknowns,

$$\begin{aligned} \mu A\boldsymbol{u}_1 - g(\Delta\rho)\lambda^{-1}G_1\gamma_1\boldsymbol{u}_1 &= \lambda(\rho_1\boldsymbol{u}_1 + \rho_2(G_1C_{11}\gamma_1\boldsymbol{u}_1 + G_1C_{12}\gamma_2\boldsymbol{u}_2)), \\ -g\rho_2\lambda^{-1}\gamma_2\boldsymbol{u}_2 &= \lambda(\rho_2C_{21}\gamma_1\boldsymbol{u}_1 + \rho_2C_{22}\gamma_2\boldsymbol{u}_2). \end{aligned} \tag{4.2}$$

In (4.2), let us perform the following substitutions,

$$A^{1/2}\boldsymbol{u}_1 = \boldsymbol{\xi}, \qquad \gamma_2\boldsymbol{u}_2 = \beta, \tag{4.3}$$

and consider that

$$\boldsymbol{\xi} \in \boldsymbol{J}_{0,S_1}(\Omega_1), \qquad \eta \in H_2 := L_2(\Gamma_2) \ominus \{1_2\}. \tag{4.4}$$

If in the first equation (4.2) we apply the operator $A^{-1/2}$ we obtain the next system of equations,

$$\begin{aligned} &\mu\boldsymbol{\xi} - g(\Delta\rho)\lambda^{-1}A^{-1/2}G_1\gamma_1A^{-1/2}\boldsymbol{\xi} \\ &\qquad = \lambda(\rho_1A^{-1/2}\boldsymbol{\xi} + \rho_2(A^{-1/2}G_1C_{11}\gamma_1A^{-1/2}\boldsymbol{\xi} + A^{-1/2}G_1C_{12}\eta)), \\ &-g\rho_2\lambda^{-1}\eta = \lambda(\rho_2C_{21}\gamma_1A^{-1/2}\boldsymbol{\xi} + \rho_2C_{22}\eta). \end{aligned} \tag{4.5}$$

This system can be written down in a vector-matrix form as follows,

$$\begin{aligned} &\mu\begin{pmatrix} I_1 & 0 \\ 0 & 0 \end{pmatrix}\begin{pmatrix} \boldsymbol{\xi} \\ \eta \end{pmatrix} - \lambda\begin{pmatrix} A^{-1/2}(\rho_1I_1 + \rho_1G_1C_{11}\gamma_1)A^{-1/2} & \rho_2A^{-1/2}G_1C_{12} \\ \rho_2C_{21}\gamma_1A^{-1/2} & \rho_2C_{22} \end{pmatrix}\begin{pmatrix} \boldsymbol{\xi} \\ \eta \end{pmatrix} \\ &\quad - g\lambda^{-1}\begin{pmatrix} (\Delta\rho)A^{-1/2}G_1\gamma_1A^{-1/2} & 0 \\ 0 & \rho_2I_2 \end{pmatrix}\begin{pmatrix} \boldsymbol{\xi} \\ \eta \end{pmatrix} = \begin{pmatrix} \boldsymbol{0} \\ 0 \end{pmatrix}. \end{aligned} \tag{4.6}$$

Let us take a closer look to the properties of the matrix coefficients that appear in equation (4.6). We are going to prove that except the unit operators I_1 and I_2 acting on $\boldsymbol{J}_{0,S_1}(\Omega_1)$ and H_2, respectively, all the other operators are compact.

In fact—as stated in Section 10.2.3—the operator $G_1C_{11}\gamma_1$ acts boundedly from $\boldsymbol{J}_{0,S_1}(\Omega_1)$ into $G_{h,S_1}(\Omega_1) \subset \boldsymbol{J}_{0,S_1}(\Omega_1)$. Therefore, the operator $A^{-1/2}(\rho_1I_1 + \rho_1G_1C_{11}\gamma_1)A^{-1/2}$ is compact as the product of a bounded operator multiplied to the left and the right with the compact operator $A^{-1/2}$. Further, C_{22} is a compact positive operator acting on H_2. This fact was often used when stating the properties of the solutions to Problems III and IV and of the operators C_{ik} in the previous chapters of the present book (see, e.g., Section 4.2).

As stated in Section 8.1.3, $\gamma_1 A^{-1/2}$—denoted there by $\gamma_n A^{-1/2}$—is a compact operator acting from $\boldsymbol{J}_{0,S_1}(\Omega_1)$ into $H_1 := L_2(\Gamma_1) \ominus \{1_1\}$. Since the operator $C_{21} : H_1 \to H_2$ is compact as well, then $C_{21}\gamma_1 A^{-1/2}$ is a compact operators acting from $\boldsymbol{J}_{0,S_1}(\Omega_1)$ into H_2. Let us remark now that $C_{12} = (C_{21})^*$ and $(A^{-1/2}G_1)^* = \gamma_1 A^{-1/2}$. The former as well as the fact that the operator matrix $(C_{ik})_{i,k=1}^2$ is positive was stated in the previous chapters of this book, whileas the latter follows from identity (2.27) applied to the elements $\psi \in H_1$ and $\boldsymbol{v} = A^{-1/2}\boldsymbol{u} \in \boldsymbol{J}^1_{0.S_1}(\Omega)$, that is,

$$(G_1\psi, A^{-1/2}\boldsymbol{u})_{\boldsymbol{J}_{0,S_1}(\Omega_1)} = (A^{-1/2}G_1\psi, \boldsymbol{u})_{\boldsymbol{J}_{0.S_1}(\Omega_1)} = (\psi, \gamma_1 A^{-1/2}\boldsymbol{u})_{L_2(\Gamma_1)}. \tag{4.7}$$

Whence it follows that $A^{-1/2}G_1C_{11}$ is a compact operator acting from H_2 into $\boldsymbol{J}_{0,S_1}(\Omega_1)$, where $(A^{-1/2}G_1C_{12})^* = C_{21}\gamma_1 A^{-1/2}$.

Finally, $A^{-1/2}G_1\gamma_1 A^{-1/2} = (\gamma_1 A^{-1/2})^*(\gamma_1 A^{-1/2})$ is a compact nonnegative operator acting on $\boldsymbol{J}_{0,S_1}(\Omega_1)$. This proves the properties of the operator coefficients in equation (4.6) that were stated previously.

We rewrite equation (4.6) as

$$(\mu\mathcal{P}_1 - \lambda\mathcal{A} - g\lambda^{-1}\mathcal{B})y = 0, \qquad y = (\boldsymbol{\xi}; \eta)^{\mathrm{t}}, \tag{4.8}$$

with

$$\begin{aligned} \mathcal{P}_1 &= \operatorname{diag}(I_1; 0) \\ \mathcal{B} &= \operatorname{diag}\left((\Delta\rho)(A^{-1/2}G_1)(\gamma_1 A^{-1/2}); \rho_2 I_2\right), \\ \mathcal{A} &= (A_{ij})_{i,j=1}^2 = \begin{pmatrix} A^{-1/2}(\rho_1 I_1 + \rho_1 G_1 C_{11}\gamma_1)A^{-1/2} & \rho_2 A^{-1/2}G_1C_{12} \\ \rho_2 C_{21}\gamma_1 A^{-1/2} & \rho_2 C_{22} \end{pmatrix}. \end{aligned} \tag{4.9}$$

Following the preceding considerations, we can state that $\mathcal{P}_1$ is the orthoprojector onto $\boldsymbol{J}_{0,S_1}(\Omega_1)$, $\mathcal{A}$ is a compact self-adjoint operator, and $\mathcal{B}$ is a bounded nonnegative operator. Note also that the operator $(A^{-1/2}G_1)(\gamma_1 A^{-1/2})$ coincides with the operator $B = (A^{1/2}T)(\gamma_n A^{-1/2})$ that occurred previously in the basic hydrodynamic problem of Chapter 8 (see Sections 8.1, 8.3, 8.5, and 8.6).

Therefore, in the sequel, we will denote

$$(A^{-1/2}G_1)(\gamma_1 A^{-1/2}) =: B \tag{4.10}$$

and notice that its eigenvalues have the asymptotic behavior (8.1.33), that is,

$$\lambda_n(B) = c_B^{1/2} n^{-1/2}[1 + \mathrm{o}(1)], \qquad n \to \infty,\ c_B = \operatorname{mes}\left(\frac{\Gamma_1}{16\pi}\right) > 0. \tag{4.11}$$

Thus, there is a self-adjoint operator pencil

$$L(\lambda) := \mu\mathcal{P}_1 - \lambda\mathcal{A} - g\lambda^{-1}\mathcal{B}, \tag{4.12}$$

with the previously mentioned properties of the operator coefficients $\mathcal{P}_1$, $\mathcal{A}$, and $\mathcal{B}$ that corresponds to the spectral problem (1.23) describing the normal oscillations of the considered partially dissipative system.

More importantly, the operator $\mathcal{A}$ in (4.9) is positive and compact. Indeed, $\mathcal{A}$ can be written as

$$\mathcal{A} = \operatorname{diag}(A^{-1/2}; I_2)\tilde{C}\operatorname{diag}(A^{-1/2}; I_2), \tag{4.13}$$

where

$$\begin{aligned}\tilde{C} :&= \begin{pmatrix} \rho_1 I_1 + \rho_2 G_1 C_{11}\gamma_1 & \rho_2 G_1 C_{12} \\ \rho_2 C_{21}\gamma_1 & \rho_2 C_{22} \end{pmatrix} \\ &= \rho_1\mathcal{P}_1 + \rho_2\begin{pmatrix} G_1 & 0 \\ 0 & I_2 \end{pmatrix}\begin{pmatrix} C_{11} & C_{12} \\ C_{21} & C_{22} \end{pmatrix}\begin{pmatrix} \gamma_1 & 0 \\ 0 & I_2 \end{pmatrix}.\end{aligned} \tag{4.14}$$

Since the properties $\mathcal{P}_1 \geq 0$, $G_1^* = \gamma_1$, $(C_{ik})_{i.k=1}^2 > 0$ are satisfied, we have that $\mathcal{A} \geq 0$. However, condition $\mathcal{A}y = 0$ leads to $y = (\boldsymbol{\xi}; \eta)^{\mathrm{t}} = 0$, and thus $\mathcal{A} > 0$.

To conclude this section, let us note that equation (4.8) generalizes equation (8.1.41) on the problem on normal oscillations of a viscous fluid in an open container (see Chapter 8) in the case of a partially dissipative hydrosystem. If in (4.8) we set $\rho_2 = 0$, then $\Delta\rho = \rho_1$, $\mu = \rho_1\nu$, the second equation turns out to be zero, and the first one coincides with equation (8.1.41) with the account of (4.10).

10.4.2 General Properties of the Spectrum

The operator pencil (4.12) is not Fredholm because $\mathcal{P}_1$ and $\mathcal{B}$ are bounded operators and none of them is compact. However, equation (4.8) can lead an equation with a Fredholm operator pencil. In fact, taking into account the previously introduced notations (4.9) and (4.10), problem (4.8) can be written in a vector-matrix form as

$$\begin{aligned}\mu\boldsymbol{\xi} - \lambda(A_{11}\boldsymbol{\xi} + A_{12}\eta) - g\lambda^{-1}(\Delta\rho)B\boldsymbol{\xi} = 0,\\ -\lambda(A_{21}\boldsymbol{\xi} + A_{22}\eta) - g\lambda^{-1}\rho_2\eta = 0,\end{aligned} \tag{4.15}$$

whence we get

$$\begin{aligned}L_1(\lambda)y :&= \begin{pmatrix} \mu I_1 & 0 \\ 0 & \rho_2 g I_2 \end{pmatrix}\begin{pmatrix} \boldsymbol{\xi} \\ \eta \end{pmatrix} + \lambda^2\begin{pmatrix} 0 & 0 \\ A_{21} & A_{22} \end{pmatrix}\begin{pmatrix} \boldsymbol{\xi} \\ \eta \end{pmatrix} \\ &\quad - \lambda\begin{pmatrix} A_{11} & A_{12} \\ 0 & 0 \end{pmatrix}\begin{pmatrix} \boldsymbol{\xi} \\ \eta \end{pmatrix} - g(\Delta\rho)\lambda^{-1}\begin{pmatrix} B & 0 \\ 0 & 0 \end{pmatrix}\begin{pmatrix} \boldsymbol{\xi} \\ \eta \end{pmatrix} \\ &= \begin{pmatrix} \mathbf{0} \\ 0 \end{pmatrix}.\end{aligned} \tag{4.16}$$

In (4.16), all the operator matrices except the first one are compact and the first matrix

$$\mathcal{I} := \operatorname{diag}(\mu I_1; \rho_2 g I_2) \tag{4.17}$$

is bounded and positive definite. Thus, after substituting $\mathcal{I}^{1/2}y = z$ and applying the operator $\mathcal{I}^{-1/2}$ in (4.17), we come to the following equation

$$L_2(\lambda)z := \mathcal{I}^{-1/2}L_1(\lambda)\mathcal{I}^{-1/2}z =: (I + \Phi(\lambda))z = 0 \tag{4.18}$$

with a Fredholm operator pencil and where $\Phi(\lambda)$ is an analytic operator function on the entire complex plane $\mathbb{C}$, with the exception of the points $\lambda = 0$ and $\lambda = \infty$, that takes compact values.

The properties of the spectrum in problem (1.23) can be deduced by studing the properties of the operator pencils (4.12), (4.16), and (4.18). Let us mention these properties and carry out their proofs.

1° The spectrum of problem (1.23) is discrete with possible limit points $\lambda = 0$ and $\lambda = \infty$, all the eigenvalues have finite multiplicity and the resolvents of the pencils (4.12), (4.16), and (4.18) are operator functions meromorphic in $\mathbb{C} \setminus \{0\}$ with poles coinciding with eigenvalues. The multiplicity of the poles coincides with the maximal multiplicity of the eigenelements corresponding to those eigenvalues.

To prove Property 1°, we use the statements in Section 1.6.3 and the previous considerations in this section. It will be enough to check that the Fredholm pencil $L_2(\lambda)$ is invertible at least in one point of the complex plane. Using (4.17) and (4.18) it is enough to check this property for $L_1(\lambda)$ in (4.16). However, for $\lambda = -1$ we have

$$L_1(-1) = \mathcal{I} + \mathcal{A} + g(\Delta\rho)\operatorname{diag}(B; 0) > \mathcal{I} \gg 0, \tag{4.19}$$

where $\mathcal{A} > 0$ is the operator matrix defined by (4.19) and $B \geq 0$ is the operator (4.10). Whence, Property 1° is proved completely.

2° Problem (1.23) has a branch of eigenvalues $\{\lambda_k^0\}_{k=1}^{\infty} \subset \mathbb{R}_+$ whose limit point is $\lambda = 0$. In this

$$\lambda_k^0 = g(\Delta\rho)\mu^{-1}\lambda_k(B)[1 + \mathrm{o}(1)], \qquad k \to \infty, \tag{4.20}$$

where $\lambda_k(B)$ are the eigenvalues of operator B and have the asymptotic behavior (4.11).

To prove Property 2°, let us eliminate the variable η from the system of equations (4.15) and consider λ to be sufficiently small, that is, $|\lambda| < r := (\rho_2 g\|A_{22}\|^{-1})^{1/2}$. Then from the second equation in (4.5) we obtain

$$\eta = -\lambda^2(\rho_2 g)^{-1}m_0(\lambda)A_{21}\boldsymbol{\xi}, \qquad m_0(\lambda) := I_2 + \lambda^2(\rho_2 g)^{-1}A_{22}. \tag{4.21}$$

Substituting η in the first equation (4.15) we come to the following equation,

$$\begin{aligned} l_1(\lambda)\boldsymbol{\xi} :&= \left(\mu I_1 - g(\Delta\rho)\lambda^{-1}B - \lambda A_{11} + \lambda^3(\rho_2 g)^{-1}A_{12}m_0^{-1}(\lambda)A_{21}\right)\boldsymbol{\xi} \\ &= \mathbf{0}, \qquad \boldsymbol{\xi} \in \boldsymbol{J}_{0.S_1}(\Omega). \end{aligned} \tag{4.22}$$

One can easily see that for $|\lambda| < r$, $l_1(\lambda)$ is a self-adjoint, holomorphic operator function, which is a Fredholm operator pencil because all the terms except the first one take compact values.

Let us now substitute the variable λ in (4.22) by $\tilde{\lambda}^{-1}$ and consider that $|\tilde{\lambda}| > r_0^{-1}$, with $r_0 < r$. Further, let us use the statement in Section 1.6.8 on the asymptotic behavior of the branches of eigenvalues corresponding to the previously introduced operator pencil

$$\begin{aligned} l_1(\tilde{\lambda}^{-1}) &= \mu I_1 - g(\Delta\rho)\tilde{\lambda}B + \Phi(\tilde{\lambda}), \\ \Phi(\tilde{\lambda}) &:= -\tilde{\lambda}^{-1}A_{11} + (\tilde{\lambda})^{-3}(\rho_2 g)^{-1}A_{12}m_0^{-1}(\tilde{\lambda}^{-1})A_{21}. \end{aligned} \tag{4.23}$$

Since $m_0^{-1}(\tilde{\lambda}^{-1}) = I_2 + \mathrm{O}(\tilde{\lambda}^{-2})$ as $\tilde{\lambda} \to \infty$, then $\Phi(\tilde{\lambda}) = \mathrm{O}(\tilde{\lambda}^{-1}) \to 0$ as $\tilde{\lambda} \to \infty$. Let us note now that $\operatorname{Ker} B = \boldsymbol{N}_0(\Omega_1) \neq \{0\}$. This particular fact and the orthogonal decomposition

$$\boldsymbol{J}_{0,S_1}(\Omega_1) = \boldsymbol{M}_0(\Omega_1) \oplus \boldsymbol{N}_0(\Omega_1), \tag{4.24}$$

appearing from the decomposition

$$\boldsymbol{J}^1_{0,S_1}(\Omega_1) = \boldsymbol{M}_1(\Omega_1) \oplus \boldsymbol{N}_1(\Omega_1), \tag{4.25}$$

were used repeatedly in Chapters 8 and 9 in the study of the problem on small oscillations of a viscous fluid in an open container.

Let us denote the orthoprojector on $\boldsymbol{M}_0(\Omega_1)$ by P, and the one on $\boldsymbol{N}_0(\Omega_1)$ by P_0. If we write down the solution $\boldsymbol{\xi}$ of equation

$$l_1(\tilde{\lambda}^{-1})\boldsymbol{\xi} = \mathbf{0}, \qquad \boldsymbol{\xi} \in \boldsymbol{J}_{0.S_1}(\Omega_1), \tag{4.26}$$

as $\boldsymbol{\xi} = P\boldsymbol{\xi} + P_0\boldsymbol{\xi}$, and project (4.26) on $\boldsymbol{M}_0(\Omega_1)$ and $\boldsymbol{N}_0(\Omega_1)$ using the orthoprojectors P and P_0, we obtain

$$\begin{aligned} \mu P\boldsymbol{\xi} - g(\Delta\rho)\tilde{\lambda}(PBP)P\boldsymbol{\xi} + P\Phi(\lambda)(P\boldsymbol{\xi} + P_0\boldsymbol{\xi}) &= \mathbf{0}, \\ \mu P_0\boldsymbol{\xi} + P_0\Phi(\tilde{\lambda})(P\boldsymbol{\xi} + P_0\boldsymbol{\xi}) &= \mathbf{0}. \end{aligned} \tag{4.27}$$

Since $\Phi(\tilde{\lambda}) = O(\tilde{\lambda}^{-1})$ as $\tilde{\lambda} \to \infty$, then from the second equation in (4.27) we have

$$P_0\boldsymbol{\xi} = -\left(\mu P_0 + P_0\Phi(\tilde{\lambda})P_0\right)^{-1}\left(P_0\Phi(\tilde{\lambda})P\right)(P\boldsymbol{\xi})$$

for sufficiently large (in absolute value) $\tilde{\lambda}$. Substituting the preceding expression into the first equation in (4.27), we finally get the following equation for $P\boldsymbol{\xi} \in \boldsymbol{M}_0(\Omega_1)$:

$$\begin{aligned} \tilde{L}(\tilde{\lambda})(P\boldsymbol{\xi}) &:= \left(\mu I - g(\Delta\rho)\tilde{\lambda}(PBP) + \Phi_1(\tilde{\lambda})\right)(P\boldsymbol{\xi}) = \mathbf{0}, \\ \Phi_1(\tilde{\lambda}) &= P\Phi(\tilde{\lambda})P - \left(P\Phi(\tilde{\lambda})P_0\right)\left(\mu P_0 + P_0\Phi(\tilde{\lambda})P_0\right)^{-1}\left(P_0\Phi(\tilde{\lambda})P\right) \\ &= O(\tilde{\lambda}^{-1}), \qquad \tilde{\lambda} \to \infty. \end{aligned} \tag{4.28}$$

Here already $\operatorname{Ker}(PBP) = \{0\}$, the numbers $\lambda_k(PBP) = \lambda_k(B) \neq 0$ have the asymptotic behavior (4.11) and $\Phi_1(\tilde{\lambda}) \to 0$ as $\tilde{\lambda} \to \infty$. Therefore, according to the statements of Section 1.6.8, problem (4.28) and consequently problem (4.26) have a branch of eigenvalues $\{\tilde{\lambda}_k^0\}_{k=1}^{\infty}$ with the asymptotic behavior

$$\tilde{\lambda}_k^0 = \mu(g(\Delta\rho))^{-1}\lambda_k^{-1}(B)[1 + o(1)], \qquad k \to \infty.$$

Whence, following the substitution $(\tilde{\lambda}_k^0)^{-1} = \lambda_k^0$ we obtain formula (4.20) and Statement 2° is proved.

3° The eigenelements $\{\boldsymbol{\xi}_k^0\}_{k=1}^{\infty}$ corresponding to the eigenvalues $\{\lambda_k^0\}_{k=1}^{\infty}$ of problem (4.22) located in the interval $(0, r)$, where $r := (\rho_2 g\|A_{22}\|^{-1})^{1/2}$, form a Riesz basis with a finite defect in $\boldsymbol{M}_0(\Omega_1)$ after being projected on $\boldsymbol{M}_0(\Omega_1)$. Moreover, that basis is a p-basis (with finite defect) for $p > p_0$, with

$$p_0 = \left(\frac{1}{2} + \frac{1}{(3/2)}\right)^{-1} = \frac{6}{7}. \tag{4.29}$$

To prove this statement, let us go back to (4.28) and multiply it by λ, then, using the definitions for $\Phi(\tilde{\lambda})$ in (4.23) and of $m_0(\lambda)$ in (4.21) we arrive at the following problem,

$$\begin{aligned} m(\lambda)(P\boldsymbol{\xi}) &:= (\lambda\mu I - g(\Delta\rho)(PBP) + \Phi_2(\lambda))(P\boldsymbol{\xi}) = \mathbf{0}, \\ \Phi_2(\lambda) &:= \lambda P\Phi(\lambda^{-1})P - \lambda\left(P\Phi(\lambda^{-1})P_0\right)\left(\mu P_0 + P_0\Phi(\lambda^{-1})P_0\right)^{-1}\left(P_0\Phi(\lambda^{-1})P_0\right), \\ \Phi(\lambda^{-1}) &= -\lambda A_{11} + \lambda^3(\rho_2 g)^{-1}A_{12}\left(\sum_{k=0}^{\infty}\left(\frac{\lambda^2 A_{22}}{\rho_2 g}\right)^k\right)A_{21}. \end{aligned} \tag{4.30}$$

Since $\Phi(\lambda^{-1}) = O(\lambda)$ as $\lambda \to 0$, then $\Phi_2(\lambda) = O(\lambda^2)$ as $\lambda \to 0$ and, therefore, according to Statement 3° in Section 1.6.10, the system of eigenelements $\{P\boldsymbol{\xi}_k^0\}_{k=1}^{\infty}$ of problem (4.30) corresponding to the eigenvalues λ_k^0 for $|\lambda| < r = (\rho_2 g\|A_{22}\|^{-1})^{1/2}$ forms a Riesz basis in $\boldsymbol{M}_0(\Omega_1)$ that has a finite defect.

Now we need to make sure that the above-mentioned basis is a p-basis (with a finite defect) in $\boldsymbol{M}_0(\Omega_1)$ for $p > p_0$, where p_0 is defined by the (4.29). To prove this we use Statement 4° in Section 1.6.10 on the property of p-basicity of a system of eigenelements for a self-adjoint holomorphic operator function $\lambda I + \sum_{k=0}^{\infty} \lambda^k \tilde{A}_k$ in the case of problem (4.30). It it obvious that $\tilde{A}_0 = g(\Delta\rho)PBP$, $\tilde{A}_1 = 0$, $\tilde{A}_2 = A_{11}$, $\tilde{A}_3 = -\mu^{-1}PA_{11}P_0A_{11}P$, $A_{11} = A^{-1/2}(\rho_1 I_1 + \rho_2 G_1 C_{11}\gamma_1)^{-1}A^{-1/2}$ [see (4.9)]. Since P and P_0 are orthoprojectors and $G_1C_{11}\gamma_1$ is a bounded operator (see Section 2.3), then using the preceding formulas and the ones on the asymptotic behavior of the eigenvalues of the operators B and A, (4.11) and (8.1.12), we get

$$\begin{aligned}
&\tilde{A}_0 \in \mathfrak{S}_{q_0}, && q_0 > \tilde{q}_0 = 2,\\
&\tilde{A}_1 \in \mathfrak{S}_{q_1}, && q_1 > \tilde{q}_1 = 0,\\
&\tilde{A}_2 \in \mathfrak{S}_{q_2}, && q_2 > \tilde{q}_2 = \frac{3}{2},\\
&\tilde{A}_3 \in \mathfrak{S}_{q_3}, && q_3 > \tilde{q}_3 = \frac{3}{4}.
\end{aligned} \tag{4.31}$$

Whence and using formula

$$\begin{aligned}
p &= \left(\min\left(\frac{1}{q_1}; \frac{1}{q_0} + \frac{1}{q_2}; \frac{2}{q_0} + \frac{1}{q_3}\right)\right)^{-1} > p_0\\
&:= \left(\min\left(\frac{1}{\tilde{q}_1}; \frac{1}{\tilde{q}_0} + \frac{1}{\tilde{q}_2}; \frac{2}{\tilde{q}_0} + \frac{1}{\tilde{q}_3}\right)\right)^{-1} = \left(\frac{1}{2} + \frac{2}{3}\right)^{-1} = \frac{6}{7},
\end{aligned}$$

which is a particular case of the general formula presented in Section 1.6.10 (for $\tilde{p}_1$), we conclude that the eigenelements $\{P\boldsymbol{\xi}_k^0\}_{k=1}^{\infty}$ corresponding to the eigenvalues λ_k^0 for $0 < \lambda < r$ form a p-basis (with a finite defect) in $\boldsymbol{M}_0(\Omega_1)$ for $p > p_0 = 6/7$.

4° If the physical properties of the studied partially dissipative hydrosystem are such that conditions

$$2(g(\Delta\rho)\|A_{11}\| \cdot \|B\|)^{1/2} < \mu < 2\|A_{11}\|\left(\frac{g}{\|C_{22}\|}\right)^{1/2} \tag{4.32}$$

are satisfied, then the system of eigenelements $\{\boldsymbol{\xi}_k^0\}_{k=1}^{\infty}$ of problem

$$\begin{aligned}
m_1(\lambda)\boldsymbol{\xi} &:= \lambda l_1(\lambda)\boldsymbol{\xi}\\
&= \left(\lambda\mu I - g(\Delta\rho)B - \lambda^2 A_{11} + \lambda^4(\rho_2 g)^{-1}A_{12}\left(I_2 + \lambda^2(\rho_2 g)^{-1}A_{22}\right)^{-1}A_{21}\right)\boldsymbol{\xi}\\
&= \boldsymbol{0},
\end{aligned} \tag{4.33}$$

corresponding to the eigenvalues λ_k^0 for $0 < \lambda < r$ together with an orthonormal basis in $\boldsymbol{N}_0(\Omega_1)$, corresponding to the infinitely multiple eigenvalue $\lambda_0 = 0$, form a Riesz basis in the space $\boldsymbol{J}_{0,S_1}(\Omega_1)$.

Let us remark that conditions (4.32) are satisfied if the density values of the two fluids, ρ_1 and ρ_2, are close enough, and the viscosity value, μ, of the first fluid is neither large nor small.

To prove Property 4° we use Statements 1° and 2° in Section 1.6.9 as well as Statements 1° and 2° in Section 1.6.10. In particular, we need to find an interval $[-\varepsilon, b] \subset \mathbb{R}$ for which conditions

$$m_1(-\varepsilon) \ll 0, \quad m_1(b) \gg 0, \quad m_1'(\lambda) \gg 0, \qquad \lambda \in [-\varepsilon, b] \tag{4.34}$$

are satisfied. Since $B \geq 0$ and $A_{11} > 0$, then

$$-m_1(-\varepsilon) = \varepsilon\mu I_1 + g(\Delta\rho)B + \varepsilon^2 A_{11} + \mathrm{O}(\varepsilon^4) \gg 0,$$

that is, $m_1(-\varepsilon) \ll 0$ for every sufficiently small $\varepsilon > 0$. Further, taking into account that $A_{22} > 0$, for $\lambda > 0$ we have

$$\begin{aligned} m_1'(\lambda) &= \mu I_1 - 2\lambda A_{11} + (\rho_2 g)^{-1} A_{12} \left(\lambda^4 \left(I_2 + \lambda^2 (\rho_2 g)^{-1} A_{22}\right)^{-1}\right)_\lambda' A_{21} \\ &\geq (\mu - 2\lambda \|A_{11}\|) I_1 \\ &\quad + (\rho_2 g)^{-1} A_{12} \Big\{ \left[4\lambda^3 I_2 + 2\lambda^5 (\rho_2 g)^{-1} A_{22}^{1/2}\right]^{1/2} \left(I_2 + \lambda^2 (\rho_2 g)^{-1} A_{22}\right)^{-2} \\ &\qquad\qquad \times \left[4\lambda^3 I_2 + 2\lambda^5 (\rho_2 g)^{-1} A_{22}\right]^{1/2} \Big\} A_{21} \\ &\geq (\mu - 2\lambda \|A_{11}\|) I_1. \end{aligned} \tag{4.35}$$

Whence, for $0 \leq \lambda < \mu/(2\|A_{11}\|) =: r_0$, we get the property $m_1'(\lambda) \gg 0$. Since the operator-function in (4.35) represented by the last term to the right takes the value $\mathrm{O}(\lambda^3)$ for $\lambda \to 0$, then the property of positive definteness of $m_1'(\lambda)$ is retained even in the case of a negative small enough $\lambda = -\varepsilon < 0$. We may now choose any $\varepsilon > 0$ that satisfies the conditions $m_1(-\varepsilon) \ll 0$, $m_1'(\lambda) \gg 0$, $\lambda \in [-\varepsilon, 0]$.

Next, we select $b > 0$ such that $b < r$ and $m_1(b) \gg 0$. We have,

$$\begin{aligned} m_1(b) &= b\mu I_1 - g(\Delta\rho)B - b^2 A_{11} + b^4 (\rho_2 g)^{-1} A_{12} \left(I_2 + b^2 (\rho_2 g)^{-1} A_{22}\right)^{-1} A_{21} \\ &\geq \left(b\mu - g(\Delta\rho)\|B\| - b^2 \|A_{11}\|\right) I_1. \end{aligned} \tag{4.36}$$

If $\mu^2 > 4g(\Delta\rho)\|B\| \cdot \|A_{11}\|$, then we have the property $m_1(b) \gg 0$ for

$$b \in (r_-, r_+), \qquad r_\pm = \frac{\mu \pm \sqrt{\mu^2 - 4g(\Delta\rho)\|B\| \cdot \|A_{11}\|}}{2\|A_{11}\|}. \tag{4.37}$$

Thus, in order to insure that conditions (4.34) and $b < r$ are satisfied, the physical and geometrical parameters of the problem should verify the relationships

$$\mu > 2(g(\Delta\rho)\|B\| \cdot \|A_{11}\|)^{1/2}, \quad b \in (r_-, r_0), \quad r_0 < r = \left(\frac{\rho_2 g}{\|A_{22}\|}\right)^{1/2},$$

Taking into account that $A_{22} = \rho_2 C_{22}$ [see (4.9)], the previous assertion leads to inequalities (4.32).

If these inequalities are satisfied, then the operator function $m_1(\lambda)$ in (4.33) admits the spectral factorization $m_1(\lambda) = m_+(\lambda)(\lambda I - Z)$, where the spectra of the operators $m_1(\lambda)$ and Z coincide on the interval $(-\varepsilon, b)$. Moreover, since $m_1(0) = -g(\Delta\rho)B$ is a compact self-adjoint operator, then according to Statement 1° in Section 1.6.10, the set of eigenelements $\{\boldsymbol{\xi}_k^0\}_{k=1}^\infty$ corresponding to the eigenvalues λ_k^0 in the interval $(0, b)$ together with an orthonormal basis in $\boldsymbol{N}_0(\Omega_1)$ corresponding to the infinitely multiple eigenvalue $\lambda_0 = 0$ in problem (4.43) form a Riesz basis in $\boldsymbol{J}_{0,S_1}(\Omega_1)$.

5° The physical meaning of the previously obtained branch of eigenvalues $\{\lambda_k^0\}_{k=1}^\infty$ that corresponds to a set of eigenelements of problem (4.1) is the following: Such eigenvalues correspond to boundary waves in a small vicinity of the surface Γ_1 that fade in deviation along a direction perpendicular to Γ_1 oriented into the depth of either the first or the second fluid.

Indeed, let us first normalize the solutions of problem (4.1) by the condition

$$\|\boldsymbol{u}_1\|^2_{\boldsymbol{J}_{0,S_1}(\Omega_1)} + \|\gamma_1 \boldsymbol{u}_1\|^2_{L_2(\Gamma_1)} + \|\zeta_1\|^2_{L_2(\Gamma_1)} + \|\zeta_2\|^2_{L_2(\Gamma_2)} = 1. \tag{4.38}$$

Since $\lambda = \lambda_k^0 \to 0$ for $k \to \infty$ and the operators in the right-hand side of (4.1) are bounded, we obtain

$$\begin{aligned} \mu A \boldsymbol{u}_{1k} + g(\Delta\rho) G_1 \zeta_{1k} &\to 0, \\ \zeta_{2k} &\to 0, \\ \gamma_1 \boldsymbol{u}_{1k} &\to 0, \\ \gamma_2 \boldsymbol{u}_{2k} &\to 0, \qquad k \to \infty. \end{aligned} \tag{4.39}$$

Multiplying the first relation by $\boldsymbol{u}_{1k}$ in the scalar product on $\boldsymbol{L}_2(\Omega_1)$ and using $G_1^* = \gamma_1$, we get

$$\mu\|A^{1/2}\boldsymbol{u}_1\|^2_{\boldsymbol{J}_{0,S_1}(\Omega_1)} + g(\Delta\rho)(\zeta_{1k}, \gamma_1\boldsymbol{u}_{1k})_{L_2(\Gamma_1)} \to \mu\|A^{1/2}\boldsymbol{u}_{1k}\|^2_{\boldsymbol{J}_{0.S_1}(\Omega_1)} \to 0, \quad k\to\infty.$$

whence it follows that $\boldsymbol{u}_{1k} \to \boldsymbol{0}$ in $\boldsymbol{J}_{0.S_1}(\Omega_1)$. Then from (4.39) and the normalization conditions we have

$$\begin{aligned}
\|\zeta_{1k}\|_{L_2(\Gamma_1)} &\to 1, \\
\|\boldsymbol{u}_{1k}\|_{\boldsymbol{J}_{0.S_1}(\Omega_1)} &\to 0, \\
\|\gamma_1\boldsymbol{u}_{1k}\|_{L_2(\Gamma_1)} &\to 0, \\
\|\gamma_2\boldsymbol{u}_{2k}\|_{L_2(\Gamma_2)} &\to 0, \qquad k\to\infty.
\end{aligned} \tag{4.40}$$

This proves Statement 5° completely.

Statements 2°–5° shed some light in the study of those solutions of problem (4.1) that correspond to a branch of eigenvalues with the limit point $\lambda = 0$. In the next section, we will study the location of the spectra of problem (4.1) in the complex plane as well as the completeness of the system of eigen- and associated elements in the case of arbitrary values attributed to the physical and gemetrical parameters of a partially dissipative hydrosystem.

10.4.3 The Theorem on Spectrum Location

Let us return now to problem (4.1) and perform in it the substitutions (4.3). Thus, we obtain the following spectral problem:

$$\mathcal{L}(\lambda)u = 0, \qquad u = (\boldsymbol{\xi};\eta;\zeta_1;\zeta_2)^{\mathrm{t}} \in \tilde{H} := \boldsymbol{J}_{0.S_1}(\Omega_1)\oplus H_2\oplus H_1\oplus H_2, \tag{4.41}$$

$$\mathcal{L}(\lambda) = \begin{pmatrix} \mu I - \lambda A_{11} & -\lambda A_{12} & g(\Delta\rho)Q^* & 0 \\ -\lambda A_{21} & -\lambda A_{22} & 0 & g\rho_2 I \\ -Q & 0 & -\lambda I & 0 \\ 0 & -I & 0 & -\lambda I \end{pmatrix}, \tag{4.42}$$

where $H_i = L_2(\Gamma_i)\ominus\{1_i\}$, $i = 1, 2$, I is the identity operator in the corresponding space, $Q := \gamma_1 A^{-1/2}$, $Q^* := A^{-1/2}G_1$, and A_{ij} are defined by (4.9). We need to get a formal expression of the matrix resolvent, $\mathcal{L}^{-1}(\lambda)$, of problem (4.41)–(4.42). With this aim in mind, we consider the following equation,

$$\mathcal{L}(\lambda)u = \tilde{u}, \qquad \tilde{u} := (\tilde{\boldsymbol{\xi}};\tilde{\eta};\tilde{\zeta}_1;\tilde{\zeta}_2)^{\mathrm{t}} \in \tilde{H}, \tag{4.43}$$

and find the specific form of its solution for any λ.

The system of equations (4.43) is equivalent to

$$\begin{aligned}\mu\boldsymbol{\xi} + g(\Delta\rho)Q^*\zeta_1 - \lambda(A_{11}\boldsymbol{\xi} + A_{12}\eta) &= \tilde{\boldsymbol{\xi}},\\ g\rho_2\zeta_2 - \lambda(A_{21}\boldsymbol{\xi} + A_{22}\eta) &= \tilde{\eta},\\ -Q\boldsymbol{\xi} - \lambda\zeta_1 &= \tilde{\zeta}_1,\\ -\eta - \lambda\zeta_2 &= \tilde{\zeta}_2.\end{aligned} \tag{4.44}$$

Then, eliminating ζ_1 and ζ_2,

$$\begin{aligned}\zeta_1 &= -\lambda^{-1}(\tilde{\zeta}_1 + Q\boldsymbol{\xi}),\\ \zeta_2 &= -\lambda^{-1}(\tilde{\zeta}_2 + \eta),\end{aligned} \tag{4.45}$$

we come to a system of two equations with two unknowns, $\boldsymbol{\xi}$ and η,

$$\begin{aligned}l_0(\lambda)\boldsymbol{\xi} &:= \left(\mu I - \lambda A_{11} - g(\Delta\rho)\lambda^{-1}B\right)\boldsymbol{\xi} = \lambda A_{12}\eta + g(\Delta\rho)\lambda^{-1}Q^*\tilde{\zeta}_1 + \tilde{\boldsymbol{\xi}},\\ m_0(\lambda)\eta &:= \left(I + \lambda^2(\rho_2 g)^{-1}A_{22}\right)\eta = -\lambda^2(\rho_2 g)^{-1}A_{21}\boldsymbol{\xi} - \tilde{\zeta}_2 - \lambda(\rho_2 g)^{-1}\tilde{\eta}.\end{aligned} \tag{4.46}$$

Assuming that the operator functions $l_0^{-1}(\lambda)$ and $m_0^{-1}(\lambda)$ exist, we get

$$\begin{aligned}l_1(\lambda)\boldsymbol{\xi} &:= \left(l_0(\lambda) + \lambda^3(\rho_2 g)^{-1}A_{12}m_0^{-1}(\lambda)A_{21}\right)\boldsymbol{\xi}\\ &= \tilde{\boldsymbol{\xi}} + g(\Delta\rho)\lambda^{-1}Q^*\tilde{\zeta}_1 - \lambda^2(\rho_2 g)^{-1}A_{12}m_0^{-1}(\lambda)\tilde{\eta} - \lambda A_{12}m_0^{-1}(\lambda)\tilde{\zeta}_2,\\ l_2(\lambda)\eta &:= \left(m_0(\lambda) + \lambda^3(\rho_2 g)^{-1}A_{21}l_0^{-1}(\lambda)A_{12}\right)\eta\\ &= -\lambda^2(\rho_2 g)^{-1}A_{21}l_0^{-1}(\lambda)\tilde{\boldsymbol{\xi}} - \lambda(\rho_2 g)^{-1}\tilde{\eta} - \left(\frac{\Delta\rho}{\rho_2}\right)\lambda A_{21}l_0^{-1}(\lambda)Q^*\tilde{\zeta}_1 - \tilde{\zeta}_2.\end{aligned} \tag{4.47}$$

We encountered the function $l_1(\lambda)$ previously; see (4.22).

Let us now assume that the operator functions $l_1^{-1}(\lambda)$ and $l_2^{-1}(\lambda)$ in (4.45)–(4.47) exist. Then the resolvent can be witten down as

$$\mathcal{L}^{-1}(\lambda) := \left(\mathcal{L}_{ik}^{-1}(\lambda)\right)_{i,k=1}^{4}, \tag{4.48}$$

where

$$\begin{aligned}\mathcal{L}_{11}^{-1}(\lambda) &= l_1^{-1}(\lambda),\\ \mathcal{L}_{12}^{-1}(\lambda) &= -\lambda^2(\rho_2 g)^{-1}l_1^{-1}(\lambda)A_{12}m_0^{-1}(\lambda),\\ \mathcal{L}_{13}^{-1}(\lambda) &= \lambda^{-1}g(\Delta\rho)l_1^{-1}(\lambda)Q^*,\end{aligned}$$

$$
\begin{aligned}
\mathcal{L}_{14}^{-1}(\lambda) &= -\lambda l_1^{-1}(\lambda)A_{12}m_0^{-1}(\lambda),\\
\mathcal{L}_{21}^{-1}(\lambda) &= -\lambda^2(\rho_2 g)^{-1}l_2^{-1}(\lambda)A_{21}l_0^{-1}(\lambda),\\
\mathcal{L}_{22}^{-1}(\lambda) &= -\lambda(\rho_2 g)^{-1}l_2^{-1}(\lambda),\\
\mathcal{L}_{23}^{-1}(\lambda) &= -\lambda\left(\frac{\Delta\rho}{\rho_2}\right)l_2^{-1}(\lambda)A_{21}l_0^{-1}(\lambda)Q^*,\\
\mathcal{L}_{24}^{-1}(\lambda) &= -l_2^{-1}(\lambda),\\
\mathcal{L}_{31}^{-1}(\lambda) &= -\lambda^{-1}Ql_1^{-1}(\lambda),\\
\mathcal{L}_{32}^{-1}(\lambda) &= \lambda(\rho_2 g)^{-1}Ql_1^{-1}(\lambda)A_{12}m_0^{-1}(\lambda),\\
\mathcal{L}_{33}^{-1}(\lambda) &= -\lambda^{-1}\left[I + Ql_1^{-1}(\lambda)Q^*g(\Delta\rho)\lambda^{-1}\right],\\
\mathcal{L}_{34}^{-1}(\lambda) &= Ql_1^{-1}(\lambda)A_{12}m_0^{-1}(\lambda),\\
\mathcal{L}_{41}^{-1}(\lambda) &= \lambda(\rho_2 g)^{-1}l_2^{-1}(\lambda)A_{21}l_0^{-1}(\lambda),\\
\mathcal{L}_{42}^{-1}(\lambda) &= (\rho_2 g)^{-1}l_2^{-1}(\lambda),\\
\mathcal{L}_{43}^{-1}(\lambda) &= \left(\frac{\Delta\rho}{\rho_2}\right)l_2^{-1}(\lambda)A_{21}l_0^{-1}(\lambda)Q^*,\\
\mathcal{L}_{44}^{-1}(\lambda) &= -\lambda^{-1}\left(I - l_2^{-1}(\lambda)\right).
\end{aligned}
\tag{4.49}
$$

From (4.48) and (4.49) it follows that $\mathcal{L}^{-1}(\lambda)$ is an analytic operator-function for $\lambda \neq 0$ and $\lambda \neq \infty$ at those points of the complex plane where the functions $m_0^{-1}(\lambda)$, $l_0^{-1}(\lambda)$, $l_1^{-1}(\lambda)$, and $l_2^{-1}(\lambda)$ are defined. Taking into account Property 1° in Section 10.4.2, we conclude that $\mathcal{L}^{-1}(\lambda)$ is a meromorphic operator function in $\mathbb{C}\setminus\{0\}$ whose poles coincide either with the eigenvalues of the operator function $l_1(\lambda)$ or, which is the same, with the eigenvaluess of the operator function $l_2(\lambda)$.

Let us show next that for small enough values of the density ρ_2 of the second fluid, the function $\mathcal{L}^{-1}(\lambda)$ is holomorphic in a certain domain of the complex plane outside some sectors adjoining the imaginary half-axes and the positive half-axis. For this purpose we introduce the notation

$$
\Lambda_{\varepsilon,R} := \{\lambda \in \mathbb{C} : \varepsilon < |\arg\lambda| < \pi/2 - \varepsilon,\ |\lambda| \geq R\},
\tag{4.50}
$$

where $\varepsilon > 0$ is an arbitrary small number and $R = R(\varepsilon)$ will be chosen in the future.

Let us recall that $\mathcal{L}^{-1}(\lambda)$ is a holomorphic function in the open left half-plane. Moreover, we can derive from (4.46) that the function $m_0^{-1}(\lambda)$ has the poles on the imaginary axis at the points $\lambda_k^{\pm} = \pm i\sqrt{\rho_2 g/\lambda_k(A_{22})} \to \infty$, $k \to \infty$, where $\lambda_k(A_{22})$ are eigenvalues of the positive compact operator A_{22}. Further, the function $l_0(\lambda)$ in (4.46) is an operator pencil of the same kind as the one studied in Chapter 8 that corresponds to the problem on normal oscillations of a viscous fluid in an open vessel. We know already from the study of the solutions to this problem that the spectrum of the pencil $l_0(\lambda)$ is situated on the positive half-axis and probably also contains no more than a finite number of nonreal eigenvalues of finite multiplicity. Whence, it follows that for any $\varepsilon > 0$ and sufficiently large R, the operator function $l_0^{-1}(\lambda)$ is defined and it is holomorphic in the domain $\Lambda_{\varepsilon,R}$ and in the open left half-plane as well.

To obtain estimates for the operator functions $l_1^{-1}(\lambda)$ and $l_2^{-1}(\lambda)$ that appear in the expressions of the matrix coefficients $\mathcal{L}_{ik}^{-1}(\lambda)$ in (4.49), we represent the functions $l_1(\lambda)$ and $l_2(\lambda)$ in (4.47) as follows:

$$\begin{aligned}
\mu l_1(\lambda) &= I - \lambda\mu^{-1}A_{11} - g(\Delta\rho)(\lambda\mu)^{-1}B + \lambda^3(\rho_2 g\mu)^{-1}A_{12}\left(I + \lambda^2(\rho_2 g)^{-1}A_{22}\right)^{-1}A_{21} \\
&= \left(I - \left(\frac{\lambda}{\mu}\right)^{1/2} A_{11}^{1/2}\right) \\
&\quad \times \left[I - g(\Delta\rho)(\lambda\mu)^{-1}\left(I - \left(\frac{\lambda}{\mu}\right)^{1/2} A_{11}^{1/2}\right)^{-1} B \left(I - \left(\frac{\lambda}{\mu}\right)^{1/2} A_{11}^{1/2}\right)^{-1}\right. \\
&\qquad + \lambda^3(\rho_2 g\mu)^{-1}\left(I - \left(\frac{\lambda}{\mu}\right)^{1/2} A_{11}^{1/2}\right)^{-1} A_{11}^{1/2}\left(A_{11}^{-1/2}A_{12}A_{22}^{-1/2}\right) \\
&\qquad \times \left(A_{22}^{1/2}\left(I + \mathrm{i}\lambda(\rho_2 g)^{-1/2}A_{22}^{1/2}\right)^{-1}\right)\left(\left(I - \mathrm{i}\lambda(\rho_2 g)^{-1/2}A_{22}^{1/2}\right)^{-1} A_{22}^{1/2}\right) \\
&\qquad \left.\times \left(A_{22}^{-1/2}A_{21}A_{11}^{-1/2}\right)\left(A_{11}^{1/2}\left(I + \left(\frac{\lambda}{\mu}\right)^{1/2} A_{11}^{1/2}\right)^{-1}\right)\right] \\
&\quad \times \left(I + \left(\frac{\lambda}{\mu}\right)^{1/2} A_{11}^{1/2}\right) \\
&=: \left(I - \left(\frac{\lambda}{\mu}\right)^{1/2} A_{11}^{1/2}\right)\left[I - g(\Delta\rho)(\lambda\mu)^{-1}T_1(\lambda) + \lambda^3(\rho_2 g\mu)^{-1}T_2(\lambda)\right] \\
&\quad \times \left(I - \left(\frac{\lambda}{\mu}\right)^{1/2} A_{11}^{1/2}\right), \qquad (4.51)
\end{aligned}$$

$$\begin{aligned}
l_2(\lambda) &= I + \lambda^2(\rho_2 g)^{-1}A_{22} + \lambda^3(\rho_2 g\mu)^{-1}A_{21}\left(I - \left(\frac{\lambda}{\mu}\right)A_{11} - g(\Delta\rho)(\lambda\mu)^{-1}B\right)^{-1}A_{12} \\
&= \left(I + \mathrm{i}\lambda(\rho_2 g)^{-1/2}A_{22}^{1/2}\right) \\
&\quad \times \left[I + \lambda^3(\rho_2 g\mu)^{-1}\left(I + \mathrm{i}\lambda(\rho_2 g)^{-1/2}A_{22}^{1/2}\right)^{-1} A_{22}^{1/2}\left(A_{22}^{-1/2}A_{21}A_{11}^{-1/2}\right)A_{11}^{1/2}\right. \\
&\qquad \times \left(I + \left(\frac{\lambda}{\mu}\right)^{1/2} A_{11}^{1/2}\right)^{-1}\left(I - g(\Delta\rho)(\lambda\mu)^{-1}\left(I - \left(\frac{\lambda}{\mu}\right)^{1/2} A_{11}^{1/2}\right)^{-1} B\right.
\end{aligned}$$

$$\times \left(I + \left(\frac{\lambda}{\mu}\right)^{1/2} A_{11}^{1/2}\right)^{-1} \left(I - \left(\frac{\lambda}{\mu}\right)^{1/2} A_{11}^{1/2}\right)^{-1} A_{11}^{1/2}$$

$$\times \left(A_{11}^{-1/2} A_{12} A_{22}^{-1/2}\right) A_{22}^{1/2} \left(I - \mathrm{i}\lambda(\rho_2 g)^{-1/2} A_{22}^{1/2}\right)^{-1} \Bigg]$$

$$\times \left(I - \mathrm{i}\lambda(\rho_2 g)^{-1/2} A_{22}^{1/2}\right)$$

$$=: \left(I + \mathrm{i}\lambda(\rho_2 g)^{-1/2} A_{22}^{1/2}\right) \left(I + \lambda^3(\rho_2 g\mu)^{-1} T_3(\lambda)\right) \left(I - \mathrm{i}\lambda(\rho_2 g)^{-1/2} A_{22}^{1/2}\right), \tag{4.52}$$

where $\operatorname{Re}\lambda^{1/2} > 0$ for $\operatorname{Re}\lambda > 0$.

We further use the following norm estimates for the nonnegative operator A:

$$\|(I - \lambda A)^{-1} A\| \le \frac{1}{|\operatorname{Im}\lambda|}, \qquad \operatorname{Re}\lambda > 0, \tag{4.53}$$

$$\|(I + \lambda A)^{-1} A\| \le \frac{1}{|\lambda|}, \qquad \operatorname{Re}\lambda > 0, \tag{4.54}$$

$$\|(I + \mathrm{i}\lambda A)^{-1} A\| \le \frac{1}{\operatorname{Re}\lambda}, \qquad \operatorname{Re}\lambda > 0,\ \operatorname{Im}\lambda > 0, \tag{4.55}$$

$$\|(I + \mathrm{i}\lambda A)^{-1} A\| \le \frac{1}{|\lambda|}, \qquad \operatorname{Re}\lambda > 0,\ \operatorname{Im}\lambda < 0 \tag{4.56}$$

that are obvious from the spectral representation $A = \int_0^\infty t \mathrm{d}P_t$ (see Section 1.1.22).

We also need to estimate the norms of the operator functions $T_i(\lambda)$, $i = 1, 2, 3$ that appear in (4.51) and (4.52). For this purpose we use the inequalities

$$\begin{aligned} &\|(I - \lambda A)^{-1}\| \le (\sin|\arg\lambda|)^{-1}, \qquad \operatorname{Re}\lambda > 0, \\ &\|(I + \lambda A)^{-1}\| \le 1, \qquad \operatorname{Re}\lambda > 0, \end{aligned} \tag{4.57}$$

and we get

$$\begin{aligned} \|T_1(\lambda)\| &= \left\| \left(I - \lambda^{1/2}\left(\mu^{-1/2} A_{11}^{1/2}\right)\right)^{-1} B \left(I + \lambda^{1/2}\left(\mu^{-1/2} A_{11}^{1/2}\right)\right)^{-1} \right\| \\ &\le (\sin|\arg\lambda^{1/2}|)^{-1} \|B\|, \qquad \operatorname{Re}\lambda > 0. \end{aligned} \tag{4.58}$$

For $\operatorname{Im}\lambda > 0$ we have

$$\|T_2(\lambda)\| \le \left\| \left(I - \left(\frac{\lambda}{\mu}\right)^{1/2} A_{11}^{1/2}\right)^{-1} A_{11}^{1/2} \right\| \left\| A_{11}^{-1/2} A_{12} A_{22}^{-1/2} \right\|$$

$$\times \left\| A_{22}^{1/2} \left(I + \mathrm{i}\lambda(\rho_2 g)^{-1/2} A_{22}^{1/2} \right)^{-1} \right\| \left\| \left(I - \mathrm{i}\lambda(\rho_2 g)^{-1/2} A_{22}^{1/2} \right)^{-1} A_{22}^{1/2} \right\|$$

$$\times \left\| A_{22}^{-1/2} A_{21} A_{11}^{-1/2} \right\| \left\| A_{11}^{1/2} \left(I + \left(\frac{\lambda}{\mu} \right)^{1/2} A_{11}^{1/2} \right)^{-1} \right\|$$

$$\leq \frac{\mu^{1/2}}{|\operatorname{Im} \lambda^{1/2}|} \left\| A_{11}^{-1/2} A_{12} A_{22}^{-1/2} \right\| \frac{(\rho_2 g)^{1/2}}{\operatorname{Re} \lambda} \left((\rho_2 g)^{1/2} |\lambda| \right)$$

$$\times \left\| A_{22}^{-1/2} A_{21} A_{11}^{-1/2} \right\| \frac{\mu^{1/2}}{|\lambda|^{1/2}}$$

$$= \left\| A_{11}^{-1/2} A_{12} A_{22}^{-1/2} \right\|^2 \rho_2 g \mu |\lambda|^{-3} (\sin \arg |\lambda^{1/2}| \cos \arg \lambda)^{-1}. \qquad (4.59)$$

A similar estimate can be found for $\operatorname{Im} \lambda < 0$.

To establish (4.59), we used the fact that operator $A_{11}^{-1/2} A_{12} A_{22}^{-1/2}$ is bounded and it is adjoint to $A_{22}^{-1/2} A_{21} A_{11}^{-1/2}$, and that their norms are equal to each other. To prove these properties we use the definitions (4.9) of the matrix elements A_{ik} and consider the operator

$$\left(A_{11}^{-1/2} A_{12} A_{22}^{-1/2} \right)^* \left(A_{11}^{-1/2} A_{12} A_{22}^{-1/2} \right)$$

$$= A_{22}^{-1/2} A_{21} A_{11}^{-1} A_{12} A_{22}^{1/2}$$

$$= \left(\rho_2^{-1/2} C_{22}^{-1/2} \right) \left(\rho_2 C_{21} \gamma_1 A^{-1/2} \right) \left(A^{1/2} (\rho_1 I_1 + \rho_2 G_1 C_{11} \gamma_1)^{-1} A^{1/2} \right)$$

$$\times \left(\rho_2 A^{-1/2} G_1 C_{12} \right) \left(\rho_2^{-1/2} C_{22}^{-1/2} \right)$$

$$= \rho_2 C_{22}^{-1/2} C_{11} \gamma_1 (\rho_1 I_1 + \rho_2 G_1 C_{11} \gamma_1)^{-1} G_1 C_{12} C_{22}^{-1/2}. \qquad (4.60)$$

From this representation and the properties of the operators C_{ik}, γ_1, and $G_1 = \gamma_1^*$, it follows that the operator defined in (4.60) is bounded from H_1 into H_2. Indeed, the following chain of bounded mappings holds true:

$$H_2 \xrightarrow{C_{22}^{-1/2}} H^{-1/2}(\Gamma_2) \xrightarrow{C_{12}} H^{1/2}(\Gamma_1) \xrightarrow{G_1} \boldsymbol{G}_{\mathrm{h},S_1}(\Omega_1) \xrightarrow{(\rho_1 I_1 + \rho_2 G_1 C_{11} \gamma_1)^{-1}} \boldsymbol{G}_{\mathrm{h},S_1}(\Omega_1)$$

$$\xrightarrow{\gamma_1} H^{-1/2}(\Gamma_1) \xrightarrow{C_{21}} H^{1/2}(\Gamma_2) \xrightarrow{C_{22}^{-1/2}} H_2.$$

Using the inequalities

$$G_1 C_{11} \gamma_1 \geq 0, \qquad (\rho_1 I_1 + \rho_2 G_1 C_{11} \gamma_1)^{-1} \leq \rho_1^{-1} I_1,$$

we obtain from (4.60) the following estimate as well:

$$\left\| A_{11}^{-1/2} A_{12} A_{22}^{-1/2} \right\|^2 \leq \frac{\rho_2}{\rho_1} \left\| G_1 C_{12} C_{22}^{-1/2} \right\|^2, \qquad (4.61)$$

where the norm squared depends only on the geometric characteristics of the domains Ω_1 and Ω_2 ocupied by the two fluids.

Let us now assume that λ belongs to $\Lambda_{\varepsilon.R}$, the domain defined in (4.50), and that $\lambda \to \infty$. Then, from (4.58), it follows that the second term in the right-hand side square brackets of (4.51) has a value of order $O(|\lambda|^{-1})$. Further, from (4.59) and (4.61), we get that the last term within square brackets has a value of order $O(1)$ as $\lambda \to \infty$, with $\lambda \in \Lambda_{\varepsilon,R}$.

We choose an arbitrary number ε, $0 < \varepsilon < \pi/4$, and introduce $R = R(\varepsilon)$, such that the operator $l_0(\lambda)$ from (4.46) is invertible in the domain $\Lambda_{\varepsilon,R(\varepsilon)}$. Since on this domain $\sin(\varepsilon/2) < \sin|(\arg\lambda)/2| = |\sin(\arg\lambda^{1/2})| < \sin((\pi/2-\varepsilon)/2)$, and $\sin\varepsilon < \cos\arg\lambda < \cos\varepsilon$, then, in virtue of (4.58), (4.59) and (4.61), we have the estimate

$$\begin{aligned}\|T(\lambda)\| &:= \left\|-g(\Delta\rho)(\lambda\mu)^{-1}T_1(\lambda) + \lambda^3(\rho_2 g\mu)^{-1}T_2(\lambda)\right\| \\ &\leq \frac{g(\Delta\rho)}{\mu}\|B\|\left(\sin\frac{\varepsilon}{2}\right)^{-1}|\lambda|^{-1} + \frac{\rho_2}{\rho_1}\left\|G_1C_{12}C_{22}^{-1/2}\right\|^2 + \left(\sin\frac{\varepsilon}{2}\sin\varepsilon\right)^{-1},\end{aligned} \tag{4.62}$$

for $\lambda \in \Lambda_{\varepsilon,R(\varepsilon)}$.

Further we will consider only the case of such hydrodynamic partially dissipative systems for which the density ρ_2 of the ideal fluid is much less than the density ρ_1 of the viscous fluid. Basically, we assume that for some ε, with $0 < \varepsilon < \pi/4$, the following condition is satisfied:

$$\rho_2 < \rho_1 \sin\frac{\varepsilon}{2}\sin\varepsilon\left\|G_1C_{12}C_{22}^{-1/2}\right\|^2. \tag{4.63}$$

Then, from (4.62) we derive that there exists such a large enough number $R = R(\varepsilon)$ for which the following condition is satisfied:

$$q := \frac{g(\Delta\rho)}{\mu}\|B\|\left(\sin\frac{\varepsilon}{2}R(\varepsilon)\right)^{-1} + \frac{\rho_2}{\rho_1}\left\|G_1C_{12}C_{22}^{-1/2}\right\|^2\left(\sin\frac{\varepsilon}{2}\sin\varepsilon\right)^{-1} < 1. \tag{4.64}$$

Thus, according to (4.62), for the newly chosen $R(\varepsilon)$ the condition

$$\|T(\lambda)\| \leq q < 1. \tag{4.65}$$

is satisfied on the region $\Lambda_{\varepsilon,R(\varepsilon)}$. This means that the operator function $I + T(\lambda)$ is invertible and, according to (4.51), in the previously mentioned domain, the operator function $l_1^{-1}(\lambda)$ exists and is holomorphic.

The reader can check independently that condition (4.63) ensures also the invertibility for the operator function $l_2(\lambda)$ defined in (4.52). In this particular case it is necessary to employ estimates similar with those in (4.58), (4.59), and (4.62)–(4.64) and to represent $l_2(\lambda)$ as in (4.52).

Thus, if condition (4.63) is satisfied and $R = R(\varepsilon)$ is chosen from condition (4.64), then in the domain $\Lambda_{\varepsilon,R(\varepsilon)}$ there exists a holomorphic operator function $\mathcal{L}^{-1}(\lambda)$ as in (4.48) with the matrix coefficients as in (4.49). We are now in a position to formulate the final result for problem (4.41), which may very well be called the theorem on spectrum localization.

If condition (4.63) *is satisfied and* $R(\varepsilon)$ *satisfies condition* (4.64), *then the spectrum of problem* (4.41), *and consequently the spectrum of the initial problem* (4.1) *on normal oscillations of a partially dissipative hydrosystem is situated in the right half-plane outside the domain* $\Lambda_{\varepsilon,R(\varepsilon)}$. *In other words, the spectrum of problem* (4.1) *consisting of finitely multiple eigenvalues may have the limit point at* $\lambda = \infty$ *and be localized (in the above-mentioned sense) in the neighborhood of the positive half-axis and the two imaginary half-axes.*

10.5 On the Completeness of the System of Modes of Normal Oscillations

This section considers the issues of the completeness of the system of eigen- and associated elements of the problem (4.41), (4.42) on normal oscillations of a partially dissipative hydrosystem, and also some heuristic arguments connected with the proof of the existence of branches of eigenvalues in the vicinity of the positive semiaxis and also of the imaginary semiaxes in the complex plane. While discussing the issues of completeness, a method of M. V. Keldysh [1, 2] proposed for linear and polynomial operator pencils will be used.

10.5.1 Auxiliary Results

Several auxiliary statements will be given here. They are necessary to prove the theorem on completeness in problem (4.41)–(4.42). Some of them are rather well known.

1. Phragmen-Lindelöff principle. Let the function $f(z)$ be holomorphic inside the angle $\Omega := \{z \in \mathbb{C} : |\arg z| < \pi/(2\alpha)\}$, $\alpha \geq 1$, and on its sides, and suppose that for some $\beta < \alpha$ one has

$$\underline{\lim}_{r\to\infty} r^{-\beta} \log \sup_{|z|=r} |f(z)| < \infty. \tag{5.1}$$

If $|f(z)| \leq M$ for $|\arg z| < \pi/(2\alpha)$, then $|f(z) \leq M$ for all $z \in \Omega$.

It should be noted that instead of the angle Ω in the Phragmen-Lindelöff principle, the region $\Omega \cap \{z : |z| \geq R\} =: \Omega_R$ may be also taken.

2. On the behavior of the resolvent of an operator function in the vicinity of an eigenvalue. Let $\lambda_0 \in \mathbb{C}$ be eigenvalue of the operator function $L(\lambda)$ acting in a Hilbert space E and analytic at the point λ_0. If $L(\lambda)$ is invertible in a punctured vicinity of the point λ_0, then a necessary condition for the function $g(\lambda) := [L^*(\bar{\lambda})]^{-1} f$ to be analytic at the point $\bar{\lambda}_0$ is the orthogonality of the element $f \in E$ to all elements of the root subspace of the function $L(\lambda)$ corresponding to the eigenvalue λ_0. If $L^{-1}(\lambda)$ has a pole at the point λ_0, then the orthogonality condition mentioned above is also sufficient for the function $g(\lambda)$ to be analytic at the point λ_0.

3. Growth order of an analytic operator function at infinity and at zero. Let the operator function $A(\lambda)$ be holomorphic in $\mathbb{C} \setminus \{0\}$. The number

$$\rho = \rho_\infty := \overline{\lim}_{|\lambda| \to \infty} \ln \ln \|A(\lambda)\| / \ln |\lambda|$$

is called *the growth order* of function $A(\lambda)$ at infinity. The growth order of function $A(\lambda)$ at zero is defined similarly as

$$\rho_0 := \overline{\lim_{\lambda \to 0}} \ln \ln \|A(\lambda)\| / \ln |\lambda|^{-1}$$

An operator function $A(\lambda)$ meromorphic in the entire complex plane is said to be a *function of finite order* ρ, provided it admits the representation

$$A(\lambda) = A_1(\lambda)/f(\lambda),$$

where the entire operator function $A_1(\lambda)$ and the scalar valued function $f(\lambda)$ are of orders ρ and the order can not be taken less than ρ.

4. An estimate of the resolvent growth of a Fredholm operator pencil. Suppose that the operator function $L(\lambda)$ has the form

$$L(\lambda) = I + \sum_{k=-m}^{n} \lambda^k A_k, \qquad A_k \in \mathfrak{S}_{p_k}, \ -m \leq k \leq n, \ m, n \geq 0, \tag{5.2}$$

and therefore is analytic in $\mathbb{C} \setminus \{0\}$. Then $L^{-1}(\lambda)$ is a meromorphic operator function in $\mathbb{C} \setminus \{0\}$ that can be represented as the ratio of two functions of orders not higher than

$$\begin{aligned} \rho_\infty &= \max_{j=1,\ldots,n} (j p_j), \qquad \lambda \to \infty, \\ \rho_0 &= \max_{j=-m,\ldots,-1} (|j| p_j), \qquad \lambda \to 0, \end{aligned} \tag{5.3}$$

and minimal types for orders ρ_∞ and ρ_0, respectively.

10.5.2 Theorem on Completeness. Keldysh Scheme Realization

Let us go back to considering the problem (4.41)–(4.42) and recall that it has a disctete spectrum situated in the right complex half-plane with a limit point at zero and probably at infinity. In the right half-plane, the spectrum is situated outside the region $\Lambda_{\varepsilon,R}$ [see (4.50)] and consists of finitely multiple eigenvalues, where the resolvent $\mathcal{L}^{-1}(\lambda)$ of problem (4.41) is a meromorphic operator function in $\mathbb{C} \setminus \{0\}$ having poles at points coinciding with eigenvalues.

Relying on these facts, the propositions of Section 10.5.1, and also the properties of coefficients of the operator function $\mathcal{L}(\lambda)$ in (4.41)–(4.42), it is possible to state the following basic result.

Theorem on Completeness. *If condition* (4.63) *is satisfied, then the system of eigen- and associated elements of problem* (4.41)–(4.42) *corresponding to eigenvalues in the region* $\{\lambda \in \mathbb{C} : \mathrm{Re}\lambda > 0\}$ *is complete and minimal in the Hilbert space*

$$\tilde{H} := \boldsymbol{J}_{0.S_1}(\Omega_1) \oplus H_2 \oplus H_1 \oplus H_2. \tag{5.4}$$

The proof of the theorem is based on the Keldysh scheme and will be carried out step by step.

(a) Suppose that the system of eigen- and associated elements of the problem (4.41)–(4.42) is incomplete in $\tilde{H}$. Then, according to the Statement 2 in Section 10.5.1 there is a nonzero element $\tilde{u} \in \tilde{H}$ such that the function

$$\tilde{\xi}(\lambda) := [\mathcal{L}^*(\bar{\lambda})]^{-1}\tilde{u} = (\boldsymbol{\xi}(\lambda);\ \eta(\lambda);\ \zeta_1(\lambda);\ \zeta_2(\lambda))^{\mathrm{t}} \tag{5.5}$$

is analytic at every point of the spectrum of the function $\mathcal{L}(\lambda)$, and therefore in the region $\mathbb{C} \setminus \{0\}$.

(b) We write down equation (5.5) in the form

$$\mathcal{L}^*(\bar{\lambda})\tilde{\xi}(\lambda) = \tilde{u} = (\boldsymbol{u}_1;\ u_2;\ u_3;\ u_4)^{\mathrm{t}} \in \tilde{H}, \tag{5.6}$$

or, by taking into account the expression (4.42) for $\mathcal{L}(\lambda)$, as a system of equations:

$$\begin{aligned}
(\mu I - \lambda A_{11})\boldsymbol{\xi}(\lambda) - \lambda A_{12}\eta(\lambda) - Q^*\zeta_1(\lambda) &= \boldsymbol{u}_1, \\
-\lambda A_{21}\boldsymbol{\xi}(\lambda) - \lambda A_{22}\eta(\lambda) - \zeta_2(\lambda) &= u_2 \\
g(\Delta\rho)Q\boldsymbol{\xi}(\lambda) - \lambda\zeta_1(\lambda) &= u_3, \\
g\rho_2\eta(\lambda) - \lambda\zeta_2(\lambda) &= u_4.
\end{aligned} \tag{5.7}$$

Hence, we get

$$\begin{aligned}\zeta_1(\lambda) &= -\lambda^{-1}(u_3 - g(\Delta\rho)Q\boldsymbol{\xi}(\lambda)),\\ \zeta_2(\lambda) &= -\lambda^{-1}(u_4 - g\rho_2\eta(\lambda)),\end{aligned} \tag{5.8}$$

and a substitution into the first two equations of (5.7) gives

$$\begin{aligned}l_0(\lambda)\boldsymbol{\xi}(\lambda) - \lambda A_{12}\eta(\lambda) &= \boldsymbol{u}_1 - \lambda^{-1}Q^*u_3,\\ m_0(\lambda)\eta(\lambda) + \frac{\lambda^2}{\rho_2 g}A_{21}\boldsymbol{\xi}(\lambda) &= -\frac{\lambda}{\rho_2 g}u_2 + \frac{1}{\rho_2 g}u_4,\end{aligned} \tag{5.9}$$

where the operator functions

$$\begin{aligned}l_0(\lambda) &:= \mu I - \lambda A_{11} - g(\Delta\rho)\lambda^{-1}B,\\ m_0(\lambda) &:= I + \frac{\lambda^2}{\rho_2 g}A_{22}\end{aligned} \tag{5.10}$$

were already met in Section 10.4.3 [see (4.46)].

From the second relation (5.9) we find

$$\eta(\lambda) = m_0^{-1}(\lambda)\Big(-\frac{\lambda}{\rho_2 g}u_2 + \frac{1}{\rho_2 g}u_4 - \frac{\lambda^2}{\rho_2 g}A_{21}\boldsymbol{\xi}(\lambda)\Big) \tag{5.11}$$

and for $\boldsymbol{\xi}(\lambda)$ we obtain the equation

$$\begin{aligned}l_1(\lambda)\boldsymbol{\xi}(\lambda) &= \boldsymbol{u}_1 - \frac{1}{\lambda}Q^*u_3 - \frac{\lambda^2}{\rho_2 g}A_{12}m_0^{-1}(\lambda)u_2 + \frac{\lambda}{\rho_2 g}A_{12}m_0^{-1}(\lambda)u_4\\ &=: \boldsymbol{f}(\lambda),\end{aligned} \tag{5.12}$$

where the function

$$l_1(\lambda) := l_0(\lambda) + \frac{\lambda^3}{\rho_2 g}A_{12}m_0^{-1}(\lambda)A_{21} \tag{5.13}$$

was studied in complete detail in Section 10.4.3.

(c) We note now that the function

$$\begin{aligned}m_0^{-1}(\lambda) &= \Big(I + \frac{\lambda^2}{\rho_2 g}A_{22}\Big)^{-1}\\ &= \Big(I - \frac{\mathrm{i}\lambda}{\sqrt{\rho_2 g}}A_{22}^{1/2}\Big)^{-1}\Big(I + \frac{\mathrm{i}\lambda}{\sqrt{\rho_2 g}}A_{22}^{1/2}\Big)^{-1},\end{aligned}$$

by virtue of the inequalities

$$\left\|\left(I+\frac{\mathrm{i}\lambda}{\sqrt{\rho_2 g}}A_{22}^{1/2}\right)^{-1}\right\| \leq \frac{1}{\cos\arg\lambda}, \qquad \mathrm{Re}\lambda>0,\ \mathrm{Im}\lambda>0,$$
$$\left\|\left(I-\frac{\mathrm{i}\lambda}{\sqrt{\rho_2 g}}A_{22}^{1/2}\right)^{-1}\right\| \leq 1, \qquad \mathrm{Re}\lambda>0,\ \mathrm{Im}\lambda>0, \tag{5.14}$$

leads to the estimate

$$\|m_0^{-1}(\lambda)\| \leq \frac{1}{\sin\varepsilon}, \qquad \lambda\in\Lambda_{\varepsilon,R}=\{\lambda\in\mathbb{C}:\varepsilon<|\arg\lambda|<\pi/2-\varepsilon,\ |\lambda|\geq R,\ \mathrm{Re}\lambda>0\}. \tag{5.15}$$

For $\mathrm{Im}\lambda<0$, the expressions on the right in (5.14) change their places.

Therefore, the function $\boldsymbol{f}(\lambda)$ in (5.12) can be estimated as follows,

$$\boldsymbol{f}(\lambda)=\mathrm{O}(|\lambda|^2), \qquad \lambda\to\infty,\ \lambda\in\Lambda_{\varepsilon,R}. \tag{5.16}$$

Since, according to the representations (4.51) and (4.62), $l_1^{-1}(\lambda)$ has the form

$$l_1^{-1}(\lambda)=\mu^{-1}\left(I+\left(\frac{\lambda}{\mu}\right)^{1/2}A_{11}^{1/2}\right)^{-1}(I-T(\lambda))^{-1}\left(I-\left(\frac{\lambda}{\mu}\right)^{1/2}A_{11}^{1/2}\right)^{-1}$$

and

$$\left\|\left(I+\left(\frac{\lambda}{\mu}\right)^{1/2}A_{11}^{1/2}\right)^{-1}\right\| \leq 1, \qquad \mathrm{Re}\lambda>0,$$
$$\left\|\left(I-\left(\frac{\lambda}{\mu}\right)^{1/2}A_{11}^{1/2}\right)^{-1}\right\| \leq (\sin|\arg\lambda|)^{-1}, \qquad \mathrm{Re}\lambda>0,$$
$$\|(I-T(\lambda))^{-1}\| \leq (1-q)^{-1}, \qquad \lambda\in\Lambda_{\varepsilon,R}, \tag{5.17}$$

[see formulas (4.57) and (4.63)–(4.65)], then the following estimate

$$\|l_1^{-1}(\lambda)\| \leq \frac{\mu^{-1}(1-q)^{-1}}{\sin\left(\frac{\varepsilon}{2}\right)}, \qquad \lambda\to\infty,\ \lambda\in\Lambda_{\varepsilon,R}, \tag{5.18}$$

takes place.

Hence from (5.16) and (5.12) we get that

$$\|\boldsymbol{\xi}(\lambda)\|=\mathrm{O}(|\lambda|^2), \qquad \lambda\to\infty,\ \lambda\in\Lambda_{\varepsilon,R}. \tag{5.19}$$

Then from (5.8), (5.11), and (5.15) we have

$$\eta(\lambda)=\mathrm{O}(|\lambda|^4), \qquad \boldsymbol{\xi}_1(\lambda)=O(|\lambda|),$$
$$\boldsymbol{\xi}_2(\lambda)=\mathrm{O}(|\lambda|^3), \qquad \lambda\to\infty,\ \lambda\in\Lambda_{\varepsilon,R}. \tag{5.20}$$

(d) Consider the behavior of the function $\tilde{\xi}(\lambda)$ from (5.5) for small λ. Since in this case $m_0^{-1}(\lambda)$ is an analytic operator function, then the function $\boldsymbol{f}(\lambda)$ in (5.12) has the following estimate,

$$\boldsymbol{f}(\lambda) = \mathrm{O}(|\lambda|^{-1}), \qquad \lambda \to 0. \tag{5.21}$$

Further, considering the estimate

$$l_1(\lambda) = \mu I - \frac{g(\Delta\rho)}{\lambda} B + \mathrm{O}(\lambda), \qquad \lambda \to 0, \tag{5.22}$$

from (5.10) and (5.13) for small λ, and also from the inequality

$$\left\| \left(I - \frac{g(\Delta\rho)}{\mu\lambda} B \right)^{-1} \right\| \leq (\sin|\arg\lambda|)^{-1}, \qquad \mathrm{Re}\lambda > 0, \tag{5.23}$$

we conclude that

$$\|l_1^{-1}(\lambda)\| \leq C_{\varepsilon,r}, \qquad \lambda \to 0,\ \lambda \in \Omega_{\varepsilon.r} := \{\lambda \in \mathbb{C} : |\arg\lambda| > \varepsilon,\ |\lambda| \leq r\}, \tag{5.24}$$

where the constant $C_{\varepsilon,r}$ depends only on the choice of ε and r.

The formulas (5.21), (5.24) and (5.8), (5.11), (5.12) show that

$$\begin{aligned}
\boldsymbol{\xi}(\lambda) &= \mathrm{O}(|\lambda|^{-1}), \\
\eta(\lambda) &= \mathrm{O}(1), \\
\zeta_1(\lambda) &= \mathrm{O}(|\lambda|^{-2}), \\
\zeta_2(\lambda) &= \mathrm{O}(|\lambda|^{-1}), \qquad \lambda \to 0,\ \lambda \in \Omega_{\varepsilon,r}.
\end{aligned} \tag{5.25}$$

(e) From (5.19), (5.20), and (5.25), and based on similar considerations, we get that, in the region

$$G_\varepsilon := \{\lambda \in \mathbb{C} \setminus \{0\} : \varepsilon < |\arg\lambda| < \pi/2 - \varepsilon,\ \pi/2 + \varepsilon < |\arg\lambda| \leq \pi\} \tag{5.26}$$

and along its boundary, the function $\tilde{\xi}(\lambda)$ in (5.5) has an estimate of the form

$$|\tilde{\xi}(\lambda)| \leq \sum_{k=-2}^{4} a_k |\lambda|^k, \qquad a_k \geq 0. \tag{5.27}$$

Here, since the operators A_{ij}, $i, j = 1, 2$, Q, and Q^* in problem (4.41)–(4.42) belong to some classes $\mathfrak{S}_q$, then, in virtue of Property 4 in Section 10.5.1, the function $\tilde{\xi}(\lambda)$ will be an analytic function in $\mathbb{C} \setminus \{0\}$ with a finite growth order at zero and at infinity.

Therefore, for the angles

$$\{\lambda \in \mathbb{C} \setminus \{0\} : |\arg \lambda| \leq \varepsilon\}, \ \{\lambda \in \mathbb{C} \setminus \{0\} : \pi/2 - \varepsilon \leq |\arg \lambda| \leq \pi/2 + \varepsilon\}$$

we may use the Phragmen-Lindelöff principle (see Statement 1 in Section 10.5.1), and conclude that the estimate (5.27) for $\boldsymbol{\xi}(\lambda)$ is realized not only in G_ε, but also in $\mathbb{C} \setminus \{0\}$.

(f) From the latter statement we get that the Laurent series expansion for $\tilde{\xi}(\lambda)$ has the form

$$\tilde{\xi}(\lambda) = \sum_{k=-2}^{4} \tilde{\xi}_k \lambda^k, \tag{5.28}$$

that is,

$$\boldsymbol{\xi}(\lambda) = \sum_{k=-2}^{4} \boldsymbol{\xi}_k \lambda^k,$$

$$\eta(\lambda) = \sum_{k=-2}^{4} \eta_k \lambda^k,$$

$$\zeta_i(\lambda) = \sum_{k=-2}^{4} \zeta_{ik} \lambda^k, \qquad i = 1, 2, \ \lambda \in \mathbb{C} \setminus \{0\} \tag{5.29}$$

We substitute the expressions (5.29) into the system of equations (5.7) and identify the coefficients of the same powers of λ. For λ^5 we obtain the following relations

$$\zeta_{14} = 0, \qquad \zeta_{24} = 0, \qquad A_{11}\boldsymbol{\xi}_4 + A_{12}\eta_4 = 0, \qquad A_{21}\boldsymbol{\xi}_4 + A_{22}\eta_4 = 0. \tag{5.30}$$

Since the operator $A = (A_{ij})_{i,j=1}^2$ is positive, then from (5.30) it follows that $\boldsymbol{\xi}_4 = \mathbf{0}$, $\eta_4 = 0$. Similarly we get that

$$\zeta_{1k} = 0, \quad \zeta_{2k} = 0, \quad \boldsymbol{\xi}_k = 0, \quad \eta_k = 0, \qquad k = 0, 1, 2, 3, 4. \tag{5.31}$$

The coefficient identification for λ^{-2} gives the relations

$$\zeta_{2,-2} = 0, \qquad \eta_{-2} = 0, \qquad \mu\boldsymbol{\xi}_{-2} - Q^*\zeta_{1,-2} = 0, \qquad g(\Delta\rho)Q\boldsymbol{\xi}_{-2} = 0. \tag{5.32}$$

From the two latter equations it follows that

$$\mu Q\boldsymbol{\xi}_{-2} = QQ^*\zeta_{1,-2} = 0,$$

and, therefore, $\|Q^*\zeta_{1,-2}\|^2 = 0$, that is, $Q^*\zeta_{1.-2} = 0$.

Thus $\boldsymbol{\xi}_{-2} = \mathbf{0}$, and from the equation $Q^*\zeta_{1.-2} = 0$ it follows that $\zeta_{1,-2} = 0$. A similar reasoning for λ^{-1} leads to the following result

$$\boldsymbol{\xi}_k = \mathbf{0}, \quad \eta_k = 0, \quad \zeta_{1k} = 0, \quad \zeta_{2k} = 0, \qquad k = -1, -2. \tag{5.33}$$

Finally, for λ^0 and taking into consideration (5.31) and (5.33), we obtain the relations

$$\mathbf{0} = \boldsymbol{u}_1, \quad 0 = u_k, \qquad k = 2, 3, 4 \tag{5.34}$$

that lead to a contradiction, since according to the initial assumption, the element $\tilde{u} = (\boldsymbol{u}_1;\ u_2;\ u_3;\ u_4)^{\mathrm{t}} \in \tilde{H}$ is nonzero.

Hence, it follows that the system of eigen- and associated elements of the problem (4.41)–(4.42) is complete in $\tilde{H}$.

(g) In conclusion, we note that the previously mentioned system consists of linearly independent elements that are the solutions of problem (4.41), which is linear relatively to the spectral parameter λ. Therefore, the system is minimal in $\tilde{H}$, that is, the deletion of any element from an arbitrary finite set of such elements contracts the space spanned by these elements.

The Theorem on Completeness is proved.

10.5.3 On the Existence of Branches of Eigenvalues With a Limit Point at Infinity. Heuristic Arguments

Let us return to considering the problem (4.1) on normal oscillations of a partially dissipative hydrosystem and turn again to its self-ajoint form (4.8), (4.9) with bounded operator coefficients:

$$L(\lambda)y := \begin{pmatrix} \mu I_1 & 0 \\ 0 & 0 \end{pmatrix} \begin{pmatrix} \boldsymbol{\xi} \\ \eta \end{pmatrix} - \lambda \begin{pmatrix} A_{11} & A_{12} \\ A_{21} & A_{22} \end{pmatrix} \begin{pmatrix} \boldsymbol{\xi} \\ \eta \end{pmatrix} - g\lambda^{-1} \begin{pmatrix} (\Delta\rho)B & 0 \\ 0 & \rho_2 I_2 \end{pmatrix} \begin{pmatrix} \boldsymbol{\xi} \\ \eta \end{pmatrix} = \begin{pmatrix} \mathbf{0} \\ 0 \end{pmatrix}, \tag{5.35}$$

where $y = (\boldsymbol{\xi};\ \eta)^{\mathrm{t}} \in \boldsymbol{J}_{0,S_1}(\Omega_1) \oplus H_2$, and the elements A_{ik} are defined in (4.9).

The Theorem of Spectrum Localization (Section 10.4.3) together with the Theorem of Completeness (Section 10.5.2) of the system of eigen- and associated elements of problem (4.41) and the directly connected with it Problem (5.35), lead to the natural assumption that in these problems there exist branches of eigenvalues situated in a vinicity of the positive semiaxis and the two imaginary semiaxes and having a limit point at infinity. In particular, a similar situation was encountered in the model problem discussed in Section 10.3.

The proof of the existence of such branches in the common situation considered now comes across serious mathematical difficulties. Thus, in stating the facts of the existence of a branch for eigenvalues ajoining the positive semiaxis and with a limit point at infinity, it is natural to consider the problem (4.22) on eigenvalues

$$
\begin{aligned}
l_1(\lambda)\boldsymbol{\xi} &:= \left(\mu I_1 - g(\Delta\rho)\lambda^{-1}B - \lambda A_{11} + \lambda^3(\rho_2 g)^{-1}A_{12}\left(I_2 + \frac{\lambda^2}{\rho_2 g}A_{22}\right)^{-1} A_{21}\right)\boldsymbol{\xi} \\
&= \mathbf{0}, \qquad \boldsymbol{\xi} \in \boldsymbol{J}_{0,S_1}(\Omega_1).
\end{aligned} \tag{5.36}
$$

Respectively, in stating the presence of eigenvalue branches ajoining the imaginary semiaxes, it is necessary to consider the problem on eigenvalues

$$
\begin{aligned}
l_2(\lambda)\eta &:= \left(I_2 + \frac{\lambda^2}{\rho_2 g}A_{22} + \frac{\lambda^3}{\rho_2 g}A_{21}(\mu I_1 - \lambda A_{11} - g(\Delta\rho)\lambda^{-1}B)^{-1}A_{12}\right)\eta \\
&= 0, \qquad \eta \in H_2.
\end{aligned} \tag{5.37}
$$

One would think that in problem (5.36), for $\lambda \to \infty$, the main terms determining the presence and asymptotic behavior of a branch of eigenvalues with a limit point at ∞ are in the contracted pencil $l_1^0(\lambda) := \mu I_1 - \lambda A_{11}$. However, for $\lambda \in \mathbb{R}_+$, $\lambda \to \infty$, the perturbation of pencil $l_1^0(\lambda)$ with additional terms turns out to be of order λ^3, that is, they could affect the existence and asymptotics of eigenvalues. A similar situation takes place in problem (5.37).

Based on these heuristic assumptions, we will consider, apart from problem (5.35), the problem on eigenvalues

$$
\begin{aligned}
L_\alpha(\lambda)y &:= \begin{pmatrix} \mu I_1 & 0 \\ 0 & 0 \end{pmatrix}\begin{pmatrix} \boldsymbol{\xi} \\ \eta \end{pmatrix} - \lambda\begin{pmatrix} A_{11} & 0 \\ 0 & A_{22} \end{pmatrix}\begin{pmatrix} \boldsymbol{\xi} \\ \eta \end{pmatrix} \\
&\quad - g\lambda^{-1}\begin{pmatrix} (\Delta\rho)B & 0 \\ 0 & \rho_2 I_2 \end{pmatrix}\begin{pmatrix} \boldsymbol{\xi} \\ \eta \end{pmatrix} - \alpha\lambda\begin{pmatrix} 0 & A_{12} \\ A_{21} & 0 \end{pmatrix}\begin{pmatrix} \boldsymbol{\xi} \\ \eta \end{pmatrix} \\
&= \begin{pmatrix} \mathbf{0} \\ 0 \end{pmatrix},
\end{aligned} \tag{5.38}
$$

corresponding to a self-adjoint operator pencil $L_\alpha(\lambda)$ that coincides with $L(\lambda)$ for $\alpha = 1$.

A peculiarity of the pencil $L_\alpha(\lambda)$ is the property that, for $\alpha = 0$, problem (5.38) splits into two independent and already studied problems on eigenvalues:

$$
\begin{aligned}
l_0(\lambda)\boldsymbol{\xi} &:= (\mu I_1 - \lambda A_{11} - g(\Delta\rho)\lambda^{-1}B)\boldsymbol{\xi} = \mathbf{0}, \qquad \boldsymbol{\xi} \in \boldsymbol{J}_{0,S_1}(\Omega_1), \\
m_0(\lambda)\eta &:= \left(I_2 + \frac{\lambda^2}{\rho_2 g}A_{22}\right)\eta = 0, \qquad \eta \in H_2.
\end{aligned} \tag{5.39}
$$

The first one was studied in detail in Chapter 8, and the second one is very simple. In the case of arbitrary viscosity values, μ, the first problem (5.39) has two branches of positive eigenvalues with limit points at zero and ∞, respectively. The spectrum of the second problem consists of two branches of eigenvalues situated on the imaginary semiaxes and having a limit point at infinity.

For the sake of simplicity, in our further considerations we will assume that the fluid viscosity value, μ, is so large that the spectrum of the first problem (5.39) is real and does not contain intermediate eigenvalues (see Section 8.3.1); then, all the problem's eigenvalues are negative or positive. We suppose also that each eigenvalue of problem (5.39) is simple, that is, of multiplicity one.

Now we note that if in equations (5.38) it is assumed that $\boldsymbol{\xi} = \boldsymbol{\xi}(\alpha) = O(1)$, $\alpha \to 0$, then $\eta = \eta(\alpha) = O(\alpha)$, $\alpha \to 0$. Further, if it is assumed that $\eta = \eta(\alpha) = O(1)$, then $\boldsymbol{\xi} = \boldsymbol{\xi}(\alpha) = O(\alpha)$, $\alpha \to 0$. Note also that under the conditions formulated above, the eigenvalues and eigenelements of problem (5.38) are analytic functions depending of the parameter α and coincide with the solutions of problem (5.39) for $\alpha = 0$. An exact analysis shows that the series expansions should be not in powers of α, but rather in powers α^2.

Taking into account what was said up until now, we will consider the solutions of problem (5.38) in the form of two classes. The first class contains the solutions for which

$$\begin{aligned} \boldsymbol{\xi}(\alpha) &= \boldsymbol{\xi}_0 + \alpha^2 \boldsymbol{\xi}_2 + O(\alpha^4), \\ \eta(\alpha) &= \alpha \eta_1 + \alpha^3 \eta_3 + O(\alpha^5), \qquad \alpha \to 0, \end{aligned} \tag{5.40}$$

and the second one, the solutions of the following kind

$$\begin{aligned} \boldsymbol{\xi}(\alpha) &= \alpha \boldsymbol{\xi}_1 + \alpha^3 \boldsymbol{\xi}_3 + O(\alpha^5), \\ \eta(\alpha) &= \eta_0 + \alpha^2 \eta_2 + O(\alpha^4), \qquad \alpha \to 0. \end{aligned} \tag{5.41}$$

For $\alpha \to 0$, the solutions of (5.40) transform into the solutions of the first problem (5.39), and the solutions of (5.41) into those of the second problem (5.39).

Let us consider now the eigenvalues $\lambda(\alpha)$ of problem (5.38) of the form

$$\lambda(\alpha) = \lambda_0 + \alpha^2 \lambda_2 + O(\alpha^4), \qquad \alpha \to 0. \tag{5.42}$$

First, let us get asymptotic solutions of the first class. Substituting the expansions (5.40) and (5.42) into (5.38) and identifying the coefficients of the same powers of α, we come to the relations

$$\begin{aligned} l_0(\lambda_0)\boldsymbol{\xi}_0 &= \mathbf{0}, \\ l_0(\lambda_0)\boldsymbol{\xi}_2 &= \lambda_2 A_{11}\boldsymbol{\xi}_0 + \lambda_0 A_{12}\eta_1 - g(\Delta\rho)\lambda_2\lambda_0^{-2} B\boldsymbol{\xi}_0, \\ m_0(\lambda_0)\eta_1 &= -\frac{\lambda_0^2}{\rho_2 g} A_{21}\boldsymbol{\xi}_0. \end{aligned} \tag{5.43}$$

As previously mentioned, the first equation in (5.43) has a countable set of solutions $\lambda_0 = \lambda_k^+ \in \mathbb{R}_+$, with $\lambda_k^+ \to +\infty$ as $k \to \infty$ and

$$\boldsymbol{\xi}_0 = \boldsymbol{\xi}_k^+ \in \boldsymbol{J}_{0.S_1}(\Omega_1),$$

for which

$$\begin{aligned}
\|\boldsymbol{\xi}_k^+\|_{\boldsymbol{L}_2(\Omega_1)} &= 1, \\
(-(\lambda l_0(\lambda))'|_{\lambda=\lambda_k^+}\boldsymbol{\xi}_k^+, \boldsymbol{\xi}_k^+)_{\boldsymbol{L}_2(\Omega_1)} &= -\mu(\boldsymbol{\xi}_k^+,\ \boldsymbol{\xi}_k^+)_{\boldsymbol{L}_2(\Omega_1)} + 2\lambda_k^+(A_{11}\boldsymbol{\xi}_k^+,\ \boldsymbol{\xi}_k^+)_{\boldsymbol{L}_2(\Omega_1)} \\
&= \lambda_k^+[(A_{11}\boldsymbol{\xi}_k^+,\ \boldsymbol{\xi}_k^+)_{\boldsymbol{L}_2(\Omega_1)} \\
&\qquad - g(\Delta\rho)(\lambda_k^+)^{-2}(B\boldsymbol{\xi}_k^+,\ \boldsymbol{\xi}_k^+)_{\boldsymbol{L}_2(\Omega_1)}] \\
&> 0. \qquad (5.44)
\end{aligned}$$

For $\lambda_0 = \lambda_k^+$ the operator $m_0(\lambda_0) \gg 0$, and the last equation in (5.43) provides the solution

$$\eta_1 = \eta_{1k}^+ = -\frac{(\lambda_k^+)^2}{\rho_2 g} m_0^{-1}(\lambda_k^+)A_{21}\boldsymbol{\xi}_k^+. \qquad (5.45)$$

Since $l_0(\lambda_k^+)$ is a bounded self-adjoint operator, then the second equation in (5.43) is solvable if and only if its right side is orthogonal to the solutions of the homogeneous equation $l_0(\lambda_k^+)\boldsymbol{\xi}_k^+ = \boldsymbol{0}$. This, together with (5.45), leads to the formula

$$\lambda_2 = \lambda_{2k} = \frac{(\lambda_k^+)^3(m_0^{-1}(\lambda_k^+)A_{21}\boldsymbol{\xi}_k^+,\ A_{21}\boldsymbol{\xi}_k^+)_{H_2}}{\rho_2 g[(A_{11}\boldsymbol{\xi}_k^+,\ \boldsymbol{\xi}_k^+)_{\boldsymbol{L}_2(\Omega_1)} - g(\Delta\rho)(\lambda_k^+)^{-2}(B\boldsymbol{\xi}_k^+,\ \boldsymbol{\xi}_k^+)_{\boldsymbol{L}_2(\Omega_1)}]} \qquad (5.46)$$

Here, the numerator of the fraction is positive by virtue of the condition $m_0^{-1}(\lambda_k^+) \gg 0$, and the denominator is also positive by virtue of (5.44).

Now let us find asymptotic solutions of the second class. The substitution of (5.41) and (5.42) into (5.38) leads to

$$\begin{aligned}
m_0(\lambda_0)\eta_0 &= 0, \\
m_0(\lambda_0)\eta_2 &= -(\rho_2 g)^{-1}[2\lambda_0\lambda_2 A_{22}\eta_0 + \lambda_0^2 A_{21}\boldsymbol{\xi}_1], \\
l_0(\lambda_0)\boldsymbol{\xi}_1 &= \lambda_0 A_{12}\eta_0. \qquad (5.47)
\end{aligned}$$

From the first equation we get

$$\begin{aligned}
\lambda_0 &= \lambda_{0k}^{\pm} := \pm\mathrm{i}\frac{\lambda_k^{-1/2}(A_{22})}{\sqrt{\rho_2 g}}, \\
\eta_0 &= \eta_k^{\pm} = \eta_k(A_{22}), \qquad k \in \mathbb{N}, \qquad (5.48)
\end{aligned}$$

where $\lambda_k(A_{22})$ are eigenvalues of the compact positive operator A_{22} and $\eta_k(A_{22})$ are its normalized eigenelements. Since in this case the operator

$$l_0(\lambda_0) = l_0(\lambda_{0k}^{\pm}) = \mu I - \lambda_{0k}^{\pm} A_{11} - g(\Delta\rho)(\lambda_{0k}^{\pm})^{-1} B \tag{5.49}$$

is invertible, then from the third equation in (5.47) we have

$$\boldsymbol{\xi}_1 = \boldsymbol{\xi}_{1k}^{\pm} = \lambda_{0k}^{\pm} l_0^{-1}(\lambda_{0k}^{\pm}) A_{12}\eta_k^{\pm}. \tag{5.50}$$

Using again the orthogonality conditions for the right-hand side of the second equation in (5.47) and taking into account (5.48)–(5.50) we obtain

$$\begin{aligned}\lambda_2 = \lambda_{2k}^{\pm} &= \frac{(\lambda_{0k}^{\pm})^4}{2\rho_2 g} \left(l_0^{-1}(\lambda_{0k}^{\pm}) A_{12}\eta_k^{\pm}, A_{12}\eta_k^{\pm}\right)_{\boldsymbol{L}_2(\Omega_1)} \\ &= \frac{\lambda_k^{-2}(A_{22})}{2(\rho_2 g)^3} \left(l_0(\lambda_{0k}^{\pm})\boldsymbol{w}_k^{\pm}, \boldsymbol{w}_k^{\pm}\right)_{\boldsymbol{L}_2(\Omega_1)}, \qquad \boldsymbol{w}_k^{\pm} = l_0^{-1}(\lambda_{0k}^{\pm}) A_{12}\eta_k^{\pm} \neq \mathbf{0}.\end{aligned} \tag{5.51}$$

We note that from this equality and from (5.49) and (5.48) it follows that

$$\operatorname{Re}\lambda_{2k}^{\pm} > 0, \quad \operatorname{Im}\lambda_{2k}^{+} < 0, \quad \operatorname{Im}\lambda_{2k}^{-} > 0. \tag{5.52}$$

All these constructions together with formulas (5.46) and (5.51) lead to the following conclusions.

(1) In problem (5.38), for small $\alpha > 0$, there is a branch of eigenvalues $\lambda_{\infty k}(\alpha)$ of the form

$$\lambda_{\infty k}(\alpha) = \lambda_k^{+} + \alpha^2\lambda_{2k} + \mathrm{O}(\alpha^4), \qquad \alpha \to 0, \tag{5.53}$$

that is situated on the positive semiaxis and has a limit point at $+\infty$ as $k \to \infty$. The eigenvalues $\lambda_{\infty k}(\alpha)$ are situated to the right of the eigenvalues λ_k^{+} of the undisturbed problem (5.38) that corresponds to the case $\alpha = 0$, that is, the first problem (5.39).

The physical meaning of the zero approximation $\lambda_{\infty k}(0)$ is the following. These eigenvalues correspond to the interior dissipative waves in the considered partially dissipative hydrodynamic system. The waves correspond to the solidified free surface Γ_2 of the ideal fluid. According to (5.46), the second term in (5.53) describes the influence of the free surface Γ_2 oscillations on the damping decrements of the interior dissipative waves.

(2) Apparently, the shift of the eigenvalues $\lambda_{\infty k}(\alpha)$ to the right of $\lambda_{\infty k}(0)$ takes place not only for small $\alpha > 0$, but also for $0 < \alpha \leq 1$. Therefore, problem (5.35), which is a a particular case of problem (5.38) for $\alpha = 1$, will also have a branch of positive eigenvalues $\{\lambda_{\infty k}(1)\}_{k=1}^{\infty}$ with a limit point at $+\infty$. This branch corresponds to interior dissipative waves with arbitrarily large damping decrements.

(3) In problem (5.38), for small $\alpha > 0$, there are two branches of eigenvalues $\lambda_k^{\pm}(\alpha)$ adjoining the imaginary semiaxes, situated in the right complex half-plane and having a limit point at infinity,

$$\lambda_k^{\pm}(\alpha) = \pm i(\rho_2 g \lambda_k(A_{22}))^{-1/2} + \alpha^2 \lambda_{2k}^{\pm} + O(\alpha^4), \qquad \alpha \to 0. \tag{5.53'}$$

The eigenvalues $\lambda_k^{\pm}(0)$ correspond to the oscillation frequencies $\omega_k(0) = (\rho_2 g \lambda_k(A_{22}))^{-1/2}$ of the ideal fluid situated under the viscous one provided that the separating boundary Γ_1 between the two fluids is not disturbed. The influence of the boundary Γ_1 motion is described by the second term in (5.53′). We deal here with oscillation damping decrements of the ideal fluid due to the motion of the viscous fluid and the separation boundary Γ_1. These decrements have the form

$$\operatorname{Re} \lambda_k^{\pm}(\alpha) = \frac{\lambda_k^{-2}(A_{22})}{2(\rho_2 g)^3} \mu \left\| \boldsymbol{w}_k^{\pm} \right\|^2_{\boldsymbol{L}_2(\Omega_1)}. \tag{5.54}$$

The influence of the viscous fluid causes the oscillation frequencies of the ideal fluid, $\omega_k(\alpha)$, to decrease in comparison with the case of a solidified viscous fluid by

$$\begin{aligned} &\omega_k(0) - \omega_k(\alpha) \\ &\quad = \alpha^2 \frac{\lambda_k^{-2}(A_{22})}{2(\rho_2 g)^3} \\ &\quad \times \left[\frac{\lambda_k^{-1/2}(A_{22})}{\sqrt{\rho_2 g}} \left(A_{11} \boldsymbol{w}_k^{\pm}, \boldsymbol{w}_k^{\pm}\right)_{\boldsymbol{L}_2(\Omega_1)} + \frac{g(\Delta\rho)\sqrt{\rho_2 g}}{\lambda_k^{-1/2}(A_{22})} \left(B \boldsymbol{w}_k^{\pm}, \boldsymbol{w}_k^{\pm}\right)_{\boldsymbol{L}_2(\Omega_1)} \right]. \end{aligned} \tag{5.55}$$

(4) Apparently, the previously mentioned physical effects and properties of the frequencies and oscillation decrements take place not only for small $\alpha > 0$ but also for $0 < \alpha \leq 1$ and, in particular, in the partially dissipative system we are studying, for $\alpha = 1$. In this case, there are two branches of eigenvalues adjoining the imaginary semiaxes and having the limit point at ∞.

(5) These asymptotic and heuristic conclusions allow us to introduce the following hypothesis.

In the problem (5.53) on normal oscillations of a partiallly dissipative hydrosystem, the following assumptions are satisfied:

(a) The solutions of this problem are three branches of eigenvalues corresponding to two types of oscillation regimens that differ by their physical meaning. The three branches have a limit point at infinity.

(b) The branch $\{\lambda_{\infty k}\}_{k=1}^{\infty} \subset \mathbb{R}_+$ corresponds to the interior dissipative waves in the viscous fluid. The modes of normal oscillations corresponding to this branch, $\{\boldsymbol{\xi}_{\infty k}\}_{k=1}^{\infty}$, describe arbitrary quickly damping aperiodic regimens, for which the boudary surface Γ_1 that separates the two fluids is almost not disturbed. The eigenvalues corresponding to this branch have the following asymptotic behavior:

$$\lambda_{\infty k} = \nu \lambda_k(A)[1 + \mathrm{o}(1)], \qquad k \to \infty,\ \nu = \frac{\mu}{\rho_1}, \tag{5.56}$$

where $\lambda_k(A)$ are the eigenvalues of operator A appearing in the auxiliary Problem I in Section 10.2.2 [see (2.9)]. The oscillation modes $\{\boldsymbol{\xi}_{\infty k}\}_{k=1}^{\infty} \subset \boldsymbol{J}_{0,S_1}(\Omega_1)$ form a system that is complete in $\boldsymbol{J}_{0,S_1}(\Omega_1)$ up to a finite defect. If the values of viscosity μ are large enough, then the system $\{\boldsymbol{\xi}_{\infty k}\}_{k=1}^{\infty}$ is complete in $\boldsymbol{J}_{0,S_1}(\Omega_1)$.

(c) The two branches $\{\lambda_k^+\}_{k=1}^{\infty}$ and $\{\lambda_k^-\}_{k=1}^{\infty}$ adjoin the two imaginary semi-axes and their corresponding solutions describe the surface waves in the upper ideal fluid. In this regard, the oscillation frequencies of the partially dissipative hydrosystem are less than the oscillation frequencies of one ideal fluid with a solidified bottom Γ_1, and the oscillation decrements are small enough. The eigenelements $\{\eta_k^+\}_{k=1}^{\infty}$ and $\{\eta_k^-\}_{k=1}^{\infty}$ corresponding to the two branches of eigenvalues form a system that is complete in H_2 up to a finite defect. If the values of the viscosity μ of the lower fluid are large enough, then the two systems of eigenelements are complete in H_2. The asymptotic behavior of the eigenvalues $\lambda_k^{\pm}$ has the form

$$\lambda_k^{\pm} = \pm \mathrm{i}\left(g\lambda_k^{-1}(C_{22})\right)^{1/2}[1 + \mathrm{o}(1)], \qquad k \to \infty, \tag{5.57}$$

where C_{22} is the positive compact operator introduced in Section 10.2.2 [see (2.16)].

We remind the reader that in problem (5.35) there exists also a branch of eigenvalues $\{\lambda_k^0\}_{k=1}^{\infty} \subset \mathbb{R}_+$ with the limit point at zero and the asymptotic behavior described in (4.20). The properties of the solutions to problem (5.35) corresponding to this branch were studied in detail in Section 10.4. The corresponding oscillation modes describe the boundary waves at the boundary surface Γ_1 that separate the two fluids.

We leave the reader the opportunity to study in full detail the properties of the interior dissipative oscillations and also the surface oscillations in a partially dissipative hydrosystem. In particular, those approaches that allow obtaining solutions to the two classes of problems for equation (5.38) in the form of series with powers of α^2 and the proof of their convergence with $\alpha = 1$. Another approach could be the use of methods from perturbation theory to the nonself-adjoint problem (5.38) and its proof in the case $\alpha = 1$.

10.5.4 Concluding Remarks

In the conclusion of this chapter, let us say some words about the problem considered here. First and foremost this problem is new both in its hydrodynamic statement and as a problem in mathematical physics. In one part of the considered region (in Ω_1) the equation is parabolic, and in the other region (in Ω_2) the equation is elliptic. Further, the boundary conditions at the boundaries Γ_1 and Γ_2 include the time derivatives with respect to the unknown functions and, therefore, a spectral parameter in the study of normal oscillations. The authors believe that problems of this kind will capture the attention of other researchers.

These are some directions in which we believe that the studies should directed in the future.

(1) In Sections 10.4 and 10.5, the theorem on spectrum localization and the theorem on the completeness of the system of root elements in problem (4.41) were stated in a rather restrictive setting, assuming condition (4.63) corresponding to sufficiently small values of ρ_2/ρ_1. From the physical point of view, however, it is natural to assume that the stability condition $\rho_2 < \rho_1$ is the only one that is satisfied. It seems that the results stated in Sections 10.4 and 10.5 might be valid even under such less restrictive conditions.

(2) In Section 10.5 we stated the theorem on the completeness of the system of root elements in problem (4.41). We believe that the Riesz basis property and even the p-basis property of the system of root elements should take place in this case.

(3) Apparently, the Riesz basis property (with a defect for arbitrary values of the viscosity and without defect for large values of viscosity of the lower fluid) should occur in the space $\boldsymbol{J}_{0,S_1}(\Omega_1)$ for the system of eigenelements $\{\boldsymbol{\xi}_{\infty k}\}_{k=1}^{\infty}$ corresponding to the eigenvalues $\{\lambda_{\infty k}\}_{k=1}^{\infty}$. A similar Riesz basis property could be expected for the solution describing the surface waves at the free surface Γ_2.

(4) Most probably, it would be useful to study in a general setting some operator pencils of the form (5.36) and (5.37), where, for $\lambda \to \infty$, the terms considered to be disturbing terms with respect to the basis problem have the order for λ greater than that of the main terms.

Chapter 11

Oscillations of Visco-Elastic and Relaxing Media

As we mentioned in the Introduction to Part IV, this chapter deals with the problem on small oscillations of a visco-elastic or relaxing fluid. In Section 11.1 we will consider the most basic problem: The oscilations of a visco-elastic fluid in an arbitrary completely filled container. Based on the connection between the tensors of viscous stresses and deformation velocities, we formulate an initial boundary-value problem. This problem leads to a differential equation of the first order in a Hilbert space. Using the dissipativity of the operator coefficient of this equation we prove the theorem on correct solvability of the initial problem. Further, we study the normal oscillations of the system, prove that the spectrum is discrete, and show the existence of its limit points, situated on the positive semiaxis and caused by the presence of visco-elastic forces.

A natural generalization of the problem in Section 11.1 is obtained in Section 11.2. Here, we study an integro-differential equation, where instead of just one operator (the Stokes operator A_0), we introduce some other operators with the same power as the basic one. In this case, the theorem on correct solvability can be also stated easily, but the problem on normal oscillations becomes more complicated. Thus, even in the case of commuting operator coefficients, the spectrum may not be discrete, and some other interesting effects, such as the appearence of segments and points in the continuous (limiting) spectrum, may occur.

Section 11.3 deals with the problem on small oscillations of a visco-elastic fluid in an open container. Both physically and mathematically, this problem is a generalization of the problems discussed in Sections 8.1–8.3. However, even in this case, the transition to a differential equation of the first order with dissipative operator coefficients enables us to state the theorem on correct solvability of the initial boundary value problem and to study the problem on normal oscillations.

In Section 11.4 we consider the problems on multiple basicity of the system of eigen- and associated elements corresponding to the problem on normal oscillations. Here, we use the generalized Laptev and Greenlee method that involves a transition to a new spectral parameter (see Section 8.2.2). This allows us to employ (with the required changes and inevitable complications) the same research scheme that was previously used in Section 8.2 and, as a result, state the property of multiple basicity of a system of eigen- and associated elements of a special kind in the spectral problem.

Another approach to the study of the spectral problem on normal oscillations of a visco-elastic fluid in an open container is used in Section 11.5. Here, based on T. Ya. Azizov's general theorem on the basicity of the system of eigen- and associated elements of a J-self-adjoint operator in a Krein space, we state the properties of Riesz and p-basicity without a complex change of the spectral parameter. We also study the existence of various branches of eigenvalues, their limit points, their location in the complex plane, etc.

Finally, in Section 11.6, we investigate the problem on oscillations of a relaxing fluid in a region with boundary. The classical statement of the problem is given for model boundary conditions. Then, we carry out a transition to a differential equation of the third order in a Hilbert space. The simplest problem on normal oscillations is studied, and the existence of two kinds of waves, acoustic-relaxing and purely relaxing, is stated. The spectral problem for variable characteristics of a relaxing medium is studied, too. We show that here, along with the branches of the discrete spectrum, there is a segment in the limiting spectrum caused by the relaxation mechanism.

11.1 Visco-Elastic Fluids in Completely Filled Containers

In this section we consider the simplest problem on small motions and normal oscillations of a visco-elastic incompressible fluid that completely fills an arbitrary container. We show that the presence of the visco-elastic forces leads to new physical effects that were not met in an ordinary viscous fluid.

11.1.1 A Model of a Visco-Elastic Fluid

There is a great deal of incompressible homogeneous viscous fluids for which the connection between the tensor of viscous stresses, $\boldsymbol{\sigma}' = (\sigma'_{ij})^3_{i,j=1}$, and the doubled tensor of deformation velocities, $\boldsymbol{\tau} = (\tau_{ij}(\boldsymbol{u}))^3_{i.j=1}$, is no longer described by the simplest Hooke law,

$$\begin{aligned} \boldsymbol{\sigma}' &= \mu \boldsymbol{\tau} \\ \sigma'_{ij} &= \sigma_{ij} + P\delta_{ij}, \\ \tau_{ij}(\boldsymbol{u}) &:= \frac{\partial u_i}{\partial x_j} + \frac{\partial u_j}{\partial x_i}, \end{aligned} \tag{1.1}$$

as it was assumed in previous chapters of this book. For such fluids, a more complex dependence that takes into account the visco-elastic properties of the medium is needed. One of such models is the Oldroit model described, for instance, in F. Airich [1]. For this model, the connection between $\boldsymbol{\sigma}'$ and $\boldsymbol{\tau}$ has the following differential character:

$$\left(1 + \sum_{j=1}^{m} \eta_j \frac{\mathrm{d}^j}{\mathrm{d}t^j}\right) \boldsymbol{\sigma}' = \left(\kappa_0 + \sum_{j=1}^{m} \kappa_j \frac{\mathrm{d}^j}{\mathrm{d}t^j}\right) \boldsymbol{\tau}, \tag{1.2}$$

where $\kappa_0, \eta_1, \dots, \eta_m$, and $\kappa_1, \dots, \kappa_m$ are some positive physical constants.

In this model it is assumed that the roots $\{\gamma_j\}_{j=1}^m$ of the polynomial

$$E(\lambda) := 1 + \sum_{j=1}^{m} \eta_j (-\lambda)^j, \tag{1.3}$$

associated with the differential operator in the left-hand side of (1.2), are distinct and positive,

$$0 < \gamma_1 < \cdots < \gamma_m < \infty. \tag{1.4}$$

The model (1.2)–(1.4) for $m = 1$ was suggested and studied by Oldroit. In this chapter we will consider the case of an arbitrary natural number m.

From $E(\lambda)$ in (1.3) and the polynomial

$$F(\lambda) := \kappa_0 + \sum_{j=1}^{m} \kappa_j (-\lambda)^j, \tag{1.5}$$

associated with the differential operator in the right-hand side of (1.2), we form the rational function

$$\frac{F(\lambda)}{E(\lambda)} =: \mu I_0(\lambda) = \mu \left(1 + \sum_{j=1}^{m} \frac{\alpha_j}{\gamma_j - \lambda}\right), \tag{1.6}$$

where $\mu = \kappa_m / \eta_m > 0$. We will additionally assume that for the considered model of a visco-elastic fluid, the following conditions are fulfilled,

$$\alpha_j > 0, \qquad j = 1, \dots, m. \tag{1.7}$$

From the physical point of view, we need to accept one more natural requirement: If, at the initial moment of time, the tensor of deformation velocities in the fluid and its derivatives in time up to the order $m-1$ are equal to zero, then the same conditions are true for the tensor of viscous stresses as well. Under this assumption, the differential connection (1.2) with the initial conditions mentioned previously leads (as it is easily seen by means of Laplace transforms) to the integral connection

$$\boldsymbol{\sigma}' = \mu \hat{I}_0(t)\boldsymbol{\tau}, \tag{1.8}$$

where, according to (1.6), the operator $\hat{I}_0(t)$ acts by the rule

$$\left(\hat{I}_0\boldsymbol{\tau}\right)(t) := \boldsymbol{\tau}(t) + \sum_{j=1}^{m} \alpha_j \int_0^t \exp(-\gamma_j(t-s))\boldsymbol{\tau}(s)\mathrm{d}s. \tag{1.9}$$

If $\alpha_j = 0$, $j = 1, \dots, m$, then we come to the model of an ordinary viscous fluid.

Let us note that in (1.8) and (1.9) the connection between the tensor $\boldsymbol{\tau}$ and the tensor of stresses $\boldsymbol{\sigma}'$ is given by means of a Volterra integral operator of the second kind. This leads to the fact that for any t the operator $\hat{I}_0(t)$ is invertible. Moreover, the inverse operator, $\hat{I}_0^{-1}(t)$, is a Volterra integral operator of the second kind as well.

11.1.2 Statement of the Initial Boundary Value Problem

Let us consider now small motions of a visco elstic incompressible fluid corresponding to the model (1.2)–(1.9) in an arbitrary region Ω. Instead of the Navier-Stokes equations

$$\frac{\partial \boldsymbol{u}}{\partial t} = -\frac{1}{\rho}\nabla p + \nu\Delta\boldsymbol{u} + \boldsymbol{f}, \quad \mathrm{div}\boldsymbol{u} = 0 \qquad \text{in } \Omega,$$

in this case, the linearized motion equations take the form

$$\frac{\partial \boldsymbol{u}}{\partial t} = -\frac{1}{\rho}\nabla p + \nu\hat{I}_0(t)(\Delta\boldsymbol{u}) + \boldsymbol{f}, \quad \mathrm{div}\boldsymbol{u} = 0 \qquad \text{in } \Omega, \tag{1.10}$$

where $\nu = \mu/\rho > 0$ is the ordinary kinematic fluid viscosity and, based on (1.9), the operator $\hat{I}_0\boldsymbol{v}$ is defined as

$$\left(\hat{I}_0\boldsymbol{v}\right)(t,x) := \boldsymbol{v}(t,x) + \sum_{j=1}^{m} \alpha_j \int_0^t \exp(-\gamma_j(t-s))\boldsymbol{v}(s,x)\mathrm{d}s, \tag{1.11}$$

for any field $\boldsymbol{v}(t,x)$.

We need to add to equations (1.10) the initial and boundary value conditions, that is, the stickiness condition on the rigid wall,

$$\boldsymbol{u}(t,x) = \mathbf{0} \qquad \text{on } S = \partial\Omega, \tag{1.12}$$

and the initial condition

$$\boldsymbol{u}(0,x) = \boldsymbol{u}^0(x) \qquad \text{in } \Omega. \tag{1.13}$$

11.1.3 On the Solvability of the Initial Boundary Value Problem

Starting with equation (1.10) and the boundary value condition (1.12), we will assume—as we did in Chapter 7—that the velocities field $\boldsymbol{u}(t,x)$ is a smooth function of variable t with values in the subspace $\boldsymbol{J}_0(\Omega)$ of the Hilbert space $\boldsymbol{L}_2(\Omega)$. Applying to both sides of equation (1.10) the orthoprojector P_0 onto the subspace $\boldsymbol{J}_0(\Omega)$, introducing the Stokes operator A_0, and taking into account that $\rho^{-1}\nabla p \perp \boldsymbol{J}_0(\Omega)$, instead of (1.10)–(1.13) we obtain the Cauchy problem for an abstract integro-differential equation of the following form,

$$\frac{\mathrm{d}\boldsymbol{u}}{\mathrm{d}t} + \nu\left(A_0\boldsymbol{u} + \sum_{j=1}^{m}\alpha_j\int_0^t \exp(-\gamma_j(t-s))A_0\boldsymbol{u}(s)\mathrm{d}s\right)$$
$$= \boldsymbol{f}_0(t) := P_0\boldsymbol{f}(t), \qquad \boldsymbol{u}(0) = \boldsymbol{u}^0. \tag{1.14}$$

We used here the fact that $\hat{I}_0(t)$ and P_0 are commuting operators.

Let us now make the transition from (1.14) to a differential equation in the orthogonal sum of spaces $(\boldsymbol{J}_0(\Omega))^{m+1} := \bigoplus_{k=1}^{m+1}\boldsymbol{J}_{0,k}(\Omega)$, where $\boldsymbol{J}_{0,k}(\Omega) := \boldsymbol{J}_0(\Omega)$. For this, we introduce the new unknown functions

$$\begin{aligned}
&\boldsymbol{u}_0(t) := \boldsymbol{u}(t),\\
&\boldsymbol{u}_j(t) := (\nu\alpha_j)^{1/2}\int_0^t \exp(-\gamma_j(t-s))A_0^{1/2}\boldsymbol{u}_0(s)\mathrm{d}s, \qquad j=1,\dots,m.
\end{aligned} \tag{1.15}$$

Whence, and from (1.14), we obtain a system of linear differential equations for the functions $\boldsymbol{u}_j(t)$, $j=0,\dots,m$, that we could better write in a vector-matrix form. For example, for $m=2$ the system looks like

$$\frac{\mathrm{d}}{\mathrm{d}t}\begin{pmatrix}\boldsymbol{u}_0\\ \boldsymbol{u}_1\\ \boldsymbol{u}_2\end{pmatrix} + \begin{pmatrix}\nu A_0 & (\nu\alpha_1)^{1/2}A_0^{1/2} & (\nu\alpha_2)^{1/2}A_0^{1/2}\\ -(\nu\alpha_1)^{1/2}A_0^{1/2} & \gamma_1 I & 0\\ -(\nu\alpha_2)^{1/2}A_0^{1/2} & 0 & \gamma_2 I\end{pmatrix}\begin{pmatrix}\boldsymbol{u}_0\\ \boldsymbol{u}_1\\ \boldsymbol{u}_2\end{pmatrix} = \begin{pmatrix}\boldsymbol{f}_0(t)\\ \boldsymbol{0}\\ \boldsymbol{0}\end{pmatrix}, \tag{1.17}$$

where I is the identity operator in $\boldsymbol{J}_0(\Omega)$. The functions $\boldsymbol{u}_j(t)$ satisfy the initial conditions

$$\boldsymbol{u}_0(0) = \boldsymbol{u}^0, \quad \boldsymbol{u}_j(0) = \boldsymbol{0}, \qquad j=1,\dots,m. \tag{1.18}$$

The system of differential equations for $m \geq 2$ has a similar form; the basic matrix operator $\mathcal{A}$ corresponding to it has a block structure,

$$\mathcal{A} := \begin{pmatrix}\nu A_0 & A_{12}\\ -A_{12}^* & A_{22}\end{pmatrix}, \tag{1.19}$$

where $A_{12} = ((\nu\alpha_1)^{1/2}A_0^{1/2}, \dots, (\nu\alpha_m)^{1/2}A_0^{1/2})$ is a row matrix, $A_{22} = \operatorname{diag}(\gamma_k I)_{k=1}^m$, and the column matrix A_{12}^* is adjoint to the row matrix A_{12}.

We consider the Cauchy problem (1.17), (1.18), that is, the problem

$$\frac{d\boldsymbol{v}}{dt} + \mathcal{A}\boldsymbol{v} = \boldsymbol{\varphi}(t), \qquad \boldsymbol{v}(0) = \boldsymbol{v}^0. \tag{1.20}$$

Here, the matrix operator $\mathcal{A}$ for $m = 2$ is given by (1.17), and for $m \geq 2$ it has the structure presented in (1.19). Other notations are obvious as well: $\boldsymbol{v}(t) := (\boldsymbol{u}_0(t), \boldsymbol{u}_1(t), \dots, \boldsymbol{u}_m(t))^{\mathrm{t}}$, $\boldsymbol{\varphi}(t) := (\boldsymbol{f}_0(t), \boldsymbol{0}, \dots, \boldsymbol{0})^{\mathrm{t}}$, and $\boldsymbol{v}^0 := (\boldsymbol{u}^0, \boldsymbol{0}, \dots, \boldsymbol{0})^{\mathrm{t}}$. It is easy to see that operator $\mathcal{A}$ is defined on the dense subset

$$\mathcal{D}(\mathcal{A}) := \left\{\boldsymbol{v} \in (\boldsymbol{J}_0(\Omega))^{m+1} : \boldsymbol{u}_0 \in \mathcal{D}(A_0), \boldsymbol{u}_i \in \mathcal{D}\left(A_0^{1/2}\right), i = 1, \dots, m\right\} \tag{1.21}$$

of the space $(\boldsymbol{J}_0(\Omega))^{m+1}$ and has the important property that $(-\mathcal{A})$ is a maximal dissipative operator. Indeed, from (1.19) we get the following representation:

$$\begin{aligned} &\mathcal{A} = \mathcal{A}_0 + \mathrm{i}\mathcal{A}_1, \\ &\mathcal{A}_1 = \mathrm{i}\begin{pmatrix} 0 & -\mathrm{i}A_{12} \\ \mathrm{i}A_{12}^* & 0 \end{pmatrix} = \mathcal{A}_1^*, \\ &\mathcal{A}_0 = \operatorname{diag}(\nu A_0, \gamma_1 I, \dots, \gamma_m I) = \mathcal{A}_0^*, \end{aligned} \tag{1.22}$$

and, therefore, for all $\boldsymbol{v} \in \mathcal{D}(\mathcal{A})$, we get

$$\begin{aligned} &\operatorname{Re}(\mathcal{A}\boldsymbol{v}, \boldsymbol{v}) = \nu \left\|A_0^{1/2}\boldsymbol{u}_0\right\|_{\boldsymbol{J}_0(\Omega)}^2 + \sum_{k=1}^m \gamma_k \|\boldsymbol{u}_k\|_{\boldsymbol{J}_0(\Omega)}^2 \geq c\|\boldsymbol{v}\|^2, \\ &c = \min\{\nu\lambda_1(A_0), \gamma_1, \dots, \gamma_m)\} > 0. \end{aligned} \tag{1.23}$$

Since $\mathcal{A}^* = \mathcal{A}_0 - \mathrm{i}\mathcal{A}_1$, we obtain similarly that

$$\operatorname{Re}(\mathcal{A}^*\boldsymbol{w}, \boldsymbol{w}) \geq c\|\boldsymbol{w}\|^2, \qquad \text{for all } \boldsymbol{w} \in \mathcal{D}(\mathcal{A}^*). \tag{1.24}$$

Thus, (1.20) is a non-homogeneous Cauchy problem for a differential equation of the first order with the maximal dissipative operator $(-\mathcal{A})$. The statements in Section 1.5.4 ensure that the semigroup $\mathcal{U}(t)$ corresponding to problem (1.20) is contractive and allows the estimate

$$\|\mathcal{U}(t)\| \leq \exp(-ct), \tag{1.25}$$

where the constant $c > 0$ is defined by (1.23). As a result, we obtain that the homogeneous problem (1.20) is uniformly correct.

Furthermore, according to the conclusion of Section 1.5.7, if $\boldsymbol{v}^0 \in \mathcal{D}(\mathcal{A})$ and $\boldsymbol{\varphi}(t)$ in $(\boldsymbol{J}_0(\Omega))^{m+1}$ is continuously differentiable, the formula

$$\boldsymbol{v}(t) = \mathcal{U}(t)\boldsymbol{v}^0 + \int_0^t \mathcal{U}(t-s)\boldsymbol{\varphi}(s)ds \tag{1.26}$$

gives the solution to problem (1.20), that is, $\boldsymbol{v}(t) \in \mathcal{D}(\mathcal{A})$ for every $t \in [0,T]$ and any $T > 0$, $\boldsymbol{v}'(t) \in (\boldsymbol{J}_0(\Omega))^{m+1}$, and equation (1.20) is satisfied for each $t \in [0,T]$.

The preceding considerations lead to the following result.

If in problem (1.14), $\boldsymbol{u}^0 \in \mathcal{D}(A_0)$ and $\boldsymbol{f}_0(t)$ is a continuously differentiable function with values in $\boldsymbol{J}_0(\Omega)$, then this problem has a solution $\boldsymbol{u}(t) \in \mathcal{D}(A_0)$ for all $t \in [0,T]$, and $\boldsymbol{u}'(t) \in \boldsymbol{J}_0(\Omega)$. If $\boldsymbol{u}^0 \in \boldsymbol{J}_0(\Omega)$ and $\boldsymbol{f}_0(t)$ is a continuous function of t with values in $\boldsymbol{J}_0(\Omega)$, then (1.26) gives a generalized solution of problem (1.20) and the first component of the function $\boldsymbol{v}(t) := (\boldsymbol{u}_0(t), \boldsymbol{u}_1(t), \dots, \boldsymbol{u}_m(t))^t$ is a generalized solution of problem (1.14).

These statements give a positive answer to the question on correct solvability of the initial boundary value problem (1.10)–(1.13) on small motions of a visco-elastic fluid in a completely filled container.

11.1.4 Normal Oscillations

Let us consider now solutions of the homogenous problem (1.20) that depend on t according to the law $\exp(-\lambda t)$. For the amplitude functions $\boldsymbol{v} = (\boldsymbol{u}_0, \boldsymbol{u}_1, \dots, \boldsymbol{u}_m)^t$ we obtain the spectral problem

$$\mathcal{A}\boldsymbol{v} = \lambda \boldsymbol{v}, \tag{1.27}$$

that is, the system of equations,

$$\begin{aligned} \nu A_0 \boldsymbol{u}_0 + \sum_{k=1}^m (\nu\alpha_k)^{1/2} A_0^{1/2} \boldsymbol{u}_k &= \lambda \boldsymbol{u}_0, \\ -(\nu\alpha_k)^{1/2} A_0^{1/2} \boldsymbol{u}_0 + \gamma_k \boldsymbol{u}_k &= \lambda \boldsymbol{u}_k, \qquad k = 1, \dots, m. \end{aligned} \tag{1.28}$$

Eliminating $\boldsymbol{u}_k$ from (1.28), we obtain the following problem,

$$\nu I_0(\lambda) A_0 \boldsymbol{u}_0 = \lambda \boldsymbol{u}_0, \tag{1.29}$$

where $I_0(\lambda)$ is defined by (1.6). Hence, it follows that the eigenelements of problem (1.29) should coincide with the eigenelements $\{\boldsymbol{u}_n(A_0)\}_{n=1}^\infty$ of the Stokes operator A_0, and the eigenvalues λ are solutions of the characteristic equations

$$I_0(\lambda) = \frac{\lambda}{\nu\lambda_n(A_0)}, \qquad n = 1, 2, \dots, \tag{1.30}$$

where the equations (1.30) are set up with regard to the multiplicity of the eigenvalues $\lambda_n(A_0)$.

Equation (1.30) can be easily studied graphically by building the graph of the function $I_0(\lambda)$ in (1.6) and the graph of the linear function $\lambda/(\nu\lambda_n(A_0))$. We now list some properties of the solutions of this equation and formulate the physical conclusions that they lead us to.

1° For any $n \geq 1$, equation (1.30) has $m+1$ roots, among which no more than two can be nonreal complex conjugate, and all the remaining ones are positive.

2° Let β_k, $k = 1, \ldots, m$, be the zeros of the function $I_0(\lambda)$, that is, of the polynomial $F(\lambda)$ in (1.5). Since $\alpha_k > 0$, as indicated in Section 11.1.1, the following inequalities are valid,

$$0 < \gamma_1 < \beta_1 < \cdots < \gamma_m < \beta_m < \infty. \tag{1.31}$$

For any $n \geq 1$ there are no more than two real roots situated to the left of the point $\lambda = \gamma_1$. For large enough n, these roots are absent. Therefore, there is only a finite number of such roots situated to the left of γ_1, for all $n \geq 1$. The remaining roots are all situated to the right of the point $\lambda = \gamma_1$.

3° For large enough n, the complex conjugate pairs of nonreal eigenvalues are absent and, therefore, the total number of nonreal eigenvalues in problem (1.30) is finite.

4° The eigenvalues of problem (1.30) for all n may be divided into $m+1$ series having the limit points β_k, $k = 1, \ldots, m$, and $+\infty$. The eigenvalues $\lambda_n^{(k)}$ corresponding to the the kth series are situated in the interval (β_k, γ_{k+1}), where γ_{m+1} is, by definition, the multiple root of equation (1.30), with the right side equal to $a\lambda$, for some $a > 0$, situated to the right of β_m. The eigenvalues $\lambda_n^{(\infty)}$ corresponding to the series whose limit point is $+\infty$ are situated in the interval (γ_{m+1}, ∞).

5° For the series $\lambda_n^{(k)}$, the following asymptotic formula takes place,

$$\lambda_n^{(k)} = \beta_k + \beta_k \left(\nu I_0'(\beta_k)\lambda_n(A_0)\right)^{-1} + \mathrm{O}\left(\lambda_n^{-2}(A_0)\right), \qquad n \to \infty, k = 1, \ldots, m. \tag{1.32}$$

6° For the series $\lambda_n^{(\infty)}$, the following formula takes place, as $n \to \infty$,

$$\lambda_n^{(\infty)} = \nu\lambda_n(A_0)\left[1 - \left(\sum_{k=1}^{m} \alpha_k\right)(\nu\lambda_n(A_0))^{-1} + \mathrm{O}\left(\lambda_n^{-2}(A_0)\right)\right]. \tag{1.33}$$

7° To the series $\lambda_n^{(\infty)}$ with the asymptotic behavior (1.33) there correspond dissipative waves in the visco-elastic fluid that are similar to the ordinary dissipative waves in an incompressible viscous homogeneous fluid (see Section 7.1.3). For these waves, the fading decrements can be as large as possible, although the presence in the hydrosystem of the elastic forces can decrease somewhat the values of the decrements, as suggested by (1.33) for $\alpha_k > 0$.

8° To the series $\{\lambda_n^{(k)}\}_{n=1}^{\infty}$, $k = 1, \ldots, m$, there corresponds a new kind of waves caused by the action of the visco-elastic forces in the system. The fading decrements corresponding to these waves are located inside finite intervals of the positive semiaxis, and have only m limit points (see (1.32)), that is, the number of integral terms in the motion equations (1.10) and (1.11). Thus, in the spectral problem (1.27) on normal oscillations of a visco-elastic fluid, we find an essential spectrum consisting of exactly m points β_k, with the properties (1.31).

9° The presence of visco-elastic forces can also account for the fact that in the given hydrodynamic system we find only a finite numbers of oscillating fading modes of the normal oscillations, together with the aperiodically fading modes.

Thus, the visco-elastic forces are the cause of new physical effects that are not characteristic for an ordinary viscous incompressible fluid filling completely a certain container. In the following sections we will consider some more complicated mathematical problems related to the problem of small oscillations in a visco-elastic fluid.

11.2 Abstract Evolution and Spectral Problems Generated by Small Motions of a Visco-Elastic Fluid

Here, the spectral and evolution problems are considered. They generalize the problem of Section 11.1 on small motions and normal oscillations of a visco-elastic fluid in a container, in the case when the operators of the problem, basic and integro-differential, include different differential operators of equal force with one and the same domain of definition. Even for commuting operators entering into the evolution equation, the spectral problem may have regions of the essential spectrum situated on the real axis.

11.2.1 Statement of the Problem. Transition to an Equation with a Dissipative Operator

In a separable Hilbert space H, let us consider the Cauchy problem for an integro-differential equation of the form

$$\frac{du}{dt} + A_0 u + \sum_{k=1}^{m} \int_0^t e^{-\gamma_k (t-\tau)} A_k u(\tau) d\tau = f(t), \qquad u(0) = u^0. \tag{2.1}$$

Here, $u = u(t)$ is an unknown function with values in H, γ_k are positive constants, $0 < \gamma_1 < \cdots < \gamma_m < \infty$, $f(t)$ is a given function with values in H, and $u^0 \in H$.

By A_k, $k = 0, \ldots, m$, in (2.1) we denote $m + 1$ unbounded positive definite operators for which

$$\mathcal{D}(A_k) = \mathcal{D}(A_0), \qquad k = 1, \ldots, m, \tag{2.2}$$

$$0 < A_k^{-1} \in \mathfrak{S}_\infty, \qquad k = 0, \ldots, m. \tag{2.3}$$

As an example of operators of this kind in the Hilbert space $H = L_2(\Omega)$, $\Omega \subset \mathbb{R}^3$, it is possible to take a set of uniformly elliptic operators,

$$A_k u := -\sum_{i,j=1}^{3} \frac{\partial}{\partial x_i} \left(a_{ij}^{(k)}(x) \frac{\partial u}{\partial x_j} \right), \qquad k = 0, \ldots, m, \tag{2.4}$$

given on one and the same domain

$$\mathcal{D}(A_k) := H_0^2(\Omega) := \left\{ u(x) \in H^2(\Omega) : u = 0 \text{ on } \partial\Omega \right\}. \tag{2.5}$$

Here, all the energetic norms $\|u\|_{A_k}$ are equivalent to the norm of the space $H_0^1(\Omega)$, and consequently equivalent to each other.

If, in particular, $H = \mathbf{J}_0(\Omega)$ and $A_k = \alpha_k A_0$, where $\alpha_k > 0$, $k = 1, \ldots, m$, then we recover the problem analyzed in Section 11.1 for the operator $A_0 = -P_0 \Delta$, that is, for the Stokes operator.

Just as in Section 11.1, we proceed from problem (2.1) to a differential equation in the orthogonal sum of spaces $\tilde{H} := \bigoplus_{k=0}^{m} H_k$, $H_k := H$, by introducing the unknown functions

$$\begin{aligned} u_0(t) &:= u(t), \\ u_k(t) &:= \int_0^t e^{-\gamma_k (t-\tau)} A_k^{1/2} u(\tau) d\tau, \qquad k = 1, \ldots, m. \end{aligned} \tag{2.6}$$

If $u(t)$ is a solution of problem (2.1), that is, a continuously differentiable function of t that for every t belongs to $\mathcal{D}(A_0) = \mathcal{D}(A_k)$, then the functions $u_k(t)$ from (2.6) are continuously differentiable and

$$\frac{\mathrm{d}u_k}{\mathrm{d}t} = A_k^{1/2} u_0(t) - \gamma_k u_k(t), \quad u_k(0) = 0, \qquad k = 1, \dots, m. \tag{2.7}$$

Together with (2.1), relations (2.7) lead to the next Cauchy problem in the space $\tilde{H}$,

$$\begin{aligned}
&\frac{\mathrm{d}\tilde{u}}{\mathrm{d}t} + \mathcal{A}\tilde{u} = \tilde{f}(t), \quad \tilde{u}(0) = \tilde{u}^0, \\
&\tilde{u}(t) := (u_0(t), u_1(t), \dots, u_m(t))^{\mathrm{t}}, \\
&\tilde{f}(t) = (f(t), 0, \dots, 0)^{\mathrm{t}}, \\
&\tilde{u}^0 = \left(u^0, 0, \dots, 0\right)^{\mathrm{t}},
\end{aligned} \tag{2.8}$$

where the $(m+1) \times (m+1)$ matrix operator $\mathcal{A}$ has the block form

$$\mathcal{A} = \begin{pmatrix} A_{11} & A_{12} \\ A_{21} & A_{22} \end{pmatrix}, \tag{2.9}$$

with

$$\begin{aligned}
A_{11} &= A_0, \\
A_{12} &= \left(A_1^{1/2}, \dots, A_m^{1/2}\right), \\
A_{21} &= -\left(A_1^{1/2}, \dots, A_m^{1/2}\right)^{\mathrm{t}}, \\
A_{22} &= \operatorname{diag}\left(\gamma_k I\right)_{k=1}^m.
\end{aligned} \tag{2.10}$$

11.2.2 On the Solvability of the Cauchy Problem for Integro-Differential Equations

The formulas (2.9), (2.10) show that the operator $\mathcal{A}$ defined on the set dense in $\tilde{H}$

$$\mathcal{D}(\mathcal{A}) = \mathcal{D}(A_0) \bigoplus \left(\bigoplus_{k=1}^m \mathcal{D}\left(A_k^{1/2}\right)\right), \qquad \mathcal{D}\left(A_k^{1/2}\right) = \mathcal{D}\left(A_0^{1/2}\right), \tag{2.11}$$

is in essence maximal dissipative, since

$$(A_{12})^* = -A_{21}, \quad (\mathcal{A}_0 \tilde{u}, \tilde{u})_{\tilde{H}} \geq c \left\|\tilde{u}\right\|_{\tilde{H}}^2, \tag{2.12}$$

where

$$\mathcal{A}_0 := \operatorname{diag}\left(A_0, A_{22}\right), c = \min\left(\lambda_1\left(A_0\right), \min_k \gamma_k\right) > 0.$$

Therefore, problem (2.8)–(2.10), for $\tilde{u}^0 \in \mathcal{D}(\mathcal{A})$ and any continuously differentiable function $\tilde{f}(t)$ with values in $\tilde{H}$, is uniquely solvable and its solution $\tilde{u}(t)$ is expressed by the formula

$$\tilde{u}(t) = \mathcal{U}(t)\tilde{u}^0 + \int_0^t \mathcal{U}(t-\tau)\tilde{f}(\tau)\mathrm{d}\tau, \tag{2.13}$$

where the semigroup $\mathcal{U}(t)$ allows the estimate

$$\|\mathcal{U}(t)\| \leq \exp(-ct), \tag{2.14}$$

with a constant c as in (2.12).

If in the original integro-differential equation (2.1) $f(t)$ is a continuously differentiable function with values in H, and $u^0 \in \mathcal{D}(A_0)$, then $\tilde{u}^0 = (u^0, 0, \dots, 0)^{\mathrm{t}} \in \mathcal{D}(\mathcal{A})$, and $\tilde{f}(t) = (f(t), 0, \dots, 0)^{\mathrm{t}}$ is a continuously differentiable function with values in $\tilde{H}$. Then, equations (2.7) are fulfilled for any t together with the equation

$$\frac{\mathrm{d}u_0}{\mathrm{d}t} + A_0 u_0(t) + \sum_{k=1}^{m} A_k^{1/2} u_k(t) = f(t), \tag{2.15}$$

where all their terms are continuous in t.

From equations (2.7) by the condition $\tilde{u}(t) \in \mathcal{D}(A)$, we get that $u_0(t) \in \mathcal{D}(A_0)$, $u_k(t) \in \mathcal{D}(A_k^{1/2})$, and thus, $\mathrm{d}u_k/\mathrm{d}t \in \mathcal{D}(A_k^{1/2})$. From (2.7) it also follows that $u_k(t)$ are expressed by the formulas (2.6). Substituting them into (2.15) we conclude that equation (2.1) is fulfilled for the function $u = u_0(t)$, where all terms in (2.1) are continuous functions.

Thus, if $u^0 \in \mathcal{D}(A_0)$ and $f(t)$ is continuously differentiable with values in H, then problem (2.1) has the solution $u(t)$, for which $u(t) \in \mathcal{D}(A_0)$ and all terms in (2.1) are continuous functions.

11.2.3 Spectral Problem. Transition to an Equation with a Bounded Operator

We consider solutions of the homogeneous problem (2.8) depending on t according to the law $\exp(-\lambda t)$. For the amplitude elements $\tilde{u} = (u_0, u_1, \dots, u_m)^{\mathrm{t}}$ we come to the spectral problem

$$\mathcal{A}\tilde{u} = \lambda\tilde{u}, \qquad \tilde{u} \in \mathcal{D}(\mathcal{A}), \tag{2.16}$$

where the operator matrix is defined by formulas (2.9) and (2.10).

We note at once that from properties (2.11) and (2.12) it follows that the spectrum of problem (2.16) is situated in the half-plane

$$\operatorname{Re}\lambda \geq c > 0. \tag{2.17}$$

From these properties we get

$$\operatorname{Re}\,(\mathcal{A}\tilde{u},\tilde{u})_{\tilde{H}} \geq c\|\tilde{u}\|^2_{\tilde{H}}, \tag{2.18}$$

and, therefore, operator $\mathcal{A}$ has a bounded inverse operator, $\mathcal{A}^{-1}$, for which

$$\|\mathcal{A}^{-1}\| \leq c^{-1}. \tag{2.19}$$

We will next derive explicit formulas for the operator $\mathcal{A}^{-1}$ when $m = 1$, and also when $m \geq 2$.

For $m = 1$, consider the equation $\mathcal{A}\tilde{u} = \tilde{v}$; we have

$$\begin{aligned} A_0 u_0 + A_1^{1/2} u_1 &= v_0, \\ -A_1^{1/2} u_0 + \gamma_1 u_1 &= v_1. \end{aligned} \tag{2.20}$$

From the second equation we find $u_1 = \gamma_1^{-1}(v_1 + A_1^{1/2}u_0)$. Substituting into the first equation we come to the relation

$$Au_0 := \left(A_0 + \gamma_1^{-1}A_1\right)u_0 = v_0 - \gamma_1^{-1}A_1^{1/2}v_1. \tag{2.21}$$

Since the operators A_0 and A_1 are positive definite and $\gamma_1 > 0$, then operator A is positive definite, and

$$u_0 = A^{-1}\left(v_0 - \gamma_1^{-1}A_1^{1/2}v_1\right).$$

Therefore, we get the next formula for the operator matrix $\mathcal{A}^{-1}$,

$$\mathcal{A}^{-1} = \begin{pmatrix} A^{-1} & -\gamma_1^{-1}\left(A_1^{1/2}A^{-1}\right)^* \\ \gamma_1^{-1}\left(A_1^{1/2}A^{-1}\right) & \gamma_1^{-1}\left(I - \gamma_1^{-1}A_1^{1/2}A^{-1}A_1^{1/2}\right) \end{pmatrix}, \qquad m = 1. \tag{2.22}$$

It can be rewritten in a more convenient form by observing that

$$\gamma_1^{-1}\left(I - \gamma_1^{-1}A_1^{1/2}A^{-1}A_1^{1/2}\right) = B^{-1}, \qquad B := \gamma_1 I + A_1^{1/2}A_0^{-1}A_1^{1/2}. \tag{2.23}$$

We prove property (2.23) assuming, if necessary, that instead of products of bounded and unbounded operators we take their closures. We first check that

$$I = \gamma_1^{-1}B\left(I - \gamma_1^{-1}A_1^{1/2}A^{-1}A_1^{1/2}\right). \tag{2.24}$$

Using the definition (2.23) for operator B we have

$$\begin{aligned}
&\gamma_1^{-1}B\left(I-\gamma_1^{-1}A_1^{1/2}A^{-1}A_1^{1/2}\right)\\
&\quad=\left(I+\gamma_1^{-1}A_1^{1/2}A_0^{-1}A_1^{1/2}\right)\left(I-\gamma_1^{-1}A_1^{1/2}A^{-1}A_1^{1/2}\right)\\
&\quad=I+\gamma_1^{-1}A_1^{1/2}A_0^{-1}A_1^{1/2}-\gamma_1^{-1}A_1^{1/2}A^{-1}A_1^{1/2}-\gamma_1^{-2}A_1^{1/2}A_0^{-1}A_1A^{-1}A_1^{1/2}\\
&\quad=:I+\gamma_1^{-1}C.
\end{aligned}$$

We next prove that $C=0$. In fact,

$$\begin{aligned}
C&=A_1^{1/2}\left(A_0^{-1}-A^{-1}-\gamma_1^{-1}A_0^{-1}A_1A^{-1}\right)A_1^{1/2}\\
&=A_1^{1/2}A_0^{-1}\left(A-A_0-\gamma_1^{-1}A_1\right)A^{-1}A_1^{1/2}=0,
\end{aligned}$$

by the definition (2.21) of operator A.

Similarly we check the property

$$I=\gamma_1^{-1}\left(I-\gamma_1^{-1}A_1^{1/2}A^{-1}A_1^{1/2}\right)B, \tag{2.25}$$

and (2.23) follows from (2.24) and (2.25).

The final expression for the operator matrix (2.22), with $m=1$, takes the form

$$\begin{aligned}
&\mathcal{A}^{-1}=\begin{pmatrix} A^{-1} & \gamma_1^{-1}\left(A_1^{1/2}A^{-1}\right)^* \\ \gamma_1^{-1}\left(A_1^{1/2}A^{-1}\right) & B^{-1}\end{pmatrix},\\
&A:=A_0+\gamma_1^{-1}A_1,\\
&B:=\gamma_1 I+A_1^{1/2}A_0^{-1}A_1^{1/2}.
\end{aligned} \tag{2.26}$$

Now we assume that $m\geq 1$ is arbitrary and use the block representation (2.9), (2.10) of the operator matrix $\mathcal{A}$. We introduce the notation $\tilde{u}:=(u_0,u_1,\ldots,u_m)^{\mathrm{t}}=:(u_0,\hat{u})^{\mathrm{t}}$, $\hat{u}:=(u_1,\ldots,u_m)^{\mathrm{t}}\in H^m$. Then, equation $\mathcal{A}\tilde{u}=\tilde{v}$ for $m\geq 1$ is equivalent to the system

$$\begin{aligned}
A_{11}u_0+A_{12}\hat{u}&=v_0,\\
-A_{12}^*u_0+A_{22}\hat{u}&=\hat{v},
\end{aligned} \tag{2.27}$$

where the first property (2.12) has been already taken into account.

Bearing in mind that the operators $A_{11}=A_0$ and $A_{22}=\operatorname{diag}(\gamma_k I)_{k=1}^m$ are invertible, from the first and second equations (2.27) we find

$$u_0=A_{11}^{-1}(v_0-A_{12}\hat{u}),\quad \hat{u}=A_{22}^{-1}(\hat{v}+A_{12}^*u_0). \tag{2.28}$$

Substituting the expression for u_0 from (2.28) into the second equation (2.27), and the expression for $\hat{u}$ into the first equation (2.27), we come to the relationships

$$\begin{aligned} Au_0 &:= \left(A_{11} + A_{12}A_{22}^{-1}A_{12}^*\right) u_0 = v_0 - A_{12}A_{22}^{-1}\hat{v}, \\ B\hat{u} &:= \left(A_{22} + A_{12}^*A_{11}^{-1}A_{12}\right) \hat{u} = \hat{v} + A_{12}^*A_{11}^{-1}v_0. \end{aligned} \tag{2.29}$$

From the definition of the operators A and B and the properties of the operators A_{11} and A_{22}, it follows that A and B are positive definite operators, and consequently the operator matrix $\mathcal{A}^{-1}$ takes the form

$$\mathcal{A}^{-1} = \begin{pmatrix} A^{-1} & -A^{-1}A_{12}A_{22}^{-1} \\ B^{-1}A_{12}^*A_{11}^{-1} & B^{-1} \end{pmatrix}. \tag{2.30}$$

Here, as in the previous calculations, we use the standard rules of multiplication of square matrices by column or row vectors.

As in the case when $m = 1$, that is, for the matrix (2.26), the matrix (2.30) may be represented in the more symmetric form

$$\mathcal{A}^{-1} = \begin{pmatrix} A^{-1} & -A^{-1}A_{12}A_{22}^{-1} \\ A_{22}^{-1}A_{12}^*A^{-1} & B^{-1} \end{pmatrix}, \tag{2.31}$$

if we note that

$$B^{-1}A_{12}^*A_{11}^{-1} = A_{22}^{-1}A_{12}^*A^{-1}. \tag{2.32}$$

This property follows from the property

$$A_{12}^*A_{11}^{-1}A = BA_{22}^{-1}A_{12}^*,$$

which is checked by using the definitions (2.29) of the operators A and B.

Based on the formulas (2.26) for $m = 1$ and (2.31) for any $m \geq 1$, from problem (2.16) we derive its equivalent spectral problem

$$\mathcal{A}^{-1}\tilde{u} = \mu\tilde{u}, \qquad \mu = 1/\lambda, \tag{2.33}$$

in which the operator $\mathcal{A}^{-1}$ is bounded. The additional study of problems (2.16) and (2.33), presented below, is based on the theory of self-adjoint operators in spaces with indefinite metrics and on the theory of self-adjoint operator functions.

11.2.4 Properties of Operator Coefficients of the Spectral Problem

Before studying problem (2.16), (2.33), we introduce some additional assumptions with respect to the operators A_k, $k = 1, \dots, m$. We will assume further that every operator $A_k \gg 0$ generates an energetic space H_{A_k} with the norm equivalent to the norm of the energetic space H_{A_0} of the operator A_0, that is,

$$0 < c_k \leq \frac{\left\| A_k^{1/2} u \right\|_H^2}{\left\| A_0^{1/2} u \right\|_H^2} \leq d_k < \infty, \qquad k = 1, \dots, m, u \in \mathcal{D}\left(A_0^{1/2}\right). \tag{2.34}$$

These properties were considered earlier in the example (2.4), (2.5).

From (2.34) we get the following important conclusions with respect to the operator elements of the matrix (2.26) and (2.31).

(a) The operators $A_k^{1/2} A_0^{-1/2}$ and $A_0^{-1/2} A_k^{1/2}$ are bounded with bounded inverse operators, and

$$c_k \leq \frac{\left\| A_k^{1/2} A_0^{-1/2} u \right\|_H^2}{\|u\|_H^2} \leq d_k, \qquad k = 1, \dots, m, \tag{2.35}$$

$$c_k \leq \frac{\left\| A_0^{-1/2} A_k^{1/2} u \right\|_H^2}{\|u\|_H^2} \leq d_k, \qquad k = 1, \dots, m. \tag{2.36}$$

Indeed, inequalities (2.35) are obtained from (2.34) by substituting the element $v = A_0^{-1/2} u$ with $u \in H$ instead of $u \in \mathcal{D}(A_0^{1/2})$. Further, the right inequalities (2.36) follow from the same inequalities (2.35) by taking into account that the norm of a bounded operator coincides with the norm of its adjoint operator. The left inequalities (2.36) are obtained from (2.34) by substituting, instead of $u \in \mathcal{D}(A_0^{1/2}) = \mathcal{D}(A_k^{1/2})$, the element $v = A_k^{-1/2} u$ with $u \in H$, estimating the norm of the operator $A_0^{1/2} A_k^{-1/2}$ and the norm of its adjoint, and then by performing the change $(A_0^{1/2} A_k^{-1/2})^* v = A_k^{-1/2} A_0^{1/2} v = w$.

(b) The operator B from (2.23) is bounded and positive definite in H,

$$(\gamma_1 + c_1) I \leq B \leq (\gamma_1 + \mathrm{d}_1) I. \tag{2.37}$$

Indeed,

$$(Bu, u)_H = \gamma_1 \|u\|_H^2 + \left\| A_0^{-1/2} A_1^{1/2} u \right\|_H^2,$$

and, therefore, inequalities (2.37) follow from (2.36).

(c) The operator $A^{-1} = (A_0 + \gamma_1^{-1} A_1)^{-1}$ is compact, positive, and has the representation

$$A^{-1} = A_0^{-1/2} K^{-1} A_0^{-1/2},$$
$$K = I + \gamma_1^{-1} A_0^{-1/2} A_1 A_0^{-1/2}, \quad \left(1 + \gamma_1^{-1} c_1\right) I \leq K \leq \left(1 + \gamma_1^{-1} d_1\right) I. \tag{2.38}$$

Formulas (2.38) follow from $A = A_0^{1/2}(I + \gamma_1^{-1} A_0^{-1/2} A_1 A_0^{-1/2}) A_0^{1/2}$ and the estimates (2.35).

(d) The off diagonal entries in the matrix (2.26) are compact operators, and the operator B^{-1} is bounded and positive definite.

In fact, by (2.38), the operator $A_1^{1/2} A^{-1} = (A_1^{1/2} A_0^{-1/2}) K^{-1} A_0^{-1/2}$ is the product of the bounded operators $A_1^{1/2} A_0^{-1/2}$ (formula (2.35)), K^{-1} (formula (2.38)), and the compact operator $A_0^{-1/2}$. The properties of boundedness and positive definiteness of operator B^{-1} follow from (2.37).

Similar properties take place for the operators A and B from the representation (2.31) of the operator matrix $\mathcal{A}^{-1}$ for any $m \geq 1$.

(e) The operator $A^{-1} = (A_{11} + A_{12} A_{22}^{-1} A_{12}^*)^{-1}$ admits the representation

$$A^{-1} = A_0^{-1/2} K^{-1} A_0^{-1/2},$$
$$K = I + \sum_{k=1}^{m} \gamma_k^{-1} A_0^{-1/2} A_k A_0^{-1/2},$$
$$\left(1 + \sum_{k=1}^{m} \gamma_k^{-1} c_k\right) \leq K \leq \left(1 + \sum_{k=1}^{m} \gamma_k^{-1} d_k\right) I. \tag{2.39}$$

Indeed, using the definitions (2.10) of the matrix block A_{ik} of the operator $\mathcal{A}$, from (2.9) we have

$$\begin{aligned} A &= A_0 + \sum_{k=1}^{m} \gamma_k^{-1} A_k \\ &= A_0^{1/2} \left(I + \sum_{k=1}^{m} \gamma_k^{-1} A_0^{-1/2} A_k A_0^{-1/2} \right) A_0^{1/2} \\ &= A_0^{1/2} K A_0^{1/2}, \end{aligned}$$

whence, by using (2.35), we get the formulas and inequalities (2.39).

(f) Operator B from (2.29) is bounded and positive definite in the space $H^m = H \oplus \ldots \oplus H$.

In fact, based on the definitions (2.9), (2.10), this operator has the representation

$$\begin{aligned} B &= \operatorname{diag}\,(\gamma_k I)_{k=1}^m + C^*C, \quad 0 < \gamma_1 < \ldots < \gamma_m, \\ C :&= \left(A_0^{-1/2}A_1^{1/2}, \ldots, A_0^{-1/2}A_m^{1/2}\right), \end{aligned} \tag{2.40}$$

where C, according to Property (a), is a bounded row matrix, and C^* is the adjoint column matrix. Therefore, from (2.40) we get the inequalities

$$(\gamma_1 + \alpha_-)\, I \le B \le (\gamma_m + \alpha_+)\, I, \tag{2.41}$$

where $\alpha_- \ge 0$ and $\alpha_+ \ge \alpha_-$ are lower and upper bounds of the matrix operator C^*C.

(g) In matrix (2.31), the off diagonal elements $A_{12}^{(-1)} := -A^{-1}A_{12}A_{22}^{-1}$ and $A_{21}^{(-1)} := A_{22}^{-1}A_{12}^*A^{-1}$ are compact operators and are related by the equation

$$A_{21}^{(-1)} = -\left(A_{12}^{(-1)}\right)^*. \tag{2.42}$$

Indeed, the connection (2.42) follows from the definition of these elements, the first formula (2.11), and the properties of the operators A_{22}^{-1} and A^{-1}. Let us prove the property of compactness, for example, for $A_{21}^{(-1)}$. We have, by the definitions (2.10) and representation (2.39),

$$\begin{aligned} A_{21}^{(-1)} &= \operatorname{diag}\,(\gamma_k^{-1}I)_{k=1}^m \left(A_1^{1/2}, \ldots, A_m^{1/2}\right)^{\mathrm{t}} A_0^{-1/2}K^{-1}A_0^{-1/2} \\ &= \operatorname{diag}\,(\gamma_k^{-1}I)_{k=1}^m \left(A_1^{1/2}A_0^{-1/2}, \ldots, A_m^{1/2}A_0^{-1/2}\right)^{\mathrm{t}} K^{-1}A_0^{-1/2}. \end{aligned} \tag{2.43}$$

Using again Properties (a) and (e), we conclude that $A_{21}^{(-1)}$ is a product of bounded operators—the first three factors in (2.43)—and the compact (positive) operator $A_0^{-1/2}$, that is, it is a compact operator.

11.2.5 Properties of Solutions of the Spectral Problem

Based on the previously stated Properties (a)–(g) of the operator elements of the matrices (2.26) and (2.31), we determine properties of the solutions of problems (2.16), (2.33), both for $m = 1$ and for any $m \geq 1$.

1° The spectrum of problem (2.16) is symmetric relatively to the real axis and situated in the half-plane $\operatorname{Re}\lambda \geq c > 0$, where c is the constant from inequality (2.12).

The location of the spectrum was mentioned previously [see formula (2.17)]. The symmetry relatively to the real axis follows from the fact that, by definitions (2.9)–(2.11), $\mathcal{A}$ is a $\mathcal{J}$-self-adjoint operator for

$$\mathcal{J} := \operatorname{diag}(I, -I, \dots, -I).$$

2° Problem (2.33) and, therefore, problem (2.16) have no more than a finite number of nonreal eigenvalues.

Indeed, for problem (2.33), the matrix

$$\mathcal{A}^{-1} = \begin{pmatrix} A^{-1} & A_{12}^{(-1)} \\ -\left(A_{12}^{(-1)}\right)^* & B^{-1} \end{pmatrix}, \tag{2.44}$$

according to Properties (e), (f), (g) has as elements the compact operators A^{-1}, $A_{12}^{(-1)}$, $-(A_{12}^{(-1)})^*$, and the bounded positive definite operator B^{-1}. Thus, according to the assumptions in Section 1.3.5, operator $\mathcal{A}^{-1}$ being $\mathcal{J}$-self-adjoint with respect to the above mentioned operator $\mathcal{J}$, has the non-negative and non-positive invariant subspaces $L_\pm \subset H^{m+1}$.

If $K_+ : H \to H^m$ is the angular operator of the subspace L_+, then

$$L_+ = \{\tilde{u} := (u_0, \hat{u})^{\mathrm{t}} \in \tilde{H} : \tilde{u} = (u_0, K_+ u_0)^{\mathrm{t}}, u_0 \in H\}.$$

For any $\tilde{u} = (u_0, K_+ u_0)^{\mathrm{t}}$ we have $\mathcal{A}^{-1}\tilde{u} \in L_+$, which leads to the equation

$$B^{-1}K_+ = K_+ A^{-1} + \left(A_{12}^{(-1)}\right)^* + K_+ A_{12}^{(-1)} K_+. \tag{2.45}$$

Since the operators A^{-1}, $A_{12}^{(-1)}$, and $(A_{12}^{(-1)})^*$ are compact, and operator B, according to Property (f), is bounded, then $K_+ \in \mathfrak{S}_\infty$. Therefore, by the same reasoning developed in Section 9.2.6 for proving Property 7°, we conclude that problem (2.33) may have no more than a finite number of nonreal eigenvalues. If $\|K_+\| < 1$, then problem (2.33) has no nonreal eigenvalues (see Property 8° in Section 9.2.6).

3° The limiting spectrum of problem (2.33) coincides with the limiting spectrum of operator B^{-1} and, therefore, the limiting spectrum of problem (2.16) coincides with the limiting spectrum of operator B.

The proof of this property is similar to the proof done in Section 6.5.7 for the limiting (essential) spectrum of the problem on eigenoscillations of an ideal rotating fluid partially filling an arbitrary container.

Namely, from (2.33) and (2.44) we have

$$\begin{aligned} A^{-1}u_0 + A_{12}^{(-1)}\hat{u} &= \mu u_0, && u_0 \in H, \mu = \lambda^{-1} \\ -\left(A_{12}^{(-1)}\right)^* u_0 + B^{-1}\hat{u} &= \mu\hat{u}, && \hat{u} = (u_1, \dots, u_m)^{\mathrm{t}} \in H^m. \end{aligned} \tag{2.46}$$

If $\mu \notin \sigma(A^{-1})$, then from the first equation we find $u_0 = -(A^{-1} - \mu I)^{-1}A_{12}^{(-1)}\hat{u}$. Substituting this relation into the second equation (2.46) we come to the problem

$$M(\mu)\hat{u} := \left[B^{-1} + \left(A_{12}^{(-1)}\right)^* \left(A^{-1} - \mu I\right)^{-1} A_{12}^{(-1)}\right]\hat{u} = \mu\hat{u}. \tag{2.47}$$

Here, the second term on the left side is a compact operator, for each fixed $\mu \notin \sigma(A^{-1})$ and by virtue of the above mentioned assertion from Section 6.5.7, the limiting spectrum of problem (2.47) coincides with the limiting spectrum of operator B.

4° In the region $\lambda \notin \sigma(B) \subset \mathbb{R}$, the spectrum of problem (2.16) is discrete and may have as limit points for (finite multiplicity) eigenvalues the point $\lambda = \infty$ and also points of the set $\sigma(B)$. In this, to the points in $\sigma(B)$, the branches of eigenvalues in the discrete spectrum may come only along the real axis (positive half-axis).

To prove this property we state a similar property for problem (2.46). If $\mu = \lambda^{-1} \notin \sigma(B^{-1})$ and $\mu \neq 0$, then by eliminating the element $\hat{u}$ from (2.46), after dividing by μ we come to the problem on eigenvalues $L(\mu)u_0 = 0$ for the operator pencil

$$\begin{aligned} L(\mu) &:= I - \Phi(\mu), \\ \Phi(\mu) &:= -\mu^{-1}\left(A^{-1} + A_{12}^{(-1)}\left(B^{-1} - \mu I\right)^{-1}\left(A_{12}^{(-1)}\right)^*\right). \end{aligned} \tag{2.48}$$

Since $\Phi(\mu)$ for $\mu \notin \sigma(B^{-1})$, $\mu \neq 0$, takes compact values, and $\Phi(\mu) \geq 0$ for $\mu < 0$, then $L(\mu)$ is invertible on the negative half-axis, that is, it is a Fredholm pencil. Therefore, according to the assertion of Section 1.6.3, the problem $L(\mu)u_0 = 0$ has a discrete spectrum with possible limit points for (finite multiplicity) eigenvalues either in the set $\sigma(B^{-1})$ or at $\mu = 0$. Hence, because of the connection $\mu = \lambda^{-1}$, for the eigenvalues of problem (2.46) or (2.16) we get the formulated above conclusion on the discreteness of the spectrum.

Since, according to Property 2°, problems (2.16), (2.46) may have no more than a finite number of nonreal eigenvalues, then the latter assertion in Property 4° immediately follows.

5° If in problem (2.16) the conditions

$$A_k = \alpha_k A_0, \qquad \alpha_k > 0, k = 1, \ldots, m, \tag{2.49}$$

are satisfied, then the problem has a discrete spectrum consisting of m series of finite multiplicity eigenvalues $\lambda_n^{(k)} \to \beta_k$ as $n \to \infty$, $k = 1, \ldots, m$, and the series $\lambda_n^{(\infty)} \to +\infty$ as $n \to \infty$. Their asymptotic behavior for $n \to \infty$ is given by formulas (1.32), (1.33). Here, β_k are infinitely multiple eigenvalues of the operator B from (2.29) in the considered case (2.49), for which the next inequalities take place

$$0 < \gamma_1 < \beta_1 < \cdots < \gamma_m < \beta_m < \infty. \tag{2.50}$$

Indeed, under condition (2.49), we get back to the situation described in Section 11.1 for $\nu = 1$ for the hydrodynamic problem on normal oscillations of a visco-elastic fluid in a vessel. There, the numbers β_k are defined as zeroes of the function $I_0(\lambda) := 1 + \sum_{k=1}^{m} \alpha_k/(\gamma_m - \lambda)$ [see (1.6)] and they have the properties (2.50). Therefore, according to Property 3°, the operator B in (2.29) has the points $\{\beta_k\}_{k=1}^{m}$ as points of the limiting spectrum. However, from formula (2.29) for operator B, definitions (2.10), and from (2.49) it follows that in this case

$$B = \operatorname{diag}(\gamma_k I)_{k=1}^{m} + (\gamma_{ik})_{i.k=1}^{m}, \qquad \gamma_{ik} = \alpha_i^{1/2}\alpha_k^{1/2} I. \tag{2.51}$$

Therefore, the problem on eigenvalues for operator B, that is, the problem

$$B\hat{u} = \lambda\hat{u}, \qquad \hat{u} = (u_1, \ldots, u_m)^{\mathrm{t}} \in H^m, \tag{2.52}$$

leads to the ordinary characteristic equation

$$\det\left((\gamma_k - \lambda)\,\delta_{ik} + \gamma_{ik}\right)_{i,k=1}^{m} = 0, \tag{2.53}$$

that has exactly m roots, which, by virtue of the previous comments coincide with the numbers β_k. Simultaneously we get that each of these roots is infinitely multiple, that is, belongs to the limiting spectrum of operator B.

Thus, in the simplest case (2.49) related to the hydrodynamic problem of Section 11.1, the spectrum of problem (2.16) is discrete and divided into $m + 1$ series of eigenvalues. It turns out that such a property of the spectrum superimposes rigid limitations on the connection of the operators A_k with the operator A_0. If the operators A_k are less closely connected with A_0, then the structure of the spectrum of problem (2.16) may be different, and, in particular, there may appear intervals in the limiting spectrum.

6° In problem (2.16) for $m = 1$ the commuting operators A_0 and A_1 may be chosen in such a way that the complete spectrum of the operator B will be the limiting spectrum and coincide with the segment $[\gamma_1 + 1, \gamma_1 + 2]$; for $m = 2$, $\gamma_1 = \gamma_2$, the commuting operators A_0, A_1, and A_2 may be chosen in such a way that the spectrum of operator B will be its limiting spectrum and coincide with the set $\{\gamma_1\} \cup [\gamma_1 + 2, \gamma_1 + 4]$.

Let us first consider the case $m = 1$. We choose the operators A_0 and A_1 such that they commute and the set of acummulation points of the spectrum of the operator $A_1^{1/2} A_0^{-1} A_1^{1/2}$ coincides with the segment $[1, 2]$. Let $\lambda_1 \geq \lambda_2 \geq \cdots > 0$ be the ordered set of eigenvalues of the operator A_1^{-1}, where every eigenvalue is simple. Since operator A_1^{-1} has to commute with operator A_0^{-1}, then A_0^{-1} has the same eigenelements as A_1^{-1}.

We choose the unknown countable set of eigenvalues of the operator $A_1^{1/2} A_0^{-1} A_1^{1/2}$ according to the following rule: $\alpha_{01} = 1$; $\alpha_{11} = 1$, $\alpha_{12} = 2$; $\alpha_{21} = 1$, $\alpha_{22} = 3/2$, $\alpha_{23} = 2$;... Here, in every following group one new value is added. That values is the midpoint of two neighboring numbers; for example, $\alpha_{31} = 1$, $\alpha_{32} = 5/4$, $\alpha_{33} = 3/2$, $\alpha_{34} = 2$, $\alpha_{41} = 1$, $\alpha_{42} = 5/4$, $\alpha_{43} = 3/2$, $\alpha_{44} = 7/4$, $\alpha_{45} = 2$, etc. We relabel these numbers as a sequence $\{\beta_n\}_{n=1}^{\infty}$, set $\mu_n := \beta_n \lambda_n$, and assume that μ_n are the eigenvalues of operator A_0^{-1} corresponding to the same eigenelements to which there correspond the eigenvalues λ_n for the operator A_1^{-1}. Then, the operator $A_1^{1/2} A_0^{-1} A_1^{1/2}$ is diagonal and its eigenvalues $\{\beta_n\}$ form a dense set in the interval $[1, 2]$, where each eigenvalue β_n by its construction is infinitely multiple.

Since, according to formula (2.23), $B = \gamma_1 I + A_1^{1/2} A_0^{-1} A_1^{1/2}$, then by virtue of the above proved facts, the spectrum of operator B coincides with its limiting spectrum and equals the segment $[\gamma_1 + 1, \gamma_1 + 2]$.

A similar contruction may be done in the case when $m = 2$ and $\gamma_2 = \gamma_1$. Here, it is necessary to put $A_2 = A_1$ and introduce, just as above, the numbers λ_n, β_n, and μ_n. Again the operator $A_1^{1/2} A_0^{-1} A_1^{1/2} = A_2^{1/2} A_0^{-1} A_2^{1/2}$ is diagonal and its eigenvalues β_n form a dense set in the segment $[1, 2]$; the operators $A_2^{1/2} A_0^{-1} A_1^{1/2}$, $A_1^{1/2} A_0^{-1} A_2^{1/2}$, coinciding with it, possess the same property. Since in this case

$$B = \operatorname{diag}(\gamma_1 I, \gamma_1 I) + (\gamma_{ik})_{i,k=1}^{2}, \qquad \gamma_{ik} = A_i^{1/2} A_0^{-1} A_k^{1/2},$$

then $B = \bigoplus_{n=1}^{\infty} B_n$, with

$$B_n = \begin{pmatrix} \gamma_1 & 0 \\ 0 & \gamma_1 \end{pmatrix} + \begin{pmatrix} \beta_n & \beta_n \\ \beta_n & \beta_n \end{pmatrix}.$$

Hence, the eigenvalues of operator B_n, for any n, are equal to γ_1 and $\gamma_1 + 2\beta_n$, respectively. Whence it follows that the complete spectrum of the operator $B = \bigoplus_{n=1}^{\infty} B_n$ is exactly its limiting spectrum and coincides with the set $\{\gamma_1\} \cup [\gamma_1 + 2, \gamma_1 + 4]$.

All these considerations show how to construct examples of problems of the form (2.16) in the case of commuting operators A_k, with a rather complicated structure of the limiting spectrum.

In connection with Properties 5° and 6°, the next natural question arises: To what extent may the operators A_k in problem (2.16) be different from the simplest operators $\alpha_k A_0$, when the general structure of the spectrum of problem (2.16) is the same as it was described in Property 5°. The answer to this question is given by some sufficient conditions formulated below in Properties 7° and 8°.

7° Let in problem (2.16) the operators A_k have the structure

$$A_k = \alpha_k A_0 + C_k, \qquad C_k = A_0^{1/2} F_k A_0^{1/2}, F_k \in \mathfrak{S}_\infty, \tag{2.54}$$

and assume that the following conditions are fulfilled:

(a) the operators

$$\tilde{A}_{0,k} := \sum_{k=1}^{m} \frac{F_k}{(\beta_k - \gamma_k)} + \beta_k A_0^{-1}, \qquad k_1 = 1, \dots, m, \tag{2.55}$$

are infinite dimensional, and

(b)

$$\tilde{A}_k := I - \left[I_0'(\beta_k)\right]^{-1} T_1 \gg 0, \qquad T_1 := A_0^{-1} - \sum_{k=1}^{m} \frac{F_k}{(\beta_k - \gamma_k)^2}. \tag{2.56}$$

Then, the spectrum of problem (2.16) is discrete and may be divided into $m+1$ series of positive eigenvalues with limit points $\{\beta_k\}_{k=1}^m$ and $+\infty$, where the numbers β_k are the same as in Property 5°.

To prove this property we derive from the system of equations (2.16) written as

$$\begin{aligned} A_0 u_0 + \sum_{k=1}^{m} A_k^{1/2} u_k &= \lambda u_0, \\ -A_j^{1/2} u_0 + \gamma_j u_j &= \lambda u_j, \qquad j = 1, \dots, m, \end{aligned} \tag{2.57}$$

a single equation in u_0. This can be done since the numbers $\lambda = \gamma_j$ are not solutions of the problem (2.57). Thus we get the spectral problem

$$A_0 u_0 + \sum_{k=1}^{m} (\gamma_k - \lambda)^{-1} A_k u_0 - \lambda u_0 = 0. \tag{2.58}$$

We next make the substitution $A_0^{1/2} u_0 = v$, apply the operator $A_0^{-1/2}$ to the left and right sides of (2.58), and use formulas (2.54); we have

$$\left(I_0(\lambda) I + \sum_{k=1}^{m} (\gamma_k - \lambda)^{-1} F_k - \lambda A_0^{-1} \right) v = 0,$$

$$I_0(\lambda) = 1 + \sum_{k=1}^{m} \frac{\alpha_k}{(\gamma_k - \lambda)}, \qquad I_0(\beta_k) = 0, \qquad k = 1, \ldots, m. \tag{2.59}$$

After dividing by $I_0(\lambda)$, we get an operator pencil of the form

$$L(\lambda) := I + I_0^{-1}(\lambda) \left(\sum_{k=1}^{m} (\gamma_k - \lambda)^{-1} F_k - \lambda A_0^{-1} \right) := I + \Phi(\lambda), \tag{2.60}$$

where $\Phi(\lambda)$ takes compact values.

It is easy to see that for each negative λ, with a large enough absolute value, the operator $L(\lambda)$ has the estimate

$$L(\lambda) \geq \left(1 - \sum_{k=1}^{m} \frac{\|F_k\|}{|\gamma_k - \lambda|} \right) I \geq cI, \qquad c > 0,$$

and, therefore, $L(\lambda)$ is a Fredholm pencil. Whence it follows that the spectrum of problem (2.57) is discrete and consists of infinitely multiple eigenvalues, with limit points in the complex plane $\mathbb{C}$ among the points where $\Phi(\lambda)$ is not analytic, that is, among the points $\{\gamma_k\}_{k=1}^m$, $\{\beta_k\}_{k=1}^m$, and $\lambda = \infty$.

We can show however that the points $\{\gamma_k\}_{k=1}^m$ are not limit points of the spectrum of problem (2.57). Assume, on the contrary, that there exists a sequence of eigenvalues $\lambda_{kn} \to \gamma_k$ as $n \to \infty$ for some k, to which correspond normalized eigenelements $\tilde{u}_n = (u_{0n}, u_{1n}, \ldots, u_{mn})^{\mathrm{t}}$ of problem (2.57). From the equation with the number $j = k$ we get

$$A_k^{1/2} u_{0n} = (\gamma_k - \lambda_{kn})\, u_{kn} \to 0, \qquad n \to \infty.$$

Whence, by inequalities (2.34), it also follows that $A_j^{1/2} u_{0n} \to 0$ for $j \neq k$, and $A_0^{1/2} u_{0n} \to 0$. Further, for $j \neq k$ we have $(\gamma_j - \lambda_{kn}) u_{jn} = A^{1/2} u_{0n} \to 0$ whence, since $\gamma_j \neq \gamma_k$, it follows that $u_{jn} \to 0$ for $j \neq k$. From the first equation (2.57) after multiplying by $A_0^{-1/2}$ we get

$$A_0^{-1/2}A_k^{1/2}u_{kn} = -A_0^{1/2}u_{0n} - \sum_{j\neq k} A_0^{-1/2}\left(A_j^{1/2}u_{jn}\right) + \lambda_{0n}A_0^{-1/2}u_{0n} \to 0, \qquad n\to\infty,$$

and since by (2.36) $A_0^{-1/2}A_k^{1/2}$ is a bounded and boundedly invertible operator, then $u_{kn} \to 0$ as $n\to\infty$. Together with the properties $u_{0n} \to 0$, and $u_{jn} \to 0$, $j\neq k$, whence it follows that $\|\tilde{u}_n\|^2_{\tilde{H}} = \|u_{0n}\|^2_H + \sum_{k=1}^m \|u_{kn}\|^2|_H \to 0$, contrary to the assumption $\|\tilde{u}\|_{\tilde{H}} = 1$.

Now we show that the point $\lambda = \infty$ is a limit point of the discrete spectrum of problem (2.57), that is, of problem (2.16). We substitute in (2.59) the spectral parameter $\lambda = \mu^{-1}$ and multiply both sides by μ; we have

$$M(\mu)v := \left(\mu I - A_0^{-1} - \mu^2 M_1(\mu)\right)v = 0,$$
$$M_1(\mu) := \sum_{k=1}^m \frac{\alpha_k I + F_k}{1-\mu\gamma_k}. \tag{2.61}$$

Since $M(0) = -A_0^{-1} \in \mathfrak{S}_\infty$, $M'(0) = I \gg 0$, then according to Statement 3° in Section 1.6.10, the eigenelements of problem (2.61) corresponding to eigenvalues from the interval $(-\varepsilon,\varepsilon)$, for any $\varepsilon > 0$, form a Riesz basis in a subspace having a finite defect in the infinite dimensional space H. By the arbitrariness of ε we get that there exists a branch of eigenvalues $\{\mu_n\}_{n=1}^\infty$ of problem (2.61) with the limit point $\mu = 0$. Therefore, problem (2.59) has a branch of eigenvalues $\lambda_n^{(\infty)} = 1/\mu_n$ with the limit point ∞. From Properties 1° and 2° it follows that this branch consists of positive numbers and therefore has the limit point $\lambda = +\infty$.

It can be similarly proved that the numbers β_k are limit points of the discrete spectrum of problem (2.16); here it is necessary to apply the conditions (a) and (b). We substitute in (2.59) $\lambda = \beta_k + \mu$ choosing any number $k = 1, 2, \ldots, m$. Taking into account that $I_0(\beta_k) = 0$ we come to the spectral problem

$$M_k(\mu)v := \left(-\tilde{A}_{0k} + \mu I_0'(\beta_k)\tilde{A}_{1k} + \tilde{A}_{2k}(\mu)\right)v = 0,$$
$$\tilde{A}_{2k}(\mu) := M_{0k}(\mu)I + \sum_{k=1}^m F_k\psi_k(\mu) = O(\mu^2), \qquad \mu\to 0,$$
$$M_{0k}(\mu) := I_0(\beta_k+\mu) - I_0'(\beta_k)\mu = O(\mu^2),$$
$$\psi_k(\mu) := (\gamma_k - \beta_k - \mu)^{-1} - (\gamma_k - \beta_k) - (\gamma_k-\beta_k)^2\mu = O(\mu^2), \tag{2.62}$$

where $\tilde{A}_{0k}$ and $\tilde{A}_k$ are the operators occuring in conditions (a) and (b) [see formulas (2.55) and (2.56)]. Taking into consideration these conditions and the property $I_0'(\beta_k) > 0$, for problem (2.62) it is possible to repeat the considerations presented above for the pencil (2.61) and to prove that to the point $\lambda = \beta_k$ in problem (2.16) there corresponds a branch of eigenvalues $\{\lambda_n^{(k)}\}_{k=1}^{\infty}$ situated on the positive half-axis and having the property $\lambda_n^{(k)} \to \beta_k$ as $n \to \infty$.

Property 7° is entirely proved. Incidentally, we also proved the Riesz basicity with a finite defect in H of each of the system of eigenelements corresponding, respectively, to the branches of eigenvalues $\{\lambda_n^{(k)}\}$, $k = 1, \dots, m$, and $\{\lambda_n^{(\infty)}\}$.

8° Let the conditions in Property 7° be fulfilled [see (2.54)–(2.56)], and assume that $\operatorname{Ker} \tilde{A}_{0k} = \{0\}$, $k = 1, \dots, m$, the eigenvalues $\lambda_j(A_0)$ of operator A_0 have the asymptotic behavior

$$\lambda_j(A_0) = a^{-1/\tilde{\alpha}_0} j^{1/\tilde{\alpha}_0}[1 + o(1)], \qquad j \to \infty, a_0 > 0\ , \tilde{\alpha}_0 > 0, \tag{2.63}$$

and the eigenvalues of the operators $\tilde{A}_{0k}$ from (2.55) have the asymptotic behavior

$$\lambda_j^{\pm}\left(\tilde{A}_{0k}\right) = \pm\left(a_k^{\pm}\right)^{1/\alpha_k^{\pm}} j^{-1/\alpha_k^{\pm}}[1 + o(1)], \qquad j \to \infty, a_k^{\pm} > 0, \alpha_k^{\pm} > 0, k = 1, \dots, m. \tag{2.64}$$

Then, the branch $\{\lambda_j^{(\infty)}\}$ has the asymptotic behavior

$$\lambda_j^{(\infty)} = \lambda_j\left(A_0\right)[1 + o(1)], \qquad j \to \infty, \tag{2.65}$$

and each of the branches $\{\lambda_n^{(k)}\}$ may be divided into two subbranches, $\lambda_j^{(k,\pm)}$, situated to the right and left of the point β_k and having the asymptotic behavior

$$\lambda_j^{(k,\pm)} = \beta_k + \lambda_j^{\pm}\left(\tilde{A}_{0k}\right)\left(I_0'(\beta_k)\right)^{-1}[1 + o(1)], \qquad j \to \infty, k = 1, \dots, m. \tag{2.66}$$

Instead of proving these properties, we note that they follow from the facts presented in Section 1.6.8, formulas (2.63) and (2.64), and also from the fact that the operator pencils (2.61) and (2.62), after substituting $\mu = \tilde{\lambda}^{-1}$, get the form of the pencil (1.7.6) and satisfy the requirements of Section 1.6.8.

We remark also that the asymptotic behavior of the branches $\lambda_j^{(k,\pm)}$ is determined not only by the properties of operator A_0, but also by properties of the compact operators F_k from (2.54). In particular, from (2.55) it follows that if $F_k \geq 0$, then, by virtue of the inequalities $\beta_k - \gamma_k > 0$, we have $\tilde{A}_{0k} > 0$, and, therefore, the branch $\lambda_j^{(k,-)}$ from (2.66) is absent. Besides, the main term in the asymptotic formula (2.64) may be defined not by the operator A_0^{-1} but by one or several operators F_k.

Let us get rid of the constraints (2.54)–(2.56) and consider again the general problem (2.16) under the assumptions (2.34). Since this time, according to Property 6°, the spectrum, generally speaking, cannot be split into $m+1$ series of eigenvalues, as it was for the conditions in 7° and 8°, then one raises the question whether it is possible to single out a branch of eigenvalues in the spectrum to which there corresponds a system of eigenelements forming a Riesz basis (p-basis) in H, with a finite defect.

9° In problem (2.16) it is possible to single out a branch of positive eigenvalues $\{\lambda_n^{(\infty)}\}$ with the limit point $\lambda = +\infty$, such that the projections onto H of the corresponding eigenelements form a Riesz basis with a finite defect in the space H. If the condition

$$A_0^{-1} \in \mathfrak{S}_{p_0} \tag{2.67}$$

is fulfilled, then the mentioned basis is a p-basis (with a finite defect) in H for $p = 2p_0$.

To prove this property, let us return to problem (2.46) with $\mu = \lambda^{-1}$, and recall that the operators A^{-1} and $A_{12}^{(-1)}$ are compact, and B^{-1} is a positive definite bounded operator. As in the proof of Property 2°, we introduce the nonnegative maximal subspace $L_+ \subset H^{m+1}$, invariant relatively to the operator $\mathcal{A}^{-1}$ in (2.33), (2.44).

Let us show that $\mathcal{A}^{-1} \mid L_+$ is a compact operator acting in L_+. Indeed, according to (2.44), the operator $\mathcal{A}^{-1} \mid L_+$ acts on any element $\tilde{u} = (u_0, K_+u_0)^{\mathrm{t}} \in L_+$ by the law

$$(\mathcal{A}^{-1} \mid L_+)\,\tilde{u} = \left(\left(A^{-1} + A_{12}^{(-1)}K_+\right)u_0, \left(\left(-A_{12}^{(-1)}\right)^* + B^{-1}K_+\right)u_0\right)^{\mathrm{t}}. \tag{2.68}$$

If $(P+ \mid L_+)$ is the restriction to L_+ of the orthoprojector P_+ acting by the law $P_+\tilde{u} := (u_0, 0)^{\mathrm{t}}$ for any element $\tilde{u} \in \tilde{H} = H^{m+1}$, then, with due account of the equation $u_0 = (P_+ \mid L_+)\tilde{u}$, from (2.69) we get, by the arbitrariness of $\tilde{u} \in \tilde{H}$,

$$(P_+ \mid L_+)\,(\mathcal{A}^{-1} \mid L_+) = \left(A^{-1} + A_{12}^{(-1)}K_+\right)(P_+ \mid L_+). \tag{2.69}$$

As it has been proved in the theory of J-spaces (see, for example, the monograph by T. Ya. Azizov and I.S. Iokhvidow [AI], Section 1.4), the operator $(P_+ \mid L_+)$ homeomorphically maps L_+ onto $H_+ = H$. Therefore, from (2.69) it follows that $(\mathcal{A}^{-1} \mid L_+)$ is similar to the operator $A^{-1} + A_{12}^{(-1)}K_+$,

$$(\mathcal{A}^{-1} \mid L_+) = (P_+ \mid L_+)^{-1}\left(A^{-1} + A_{12}^{(-1)}K_+\right)(P_+ \mid L_+). \tag{2.70}$$

Since by the compactness of A^{-1}, $A_{12}^{(-1)}$, and the boundedness of K_+, the operator $A^{-1} + A_{12}^{(-1)}K_+$ is compact, then from $\|P_+ \mid L_+\| \le 1$ and $\|(P_+ \mid L_+)^{-1}\| \le \sqrt{2}$ we conclude that the operator $(\mathcal{A}^{-1} \mid L_+)$ is also compact.

Further, as it is proved in the theory of operators in spaces with indefinite metrics, the operator $(\mathcal{A}^{-1} \mid L_+)$ is similar to a self-adjoint operator, and, therefore, has a countable set of positive eigenvalues $\mu_k^0 \to 0$ as $k \to \infty$. In problem (2.16), to them there correspond eingenvalues $\lambda_n^{(\infty)}$ and $\mathcal{J}$-orthonormal eigenelements forming almost a $\mathcal{J}$-orthonormal basis in L_+ for $\mathcal{J} := \operatorname{diag}(I, -I, \ldots, -I)$. The projections of these elements on $\tilde{H}_+ = (P_+ \mid L_+)\tilde{H} = H$ give a p-basis with a finite defect in H if and only if the angular operator K_+ of the invariant subspace L_+ belongs to the class $\mathfrak{S}_{2p}$; this fact was discussed in T. Ya. Azizov's dissertation [1] (consequence 1.21, p.55).

To conclude the proof, we consider equation (2.45) satisfied by the angular operator K_+. As it follows from (2.67) and formula (2.39) for the operator A^{-1}, this operator belongs to the class $\mathfrak{S}_{p_0}$. Analogously, from (2.43) we get that the operators $A_{12}^{(-1)}$ and $A_{21}^{(-1)} = -(A_{12}^{(-1)})^*$ belong to the class $\mathfrak{S}_{p_0}$. Therefore, by virtue of equation (2.45) and the boundedness of B, we have $K_+ \in \mathfrak{S}_{2p_0}$. Statement 9° is proved.

11.3 Small Motions and Normal Oscillations of a Visco-Elastic Fluid in an Open Container

The problem discussed in this section generalizes in the case of a visco-elastic fluid the problem on oscillations of an ordinary viscous fluid in an open vessel (see Sections 8.1–8.3) and mathematically it is not as simple as the problem from Section 11.1. The spectrum of this problem has a rather complicated structure and consists of several spectral series of eigenvalues corresponding to various kinds of normal oscillation waves.

11.3.1 Mathematical Statement of the Problem

We consider small motions of a visco-elastic fluid partially filling a container. Usually, at rest, the domain filled with fluid is denoted by Ω, the rigid container wall is denoted by S, and the horizontal free surface is denoted by Γ.

Let us assume that Γ has the equation $x_3 = 0$ where the Ox_3-axis along the unit vector $\boldsymbol{e}_3$ is directed vertically upward, that is, opposite to the acceleration vector of the gravity force $\boldsymbol{g} = -g\boldsymbol{e}_3$. Then, the equilibrium pressure in fluid is given by

$$P_0 = P_0(x_3) = p_a - \rho g x_3, \tag{3.1}$$

where p_a is the external constant pressure.

For the velocity field $\boldsymbol{u}(t,x)$ and the deviation $p(t,x)$ of the pressure field $P(t,x)$ from the equilibrium field $P_0(x_3)$ we obtained, just as in Section 11.1.2, the linearized equation (1.10)–(1.11) in the domain Ω and the stickiness condition (1.12) on a rigid wall S,

$$\frac{\partial \boldsymbol{u}}{\partial t} = -\frac{1}{\rho}\nabla p + \nu \hat{I}_0(t)\Delta \boldsymbol{u} + \boldsymbol{f}(t,x), \operatorname{div} \boldsymbol{u} = 0 \quad \text{in } \Omega, \tag{3.2}$$

$$\boldsymbol{u} = \boldsymbol{0} \quad \text{on } S. \tag{3.3}$$

On the equilibrium surface Γ, just as for an ordinary viscous fluid, the kinematic and dynamic conditions should be fulfilled. If the equation of the moving surface $\Gamma(t)$ has the form $x_3 = \zeta(t, x_1, x_2)$, $(x_1, x_2) \in \Gamma$, then the kinematic condition is the following,

$$u_n \,(= u_3 = \boldsymbol{u}\cdot\boldsymbol{n}) = \frac{\partial\zeta}{\partial t} \qquad \text{on } \Gamma. \tag{3.4}$$

As for the three dynamic conditions on Γ, they are obtained by linearization and a drift on Γ from the conditions of equality to zero of the viscous stresses on $\Gamma(t)$. In the chosen coordinate system $Ox_1x_2x_3$, we get the conditions $\sigma_{13} = 0$, $\sigma_{23} = 0$, $\sigma_{33} = 0$ (on $\Gamma(t)$). Recalling the connections (1.1), (1.8) between the tensors $\boldsymbol{\sigma}$, $\boldsymbol{\sigma}'$, and $\boldsymbol{\tau}$ in a visco-elastic fluid, and also the relationship $P = P_0 + p$ and formula (3.1), we get the following dynamic conditions

$$\begin{aligned} \nu\hat{I}_0(t)\tau_{i3}(\boldsymbol{u}) := \nu\hat{I}_0(t)\left(\frac{\partial u_i}{\partial x_3} + \frac{\partial u_3}{\partial x_i}\right) &= 0, \qquad i = 1, 2, \\ -\rho^{-1}p + \nu\hat{I}_0(t)\tau_{33}(\boldsymbol{u}) &= -g\zeta \qquad \text{on } \Gamma. \end{aligned} \tag{3.5}$$

Thus, the initial boundary value problem on small motions of a visco-elastic fluid in an open vessel consists in finding the velocity field $\boldsymbol{u}(t, x)$, the pressure field $p(t, x)$, and the function $\zeta(t, x_1, x_2)$ of deviation of the moving surface $\Gamma(t)$ from the equilibrium surface Γ, from equation (3.2), the boundary value conditions (3.3)–(3.5), and the following initial conditions,

$$\begin{aligned} \boldsymbol{u}(0, x) &= \boldsymbol{u}^0(x) \qquad \text{in } \Omega, \\ \zeta(0, x_1, x_2) &= \zeta^0(x_1, x_2) \qquad \text{on } \Gamma. \end{aligned} \tag{3.6}$$

We recall that $\hat{I}_0(t)$ in (3.2) and (3.5) is an integral Volterra operator in the variable t which on vector fields is defined by formula (1.11) and on tensor fields and their components by the similar formula (1.9).

11.3.2 Transition to a System of Operator Equations

Let us consider the velocity field $\boldsymbol{u}(t, x)$ for every t to be an element of the Hilbert space $\boldsymbol{L}_2(\Omega)$. Then, by the condition of solenoidality and from (3.3) we get that $\boldsymbol{u}(t, x) \in \boldsymbol{J}_{0,S}(\Omega)$. Introducing the orthoprojector $P_{0,S}$ onto the subspace $\boldsymbol{J}_{0,S}(\Omega)$ and applying it to the first equation (3.2) we get

$$\begin{aligned} \frac{\partial \boldsymbol{u}}{\partial t} &= -\frac{1}{\rho}\nabla\tilde{p} + \nu\hat{I}_0(t)P_{0,S}\Delta\boldsymbol{u} + \boldsymbol{f}_{0,S}, \\ \nabla\tilde{p} : &= P_{0,S}\nabla p \in \boldsymbol{G}_{h,S}(\Omega) \subset \boldsymbol{J}_{0,S}(\Omega), \boldsymbol{f}_{0,S} := P_{0,S}\boldsymbol{f}(t, x). \end{aligned} \tag{3.7}$$

We represent the field $\nabla\tilde{p}$, just as in the problem of Section 10.2, as a sum of two terms,

$$\nabla\tilde{p} = \nabla\varphi_1 + \nabla\varphi_2, \nabla\varphi_i \in \boldsymbol{G}_{h,S}(\Omega), \qquad i = 1,2, \tag{3.8}$$

and assume that the field $\boldsymbol{v}(t,x) := \hat{I}_0(t)\boldsymbol{u}(t,x)$ and $\nabla\varphi_1(t,x)$ for every t are solutions of the first auxiliary problem considered in Section 10.2.2, that is, the problem

$$\begin{aligned} &-\nu P_{0,S}\Delta\boldsymbol{v} + \nabla\varphi_1 =: \nu A\boldsymbol{v} = \boldsymbol{\eta} := -\frac{\partial\boldsymbol{u}}{\partial t} - \nabla\varphi_2 + \boldsymbol{f}_{0,S}, \\ &\operatorname{div}\boldsymbol{v} = 0 \quad \text{in } \Omega, \qquad \boldsymbol{v} = 0 \qquad \text{on } S, \\ &\nu\tau_{i3}(\boldsymbol{v}) = 0, \quad i = 1,2, \qquad \nu\tau_{33}(\boldsymbol{v}) = -\varphi_1 + 2\nu\frac{\partial v_3}{\partial x_3} = 0 \quad \text{on } \Gamma. \end{aligned} \tag{3.9}$$

In (3.7) and (3.9) we used the commuting property of the operator $\hat{I}_0(t)$ and the operators acting on space variables.

If $\boldsymbol{\eta} = \boldsymbol{\eta}(t) \in \boldsymbol{J}_{0,S}(\Omega)$, then from (3.9) we get that

$$\nu\boldsymbol{v} := \nu\hat{I}_0(t)\boldsymbol{u}(t) = A^{-1}\left(-\frac{\mathrm{d}\boldsymbol{u}}{\mathrm{d}t} - \nabla\varphi_2 + \boldsymbol{f}_{0,S}\right), \tag{3.10}$$

where the operator A has been already studied in detail in the previous chapters.

Now we consider the second auxiliary problem for the field $\varphi_2(t,x)$, similar to the problem (10.2.11),

$$\begin{aligned} &\Delta\varphi_2 = 0 \quad \text{in } \Omega, \qquad \frac{\partial\varphi_2}{\partial n} = 0 \quad \text{on } S, \\ &\varphi_2 = \psi := g\zeta \quad \text{on } \Gamma, \qquad \int_\Gamma \zeta d\Gamma = 0. \end{aligned} \tag{3.11}$$

This problem—a Zaremba Problem—has a unique solution $\varphi_2(t,x) \in H^1_\Gamma(\Omega)$ for any $\psi(t,x_1,x_2)$ taking, for fixed t, values in $H^{1/2}_\Gamma$, and

$$\nabla\varphi_2 = G\psi = gG\zeta, \tag{3.12}$$

where G is a bounded operator acting from $H^{1/2}_\Gamma$ in $\boldsymbol{G}_{h,S}(\Omega)$ and obtained as the adjoint to the operator that gives the normal component $\gamma_n\boldsymbol{u}$ for any field $\boldsymbol{u} \in \boldsymbol{J}_{0,S}(\Omega)$.

Based on the solutions of the auxiliary problems (3.9) and (3.11) and on the relations (3.8), (3.10), (3.12), we conclude that the equations, boundary value and initial conditions (3.2)–(3.6) give the following Cauchy problem for a system of operator equations relatively to two unknown functions (the function $\boldsymbol{u}(t)$, with values in the space $\boldsymbol{J}_{0,S}(\Omega)$, and the function $\zeta(t)$, with values in the space $H_\Gamma := L_2(\Gamma) \ominus \{1\}$):

$$\begin{aligned} &\frac{\mathrm{d}\boldsymbol{u}}{\mathrm{d}t} = -\nu\hat{I}_0(t)A\boldsymbol{u} - gG\zeta + \boldsymbol{f}_{0,S}(t), \\ &\frac{\mathrm{d}\zeta}{\mathrm{d}t} = \gamma_n\boldsymbol{u}, \boldsymbol{u}(0) = \boldsymbol{u}^0, \zeta(0) = \zeta^0. \end{aligned} \tag{3.13}$$

11.3.3 On the Solvability of the Initial Boundary Value Problem

Let us rewrite problem (3.13) by taking into account the definition (1.11) of operator $\hat{I}_0(t)$ and by performing the substitution

$$g^{1/2}\zeta =: \eta. \tag{3.4}$$

We obtain the Cauchy problem

$$\begin{aligned}
&\frac{d\boldsymbol{u}}{dt} + \nu\left(A\boldsymbol{u} + \sum_{j=1}^{m}\alpha_j\int_0^t \exp\left(-\gamma_j(t-s)\right)A\boldsymbol{u}(s)ds\right) + g^{1/2}G\eta = \boldsymbol{f}_{0,S}(t),\\
&\frac{d\eta}{dt} = g^{1/2}\gamma_n\boldsymbol{u},\\
&\boldsymbol{u}(0) = \boldsymbol{u}^0, \qquad \eta(0) := g^{1/2}\zeta^0 =: \eta^0.
\end{aligned} \tag{3.15}$$

As in Section 11.1.3, from problem (3.15) we switch to a differential equation in the space $H := (\boldsymbol{J}_{0,S}(\Omega))^{m+1} \oplus H_\Gamma$. For this we introduce the new unknown functions (1.15) and use the relations (1.16). Then from (3.15) we have the following problem in a vector-matrix form, which for simplicity is written down for $m = 2$:

$$\begin{aligned}
&\frac{d}{dt}\begin{pmatrix}\boldsymbol{u}_0\\ \boldsymbol{u}_1\\ \boldsymbol{u}_2\\ \eta\end{pmatrix} + \begin{pmatrix}\nu A & (\nu\alpha_1)^{1/2}A^{1/2} & (\nu\alpha_2)^{1/2}A^{1/2} & g^{1/2}G\\ -(\nu\alpha_1)^{1/2}A^{1/2} & \gamma_1 I & 0 & 0\\ -(\nu\alpha_2)^{1/2}A^{1/2} & 0 & \gamma_2 I & 0\\ -g^{1/2}G^* & 0 & 0 & 0\end{pmatrix}\begin{pmatrix}\boldsymbol{u}_0\\ \boldsymbol{u}_1\\ \boldsymbol{u}_2\\ \eta\end{pmatrix}\\
&\quad = \begin{pmatrix}\boldsymbol{f}_{0,S}\\ \boldsymbol{0}\\ \boldsymbol{0}\\ \boldsymbol{0}\end{pmatrix},\\
&\boldsymbol{u}_0(0) = \boldsymbol{u}^0, \qquad \boldsymbol{u}_j(0) = \boldsymbol{0} \quad j = 1, 2, \qquad \eta(0) = \eta^0,
\end{aligned} \tag{3.16}$$

where it has been already taken into account that the operator γ_n is adjoint to G.

The matrix operator in (3.16) has the same block structure (1.19) as the operator $\mathcal{A}$ from (1.17), with the substitution of the operator A_0 by A, and with the difference that the block A_{22} in (3.16) has the form $\mathrm{diag}(\gamma_1 I, \dots, \gamma_m I, 0)$, and, therefore, is nonnegative but not positive definite as in (1.17), where $A_{22} = \mathrm{diag}(\gamma_1 I, \dots, \gamma_m I)$, $\gamma_k > 0$, $k = 1, \dots, m$.

Whence it follows that the operator $\mathcal{A}$ in (3.16) is again representable in the form (1.22), that is, $\mathcal{A} = \mathcal{A}_0 + \mathrm{i}\mathcal{A}_1$, where $\mathcal{A}_0 := \mathrm{diag}(\nu A, \gamma_1 I, \dots, \gamma_m I, 0)$ is a nonnegative operator. Therefore, we get inequality (1.23) with the constant $c = 0$, and also the similar inequality (1.24).

Thus, the problem (3.16) is a nonhomogeneous Cauchy problem for a differential equation of the first order with a maximal dissipative operator $(-\mathcal{A})$. Therefore, the semigroup $\mathcal{U}(t)$ corresponding to problem (3.16) is contractive. Whence, just as in Section 11.1, we get the following conclusion.

If in problem (3.2)–(3.6), $\boldsymbol{f}(t,x)$ is a continuously differentiable function in the variable t with values in $\boldsymbol{L}_2(\Omega)$, $\boldsymbol{u}^0(x) \in \mathcal{D}(A)$, $\zeta^0(x_1,x_2) \in \mathcal{D}(G) = H_\Gamma^{1/2}$, then this problem has a solution $\{\boldsymbol{u}(t,x), \zeta(t,x)\}$ such that $\boldsymbol{u}(t,x) \in \mathcal{D}(A)$ for any $t \in [0,T]$, with $T > 0$ arbitrary, $\zeta(t,x_1,x_2) \in H_\Gamma^{1/2}$ for any $t \in [0,T]$, the functions $\partial \boldsymbol{u}/\partial t$ and $\partial\zeta/\partial t$ are continuous in $\boldsymbol{J}_{0,S}(\Omega)$ and $H_\Gamma^{1/2}$, respectively, and the equations (3.13) are satisfied. The homogeneous problem (3.13) is uniformly correct.

If $\boldsymbol{f}(t,x)$ is a continuous function in t with the values in $\boldsymbol{L}_2(\Omega)$, $\boldsymbol{u}^0(x) \in \boldsymbol{J}_{0,S}(\Omega)$, $\zeta^0(x_2,x_2) \in H_\Gamma$, then formula (1.26) gives a generalized solution of problem (3.13).

11.3.4 Normal Oscillations. Main Operator Pencil

We consider the spectral problem on normal oscillations corresponding to the evolution problem (3.16) (for arbitrary values of m). Assuming $\boldsymbol{f}_{0,S}(t) \equiv \boldsymbol{0}$, a dependence on time for the unknown functions in the form $\exp(-\lambda t)$, and denoting the amplitude elements by the previous symbols, we have

$$\begin{aligned}
\nu A\boldsymbol{u}_0 + \sum_{k=1}^{m} (\nu\alpha_k)^{1/2} A^{1/2}\boldsymbol{u}_k + g^{1/2} G\eta &= \lambda \boldsymbol{u}_0, \\
-(\nu\alpha_k)^{1/2} A^{1/2}\boldsymbol{u}_0 + \gamma_k \boldsymbol{u}_k &= \lambda \boldsymbol{u}_k, \qquad k = 1,\dots,m, \\
-g^{1/2} G^* \boldsymbol{u}_0 &= \lambda\eta. \qquad (3.17)
\end{aligned}$$

First we observe some simple properties of the spectrum of problem (3.17).

1° The number $\lambda = 0$ is not an eigenvalue of problem (3.17).

Indeed, assuming in (3.17) that $\lambda = 0$, multiplying the first equation scalarly by $\boldsymbol{u}_0$, the following m equations scalarly by $\boldsymbol{u}_k$, $k = 1,\dots,m$, and the last one by η (the first in $\boldsymbol{J}_{0,S}(\Omega)$, and the last in H_Γ), by taking the real part we get

$$\begin{aligned}
0 &= \nu \left\| A^{1/2}\boldsymbol{u}_0 \right\|^2 + \sum_{k=1}^{m} \gamma_k \|\boldsymbol{u}_k\|^2 \\
&\geq \min\{\nu\lambda_1(A), \gamma_1, \dots, \gamma_m\} \left(\|\boldsymbol{u}_0\|^2 + \sum_{k=1}^{m} \|\boldsymbol{u}_k\|^2 \right). \qquad (3.18)
\end{aligned}$$

Hence, it follows that $\boldsymbol{u}_0 = \boldsymbol{u}_1 = \cdots = \boldsymbol{u}_m = \mathbf{0}$. Then from the first equation (3.17) (for $\lambda = 0$) we get that $G\eta = 0$ and therefore $\eta = 0$. Thus, for $\lambda = 0$, problem (3.17) has the trivial solution.

2° The eigenvalues of problem (3.17) are situated in the right half-plane.

Indeed, if $\lambda \neq 0$ is an eigenvalue, then calculations similar to the ones presented previously [see (3.18)] lead to the relation

$$\nu \left\| A^{1/2}\boldsymbol{u}_0 \right\|^2 + \sum_{k=1}^{m} \gamma_k \|\boldsymbol{u}_k\|^2 = \operatorname{Re}\lambda \left(\|\boldsymbol{u}_0\|^2 + \sum_{k=1}^{m} \|\boldsymbol{u}_k\|^2 + \|\eta\|^2 \right). \tag{3.19}$$

whence it follows that $\operatorname{Re}\lambda > 0$.

3° The points $\lambda = \gamma_k$, $k = 1, \dots, m$, are not eigenvalues of problem (3.17).

Indeed, if $\lambda = \gamma_j$, then from the equation with $k = j$ in (3.17) we get $A^{1/2}\boldsymbol{u}_0 = 0$, and, therefore, $\boldsymbol{u}_0 = \mathbf{0}$. Then, from the equations with $k \neq j$ we have $(\gamma_k - \gamma_j)\boldsymbol{u}_k = \mathbf{0}$, that is, $\boldsymbol{u}_k = \mathbf{0}$, $k \neq j$, and from the last one we get $\eta = 0$. At last, from the first equation (3.17) we have $(\nu\alpha_j)^{1/2}A^{1/2}\boldsymbol{u}_j = \mathbf{0}$, whence it follows that $\boldsymbol{u}_j = \mathbf{0}$. Thus, for any $\lambda = \gamma_k$, $k = 1, \dots, m$, problem (3.17) has the trivial solution.

Based on these properties and eliminating in (3.17) all the unknown elements except $\boldsymbol{u}_0$, we obtain the next problem,

$$\nu \left(1 + \sum_{k=1}^{m} \frac{\alpha_k}{\gamma_k - \lambda} \right) A\boldsymbol{u}_0 - g\lambda^{-1}GG^*\boldsymbol{u}_0 = \lambda\boldsymbol{u}_0, \tag{3.20}$$

with unbounded operator coefficients. Performing in (3.20) the substitution

$$A^{1/2}\boldsymbol{u}_0 = \boldsymbol{\xi} \in \boldsymbol{J}_{0.S}(\Omega), \tag{3.21}$$

applying on the left the operator $A^{-1/2}$, and recalling the definitions (1.6) of the function $I_0(\lambda)$ and also of the operator $B = (A^{-1/2}G)(\gamma_n A^{-1/2})$, $\gamma_n = G^*, A^{-1/2}G = A^{1/2}T$, which has been already met in Chapter 8 and Section 10.4 [see Section 10.4.1 and formula (10.4.10)], we arrive at the spectral problem

$$L(\lambda)\boldsymbol{\xi} := \left(\nu I_0(\lambda)I - \lambda A^{-1} - g\lambda^{-1}B\right)\boldsymbol{\xi} = \mathbf{0} \tag{3.22}$$

for what in this section will be called the main operator pencil.

Its study enables us to reveal properties of the spectrum and the eigen- and associated elements of the problem on normal oscillations of a visco-elastic fluid in a partially filled container. We remark that if $\alpha_k = 0$, $k = 1, \ldots, m$, the pencil (3.22) transforms into the main operator pencil of the problem on normal oscillations of an ordinary viscous fluid [see (8.1.41)]. We recall also that A^{-1} in (3.22) is a compact positive operator, and B is a compact nonnegative operator. Properties of these operators and the asymptotics of their eigenvalues were studied in detail in Chapter 8. At last, we notice one more fact: If in (3.22) it is formally assumed that $g = 0$ and $A = A_0$, where A_0 is the Stokes operator, then problem (3.22) transforms [after the inverse substitution (3.21)] into problem (1.29) on normal oscillations of a visco-elastic fluid in a completely filled container.

We mention another property of the solution of problem (3.22), and, therefore, of problem (3.17).

4° The numbers β_k, $k = 1, \ldots, m$, that is, the zeroes of the function $I_0(\lambda)$, are not eigenvalues of problem (3.22).

Indeed, for $\lambda = \beta_k$, from (3.22) we get $(\beta_k A^{-1} + g\beta_k^{-1} B)\boldsymbol{\xi} = 0$, and since $\beta_k > 0$, $A^{-1} > 0$, $B \geq 0$, it follows that $\boldsymbol{\xi} = \mathbf{0}$.

Based on this property it is now easy to state the following important fact.

5° The spectrum of problem (3.22) is discrete, with possible limit points $\lambda = 0$, $\lambda = \beta_k$, $k = 1, \ldots, m$, and $\lambda = \infty$.

To prove this property, we observe that the function $I_0(\lambda)$ in (3.22) equals zero at points that are not eigenvalues, so we can divide both sides of (3.22) by $\nu I_0(\lambda)$. In effect we get the problem

$$l(\lambda)\boldsymbol{\xi} := \left(I - (\nu I_0(\lambda))^{-1} \left(\lambda A^{-1} + g\lambda^{-1} B\right)\right)\boldsymbol{\xi} = \mathbf{0}, \tag{3.33}$$

for a Fredholm operator pencil of the form $I + \Phi(\lambda)$, where $\Phi(\lambda)$ takes compact values and is an analytic operator-function in all the complex plane $\mathbb{C}$, except the points $\lambda = 0$, $\lambda = \beta_k$, $k = 1, \ldots, m$, and $\lambda = \infty$. Since for each negative λ the function

$$I_0(\lambda) := 1 + \sum_{k=1}^{m} \frac{\alpha_k}{\gamma_k - \lambda}, \qquad \alpha_k > 0, k = 1, \ldots, m, \tag{3.34}$$

takes positive values (> 1), then $\Phi(\lambda)$ for $\lambda < 0$ is a positive operator-function and therefore $I + \Phi(\lambda)$ is invertible on the half-axis $(-\infty, 0)$.

Thus, the operator-function $l(\lambda)$ from (3.33) is a regular Fredholm pencil and, according to the statements of Section 1.6.3, its spectrum consists of finitely multiple eigenvalues with the possible limit points $\lambda = 0$, $\lambda = \infty$, and $\lambda = \beta_k$, $k = 1, \ldots, m$. The resolvent $l^{-1}(\lambda)$ is a meromorphic operator-function having its poles at points coinciding with the eigenvalues, where the pole multiplicity coincides with the maximal multiplicity of eigenelements corresponding to the given eigenvalue.

11.4 Multiple Basicity of the System of Eigen- and Associated Elements for the Problem on Normal Oscillations of a Visco-Elastic Fluid in an Open Container

In this section we study the problems on multiple completeness and basicity of the system of eigenelements of problem (3.33). In studying them, we use the method of reduction to an operator pencil linear relatively to a new spectral parameter, thus generalizing the approach discussed in Section 8.2.2 for a more simple problem on normal oscillations of an ordinary viscous fluid. The schemes and constructions from Sections 8.2.2–8.2.5 are preserved in what follows.

11.4.1 The Linearization of the Pencil

We perform in (3.33) two formal substitutions, redenoting, for the convenience of further recordings,

$$\nu^{-1}A^{-1} \to A, \qquad g\nu^{-1}B \to B. \tag{4.1}$$

We have

$$l(\lambda) := I - I_0^{-1}(\lambda)(\lambda A + \lambda^{-1}B), \tag{4.2}$$

$$I_0(\lambda) := 1 + \sum_{k=1}^{m} \frac{\alpha_k}{\gamma_k - \lambda} \equiv \prod_{k=1}^{m}(\beta_k - \lambda) / \prod_{k=1}^{m}(\gamma_k - \lambda),$$

$$0 < \gamma_1 < \beta_1 < \cdots < \gamma_m < \beta_m < \infty. \tag{4.3}$$

We decompose the functions

$$a(\lambda) := -\lambda I_0^{-1}(\lambda), \qquad b(\lambda) := -(\lambda I_0(\lambda))^{-1}, \tag{4.4}$$

as sums of simple fractions.

For $a(\lambda)$ we get

$$a(\lambda) = -\lambda - c + \sum_{k=1}^{m} \frac{a_k}{\beta_k - \lambda} = -\lambda \cdot \frac{\prod_{i=1}^{m}(\gamma_i - \lambda)}{\prod_{i=1}^{m}(\beta_i - \lambda)}, \tag{4.5}$$

where c and a_k are to be defined. Multiplying (4.5) by $\prod_{i=1}^{m}(\beta_i - \lambda)$ we have the identity

$$-(\lambda + c)\prod_{i=1}^{m}(\beta_i - \lambda) + \sum_{k=1}^{m} a_k \prod_{i\neq k}(\beta_i - \lambda) \equiv -\lambda \prod_{i=1}^{m}(\gamma_i - \lambda). \tag{4.6}$$

Assuming $\lambda = \beta_k$ we get

$$a_k \prod_{i\neq k}(\beta_i - \beta_k) = -\beta_k \prod_{i=1}^{m}(\gamma_i - \beta_k),$$

whence

$$a_k = -\beta_k \cdot \frac{\prod_{i=1}^{m}(\gamma_i - \beta_k)}{\prod_{i\neq k}(\beta_i - \beta_k)} > 0, \qquad k = 1, \dots, m. \tag{4.7}$$

[The positiveness of the coefficients a_k is obtained by using inequality (4.3)].

To get a formula for the constant c from (4.5), we equate in (4.6) the coefficients of λ^m on the left and the right sides, and have

$$c = \sum_{i=1}^{m}(\beta_i - \gamma_i) > 0. \tag{4.8}$$

Similarly, for $b(\lambda)$, from (4.4) we have the decomposition

$$b(\lambda) = -\frac{d}{\lambda} + \sum_{k=1}^{m} \frac{b_k}{\beta_k - \lambda} = -\frac{\prod_{i=1}^{m}(\gamma_i - \lambda)}{\lambda \prod_{i=1}^{m}(\beta_i - \lambda)}, \tag{4.9}$$

and, therefore,

$$d \cdot \prod_{i=1}^{m}(\beta_i - \lambda) - \lambda \sum_{k=1}^{m} b_k \prod_{i\neq k}(\beta_i - \lambda) \equiv \prod_{i=1}^{m}(\gamma_i - \lambda). \tag{4.10}$$

Whence, for $\lambda = 0$ we get that

$$d = \frac{\prod_{i=1}^{m}\gamma_i}{\prod_{i=1}^{m}\beta_i} > 0. \tag{4.11}$$

Putting $\lambda = \beta_k$ into (4.10), we have

$$-\beta_k b_k \prod_{i\neq k}(\beta_i - \beta_k) = \prod_{i=1}^{m}(\gamma_i - \beta_k),$$

and, therefore,

$$b_k = -\beta_k \cdot \frac{\prod\limits_{i-1}^{m}(\gamma_i - \beta_k)}{\prod\limits_{i\neq k}(\beta_i - \beta_k)} = \frac{a_k}{\beta_k^2} > 0, \qquad k = 1, \dots, m. \tag{4.12}$$

Based on the decompositions (4.5) and (4.9), with the coefficients (4.7), (4.8), (4.11), and (4.12), we perform a linearization on the spectral parameter in the problem

$$l(\lambda)\boldsymbol{\xi} := (I + a(\lambda)A + b(\lambda)B)\boldsymbol{\xi} = \mathbf{0}, \qquad \boldsymbol{\xi} \in \boldsymbol{J}_{0,S}(\Omega); \tag{4.13}$$

here we use the method applied in Sections 8.2.2 and 8.4.5 to a more simple situation, and based on the introduction of a new spectral parameter and new unknown elements.

Let us introduce the notation

$$\mu := \lambda - \lambda^{-1} + \sum_{k=1}^{m}(\beta_k - \lambda)^{-1}, \qquad \boldsymbol{\eta}_0 := -\frac{\boldsymbol{\xi}}{\lambda},$$
$$\boldsymbol{\eta}_k := \frac{\boldsymbol{\xi}}{\beta_k - \lambda}, \qquad k = 1, \dots, m. \tag{4.14}$$

We rewrite equation (4.13) in the form

$$(I - cA)\boldsymbol{\xi} + \left(-\lambda + \lambda^{-1} - \sum_{k=1}^{m}\frac{1}{\beta_k - \lambda} - \lambda^{-1} + \sum_{k=1}^{m}\frac{a_k + 1}{\beta_k - \lambda}\right)A\boldsymbol{\xi}$$
$$+ \left(-\frac{d}{\lambda} + \sum_{k=1}^{m}\frac{b_k}{\beta_k - \lambda}\right)B\boldsymbol{\xi} = 0.$$

With regard to the notations introduced in (4.14) we have

$$(I - cA)\boldsymbol{\xi} - \mu A\boldsymbol{\xi} + A\boldsymbol{\eta}_0 + \sum_{k=1}^{m}(a_k + 1)A\boldsymbol{\eta}_k + d \cdot B\boldsymbol{\eta}_0 + \sum_{k=1}^{m} b_k B\boldsymbol{\eta}_k = \mathbf{0}. \tag{4.15}$$

Now we divide (4.13) by $(-\lambda)$. Using the identities

$$\frac{1}{\lambda(\beta-\lambda)}=\frac{1}{\beta}\left(\frac{1}{\lambda}+\frac{1}{\beta-\lambda}\right),\qquad -\frac{\lambda}{\beta-\lambda}=1-\frac{\beta}{\beta-\lambda},$$

we rewrite the obtained equation in the form

$$(I-cA)\boldsymbol{\eta}_0+\left(1-\sum_{k-1}^{m}\frac{a_k}{\beta_k}\left(\frac{1}{\lambda}+\frac{1}{\beta_k-\lambda}\right)\right)A\boldsymbol{\xi}$$

$$+\left(\left(\lambda-\frac{1}{\lambda}+\sum_{k=1}^{m}\frac{1}{\beta_k-\lambda}\right)-\lambda+\sum_{k=1}^{m}\frac{\frac{b_k}{d}-1}{\beta_k-\lambda}\right)d\cdot B\left(-\frac{\boldsymbol{\xi}}{\lambda}\right)=\mathbf{0}.$$

In the notations of (4.14) we have

$$(I-cA)\boldsymbol{\eta}_0+A\boldsymbol{\xi}+\sum_{k=1}^{m}\frac{a_k}{\beta_k}(A\boldsymbol{\eta}_0-A\boldsymbol{\eta}_k)+\mu d\cdot B\boldsymbol{\eta}_0+d\cdot B\boldsymbol{\xi}$$

$$+\sum_{k=1}^{m}\frac{b_k-d}{\beta_k}(B\boldsymbol{\eta}_0-B\boldsymbol{\eta}_k)=\mathbf{0}. \tag{4.16}$$

Finally, we divide (4.13) by $\beta_k-\lambda$ and using the previous identities and also the identity

$$\frac{1}{(\beta_k-\lambda)(\beta_i-\lambda)}=\frac{1}{(\beta_i-\beta_k)}\left(\frac{1}{\beta_k-\lambda}-\frac{1}{\beta_i-\lambda}\right),$$

we have

$$(I-cA)\boldsymbol{\eta}_k$$

$$+\left(\left(\lambda-\frac{1}{\lambda}+\sum_{k=1}^{m}\frac{1}{\beta_k-\lambda}\right)+\frac{1}{\lambda}-\left(\frac{1}{a_k}+1\right)\lambda+\sum_{i\neq k}\frac{\frac{a_i}{a_k}-1}{\beta_i-\lambda}\right)a_kA\left(\frac{\boldsymbol{\xi}}{\beta_k-\lambda}\right)$$

$$+\left(\left(\lambda-\frac{1}{\lambda}+\sum_{k=1}^{m}\frac{1}{\beta_k-\lambda}\right)-\lambda-\frac{1}{\lambda}\left(\frac{d}{b_k}-1\right)+\sum_{i\neq k}\frac{\frac{b_i}{b_k}-1}{\beta_i-\lambda}\right)b_kB\left(\frac{\boldsymbol{\xi}}{\beta_k-\lambda}\right)$$

$$=\mathbf{0},$$

or

$$(I-cA)\boldsymbol{\eta}_k+\mu(a_kA+b_kB)\boldsymbol{\eta}_k$$

$$+\left((a_k+1)A\boldsymbol{\xi}-\beta_k(a_k+1)A\boldsymbol{\eta}_k-\frac{a_k}{\beta_k}A\boldsymbol{\eta}_0+\frac{a_k}{\beta_k}A\boldsymbol{\eta}_k+\sum_{i\neq k}\frac{a_i-a_k}{\beta_i-\beta_k}A(\boldsymbol{\eta}_k-\boldsymbol{\eta}_i)\right)$$

$$+\left(\frac{d-b_k}{\beta_k}(B\boldsymbol{\eta}_0-B\boldsymbol{\eta}_k)+\sum_{i\neq k}\frac{b_i-b_k}{\beta_i-\beta_k}B(\boldsymbol{\eta}_k-\boldsymbol{\eta}_i)+b_kB(\boldsymbol{\xi}-\beta_k\boldsymbol{\eta}_k)\right)$$

$$=\mathbf{0}. \tag{4.17}$$

We will treat the collection of $m+2$ equations (4.15)–(4.17) containing the new spectral parameter μ as one linear equation with respect to μ in the orthogonal sum of spaces $\hat{\boldsymbol{H}} := (\boldsymbol{J}_{0,S}(\Omega))^{m+2}$, relatively to the vector-colomn $\hat{\boldsymbol{y}} := (\boldsymbol{\xi}, \boldsymbol{\eta}_0, \boldsymbol{\eta}_1, \ldots, \boldsymbol{\eta}_m)^{\mathrm{t}} \in \hat{\boldsymbol{H}}$ with the components (4.14). Then, the mentioned system of equations may be written in the vector-matrix form

$$\mathcal{L}(\mu)\hat{\boldsymbol{y}} := (\mathcal{I} - \mu\mathcal{H} + \mathcal{F})\hat{\boldsymbol{y}} = \boldsymbol{0}. \tag{4.18}$$

Here $\mathcal{I}$ is the identity operator-matrix in $\hat{\boldsymbol{H}}$, the matrix $\mathcal{H}$ has the form

$$\mathcal{H} := \operatorname{diag}(A, -d \cdot B, -a_1 A - b_1 B, \ldots, -a_m A - b_m B), \tag{4.19}$$

and the operator-matrix $\mathcal{F} = (F_{ij})_{i,j=1}^{m+2}$ has the following components

$$\begin{aligned}
F_{11} &= -cA,\\
F_{12} = F_{21} &= A + d \cdot B,\\
F_{1,k+2} = F_{k+2,1} &= (a_k + 1)A + b_k B,\\
F_{22} &= -cA + \left(\sum_{k=1}^{m} \frac{a_k}{\beta_k}\right) A + \left(\sum_{k=1}^{m} \frac{b_k - d}{\beta_k}\right) B,\\
F_{2,k+2} = F_{k+2,2} &= -\frac{a_k}{\beta_k} A - \frac{b_k - d}{\beta_k} B,\\
F_{k+2,k+2} &= -cA + \frac{a_k}{\beta_k} A - \beta_k (a_k + 1)A + \sum_{i \neq k} \frac{a_i - a_k}{\beta_i - \beta_k} A + \frac{b_k - d}{\beta_k} B - \beta_k b_k B\\
&\quad + \sum_{i \neq k} \frac{b_i - b_k}{\beta_i - \beta_k} B,\\
F_{k+2,l+2} &= -\frac{a_l - a_k}{\beta_l - \beta_k} A - \frac{b_l - b_k}{\beta_l - \beta_k} B, \qquad k, l = 1, \ldots, m, k \neq l.
\end{aligned} \tag{4.20}$$

From formulas (4.19) and (4.20), it follows that the operators $\mathcal{H}$ and $\mathcal{F}$ are self-adjoint and compact, since the operators A and B possess these properties. Note also that from (4.4) we have $a(0) = 0$, and, therefore,

$$-c + \sum_{k=1}^{m} \left(\frac{a_k}{\beta_k}\right) = 0; \tag{4.21}$$

using this equality, the formula for F_{22} may be rewritten in the form

$$F_{22} = \left(\sum_{k=1}^{m} \frac{b_k - d}{\beta_k}\right) B = \left(c - d \cdot \sum_{k=1}^{m} \beta_k^{-1}\right) B. \tag{4.22}$$

11.4.2 The Theorem on Basicity of the System of Root Elements of the Linear Pencil

We consider the spectral problem (4.18) and carry out the same methods already presented in Sections 8.2.4–8.2.5 in addressing the pencil structure problem (8.2.10). First of all, we mention that operator $\mathcal{H}$ from (4.19) has a nonzero kernel,

$$\operatorname{Ker}\mathcal{H} = \{\hat{\boldsymbol{y}} : \hat{\boldsymbol{y}} = (\mathbf{0}, \boldsymbol{\eta}_{00}, \mathbf{0}, \ldots, \mathbf{0})^{\mathrm{t}}, \qquad \boldsymbol{\eta}_{00} \in \operatorname{Ker} B = \boldsymbol{N}_0(\Omega) \subset \boldsymbol{J}_{0,S}(\Omega)\}, \quad (4.23)$$

whence $\dim \operatorname{Ker}\mathcal{H} = \infty$. We denote by $\mathcal{P}_0$ the orthoprojector onto $\operatorname{Ker}\mathcal{H}$, $\mathcal{P}_1 = \mathcal{I} - \mathcal{P}_0$, and set $\hat{\boldsymbol{y}} = \hat{\boldsymbol{y}}_0 + \hat{\boldsymbol{y}}_1$, $\hat{\boldsymbol{y}}_i = \mathcal{P}_i\hat{\boldsymbol{y}}$, $i = 0, 1$.

Applying to (4.19) the orthoprojectors $\mathcal{P}_i$, we obtain the system

$$\begin{aligned} \hat{\boldsymbol{y}}_1 - \mu\mathcal{H}_1\hat{\boldsymbol{y}}_1 + \mathcal{P}_1\mathcal{F}\mathcal{P}_0\hat{\boldsymbol{y}}_0 + \mathcal{P}_1\mathcal{F}\mathcal{P}_1\hat{\boldsymbol{y}}_1 &= \hat{\mathbf{0}}, \\ \hat{\boldsymbol{y}}_0 + \mathcal{P}_0\mathcal{F}\mathcal{P}_0\hat{\boldsymbol{y}}_0 + \mathcal{P}_0\mathcal{F}\mathcal{P}_1\hat{\boldsymbol{y}}_1 &= \hat{\mathbf{0}}. \end{aligned} \quad (4.24)$$

Taking into account the structure of operator $\mathcal{F}$ from (4.21) and formula (4.23) for F_{22}, we have

$$\begin{aligned} \mathcal{P}_0\mathcal{F}\mathcal{P}_0 &= \operatorname{diag}(0, P_0, 0, \ldots, 0) \cdot \operatorname{diag}(0, F_{22}, 0, \ldots, 0) \\ &\quad \cdot \operatorname{diag}(0, P_0, 0, \ldots, 0) \\ &= \operatorname{diag}(0, P_0F_{22}P_0, 0, \ldots, 0), \end{aligned} \quad (4.25)$$

where P_0 is the orthoprojector onto $\operatorname{Ker} B = \boldsymbol{N}_0(\Omega)$. Therefore, from the second equation (4.25), we get that $\hat{\boldsymbol{y}}_0 = -\mathcal{P}_0\mathcal{F}\mathcal{P}_1\hat{\boldsymbol{y}}_1$, and the first equation (4.25) changes into the next equation for $\hat{\boldsymbol{y}}_1$,

$$\begin{aligned} \mathcal{L}_1(\mu)\hat{\boldsymbol{y}}_1 &:= (\mathcal{I}_1 - \mu\mathcal{H}_1 + \mathcal{F}_1)\hat{\boldsymbol{y}}_1 = \hat{\mathbf{0}}, \\ \hat{\boldsymbol{y}}_1 \in \hat{\boldsymbol{H}}_1 &:= \mathcal{P}_1\hat{\boldsymbol{H}}, \\ \mathcal{H}_1 &:= \mathcal{P}_1\mathcal{H}\mathcal{P}_1, \\ \mathcal{F}_1 &:= \mathcal{P}_1\mathcal{F}\mathcal{P}_1 - \mathcal{P}_1\mathcal{F}\mathcal{P}_0\mathcal{P}\mathcal{F}_1, \\ \mathcal{I}_1 &= \mathcal{P}_1. \end{aligned} \quad (4.26)$$

Here, the operator $\mathcal{H}_1$ is complete, that is, $\operatorname{Ker}\mathcal{H}_1 = \{\hat{\mathbf{0}}\}$, and $\mathcal{F}_1$ is a compact self-adjoint operator. Since, in addition, $\mathcal{H}_1$ is compact, self-adjoint, and belongs to a certain class $\mathfrak{S}_p$, $p > 0$, then, according to the second Keldysh theorem (Section 1.6.4), the system of eigen- and associated elements of problem (4.26) is complete in the space

$$\mathcal{P}_1\hat{\boldsymbol{H}} := \hat{\boldsymbol{M}}(\Omega) := \boldsymbol{J}_{0,S}(\Omega) \oplus \mathbf{M}_0(\Omega) \oplus (\boldsymbol{J}_{0,S}(\Omega))^m. \quad (4.27)$$

Moreover, the spectrum of this problem is discrete and has the limit point $\mu = \infty$, where for any $\varepsilon > 0$, all eigenvalues μ, except maybe a finite number, are situated in the sectors

$$|\arg \mu| < \varepsilon, \qquad |\pi - \arg \mu| < \varepsilon.$$

Therefore, we can find a real number $\mu = a$ such that the operator

$$\mathcal{L}_1(a) := \mathcal{I} - a\mathcal{H}_1 + \mathcal{F}_1$$

is invertible. Then, performing in (4.26) a substitution of the spectral parameter according to the formula $\mu = a + \tilde{\mu}$, we get the problem

$$\hat{\boldsymbol{u}}_1 = \tilde{\mu}\mathcal{L}_1^{-1}(a)\mathcal{H}_1\hat{\boldsymbol{y}}_1 \tag{4.28}$$

that concerns the definition of the characteristic numbers of the compact $\mathcal{J}$-self-adjoint operator $\mathcal{L}_1^{-1}(a)\mathcal{H}_1$, where

$$\mathcal{J} = \mathcal{L}_1(a) = \mathcal{I} + \mathcal{K}, \qquad \mathcal{K} \in \mathfrak{S}_\infty.$$

Since the root subspace $L_0(\mathcal{L}_1^{-1}(a)\mathcal{H}_1)$ corresponding to the zero eigenvalue of the operator $\mathcal{L}_1^{-1}(a)\mathcal{H}_1$ is not degenerated according to the constructions mentioned above, then, in virtue of the statements in Section 1.4.7, the system of root elements of problem (4.26) forms a Riesz basis in $\hat{\boldsymbol{M}}(\Omega)$. In addition, the nonreal spectrum of problem (4.28), and, therefore, of the problems (4.26), (4.18), consists of no more than a finite number of normal eigenvalues that are symmetric relatively to the real axis. There exists no more than a finite number of real eigenvalues to which there correspond nontrivial Jordan chains of problem (4.18).

11.4.3 Connection between the System of Root Elements of the Two Problems

Let us use the previous results related to problem (4.18) to study properties of the solutions of the spectral problem (4.13) and the pencil $l(\lambda)$. We will subsequently assume that the following conditions are fulfilled:

$$1^\circ \ \kappa(\lambda) := \mu'(\lambda) = 1 + \lambda^{-2} + \sum_{k=1}^{m}(\beta_k - \lambda)^2 \neq 0, \qquad \mu(\lambda) \in \sigma(\mathcal{L}(\mu)). \tag{4.29}$$

2° All the roots of the equation $\kappa(\lambda) = 0$ are different in pairs.

We introduce, for $\lambda \in \mathbb{C}$, the subspace $\hat{\boldsymbol{H}}_0(\lambda) \subset \hat{\boldsymbol{H}}$ made up of the vector-columns

$$\hat{\boldsymbol{y}}_0 := \left(\boldsymbol{\xi}, -\frac{\boldsymbol{\xi}}{\lambda}, \frac{\boldsymbol{\xi}}{\beta_1 - \lambda}, \ldots, \frac{\boldsymbol{\xi}}{\beta_m - \lambda}\right)^{t}, \tag{4.30}$$

where $\boldsymbol{\xi}$ is an arbitrary element running through the entire space $\boldsymbol{J}_{0,S}(\Omega)$. From the relations (4.15)–(4.18), we get the formula

$$\mathcal{L}(\mu(\lambda))\boldsymbol{y}_0 = \left(l(\lambda)\boldsymbol{\xi}, -l(\lambda)\frac{\boldsymbol{\xi}}{\lambda}, l(\lambda)\frac{\boldsymbol{\xi}}{\beta_1 - \lambda}, \dots, l(\lambda)\frac{\boldsymbol{\xi}}{\beta_m - \lambda}\right)^{\mathrm{t}}, \tag{4.31}$$

indicating in particular that

$$\mathcal{L}(\mu(\lambda))\hat{\boldsymbol{H}}_0(\lambda) \subset \hat{\boldsymbol{H}}_0(\lambda). \tag{4.32}$$

Now we note that the subspace $\hat{\boldsymbol{H}}'(\lambda) := \hat{\boldsymbol{H}} \ominus \hat{\boldsymbol{H}}_0(\lambda)$ is invariant relatively to the operator $\mathcal{L}^*(\mu(\lambda)) = \mathcal{L}(\mu(\bar{\lambda}))$, and consists of the elements

$$\hat{\boldsymbol{z}} = (\boldsymbol{z}, \boldsymbol{z}_0, \boldsymbol{z}_1, \dots, \boldsymbol{z}_m)^{\mathrm{t}} \in \hat{\boldsymbol{H}},$$

for which

$$\boldsymbol{z} - \frac{\boldsymbol{z}_0}{\bar{\lambda}} + \sum_{k=1}^{m} \frac{\boldsymbol{z}_k}{\beta_k - \bar{\lambda}} = \boldsymbol{0}.$$

Substituting here λ by $\bar{\lambda}$, we get that $\mathcal{L}(\mu(\lambda))\hat{\boldsymbol{H}}'(\lambda) \subset \hat{\boldsymbol{H}}'(\lambda)$, where,

$$\hat{\boldsymbol{H}}'(\lambda) := \left\{\hat{\boldsymbol{z}} \in \hat{\boldsymbol{H}} : \hat{\boldsymbol{z}} = (\boldsymbol{z}, \boldsymbol{z}_0, \boldsymbol{z}_1, \dots, \boldsymbol{z}_m)^{\mathrm{t}}, \boldsymbol{z} - \frac{\boldsymbol{z}_0}{\lambda} + \sum_{k=1}^{m} \frac{\boldsymbol{z}_k}{\beta_k - \lambda} = \boldsymbol{0}\right\}. \tag{4.33}$$

Let us prove that the direct sum of the subspaces $\hat{\boldsymbol{H}}_0(\lambda)$ and $\hat{\boldsymbol{H}}'(\lambda)$ equals the entire space $\hat{\boldsymbol{H}}$. In fact, let $\hat{\boldsymbol{x}} = (\boldsymbol{x}, \boldsymbol{x}_0, \boldsymbol{x}_1, \dots, \boldsymbol{x}_m)^{\mathrm{t}}$ be an arbitrary element from $\hat{H}$. Consider the system of equations

$$\begin{aligned}
\boldsymbol{\xi} + \boldsymbol{z} &= \boldsymbol{x}, \\
-\frac{\boldsymbol{\xi}}{\lambda} + \boldsymbol{z}_0 &= \boldsymbol{x}_0, \\
\frac{\boldsymbol{\xi}}{\beta_k - \lambda} + \boldsymbol{z}_k &= \boldsymbol{x}_k, \qquad k = 1, \dots, m, \\
\boldsymbol{z} - \frac{\boldsymbol{z}_0}{\lambda} + \sum_{k=1}^{m} \frac{\boldsymbol{z}_k}{\beta_k - \lambda} &= \boldsymbol{0},
\end{aligned} \tag{4.34}$$

relatively to the variables $\boldsymbol{\xi}, \boldsymbol{z}, \boldsymbol{z}_0, \boldsymbol{z}_1, \dots, \boldsymbol{z}_m$. This linear system of $m+3$ equations with $m+3$ unknowns has the unique solution

$$\begin{aligned}
\boldsymbol{\xi} &= \kappa^{-1}(\lambda)\left[\boldsymbol{x} - \frac{\boldsymbol{x}_0}{\lambda} + \sum_{k=1}^{m} \frac{\boldsymbol{x}_k}{\beta_k - \lambda}\right], \\
\boldsymbol{z} &= \boldsymbol{x} - \boldsymbol{\xi}, \\
\boldsymbol{z}_0 &= \boldsymbol{x}_0 + \frac{\boldsymbol{\xi}}{\lambda}, \\
\boldsymbol{z}_k &= \boldsymbol{x}_k - \frac{\boldsymbol{\xi}}{\beta_k - \lambda}, \qquad k = 1, \dots, m.
\end{aligned} \tag{4.35}$$

Since $(\boldsymbol{x}, \boldsymbol{x}_0, \boldsymbol{x}_1, \ldots, \boldsymbol{x}_m)^{\mathrm{t}}$ in (4.34) is an arbitrary element of the space $(\boldsymbol{J}_{0,S}(\Omega))^{m+2}$, then from the just proved fact it follows that

$$\hat{\boldsymbol{H}} = \hat{\boldsymbol{H}}_0(\lambda) \dotplus \hat{\boldsymbol{H}}'(\lambda). \tag{4.36}$$

To continue the study of the problem, we need several auxiliary statements that are presented below.

(a) We introduce the operator

$$D_0(\lambda) := \frac{1}{\kappa(\lambda)} \begin{pmatrix} 1 & -\lambda^{-1} & \cdots & (\beta_m - \lambda)^{-1} \\ -\lambda^{-1} & \lambda^{-2} & \cdots & -\lambda^{-1}(\beta_m - \lambda)^{-1} \\ (\beta_1 - \lambda)^{-1} & \cdots & \cdots & (\beta_1 - \lambda)^{-1}(\beta_m - \lambda)^{-1} \\ \cdots & \cdots & \cdots & \cdots \\ (\beta_m - \lambda)^{-1} & \cdots & \cdots & (\beta_m - \lambda)^{2} \end{pmatrix}. \tag{4.37}$$

A straightforward calculation shows that $D_0(\lambda)$ projects the space $\hat{\boldsymbol{H}}$ onto $\hat{\boldsymbol{H}}_0(\lambda)$ parallel to $\hat{\boldsymbol{H}}'(\lambda)$.

(b) We denote by $\xi_j = \xi_j(\lambda)$, $j = 0, \ldots, m+1$, $\xi_0(\lambda) \equiv \lambda$, the roots of the equation

$$\mu(\xi) = \mu(\lambda) := \lambda - \lambda^{-1} + \sum_{k=1}^{m} (\beta_k - \lambda)^{-1}. \tag{4.38}$$

We note the following important property of the roots $\xi_j(\lambda)$: If the point μ belongs to the spectrum of the operator pencil $\mathcal{L}(\mu)$ from (4.18), then all the roots of equation (4.38) are different.

Indeed, if $\xi_i = \xi_j = \xi$, $i \neq j$, then $\mu'(\xi) = 1 + \xi^{-2} + \sum_{k=1}^{m} (\beta_k - \xi)^2 = \kappa(\xi) = 0$, and, therefore, the pencil $\mathcal{L}(\mu)$ has for this value of ξ the eigenvalue $\mu(\xi)$ for which $\kappa(\xi) = 0$, a conclusion that contradicts Property 2° from (4.29).

Hence, it follows that the functions $\xi_j(\lambda)$, $j = 0, \ldots, m+1$, are analytic in neighbourhoods of those points λ for which $\mu(\lambda) \in \sigma(\mathcal{L}(\mu))$.

(c) The next direct sum decomposition takes place,

$$\begin{aligned} \hat{\boldsymbol{H}} &= \hat{\boldsymbol{H}}_0(\lambda) \dotplus \hat{\boldsymbol{H}}_1(\lambda) \dotplus \cdots \dotplus \hat{\boldsymbol{H}}_m(\lambda) \dotplus \hat{\boldsymbol{H}}_{m+1}(\lambda), \\ \hat{\boldsymbol{H}}_j(\lambda) &:= \left\{ \hat{\boldsymbol{z}} : \hat{\boldsymbol{z}} = \left(\boldsymbol{z}, -\frac{\boldsymbol{z}}{\xi_j}, \frac{\boldsymbol{z}}{\beta_1 - \xi_j}, \ldots, \frac{\boldsymbol{z}}{\beta_m - \xi_j} \right)^{\mathrm{t}} \right\}, \\ & j = 0, \ldots, m+1, \boldsymbol{z} \in \boldsymbol{J}_{0.S}(\Omega). \end{aligned} \tag{4.39}$$

Let us prove this statement. To this end we need to make sure that the equation

$$\sum_{j=0}^{m+1} \hat{\boldsymbol{f}}_j = \hat{\boldsymbol{g}} \tag{4.40}$$

has a unique solution $\boldsymbol{f}_j \in \hat{\boldsymbol{H}}_j(\lambda)$, $j = 0, \dots, m+1$, for any $\hat{\boldsymbol{g}} \in \hat{\boldsymbol{H}}$. Put $\hat{\boldsymbol{g}} := (\boldsymbol{g}, \boldsymbol{g}_0, \boldsymbol{g}_1, \dots \boldsymbol{g}_m)^{\mathrm{t}}$, $\hat{\boldsymbol{f}}_j = (\boldsymbol{f}^j, \boldsymbol{f}_0^j, \boldsymbol{f}_1^j, \dots, \boldsymbol{f}_m^j)^{\mathrm{t}}$, $j = 0, \dots, m+1$. Breaking down equation (4.40) by coordinates, we easily see that for the uniqueness of its solution it is necessary and sufficient to fulfill the condition

$$\Delta(\lambda) \neq 0, \tag{4.41}$$

where $\Delta(\lambda)$ is the determinant of system (4.40),

$$\Delta(\lambda) := \begin{vmatrix} 1 & 1 & \cdots & 1 \\ \xi_0^{-1} & \xi_1^{-1} & \cdots & \xi_m^{-1} \\ (\beta_1 - \xi_0)^{-1} & (\beta_1 - \xi_1)^{-1} & \cdots & (\beta_1 - \xi_m)^{-1} \\ \cdots & \cdots & \cdots & \cdots \\ (\beta_m - \xi_0)^{-1} & (\beta_m - \xi_1)^{-1} & \cdots & (\beta_m - \xi_m)^{-1} \end{vmatrix}. \tag{4.42}$$

By induction on m it may be checked that

$$\Delta(\lambda) = \prod_{0 \le i < j \le m+1} (\xi_i - \xi_j) \cdot \prod_{0 \le i < j \le m} (\beta_i - \beta_j) \cdot \prod_{\substack{i=0,\dots,m+1 \\ j=0,\dots,m}} (\beta_j - \xi_i)^{-1}, \qquad \beta_0 := 0, \tag{4.43}$$

whence (4.41) follows since $\xi_i \neq \xi_j$, $i \neq j$, $\beta_i \neq \beta_j$, $i \neq j$, and $\xi_i \neq \beta_j$, for all i, j.

(d) We introduce the notations

$$D_j(\lambda) := D_0(\xi_j(\lambda)), \qquad j = 0, \dots, m+1. \tag{4.44}$$

As previously, it can be checked that the next relations are true,

$$D_j \hat{\boldsymbol{H}}_j(\lambda) \subset \hat{\boldsymbol{H}}_j(\lambda), \qquad D_j \hat{\boldsymbol{H}}'(\lambda) = \{\hat{\boldsymbol{0}}\}. \tag{4.45}$$

Now we show that

$$\hat{\boldsymbol{H}}_j(\lambda) \subset \hat{\boldsymbol{H}}_i'(\lambda), \qquad i \neq j. \tag{4.46}$$

With this goal in mind it is enough to check the property $\hat{\boldsymbol{H}}_j(\lambda) \perp \hat{\boldsymbol{H}}_i(\lambda)$ for $i \neq j$, that is, the property

$$1 + \frac{1}{\xi_i \xi_j} + \sum_{k=1}^{m} \frac{1}{(\beta_k - \xi_i)(\beta_k - \xi_j)} = 0, \qquad i \neq j.$$

To prove this fact, we successively set $\xi = \xi_i(\lambda)$ and $\xi = \xi_j(\lambda)$ in (4.38), and then subtract one equation from the other and get,

$$\begin{aligned} 0 &= \mu(\lambda) - \mu(\lambda) \\ &= \mu(\xi_i) - \mu(\xi_j) \\ &= \xi_i - \xi_j - \frac{1}{\xi_i} + \frac{1}{\xi_j} + \sum_{k=1}^{m} \left(\frac{1}{\beta_k - \xi_i} - \frac{1}{\beta_k - \xi_j} \right) \\ &= (\xi_i - \xi_j) \left(1 + \frac{1}{\xi_i \xi_j} + \sum_{k=1}^{m} \frac{1}{(\beta_k - \xi_i)(\beta_k - \xi_j)} \right). \end{aligned}$$

Since $\xi_i \neq \xi_j$, the required equation and statement (4.46) follow.

As a corollary of the relations (4.45), (4.46), we get the formulas

$$D_i(\lambda) D_j(\lambda) = D_j(\lambda) D_i(\lambda) = 0, i \neq j, \text{ and } \sum_{i=0}^{m+1} D_i(\lambda) = I. \tag{4.47}$$

Substituting λ in formula (4.31) by $\xi_i(\lambda)$ and taking into account (4.38) we get, for each $j = 0, \ldots, m+1$, the properties

$$\mathcal{L}(\mu(\lambda)) \hat{\boldsymbol{H}}_j(\lambda) \subset \hat{\boldsymbol{H}}_j(\lambda), \qquad \mathcal{L}(\mu(\lambda)) D_j(\lambda) = D_j(\lambda) \mathcal{L}(\mu(\lambda)). \tag{4.48}$$

We employ now the concept of a root function of an operator pencil (Section 1.6.2). Let $\boldsymbol{\xi}(\lambda)$ be a root function of order n of the operator pencil $l(\lambda)$ of problem (4.13), corresponding to the point $\lambda = \lambda_0$. It means that $l(\lambda)\boldsymbol{\xi}(\lambda)$ has at the point λ_0 a zero of the order $n+1$.

For the operator

$$C(\lambda) := \operatorname{diag} \left(I, -\frac{I}{\lambda}, \frac{I}{\beta_1 - \lambda}, \ldots, \frac{I}{\beta_m - \lambda} \right)$$

the next relationship takes place,

$$\begin{aligned} &\mathcal{L}(\mu(\lambda)) \left(\boldsymbol{\xi}(\lambda), -\frac{\boldsymbol{\xi}(\lambda)}{\lambda}, \frac{\boldsymbol{\xi}(\lambda)}{\beta_1 - \lambda}, \ldots, \frac{\boldsymbol{\xi}(\lambda)}{\beta_m - \lambda} \right)^{\mathrm{t}} \\ &\quad = \mathcal{L}(\mu(\lambda)) C(\lambda) (\boldsymbol{\xi}(\lambda), \boldsymbol{\xi}(\lambda), \ldots, \boldsymbol{\xi}(\lambda))^{\mathrm{t}} =: \mathcal{L}(\mu(\lambda)) C(\lambda) \hat{\boldsymbol{\xi}}(\lambda) \\ &\quad = \left(l(\lambda)\boldsymbol{\xi}(\lambda), -l(\lambda)\frac{\boldsymbol{\xi}(\lambda)}{\lambda}, l(\lambda)\frac{\boldsymbol{\xi}(\lambda)}{\beta_1 - \lambda}, \ldots, l(\lambda)\frac{\boldsymbol{\xi}(\lambda)}{\beta_m - \lambda} \right)^{\mathrm{t}}. \end{aligned} \tag{4.49}$$

Whence it follows that $C(\lambda)\hat{\boldsymbol{\xi}}(\lambda) =: \hat{\boldsymbol{z}}(\lambda)$ is a root function of order n for the operator pencil $\mathcal{L}(\mu(\lambda))$ at the point λ_0. We calculate its corresponding eigen- and associated elements,

$$\begin{aligned}
\hat{\boldsymbol{z}}_k &= \frac{1}{k!}\frac{\mathrm{d}^k\hat{\boldsymbol{z}}(\lambda)}{\mathrm{d}\lambda^k}\bigg|_{\lambda=\lambda_0} \\
&= \frac{1}{k!}\left(\frac{\mathrm{d}^k\boldsymbol{\xi}(\lambda)}{\mathrm{d}\lambda^k}\bigg|_{\lambda=\lambda_0}, -\frac{\mathrm{d}^k}{\mathrm{d}\lambda^k}\left(\frac{\boldsymbol{\xi}(\lambda)}{\lambda}\right)\bigg|_{\lambda=\lambda_0}, \frac{\mathrm{d}^k}{\mathrm{d}\lambda^k}\left(\frac{\boldsymbol{\xi}(\lambda)}{\beta_1-\lambda}\right)\bigg|_{\lambda=\lambda_0}, \dots,\right. \\
&\qquad \left.\frac{\mathrm{d}^k}{\mathrm{d}\lambda^k}\left(\frac{\boldsymbol{\xi}(\lambda)}{\beta_m-\lambda}\right)\bigg|_{\lambda=\lambda_0}\right)^{\mathrm{t}} \\
&= \frac{1}{k!}\left(\boldsymbol{\xi}_k, \sum_{j=1}^{k}\frac{(-1)^j\boldsymbol{\xi}_{k-j}}{(-\lambda_0)^{j+1}}, \sum_{j=0}^{k}\frac{(-1)^j\boldsymbol{\xi}_{k-j}}{(\beta_1-\lambda_0)^{j+1}}, \dots, \sum_{j-0}^{k}\frac{(-1)^j\boldsymbol{\xi}_{k-j}}{(\beta_m-\lambda_0)^{j+1}}\right)^{\mathrm{t}}, \\
&\qquad k = 0, \dots, n. \qquad (4.50)
\end{aligned}$$

The elements in the form (4.50), where $\boldsymbol{\xi}_0$ is eigen- element and $\boldsymbol{\xi}_k$, $k = 1, \dots, n$, are associated elements of $l(\lambda)$ corresponding to the eigenvalue λ_0 will be called *elements of special form* for $\mathcal{L}(\mu(\lambda))$ at the point $\mu_0 = \mu(\lambda_0)$.

(e) Let $\mu_0 \in \sigma(\mathcal{L}(\mu))$. Consider the root function $\hat{\boldsymbol{u}}(\mu)$ of order n for the operator-function $\mathcal{L}(\mu)$ relatively to the point $\mu_0 = \mu(\xi_j^0)$; then $\mathcal{L}(\mu)\hat{\boldsymbol{u}}(\mu) = \boldsymbol{O}((\mu-\mu_0)^{n+1})$ as $\mu \to \mu_0$. By virtue of the equality $\mu - \mu_0 = \mu'(\xi_j^0)(\lambda - \xi_j^0) + \cdots$, where $\mu'(\xi) = \kappa(\xi) \neq 0$ for $\xi = \xi_j^0$, the function $\hat{\boldsymbol{u}}(\mu(\lambda))$ is a root function of order n of the operator-function $\mathcal{L}(\mu(\lambda))$ relatively to the point ξ_j^0, $j = 0, \dots, m+1$. We introduce the functions $D_j(\lambda)\hat{\boldsymbol{u}}(\mu(\lambda))$; if $D_j(\lambda_0)\hat{\boldsymbol{u}}(\mu_0) \neq \boldsymbol{0}$, then they are root functions of the pencil $\mathcal{L}(\mu(\lambda))$ of order not lower than n, relatively to the points ξ_j^0, $j = 0, \dots, m+1$.

Let us recall that since the spectum of the pencil $\mathcal{L}(\mu)$ is discrete, there is a real number a such that $\mathcal{L}(a)$ is invertible. Then, by the identity,

$$\mathcal{L}(\mu) = \mathcal{L}(a) - \mu\mathcal{H} = \mathcal{L}(a)(\mathcal{I} - \mu\mathcal{L}^{-1}(a)\mathcal{H}),$$

the spectrum of the pencil $\mathcal{L}(\mu)$ coincides with the set of characteristic numbers of the operator $\mathcal{L}^{-1}(a)\mathcal{H}$.

(f) In every root subspace of the operator $\mathcal{L}^{-1}(a)\mathcal{H}$ one can choose a basis consisting of elements of special form.

We carry out the proof of this statement as follows. Let μ_0 be a characteristic number of the operator $\mathcal{L}^{-1}(a)\mathcal{H}$, $\hat{\boldsymbol{M}}_0$ an eigensubspace, and $\hat{\boldsymbol{N}}_0$ a root subspace ($\hat{\boldsymbol{M}}_0 \subset \hat{\boldsymbol{N}}_0$). Denote a basis of eigen- and associated elements in $\hat{\boldsymbol{N}}_0$ by $\hat{\boldsymbol{y}}_k^{(i)}$, $i = 1, \dots, r$, $k = 0, \dots, n$. These elements are eigen- and associated elements for the operator-function $\mathcal{L}(\mu)$, therefore, we can construct r root functions, $\hat{\boldsymbol{u}}^i(\mu)$, $i = 1, \dots, r$, of orders n_i for $\mathcal{L}(\mu)$ relatively to the point μ_0.

Every eigenelement $\hat{\boldsymbol{y}}_0^{(i)}$, $i = 1, \dots, r$, according to formulas (4.47), may be represented in the form

$$\begin{aligned}\hat{\boldsymbol{y}}_0^{(i)} &= D_0(\lambda_0)\hat{\boldsymbol{y}}_0^{(i)} + D_1(\lambda_0)\hat{\boldsymbol{y}}_0^{(i)} + \dots + D_{m+1}(\lambda_0)\hat{\boldsymbol{y}}_0^{(i)} \\ &=: \hat{\boldsymbol{y}}_{0,0}^{(i)} + \hat{\boldsymbol{y}}_{0,1}^{(i)} + \dots + \hat{\boldsymbol{y}}_{0,m+1}^{(i)} \end{aligned} \tag{4.51}$$

As it will be proved below [see Property (g)], in the subspace $\hat{\boldsymbol{M}}_0$ one can choose a basis such that, after relabeling, it consists of the elements

$$\hat{\boldsymbol{y}}_{0,0}^{(1)}, \dots, \hat{\boldsymbol{y}}_{0,0}^{(r_0)}, \hat{\boldsymbol{y}}_{0,1}^{(r_0+1)}, \dots, \hat{\boldsymbol{y}}_{0,1}^{(r_1)}, \hat{\boldsymbol{y}}_{0,2}^{(r_1+1)}, \dots, \hat{\boldsymbol{y}}_{0,2}^{(r_2)}, \dots, \hat{\boldsymbol{y}}_{0,m+1}^{(r_m+1)}, \dots, \boldsymbol{y}_{0,m+1}^{(r_{m+1})}. \tag{4.52}$$

Based on this fact, consider the functions

$$\begin{aligned} &D_0(\mu)\hat{\boldsymbol{u}}^{(i)}(\mu), && 0 \le i \le r_0, \\ &D_1(\mu)\hat{\boldsymbol{u}}^{(i)}(\mu), && r_0 + 1 \le i \le r_1, \\ &\dots \\ &D_{m+1}(\mu)\hat{\boldsymbol{u}}^{(i)}(\mu), && r_m + 1 \le i \le r_{m+1}; \end{aligned} \tag{4.53}$$

we recall that $D_j(\mu) = D_0(\xi_j(\mu))$ where $\xi_j(\mu)$ are branches of the solution of the equation $\mu(\xi) = \mu$, $\xi_j(\mu_0) = \xi_j^0$. The functions (4.53) are root functions of the operator $\mathcal{L}(\mu)$. The eigenelements corresponding to them form a basis in $\hat{\boldsymbol{M}}_0$ and together with the corresponding associated elements, form a linearly independent system in $\hat{\boldsymbol{N}}_0$. Since the length of the ith chain is not less than $n_i + 1$, then it is exactly equal to $n_i + 1$.

It remains to show that the functions (4.53) define eigen- and associated elements of special form. Differentiating a composite function we have

$$\begin{aligned} \hat{\boldsymbol{z}}_0^{(i)} &:= D_p(\mu_0)\hat{\boldsymbol{y}}_0^{(i)} = D_0(\xi_p^0)\hat{\boldsymbol{y}}_0^{(i)}, \\ \hat{\boldsymbol{z}}_k^{(i)} &= \frac{1}{k!}\frac{\mathrm{d}^k}{\mathrm{d}\mu^k}(D_p(\mu)\hat{\boldsymbol{u}}^{(i)}(\mu))\bigg|_{\mu=\mu_0} = \frac{1}{k!}\sum_{j=1}^{m} \frac{d^j D_0(\xi_p)\hat{\boldsymbol{u}}(\mu(\xi_p))}{d\xi_p^j}\bigg|_{\xi_p=\xi_p^0} \cdot \varphi_j^{(p)}(\mu_0), \end{aligned}$$

where $p = p(i)$ and $\varphi_j^{(p)}(\mu_0)$ are some constants appearing due to the change of the spectral parameter (see Section 1.6.2). Now we note that the next relationships are fulfilled,

$$\begin{aligned} D_0(\xi_p)\hat{\boldsymbol{u}}^{(i)}(\mu(\xi_p)) &= \frac{1}{\kappa(\xi_p)}C(\xi_p)\begin{pmatrix} I & -\dfrac{I}{\xi_p} & \cdots & \dfrac{I}{\beta_m - \xi_p} \\ I & -\dfrac{I}{\xi_p} & \cdots & \dfrac{I}{\beta_m - \xi_p} \\ \cdots & \cdots & \cdots & \cdots \\ I & -\dfrac{I}{\xi_p} & \cdots & \dfrac{I}{\beta_m - \xi_p} \end{pmatrix} \cdot \hat{\boldsymbol{u}}^{(i)}(\mu(\xi_p)) \\ &= C(\xi_p)(\boldsymbol{\xi}^{(i)}(\xi_p), \dots, \boldsymbol{\xi}^{(i)}(\xi_p))^{\mathrm{t}}. \end{aligned} \tag{4.54}$$

By (4.49), the functions $\boldsymbol{\xi}^{(i)}(\xi_p)$ are root functions of the operator $l(\xi_p)$. It means that the elements

$$\left.\frac{\mathrm{d}^j D_0(\xi_p)\hat{\boldsymbol{u}}^{(i)}(\mu(\xi_p))}{\mathrm{d}\xi_p^j}\right|_{\xi_p=\xi_p^0} = \frac{\mathrm{d}^j}{\mathrm{d}\xi_p^j}(C(\xi_p)(\boldsymbol{\xi}^{(i)}(\xi_p),\ldots,\boldsymbol{\xi}^{(i)}(\xi_p))^{\mathrm{t}})_{\xi_p=\xi_p^0}$$

are elements of special form relatively to the point ξ_p^0. Statements (f) is proved.

(g) Now we give an exact formulation and a proof of the fact mentioned and used previously with regard to the set (4.52).

Suppose that the n-dimensional vector space R is expressed as a direct sum

$$R = R^{(1)} \dot{+} R^{(2)} \dot{+} \cdots \dot{+} R^{(q)}, \qquad n = n_1 + n_2 + \cdots + n_q.$$

Let the system of vectors $\{x_k\}_{k=1}^n, x_k = x_k^{(1)} + x_k^{(2)} + \cdots + x_k^{(q)}$, form a basis in R. Then in R one can choose a basis in such a way that from each group of vectors $\{x_k^{(1)}, x_k^{(2)}, \ldots, x_k^{(q)}\}$ one and only one element enters it.

Proof. Let $n = q$. Then all the spaces $R^{(k)}$ are one-dimensional; we choose in each of them a nonzero vector r_k. It is obvious that the vectors r_k, $k = 1, \ldots, n$, form a basis in R. Express vectors x_k in the basis $\{r_k\}$,

$$x_k = \sum_{j=1}^{q} x_k^{(j)} = \sum_{j=1}^{n} a_{kj} r_j, \qquad k = 1, \ldots, n.$$

Since the determinant of the matrix $A = (a_{ij})_{i.j=1}^n$ associated with the transition from one basis in R to another differs from zero, then a nonzero product $a_{1d_1}a_{2d_2}\ldots a_{nd_n}$ can be found, where the indexes $d_1, d_2, \ldots, d_n$ give a permutation of the numbers 1, 2, $\ldots, n$. As a required basis we can choose the basis $x_{d_j}^{(j)}$, $j = 1, \ldots, n$.

For $n > q$ we give a proof by induction on n. Decompose the vectors $x_n^{(j)}$ relatively to the basis x_k,

$$x_n^{(j)} = \sum_{m=1}^{n} d_m^{(j)} x_m, \qquad j = 1, \ldots, q. \tag{4.55}$$

Here, not all the numbers $d_n^{(j)}$, $j = 1, \ldots, q$ are equal to zero, since in the opposite case, after adding up the equalities (4.55) we get a linear dependence of x_n on the other vectors x_k, $k = 1, \ldots, n-1$. For definiteness, let us assume that $d_n^{(q)} \neq 0$, and decompose the space $R^{(q)}$ into a direct sum of two subspaces,

$$R^{(q)} = \{\lambda x_n^{(q)}\} \dot{+} R_1^{(q)}.$$

Denote by P the projector onto the subspace

$$R_1 = R^{(1)} \dot{+} R^{(2)} \dot{+} \cdots \dot{+} R_1^{(q)}$$

parallel to the one-dimensional subspace $\{\lambda x_n^{(q)}\}$; then $Px_n^{(q)} = 0$. The vectors Px_k, $k = 1, \ldots, n-1$, are linearly independent and form a basis in R_1. Indeed, in the opposite case,

$$0 = \sum_{k=1}^{n-1} P(a_k x_k) \Leftrightarrow \sum_{k=1}^{n-1} a_k x_k = \lambda x_n^{(q)}.$$

For $\lambda \neq 0$ this relationship contradicts (4.55) with $j = q$, since $d_n^{(q)} \neq 0$, and for $\lambda = 0$ it contradicts the basicity of the vectors $\{x_k\}$.

By the induction assumption, in the space R_1 one can choose a basis in such a way that from each group of vectors $\{x_k^{(1)}, x_k^{(2)}, \ldots, x_k^{(q-1)}, Px_k^{(q)}\}$ only one vector will enter that basis. Changing the labeling it is possible to think that the vectors

$$x_1^{(1)}, \ldots, x_{r_1}^{(1)}, x_{r_1+1}^{(2)}, \ldots, x_{r_2}^{(2)}, \ldots, Px_{r_{q-1}+1}^{(q)}, \ldots, Px_{r_q}^{(q)} \tag{4.56}$$

form a basis in R_1. Then, the desired basis in R is formed by the vectors

$$x_1^{(1)}, \ldots, x_{r_1}^{(1)}, x_{r_1+1}^{(2)}, \ldots, x_{r_2}^{(2)}, \ldots, x_{r_{q-1}+1}^{(q)}, \ldots, x_{r_q}^{(q)}, \tag{4.57}$$

since they are linearly independent.

Indeed, in the opposite case we have a nontrivial equation of the form

$$\sum_k a_k^{(1)} x_k^{(1)} + \sum_k a_k^{(2)} x_k^{(2)} + \cdots + \sum_k a_k^{(q)} x_k^{(q)} = 0. \tag{4.58}$$

Applying the operator P to the left and using the property of linear independence of the vectors (4.56) we get that $a_k^{(j)} = 0$ for all k and j, except $j = q$, $k = n$. Then from (4.58) it follows that $a_n^{(q)} = 0$, in contradiction to the previous assumption. Statement (g) is proved.

11.4.4 The Theorem on Basicity of Special Form Elements

The conclusions obtained in the previous subsections allude to the basic result of this section:

The elements of special form (4.50), where $\lambda_0 \in \sigma(l(\lambda))$, form a Riesz basis in the space $\hat{\boldsymbol{M}}(\Omega) = \boldsymbol{J}_{0,S}(\Omega) \oplus \boldsymbol{M}_0(\Omega) \oplus (\boldsymbol{J}_{0.S}(\Omega))^m$.

The proof of the theorem will be done in several steps.

Step 1. Taking into account the results presented in Section 11.4.3, to prove the theorem it is necessary to establish the substantiality of the requirements (4.29).

First, we prove that from problem (4.13) for the pencil $l(\lambda)$ it is possible to switch to a problem of the same kind with a pencil $l_\varepsilon(\lambda)$ for whose spectrum points both Conditions 1° and 2° from (4.29) are fulfilled. For $\varepsilon > 0$, we put

$$\begin{aligned} l_\varepsilon(\nu) &:= l(\varepsilon\nu) \\ &= I - \nu \left(\frac{\prod_{k=1}^{m} \left(\frac{\gamma_k}{\varepsilon} - \nu \right)}{\prod_{k=1}^{m} \left(\frac{\beta_k}{\varepsilon} - \nu \right)} \right) \varepsilon A - \nu^{-1} \left(\frac{\prod_{k=1}^{m} \left(\frac{\gamma_k}{\varepsilon} - \nu \right)}{\prod_{k=1}^{m} \left(\frac{\beta_k}{\varepsilon} - \nu \right)} \right) \varepsilon^{-1} B \\ &=: I + a(\nu;\varepsilon) A + b(\nu;\varepsilon) B. \end{aligned}$$

The pencil $l_\varepsilon(\nu)$ has the form of the pencil $l(\lambda)$ from (4.2)–(4.4), where the next substitutions have been done,

$$A \to \varepsilon A, \quad B \to \varepsilon B, \quad \gamma_k \to \frac{\gamma_k}{\varepsilon}, \quad \beta_k \to \frac{\beta_k}{\varepsilon}, \qquad k = 1, \dots, m. \tag{4.59}$$

The first condition (4.29) for the pencil $l_\varepsilon(\nu)$ takes the form

$$\kappa(\nu;\varepsilon) := 1 + \nu^{-2} \sum_{k=1}^{m} \left(\frac{\beta_k}{\varepsilon} - \nu \right)^{-2} \neq 0, \qquad \nu \in \sigma(l_\varepsilon(\nu)). \tag{4.60}$$

For $\varepsilon \to 0$, the equation $\kappa(\nu;\varepsilon) = 0$ has a pair of bounded roots

$$\nu_0^{\pm}(\varepsilon) = \pm \mathrm{i} \mp \frac{\mathrm{i}\varepsilon^2}{2} \left(\sum_{k=1}^{m} \beta_k^2 \right) + \mathrm{O}(\varepsilon^3). \tag{4.61}$$

Based on (4.61) and assuming in formula (4.58) for $a(\nu;\varepsilon)$ that

$$\varepsilon\nu = \varepsilon\nu_0^{\pm}(\varepsilon) =: \mathrm{i}\delta, \qquad \text{where } \delta = \delta(\varepsilon) = \mathrm{O}(\varepsilon) \text{ as } \varepsilon \to 0,$$

we have

$$
\begin{aligned}
a(\nu;\varepsilon)
&= -\mathrm{i}\delta\cdot\frac{\prod_{k=1}^{m}(\gamma_k-\mathrm{i}\delta)}{\prod_{k=1}^{m}(\beta_k-\mathrm{i}\delta)}\\
&= -\mathrm{i}\delta\cdot\frac{\prod_{k=1}^{m}\gamma_k}{\prod_{k=1}^{m}\beta_k}-\delta^2\cdot\frac{\left\{\left(\sum_{k=1}^{m}\prod_{i\neq k}\gamma_i\right)\left(\prod_{k=1}^{m}\beta_k\right)-\left(\sum_{k=1}^{m}\prod_{i\neq k}\beta_i\right)\left(\prod_{k=1}^{m}\gamma_k\right)\right\}}{\prod_{k=1}^{m}\beta_k^2}+\mathrm{O}(\delta^3)\\
&=: -\mathrm{i}\delta\cdot\frac{\prod_{k=1}^{m}\gamma_k}{\prod_{k=1}^{m}\beta_k}-\alpha\delta^2+\mathrm{O}(\delta^3).
\end{aligned}
\tag{4.62}
$$

A similar calculation for $b(\nu;\varepsilon)$ gives

$$
b(\nu;\varepsilon) = -(\mathrm{i}\delta)^{-1}+\alpha+\mathrm{O}(\delta), \qquad \text{as } \varepsilon\to 0. \tag{4.63}
$$

Based on the inequalities $0<\gamma_1<\beta_1<\cdots<\gamma_m<\beta_m<\infty$ we conclude that

$$
\alpha = \frac{\prod_{k=1}^{m}\gamma_k\beta_k^{-1}}{\sum_{k=1}^{m}(\gamma_k^{-1}-\beta_k^{-1})} > 0. \tag{4.64}
$$

Therefore, from (4.62), (4.63), for the indicated values of $\varepsilon\nu$ and a small enough $\varepsilon>0$ we have

$$
\operatorname{Re} l_\varepsilon(\nu) = I-\varepsilon^2[\alpha+\mathrm{O}(\varepsilon)]A+[\alpha+\mathrm{O}(\varepsilon)]B \gg 0. \tag{4.65}
$$

Whence it follows that the operator $l_\varepsilon(\nu)$ for $\nu=\nu_0^{\pm}(\varepsilon)$ and a small $\varepsilon>0$ is invertible and, therefore, in a neighbourhood of this point there is no spectrum.

Now we consider the remaining roots of the equation $\kappa(\nu;\varepsilon)=0$. For a small $\varepsilon>0$ they are divided into m pairs,

$$
\nu_k^{\pm}(\varepsilon) = \pm\mathrm{i}-\frac{\beta_k}{\varepsilon}+\mathrm{O}(\varepsilon), \qquad k=1,\dots,m, \text{ as } \varepsilon\to 0.
$$

Substituting $\varepsilon\nu=\mathrm{i}\delta-\beta_k+\mathrm{O}(\varepsilon^2)$, where $\delta=\delta(\varepsilon)=\mathrm{O}(\varepsilon)$, in the functions $a(\nu;\varepsilon)$

and $b(\nu;\varepsilon)$ we have

$$\operatorname{Im} a(\nu;\varepsilon) = \rho_k \operatorname{Im}(\mathrm{i}\delta)^{-1} + \mathrm{O}(\varepsilon), \qquad \rho_k := -\beta_k \cdot \frac{\prod_{i\neq k}(\gamma_i - \beta_k)}{\prod_{i\neq k}(\beta_i - \beta_k)},$$

$$\operatorname{Im} b(\nu;\varepsilon) = \frac{\rho_k}{\beta_k^2} \operatorname{Im}(\mathrm{i}\delta)^{-1} + \mathrm{O}(\varepsilon), \qquad \text{as } \varepsilon \to 0. \tag{4.66}$$

From (4.66) it follows that for any small $\varepsilon > 0$ the operator

$$\operatorname{Im} l_\varepsilon(\nu) = (\rho_k \operatorname{Im}(\mathrm{i}\delta)^{-1} + \mathrm{O}(\varepsilon))A + \left(\frac{\rho_k}{\beta_k^2}\operatorname{Im}(\mathrm{i}\delta)^{-1} + \mathrm{O}(\varepsilon)\right) B \tag{4.67}$$

is of a fixed sign, and, therefore, from the equation $\operatorname{Im}(l_\varepsilon(\nu)\boldsymbol{\xi},\boldsymbol{\xi}) = 0$ it follows that $\boldsymbol{\xi} = \mathbf{0}$. Thus the operator $l_\varepsilon(\nu)$ is invertible in neighbourhoods of the roots $\nu_k^{\pm}(\varepsilon)$.

Thus, for small $\varepsilon > 0$, Condition 1° from (4.29) is fulfilled for the functions $\kappa(\nu;\varepsilon)$. The previous considerations showed also that, for small ε, all the roots of the equation $\kappa(\nu;\varepsilon) = 0$ are different in pairs, that is, Condition 2° from (4.29) is also fulfilled. With this, the proof of Step 1 is concluded.

Step 2. We now assume that the transition indicated in Step 1 has already been done, and we again return to pencil $l(\lambda)$ supposing that condtions (4.29) are fulfilled for it.

Consider the equation

$$\mu(\lambda) := \lambda - \lambda^{-1} + \sum_{k=1}^{m}(\beta_k - \lambda)^{-1} = \mu_0 \tag{4.68}$$

for any $\mu_0 \in \mathbb{C}$. Since acccording to the second condition (4.29) the roots of the equation $\mu'(\lambda) = \kappa(\lambda) = 0$ are different in pairs, then equation (4.68) has no roots of multiplicity greater than two. Let it have r two-multiple roots, $\lambda_1, \dots, \lambda_r$, and s simple roots, $\xi_1, \dots, \xi_s$, $2r + s = m + 2$.

Subtracting the left and right sides of the equations $\mu(\lambda_p) = \mu_0$, $\mu(\xi_q) = \mu_0$, $1 \le p \le r$, $1 \le q \le s$, we get the identities

$$1 + \frac{1}{(\lambda_p \xi_q)} + \sum_{k=1}^{m}((\beta_k - \lambda_p)(\beta_k - \xi_q))^{-1} = 0, \qquad 1 \le p \le r, 1 \le q \le s, \tag{4.69}$$

$$1 + \frac{1}{(\lambda_i \lambda_j)} + \sum_{k=1}^{m}((\beta_k - \lambda_i)(\beta_k - \lambda_j))^{-1} = 0, \qquad i \neq j, 1 \le i, j \le r, \tag{4.70}$$

$$1 + \frac{1}{(\xi_i \xi_j)} + \sum_{k=1}^{m}((\beta_k - \xi_i)(\beta_k - \xi_j))^{-1} = 0, \qquad i \neq j, 1 \le i, j \le s, \tag{4.71}$$

The two-multiple roots λ_p satisfy the equation

$$\kappa'(\lambda_p) = 1 + \lambda_p^{-2} + \sum_{k=1}^{m}(\beta_k - \lambda_p)^{-2} = 0, \qquad 1 \le p \le r. \tag{4.72}$$

Subtracting from the left and right sides of (4.72) the corresponding sides of (4.69) and (4.70) we get the relations

$$-\frac{1}{(\lambda_p^2\xi_q)} + \sum_{k=1}^{m}(\beta_k - \lambda_p)^{-2}(\beta_k - \xi_q)^{-1} = 0, \qquad 1 \le p \le r, 1 \le q \le s, \tag{4.73}$$

$$-\frac{1}{(\lambda_i^2\lambda_j)} + \sum_{k=1}^{m}(\beta_k - \lambda_i)^{-2}(\beta_k - \lambda_j)^{-1} = 0, \qquad 1 \ne j, 1 \le i, j \le r. \tag{4.74}$$

Now, differentiate by λ the identity (4.31) and write it at the points $\lambda = \lambda_p$, $p = 1, \dots, r$. Since $\mathrm{d}\mathcal{L}(\mu(\lambda))/\mathrm{d}\lambda = \mathcal{L}'(\mu)\mu'(\lambda) = 0$ for $\lambda = \lambda_p$, whence we get

$$\begin{aligned}\mathcal{L}(\mu(\lambda))&\left(\mathbf{0}, \frac{\boldsymbol{\xi}}{\lambda^2}, \frac{\boldsymbol{\xi}}{(\beta_1-\lambda)^2}, \dots, \frac{\boldsymbol{\xi}}{(\beta_m-\lambda)^2}\right)^{\mathrm{t}} \\ &= \left(l'(\lambda)\boldsymbol{\xi},\ -l'(\lambda)\frac{\boldsymbol{\xi}}{\lambda} + l(\lambda)\frac{\boldsymbol{\xi}}{\lambda^2},\ l'(\lambda)\frac{\boldsymbol{\xi}}{(\beta_1-\lambda)} + l(\lambda)\frac{\boldsymbol{\xi}}{(\beta_1-\lambda)^2}, \dots,\right. \\ &\qquad \left. l'(\lambda)\frac{\boldsymbol{\xi}}{(\beta_m-\lambda)} + l(\lambda)\frac{\boldsymbol{\xi}}{(\beta_m-\lambda)^2}\right)^{\mathrm{t}}, \qquad \lambda = \lambda_p.\end{aligned} \tag{4.75}$$

From (4.75) and (4.32) we get the invariance relatively to $\mathcal{L}(\mu(\lambda))$ for $\lambda = \lambda_p$ of the subspace of elements in the form

$$\begin{aligned}\hat{\boldsymbol{H}}^1(\lambda) := \Bigg\{\hat{\boldsymbol{x}} \in \hat{\boldsymbol{H}} : \hat{\boldsymbol{x}} = &\left(\boldsymbol{y}, -\frac{\boldsymbol{y}}{\lambda} + \frac{\boldsymbol{z}}{\lambda^2}, \frac{\boldsymbol{y}}{(\beta_1-\lambda)} + \frac{\boldsymbol{z}}{(\beta_1-\lambda)^2}, \dots,\right. \\ &\left.\frac{\boldsymbol{y}}{(\beta_m-\lambda)} + \frac{\boldsymbol{z}}{(\beta_m-\lambda)^2}\right)^{\mathrm{t}}, \ \boldsymbol{y}, \boldsymbol{z} \in \boldsymbol{J}_{0,S}(\Omega)\Bigg\}.\end{aligned} \tag{4.76}$$

Step 3. The following relationships take place,

$$\hat{\boldsymbol{H}}^1(\lambda_i) \cap \hat{\boldsymbol{H}}^1(\lambda_j) = \{\hat{\mathbf{0}}\}, \qquad 1 \le i, j \le r,\ i \ne j. \tag{4.77}$$

We continue with a proof of the properties (4.77). Let $\hat{\boldsymbol{x}} \in \hat{\boldsymbol{H}}^1(\lambda_i) \cap \hat{\boldsymbol{H}}^1(\lambda_j)$. Then by (4.76) we have

$$\begin{aligned}\hat{\boldsymbol{x}} &= \left(\boldsymbol{\xi}_i, -\frac{\boldsymbol{\xi}_i}{\lambda_i} + \frac{\boldsymbol{\eta}_i}{\lambda_i^2}, \frac{\boldsymbol{\xi}_i}{(\beta_1-\lambda_i)} + \frac{\boldsymbol{\eta}_i}{(\beta_1-\lambda_i)^2}, \dots, \frac{\boldsymbol{\xi}_i}{(\beta_m-\lambda_i)} + \frac{\boldsymbol{\eta}_i}{(\beta_m-\lambda_i)^2}\right)^{\mathrm{t}} \\ &= \left(\boldsymbol{\xi}_j, -\frac{\boldsymbol{\xi}_j}{\lambda_j} + \frac{\boldsymbol{\eta}_j}{\lambda_j^2}, \frac{\boldsymbol{\xi}_j}{(\beta_1-\lambda_j)} + \frac{\boldsymbol{\eta}_j}{(\beta_1-\lambda_j)^2}, \dots, \frac{\boldsymbol{\xi}_j}{(\beta_m-\lambda_j)} + \frac{\boldsymbol{\eta}_j}{(\beta_m-\lambda_j)^2}\right)^{\mathrm{t}}.\end{aligned}$$

Comparing the first components we get $\boldsymbol{\xi}_i = \boldsymbol{\xi}_j =: \boldsymbol{\xi}$, and then comparing the remaining ones, we obtain the equations

$$-\frac{1}{\lambda_i}\boldsymbol{\xi} + \frac{1}{\lambda_i^2}\boldsymbol{\eta}_i = -\frac{1}{\lambda_j}\boldsymbol{\xi} + \frac{1}{\lambda_j^2}\boldsymbol{\eta}_j, \tag{4.78}$$

$$\frac{\boldsymbol{\xi}}{(\beta_k - \lambda_i)} + \frac{\boldsymbol{\eta}_i}{(\beta_k - \lambda_i)^2} = \frac{\boldsymbol{\xi}}{(\beta_k - \lambda_j)} + \frac{\boldsymbol{\eta}_j}{(\beta_k - \lambda_j)^2}, \qquad k = 1, \ldots, m. \tag{4.79}$$

We divide (4.78) by $(-\lambda_j)$, and each of the equations (4.79) by $(\beta_k - \lambda_j)$; after adding the left and right sides of the obtained equations we have the relations

$$\boldsymbol{\xi}\left(\frac{1}{\lambda_i\lambda_j} + \sum_{k=1}^{m} \frac{1}{(\beta_k - \lambda_i)(\beta_k - \lambda_j)}\right) - \boldsymbol{\eta}_i\left(-\frac{1}{\lambda_i^2\lambda_j} + \sum_{k=1}^{m} \frac{1}{(\beta_k - \lambda_i)^2(\beta_k - \lambda_j)}\right)$$
$$= \boldsymbol{\xi}\left(\frac{1}{\lambda_j^2} + \sum_{k=1}^{m} \frac{1}{(\beta_k - \lambda_j)^2}\right) - \boldsymbol{\eta}_j\left(-\frac{1}{\lambda_j^3} + \sum_{k=1}^{m} \frac{1}{(\beta_k - \lambda_j)^3}\right). \tag{4.80}$$

Using the identities (4.70), (4.72), (4.74), from (4.80) we deduce the relation

$$\boldsymbol{\eta}_j\left(-\lambda_j^{-3} + \sum_{k=1}^{m}(\beta_k - \lambda_j)^{-3}\right) = \kappa'(\lambda_j)\boldsymbol{\eta}_j = \mathbf{0}. \tag{4.81}$$

Since the root λ_j from equation (4.68) could not be three-multiple, that is, $\mu''(\lambda_j) = \kappa'(\lambda_j) \neq 0$, then from (4.81) it follows that $\boldsymbol{\eta}_j = \mathbf{0}$. Similarly, using a division of (4.79) by $(\beta_k - \lambda_i)$ and the same considerations, it follows that $\boldsymbol{\eta}_i = \mathbf{0}$. Then, from (4.78) it follows that $\boldsymbol{\xi} = \mathbf{0}$, that is, $\hat{\boldsymbol{x}} = \mathbf{0}$, and the statements (4.77) are completely proved.

Step 4. Based on (4.77), we form the subspace

$$\hat{\boldsymbol{H}}^r := \sum_{p=1}^{r} (\dot{+}) \hat{\boldsymbol{H}}^1(\lambda_p), \tag{4.82}$$

where $\hat{\boldsymbol{H}}^1(\lambda_p)$ is the subspace of the form (4.76) corresponding to the multiple root λ_p, $p = 1, \ldots, r$, of equation (4.68). As it has been proved previously, $\hat{\boldsymbol{H}}^r$ is invariant with respect to the operator $\mathcal{L}(\mu_0)$.

Similarly to the way we proved the existence of the subspace $\hat{\boldsymbol{H}}'(\lambda)$ from (4.33), for $\mathcal{L}(\mu_0)$ one more invariant subspace can be obtained along with $\hat{\boldsymbol{H}}^r$, namely,

$$\hat{\boldsymbol{H}}_r := \left\{\hat{\boldsymbol{x}} := (\boldsymbol{x}, \boldsymbol{x}_0, \boldsymbol{x}_1, \ldots, \boldsymbol{x}_m)^{\mathrm{t}} : \boldsymbol{x} - \frac{\boldsymbol{x}_0}{\lambda_p} + \sum_{k=1}^{m} \frac{\boldsymbol{x}_k}{(\beta_k - \lambda_p)} = \mathbf{0},\right.$$
$$\left.\frac{\boldsymbol{x}_0}{\lambda_p^2} + \sum_{k=1}^{m} \frac{\boldsymbol{x}_k}{(\beta_k - \lambda_p)^2} = \mathbf{0}, 1 \le p \le r\right\}. \tag{4.83}$$

The next direct decomposition takes place,

$$\hat{\boldsymbol{H}} = \hat{\boldsymbol{H}}^r (\dot{+}) \hat{\boldsymbol{H}}_r. \tag{4.84}$$

To prove this fact, we need to show that the equation

$$\hat{\boldsymbol{x}} + \hat{\boldsymbol{z}} = \hat{\boldsymbol{y}} = (\boldsymbol{y}, \boldsymbol{y}_0, \boldsymbol{y}_1, \dots, \boldsymbol{y}_m)^{\mathrm{t}}, \tag{4.85}$$

for any $\hat{\boldsymbol{y}} \in \hat{\boldsymbol{H}}$ has a unique solution $\hat{\boldsymbol{x}} \in \hat{\boldsymbol{H}}_r$, and

$$\hat{\boldsymbol{z}} = \Bigg(\sum_{p=1}^{r} \boldsymbol{z}_p, \sum_{p=1}^{r} \Big(-\frac{\boldsymbol{z}_p}{\lambda_p} + \frac{\boldsymbol{w}_p}{\lambda_p^2} \Big), \sum_{p=1}^{r} \Big(\frac{\boldsymbol{z}_p}{\beta_1 - \lambda_p} + \frac{\boldsymbol{w}_p}{(\beta_1 - \lambda_p)^2} \Big), \dots, \\ \sum_{p=1}^{r} \Big(\frac{\boldsymbol{z}_p}{\beta_m - \lambda_p} + \frac{\boldsymbol{w}_p}{(\beta_m - \lambda_p)^2} \Big) \Bigg)^{\mathrm{t}} \in \hat{\boldsymbol{H}}^r.$$

Taking into account the equations (4.83) for any element $\hat{\boldsymbol{x}} \in \hat{\boldsymbol{H}}^r$, from (4.85) we get the system of equations

$$\sum_{p=1}^{r} \boldsymbol{z}_p + \boldsymbol{x} = \boldsymbol{y}, \tag{4.86}$$

$$\sum_{p=1}^{r} \Big(-\frac{\boldsymbol{z}_p}{\lambda_p} + \frac{\boldsymbol{w}_p}{\lambda_p^2} \Big) + \boldsymbol{x}_0 = \boldsymbol{y}_0, \tag{4.87}$$

$$\sum_{p=1}^{r} \Big(\frac{\boldsymbol{z}_p}{\beta_k - \lambda_p} + \frac{\boldsymbol{w}_p}{(\beta_k - \lambda_p)^2} \Big) + \boldsymbol{x}_k = \boldsymbol{y}_k, \qquad k = 1, \dots, m, \tag{4.88}$$

$$\boldsymbol{x} - \frac{\boldsymbol{x}_0}{\lambda_p} + \sum_{k=1}^{m} \frac{\boldsymbol{x}_k}{\beta_k - \lambda_p} = \boldsymbol{0},$$

$$\frac{\boldsymbol{x}_0}{\lambda_p^2} + \sum_{k=1}^{m} \frac{\boldsymbol{x}_k}{(\beta_k - \lambda_p)^2} = \boldsymbol{0}, \qquad 1 \le p \le r. \tag{4.89}$$

We divide (4.87) by $(-\lambda_i)$ and (4.88) by $(\beta_k - \lambda_i)$, respectively, add them together with (4.86), and get

$$\sum_{p=1}^{m} \boldsymbol{z}_p \Big(1 + \frac{1}{\lambda_i \lambda_p} + \sum_{k=1}^{m} \frac{1}{(\beta_k - \lambda_i)(\beta_k - \lambda_p)} \Big) - \boldsymbol{w}_i \Big(-\frac{1}{\lambda_i^3} + \sum_{k=1}^{m} \frac{1}{(\beta_k - \lambda_i)^3} \Big) \\ + \sum_{p \ne i} \boldsymbol{w}_p \Big(-\frac{1}{\lambda_p^2 \lambda_i} + \sum_{k=1}^{m} \frac{1}{(\beta_k - \lambda_p)^2 (\beta_k - \lambda_i)} \Big) + \boldsymbol{x} - \frac{\boldsymbol{x}_0}{\lambda_i} + \sum_{k=1}^{m} \frac{\boldsymbol{x}_k}{\beta_k - \lambda_i} \\ = \boldsymbol{y} - \frac{\boldsymbol{y}_0}{\lambda_i} + \sum_{k=1}^{m} \frac{\boldsymbol{y}_k}{\beta_k - \lambda_i}, \qquad 1 \le i \le r. \tag{4.90}$$

By the identities (4.70) and (4.74) and the first relationship (4.89), using again the property $\kappa'(\lambda_i) \neq 0$ we get that $\boldsymbol{w}_i$, $i = 1, \dots, r$, are uniquely defined by the components $\hat{\boldsymbol{y}}$.

Dividing (4.87) by λ_i^2 and (4.88) by $(\beta_k - \lambda_i)^2$ and then adding the resulting equations we obtain the relations

$$\begin{aligned}
&\boldsymbol{z}_i\left(-\frac{1}{\lambda_i^3} + \sum_{k=1}^{m} \frac{1}{(\beta_k - \lambda_i)^3}\right) + \sum_{p \neq i} \boldsymbol{z}_p\left(-\frac{1}{\lambda_i^2 \lambda_p} + \sum_{k=1}^{m} \frac{1}{(\beta_k - \lambda_i)^2 (\beta_k - \lambda_p)}\right) \\
&\quad - \sum_{p=1}^{m} \boldsymbol{w}_p\left(\frac{1}{\lambda_p^2 \lambda_i^2} + \sum_{k=1}^{m} \frac{1}{(\beta_k - \lambda_i)^2 (\beta_k - \lambda_p)^2}\right) + \frac{\boldsymbol{x}_0}{\lambda_i^2} + \sum_{k=1}^{m} \frac{\boldsymbol{x}_k}{(\beta_k - \lambda_i)^2} \\
&\quad = \frac{\boldsymbol{y}_0}{\lambda_i^2} + \sum_{k=1}^{m} \frac{\boldsymbol{y}_k}{(\beta_k - \lambda_i)^2}, \qquad i = 1, \dots, r. \qquad (4.91)
\end{aligned}$$

Hence, by using (4.74) and the second equation in (4.89), we get that the components $\boldsymbol{z}_i$, $i = 1, \dots, r$, are also uniquely defined by the components $\hat{\boldsymbol{y}}$ and $\hat{\boldsymbol{w}}$, $i = 1, \dots, r$.

Finally, we observe that $\boldsymbol{x}, \boldsymbol{x}_0$, and $\boldsymbol{x}_k$, $k = 1, \dots, m$, are uniquely determined from (4.86)–(4.88). Hence the decomposition (4.84) follows.

Step 5. For the simple roots ξ_q, $q = 1, \dots, s$, of equation (4.68), we introduce the subspaces of the form (4.30),

$$\hat{\boldsymbol{H}}(\xi_q) := \left\{ \hat{\boldsymbol{y}} \in \hat{\boldsymbol{H}} : \hat{\boldsymbol{y}} = \left(\boldsymbol{y}, -\frac{\boldsymbol{y}}{\xi_q}, \frac{\boldsymbol{y}}{\beta_1 - \xi_q}, \dots, \frac{\boldsymbol{y}}{\beta_m - \xi_q} \right)^{\mathrm{t}} \right\}. \qquad (4.92)$$

We prove that the next formula is valid,

$$\hat{\boldsymbol{H}} = \hat{\boldsymbol{H}}^{\mathrm{r}} (\dot{+}) \left(\sum_{q=1}^{s} (\dot{+}) \hat{\boldsymbol{H}}(\xi_q) \right). \qquad (4.93)$$

By the decomposition (4.84),it is enough to show that

$$\hat{\boldsymbol{H}}_r = \sum_{q=1}^{s} (\dot{+}) \hat{\boldsymbol{H}}(\xi_q). \qquad (4.94)$$

This is equivalent to checking that the system of equations

$$\begin{aligned}
\sum_{q=1}^{s} \boldsymbol{z}_q &= \boldsymbol{x}, \\
-\sum_{q=1}^{s} \frac{\boldsymbol{z}_q}{\xi_q} &= \boldsymbol{x}_0, \qquad (4.95) \\
\sum_{q=1}^{s} \frac{\boldsymbol{z}_q}{\beta_k - \xi_q} &= \boldsymbol{x}_k, \qquad k = 1, \dots, m, \qquad (4.96)
\end{aligned}$$

with the constraints (4.89), has a unique solution.

The system consisting of the equations (4.95) and the first $s-2$ equations (4.96) has a unique solution, which by Cramer's formulas is representable as

$$\boldsymbol{z}_q = \frac{\boldsymbol{\Delta}_q}{\Delta}, \qquad 1 \le q \le s. \tag{4.97}$$

Here, the determinant $\Delta = \Delta(\lambda)$ is expressed by formula (4.42), it involves the coefficients for $\boldsymbol{z}_q$, $q = 1, \dots, s$, and is not equal to zero, as it has been proved in Section 11.4.3. The determinants $\boldsymbol{\Delta}_q$ are obtained from Δ by substituting column q in (4.42) by the column of the right sides from (4.95) and (4.96). Expanding $\boldsymbol{\Delta}_q$ by column q we have

$$\boldsymbol{z}_q = \boldsymbol{x}\frac{\Delta_{1q}}{\Delta} + \boldsymbol{x}_0\frac{\Delta_{2q}}{\Delta} + \sum_{k=1}^{s-2} \boldsymbol{x}_k \frac{\Delta_{kq}}{\Delta}, \tag{4.98}$$

where Δ_{ij} is the algebraic complement of the element situated in row i and column j of the determinant Δ. Substituting $\boldsymbol{z}_q$ from (4.97) in the last $m-(s-2) = 2r$ equations in (4.96) we get

$$\boldsymbol{x}_k = \frac{1}{\Delta}\sum_{q=1}^{s} \frac{\boldsymbol{\Delta}_q}{(\beta_k - \xi_q)}, \qquad s-1 \le k \le m. \tag{4.99}$$

The equations (4.89) allow to express $\boldsymbol{x}_k$, $s-1 \le k \le m$ in terms of $\boldsymbol{x}, \boldsymbol{x}_0, \boldsymbol{x}_k$, $1 \le k \le s-2$, since the determinant Δ_0 formed by coefficients of $\boldsymbol{x}_k$, $s-1 \le k \le m$, is different from zero. Let us prove the fact.

The odd rows of this determinant have the form

$$\left(\frac{1}{\beta_{s-1} - \lambda_p}, \dots, \frac{1}{\beta_m - \lambda_p}\right),$$

and the even rows are of the form

$$\left(\frac{1}{(\beta_{s-1} - \lambda_p)^2}, \dots, \frac{1}{(\beta_m - \lambda_p)^2}\right), \qquad 1 \le p \le r.$$

Consider the determinant Δ' differing from Δ_0 by even rows of the form

$$\left(\frac{1}{\beta_{s-1} - \lambda'_p}, \dots, \frac{1}{\beta_m - \lambda'_p}\right).$$

Applying a formula presented by D. K. Faddeev and I. S. Sominsky on p. 48 of their book entitled "Problems and Exercises in Algebra" [1] we get

$$\Delta' = \prod_{i<j}(\lambda_i - \lambda_j) \prod_{i<j}(\lambda_i - \lambda'_i) \prod_{i \le j}(\lambda_i - \lambda'_j) \prod_{i<j}(\lambda'_i - \lambda_j)$$
$$\cdot \prod_{i<j}(\beta_i - \beta_j) \prod_{i.j}(\beta_j - \lambda_i)^{-1} \prod_{i.j}(\beta_j - \lambda'_i)^{-1}.$$

Assuming now that $\lambda'_p = \lambda_p + \delta$, subtracting from each even row the previous one, and dividing Δ' by $(-\delta)^r$, we get

$$\Delta_0 = \lim_{\delta\to 0} \frac{\Delta'(\delta)}{(-\delta)^r} = \prod_{i<j}(\lambda_i - \lambda_j)^2 \prod_{i<j}(\beta_i - \beta_j) \prod_{i,j}(\beta_j - \lambda_i)^2 \neq 0.$$

To finish the proof of the unique solvability of the system of the equations (4.95), (4.96), (4.89) it remains to check that the expressions (4.99) identically satisfy the relationships (4.89). Substituting (4.99) into the first equation (4.89) we have

$$\boldsymbol{x} - \frac{\boldsymbol{x}_0}{\lambda_p} + \sum_{k=1}^{s-2} \frac{\boldsymbol{x}_k}{\beta_k - \lambda_p} + \sum_{k=s-1}^{m} \sum_{q=1}^{s} \frac{\frac{\boldsymbol{\Delta}_q}{\Delta}}{(\beta_k - \xi_p)(\beta_k - \lambda_p)} = \boldsymbol{0}. \tag{4.100}$$

Changing the order of summation and using identity (4.69) we get

$$\boldsymbol{x} - \frac{\boldsymbol{x}_0}{\lambda_p} + \sum_{j=1}^{s-2} \frac{\boldsymbol{x}_j}{\beta_j - \lambda_p} - \sum_{q=1}^{s} \frac{\boldsymbol{\Delta} q}{\Delta}\left(1 + \frac{1}{\lambda_p \xi_q} + \sum_{k=1}^{s-2} \frac{1}{(\beta_k - \lambda_p)(\beta_k - \xi_q)}\right) = \boldsymbol{0} \tag{4.101}$$

Using formula (4.98), we write the coefficients for $\boldsymbol{x}, \boldsymbol{x}_0$ and $\boldsymbol{x}_k$ respectively; we get

$$1 - \sum_{q=1}^{s} \frac{\Delta_{1q}}{\Delta}\left(1 + \frac{1}{\xi_q \lambda_p} + \sum_{k=1}^{s-2} \frac{1}{(\beta_k - \lambda_p)(\beta_k - \xi_q)}\right) = 0,$$

$$-\frac{1}{\lambda_p} - \sum_{q=1}^{s} \frac{\Delta_{2q}}{\Delta}\left(1 + \frac{1}{\lambda_p \xi_q} + \sum_{k=1}^{s-2} \frac{1}{(\beta_k - \lambda_p)(\beta_k - \xi_q)}\right) = 0,$$

$$\frac{1}{\beta_j - \lambda_p} - \sum_{q=1}^{s} \frac{\Delta_{jq}}{\Delta}\left(1 + \frac{1}{\lambda_p \xi_q} + \sum_{k=1}^{s-2} \frac{1}{(\beta_k - \lambda_p)(\beta_k - \xi_q)}\right) = 0. \tag{4.102}$$

To check the relations (4.102) it is enough to open the brackets and use the equalities

$$\sum_{q=1}^{s} \frac{\Delta_{1q}}{\Delta} = 1,$$

$$\sum_{q=1}^{s} \frac{\Delta_{1q}}{\Delta} \cdot \frac{1}{\xi_q} = 0,$$

$$\sum_{q=1}^{m} \frac{\Delta_{1q}}{\Delta} \cdot \frac{1}{(\beta_k - \xi_q)} = 0, \tag{4.103}$$

known from matrix theory.

The second group of relations (4.89) for $p = 1, \dots, r$, is similarly checked. This completes the proof of the decomposition (4.94).

Step 6. Coming to the last stage of the proof of the theorem on the basicity of elements of special form formulated at the beginning of Section 11.4.4, we use the decompositions (4.82), (4.93). The problem

$$\mathcal{L}(\mu_0)\hat{\boldsymbol{y}} = \hat{\boldsymbol{0}}, \qquad \hat{\boldsymbol{y}} \in \hat{\boldsymbol{H}}, \tag{4.104}$$

with the spectral parameter μ_0, is equivalent to a system of two problems corresponding to the invariant subspaces (4.82) and (4.94) for the operator $\mathcal{L}(\mu_0)$, namely,

$$\begin{aligned} &\mathcal{L}(\mu_0(\xi_q))\hat{\boldsymbol{y}}_q = \boldsymbol{0}, \\ &\qquad \hat{\boldsymbol{y}}_q = \left(\boldsymbol{y}_q,\ -\frac{\boldsymbol{y}_q}{\xi_q},\ \frac{\boldsymbol{y}_q}{\beta_1 - \xi_q}, \dots,\ \frac{\boldsymbol{y}_q}{\beta_m - \xi_q}\right)^{\mathrm{t}} \in \hat{\boldsymbol{H}}_r, \qquad 1 \le q \le s, \end{aligned} \tag{4.105}$$

and

$$\begin{aligned} &\mathcal{L}(\mu_0(\lambda_p))\hat{\boldsymbol{z}}_p = \hat{\boldsymbol{0}}, \\ &\qquad \hat{\boldsymbol{z}}_p = \left(\boldsymbol{z}_p,\ -\frac{\boldsymbol{z}_p}{\lambda_p} + \frac{\boldsymbol{w}_p}{\lambda_p^2},\ \frac{\boldsymbol{z}_p}{\beta_1 - \lambda_p} + \frac{\boldsymbol{w}_p}{(\beta_1 - \lambda_p)^2}, \dots, \right. \\ &\qquad\qquad \left. \frac{\boldsymbol{z}_p}{\beta_m - \lambda_p} + \frac{\boldsymbol{w}_p}{(\beta_m - \lambda_p)^2}\right)^{\mathrm{t}} \in \hat{\boldsymbol{H}}^r, \qquad 1 \le p \le r. \end{aligned} \tag{4.106}$$

We use for the equation (4.106) the relations (4.49) and (4.75) that lead to the following equations

$$\begin{aligned} l(\lambda_p)\boldsymbol{w}_p &= \boldsymbol{0}, \\ l(\lambda_p)\boldsymbol{z}_p + l'(\lambda_p)\boldsymbol{w}_p &= \boldsymbol{0}, \qquad 1 \le p \le r. \end{aligned} \tag{4.107}$$

We recall now that at the points λ_p, which, as two-multiple roots of the equation (4.68) are roots of the equation $\mu'(\lambda) = \kappa(\lambda) = 0$, the operator $l(\lambda)$ is invertible. This fact was ascertained in the proof of Statement 1° [see formula (4.65) and the considerations immediately following]. Therefore, from the first equation (4.107) it follows that $\boldsymbol{w}_p = \boldsymbol{0}$, and then, from the second one, also $\boldsymbol{z}_p = \boldsymbol{0}$, $1 \le p \le r$.

Thus, problem (4.106) has only the trivial solution. Therefore, any solution $\hat{\boldsymbol{y}}$ of the spectral problem (4.104) belongs to the subspace $\hat{\boldsymbol{H}}_r$, and by (4.94) is a solution of one of the problems (4.105). As it was shown in Section 11.4.3, the solutions of problem (4.104) are elements of special form (4.50), and they form a basis in the space $\hat{\boldsymbol{M}}(\Omega) = \boldsymbol{J}_{0,S}(\Omega) \oplus \boldsymbol{M}_0(\Omega) \oplus (\boldsymbol{J}_{0.S}(\Omega))^m$, if λ_0 runs through all points of the spectrum of the operator pencil $l(\lambda)$.

The theorem formulated at the beginning of Section 11.4.4 has been completely proved.

An important consequence of the results obtained in this section is summarized in the next statement.

Except for no more than a finite number of eigenvalues, the spectrum of the pencil $l(\lambda)$ in problem (3.33) is real, and to no more than a finite number of eigenvalues there correspond nontrivial chains of associated elements.

11.5 Additional Properties of Solutions of the Spectral Problem

In this section we study additional properties of the solutions of the spectral problem (3.22), or of the equivalent problem (3.33), that were briefly mentioned at the end of Section 11.3. Moreover, in studying the properties of multiple basicity, we use an approach that is not based on the change (4.14) of the spectral parameter and the theory of self-adjoint operators in a Pontryagin space. The approach that we will develop next is based on a direct study of the properties of the spectral problem (3.17) in a Krein space, and an application of the basis criterion presented in Section 1.3.7.

11.5.1 On the Existence of Different Branches of Eigenvalues

As we mentioned at the end of Section 11.3, the spectrum of problem (3.22),

$$L(\lambda)\boldsymbol{\xi} := \left(\nu I_0(\lambda)I - \lambda A^{-1} - \lambda^{-1}gB\right)\boldsymbol{\xi} = \mathbf{0}, \qquad \boldsymbol{\xi} \in \boldsymbol{J}_{0,S}(\Omega),$$
$$I_0(\lambda) := 1 + \sum_{k=1}^{m} \frac{\alpha_k}{\gamma_k - \lambda}, \tag{5.1}$$

is discrete, situated in the right half-plane, and may have as limit points the points $\lambda = 0$, $\lambda = \infty$, and $\lambda = \beta_k$, $k = 1, \ldots, m$, $I_0(\beta_k) = 0$ (see Property 5° in Section 11.3). We show that all these points are really limit points of the discrete spectrum in problem (5.1). The corresponding propositions are formulated as distinct properties.

(1) To the point $\lambda = \infty$ there corresponds a branch of eigenvalues $\{\lambda_n^{(\infty)}\}_{n=1}^{\infty}$ situated on the positive semiaxis in the interval (γ_{m+1}, ∞) (where the number $\gamma_{m+1} > \beta_m$ was introduced in Property 4° of Section 11.1.4), and having the asymptotic behavior

$$\lambda_n^{(\infty)} = \nu\lambda_n(A)[1 + \mathrm{o}(1)], \qquad n \to \infty. \tag{5.2}$$

The eigenelements $\{\boldsymbol{\xi}_n^{(\infty)}\}_{n=1}^{\infty}$ of problem (5.1) corresponding to the branch $\{\lambda_n^{(\infty)}\}$ form a Riesz basis with defect, and even a p-basis with defect for $p > 3/2$.

To prove Property (1), we substitute in (5.1) the spectral parameter $\lambda = 1/\tilde{\lambda}$ and consider the problem

$$\tilde{L}(\tilde{\lambda})\boldsymbol{\xi} := \tilde{\lambda} L\left(\tilde{\lambda}^{-1}\right)\boldsymbol{\xi} := \left(\tilde{\lambda}\nu I_0(\tilde{\lambda}^{-1})I - A^{-1} - \tilde{\lambda}^2 B\right)\boldsymbol{\xi} = \mathbf{0}. \tag{5.3}$$

Since for the operator pencil $\tilde{L}(\tilde{\lambda})$ the next conditions

$$\tilde{L}(0) = -A^{-1} \in \mathfrak{S}_\infty, \qquad \tilde{L}'(0) = \nu I \gg 0 \tag{5.4}$$

are fulfilled, then according to Statement 3° in Section 1.6.10, for any $\varepsilon > 0$ problem (5.3) has a branch of eigenvalues $\{\tilde{\lambda}_n^0\}_{n=1}^{\infty}$ with limit point at zero situated in the interval $(0, \varepsilon)$, and the corresponding eigenelements $\{\boldsymbol{\xi}_n^{(\infty)}\}_{n=1}^{\infty}$ form a Riesz basis with defect in the space $\boldsymbol{J}_{0,S}(\Omega)$. Further, since the eigenvalues $\lambda_n(A^{-1})$ of the operator A^{-1} have power asymptotics [see (8.1.12)], then, by the statement of Section 1.6.8, the asymptotic behavior of the eigenvalues of problem (5.1) with the limit point $\lambda = \infty$ or, equivalently, of problem (3.33),

$$l(\lambda)\boldsymbol{\xi} := \left(I - \lambda\left(\nu I_0(\lambda)\right)^{-1} A^{-1} - g\lambda^{-1}\left(\nu I_0(\lambda)\right)^{-1} B\right)\boldsymbol{\xi} = \mathbf{0}, \tag{5.5}$$

and the eigenvalues of the problem

$$l_0(\lambda)\boldsymbol{\xi} := \left(I - \lambda\nu^{-1}A^{-1}\right)\boldsymbol{\xi} = \mathbf{0}$$

for the shortened pencil $l_0(\lambda)$, is similar, that is, formula (5.2) takes place.

We now note that the property of defect basicity of the elements $\{\boldsymbol{\xi}_n^{(\infty)}\}_{n=1}^{\infty}$ holds, if instead of the interval $(\varepsilon^{-1}, \infty)$ for any $\varepsilon > 0$, we take the interval (γ_{m+1}, ∞) and consider that the eigenvalues $\{\lambda_n^{(\infty)}\}_{n=1}^{\infty}$ are situated only on it.

At last, the property of p-basicity (with defect) for $p > 3/2$ of eigenelements $\{\boldsymbol{\xi}_n^{(\infty)}\}_{n=1}^{\infty}$ corresponding to eigenvalues from the interval (γ_{m+1}, ∞) follows from Statements 4° and 5° of Section 1.6.10, if we note that

$$\begin{aligned} A^{-1} &\in \mathfrak{S}_p, \qquad p > 3/2, \\ B &\in \mathfrak{S}_q, \qquad q > 2, \end{aligned} \tag{5.6}$$

and apply these statements to the pencil (5.3) after its division by $\nu I_0(\tilde{\lambda}^{-1})$. Properties (1) are entirely proved.

(2) To the point $\lambda = 0$ there corresponds a branch of eigenvalues $\{\lambda_n^{(0)}\}_{n=1}^{\infty}$ situated in the interval $(0, \gamma_1)$, and having the asymptotic behavior

$$\lambda_n^{(0)} = \nu^{-1} g \lambda_n(B)[1 + \mathrm{o}(1)], \qquad n \to \infty. \tag{5.7}$$

The eigenelements $\{\boldsymbol{\xi}_n^{(0)}\}_{n=1}^{\infty}$ corresponding to the nonzero eigenvalues $\lambda_n^{(0)}$ in conjunction with the basis from $\operatorname{Ker} B = \boldsymbol{N}_0(\Omega)$ (see the expansion (8.3.24)) form a Riesz basis with defect in the space $\boldsymbol{J}_{0,S}(\Omega)$. The projections of the elements $\{\boldsymbol{\xi}_n^{(0)}\}_{n=1}^{\infty}$ on the subspace $\boldsymbol{M}_0(\Omega) \subset \boldsymbol{J}_{0,S}(\Omega)$ form a defect Riesz basis and even a defect p-basis in $\boldsymbol{M}_0(\Omega)$ for $p > 2$.

Let us prove Property (2). Use the representation (4.3)

$$I_0(\Omega) = \frac{\prod_{k=1}^{m} (\beta_k - \lambda)}{\prod_{k=1}^{m} (\gamma_k - \lambda)} \tag{5.8}$$

for the function $I_0(\lambda)$ and consider instead of (5.1) the equivalent problem

$$\begin{aligned} M(\lambda)\boldsymbol{\xi} := \Bigg\{ \lambda\nu \left(\prod_{k=1}^{m} (\beta_k - \lambda) \right) I - g \left(\prod_{k=1}^{m} (\gamma_k - \lambda) \right) B \\ -\lambda^2 \left(\prod_{k=1}^{m} (\gamma_k - \lambda) \right) A^{-1} \Bigg\} \boldsymbol{\xi} = \mathbf{0}. \end{aligned} \tag{5.9}$$

Since the next conditions are fulfilled,

$$\begin{aligned} M(0) &= -g \left(\prod_{k=1}^{m} \gamma_k \right) B \in \mathfrak{S}_\infty, \\ M'(0) &= \nu \left(\prod_{k=1}^{m} \beta_k \right) I + g \sum_{m=1}^{m} \left(\prod_{j \neq k} \gamma_j \right) B \gg \nu \left(\prod_{k=1}^{m} \beta_k \right) I \gg 0, \end{aligned} \tag{5.10}$$

then, by repeating the same arguments presented above in proving Property (1), we conclude that in problem (5.1) there is a branch of eigenvalues $\{\lambda_n^{(0)}\}_{n=1}^{\infty}$ with a limit point at zero and situated in the interval $[0, \gamma_1)$, where the eigenelements $\{\boldsymbol{\xi}_n^{(0)}\}_{n=1}^{\infty}$ corresponding to the nonzero eigenvalues $\{\lambda_n^{(0)}\}_{n=1}^{\infty}$ in conjuction with the basis of the kernel $\operatorname{Ker} B = \boldsymbol{N}_0(\Omega)$ form a Riesz basis with defect in the space $\boldsymbol{J}_{0,S}(\Omega)$.

Since in the considered case $\operatorname{Ker} B \neq \{\mathbf{0}\}$, then to problem (5.9) there corresponds a degenerate Fredholm pencil, and considerations similar to those presented in Section 1.6.4 for a Keldysh pencil lead to the conclusion that the projections of the elements $\{\boldsymbol{\xi}_n^{(0)}\}_{n=1}^{\infty}$ on the subspace $\boldsymbol{M}_0(\Omega) = \boldsymbol{J}_{0.S}(\Omega) \ominus \boldsymbol{N}_0(\Omega)$ form a Riesz basis (with defect) in $\boldsymbol{M}_0(\Omega)$.

To derive the asymptotic formula (5.7) it is sufficient to use the form (5.5) of the spectral problem, where $I_0^{-1}(\lambda)$ is represented as a Taylor series with respect to λ, the asymptotic formula (8.1.33) for the eigenvalues of operator B, and the statements of Section 1.6.8.

At last, the assertion on p-basicity with defect in $\boldsymbol{M}_0(\Omega)$ for $p > 2$ follows from (5.6) and Statements 4° and 5° of Section 1.6.10 applied the pencil $\lambda l(\lambda)$ [see (5.5)].

11.5.2 On the Location of Nonreal Eigenvalues in the Complex Plane

As it was stated in Section 11.4, problem (5.1) may have no more than a finite number of nonreal eigenvalues λ. In this section we determine the zones of the complex right half-plane, where such eigenvalues could be located. Simultaneously, we will point out some properties of the real spectrum of problem (5.1).

Let λ be a nonreal eigenvalue of problem (5.1), and $\boldsymbol{\xi} \in \boldsymbol{J}_{0.S}(\Omega)$ a normalized eigenelement. Then, by (5.9), we have

$$\psi(\lambda) := (M(\lambda)\boldsymbol{\xi}, \boldsymbol{\xi})_{\boldsymbol{L}_2(\Omega)} = \lambda\nu \prod_{j=1}^{m} (\beta_j - \lambda) - (p\lambda^2 + gq) \prod_{j=1}^{m} (\gamma_j - \lambda) = 0,$$
$$p := \left(A^{-1}\boldsymbol{\xi}, \boldsymbol{\xi}\right)_{\boldsymbol{L}_2(\Omega)} > 0,$$
$$q := (B\boldsymbol{\xi}, \boldsymbol{\xi})_{\boldsymbol{L}_2(\Omega)} \geq 0. \tag{5.11}$$

We next indicate some common properties of the solutions of equation (5.11) under the assumption that $\psi(\lambda)$ has nonreal roots.

(1) The function $\psi(\lambda)$ has exactly m real roots $\{\xi_j\}_{j=1}^m$ satisfying the inequalities

$$0 < \xi_1 < \gamma_1 < \beta_1 < \xi_2 < \gamma_2 < \beta_2 < \cdots < \xi_m < \gamma_m < \beta_m. \tag{5.12}$$

Indeed, $\psi(\lambda) < 0$ for $\lambda < 0$, and $\psi(\gamma_k) = (-1)^{k+1}c_k$, $c_k := \gamma_k \prod_{j=1}^m |\beta_j - \gamma_k| > 0$, $k = 1, \ldots, m$. Therefore, in the interval $[0, \gamma_m]$, function $\psi(\lambda)$ has m real roots $\{\xi_j\}_{j=1}^m$, for which $0 \leq \xi_1 < \gamma_1 < \xi_2 < \gamma_2 < \cdots < \xi_m < \gamma_m$. Since $\psi(\lambda)$ is a polynomial of degree $m+2$ with real coefficients [see (5.11)] and, according to the assumption, it has nonreal roots, then the number of real roots exactly equals m. We note also that since all the eigenvalues of problem (5.1) are situated in the right complex half-plane, then $\xi_1 > 0$. Further, according to Properties 3° and 4° proved in Section 11.3.4, the points γ_k and β_k, $k = 1, \ldots, m$, are not eigenvalues of problem (5.1).

We show that in the intervals (γ_k, β_k), $k = 1, \ldots, m$, problem (5.1) also has no eigenvalues and therefore the properties (5.12) are fulfilled.

Indeed, if $\lambda = \beta_k - \varepsilon$, $0 < \varepsilon < (\beta_k - \gamma_k)$, then the operator $M(\lambda)$ from (5.9) has the properties

$$M(\beta_k - \varepsilon)(-1)^{k+1} \gg 0, \qquad k = 1, \ldots, m. \tag{5.13}$$

In fact,

$$M(\beta_k - \varepsilon) = \nu(\beta_k - \varepsilon)\prod_{j=1}^{m}(\beta_j - \beta_k + \varepsilon)I - \prod_{j=1}^{m}(\gamma_j - \beta_k + \varepsilon)\left(g(\beta_k - \varepsilon)^2 B + A^{-1}\right),$$

and since $0 < \gamma_1 < \beta_1 < \cdots < \gamma_m < \beta_m$, $B \geq 0$, $A^{-1} > 0$, then (5.13) takes place.

Let us denote the complex conjugate pair of nonreal roots by $\lambda_0 = \xi_0 \pm i\eta_0$, $\eta_0 \neq 0$, and get some estimates for ξ_0 and η_0.

(2) For the nonreal eigenvalues λ_0 of problem (5.1) the inequalities

$$\nu\left(2\left\|A^{-1}\right\|\right)^{-1} \leq \operatorname{Re}\lambda_0 = \xi_0 \leq 2\sum_{k=1}^{m}\beta_k - \frac{1}{2}\sum_{k=1}^{m}\gamma_k + 2g\frac{\|B\|}{\nu} \tag{5.14}$$

are fulfilled.

To prove properties (5.14) we represent the polynomial $\psi(\lambda)$ from (5.11) with the roots $\{\xi_j\}_{j=1}^{\infty}$ and $\lambda_0^{\pm} = \xi_0 \pm i\eta_0$ in the form

$$\psi(\lambda) = (-p)(-1)^m \prod_{j=1}^{m}(\lambda - \xi_j)\left[(\lambda - \xi_0)^2 + \eta_0^2\right]. \tag{5.15}$$

Comparing the coefficients of the powers λ^{m+1} and λ^m in (5.11) and (5.15) we get the relations

$$2\xi_0 + \sum_{j=1}^{m}\xi_j = \sum_{j=1}^{m}\gamma_j + \frac{\nu}{p}, \tag{5.16}$$

$$\xi_0^2 + \eta_0^2 + 2\xi_0\sum_{j=1}^{m}\xi_j + \sum_{i<j}\xi_i\xi_j = \frac{gq}{p} + \sum_{k<j}\gamma_i\gamma_j + \frac{\nu}{p}\sum_{j=1}^{m}\beta_j. \tag{5.17}$$

By the inequalities (5.12) we have $\sum_{j=1}^{m}(\gamma_j - \xi_j) > 0$, and, therefore, from (5.16) we get

$$2\xi_0 > \frac{\nu}{p} \geq \frac{\nu}{\|A^{-1}\|},$$

whence the left inequality (5.14) follows.

Further, from (5.17) with regard to (5.16) we have

$$\begin{aligned}\eta_0^2 &= -\xi_0^2 - 2\xi_0\left(-2\xi_0 + \sum_{j=1}^{m}\gamma_j + \frac{\nu}{p}\right) + \sum_{i<j}(\gamma_i\gamma_j - \xi_i\xi_j) + \frac{gq}{p} + \frac{\nu}{p}\sum_{j=1}^{m}\beta_j \\ &= 3\xi_0^2 - 2\xi_0\left(\sum_{j=1}^{m}\gamma_j + \frac{\nu}{p}\right) + \sum_{i<j}(\gamma_i\gamma_j - \xi_i\xi_j) + \frac{gq}{p} + \frac{\nu}{p}\sum_{j=1}^{m}\beta_j. \qquad (5.18)\end{aligned}$$

Let us introduce the notations

$$2\omega := \frac{\nu}{p}, \qquad 2\delta := \sum_{j=1}^{m}(\gamma_j - \xi_j)\,, \qquad (5.19)$$

and observe that by (5.16)

$$\xi_0 = \omega + \delta. \qquad (5.20)$$

Substituting ξ_0 from (5.20) into (5.18) we get

$$\eta_0^2 = -\omega^2 + 2\omega\left[\delta - \sum_{j=1}^{m}\gamma_j + \frac{gq}{\nu} + \sum_{j=1}^{m}\beta_j\right] + 3\delta^2 - 2\delta\sum_{j=1}^{m}\gamma_j + \sum_{i<j}(\gamma_i\gamma_j - \xi_i\xi_j)\,. \qquad (5.21)$$

Let us show that the expression not containig ω in (5.21) is negative, that is,

$$3\delta^2 < 2\delta\sum_{j=1}^{m}\gamma_j - \sum_{i<j}(\gamma_i\gamma_j - \xi_i\xi_j)\,. \qquad (5.22)$$

Taking into account the notations (5.19) for δ, inequality (5.22) gets the form

$$\begin{aligned}\frac{3}{4}\left(\sum_{j=1}^{m}(\gamma_j - \xi_j)\right)^2 &< \left(\sum_{j=1}^{m}(\gamma_j - \xi_j)\right)\left(\sum_{i=1}^{m}\gamma_i\right) - \sum_{i<j}(\gamma_i\gamma_j - \xi_i\xi_j) \\ &= \sum_{j=1}^{m}(\gamma_j - \xi_j)\left[\sum_{i=1}^{m}(\gamma_i - \xi_i) + \sum_{i=1}^{m}\xi_i\right] - \sum_{i<j}(\gamma_i\gamma_j - \xi_i\xi_j)\,. \qquad (5.23)\end{aligned}$$

It is clear that in order to prove inequality (5.23) it is sufficient to show that

$$\sum_{j=1}^{m}(\gamma_j-\xi_j)\left(\sum_{i=1}^{m}\xi_i\right)>\sum_{i<j}(\gamma_i\gamma_j-\xi_i\xi_j)\,. \tag{5.24}$$

Let us check this fact. Using for $i<j$ the identity

$$\gamma_i\gamma_j-\xi_i\xi_j=\gamma_i(\gamma_j-\xi_j)+\xi_j(\gamma_i-\xi_i)\,, \tag{5.25}$$

then the right side in (5.24) takes the form

$$\begin{aligned}\sum_{i<j}(\gamma_i\gamma_j-\xi_i\xi_j)&=\sum_{i<j}\left[\gamma_i(\gamma_j-\xi_j)+\xi_j(\gamma_i-\xi_i)\right]\\&=\sum_{j=1}^{m}(\gamma_j-\xi_j)\left[\sum_{j<i}\xi_i+\sum_{j>i}\gamma_i\right].\end{aligned} \tag{5.26}$$

Since the inequalities (5.12) are fulfilled, then the right side of (5.26) is smaller than the left side of (5.24), and inequality (5.23) is proved.

Equation (5.21) together with inequality (5.22) allows us to estimate ω. We have

$$\begin{aligned}\eta_0^2&-\left\{3\delta^2-2\delta\sum_{j=1}^{m}\gamma_m+\sum_{i<j}(\gamma_i\gamma_j-\xi_i\xi_j)\right\}\\&=-\omega^2+2\omega\left[\delta+\frac{gq}{\nu}+\sum_{j=1}^{m}(\beta_j-\gamma_j)\right]>0,\end{aligned}$$

whence, by the property $\omega>0$, we get

$$\begin{aligned}\omega&<2\left[\delta+gq\nu^{-1}+\sum_{j=1}^{m}(\beta_j-\gamma_j)\right]\\&<2\left[\frac{1}{2}\sum_{j=1}^{m}(\gamma_j-\xi_j)+gq\nu^{-1}+\sum_{j=1}^{m}(\beta_j-\gamma_j)\right]\\&<2\left[\sum_{j=1}^{m}\beta_j-\frac{1}{2}\sum_{j=1}^{m}\gamma_j+g\nu^{-1}\|B\|\right].\end{aligned} \tag{5.27}$$

Hence

$$\xi_0 = \omega + \delta < 2\left[\sum_{j=1}^{m}\beta_j - \frac{1}{2}\sum_{j=1}^{m}\gamma_j + g\nu^{-1}\|B\|\right] + \frac{1}{2}\sum_{j=1}^{m}(\gamma_j - \xi_j)$$
$$< 2\sum_{j=1}^{m}\beta_j - \frac{1}{2}\sum_{j=1}^{m}\gamma_j + 2g\nu^{-1}\|B\|;$$

the right inequality (5.14) is proved.

(3) For the nonreal eigenvalues λ_0 of problem (5.1) we have the estimate

$$|\lambda_0|^2 < 2\omega\left(q\nu^{-1} + \sum_{j=1}^{m}\beta_j\right)$$
$$\leq 4\left(g\nu^{-1}\|B\| + \sum_{j=1}^{m}\beta_j\right)\left(g\nu^{-1}\|B\| + \sum_{j=1}^{m}\beta_j - \frac{1}{2}\sum_{j=1}^{m}\gamma_j\right). \tag{5.28}$$

Let us prove property (5.28). From (5.17) with regard to (5.20), (5.19) we obtain

$$|\lambda_0|^2 = \xi_0^2 + \eta_0^2 = 2\omega\left(gq\nu^{-1} + \sum_{j=1}^{m}\beta_j\right) - 2\omega\sum_{j=1}^{m}\xi_j - F, \tag{5.29}$$

$$F := \sum_{j=1}^{m}(\gamma_j - \xi_j)\left(\sum_{i=1}^{m}\xi_i\right) - \sum_{i<j}(\gamma_i\gamma_j - \xi_i\xi_j). \tag{5.30}$$

Let us show that $F > 0$. Using relation (5.26) we have

$$F = \sum_{j=1}^{m}(\gamma_j - \xi_j)\left[\sum_{k=1}^{m}\xi_k - \sum_{j<k}\xi_k - \sum_{j>k}\gamma_k\right]$$
$$= \sum_{j=1}^{m}(\gamma_j - \xi_j)\left[\sum_{j\geq k}\xi_k - \sum_{k<j}\gamma_k\right] > 0, \tag{5.31}$$

since, by (5.12), $(\gamma_j - \xi_j) > 0$, $(\xi_{j+1} - \gamma_j) > 0$. Since $\omega > 0$, $\xi_k > 0$, $k = 1, \ldots, m$, then inequality (5.28) follows from (5.29), (5.31), and (5.27).

(4) The nonreal eigenvalues λ of problem (5.1) are separated from the points β_k and γ_k, $k = 1, \ldots, m$, in such a way that the following inequalities take place,

$$|\lambda - \beta_k|^2 \geq a_k, \quad |\lambda - \gamma_k|^2 \geq \alpha_k\frac{\nu}{\|A^{-1}\|}, \qquad k = q, \ldots, m, \tag{5.32}$$

where a_k and b_k are the positive constants from the expansions (4.5), (4.9) [see (4.7), (4.12)].

Next, let us prove the first inequalities (5.32). From (5.5), (4.4), (4.9), (4.1) for the normalized solutions of problem (5.1) we have

$$1+\left(-\lambda-c+\sum_{k=1}^{m}\frac{a_k}{\beta_k-\lambda}\right)\frac{p}{\nu}+\left(-\frac{d}{\lambda}+\sum_{k=1}^{m}\frac{b_k}{\beta_k-\lambda}\right)\frac{g}{\nu}q=0$$

where p and q are defined in (5.11). Separating here the imaginary part and taking into account that $\operatorname{Im}\lambda\neq 0$ we get

$$\left(1-\sum_{k=1}^{m}\frac{a_k}{|\lambda-\beta_k|^2}\right)\frac{p}{q}=\left(\frac{d}{|\lambda|^2}+\sum_{k=1}^{m}\frac{b_k}{|\lambda-\beta_k|^2}\right)\frac{g}{\nu}q\geq 0. \tag{5.33}$$

Since $p>0$, whence it follows that the first inequalities (5.32) are valid.

To prove the second inequalities (5.32) we use the relation

$$\nu\left(1+\sum_{k=1}^{m}\frac{\alpha_k}{\gamma_k-\lambda}\right)-\lambda p-\lambda^{-1}gq=0, \tag{5.34}$$

which is valid for the normalized solutions of problem (5.1). Separating the imaginary part in (5.34) and using the property $\operatorname{Im}\lambda\neq 0$ we get

$$\nu\sum_{k=1}^{m}\frac{\alpha_k}{|\lambda-\gamma_k|^2}+gq\frac{1}{|\lambda|^2}=p\leq \left\|A^{-1}\right\|. \tag{5.35}$$

Whence the second group of inequalities (5.32) follows.

Now we consider those real eigenvalues of problem (5.1) to which, along with their eigenelements, there also correspond associated elements. As it has been mentioned in Section 1.6.10, for an operator pencil $A(\lambda)$ with eigenvalue $\lambda_0\in\mathbb{R}$ and eigenelement $\varphi_0\in E$, in this case the conditions

$$A(\lambda_0)\varphi_0=0,\qquad (A'(\lambda_0)\varphi_0,\varphi_0)=0, \tag{5.36}$$

are fulfilled. Such eigenvalues were earlier called degenerate (Section 8.2.8), that is, not definite (neither positive, nor negative).

(5) Degenerate eigenvalues $\lambda_0=\xi_0\in\mathbb{R}$ of problem (5.1) satisfy the inequalities (5.14), where the intervals

$$|\xi_0-\beta_k|<\sqrt{a_k},\quad |\xi_0-\gamma_k|<\left(\frac{\nu\alpha_k}{\|A^{-1}\|}\right)^{1/2},\qquad k=1,\dots,m, \tag{5.37}$$

do not contain degenerate eigenvalues.

We prove the last statement and properties (5.37). Let $\boldsymbol{\xi}$ be a normalized eigenelement of problem (5.1), corresponding to the degenerate eingenvalue ξ_0. Then according to (5.35) the relation

$$\left(L'\left(\xi_0\right)\boldsymbol{\xi},\boldsymbol{\xi}\right) = \left(\left(\nu I_0'\left(\xi_0\right)I - A^{-1} + \xi_0^{-2}gB\right)\boldsymbol{\xi},\boldsymbol{\xi}\right) = 0$$

is fulfilled, whence the inequality (5.35) and thus the second group of inequalities (5.32) follow. It means that if the second conditions (5.37) are fulfilled then the corresponding eigenvalues $\lambda_0 = \xi_0 \in \mathbb{R}$ could not be degenerate.

To prove the first inequalities (5.37) for a degenerate eigenvalue $\xi_0 \in \mathbb{R}$ it is necessary to consider the pencil $l(\lambda)$ from (5.5). The condition $(l'(\xi_0)\boldsymbol{\xi},\boldsymbol{\xi}) = 0$ leads to the relation (5.33) for $\lambda = \xi_0$, whence the first inequalities (5.32) follow.

Further, according to Property (1), function $\psi(\lambda)$ has not less than m different real eigenvalues. If, in addition, it has a two-multiple real root, then there are no nonreal roots. Such a two-multiple root may be situated in the interval (γ_m,∞) or in any of the intervals (γ_{k-1},γ_k), $k=1,\ldots,m$, $\gamma_0 := 0$. In the latter case, the root could be three-multiple and coincide with ξ_k, however only one two-multiple root could not be in the interval (γ_{k-1},γ_k).

It is easy to see that, in all the mentioned cases, the relations (5.16), (5.17) for $\eta_0 = 0$ were fulfilled (since $\lambda_0 = \xi_0 \in \mathbb{R}$). Just as in the proof of Properties (2) we get that the inequalities (5.14) are fulfilled and Properties (5) are proved.

(6) As consequences of the previously proved properties of the solutions of problem (5.1) we have the next results.

(a) The collection of intermediate eigenvalues of the problem on normal oscillations of a viscous fluid in an open container, that is, nonreal and degenerate real eigenvalues of problem (5.1), forms a finite set in the right complex half-plane, every point λ_0 of which is satisfying inequalities (5.14), (5.28), (5.32). The circles

$$|\lambda_0 - \beta_k| < \sqrt{a_k}, \quad |\lambda_0 - \gamma_k| < \left(\frac{\nu\alpha_k}{\|A^{-1}\|}\right)^{1/2}, \qquad k=1,\ldots,m,$$

do not contain intermediate eigenvalues and the intervals $[\gamma_j,\beta_j]$, $j=1,\ldots,m$, do not contain eigenvalues of problem (5.1).

(b) If the rough condition of strong damping is fulfilled,

$$\nu^2 > 4g\left\|A^{-1}\right\|\cdot\|B\| + 2\nu\left(\sum_{k=1}^{m}\beta_k - \frac{1}{2}\sum_{k=1}^{m}\gamma_k\right)\left\|A^{-1}\right\|, \tag{5.38}$$

then the intermediate eigenvalues are absent, every eigenvalue of problem (5.1) is real, and no associated elements correspond to it.

(c) If condition (5.38) is fulfilled, then the spectrum of problem (5.1) is divided into $m+2$ branches of real eigenvalues $\{\lambda_n^{(0)}\}_{n=1}^{\infty}$, $\{\lambda_n^{(\infty)}\}_{n=1}^{\infty}$, $\{\lambda_n^{(k)}\}_{n=1}^{\infty}$, $k = 1, \dots, m$, with the limit points 0, $+\infty$ and $\{\beta_k\}_{k=1}^{m}$, respectively [and asymptotic behavior (5.7), (5.2)]. In addition, the branches $\{\lambda_n^{(0)}\}_{n=1}^{\infty}$ and $\{\lambda_n^{(k)}\}_{n=1}^{\infty}$ approach the points 0, β_k, $k = 1, \dots, m$, from the right.

It is necessary to prove only Assertion (b) This assertion follows from the fact that under condition (5.38) the inequalities (5.14) are impossible, and, therefore, by Properties (2) and (5) the intermediate eigenvalues form the empty set.

In conclusion, we note that condition (5.38) is fulfilled for sufficiently large values of the kinematic viscosity ν of a visco-elastic fluid. It is a generalization of the rough condition of strong damping

$$\nu^2 > 4g \left\|A^{-1}\right\| \cdot \|B\|$$

[see (1.3.6)] corresponding to an ordinary viscous homogeneous fluid and turns into it for $\alpha_k = 0$, $k = 1, \dots, m$ [see (5.1)].

11.5.3 Multiple Basicity and p-Basicity. An Indefinite Metric Approach Using Krein Space Theory

Let us return to the study of the spectral problem (3.17) on normal oscillations of a visco-elastic fluid in an open container and apply to it an approach based not on operator theory in a Pontryagin space and complex changes of the spectral parameter, as it has been done in Sections 11.4.1, 11.4.2, and further, but on constructions that make it possible to consider the problem in a Krein space.

Just for the sake of simplicity, we limit ourselves to the case $m = 1$. For any $m \geq 1$ the constructions are similar.

If $m = 1$, problem (3.17) takes the form

$$\mathcal{A}u = \lambda u, \qquad u = (\boldsymbol{u}_0, \boldsymbol{u}_1, \eta)^{\mathrm{t}} \in \boldsymbol{J}_{0,S}(\Omega) \oplus \boldsymbol{J}_{0,S}(\Omega) \oplus H_\Gamma, \tag{5.39}$$

$$\mathcal{A} = \begin{pmatrix} \nu A & (\nu\alpha_1)^{1/2} A^{1/2} & g^{1/2} G \\ -(\nu\alpha_1)^{1/2} A^{1/2} & \gamma_1 I & 0 \\ -g^{1/2} G^* & 0 & 0 \end{pmatrix}. \tag{5.40}$$

In (5.39) we substitute the spectral parameter according to the formula

$$\lambda = -a + \mu^{-1}, \qquad a > 0, \tag{5.41}$$

and get from (5.39) the equivalent problem

$$\mathcal{M}_a u := \mathcal{A}_a^{-1} u = \mu u, \qquad \mathcal{A}_a := \mathcal{A} + a\mathcal{I}. \tag{5.42}$$

We recall that, first, $\lambda = 0$ is not an eigenvalue of problem (5.39) (see Property 1° in Section 11.3.4) and, second, for operator $\mathcal{A}$ we have $\mathrm{Re}\mathcal{A} \geq 0$. Therefore, $\mathcal{A}_a^{-1}$ is a bounded operator.

We write down an explicit formula for the matrix operator $\mathcal{A}_a^{-1} = \mathcal{M}_a$,

$$\begin{aligned}
&\mathcal{M}_a = (\mathcal{M}_{jk})_{j,k=1}^{3}\\
&\mathcal{M}_{11} = A^{-1/2}L^{-1}(-a)A^{-1/2} = \mathcal{M}_{11}^*, \quad L(\lambda) := \nu\left(1 + \frac{\alpha_1}{\gamma_1 - \lambda}\right)I - \lambda A^{-1} - g\lambda^{-1}B,\\
&\mathcal{M}_{12} = -\left(\nu\alpha_1\right)^{1/2}\left(\gamma_1 + a\right)^{-1}A^{-1/2}L^{-1}(-a) = -\mathcal{M}_{21}^*,\\
&\mathcal{M}_{13} = -a^{-1}g^{1/2}A^{-1/2}L^{-1}(-a)Q^* = -\mathcal{M}_{31}^*, \quad Q := G^*A^{-1/2} = \gamma_n A^{-1/2},\\
&\mathcal{M}_{22} = (\gamma_1 + a)^{-1}\left[I - \nu\alpha_1\left(\gamma_1 + a\right)^{-1}L^{-1}(-a)\right] = \mathcal{M}_{22}^*,\\
&\mathcal{M}_{23} = -\left(\gamma_1 + a\right)^{-1}\left(\nu\alpha_1\right)^{1/2}g^{1/2}a^{-1}L^{-1}(-a)Q^* = \mathcal{M}_{32}^*,\\
&\mathcal{M}_{33} = a^{-1}\left[I - a^{-1}gQL^{-1}(-a)Q^*\right] = \mathcal{M}_{33}^*,
\end{aligned} \tag{5.43}$$

and describe some properties of its entries $\mathcal{M}_{ij}$.

(1) Operator $\mathcal{M}_{11}$ is positive, compact, and belongs to the class $\mathfrak{S}_{p_{11}}$, where $p_{11} = p(A^{-1}) > 3/2$.

Indeed, the operator $L(-a) = \nu(1 + \alpha_1/(\gamma_1 + a))I + aA^{-1} + ga^{-1}B$ is bounded and positive definite, therefore $L^{-1}(-a)$ is a bounded positive definite operator acting in $\boldsymbol{J}_{0,S}(\Omega)$. Whence from the formula for $\mathcal{M}_{11}$ and also from the properties of operators of the classes $\mathfrak{S}_p$ we get that $p_{11} = p(A^{-1})$. At last, the property $p(A^{-1}) > 3/2$ follows from the asymptotic formula (8.1.12) for the eigenvalues of operator A.

(2) The operators $\mathcal{M}_{12}$ and $\mathcal{M}_{21}$ are compact operators of the class $\mathfrak{S}_{p_{12}}$, $p_{12} = p((A^{-1/2}) > 3$.

These assertions are based on the formulas for $\mathcal{M}_{12}$ and $\mathcal{M}_{21}$, some arguments similar to the ones mentioned above, and also on the fact that $p(A^{-1/2}) = 2p(A^{-1})$.

(3) The operators $\mathcal{M}_{13}$ and $\mathcal{M}_{31}$ are compact operators of the same class $\mathfrak{S}_{p_{13}}$ as the operators $A^{-1/2}B^{1/2}$ or $B^{1/2}A^{-1/2}$, and, therefore, $p_{13} > 12/7$.

Indeed, it is easy to see according to the formula $B = Q^*Q$ that Q and Q^* are compact operators of the class $\mathfrak{S}_{p(B^{1/2})}$. Since from the asymptotic formula (8.1.33) it follows that $p(B) > 2$, then $p(Q) = p(Q^*) = p(B^{1/2}) > 4$. Therefore, from the formulas for $\mathcal{M}_{13}$ and $\mathcal{M}_{31}$ we get that $p_{13} > \tilde{p}_{13}$, $\tilde{p}_{13}^{-1} = (p(A^{-1/2}))^{-1} + (p(B^{-1/2}))^{-1} = 1/3 + 1/4 = 7/12$.

(4) The operators $\mathcal{M}_{23}$ and $\mathcal{M}_{32}$ are compact operators of the class $\mathfrak{S}_{p_{23}}$, $p_{23} = p(Q^*) > 4$.

(5) Operator $\mathcal{M}_{22}$ is a bounded positive definite operator acting in $\boldsymbol{J}_{0,S}(\Omega)$. Indeed,

$$\begin{aligned}
(\gamma_1 + a)\,\mathcal{M}_{22}
&= I - \frac{\nu\alpha_1}{\gamma_1 + a} L^{-1}(-a) \\
&= L^{-1/2}(-a)\left[L(-a) - \frac{\nu\alpha_1}{\gamma_1 + a} I\right] L^{-1/2}(-a) \\
&= L^{-1/2}(-a)\left[\nu\left(1 + \frac{\alpha_1}{\gamma_1 + a}\right) I + aA^{-1} + ga^{-1}B - \frac{\nu\alpha_1}{\gamma_1 + a} I\right] L^{-1/2}(-a) \\
&= L^{-1/2}(-a) L_0(-a) L^{-1/2}(-a),
\end{aligned}$$

$$L_0(\lambda) := \nu I - \lambda A^{-1} - g\lambda^{-1} B. \tag{5.44}$$

Since $L_0(-a) \gg 0$, $L^{-1/2}(-a) \gg 0$, the property to be proved follows from (5.44).

(6) Operator $\mathcal{M}_{33}$ has the structure

$$\mathcal{M}_{33} = a^{-1} I - a^{-2} \mathcal{M}_{33}^0, \quad 0 \le \mathcal{M}_{33}^0 \in \mathfrak{S}_{p_{33}}, \qquad p_{33} = p(B) > 2.$$

In fact, the operator $QL^{-1}(-a)Q^*$ is a nonnegative compact operator of the same class as the operator QQ^*, that is, of the class $\mathfrak{S}_{p(B)}$.

As a consequence of the properties (1)–(6) we have the next important conclusion.

(7) In a block form, the matrix operator $\mathcal{M}_a$ has the structure

$$\mathcal{M}_a = \begin{pmatrix} \tilde{\mathcal{M}}_{11} & \tilde{\mathcal{M}}_{12} \\ -\tilde{\mathcal{M}}_{12}^* & \tilde{\mathcal{M}}_{22} \end{pmatrix}, \tag{5.45}$$

$$\begin{aligned}
&\tilde{\mathcal{M}}_{11} := \mathcal{M}_{11} = \tilde{\mathcal{M}}_{11}^* \in \mathfrak{S}_p, \qquad p > 3/2, \\
&\tilde{\mathcal{M}}_{12} := (\mathcal{M}_{12}, \mathcal{M}_{13}) = -\tilde{\mathcal{M}}_{12}^* \in \mathfrak{S}_p, \qquad p > 3, \\
&\tilde{\mathcal{M}}_{22} := T_{22} - \tilde{\mathcal{M}}_{22}^0, \\
0 \ll\ &T_{22} := \operatorname{diag}\left(\mathcal{M}_{22}, a^{-1} I\right) \\
&\tilde{\mathcal{M}}_{22}^0 := \begin{pmatrix} 0 & \mathcal{M}_{23} \\ \mathcal{M}_{23}^* & a^{-2}\mathcal{M}_{33}^0 \end{pmatrix} = \left(\tilde{\mathcal{M}}_{22}^0\right)^* \in \mathfrak{S}_p, \qquad p > 4,
\end{aligned} \tag{5.46}$$

and it is $\mathcal{J}$-self-ajoint in the space

$$\boldsymbol{H} := \boldsymbol{J}_{0.S}(\Omega) \oplus \boldsymbol{J}_{0.S}(\Omega) \oplus H_{\Gamma},$$
$$\mathcal{J} := \operatorname{diag}(I, -I, -I_{\Gamma}). \tag{5.47}$$

Based on the just stated properties of operator $\mathcal{M}_a$ and the fundamental theorem in Section 1.3.7 on completeness and basicity of the system of root elements of a $\mathcal{J}$-self-ajoint operator, it is now possible to prove the theorem on basicity and p-basicity of the root elements of problem (5.39), (5.40), that is, the multiple basicity of the system of eigen- and associated elements of problem (3.22). We realize this program step by step, by proving additional properties of problem (5.40).

(8) Operator $\mathcal{M}_a$ of problem (5.42) belongs to the class $\mathcal{H}$ (see Section 1.3.7), that is, there is at least one maximal nonnegative invariant subspace, and every such subspace decomposes into a direct sum of a finite-dimensional neutral and an uniformly positive subspace.

To prove this property it is sufficient to check (see Section 1.3.7) that for the operator $\mathcal{M}_a$ there exists at least one maximal nonnegative subspace, and the angular operator of that subspace is compact. Let us verify these properties.

Since the element $\tilde{\mathcal{M}}_{12}$ of the matrix $\mathcal{M}_a$ according to (5.45), (5.46) is compact, then according to a theorem of Langer (see Section 1.3.5), for the operator $\mathcal{M}_a$ there exists a maximal nonnegative invariant subspace $L_+(\mathcal{M}_a)$. We denote the angular operator of this subspace by K. It is easy to deduce, as it was done in Section 9.2.4, for example, that K is the solution in the unit operator ball of the operator equation

$$K\left(\tilde{\mathcal{M}}_{11} + \tilde{\mathcal{M}}_{12}K\right) = -\tilde{\mathcal{M}}_{12}^{*} + \tilde{\mathcal{M}}_{22}K, \tag{5.48}$$

whence it follows that

$$K = T_{22}^{-1}\left(K\tilde{\mathcal{M}}_{11} + K\tilde{\mathcal{M}}_{12}K + \tilde{\mathcal{M}}_{12}^{*} + \tilde{\mathcal{M}}_{22}^{0}K\right). \tag{5.49}$$

By formulas (5.46) and the boundedness of operator T_{22}^{-1}, from (5.49) it follows that K is a compact operator and, moreover,

$$K \in \mathfrak{S}_p, \qquad p > 4. \tag{5.50}$$

Assertion (8) is proved.

(9) Operator $\mathcal{A}$ of problem (5.40) is an unbounded $\mathcal{J}$-self-ajoint operator of the class $\mathcal{K}(\mathcal{H})$ (see Section 1.3.7), that is, there exists a $\mathcal{J}$-self-ajoint operator of class $\mathcal{H}$ commuting with $\mathcal{A}$.

The proof of this property is trivial, since $\mathcal{M}_a = \mathcal{A}_a^{-1} = (\mathcal{A}+a\mathcal{I})^{-1}$ commutes with $\mathcal{A}$ and $\mathcal{M}_a \in \mathcal{H}$.

(10) The number $a > 0$ in problem (5.41), (5.42) may be chosen in such a way that the angular operator K from (5.48) will have, instead (5.50), the property

$$K \in \mathfrak{S}_p, \qquad p > 3. \tag{5.51}$$

The proof of this property is based on the fact that if the operator $\tilde{\mathcal{M}}_{22}$ in (5.48) is invertible, then by virtue of its structure (5.46) the inverse operator $\tilde{\mathcal{M}}_{22}^{-1}$ will be bounded. Then from (5.48) we will have

$$K = \tilde{\mathcal{M}}_{22}^{-1}\left(\tilde{\mathcal{M}}_{12}^{*} + K\mathcal{M}_{11} + K\tilde{\mathcal{M}}_{12}K\right) \tag{5.52}$$

and from the properties (5.46) of the matrix elements $\tilde{\mathcal{M}}_{jk}$ we get that property (5.51) takes place.

Now we show that there exists a sufficiently wide set of values $a > 0$ such that the operator $\tilde{\mathcal{M}}_{22} = \tilde{\mathcal{M}}_{22}(a)$ is invertible. With this purpose in mind, we consider the equation

$$\tilde{\mathcal{M}}_{22}(a)\boldsymbol{u} = \boldsymbol{0}, \qquad \boldsymbol{u} = (\boldsymbol{u}_1, u_2)^{\mathrm{t}} \in \boldsymbol{J}_{0,S}(\Omega) \oplus H_\Gamma. \tag{5.53}$$

Taking into account (5.46) and (5.43)–(5.45) we have

$$\begin{aligned}
&\frac{1}{(\gamma_1+a)}L^{-1/2}(-a)L_0(-a)L^{-1/2}(-a)\boldsymbol{u}_1 - \frac{1}{(\gamma_1+a)}(\nu\alpha_1)^{1/2}g^{1/2}a^{-1}L^{-1}(-a)Q^*u_2 = \boldsymbol{0},\\
&-\frac{1}{(\gamma_1+a)}(\nu\alpha_1)^{1/2}g^{1/2}a^{-1}QL^{-1}(-a)\boldsymbol{u}_1 + \left(a^{-1}I_\Gamma - a^{-2}gQL^{-1}(-a)Q^*\right)u_2 = 0.
\end{aligned} \tag{5.54}$$

Since $L^{-1/2}(-a)$ and $L_0(-a)$ are bounded positive operators, then eliminating $\boldsymbol{u}_1$ from (5.54) we come to the equation

$$\begin{aligned}
&\left[I_\Gamma - ga^{-1}\left(QL^{-1}(-a)Q^* + \frac{\nu\alpha_1}{(\gamma_1+a)}QL^{-1/2}(-a)L_0^{-1}(-a)L^{-1/2}(-a)Q^*\right)\right]u_2\\
&=: \left[I_\Gamma - T(a)\right]u_2 = 0, \qquad u_2 \in H_\Gamma,
\end{aligned} \tag{5.55}$$

where $T(a)$ is a compact (by the compactness of Q and Q^*) and nonnegative operator (for $a > 0$).

We now note that the operator $T(a)$ takes compact values and is a holomorphic operator function with respect to the parameter a through all the complex plane except the points $a = 0$, $a = \infty$, and $a = -\gamma_1$. Therefore, the operator pencil $I - T(a)$ is Fredholm, and problem (5.55) has a discrete spectrum with the possible limit points $a = 0$, $a = \infty$, and $a = -\gamma_1$. Hence, on the positive semiaxis, any point

a not coinciding with some point of the discrete spectrum $\{a_j\}_{j=1}^{\infty}$ of problem (5.55) has the property that the operator $I_\Gamma - T(a)$, and along with it, the operator $\tilde{\mathcal{M}}_{22}(a)$ have bounded inverses.

Property (10) is completely proved.

At last, let us formulate the basic statement of this section.

(11) Problem (5.40) has a discrete spectrum consisting of normal eigenvalues, with the limit points 0, β_k, $k = 1, \ldots, m$, and $+\infty$. All the eigenvalues of problem (5.40) except probably a finite number of them are positive, and their corresponding eigenelements do not have associated elements.

For any arbitrary value of the viscosity ν, the set of the root elements of problem (5.40) forms an almost $\mathcal{J}$-orthonormal basis in $\boldsymbol{H}$, with the operator $\mathcal{J}$ from (5.47), and this basis is a p-basis for $p > 3$.

If the rough condition of strong damping (5.38) is fulfilled, that is, the viscosity ν of the considered hydrodynamic system is sufficiently large, then problem (5.40) has no intermediate eigenvalues (see Section 11.5.2). In this case, the set of eigenelements of problem (5.40) forms a $\mathcal{J}$-orthonormal basis in $\boldsymbol{H}$, which is also a p-basis for $p > 3$.

We prove only the second and third statements of (11), since the first one has been already proved in Section 11.5.2. By Property (9), the operator $\mathcal{A}$ of problem (5.40) belongs to the class $\mathcal{K}(\mathcal{H})$. We recall that a point $\lambda \in \mathbb{R}$ for the $\mathcal{J}$-self-ajoint operator $\mathcal{A}$ is called degenerate, if $\mathrm{Ker}(\mathcal{A} - \lambda\mathcal{I})$ is degenerate, that is, it contains nonzero isotropic elements. The set $s(\mathcal{A})$ of the critical points in problem (5.40) obviously is contained in the set $\{0;\, \beta_k, k = 1, \ldots, m\}$ of all finite limit points of the spectrum of operator $\mathcal{A}$, since the normal eigenvalues of a $\mathcal{J}$-self-ajoint operator generate a nondegenerate root subspace (the last fact is stated, for example, in the book of T. Ya. Azizov and I.S. Iokhvidov [AI], Section 2.2, Consequence 2.25).

Thus, the operator $\mathcal{A}$ of problem (5.40) satisfies the conditions of the basis criterion in a Krein space presented in Section 1.3.7. In fact, the set of accumulation points of the spectrum of operator $\mathcal{A} \in \mathcal{K}(\mathcal{H})$ is finite, and $s(\mathcal{A}) = \emptyset$, since the numbers 0 and β_k, $k = 1, \ldots, m$, are not eigenvalues of operator $\mathcal{A}$ (see Properties 1° and 4° in Section 11.3.4). Therefore, according to Assertion (c) from the criterion in Section 1.3.7, we get that for any arbitrary value of the viscosity ν the set of root elements of problem (5.40) is complete in $\boldsymbol{H}$. Further, according to Assertion (d) from the mentioned criterion, we get that the root elements of operator $\mathcal{A}$ form an almost $\mathcal{J}$-orthonormal basis in $\boldsymbol{H}$. That basis is also a p-basis for $p > 3$. Indeed, by choosing $a > 0$ according to Property (10), the angular operator K corresponding to operator $\mathcal{M}_a$, and, therefore, to operator $\mathcal{A}$, belongs to the class $\mathfrak{S}_p$, $p > 3$. By this remark, the second statement in Property (11) is proved.

At last, if the condition (5.38) is fulfilled, then according to Assumption (6) in Section 11.5.2, problem (5.40) has only a real (positive) spectrum and has no associated elements. In this case, according to the criterion in Section 1.3.7, the system of eigenelements of problem (5.40) [Assertions (b), (d) and (e)] is not only complete in $\boldsymbol{H}$, but also a $\mathcal{J}$-orthonormal basis and a p-basis for $p > 3$.

Properties (11) are proved.

In conclusion, we note that in Sections 11.1–11.5 we studied in rather sufficient detail the properties of solutions of various problems on small oscillations of a visco-elastic fluid in a completely or partially filled container, and some of their abstract generalizations. Further, we will investigate the problem on small motions of a somewhat complex hydrosystem containing not a viscous but an ideal fluid.

11.6 Oscillations of Relaxing Fluids

In this section, a new class of problems is considered, where the so-called phenomenon of relaxation is displayed. The motion of a fluid with the property of relaxation is determined not only by its velociy, pressure, and density fields at the present moment of time, but also by the behavior of these fields during the complete motion until the indicated moment. In other words, relaxing fluids possess "memory" taking into account previous motions.

Here, the first simplest problem of this kind, a model example, is considered. General considerations on the structure of the spectrum of normal oscillations of a relaxing ideal fluid are given for some special cases.

11.6.1 Classical Statement of the Problem on Small Motions of a Relaxing Fluid

We consider an ideal nonhomogeneous fluid completely filling some domain Ω from $\mathbb{R}^3$. At rest, the fluid has density $\rho_0(x), x \in \Omega$, and the equilibrium pressure $P_0(x)$ is obtained from the relation

$$-\nabla P_0 - \rho_0(x) g \boldsymbol{e}_3 = \boldsymbol{0}, \tag{6.1}$$

where $g > 0$ is the acceleration of the gravitational field, and $\boldsymbol{e}_3$ is the unit vector of the axis Ox_3 directed upward, that is, opposite the vector of gravity. From (6.1) it follows that

$$P_0(x) = -g \int_0^{x_3} \rho_0(x_1, x_2, \xi) \mathrm{d}\xi + P_0(x_1, x_2, 0), \tag{6.2}$$

where $P_0(x_1, x_2, 0)$ is the pressure distribution in the plane Ox_1x_2.

Let us consider small fluid motions close to the state of rest (6.1), (6.2). We will assume as a rule that the complete pressure in fluid is equal to $P(t,x) = P_0(x)+p(t,x)$, where $p(t,x)$ is the dynamical pressure, and the density field is equal, respectively, to the sum $\rho_0(x) + \rho(t,x)$, where $\rho(t,x)$ is the density perturbation in the process of small oscillations. For a relaxing fluid, the following equation of state, connecting the pressure perturbation and density with each other, is used:

$$p(t,x) = a_\infty^2(x)\rho(t,x) - \int_0^t K(t-s,x)\rho(s,x)ds, \tag{6.3}$$

where $K(t,x)$ is a function defining the kernel of an integral Volterra operator, and $a_\infty^2(x)$ is the square of sound velocity spreading in the nonhomogeneous fluid. An important and typical case for a model of relaxing fluid is given by

$$K(t,x) = K_0(x)\exp(-b(x)t), \tag{6.4}$$

where $K_0(x)$ and $b(x)$ are positive functions in the domain Ω.

For deriving the linearized motion equations and linear continuity equation we use the nonlinear Euler equation for an ideal nonhomogeneous fluid and the nonlinear continuity equation. With the account of the scalar fields introduced above, these equations will take the form

$$(\rho_0(x)+\rho(t,x))\left(\frac{\partial \boldsymbol{u}}{\partial t} + (\boldsymbol{u}\cdot\nabla)\boldsymbol{u}\right) = -\nabla P_0 - \nabla p(t,x) - (\rho_0(x)+\rho(t,x))g\boldsymbol{e}_3,$$

$$\frac{\partial}{\partial t}(\rho_0(x)+\rho(t,x)) + \operatorname{div}((\rho_0(x)+\rho(t,x))\boldsymbol{u}(t,x)) = 0, \tag{6.5}$$

where $\boldsymbol{u}(t,x)$ is the velocity field.

Considering small first order infinitesimals of the functions $\boldsymbol{u}(t,x), p(t,x)$ and $\rho(t,x)$, and linearizing the equations with account of (6.1), we have

$$\rho_0(x)\frac{\partial \boldsymbol{u}}{\partial t} = -\nabla p, \qquad \frac{\partial \rho}{\partial t} + \operatorname{div}(\rho_0(x)\boldsymbol{u}) = 0 \quad \text{in } \Omega. \tag{6.6}$$

As for boundary value conditions, for an ideal fluid one such condition is obvious, that is, the nonleaking condition on the boundary of the domain Ω:

$$\boldsymbol{u}\cdot\boldsymbol{n} =: u_n = 0 \quad \text{on } \partial\Omega. \tag{6.7}$$

We also consider that the deviation of the density field on $\partial\Omega$ from the initial field $\rho_0(x)$ is equal to zero:

$$\rho(t,x) = 0 \quad \text{on } \partial\Omega. \tag{6.8}$$

Thus, the problem on small motions of an ideal relaxing fluid consists of finding the fields $\boldsymbol{u}(t,x), p(t,x)$ and $\rho(t,x)$ from the equations (6.3), (6.6), boundary value conditions (6.7), (6.8), and also the initial conditions

$$\boldsymbol{u}(0,x) = \boldsymbol{u}^0(x), \qquad \rho(0,x) = \rho^0(x). \tag{6.9}$$

11.6.2 Transition to an Initial Boundary Value Problem for One Scalar Function

In problem (6.3), (6.6)–(6.9), the unknown functions $p(t, x)$, $u_i(t, x)$, $i = 1, 2, 3$, may be expressed by just one scalar function, the deviation of the density field from the equilibrium field $\rho_0(x)$. For this function, $\rho(t, x)$, we formulate below an initial-boundary value problem for an integro-differential equation.

From the equation of the state (6.3) we have

$$\nabla p = \nabla(a_\infty^2(x)\rho(t, x)) - \nabla\left(\int_0^t K(t - s, x)\rho(s, x)\mathrm{d}s\right), \tag{6.10}$$

and hence

$$\operatorname{div}\nabla p \equiv \Delta p = \Delta(a_\infty^2(x)\rho(t, x)) - \Delta\left(\int_0^t K(t - s, x)\rho(s, x)\mathrm{d}s\right). \tag{6.11}$$

On the other hand, from the first and second equations (6.6) we get

$$\Delta p = -\operatorname{div}\left(\rho_0(x)\frac{\partial \boldsymbol{u}}{\partial t}\right) = -\frac{\partial}{\partial t}\operatorname{div}(\rho_0 \boldsymbol{u}) = \frac{\partial^2 \rho}{\partial t^2}. \tag{6.12}$$

Equalizing the right sides of relations (6.11) and (6.12) we obtain the integro-differential equation

$$\frac{\partial^2 \rho}{\partial t^2} - \Delta(a_\infty^2(x)\rho) + \Delta\left(\int_0^t K(t - s, x)\rho(s, x)\mathrm{d}s\right) = 0 \tag{6.13}$$

for the function $\rho(t, x)$.

For $t = 0$, the initial conditions

$$\rho(0, x) = \rho^0(x), \qquad \frac{\partial \rho}{\partial t}(0, x) = -\operatorname{div}(\rho_0(x)\boldsymbol{u}^0(x)), \tag{6.14}$$

should be fulfilled, which follow from the second equation (6.6) and from (6.9).

We make some comments on the choice of the boundary value condition on $\partial\Omega$ for function $\rho(t, x)$. If $\boldsymbol{n}$ is the external normal to $\partial\Omega$, then from (6.10) we have

$$\frac{\partial p}{\partial n} = \nabla p \cdot \boldsymbol{n} = \frac{\partial}{\partial n}(a_\infty^2(x)\rho) - \frac{\partial}{\partial n}\left(\int_0^t K(t - s, x)\rho(s, x)\mathrm{d}s\right).$$

On the other hand, from the first relation (6.6) and from (6.7) we get $\partial p/\partial n = -\rho_0(x)(\partial u_n/\partial t) = 0$ on $\partial\Omega$, and, therefore, on $\partial\Omega$ the function $\rho(t, x)$ should satisfy the boundary value condition

$$\frac{\partial}{\partial n}(a_\infty^2(x)\rho(t, x)) - \int_0^t \frac{\partial}{\partial n}(K(t - s, x)\rho(s, x))\mathrm{d}s = 0 \quad \text{on } \partial\Omega. \tag{6.15}$$

In particular, if the model (6.4) is accepted and

$$\begin{aligned} a_\infty^2(x) &= a_\infty^2 = \text{const} > 0, \\ K_0(x) &= K_0 = \text{const} > 0, \\ b(x) &= b_0 = \text{const} > 0, \end{aligned} \tag{6.16}$$

then from (6.15) we come to the Neuman condition

$$\frac{\partial \rho}{\partial n} = 0 \quad \text{on } \partial\Omega. \tag{6.17}$$

Further, for the sake of simplicity and also for obtaining a qualitative picture of the phenomena studied in problem (6.13), (6.14), the condition (6.8) will be accepted but not the more bulky condition (6.15).

For a relaxing fluid with the function $K(t,x)$ as in (6.4), it is possible to switch from the integro-differential equation (6.13) to a differental equation for the function

$$v(t,x) := \int_0^t K_0(x)\exp(-b(x)(t-s))\rho(s,x)ds. \tag{6.18}$$

Indeed, from (6.18) we have

$$\frac{\partial v}{\partial t} = K_0(x)\rho(t,x) - b(x)v(x), \tag{6.19}$$

and, therefore,

$$\begin{aligned} \rho(t,x) &= K_0^{-1}(x)\left(\frac{\partial v}{\partial t} + b(x)v\right), \\ \frac{\partial^2 \rho}{\partial t^2} &= K_0^{-1}(x)\left(\frac{\partial^3 v}{\partial t^3} + b(x)\frac{\partial^2 v}{\partial t^2}\right). \end{aligned} \tag{6.20}$$

From (6.18), (6.20), and from (6.13) we now get the problem

$$K_0^{-1}(x)\left(\frac{\partial^3 v}{\partial t^3} + b(x)\frac{\partial^2 v}{\partial t^2}\right) - \Delta\left(a_\infty^2(x)K_0^{-1}(x)\left(\frac{\partial v}{\partial t} + b(x)v\right)\right) + \Delta v = 0, \tag{6.21}$$

with the initial conditions

$$\begin{aligned} v(0,x) &= 0, \\ \frac{\partial v}{\partial t}(0,x) &= K_0(x)\rho^0(x), \\ \frac{\partial^2 v}{\partial t^2}(0,x) &= -K_0(x)\left[\operatorname{div}(\rho_0(x)\boldsymbol{u}^0(x) + b(x)\rho^0(x))\right], \end{aligned} \tag{6.22}$$

following from (6.18), (6.19), and (6.14).

By the mentioned above assumption on the choice of the boundary value condition for $\rho(t,x)$, we will consider that for $v(t,x)$ on $\partial\Omega$ the next model condition is fulfilled,

$$v(t,x) = 0 \quad \text{on } \partial\Omega. \tag{6.23}$$

11.6.3 The Simplest Problem on Oscilations of a Relaxing Fluid

In order to understand on a simple example the physical effects that are possible in an ideal relaxing fluid, we consider the special case (6.16) for problem (6.21)–(6.23) when all the medium characteristics are constant. From (6.21) we obtain the equation

$$b_0^{-1}\frac{\partial}{\partial t}\left(\frac{\partial^2 v}{\partial t^2} - a_\infty^2 \Delta v\right) + \left(\frac{\partial^2 v}{\partial t^2} - (a_\infty^2 - K_0 b_0^{-1})\Delta v\right) = 0 \tag{6.24}$$

considered in the domain Ω, with the boundary value condition (6.23) and the initial conditions (6.22). If in (6.24) one takes the limit $b_0 \to +\infty$, then we come to the ordinary wave equation for $v(t, x)$. The parameter b_0^{-1} is called the *time of relaxation.* It is usually small in comparison to other time intervals during which significant changes occur in the hydrodynamic system.

We introduce new parameters in (6.24) by the formulas

$$\tau := a_\infty / b_0, \qquad \gamma := K_0 / a_\infty^2, \qquad \tilde{t} := a_\infty t, \tag{6.25}$$

and denote the new time variable $\tilde{t}$ by the previous letter t. Then, instead of (6.24) we get the equation

$$\tau\frac{\partial}{\partial t}\left(\frac{\partial^2 v}{\partial t^2} - \Delta v\right) + \left(\frac{\partial^2 v}{\partial t^2} - \alpha(\tau)\Delta v\right) = 0. \tag{6.26}$$

Here, τ is the new relaxation time, and for $\alpha(\tau)$ we assume the conditions

$$0 < \alpha(\tau) := 1 - \gamma\tau < 1 \tag{6.27}$$

to be fulfilled.

We proceed from problem (6.26), (6.22), (6.23) to a differential equation in the Hilbert space $L_2(\Omega)$, with the standard scalar product. To this end, we consider on the set

$$\mathcal{D}(A) := \{v(x) \in C^2(\Omega) : v(x) = 0 \text{ on } \partial\Omega\} \subset L_2(\Omega)$$

the differential operator A acting according to the law $Av := -\Delta v$. This operator is symmetric and positive definite on $\mathcal{D}(A)$. Let us next take its Friedrichs extension, and denote that extension again by A. Then we have $A \gg 0$, $\mathcal{D}(A^{1/2}) = H_0^1(\Omega)$, $0 < A^{-1} \in \mathfrak{S}_\infty$, and the eigenvalues $\{\lambda_k(A)\}_{k=1}^\infty$ have the asymptotic behavior

$$\lambda_k(A) = \left(\frac{\operatorname{mes}\Omega}{6\pi^2}\right)^{-2/3} k^{2/3}[1 + o(1)], \qquad k \to \infty. \tag{6.28}$$

By means of operator A, taking also into account the boundary value condition (6.23), problem (6.22), (6.23), (6.26) may be rewritten in the form of a Cauchy problem

$$\begin{aligned}\tau\frac{d}{dt}\left(\frac{d^2v}{dt^2}+Av\right)+\left(\frac{d^2v}{dt^2}+\alpha(\tau)Av\right)&=0,\\ v(0)&=0,\\ v'(0)&=v^1,\\ v''(0)&=v^2,\end{aligned} \tag{6.29}$$

for the function $v=v(t)$ with values in $L_2(\Omega)$.

Based on (6.29), we consider normal motions of the relaxing fluid assuming $v(t)=v\exp(-\lambda t)$. Then, to find the amplitude elements v, we get the spectral problem

$$(-\tau\lambda+\alpha(\tau))Av=(\tau\lambda^3-\lambda^2), \tag{6.30}$$

which obviously amounts to a sequence of characteristic equations,

$$(-\tau\lambda+\alpha(\tau))\lambda_k=\tau\lambda^3-\lambda^2, \qquad k=1,2,\dots, \tag{6.31}$$

where $\lambda_k=\lambda_k(A)$ are the eigenvalues of operator A.

The study of the equations (6.31) leads to the following conclusions.

1° Problem (6.30) has a discrete spectrum situated in the right complex half-plane symmetrically relatively to the real axis, with limit points $\lambda_0=\alpha(\tau)/\tau>0$ and $\lambda=\infty$.

2° All the eigenvalues are divided into three branches,

$$\{\lambda_k^0\}_{k=1}^\infty, \quad \{\lambda_k^+\}_{k=1}^\infty, \quad \text{and} \quad \{\lambda_k^-\}_{k=1}^\infty.$$

The branch $\{\lambda_k^0\}_{k=1}^\infty$ is situated in the range $[\alpha(\tau)/\tau, 1/\tau]$, and the branch $\{\lambda_k^+\}_{k=1}^\infty$ consists of complex numbers situated in the upper half-plane, where $\mathrm{Re}\lambda_k^+\in(0,\underline{\gamma/2)}$, and $\mathrm{Im}\,\lambda_k^+\to+\infty$ as $k\to\infty$. The branch $\{\lambda_k^-\}_{k=1}^\infty$ consists of the numbers $\lambda_k^-=\overline{\lambda_k^+}$.

3° For $k\to\infty$ we have the following asymptotic formulas,

$$\begin{aligned}\lambda_k^0&=\frac{\alpha}{\tau}+\alpha^2(1-\tau)\tau^{-3}\lambda_k^{-1}(A)+\alpha^3(1-\alpha)(2-3\alpha)\tau^{-5}\lambda_k^{-2}(A)\\&\quad+3\alpha^4(1-\alpha)(1-3\alpha^2)\tau^{-7}\lambda_k^{-3}(A)+O(\lambda_k^{-4}(A)),\end{aligned} \tag{6.32}$$

$$\begin{aligned}\lambda_k^\pm&=\pm\mathrm{i}\lambda_k^{1/2}(A)+\frac{1-\alpha}{2\tau}\mp\mathrm{i}(1+3\alpha)\frac{1-\alpha}{8\tau^2\lambda_k^{1/2}(A)}\\&\quad-\alpha^2\frac{1-\alpha}{2\tau^3\lambda_k(A)}+O(\lambda_k^{-3}(A)), \qquad \alpha=\alpha(\tau)=1-\gamma\tau,\end{aligned} \tag{6.33}$$

and for $\tau \to 0$, the formulas

$$\lambda_k^{\pm} = \pm \mathrm{i}\lambda_k^{1/2}(A) \pm \mathrm{i}\gamma\lambda_k^{1/2}(A)\frac{\tau}{2} + \left(\gamma\frac{\lambda_k(A)}{2} \mp \mathrm{i}\gamma^2\frac{\lambda_k^{1/2}(A)}{8}\right)\tau^2$$
$$\pm \mathrm{i}\left(8\gamma\lambda_k^{3/2}(A) - \gamma^3\frac{\lambda_k^{1/2}(A)\tau^3}{16}\right) + \mathrm{O}(\tau^4). \tag{6.34}$$

4° From Properties 1°–3° it follows that in an ideal relaxing fluid there are two types of wave motions: *acoustic-relaxing waves*, similar to ordinary acoustic waves spreading in a bounded domain Ω, and also a new type of waves, *purely relaxing normal oscillations*, that are absent in an ordinary (not relaxing) fluid.

5° To the acoustic-relaxing waves there correspond two branches of eigenvalues, $\{\lambda_k^{\pm}\}_{k=1}^{\infty}$, $\lambda_k^- = \overline{\lambda_k^+}$, that are close to the imaginary half-axis. The presence of this relation in the given hydrodynamic system leads to the fact that the acoustic-gravitational waves, opposite to ordinary acoustic waves, have the fading decrements $\mathrm{Re}\lambda_k^{\pm} > 0$ different from zero that, for $k \to \infty$, approach asymptotically the constant value $(1-\alpha(\tau))/(2\tau) = \gamma/2$ from the left.

6° To purely relaxing waves there correspond aperiodic fading modes of normal oscillations with decrements λ_k^0 situated in the range $[1/\tau - \gamma, 1/\tau]$ of width γ. As $k \to \infty$, these decrements are asymptotically equal to $1/\tau - \gamma$.

7° If the relaxation time $\tau \ll 1$ tends to zero, then the acoustic-relaxation waves pass into ordinary acoustic waves [see (6.34)], and the fading decrements of purely relaxing waves become as large as possible and in the limit go to $+\infty$.

Based on these properties of the solutions of the spectral problem (6.30), we return to the Cauchy problem (6.29) and represent its generalized solution as a superposition of acoustic-relaxing and purely relaxing waves,

$$v(t) = \sum_{k=1}^{\infty} \left(c_k^0 \exp(-\lambda_k^0 t) + c_k^+ \exp(-\lambda_k^+ t) + c_k^- \exp(-\lambda_k^- t)\right) u_k(A), \tag{6.35}$$

with some coefficients c_k^0, c_k^+, and $c_k^- = \overline{c_k^+}$. Expanding the initial elements of (6.29) into series on the basis $\{u_k(A)\}_{k=1}^{\infty}$ of eigenelements of operator A that are considered to be orthonormalized in $L_2(\Omega)$ and satisfying the initial conditions (6.29), we come to the system of linear equations

$$\begin{aligned} c_k^0 + c_k^+ + c_k^- &= 0, \\ \lambda_k^0 c_k^0 + \lambda_k^+ c_k^+ + \lambda_k^- c_k^- &= -(v^1, u_k)_{L_2(\Omega)}, \\ (\lambda_k^0)^2 c_k^0 + (\lambda_k^+)^2 c_k^+ + (\lambda_k^-)^2 c_k^- &= (v^2, u_k)_{L_2(\Omega)}, \qquad k = 1, 2, \ldots. \end{aligned} \tag{6.36}$$

Here, for every k, the determinant

$$\Delta_k = \begin{vmatrix} 1 & 1 & 1 \\ \lambda_k^0 & \lambda_k^+ & \lambda_k^- \\ (\lambda_k^0)^2 & (\lambda_k^+)^2 & (\lambda_k^-)^2 \end{vmatrix}$$

is a Vandermonde determinant equal to

$$\Delta_k = -2\mathrm{i}\beta_k((\alpha_k - \lambda_k^0)^2 + \beta_k^2), \qquad \lambda_k^\pm = \alpha_k \pm \mathrm{i}\beta_k. \tag{6.37}$$

If $\beta_k \neq 0$, then $\Delta_k \neq 0$ and the system (6.36) has a unique solution allowing to uniquely obtain the generalized solution (6.35) by the initial conditions (6.29).

In addition to Properties 1°–7° connected with problem (6.30) and the solutions of equations (6.31), we note that for some combinations of physical parameters a finite number of orders k in (6.31), for which all roots are real, is possible, and then they could be two-multiple or three-multiple. Thus, a three-multiple root is possible only under the condition $1 - \gamma\tau = 1/9$, $\tau^2\lambda_k(A) = 1/3$, and then it is equal to $\lambda = 1/(3\tau)$. A two-multiple root may take place under the condition $1 - \gamma\tau \leq 1/3$, that is, for a large enough parameter of relaxation τ.

In these cases, to represent the generalized solution in the form (6.35) it is necessary to choose the dependence on t for a term with the mentioned values of k in the form $c_k^0 \exp(-\lambda_k^0 t) + (c_{1k} + c_{2k}t)\exp(-\beta t)$ for a two-multiple root β ($\neq \lambda_k^0$), or in the form $\exp(-\beta t)(c_{0k} + c_{1k}t + c_{2k}t^2)$ for a three-multiple root $\beta > 0$. It is possible to check that, in this case, the determinant in the system of linear equations (6.36) for such particular situations is different from zero.

We give the reader a possibility to study in detail the generalized solution (6.35) of problem (6.29), to formulate the requirement for the initial data v^1 and v^2 guaranteeing the convergence of series (6.35) to the solution of problem (6.29), and also to state the property of asymptotic (exponential) stability of this equation.

11.6.4 On the Solvability of the Cauchy Problem for an Abstract Integro-Differential Equation Connected with Small Motions of a Relaxing Fluid

Let us go back to the problem (6.13), (6.8), (6.14) and consider that

$$K(t, x) = \exp(-\gamma t)b(x), \qquad \gamma > 0. \tag{6.38}$$

If we reintroduce the operator A from Section 11.6.3 defined on $\mathcal{D}(A)$ and acting in $L_2(\Omega)$, then the problem (6.13), (6.8), (6.14), (6.38) may be rewritten in the form

$$\begin{aligned} &\frac{\mathrm{d}^2\rho}{\mathrm{d}t^2} + A(a_\infty^2(x)\rho) - \int_0^t \exp(-\gamma(t-s))A(b(x)\rho(s,x))\mathrm{d}s = 0, \\ &\rho(0) = \rho^0, \\ &\rho'(0) = \rho^1. \end{aligned} \tag{6.39}$$

It may be reduced to a more symmetric form if we introduce the new unknown function

$$u(t,x) = a_\infty(x)\rho(t,x)$$

with values in $L_2(\Omega)$, and multiply both sides of equation (6.39) by $a_\infty(x) > 0$. We obtain the problem

$$\begin{aligned}
&\frac{d^2u}{dt^2} + A_1u + \int_0^t \exp(-\gamma(t-s))A_2u(s)ds = 0,\\
&u(0) = u^0, \qquad u'(0) = u^1,\\
&A_1u := a_\infty(x)A(a_\infty(x)u),\\
&A_2u := -a_\infty(x)A(b(x)a_\infty^{-1}(x)u),
\end{aligned} \tag{6.40}$$

with the operator $A_1 \gg 0$, $0 < A_1^{-1} \in \mathfrak{S}_\infty$, and the operator A_2 such that $\mathcal{D}(A_2) = \mathcal{D}(A_1)$. In this setting, the operator $(A_1 + \gamma^2 I)^{-1}A_2$ may be extended by taking the closure to a bounded operator acting in $L_2(\Omega)$.

Generalizing further problem (6.40), we consider in a Hilbert space H the Cauchy problem for the integro-differential equation

$$\frac{\mathrm{d}^2u}{\mathrm{d}t^2} + A_1u + \sum_{k=2}^{m} \int_0^t \exp(-\gamma_k(t-s))A_ku(s)\mathrm{d}s = f(t),$$

$$u(0) = u^0, \quad u'(0) = u^1, \quad \gamma_k > 0, \qquad k = 2,\dots,m, \tag{6.41}$$

$$A_1 \gg 0, \quad 0 < A_1^{-1} \in \mathfrak{S}_\infty, \quad \mathcal{D}(A_k) \supset \mathcal{D}(A_1), \qquad k = 2,\dots,m, \tag{6.42}$$

where after taking the closure,

$$(A_1 + \gamma^2 I)^{-1}A_k \in \mathcal{L}(H), \qquad k = 2,\dots,m. \tag{6.43}$$

If all the integral terms in (6.41) disappear, then this problem reduces to the Cauchy problem for an abstract hyperbolic equation. Let us show that the conditions of existence of a generalized solution of problem (6.41)–(6.43) have the same form as for a hyperbolic equation.

We introduce the function

$$\tilde{f}(t) := f(t) - \sum_{k=2}^{m} \int_0^t \exp(-\gamma_k(t-s))A_ku(s)\mathrm{d}s \tag{6.44}$$

and consider it to be known. Then, problem (6.41) has the generalized solution

$$\begin{aligned}
&u(t) = \varphi_0(t) + \int_0^t \sin[A_1^{1/2}(t-s)](A_1^{-1/2}\tilde{f}(s))\mathrm{d}s,\\
&\varphi_0(t) = \cos(A_1^{1/2}t)u^0 + \sin(A_1^{1/2}t)(A_1^{-1/2}u^1).
\end{aligned} \tag{6.45}$$

If we substitute the function $\tilde{f}(t)$ from (6.44) into (6.45), then we obtain the Volterra integral equation

$$\begin{aligned} u(t) &= \varphi_0(t) + \varphi_1(t) \\ &\quad - \sum_{k=2}^{m} \int_0^t \sin(A_1^{1/2}(t-s))A_1^{-1/2}\Big(\int_0^s \exp(-\gamma_k(s-\xi))A_k u(\xi)\mathrm{d}\xi\Big)\mathrm{d}s, \\ \varphi_1(t) &= \int_0^t \sin(A_1^{1/2}(t-s))(A_1^{-1/2}f(s))\mathrm{d}s. \end{aligned} \tag{6.46}$$

We transform the integral operators by changing the order of integration and have

$$\begin{aligned} &\int_0^t \sin(A_1^{1/2}(t-s))A_1^{-1/2}\Big(\int_0^s \exp(-\gamma_k(s-\xi))A_k u(\xi)\mathrm{d}\xi\Big)\mathrm{d}s \\ &\quad = \int_0^t \Big(\int_\xi^t \sin(A_1^{1/2}(t-s))A_1^{-1/2}\exp(-\gamma_k(s-\xi))\mathrm{d}s\Big)A_k u(\xi)\mathrm{d}\xi \\ &\quad =: \int_0^t U_k(t,\xi)A_k u(\xi)\mathrm{d}\xi. \end{aligned} \tag{6.47}$$

A direct calculation shows that

$$\begin{aligned} U_k(t,\xi) &= (\exp(-\gamma_k(t-\xi))I - \cos(A_1^{1/2}(t-\xi)) \\ &\quad + \gamma_k \sin(A_1^{1/2}(t-\xi))A_1^{-1/2})(A_1+\gamma_k^2 I)^{-1}, \qquad k=2,\dots,m. \end{aligned} \tag{6.48}$$

The relations (6.46)–(6.48) lead to a Volterra integral equation of the second kind,

$$\begin{aligned} u(t) &= \varphi_0(t) + \varphi_1(t) + \int_0^t U(t,s)u(s)\mathrm{d}s, \\ U(t,s) &:= -\sum_{k=2}^{m} U_k(t,s)A_k, \end{aligned} \tag{6.49}$$

with a continuous kernel $U(t,s)$, since the first factors in (6.48) are continuous and conditions (6.42) are fulfilled.

Let us formulate the final result obtained by studying problem (6.41)–(6.43) based on the transition to equation (6.49) and the statements of Sections 1.5.8 and 1.5.9.

If the function $f(t)$ in (6.41) is continuously differentiable, $u^0 \in \mathcal{D}(A_1)$, $u^1 \in \mathcal{D}(A_1^{1/2})$, then there exists a unique solution $u(t)$ of problem (6.41) that may be obtained by applying the method of successive approximations to the equation (6.49). If $f(t)$ is a continuous function with values in H, $u^0 \in H$, $u^1 \in H$, then the solution of the Volterra equation (6.49) gives a generalized solution of problem (6.41).

Based on the just obtained general result, the reader can formulate without difficulty the statement on the solvability of the integro-differential equation (6.39) arising in the problem on small oscillations of a relaxing fluid (assuming that the function $a_\infty(x)$ is smooth enough). This statement will not be included here in more detail.

11.6.5 Normal Oscillations of a Relaxing Fluid with Variable Medium Characteristics

Let us go back to problem (6.21)–(6.23) on small motions of a relaxing fluid and consider that $K_0(x)$ and $a_\infty^2(x)$ are some functions of the point $x \in \Omega$, and $b(x) = b_0 = \text{const} > 0$. As in Section 11.6.3, we introduce the operator A and we consider the normal oscillations $v(t,x) = v(x)\exp(-\lambda t)$. For the amplitude elements $v(x) \in L_2(\Omega)$ we arrive at the following problem,

$$\begin{aligned} &-\tau\lambda^3\left(K_0^{-1}(x)v(x)\right) + \lambda^2\left(K_0^{-1}(x)v(x) - \tau\lambda A\left(a_\infty^2(x)\left(K_0^{-1}(x)v(x)\right)\right)\right) \\ &\quad + A\left(a_\infty^2(x)\left(K_0^{-1}(x)v(x)\right)\right) - \tau A v(x) = 0, \qquad \tau = b_0^{-1} > 0. \end{aligned} \tag{6.50}$$

Applying the operator $A^{-1} > 0$ to the left-hand side of (6.50) we obtain

$$\left(\lambda^2(1-\tau\lambda)A^{-1} - \tau a_\infty^2(x) + \left[a_\infty^2(x) - \tau K_0(x)\right]\right)\left(K_0^{-1}(x)v(x)\right) = 0. \tag{6.51}$$

To simplify this problem and to switch to a self-adjoint pencil with bounded operator coefficients, we make the transition to the following unknown variable

$$a_\infty(x)K_0^{-1}(x)v(x) =: u(x). \tag{6.52}$$

Then, multiplying on the left by the function $a_\infty^{-1}(x) > 0$, we finally get the following problem on eigenvalues,

$$\begin{aligned} L(\lambda)u &:= \left[\lambda^2(1-\tau\lambda)A_0 - \lambda\tau I + B\right]u = 0, \\ A_0 &:= a_\infty^{-1}(x)A^{-1}\left(a_\infty^{-1}(x)\right), \\ Bu &:= \alpha(\tau,x)u, \qquad\qquad (6.53) \\ \alpha(\tau,x) &:= 1 - \tau\gamma(x), \\ \gamma(x) &:= \frac{K_0(x)}{a_\infty^2(x)}. \qquad\qquad (6.54) \end{aligned}$$

Let us discuss now the properties of the operator coefficients A_0 and B in the pencil $L(\lambda)$. Because of the operator A (obtained as the Friedrichs extension of the operator $(-\Delta)$ under the Dirichlet condition on $\partial\Omega$) and from equation (6.50), it is only natural to assume that the functions $K_0(x)$ and $a_\infty^2(x)$ are twice continuously differentiable in $\bar{\Omega}$. Further, as in Section 11.6.3, we assume that the relaxation time

$\tau = b_0^{-1}$ is small enough, and, therefore, similarly to condition (6.27), here the following condition is satisfied

$$0 < \alpha(\tau, x) < 1. \tag{6.55}$$

Let

$$\gamma_- = \min_{x \in \bar{\Omega}} \gamma(x), \qquad \gamma_+ = \max_{x \in \bar{\Omega}} \gamma(x). \tag{6.56}$$

Then from (6.55) it follows that

$$0 < \alpha_-(\tau) := 1 - \tau\gamma_+ \leq \alpha(\tau, x) \leq 1 - \tau\gamma_- =: \alpha_+(\tau) < 1. \tag{6.57}$$

Because the function $\gamma(x)$ is continuous in $\bar{\Omega}$, we also get that the interval in which the function $\alpha(\tau, x)$ takes values is the segment $[\alpha_-(\tau), \alpha_+(\tau)]$, and, therefore, the operator B in (6.54) being the operator of multiplication by the function $\alpha(\tau, x)$ is a self-adjoint bounded operator acting in $L_2(\Omega)$, whose spectrum is continuous and situated in the segment $[\alpha_-(\tau), \alpha_+(\tau)]$. If, in particular, the characteristics of the relaxing medium—as in Section 11.6.3—are constant, that is, $K_0(x) = K_0 > 0$ and $a_\infty^2(x) = a_\infty^2 > 0$, then $\gamma(x) = \gamma = \text{const}$, and $\alpha_-(\tau) = \alpha_+(\tau) = 1 - \gamma\tau$, that is, the segment turns into a point of continuous (limiting) spectrum.

Next, we consider the properties of operator A_0 in (6.54). Because A^{-1} is a positive compact operator and the function $a_\infty^{-1}(x)$ is positive and continuous in $\bar{\Omega}$, the operator A_0 is compact and positive in $L_2(\Omega)$. Its eigenvalues $\lambda_k(A_0)$ are found from the equation $a_\infty^{-1}(x)A^{-1}a_\infty^{-1}(x)u = \lambda u$, which, after the substitution $A^{-1}a_\infty^{-1}(x)u = v$, becomes $a_\infty^2(x)v = \lambda Av$. Whence, it follows that the numbers $\lambda_k(A_0)$ are consecutive maxima of the variational ratio

$$\frac{\int_\Omega a_\infty^{-2}(x)|v|^2 \mathrm{d}\Omega}{\|A^{1/2}v\|_{L_2(\Omega)}^2} = \frac{\int_\Omega a_\infty^{-2}(x)|v|^2 \mathrm{d}\Omega}{\int_\Omega |\nabla v|^2 \mathrm{d}\Omega} \tag{6.58}$$

considered on the elements $v(x) \in H_0^1(\Omega)$. This fact allows us to state the character of the asymptotic behavior of the numbers $\lambda_k(A_0)$ as $k \to \infty$,

$$\lambda_k(A_0) = \left(\frac{1}{6\pi}\int_\Omega a_\infty^{-3}(x)\mathrm{d}\Omega\right)^{2/3} k^{-2/3}[1 + \mathrm{o}(1)]. \tag{6.59}$$

Based on these properties of the coefficients A_0 and B in problem (6.53), we can prove next several statements related to properties of its solutions.

1° Outside the segment $[\alpha_-(\tau)/\tau, \alpha_+(\tau)/\tau]$ problem (6.53) has a discrete spectrum consisting of finitely multiple eigenvalues situated symmetrically with respect to the real axis and with possible limit points $\lambda \in [\alpha_-(\tau)/\tau, \alpha_+(\tau)/\tau]$ and $\lambda = \infty$.

The spectrum's symmetry follows from the self-adjointness of the operator pencil $L(\lambda)$, which takes place in virtue of the fact that A_0 and B are self-adjoint operators. Since $\sigma(B) = [\alpha_-(\tau), \alpha_+(\tau)]$, then outside the segment $[\alpha_-(\tau)/\tau, \alpha_+(\tau)/\tau]$ the operator $B - \lambda\tau I$ is invertible and the resolvent $(B - \lambda\tau I)^{-1}$ is a holomorphic operator function. Applying it to both sides of equation (6.53) for the previously mentioned λ we arrive at the following problem,

$$(I + \Phi(\lambda))u = 0, \qquad \Phi(\lambda) := \lambda^2(1 - \tau\lambda)(B - \lambda\tau I)^{-1}A_0, \tag{6.60}$$

where $\Phi(\lambda)$ is an operator function taking compact values because A_0 is a compact operator. Since $\Phi(0) = 0$, then $(I + \Phi(\lambda))_{\lambda=0} \gg 0$, and, therefore, $I + \Phi(\lambda)$ is a Fredholm holomorphic operator function outside the previously mentioned segment. Thus, according to the statement in Section 1.6.3, the spectrum of problem (6.60) is discrete with possible limit points at those points where the holomorphicity property disappears, that is, points of the segment $[\alpha_-(\tau)/\tau, \alpha_+(\tau)/\tau]$ and the point $\lambda = \infty$.

2° The segment $[\alpha_-(\tau)/\tau, \alpha_+(\tau)/\tau]$ is the limiting spectrum of problem (6.53).

The proof of this property goes along the same lines as the proof of a similar property in Section 6.5.7, in which we considered the question on the existence of internal inertia waves for a rotating fluid.

Let us perform in (6.53) the trasformation $\lambda\tau =: \tilde{\lambda}$. Then we come across the following problem,

$$M(\tilde{\lambda})u := \left(B + \tilde{\lambda}^2(1 - \tilde{\lambda})\tau^{-2}A_0\right) u = \tilde{\lambda}u. \tag{6.61}$$

For any fixed $\tilde{\lambda}_1 \in [\alpha_-(\tau), \alpha_+(\tau)]$ we consider the problem

$$M(\tilde{\lambda}_1)u = \tilde{\lambda}u. \tag{6.62}$$

Since A_0 is a compact operator and the spectrum of operator B is the segment $[\alpha_-(\tau), \alpha_+(\tau)]$, then according to Weyl theorem, the limiting spectrum of the operator $M(\tilde{\lambda}_1)$ coincide with the seqment $[\alpha_-(\tau), \alpha_+(\tau)]$. This means that for any $\tilde{\lambda}_2 \in [\alpha_-(\tau), \alpha_+(\tau)]$ we can find a sequence of elements $\{u_k\}_{k=1}^{\infty}$ (Weyl sequence) such that

$$\lim_{k\to\infty} \left(M(\tilde{\lambda}_1)u_k - \tilde{\lambda}_2 u_k\right) = 0.$$

All we have to do now is to choose $\tilde{\lambda}_2 = \tilde{\lambda}_1$ and consider it to be any point in the seqment $[\alpha_-(\tau), \alpha_+(\tau)]$. Property 2° is proved.

3° The real spectrum of problem (6.53) is situated on the interval $[\alpha_-(\tau)/\tau, 1/\tau)$, and, therefore, in virtue of Properties 1° and 2°, in the interval $(\alpha_+(\tau)/\tau, 1/\tau)$ there is no more than a countable set of eigenvalues $\{\lambda_k^0\}$ that have limit points situated in the interval $[\alpha_-(\tau)/\tau, \alpha_+(\tau)/\tau]$.

Let $\lambda_0 < \alpha_-(\tau)/\tau$. Then $B - \lambda_0 \tau I \geq (\alpha_-(\tau) - \lambda_0\tau)I \gg 0$, $\lambda_0^2(1-\lambda_0\tau)A_0 > 0$ because by (6.57), $1 - \lambda_0\tau > 0$ and $A_0 > 0$. Therefore, for the previously mentioned $\lambda_0 \in \mathbb{R}$, the property $L(\lambda_0) \gg 0$ is satisfied for the operator pencil $L(\lambda)$, and problem (6.53) has a trivial solution for $\lambda = \lambda_0$.

If $\lambda_0 \geq 1/\tau$, then similarly we get that $\lambda_0^2(1-\tau\lambda_0)A_0 \leq 0$,

$$B - \lambda_0\tau I \leq (\alpha_+(\tau) - \lambda_0\tau)I = [(1-\lambda_0\tau) + (\alpha_+(\tau) - 1)]I \leq -(1-\alpha_+(\tau))I \ll 0,$$

and hence $L(\lambda_0) \ll 0$.

4° If the relaxation time τ is small enough so that

$$0 < \tau \leq \frac{1}{3\gamma_+}, \tag{6.63}$$

then the nonreal eigenvalues of problem (6.53) can not have their limit points situated in the interval $[\alpha_-(\tau)/\tau, \alpha_+(\tau)/\tau]$, that is, the limiting spectrum of problem (6.53).

The proof of this property relies on the posibility of a spectral factorization of the pencil (6.53) due to condition (6.63). Let us remark that this condition is quite natural from a physical point of view because in real problems the relaxation time is often considered to be a small parameter.

We consider now the interval $[\alpha_-(\tau)/\tau - \varepsilon, 1/\tau]$, where $\varepsilon > 0$ is a small enough number. As it was stated in the proof of Property 3°,

$$L\left(\frac{\alpha_-(\tau)}{\tau} - \varepsilon\right) \gg 0, \qquad L\left(\frac{1}{\tau}\right) \ll 0. \tag{6.64}$$

From (6.63) it follows that

$$L'(\lambda) = (2\lambda - 3\tau\lambda^2)A_0 - \tau I \ll 0, \qquad \lambda \in \left[\frac{\alpha_-(\tau)}{\tau} - \varepsilon, \frac{1}{\tau}\right]. \tag{6.65}$$

Indeed, the function $2\lambda - 3\tau\lambda^2$ takes positive values in the interval $(0, 2/(3\tau))$ and nonpositive values at all the other points of the real axis. Therefore, if $\alpha_-(\tau)/\tau = (1-\gamma_+\tau)/\tau \geq 2/(3\tau)$, that is, if inequality (6.63) is satisfied for all points of the interval $[\alpha_-(\tau)/\tau, 1/\tau]$, then we have $(2\lambda - 3\tau\lambda^2)A_0 \leq 0$, because A_0 is a positive operator. Hence, for small enough ε, property (6.65) is satisfied as well.

From (6.64) and (6.65), it follows that conditions (1.6.9) are sufficient for the spectral factorization of the pencil $(-L(\lambda))$. Thus, according to Statement 1° in Section 1.6.9, we have

$$L(\lambda) = L_+(\lambda)(\lambda I - Z) := (L_0 + \lambda L_1 + \lambda^2 L_2)(\lambda I - Z), \tag{6.66}$$

where $L_+(\lambda)$ is holomorphic and holomorphically invertible in a neighborhood U of the interval $[\alpha_-(\tau)/\tau - \varepsilon, 1/\tau]$. The spectrum of problem (6.53) in U coincides with the spectrum of operator Z, which is similar to a self-adjoint operator. It means that in the domain U the spectrum of problem (6.53) is situated in the interval $[\alpha_-(\tau)/\tau - \varepsilon, 1/\tau]$, and, therefore, Statement 1° holds true. At the same time; from (6.66) we get the property $\tau Z = B + K$, where K is a compact operator. Indeed, it can be proved that the factor Z satisfies the equation $B - \tau Z + A_0 Z^2 - \tau A_0 Z^3 = 0$ and thus, by the fact that A_0 is compact and Z is bounded, we get the previously mentioned property.

5° Problem (6.23) has two branches of eigenvalues $\{\lambda_k^\pm\}_{k=1}^\infty$ adjacent to the imaginary half-axis $\arg\lambda = \pm\pi/2$ and having the asymptotic behavior

$$\lambda_k^\pm = \pm \mathrm{i}(\lambda_k(A_0))^{-1/2}[1 + o(1)], \qquad k \to \infty, \tag{6.67}$$

where $\lambda_k(A_0)$ are the eigenvalues of operator A_0 in (6.54) with the asymptotic behavior (6.59).

To prove this property, let us divide both sides of (6.53) by $-\lambda\tau$ and then perform the substitution $v = \mathrm{i}\lambda A_0^{1/2} u$ in the thus obtained equation. This equation together with the relation between u and v lead to the vector-matrix equation

$$\begin{pmatrix} I & \mathrm{i}\tau^{-1}A_0^{1/2} \\ 0 & I \end{pmatrix}\begin{pmatrix} u \\ v \end{pmatrix} - \tilde{\lambda}\begin{pmatrix} 0 & A_0^{1/2} \\ A_0^{1/2} & 0 \end{pmatrix}\begin{pmatrix} u \\ v \end{pmatrix} - \frac{\mathrm{i}}{\tau\tilde{\lambda}}\begin{pmatrix} B & 0 \\ 0 & 0 \end{pmatrix}\begin{pmatrix} u \\ v \end{pmatrix}$$
$$= \begin{pmatrix} 0 \\ 0 \end{pmatrix}, \qquad \tilde{\lambda} := \mathrm{i}\lambda, \tag{6.68}$$

considered in the Hilbert space $(L_2(\Omega))^2$.

Because $A_0^{1/2}$ is a positive compact operator, whose eigenvalues, according to formula (6.59), have power asymptotics, then to problem (6.68) there corresponds an operator pencil as the one given by (1.6.6) in Section 1.6.8. According to the conclusions of that section, problem (6.68) has two branches of eigenvalues $\{\tilde{\lambda}_k^{\pm}\}_{k=1}^{\infty}$, whose asymptotic behavior for $k \to \infty$ is similar to the one of the shortened operator pencil

$$\begin{pmatrix} I & 0 \\ 0 & I \end{pmatrix} - \tilde{\lambda} \begin{pmatrix} 0 & A_0^{1/2} \\ A_0^{1/2} & 0 \end{pmatrix},$$

that is, it has the form

$$\tilde{\lambda}_k^{\pm} = \pm(\lambda_k(A_0))^{-1/2}[1 + \mathrm{o}(1)], \qquad k \to \infty. \tag{6.69}$$

Whence, formulas (6.67) follow.

Coming to the study of other properties of the solutions of problem (6.53), subsequently, we will refer to its nonreal eigenvalues that are situated in the upper and lower half-plane, respectively, and counted in the increasing order of their absolute values.

6° The eigenvalues $\{\lambda_k^{\pm}\}_{k=1}^{\infty}$ are situated in the right complex half-plane in the band

$$\frac{\tau}{2\lambda_1(A_0)} < \operatorname{Re}\lambda < \frac{2}{3\tau}. \tag{6.70}$$

Indeed, multiplying (scalarly in $L_2(\Omega)$) both sides of (6.53) by u, and taking its imaginary part we get

$$\begin{aligned} &- \tau \operatorname{Im}\lambda((\operatorname{Re}\lambda)^2 - (\operatorname{Im}\lambda)^2)(A_0 u, u)_{L_2(\Omega)} - 2\tau(\operatorname{Re}\lambda)^2 \operatorname{Im}\lambda(A_0 u, u)_{L_2(\Omega)} \\ &\quad - \tau(\operatorname{Im}\lambda)(u,u)_{L_2(\Omega)} + 2(\operatorname{Im}\lambda)(\operatorname{Re}\lambda)(A_0 u, u)_{L_2(\Omega)} = 0. \end{aligned}$$

Because here $\operatorname{Im}\lambda \neq 0$ we obtain the following inequality,

$$\begin{aligned} &\tau(\operatorname{Im}\lambda)^2(A_0 u, u)_{L_2(\Omega)} \\ &\quad = 3\tau(\operatorname{Re}\lambda)^2(A_0 u, u)_{L_2(\Omega)} + \tau(u,u)_{L_2(\Omega)} - 2\operatorname{Re}\lambda(A_0 u, u)_{L_2(\Omega)} \\ &\quad > 0, \end{aligned}$$

whence it follows that

$$\operatorname{Re}\lambda > \frac{\tau}{2}\left(3(\operatorname{Re}\lambda)^2 + \frac{(u,u)_{L_2(\Omega)}}{(A_0 u, u)_{L_2(\Omega)}}\right) > 0,$$

since $A_0 > 0$. Hence, we also get the following relations

$$\operatorname{Re}\lambda > \frac{\tau}{2}\frac{(u,u)_{L_2(\Omega)}}{(A_0 u, u)_{L_2(\Omega)}} \geq \frac{\tau}{2\lambda_1(A_0)}, \qquad \operatorname{Re}\lambda > \frac{3\tau}{2}(\operatorname{Re}\lambda)^2,$$

that is, inequalities (6.70) follow.

As a consequence of Properties 3° and 6° we have the following.

7° If condition (6.57) is fulfilled, then the spectrum of problem (6.53) is situated in the right complex half-plane and may be divided into three sets: two branches of nonreal eigenvalues $\{\lambda_k^{\pm}\}_{k=1}^{\infty}$ adjacent to the imaginary half-axis, and a spectral set situated in the interval $[\alpha_-(\tau)/\tau, 1/\tau)$ that consists of the limiting spectrum $[\alpha_-(\tau)/\tau, \alpha_+(\tau)/\tau)]$ and a finite or countable set of finitely multiple eigenvalues $\{\lambda_k^0\}$.

8° The system of eigen- and associated elements of problem (6.68) corresponding to the eigenvalues $\{\lambda_k^{\pm}\}_{k=1}^{\infty}$ has no more than a finite defect in the space $(L_2(\Omega))^2$.

In order to prove this property, let us perform in (6.68) the substitution $\tilde{\lambda} = \mu^{-1}$, multiply both sides of this equation by μ, and get

$$\begin{pmatrix} 0 & A_0^{1/2} \\ A_0^{1/2} & 0 \end{pmatrix}\begin{pmatrix} u \\ v \end{pmatrix} - \mu\begin{pmatrix} I & 0 \\ 0 & I \end{pmatrix}\begin{pmatrix} u \\ v \end{pmatrix} - \mu i\tau^{-1}\begin{pmatrix} \mu B & -A_0^{1/2} \\ 0 & 0 \end{pmatrix}\begin{pmatrix} u \\ v \end{pmatrix} = \begin{pmatrix} 0 \\ 0 \end{pmatrix}. \tag{6.71}$$

Thus, we come to a pencil similar to the one discussed in Section 1.6.7.

Here the first operator matrix is self-adjoint, compact, belongs—by formula (6.59)—to the class $\mathfrak{S}_p$ for $p > 3$, and has a zero kernel. Further, the last term is a holomorphic function in the circle of any radius $|\mu| < r$ that takes bounded values. Therefore, according to the conclusions of Section 1.6.7, the system of eigen- and associated elements of problem (6.71) corresponding to eigenvalues μ in the circle $|\mu| < r$ has no more than a finite defect in $(L_2(\Omega)))^2$. The property is satisfied for problem (6.68), after we perform the inverse substitution, in the circle $|\lambda| = |\mu|^{-1} > r^{-1}$, with any $r > 0$.

9° If the parameters of the considered physical system are such that

$$\begin{gathered} \lambda_1^{1/2}(A_0) < \tau, \\ 1 - \tau\gamma_- < 0, \\ \tau^2 + \lambda_1(A_0) - 6\tau\lambda_1^{1/2}(A_0) + 4\tau^2\lambda_1^{1/2}(A_0)\gamma_- > 0, \end{gathered} \tag{6.72}$$

then the system of eigen- and associated elements of problem (6.71) corresponding to eigenvalues in the circle $|\mu| < r_-$, where r_- is defined by formula (6.73), is complete in the space $(L_2(\Omega))^2$.

To prove this statement, let us use the condition in Section 1.6.6 that is sufficient to factorize the operator-function (6.71). Since the norm of the first matrix operator in (6.7) is equal to $\lambda_1^{1/2}(A_0)$ and $\|B\| = 1 - \tau\gamma-$, then the previously mentioned condition can be reduced to the following inequality for some $r > 0$,

$$\frac{\lambda_1^{1/2}(A_0)}{r} + \frac{\lambda_1^{1/2}(A_0)}{\tau} + r\frac{(1-\tau\gamma-)}{\tau} < 1.$$

Whence it follows that $r \in (r_-, r_+)$, where

$$r_\pm = \frac{\left(\tau - \lambda_1^{1/2}(A_0)\right) \pm \sqrt{\left(\tau - \lambda_1^{1/2}(A_0)\right)^2 - 4\tau\lambda_1^{1/2}(A_0)(1-\tau\gamma-)}}{2(1-\tau\gamma-)}, \tag{6.73}$$

and, if inequalities (6.72) are satisfied, then $0 < r_- < r_+ < \infty$.

We introduce the notations

$$\frac{\lambda_1^{1/2}(A_0)}{\tau} =: \xi,$$
$$\lambda_1^{1/2}(A_0)\gamma_- =: \delta.$$

Then, inequalities (6.72) lead to the following relations,

$$0 < \delta < \xi < 1, \qquad \xi^2 - 6\xi + 1 + 4\delta > 0,$$

that are satisfied for small enough $0 < \delta < \xi$ like, for example, for $\xi = 1/3$, $2/9 < \delta < 1/3$.

Since, under conditions (6.72), the operator-function of problem (6.71) allows a spectral factorization relative to the circle $|\mu| = r \in (r_-, r_+)$, then we get the following factorization,

$$\mathcal{A} - \mu\mathcal{I} - \mu\mathcal{S} + \mu^2\mathcal{B} = (\mathcal{D} - \mu\mathcal{B})(\mu\mathcal{I} - \mathcal{F}), \tag{6.74}$$

where the operator matrices $\mathcal{A}$, $\mathcal{I}$, $\mathcal{S}$, and $\mathcal{B}$ are defined by the left-hand side of (6.71). In particular, $\mathcal{I}$ is the identity operator in $(L_2(\Omega))^2$, $\mathcal{D} - \mu\mathcal{B}$ is holomorphically invertible in the circle $|\mu| < r_+$, and the spectrum of problem (6.71) coincides with the spectrum of operator $\mathcal{F}$ and is situated in the circle $|\mu| \leq r_-$.

Comparing the coefficients of the zero and first degrees of μ in (6.74) we arrive to the following relations:

$$\mathcal{A} = -\mathcal{D}\mathcal{F},$$
$$\mathcal{I} + \mathcal{S} = \mathcal{D} - \mathcal{B}\mathcal{F}.$$

Hence, it follows that $\mathcal{F} = -\mathcal{D}^{-1}\mathcal{A}$ is a compact operator and since $-\mathcal{D} = \mathcal{I}+\mathcal{S}+\mathcal{B}\mathcal{F}$, the sum of the identity operator and of two compact operators, then $-\mathcal{D}^{-1} = \mathcal{I}+\mathcal{H}$, where $\mathcal{H}$ is compact. Thus, in the circle $|\mu| \le r_-$, problem (6.71) can be replaced with the following problem on eigenvalues

$$(\mathcal{I}+\mathcal{H})\mathcal{A}w = \mu w, \qquad w = (u;v)^{\mathrm{t}} \in (L_2(\Omega))^2, \tag{6.75}$$

for the weakly perturbed compact operator $(\mathcal{I}+\mathcal{H})\mathcal{A}$. Since, in this case, $\mathcal{A} \in \mathfrak{S}_p$, $p > 3$, $\mathrm{Ker}[(\mathcal{I}+\mathcal{H})\mathcal{A}] = \{0\}$, then, according to Keldysh first theorem formulated in Section 1.6.4, the system of eigen- and associated elements of problem (6.75), that is, problem (6.71) for $|\mu| \le r_-$, is complete in $(L_2(\Omega))^2$ and thus Statement 9° is proved.

10° For the real eigenvalues λ_k^0 of problem (6.53) that are situated in the interval $(\alpha_+(\tau)/\tau, 1/\tau)$ there exist the two-sided estimates

$$\frac{\delta_k^-}{\tau} \le \lambda_k^0 \le \frac{\delta_k^+}{\tau}, \qquad k = 1, 2, \ldots, \tag{6.76}$$

where $\delta_k^\pm$ are the roots of the equations

$$\lambda_k(A_0)x^2(1-x) = (x - \alpha_\pm(\tau))\tau^2, \tag{6.77}$$

situated in the interval $(\alpha_+(\tau), 1)$.

The proof of (6.76) and (6.77) is based on the introduction of a new spectral parameter in problem (6.61), on a two-sided estimate of the variation ratio of a linear spectral problem, and also on the minimax principle.

Going back to problem (6.61) with the parameter $\tilde\lambda = \tau\lambda$, let us rewrite it in the form

$$\begin{aligned} P(\tilde\lambda)u &= \beta(\tilde\lambda)\tau^{-2}A_0u, \\ P(\tilde\lambda) &:= \tilde\lambda I - B, \\ \beta(\tilde\lambda) &:= \tilde\lambda^2(1-\tilde\lambda), \end{aligned} \tag{6.78}$$

where β is a spectral parameter, and $\tilde\lambda \in (\alpha_+(\tau), 1]$ is an ordinary parameter that the solutions of problem (6.78) are depending on. Since $\sigma(B) = [\alpha_-(\tau), \alpha_+(\tau)]$, then for the operator $P(\tilde\lambda)$ we have the two-sided estimate

$$0 \le \left(\tilde\lambda - \alpha_+(\tau)\right) I \le P(\tilde\lambda) \le \left(\tilde\lambda - \alpha_-(\tau)\right) I,$$

and, therefore,

$$\frac{\left(\tilde{\lambda}-\alpha_+(\tau)\right)}{\tau^{-2}}\cdot\frac{(u,u)_{L_2(\Omega)}}{(A_0u,u)_{L_2(\Omega)}}\leq\frac{\left(P(\tilde{\lambda})u,u\right)_{L_2(\Omega)}}{\tau^{-2}(A_0u,u)_{L_2(\Omega)}}\leq\frac{\left(\tilde{\lambda}-\alpha_-(\tau)\right)}{\tau^{-2}}\cdot\frac{(u,u)_{L_2(\Omega)}}{(A_0u,u)_{L_2(\Omega)}}. \tag{6.79}$$

Since in (6.78) the operator $P(\tilde{\lambda})$ is positive definite and bounded, then, in virtue of the results in Section 1.4.2, this problem has a discrete positive spectrum $\{\beta_k(\tilde{\lambda})\}_{k=1}^{\infty}$ with a limit point at $+\infty$, and we can apply the minimax principle to the numbers $\beta_k^{-1}(\tilde{\lambda})$. Then, a maximin principle takes place for $\beta_k(\tilde{\lambda})$, which applied to the variational ratio (6.79) gives the inequalities

$$\frac{\tau^2\left(\tilde{\lambda}-\alpha_+(\tau)\right)}{\lambda_k(A_0)}\leq\beta_k(\tilde{\lambda})\leq\frac{\tau^2\left(\tilde{\lambda}-\alpha_-(\tau)\right)}{\lambda_k(A_0)}. \tag{6.80}$$

Let us note now that problem (6.61) can have nontrivial solutions $\tilde{\lambda}_k^0$ in the interval $(\alpha_+(\tau),1)$ if the roots of the equations

$$\beta_k(\tilde{\lambda})=\tilde{\lambda}^2(1-\tilde{\lambda}),\qquad k=1,2,\ldots \tag{6.81}$$

are found in the same interval. However, because $\beta_k(\tilde{\lambda})$ depends continuously on and increases monotonously with the parameter $\tilde{\lambda}$, taking into consideration the two-sided estimates (6.80) and the graph of equation (6.81), and also majorizing the equations

$$\frac{\tau^2\left(\tilde{\lambda}-\alpha_\pm(\tau)\right)}{\lambda_k(A_0)}=\tilde{\lambda}^2(1-\tilde{\lambda}),\qquad k=1,2,\ldots, \tag{6.82}$$

reducing to equations (6.77) it follows that

$$\delta_k^-\leq\tilde{\lambda}_k^0\leq\delta_k^+,\qquad k=1,2,\ldots, \tag{6.83}$$

where $\delta_k^\pm$ are the roots of equation (6.77). Since $\tilde{\lambda}_k^0=\tau\lambda_k^0$, then the two-sided estimate (6.76) follows from (6.83), and, thus, Statement 10° is proved.

Let us note that, at the same time, from the graphic solution of equation (6.82) we get $\delta_k^+\to\alpha_+(\tau)$ and $\delta_k^-\to\alpha_-(\tau)$ for $k\to\infty$. Therefore, in principle eigenvalues λ_k^0 of problem (6.53) that are not situated inside the interval $(\alpha_+(\tau)/\tau,1/\tau)$, that is, outside the limiting spectrum of this problem, but on this very limiting spectrum, are possible. Hence, it is also clear that the estimates (6.76) can be effective only for small values of k.

11.6.6 Physical Conclusions

The properties of the solutions to problem (6.53) that were proved in Section 11.6.5 allow us to draw the following physical conclusions.

1° In a relaxing fluid with variable medium characteristics, when in the chosen model (6.3), (6.4) the square of sound velocity $a^2_\infty(x)$ and the function $K_0(x)$ are variable, the relaxation time $\tau = 1/b_0$ ($b_0 \equiv b_0(x)$) is constant, and conditions (6.57) are satisfied, there exist two types of wave motions: acoustic-relaxing waves and purely relaxing waves.

2° To acoustic-relaxing waves there correspond nonreal eigenvalues $\{\lambda_k^\pm\}_{k=1}^\infty$ situated in the band (6.70) and having the asymptotic behavior (6.67). The eigen-elements and associated elements of problem (6.53) that correspond to them form a two-multiple complete system in $L_2(\Omega)$ with finite defect, in the sense that the system of eigen- and associated elements of problem (6.71) has no more than a finite defect in $(L_2(\Omega))^2$. If conditions (6.72) are satisfied, then the property of two-multiple completeness with a defect is changed into the property of two-multiple completeness.

3° To purely relaxing waves there corresponds a real spectrum situated on the positive half-axis in the interval $[\alpha_-(\tau)/\tau, 1/\tau)$ and consisting of two sets: the limiting spectrum in the form of the segment $[\alpha_-(\tau)/\tau, \alpha_+(\tau)/\tau]$ and a finite or a countable set of finitely multiple eigenvalues λ_k^0 with possible limit points situated on the limiting spectrum. Thus, to the numbers λ_k^0 there correspond aperiodic fading modes of normal oscillations with decrements bounded both upward and downward. For these decrements λ_k^0, the two-sided estimates (6.76) are satisfied.

4° If the medium charascteristics tend to constants, that is, if $\gamma(x) \to \gamma =$ const, then the limiting spectrum of problem (6.53) reduces to the point $\alpha(\tau) = 1-\gamma\tau$, the number of eigenvalues λ_k^0 becomes countable, and Properties 1°–7° obtained in Section 11.6.3 are satisfied.

In the conclusion of this section we should note that in all the problems considered here, we take into account the Dirichlet model condition (6.8) on the boundary $\partial\Omega$ of the considered domain Ω filled with a relaxing ideal fluid. Similar results take place if we choose the Neuman condition (6.17) and even in the case of the more complicated condition (6.15), that is, the qualitative pattern of the physical phenomena studied here will be similar to the one obtained under condition (6.8).

Appendix D

Remarks and Reference Comments to Part IV

D.1 Chapter 10

10.1. Problems on oscillations of a two layer hydrosystem consisting of an ideal and a viscous fluid layer have not been systematically studied until now. In this regard, we can mention only the work of E. Hasegawa [1], where the problem on running waves at the boundary region between two horizontal layers has been studied: an upper layer of a viscous fluid and a bottom layer of an ideal fluid. At the top and at the bottom, both layers were limited by horizontal solid walls. In recent years, the problems on oscillations of partially dissipative hydrosystems have been studied by N. D. Kopachevsky, T. P. Temchenko, and B. M. Vronskii [1, 2].

The statement of the problem on small oscillations of a hydrosystem consisting of a viscous fluid and an ideal one, and filling an arbitrary container, is due to N. D. Kopachevsky. N. D. Kopachevsky also obtained the balance law of full energy and the equations of the problem on normal oscillations.

10.2. The study of the initial boundary value problem (10.1.2)–(10.1.11) based on an operator approach, and using the additional boundary value problems I–V, was carried out by N. D. Kopachevsky. In this process, instead of the scheme used earlier in Chapter 8, in particular, the second additional problem of Section 8.1 [see problem (8.1.13)–(8.1.14)], we developed a new approach, with a new additional Problem II (see Section 10.2.2), in which the velocity field in the viscous fluid is not decomposed into two new unknown fields. This approach allows to a certain extent to simplify the transition to the system of evolution operator equations. The investigation of the properties of matrix blocks (Section 10.2.3) and, relying on them, the proof of the theorem on correct solvability of the original initial boundary value problem, was completed by N. D. Kopachevsky.

10.3. The statement of the model problem on normal oscillations of a partially dissipative hydrosystem (Section 10.3) is due to N. D. Kopachevsky, and its study was carried out by I. M. Klinchin and N. D. Kopachevsky [1]. The scheme of analyzing the characteristic equation (10.3.23) is close to that applied by N. D. Kopachevsky and A. D. Mychkis [1, 2] in problems on normal oscillations of a capillary viscous fluid. The derivation of the asymptotic formulas (10.3.38), (10.3.39) is a joint work with A. D. Myshkis. The general conclusions addressing the model problem, and the hypotheses arising from it, were deduced by N. D. Kopachevsky.

10.4. The study of normal oscillations of a partially dissipative hydrosystem in an arbitrary region (Section 10.4) was carried out by N. D. Kopachevsky and is presented here for the first time. We should stress that the properties of the spectrum and the system of eigen- and associated elements of the operator pencil (10.4.8), (10.4.9) deserve an additional, more detailed, research. In particular, close attention should be paid to studying these properties when the viscosity μ and the ratio ρ_2/ρ_1 of densities of the ideal and viscous fluids are changing. The spectrum properties stated in Section 10.4.2 were obtained by means of the same methods applied to the problems of Chapter 8. The theorem on spectrum localization was proved following the scheme applied earlier by M. B. Orazov [1–3] in problems of hydrodynamics and hydroelasticity, and goes back to M. V. Keldysh [1, 2], and G. V. Radzievskii [1–4].

10.5. The theorem on multiple completeness of the system of eigen- and associated elements of the problem on normal oscillations of a partially dissipative hydrosystem was proved by N. D. Kopachevsky, and is explained here for the first time. In this proof, a linearization on the spectal parameter was first done, and then Keldysh classical scheme [1, 2] has been used. As an interesting and not completely clarified problem, we should call attention to the existence of three branches of eigenvalues localized in a neighborhood of the imaginary semiaxes and the positive semiaxis, and having a common limit point at infinity. In the form presented in Section 10.5.3, this study was done by N. D. Kopachevsky, and it is also outlined for the first time here.

D.2 Chapter 11

11.1. The first five sections of this Chapter were motivated by the work of A. I. Miloslavskii [1, 4], where the problems on small oscillations of a viscous fluid in a completely or partially filled rigid container have been studied.

In describing small motions of a visco-elastic fluid, as a basic paradigm we have chosen Oldroite generalized model (11.1.2) of the connection between the tensors of tensions and deformation velocities. The physical and mathematical statements of the problem on motion of a visco-elastic fluid in a bounded region (Section 11.1.1, 11.1.2) were taken from the work of A. I. Miloslavskii [2, 3].

The theorem on correct solvability of the initial boundary value problem (Section 11.1.3) was proved by L. D. Bolgova and N. D. Kopachevsky [1]. The problem on normal oscillations (Section 11.1.4) has been also studied by L. D. Bolgova and N. D. Kopachevsky [1].

11.2. Abstract evolution and spectral problems (Sections 11.2) have been studied by T. Ya. Azizov, N. D. Kopachevsky, and L. D. Orlova (Bolgova) [1], and also by L. D. Bolgova (Orlova) and N. D. Kopachevsky [2]. In the investigation of the spectral problem carried out by T. Ya. Azizov, N. D. Kopachevsky, and L. D. Orlova, the block method [see (11.2.27)–(11.2.31)] was proposed by T. Ya. Azizov. The example formulated in Property 6° (for $m = 2$) from Section 11.2.5 was constructed by T. Ya. Azizov. Property 9° has also been proved by T. Ya. Azizov. The other properties of solutions of the spectral problem have been stated and proved by N. D. Kopachevsky and L. D. Orlova (Bolgova).

11.3. The problem on small motions and normal oscillations of a visco-elastic fluid in an open container was studied by A. I. Miloslavskii [3]. Here, the presentation is based on the general scheme stated in Section 1.8, and also on previous examinations of the problem done in Chapter 8. The theorem on correct solvability of the initial boundary-value problem (Section 11.3.3) has been proved by N. D. Kopachevsky and L. D. Orlova (Bolgova).

11.4. The study of the properties of multiple basicity of the system of eigen- and associated functions in the problem on normal oscillations of a visco-elastic fluid in an open container (Section 11.4) was carried out following the scheme of A. I. Miloslavskii [3].

11.5. The study of the properties of the spectrum in the problem on normal oscillations of a visco-elastic fluid in an open container is due to A. I. Miloslavskii [3]. However, the presentation here was adapted by N. D. Kopachevsky. In particular, the proof of some spectrum properties, and the property of p-basicity of the system of eigen- and associated elements, was done without changing the spectral parameter. These constructions belong to N. D. Kopachevsky and L. D. Orlova (Bolgova).

11.6. As it has been already mentioned the problem on small motions and normal oscillations of an ideal relaxing fluid is the only problem for ideal fluids not addressed in Volume I, and deferred to Volume II of the present monograph. The twofold reason is that, first, by its physical effects the problem is close to hydrodynamic systems containing a viscous fluid and, second, it is a rather complex nonself-adjoint problem.

The first studies of similar problems using the spectral theory of operator pencils were done by L. D. Bolgova and N. D. Kopachevsky [2, 3], and also N. D. Kopachevsky and L. D. Orlova (Bolgova) [1]. Results of these investigations are presented also in the report by N. D. Kopachevsky, L. D. Orlova (Bolgova), and Yu. S. Pashkova [1].

The exposition in Section 11.6 is a version elaborated by N. D. Kopachevsky.

Standard Reference Texts

[AI] Azizov, T. Y. and Iokhvidov, I. S. (1986). *Foundations of the Theory of Linear Operators in Spaces with Indefinite Metrics*, pp. 352. Moscow: Nauka.

[AG] Akhiezer, N. I. and Glasman, I. M. (1966). *Theory of Linear Operators in Hilbert Spaces*, pp. 554. Moscow: Nauka.

[BS] Birman, M. S. and Solomyak, M. Z. (1980). *Spectral Theory of Self-adjoint Operators in Hilbert Spaces*, pp. 264. Leningrad: Izd. LGU.

[F] Krein, S. G., ed. (1972). *Functional Analysis*, 2nd edition, pp. 544. Moscow: Nauka.

[GG] Gohberg, I. and Goldberg, S. (1981). *Operator Theory*, pp. 285. Basel: Birkhäuser Verlag.

[GGK] Gohberg, I., Goldberg, S., and Kaashoek, M. A. (1990; 1993). *Classes of Linear Operators*, Vol. 1, pp. 468; Vol. 2, pp. 556. Basel: Birkhäuser Verlag.

[GK] Gohberg, I. and Krein, M. G. (1965). *Introduction to the Theory of Nonselfadjoint Operators in Hilbert Spaces*, pp. 448. Moscow: Nauka.

[KKN] Kopachevsky, N. D., Krein, S. G., and Ngo Zui Kan (1989). *Operator Methods in Linear Fluid Dynamics: Evolution and Spectral Problems*, pp. 416. Moscow: Nauka.

[K] Krein, S. G. (1967). *Linear Differential Equations in Banach Spaces*, pp. 464. Moscow: Nauka.

[M] Markus, A. S. (1986). *Introduction to the Spectral Theory of Polynomial Operator Pencils*, pp. 260. Kishinev: Ştiinţa.

Bibliography

AIRICH, F.

[1] *Reology.* Moscow: Izd. Inost. Lit., 1962, pp. 824.

ASKEROV, N. K., KREIN, S. G., AND LAPTEV, G. I.

[1] On a class of nonself-adjoint boundary value problems. *DAN SSSR,* 155(3), 1964.

[2] The problem on oscillations of a viscous fluid and related operator equations. *Functional Analysis and Its Applications,* 2(2), 1968, 21–32.

AZIZOV, T. Y.

[1] *Spectral Theory and the Theory of Extensions of Operators in Spaces with Indefinite Metrics.* Dissertation for the Second Doctoral Degree (Professorship), Voronezh University, Voronezh, 1985, pp. 275.

AZIZOV, T. Y. AND KOPACHEVSKY, N. D.

[1] On the basicity of the system of eigen- and associated elements of the S. G. Krein's problem on normal oscillations of a viscous fluid in an open container. In *Spectral and Evolution Problems. Summaries of Lectures and Papers of the Third Crimean Fall Math. School—Symp. (CFMS-III), 1992,* Sevastopol, Laspi, Izd. SGU, 3, 1992, 43.

AZIZOV, T. Y., KOPACHEVSKY, N. D., AND ORLOVA, L. D. (BOLGOVA)

[1] Evolution and spectral problems generated by the problem on small movements of a visco-elastic fluid. *Spectral and Evolution Problems. Proceedings of the Fourth Crimean Fall Math. School—Symp. (CFMS-IV), 1993,* Sevastopol, Laspi, Izd. SGU, 4, 1993, 85–89.

AZIZOV, T. Y. AND SUKHOCHEVA, L. I.

[1] On the completeness of the root vector system of S. G. Krein's pencil in a Pontryagin space. In *Spectral and Evolution Problems. Summaries of Lectures

and Papers of the Second Crimean Fall Math. School—Symp. (CFMS-II), 1991, Sevastopol, Laspi, Izd. SGU, 2, 1991, 69–70.

BABSKII, V. G., KOPACHEVSKY, N. D., MYSHKIS, A. D., SLOBOZHANIN, L. A., AND TYUPTSOV, A. D.

[1] *Gidromehanika Nevesomosty*. Moscow: Nauka, 1976, pp. 504.

BARKAR, S. M.

[1] On the spectrum of a quadratic operator pencil in the open right half-plane. In *Spectral and Evolution Problems. Summaries of Lectures and Papers of the Second Crimean Fall Math. School—Symp. (CFMS-II), 1991,* Sevastopol, Laspi, Izd. SGU, 2, 1991, 70–71.

[2] On the eigenvalues of a boundary value problem in the open right half-plane. *Mat. Zametki*, 48(5), 1990, 138–140.

[3] On the spectrum of a boundary value problem in the open right half-plane. *Izv. AN MoldSSR. Mat.*, 1(4), 1991, 83–88.

BARNIAK, M. YA.

[1] Projection method of solving the problem on small free oscillations of a viscous incompressible fluid in a vessel. In *Direct Methods in Problems of Dynamics and Stability of Multidimensional Systems*, Kiev: Math. Institute AN USSR, 1986, 36–42.

[2] Investigation of small oscillations of a spherical pendulum with a cavity filled with a viscous incompressible fluid. In *Applied Problems of Dynamics and Stability of Mechanical Systems*, Kiev: Math. Institute AN USSR, 1987, 27–33.

[3] Small transverse oscillations of a viscous incompressible fluid in a horizontal channel. In *Mathematical Simulation of Dynamic Processes in a Systems of Bodies with Fluids*, Kiev: Math. Institute AN USSR, 1988, 14–23.

[4] Small oscillations of a viscous incompressible fluid in a container, Kiev, 1989, Math. Institute AN USSR, Preprint no. 89.43, pp. 60.

BARNIAK, M. Y. AND TSEBRII, R. I.

[1] Approximate method of investigating proper oscillations of a pendulum with a cylindrical cavity filled with a viscous incompressible fluid. In *Approximate Methods of Investigation of Dynamics and Stability of Multidimensional Systems*, Kiev: Math. Institute AN USSR, 1983, 85–92.

[2] Approximate method of solving problems on free oscillations of a pendulum with a cavity filled with an incompressible fluid. In *Direct Methods in Problems of Dynamics and Stability of Multidimensional Systems*, Kiev: Math. Institute AN USSR, 1986, 43–52.

BOLGOVA, L. D. AND KOPACHEVSKY, N. D.

[1] On a problem of mathematical physics connected with small and normal movements of a visco-elastic fluid in an arbitrary container. In *Spectral and Evolution Problems. Summaries of Lectures and Papers of the Third Crimean Fall Math. School—Symp. (CFMS-III), 1992*, Sevastopol, Laspi, Izd. SGU, 3, 1992, 44–46.

[2] Boundary value problems on small oscillations of an ideal relaxing fluid and generalizations. In *Spectral and Evolution Problems. Summaries of Lectures and Papers of the Third Crimean Fall Math. School—Symp. (CFMS-III), 1992*, Sevastopol, Laspi, Izd. SGU, 3, 1992, 46–47.

[3] Model problem on small movements and normal oscillations of an ideal relaxing incompressible fluid. In *Spectral and Evolution Problems. Summaries of Lectures and Papers of the Second Crimean Fall Math. School—Symp. (CFMS-II), 1991*, Sevastopol, Laspi, Izd. SGU, 2, 1991, 80–81.

CHAN, T. H.

[1] On a nonself-adjoint operator pencil. *Mat. Zametki*, 1990, 47(3), 133–135.

[2] The problem on flowing of a nonuniformly heated viscous fluid. *J. Comp. Math. and Math. Phys.*, 1989, 29(8), 1153–1158.

[3] *On the Convection Spectrum Problem for a System of Viscous Incompressible Fluids.* Voronezh University, pp. 26.

[4] *The Convection Problem for a Viscous Incompressible Fluid Partially Filling a Container.* Voronezh University, pp. 15.

CHANDRASEKHAR, S.

[1] The oscillations of a viscous liquid globe. *Proc. London Math. Soc.*, 1959, 9(33), 141–149.

CHERNOUSKO, F. L.

[1] Motion of a rigid body with cavities filled with a viscous fluid with small Reinolds numbers. *J. Comp. Math. and Math. Phys.*, 1965, 5(6), 1049–1070.

[2] On the motion of body with a cavity partially filled with a viscous fluid. *PMM*, 1966, 30(6), 977–992.

[3] Motion of a body with a cavity filled with a viscous fluid with large Reinolds numbers. *PMM*, 1966, 30(3), 476–494.

[4] On free oscilations of a viscous fluid in a container. *PMM*, 1966, 30(5), 836–847.

[5] Oscillations of a container with a viscous fluid. *Izv. AN SSSR, MJG*, 1967, 1, 58–66.

[6] *Motions of a Rigid Body with Cavities Filled with a Viscous Fluid.* Moscow: Izd. Vychislit. Tsentra AN SSSR, 1968, pp. 230.

FADDEEV, D. K. AND SOMINSKII, I. S.

[1] *Problems and Exercises in Higher Algebra.* Moscow: Nauka., 1964, pp. 304.

GARADZHAEV, A.

[1] On the problem on normal oscillations of a heavy viscous fluid in a container. *Sibirsk. Math. J.*, 1984, 25(2), 213–216.

[2] Some model problems connected with the problem on oscillations of a viscous fluid in an elastic container. *Diff. Equations*, 1984, 20(1), 14–19.

GELFAND, I. M., MINLOS, R. A., AND SHAPIRO, Z. Y.

[1] *Representations of the Rotation and Lorentz Groups.* Moscow: Fizmatgiz, 1958, pp. 368.

GREENLEE, W. H.

[1] Double unconditional basis associated with a quadratic characteristic parameter problem. *J. Funct. Anal.*, 1974, 15, 306–339.

[2] Linearized hydrodynamic stability of a viscous liquid in an open container. *J. Funct. Anal.*, 1976, 22, 106–129.

GREENSTEIN, V. A.

[1] *On p-Basicity of the Eigen- and Associated Vector System of a Polymonial Self-Adjoint Operator Pencil.* Simferopol, SGU, Report no. 890, Uk. 90, Ukraine, 1990, pp. 8.

[2] Basicity of a part of the eigenvector system of a holomorphic operator function. *Mat. Zametki*, 1991, 50(1), 142–143.

GREENSTEIN, V. A. AND KOPACHEVSKY, N. D.

[1] *On the Spectrum Problem of a Bordered Self-Adjoint Operator.* Report no. 109, Uk. 89, Ukraine, NIINTI, 1989, pp. 25.

[2] On p-basicity of the eigenelement system of a self-adjoint operator function. In *Proceedings of the XVth All-Union School on Operator Theory in Functional Spaces. Part I*, Ulianovsk, 1990, 72.

HASEGAWA, E.

[1] Waves on the interface of two-liquid layers in a vertical periodic motion. *Bulletin of the JSME*, 1983, 26(211), 51–56.

IEVLEVA, O. B.

[1] On oscillations of a pendulum with a spherical cavity filled with a viscous fluid. *PMM*, 1964, 28(6), 1132–1135.

[2] On oscillations of a body filled with a viscous fluid. *J. PMTF*, 1966, 6, 27–34.

KARAZEEVA, N. A. AND SOLOMYAK, M. Z.

[1] Spectral asymptotics of the contact problem for ellyptic equations of the second order. In *Problems in Mathematical Analysis*, Izd. Leningrad Univ., 1981, 8, 36–48.

KELDYSH, M. V.

[1] On eigenvalues and eigenfunctions of some classes of nonself-adjoint operators. *DAN SSSR*, 1951, 77(1), 11–14.

[2] On completeness of eigenfunctions of some classes of nonself-adjoint linear operators. *Uspehi Mat. Nauk*, 1971, 24(4), 15–41.

KISELEV, A. A.

[1] Solving the linearized equations of nonstationary flow of a viscous incompressible fluid in a bounded region. *DAN SSSR*, 1955, 101(1), 43–46.

KLINCHIN, I. M. AND KOPACHEVSKY, N. D..

[1] Model problem on oscillations of a partially dissipative hydrosystem. In *Wave Motions of Fluids and Related Questions.* Izd. Kubansk Univ., Krasnodar, 1991, 51–67.

KOBRIN, A. G.

[1] On the problem of motion of a body with a cavity filled with a viscous fluid relatively to the mass center in a potential field of mass forces. *PMM*, 1969, 35(3), 431–446.

KOPACHEVSKY, N. D.

[1] On oscillations of a capillary viscous rotating fluid. *DAN SSSR*, 1974, 219(5), 1065–1968.

[2] On oscillations of a capillary viscous fluid (model problem). In *Works of the Fourth Winter School in Mathematical Programing and Adjacent Questions, Drogobych, 1973*, Functional Analysis and its Applications, M.: TEMI AN SSSR, 1975, 197–214.

[3] Inversion of Lagrange theorem on stability of small oscillations of a capillary viscous fluid. *DAN SSSR*, 314(1), 1990, 71–73.

[4] Small Movements and Normal Oscillations of a System of Heavy Viscous Rotating Fluids. Preprint FTINT AN SSSR, Kharkov, 1978, pp. 60.

[5] Normal oscillations of a system of heavy viscous rotating fluids. *DAN UkSSR, Series A*, 1978, 7, 586–590.

[6] On the spectral theory of self-adjoint operator functions and its applications in hydrodynamics. In *Summary of Reports of the Seventh School on Operator theory in Functional Spaces*, Minsk, 1982, 89–90.

[7] On p-basicity of the system of root vectors of the self-adjoint operator pencil $I-\lambda A-\lambda^{-1}B$. In *Functional Analysis and Applied Mathematics*, Kiev: Naukova Dumka, 1982, 55–70.

[8] Gravitational convection in an arbitrary partially filled container. In *Proceedings of the Fourth All-Union Seminar on Hydromechanics and Heat-Mass Transfer in Zero-Gravity*, Novosibirsk, 1987, 33–34.

[9] Inversion of Lagrange theorem on stability of small oscillations of a capillary fluid. In *Proceedings of the All-Union Conference*, Ukraine, Sevastopol, 1989, 99–100.

[10] On stability of small movements of a capillary viscous rotating fluid. In *Summaries of Reports of the International Mathematical Conference in Honor of A. M. Liapunov*, Kharkov, 1992, 87–88.

[11] On basicity properties of the system of eigen- and associated vectors of the self-adjoint operator pencil $I-\lambda A-\lambda^{-1}B$. *Functional Analysis and its Applications*, 1981, 15(2), 77–78.

[12] The Cauchy problem for small oscillations of a viscous fluid in a weak field of mass forces. *J. Comp. Math. and Math. Phys.*, 1967, 7(1), 128–146.

KOPACHEVSKY, N. D. AND KREIN, S. G.

[1] On some flowing problem for a viscous fluid. *Zeitshrift für Analysis und Ihre Anwendungen*, 1989, 8(6), 557–561.

KOPACHEVSKY, N. D. AND MYSHKIS, A. D.

[1] Fluid dynamics in weak force fields: On small oscillations of a viscous fluid in a potential field of mass forces. *J. Comp. Math. and Math. Phys.*, 1966, 6(6), 1054–1063.

[2] On free oscillations of a fluid self-gravitating sphere with regard to capillary and viscous forces. *J. Comp. Math. and Math. Phys.*, 1968, 8(4), 1291–1305.

KOPACHEVSKY, N. D. AND NGO ZUI KAN

[1] On a problem in the theory of free convection. *DAN SSSR*, 1980, 251(6), 1334–1337.

KOPACHEVSKY, N. D. AND ORLOVA (BOLGOVA), L. D.

[1] Integro-differential equations connected with the problem on small oscillations of a relaxing fluid. In *Spectral and Evolution Problems. Proceedings of the Fourth Crimean Fall Math. School—Symp. (CFMS-IV), 1995*, Sevastopol, Laspi, Izd. SGU, 4, 98–103.

KOPACHEVSKY, N. D., ORLOVA (BOLGOVA), L. D., AND PASHKOVA, J. S.

[1] Small oscillations and stability of hydrodynamical systems. In *Abstracts of the International Conference on Nonlinear Differential Equations*, Ukraine, Kiev, 1995, 80.

KOPACHEVSKY, N. D. AND PIVOVARCHIK, V. N.

[1] Sufficient conditions of instability of normal convective motions of a fluid in an open container. In *Spectral and Evolution Problems. Summaries of Lectures and Papers of the First Crimean Fall Math. School—Symp. (CFMS-I), 1991*, Sevastopol, Laspi, Izd. SGU, 2, 1991, 32.

[2] Sufficient conditions of instability of convective fluid movements in an open container. *J. Comp. Math. and Math. Phy.*, 1993, 33(1), 101–118.

KOPACHEVSKY, N. D. AND RADYAKIN, N. K.

[1] Two problems on normal oscillations of a system of slightly viscous capillary fluids. In *Topics in Mathematical Physics and Functional Analysis: Materials of Scientific Seminars FTINT AN UkSSR*, Kiev: Naukova Dumka, 1976, 93–110.

KOPACHEVSKY, N. D., TEMCHENKO, T. P., AND VRONSKII, B. M.

[1] Oscillations of partially dissipative hydrosystems. In *Proceedings of the XIIth School on Operator Theory in Functional Spaces. Part I*, Tambov, Russia,1987, 40.

[2] Spectral problems on oscillations of partially dissipative hydrosystems. In *Proceedings of the XIVth School on Operator Theory in Functional Spaces. Part I*, Novgorod, Russia, 1989, 52.

KOPACHEVSKY, N. D. AND TEMNOV, A. N.

[1] *Small Oscillations of a Viscous Stratified Fluid in a Container.* Report no. 4531-83, Moscow, 1983, pp. 45.

[2] Oscillations of a stratified fluid in a container of arbitrary form. *J. Comp. Math. and Math. Phys.*, 1986, 26, 734–755.

KOPACHEVSKY, N. D. AND VADIAA, A.

[1] Oscillations of a plane pendulum with a cavity containing a system of heavy nonmixing ideal fluids. In *Spectral and Evolution Problems: Proceedings of the Fourth Crimean Fall Math. School—Symp. (CFMS-IV)*, 1995, Sevastopol, Laspi, Izd. SGU, 4, 1995, 90–94.

[2] Small oscillations of a plane pendulum with a cavity partially filled with a viscous capillary fluid. In *Spectral and Evolution Problems: Proceedings of the Fourth Crimean School—Symp. (CFMS-IV)*, 1995, Sevastopol, Laspi, Izd. SGU, 4, 1995, 94–98.

KOPACHEVSKY, N. D. AND VOLODKOVICH, E. D.

[1] Small oscillations of a plane pendulum with a cavity completely filled with a viscous fluid. In *Spectral and Evolution Problems: Summaries of Lectures and Papers of the Third Crimean Fall Math. School—Symp. (CFMS-III), 1992*, Sevastopol, Laspi, Izd. SGU, 3, 1992, 52–54.

[2] Small movements and normal oscillations of a plane pendulum with a cavity partially filled with a viscous fluid. In *Spectral and Evolution Problems: Summaries of Lectures and Papers of the Third Crimean Fall Math. School—Symp. (CFMS-III), 1992*, Sevastopol, Laspi, Izd. SGU, 3, 1992, 55–58.

[3] Inversion of Lagrange theorem on stability of small motions of a capillary viscous rotating fluid. In *Spectral and Evolution Problems: Summaries of Lectures and Papers of the Second Crimean Fall Math. School-Symp. (CFMS-II), 1991*, Sevastopol, Laspi, Izd. SGU, 2, 1991, 73–34.

KORENEV, N. K.

[1] On some problems of convection in a viscous incompressible fluid. *Vestnik Leningrad Univ.*, 1971, 7(2), 29–39.

KOSTYUCHENKO, A. G. AND ORAZOV, M. B.

[1] On some properties of the roots of a self-adjoint quadratic pencil. *Functional Analysis and its Applications*, 9(4), 1975, 28–40.

[2] The problem on oscillations of an elastic half-cylinder and the related self-adjoint quadratic pencil. In *Proceedings of a Seminar in Honor of I. G. Petrovskii, 1981*, 6, 97–146.

KOSTYUCHENKO, A. G. AND SHKALIKOV, A. A.

[1] Self-adjoint quadratic pencils of operators and elliptic problems. *Functional Analysis and its Applications*, 1983, 17(2), 38–61.

[2] On the theory of self-adjoint pencils of operators. *Vestnik MGU, Series 1: Mathematics and Mechanics*, 1983, 6, 40–51.

KOTKO, L. A. AND KREIN, S. G.

[1] On completeness of the system of eigen- and associated functions of boundary value problems with a parameter in boundary conditions. In *Selection of Scientific Articles on Applications of Functional Analysis*, S. G. Krein (ed.), Voronezh, Voronezhskii Technologicheskii Institut, 1975, 71–89.

KOSHEVNIKOV, A. N.

[1] On the operator of the linearized stationary Navier–Stokes problem. *Mat. Zametki*, 1984, 125(1), 3–18.

[2] Separate asymptotics of two branches of eigenvalues of an ellyptic boundary value problem. *Mat. Zametki*, 1977, 22(5), 699–709.

KRASNOSCHYOKOV, P. S.

[1] On oscillations of a physical pendulum that has cavities filled with a viscous fluid. *PMM*, 27(2), 1063, 193–202.

[2] Small oscillations of a rigid body that has cavities filled with a viscous fluid. In *Numerical Methods of Solving Problems of Mathematical Physics*, N. N. Moiseev (ed.), Moscow: Nauka, 1966, 258–266.

KREIN, M. G. AND LANGER, G. K.

[1] On the theory of quadratic pencils of nonself-adjoint operators. *DAN SSSR*, 154(6), 1964, 1258–1261.

[2] On some mathematical principles of linear theory of damped continuum oscillations. In *Proceedings of an International Symposium on the Applications of Complex Analysis in the Mechanics of Continuous Media, Vol. 2*, Moscow: Nauka, 1965, pp. 283–322.

KREIN, S. G.

[1] Differential equations in Banach spaces and applications to hydrodynamics. *Uspekhi Math. Nauk.*, 12(1), 1957, 208–211.

[2] On functional properties of operators of vector analysis and hydrodynamics. *DAN SSSR*, 93(6), 1953, 969–972.

[3] On oscillations of a viscous fluid in a container. *DAN SSSR*, 159(2), 1964, 262–265.

[4] On the theory of viscous fluid oscillations. In *Abstracts of the 2nd All-Union Congress on Theoretical and Applied Mechanics*, Moscow: Nauka, 1964, pp. 121.

KREIN, S. G. AND LAPTEV, G. I.

[1] On the problem of viscous fluid movement in an open container. *Functional Analysis and Its Applications*, 2(1), 1968, 40–50.

KREIN, S. G. AND NGO, Z. K.

[1] On the motion of a rigid body with cavities filled with a viscous fluid. In *The 3rd All-Union Congress on Theoretical and Applied Mechanics, Annotated Reports*, Moscow: Nauka, 1968, pp. 177.

[2] Problems on small movements of a body with a cavity partially filled with a viscous fluid. *PMM*, 33(1), 1969, 117–123.

[3] Asymptotic method in the problem on oscillations of a strongly viscous fluid. *PMM*, 33, 1969, 456–466.

[4] Motion of a rigid body with a cavity filled with several nonmixing viscous incompressible fluids. In *The All-Union Conference in the Mechanics of Continuous Media, Annotated Reports*, Tashkent: Izd. Fan UzSSR, 1979, pp. 20.

[5] Asymptotic solutions of the problem on the rotating motion of a body with a cavity filled with several nonmixing strongly viscous incompressible fluids. In *Abstracts of the All-Union Conference on Asymptotic Methods in the Theory of Singularly Perturbed Equations, Alma-Ata, Part 2*, Alma-Ata: Izd. Nauka KazSSR, 1979, 89–90.

KRUSHINSKAYA, S. I.

[1] Oscillations of a viscous heavy fluid in a moving container. *J. Comp. Math. and Math. Phys.*, 5(3), 1965, 519–563.

LADYZHENSKAYA, O. A.

[1] *Mathematical Issues on the Dynamics of a Viscous Incompressible Fluid.* Moscow: Nauka, 1970, pp. 288.

LARIONOV, E. A.

[1] On bases made of root vectors of an operator pencil. *DAN SSSR*, 206(2), 1972, 283–286.

MARKUS, A. S. AND MATSAEV, V. I.

[1] On the basicity of a part of the eigen- and associated vectors of a self-adjoint operator pencil. *Mat. Sbornik*, 133(3), 1987, 293–313.

[2] Theorems on spectra comparison of linear operators and spectral asymptotics. *Proc. Moscow Math. Soc.*, 45, 1982, 133–181.

[3] A theorem on spectra comparison and spectral asymptotics due to M.V. Keldysh. *Mat. Sbornik*, 123(165), N3, 1984, 391–406.

[4] The basicity of a subsystem of eigen- and associated vectors of a self-adjoint operator pencil. *J. Funct. Anal.*, 21(1), 1987, 82–83.

[5] Theorems on spectra comparison of linear operators and Keldysh asymptotics for pencils. Preprint, Math. Institute AN MoldSSR, 1983, pp. 38.

MARKUS, A. S., MATSAEV, V. I., AND RUSSU, G. I.

[1] On some generalizations of the theory of overdamped pencils in the case of pencils of an arbitrary order. *Acta Sci. Math. Szeged*, 34, 1973, 245–271.

METIVIER, G.

[1] Valeurs propres d'operateurs definis par le restriction de systemes variationalles a des sousespaces. *J. Math. Pures et Appl.* 57(2), 1978, 133–156.

MILOSLAVSKII, A. I.

[1] On the stability of some classes of evolution equations. *Sibirsk. Math. J.*, 36(5), 1985, 118–132.

[2] On the stability of a visco-elastic isotropic medium. *DAN SSSR*, 299(6), 1988, 1341–1343.

[3] *Spectral Analysis of Small Oscillations of a Visco-elastic Fluid in an Open Container.* Report no. 1221, Uk 89, Ukraine, 1989, pp. 78.

[4] Spectral properties of a class of quadratic operator pencils. *Functional Analysis and Its Applications* 15(2), 1981, 81–82.

MOGULSKII, E. Z.

[1] On the completeness of the eigen- and associated vector system of an operator pencil. *DAN SSSR*, 183(4), 1968, 775–778.

MOISEEV, N. N.

[1] Boundary value problems for linearized Navier–Stokes equations in the case of small viscosity. *J. Comp. Math. and Math. Phys.* 1(3), 1961, 543–550.

MYSHKIS, A. D., BABSKII, V. G., KOPACHEVSKY, N. D., SLOBOZHANIN, L. A., AND TYUPTSOV, A. D.

[1] *Low-Gravity Fluid Mechanics.* Springer-Verlag, 1987, pp. 583.

MYSHKIS, A. D., BABSKII, V. G., ZHUKOV, M. F., KOPACHEVSKY, N. D., SLOBOZHANIN, L. A., AND TYUPTSOV, A. D.

[1] *Methods of Solving Problems in Hydrodynamics for Zero-Gravity Conditions.* Kiev: Naukova Dumka, 1992, pp. 592.

NGO, Z. K.

[1] On the motion of a rigid body with cavities filled with an incompressible viscous fluid. *J. Comp. Math. and Math. Phys.*, 8(4), 1968, 914–917.

[2] On the motion of a pendulum containing dampers. *Izv. AN SSSR, MTT*, 1, 1969, 23–25.

[3] On the rotating motion of a rigid body with a cavity filled with a viscous fluid. *J. Comp. Math. and Math. Phys.*, 11(5), 1971, 1488–1497.

[4] On the motion of a rigid body with a cavity partially filled with a viscous fluid in complete low-gravity. *PMM*, 43(2), 1979, 267–273.

[5] On the free rotation of a rigid body with a cavity filled with viscous fluids. *DAN SSSR*, 246(4), 1979, 823–827.

[6] On the rotating motion of a rigid body with a cavity partially filled with a nonuniformly heated viscous incompressible fluid. *J. Comp. Math. and Math. Phys.*, 20(6), 1980, 1564–1579.

[7] On the influence of surface tension on fading oscillations of a viscous incompressible fluid in a cylindrical channel. *Izv. AN SSSR MJG*, 3, 1980, 120–124.

ORAZOV, M. B.

[1] On the problem of normal oscillations of a viscous fluid partially filling a cylindrical hull. *DAN SSSR*, 259(1), 1981, 79–82.

[2] On the problem of normal oscillations of a viscous capillary fluid in a container. *DAN SSSR*, 271(2), 1983.

[3] On location of the spectrum in the problem on normal oscillations of an elastic hull filled with a viscous incompressible fluid. *J. Comp. Math. and Math. Phys.* 25(3), 1985, 403–412.

ORAZOV, M. B. AND SHUKUROV, K. A.

[1] On some problems of the visco-elastic theory. In *Spectral and Evolution Problems: Summaries of Lectures and Papers of the Second Crimean Fall Math. School—Symp. (CFMS-II)*, Sevastopol, Laspi, Izd. SGU, 2, 1991, 88–89.

PIVOVARCHIK, V. N.

[1] On the spectrum of quadratic operator pencils in the right half-plane. *Mat. Zametki*, 45(6), 1989, 101–103.

[2] On the right half-plane spectrum of a quadratic pencil describing systems under the action of linear nonconservative positional and dissipative forces. In *Spectral and Evolution Problems: Proceedings of the Fourth Crimean Fall Math. School—Symp. (CFMS-IV)*, Sevastopol, Laspi, Izd. SGU, 4, 1995, 112–113.

RADYAKIN, N. K.

[1] Normal oscillations of a slightly viscous rotating drop. *Izv. AN SSSR MJG*, 4, 1979, 108–117.

[2] On the calculation of frequencies of oscillations of a slightly viscous fluid rotating in an axially symmetrical container. *Izv. AN SSSR MJG* 2, 1980, 128–133.

RADZIEVSKII, G. V.

[1] Multiple completeness of root vectors of the Keldysh pencil perturbed by an operator function analytic on a circle. *Mat. Sbornik*, 91(133), N 3, 1973.

[2] On the basicity of derived chains. *Izv AN SSSR, Series Math.*, 39(5). (English translation: *Math. USSR Izv.*, 9(5), 1975, 1119–1154.)

[3] On minimality, basicity and completeness of a part of the root vectors of a quadratic operator pencil. *DAN SSSR*, 283(1), 1985, 53–57.

[4] The problem on root vector completeness in the operator function spectral theory. *Uspekhi Mat. Nauk*, 37(2), 1982, 81–145.

RUMYANTSEV, B. N.

[1] On the motion of a rigid body containing cavities filled with a viscous fluid. *PMM*, 28(6), 1964, 1127–1132.

SHKALIKOV, A. A.

[1] The property of being a basis for the eigenvectors of a quadratic operator pencil. *Mat. Zametki*, 30(3), 1981, 371–385. (English translation: *Math. Notes*, 30(3), 1981, 676–684.)

[2] On properties of a part of the eigen- and associated elements of self-adjoint quadratic operator pencils. *DAN SSSR*, 283(5), 1985, 1101–1105.

[3] On the minimality property of derived chains corresponding to a part of the eigen- and associated elements of self-adjoint operator pencils. *Vestnik Moskovskovo Univ. Series Math.*, 6, 1985, 10–20. (English translation: *Moscow Univ. Math. Bull.*, 40(6), 1985.)

[4] On minimality and completeness of systems constructed from eigen- and associated elements of a quadratic operator pencil. *DAN SSSR*, 285(6), 1985, 1334–1339. (English translation: *Soviet Math. Dokl.*, 32(3), 1985, 902–907.)

[5] Selection principles and properties of a part of the eigen- and associated elements of an operator pencil. *Vestnik Moskovskovo Univ. Series Math.*, 4, 1988, 16–25. (English translation: *Moscow Univ. Math. Bull.*, 43(4), 1988, 16–28.)

[6] Strongly damped pencils of operators and the solvability of the corresponding operator-differential equations. *Mat. Sbornik*, 135(1), 1988, 96–118. (English translation: *Math. USSR Sbornik*, 63(1), 1989, 97–119.)

[7] *Operator Pencils Arising in Elasticity and Hydrodynamics. The Instability Index Formula.* Operator Theory: Advances and Applications, Birkhäuser Verlag (to appear).

SHKALIKOV, A. A. AND PLIEV, V. T.

[1] Compact perturbations of overdamped operator pencils. *Mat. Zametki*, 45(2), 1989, 118–128.

SHMIDT, A. G.

[1] Gravitational and capillary waves on the surface of a sphere layer of a viscous gravitating fluid. *J. Comp. Math. and Math. Phys.*, 4(1), 1964, 183–189.

SMIRNOVA, E. P.

[1] Motion of a fluid with high viscosity in a rotating torus. *PMM*, 39(1), 1975, 177–182.

SOLONNIKOV, V. A.

[1] On the differential properties of solutions of the first boundary value problem for a nonstationary system of Navier–Stokes equations. *Works of MI AN SSSR*, 73, 1964, 221–291.

[2] On general boundary value problems for systems ellyptic in the sense of Duglis–Nirenberg. *Izv. AN SSSR, Series Math.* 28, 1964, 665–706; *Works of MI AN SSSR*, 92, 1966, 233–297.

[3] The solvability of the problem on plane motion of a heavy viscous incompressible and capillary fluid. *Izv. AN SSSR, Series Math.*, 43(1), 1979, 203–236.

SOLONNIKOV, V. A. AND SCHADILOV, V. E.

[1] On a boundary value problem for a stationary system of Navier–Stokes equations. *Works of MI AN SSSR*, 125, 1973, 196–210.

STIOPIN, S. A.

[1] *On the Frequencies of Small Oscillations of a Capillary Viscous Fluid Ring.* Report no. 2889-B90, Moscow Univ., 1990, pp. 12.

[2] On the Laplace operator with a spectral parameter in the boundary condition. In *Spectral and Evolution Problems: Summaries of Lectures and Papers of the Third Crimean Fall Math. School—Symp. (CFMS-III)*, Sevastopol, Laspi, Izd. SGU, 3, 1992, 68–69.

SUSLINA, T. A.

[1] Asymptotics of the spectrum of variational problems for solutions of a homogeneous elliptic equation with bindings on a part of the boundary. In *Problems in Mathematical Analysis*, Leningrad: Izd. LGU, 9, 1984, 84–87.

[2] Asymptotics of the spectrum of variational problems for solutions of an elliptic equation in a region with a piecewise smooth boundary. *Zap. Nauk Sem. LOMI AN SSSR*, 147, 1985, 179–183.

[3] *Asymptotics of the Spectrum of Some Problems Connected with Oscillations of Fluids.* Leningrad Electromech. Inst. Svyazi, Leningrad, VITINI, Report no. 8058-B85, 1985, pp. 79.

[4] On asymptotics of the spectrum of a model problem of fluid oscillation theory. In *Spectral and Evolution Problems: Summaries of Lectures and Papers of the First Crimean Fall Math. School—Symp. (CFMS-I), 1990*, Sevastopol, Laspi, Izd. SGU, 1, 1990, 43–44.

[5] On asymptotics of the spectrum of some problems related to oscillations of fluids. *Zap. Nauk Sem. LOMI AN SSSR*, 152, 1986, 158–164.

[6] On asymptotics of the spectrum of some model problems of oscillation theory for a capillary viscous fluid. In *Spectral and Evolution Problems: Summaries of Lectures and Papers of the Second Crimean Fall Math. School—Symp. (CFMS-II), 1991*, Sevastopol, Laspi, Izd. SGU, 2, 1991, 95–96.

[7] Asymptotics of the spectrum of a variational problem with an elliptic connection in a domain with an edge. In *Spectral and Evolution Problems: Summaries of Lectures and Papers of the Second Crimean Fall Math. School—Symp. (CFMS-II), 1991*, Sevastopol, Laspi, Izd. SGU, 2, 1991, 96–97.

[8] Asymptotics of the spectrum of a "nonlocal" problem of Steklov type. In *Spectral and Evolution Problems: Summaries of Lectures and Papers of the Third Crimean Fall Math. School—Symp. (CFMS-III), 1992*, Sevastopol, Laspi, Izd. SGU, 2, 1992, 69–70.

[9] Spectrum asymptotics of two model problems of fluid oscillation theory. *Izv. St. Petersburg. Elektrotech. Inst.*, 449, 1992, 82–88.

[10] Asymptotics of the spectrum of a variational quotient arising in the fluid convection problem. In *Spectral and Evolution Problems: Summaries of Lectures and Papers of the Fourth Crimean Fall Math. School—Symp. (CFMS-IV), 1993*, Sevastopol, Laspi, Izd. SGU, 4, 1993, 132–134.

[11] Asymptotics of the spectrum of variational problems with pseudo-differential restrictions. In *Spectral and Evolution Problems: Summaries of Lectures and Papers of the Fourth Crimean Fall Math. School—Symp. (CFMS-IV), 1993*, Lapsi, Izd. SGU, 4, 1993, 134–136.

TARUNIN, E. L.

[1] Nonstationary convection of a fluid in a closed cavity. *Izv. AN SSSR MJG*, 6, 1968, 83–88.

TEMAM, R.

[1] *Navier-Stokes Equations.* Moscow: Mir, 1981, pp. 408.

TEMNOV, A. N.

[1] Small movements of a viscous self-gravitating fluid. In *Spectral and Evolution Problems: Summaries of Lectures and Papers of the Third Crimean Fall Math. School—Symp. (CFMS-III), 1992*, Sevastopol, Laspi, Izd. SGU, 3, 1992, 70–72.

[2] Small oscillations of a rigid body with a nonhomogeneous viscous fluid. In *Spectral and Evolution Problems: Summaries of Lectures and Papers of the Fourth Crimean Fall Math. School—Symp. (CFMS-IV), 1993*, Sevastopol, Laspi, Izd. SGU, 4, 1993, 136–128.

TRUBACHOV, A. V.

[1] Spectrum location of the oscillations of a viscous capillary fluid. *Mat. Zametki*, 47(5), 1990, 116–126.

TSEBRII, R. I.

[1] Investigation of free oscillations of a physical pendulum with a cylindrical cavity filled with a viscous incompressible fluid. In *Numerical Analysis Methods of Investigation of the Dynamics and Stability of Complex Systems*, Kiev: Math. Institute AN UkSSR, 1984, 48–60.

[2] Investigation of free oscillations of a pendulum with a cylindrical cavity filled with a viscous incompressible fluid. In *Investigations on Theoretical and Practical Topics in Mathematics*, Kiev: Math. Institute AN UkSSR, 1986, 83.

VADIAA, A.

[1] *Small Oscillations of a Spherical Pendulum with a Cavity Filled by a System of Nonmixing Ideal Fluids.* Report no. 100, Uk 94, Simferopol Univ., Simferopol, 1994, pp. 34.

[2] *Small Oscillations of a Plane Pendulum with a Cavity Filled by a System of Nonmixing Ideal Fluids.* Report no. 2113, Uk 93, Simferopol Univ., Simferopol, 1993, pp. 40.

[3] *Small Oscillations of a Plane Pendulum with a Cavity Filled by an Ideal Capillary Fluid.* Report no. 98, Uk 94, Simferopol Univ., Simferopol, 1994, pp. 21.

[4] Proper oscillations of a plane pendulum with a cavity filled by a system of nonmixing viscous fluids. In *Spectral and Evolution Problems: Proceedings of the Fourth Crimean Fall Math. School—Symp. (CFMS-IV), Sept. 29–Oct. 10, 1993*, Sevastopol, Laspi, Izd. SGU, 4, 68–70.

VASILIEVA, A. B. AND BUTUZOV, V. F.

[1] *Asymptotic Expansions of Solutions of Singular Perturbed Equations.* Moscow: Nauka, 1973, pp. 272.

VRONSKII, B. M.

[1] Oscillations of a system "fluid–gas" in a rotating container. In *Spectral and Evolution Problems: Summaries of Lectures and Papers of the Third Crimean Fall Math. School—Symp. (CFMS-III), 1992*, Sevastopol, Laspi, Izd. SGU, 3, 1992, 58–59.

[2] Oscillations of a partially dissipative hydrosystem "fluid–gas". In *Spectral and Evolution Problems: Summaries of Lectures and Papers of the Third Crimean Fall Math. School—Symp. (CFMS-III), 1992*, Sevastopol, Laspi, Izd. SGU, 3, 1992, 59.

YAKUBOV, S. Y.

[1] On two-multiple completeness of eigen- and associated elements of a quadratic operator pencil. *Functional Analysis and Its Applications*, 7(1), 1973, 92–94.

YUDOVICH, V. I.

[1] *Method of Linearization in Hydrodynamic Theory of Stability.* Rostov-on-Don: Izd. RGU, 1984, pp. 192.

[2] *On the Spectrum of Some Bordered Self-Adjoint Operators.* Rostov-on-Don: Izd. RGU, Report no. 3632-80, 1980, pp. 15.

ZAGNIBEDA, I. G.

[1] On the solvability of the problem on dynamics of a rigid body with a cavity containing a viscous fluid. *DAN SSSR*, 238(5), 1978, 1075–1078.

ZARUBIN, A. G.

[1] The problem on stationary free convection. *J. Comp. Math. and Math. Phys.*, 8(6), 1968, 1378–1383.

ZARUBIN, A. G. AND NGO, Z. K.

[1] On the motion of a rigid body with a cavity filled with a heated viscous incompressible fluid. *Scientific Reports of Far-Eastern Univ., Series Phys.-Mat. Sciences*, Vladivostok: Izd. DGU, 17, 1969, 82–95.

[2] On convection in a fluid completely filling the cavity of a moving rigid body. *PMM*, 44(6), 1980, 1027–1034.

Z[illegible]

[1] The problem [illegible] the solvability [illegible] [illegible], 1972, 129[illegible]

Z[illegible]

[1] On the problem of a rigid body with [illegible] incompressible fluid. Soviet [illegible] Sov. Phys. Dokl. [illegible] 1970, 1[illegible]

[2] On convection in a fluid [illegible] PMM, 44(6), 1980, 1027-1034.

List of Symbols

$\emptyset$, $\mathbb{N}$, $\mathbb{R}$, $\mathbb{C}$	the empty set and the sets of natural, real, and complex numbers
$\in$, $\subset$, $\cup$, $\cap$	membership, inclusion, union, and intersection
$A := B$ $(B =: A)$	equality by definition; the colon sign is placed on the side of the object to be defined
E, F, G	abstract Hilbert or Banach spaces (Chapter 1)
$\dim E$	dimension of space E
$(\cdot, \cdot)_E$ $([\cdot, \cdot]_E)$	definite (indefinite) scalar product in E (Sections 1.1, 1.3)
$\|\cdot\|_E$	norm in E (Section 1.1)
$L\dot{+}M$	direct sum of subspaces L and M (Section 1.8.14)
$L \oplus M$ $(L \ominus M)$	orthogonal sum (difference) of subspaces L and M (Sections 1.1.3, 1.1.14)
$F \hookrightarrow E$	embedding of space F into space E (Section 1.1.6)
F^*	space dual to space F (Section 1.1.6)
Π_κ	Pontryagin space (Section 1.3.9)
$M[+]N$	direct J-orthogonal sum of subspaces (Section 1.3.1)
E^α $(-\infty < \alpha < \infty)$	Hilbert scale of spaces (Section 1.8.3)
$N^\perp$ $(N^{[\perp]})$	orthogonal (J-orthogonal) complement of N in E (Sections 1.1.3, 1.3.1)
A, B, C	linear operators acting from E into F, or from E into E (Sections 1.1.3, 1.3.1)
$\mathcal{D}(A)$, $\mathcal{R}(A)$	domain (range) of operator A (Section 1.1.8)
A^*	adjoint of operator A in a space with a definite scalar product (Section 1.1.9)

A^+	adjoint of the operator A in a space with an indefinite scalar product (Section 1.3.4)
$A \geq 0$ ($A > 0$, $A \gg 0$)	nonnegativeness, positiveness, and positive definiteness of operator A, respectively (Section 1.1.10)
$A, L, \gamma, T = \gamma^* = \partial^{-1}$, $C = \gamma T$	operators from the general scheme for solving boundary value problems (Section 1.8)
$\mathcal{L}(E, F)$ ($\mathcal{L}(E)$)	linear bounded operators from E into F (from E into E) (Section 1.1.8)
$\mathfrak{S}_\infty(E, F)$, $\mathfrak{S}_\infty$	compact operators from E into F, or from E into E (Section 1.1.12)
$\mathfrak{S}_p$	operators of a finite order (Section 1.1.12)
P, P_0, P_Γ, P_L	orthoprojectors on subspaces from E (Section 1.1.10)
$\operatorname{diag}(A_1, A_2)$	diagonal matrix with operator entries on the main diagonal (Section 1.1.14)
$A_1 \oplus A_2$	orthogonal sum of operators A_1 and A_2 (Section 1.1.14)
$\operatorname{Ker} A$	kernel (null space) of operator A (Section 1.6.5)
$W(A)$	numerical range of operator A (Section 1.1.17)
$\lambda_j(A)$, $j \in \mathbb{N}$	eigenvalues of operator A (Section 1.1.15)
$s_j(A)$, $j \in \mathbb{N}$	s-numbers of operator A (Section 1.1.12)
$r(A)$	spectral radius of operator A (Section 1.1.17)
$A(\lambda)$ ($L(\lambda)$, $M(\lambda)$)	operator pencil, operator-valued function (Section 1.6.2)
$\sigma(A)$ ($\sigma(A(\lambda))$)	spectrum of operator A (operator-valued function $A(\lambda)$) (Sections 1.1.17, 1.6.2)
$\sigma_p(A)$, $\sigma_c(A)$, $\sigma_r(A)$, $\sigma_e(A)$, $\sigma_d(A)$	subsets of $\sigma(A)$ (Sections 1.1.18, 1.1.9)
$\rho(A)$, $\rho(A(\lambda)))$	resolvent set of operator A (of operator pencil $A(\lambda)$) (Sections 1.1.17, 1.6.2)
$n(r, A)$, $n^{\pm}(r, A)$ ($n(r, A(\lambda))$, $n^{\pm}(A(\lambda))$)	distribution functions of eigenvalues of operator A (operator pencil $A(\lambda)$) (Section 1.6.11)
$\mathcal{Z}_{\lambda_0}(A) = \operatorname{Ker}(A - \lambda_0 I)$	eigensubspace of operator A corresponding to eigenvalue λ_0 (Section 1.1.15)
α_{λ_0} (ν_{λ_0})	eigen (algebraic) multiplicity of operator A corresponding to eigenvalue λ_0 (Section 1.1.15)
$\varphi_0, \varphi_1, \ldots, \varphi_{p-1}$	chain of eigen- and associated elements corresponding to eigenvalue λ_0 (Sections 1.1.15, 1.6.2)
$\varphi(\lambda)$	root function corresponding to an eigenvalue of pencil $A(\lambda)$ (Section 1.6.3)
e^{tA} ($\exp(tA)$)	exponential function of operator A (Section 1.5.1)

$\mathcal{U}(t)$	semigroup of operators (Section 1.5.2)
$\cos(tA)$, $\sin(tA)$	trigonometric functions of operator A (Section 1.5.8)
$L_2(\Omega)$, $H^l(\Omega)$, $L_2^1(\Omega)$, $H_0^1(\Omega)$, $H_h^1(\Omega)$, $H_{\partial\Omega}^1(\Omega)$, $H_\Gamma^1(\Omega)$, H_Ω^1, $H_{0,\Gamma}^1(\Omega)$, $L_2(\Gamma)$, $L_{2,\Gamma}$, $H_{h,S}^1(\Omega)$, $H_{\partial\Omega}^{1/2}$, ...	the Hilbert spaces of square-integrable scalar-valued functions and their subspaces (Sections 1.1, 1.8)
$\boldsymbol{L}_2(\Omega)$, $\boldsymbol{G}_0(\Omega)$, $\boldsymbol{J}(\Omega)$, $\boldsymbol{J}_0(\Omega)$, $\boldsymbol{G}(\Omega)$, $\boldsymbol{J}_{0,S}(\Omega)$, $\boldsymbol{G}_{0,\Gamma}(\Omega)$, ...	spaces and subspaces of hydrodynamics for an ideal fluid (Section 2.1)
$\boldsymbol{H}^1(\Omega)$, $\boldsymbol{J}^1(\Omega)$, $\boldsymbol{J}_0^1(\Omega)$, $\boldsymbol{M}_1(\Omega)$, $\boldsymbol{N}_1(\Omega)$	spaces and subspaces of hydrodynamics for a viscous fluid (Section 2.2)
$\boldsymbol{u} = \boldsymbol{u}(t,x)$	velocity field of fluid (Section 2.2.1)
ρ, ν, $\mu = \rho\nu$	density and the kinematic and dynamic viscosities of a fluid (Sections 2.1.1, 2.2.1)
$p = p(t,x)$, $P = P(t,x)$	dynamic and full pressure (Section 3.1.1)
σ	surface tension coefficient of a capillary fluid (Section 3.4.1)
$\boldsymbol{g}$	acceleration of the gravitational field (Section 3.1.5)
$\operatorname{div}\boldsymbol{u}$	divergence of field $\boldsymbol{u}$ (Section 2.1.3)
$\Delta p := \operatorname{div}\nabla p$	Section 2.1.5
$\gamma_n\boldsymbol{u} = u_n = \boldsymbol{u}\cdot\boldsymbol{n}$	normal component of field $\boldsymbol{u}$ on the boundary (Section 2.1.6)
$\tau_{ij}(\boldsymbol{u}) := \dfrac{\partial u_i}{\partial x_j} + \dfrac{\partial u_j}{\partial x_i}$	Section 2.2.1
$\tilde{\tau}_{ij}(\boldsymbol{u}) := -P\delta_{ij} + \rho\nu\tau_{ij}(\boldsymbol{u})$	tensor of tensions (Section 2.2.1)
$\boldsymbol{J}$	inertia tensor (Section 3.1.3)

Subject Index

Sections

A

absolute velocity 3.1.1
abstract Cauchy problem 1.5.1
abstract Green formula 1.8.7
acceleration 3.1.1
acoustic-relaxing waves 11.6.3
adjoint operator 1.1.9
adjoint space 1.1.5
algebraic multiplicity of an eigenvalue
  for operators 1.1.15
  for operator-valued functions 1.6.1
angular operator 1.3.1
angular velocity 3.1.1
associated elements 1.1.15, 1.6.1
asymptotic expansion 1.7.4
asymptotically stable solution 1.5.6

B

basis
  canonical 1.6.1
  orthonormal 1.1.7
  Riesz (p-basis) 1.1.13
bounded
  functional 1.1.5
  operator 1.1.8

C

canonical factorization 1.6.6
Cartesian coordinate system
  stationary 3.1.1
  nonstationary 3.1.1
  semistationary 3.1.1
Cauchy-Lagrange integral 3.3.2
Cauchy problem 1.5.1
closable operator 1.1.8
closed
  extension 1.1.16
  operator 1.1.16
codimension 1.1.3
compact operator 1.1.8
  of finite order 1.1.12
complete

space 1.1.1
system of eigen- and associated elements 1.6.4
completeness
with a finite defect 1.6.7
continuous spectrum 1.1.16
contractive semigroup 1.5.4
connective movements of fluid 8.9.1
Coriolis acceleration 3.1.1

D

defect 1.1.2
definite subspace 1.3.1
degenerate pencil 1.6.4
densely embedded space 1.8.1
dimension 1.1.3
direct decomposition 1.1.14
Dirichlet
problem 1.8.2
norm 1.2.4
discrete spectrum 1.1.8
displacement
acceleration 3.1.1
velocity 3.1.1
dissipation of the energy 2.2.1
dissipative
operator 1.5.4
waves 7.5.4, 7.5.5, 8.3.1
dissipatively thermal waves 8.9.5
distribution function of eigenvalues 1.6.8
divergence of a field 2.1.3
domain of definition 1.1.16
Dupré-Young condition 4.1.1

E

eigenelement
for an operator 1.1.15
for an operator-valued function 1.6.1
eigenmultiplicity 1.1.15
eigenvalue
for an operator 1.1.15
for an operator-valued function 1.6.1
embedding
compact 1.2.6, 1.8.1
continuous 1.2.6, 1.8.1
energetic space 4.1.5
equation
of hyperbolic type 1.5.8
Navier-Stokes 3.1.2
Poisson 1.8.2
Sophie Germaine 4.2.2
Volterra 1.5.9
equipped spaces 1.2.5
equivalent norms 1.1.4
exponential stability 1.5.6
essential spectrum 1.1.19
extended spectrum 1.1.17
extension of a linear operator 1.1.16

F

factor-space 1.1.23
field of mass external forces 3.1.2
field with a finite kinetic energy 2.1.1
finite function 1.2.1
Fisher-Courant-Weyl principle 1.6.11
fluid
capillary 3.1.3
ideal 2.1.1
incompressible 2.1.1
heavy (noncapillary) 3.1.3
homogeneous 2.1.1
viscous 2.2.1

form of an operator 1.1.10
Fourier coefficient 1.1.7
Fredholm
  holomorphic operator-valued
    function 1.6.3
  pencil 1.6.3
Friedrichs extension 1.8.4
function of
  operator 1.1.20
  self-adjoint operator 1.1.22

G

generalized
  derivative 1.2.2
  solution 1.8.4
generating operator 1.8.1
gyroscopic operator 6.3.4
gyrostate 3.1.4
gyrostatical moment 3.1.4
Green's formula 2.1.7, 2.2.3

H

harmonic field 2.1.7
Hilbert
  pair 1.8.2
  scale 1.8.3
  space 1.1.1
Hilbert-Schmidt operator 1.1.12
holomorphic operator-valued
  function 1.6.1

I

indefinite metrics 1.3.1
inequality of
  Bessel 1.1.7
  Buniakowsky-Schwarz 1.1.1
  Korn 2.2.4
  Poincare 1.2.4
inertia tensor 3.1.4
integral Volterra equation 1.5.9
interlacing operator 1.7.2
intermediate waves 8.3.1, 8.3.3
internal waves 8.9.6
invariant subspace 1.1.14
inverse operator 1.1.8
isometric operator 1.1.8
isomorphic spaces 1.1.8
isotropic
  element 1.3.1
  subspace 1.3.1

J

Jordan chain 1.6.1

K

kernel
  function 1.6.2
  lineal 1.1.15
  of an operator 1.1.15
  subspace 1.1.15
kinetic energy 2.1.1
  energy operator 3.3.3
  moment 3.1.4
Korn identity 2.2.4
  inequality 2.2.4

L

Laplace-Beltrami operator 4.1.2
Laplace operator 2.1.5
length of chain 1.6.1
limiting spectrum 1.1.19
lineal
  degenerate 1.3.1
  isotropic 1.3.1
  negative 1.3.1
  neutral 1.3.1

nonnegative 1.3.1
nonpositive 1.3.1
positive 1.3.1
linear
functional 1.1.5
hull 1.1.3
operator 1.1.8
pencil 1.6.4

M

maximal
dissipative operator 1.5.4
subspace 1.3.1
minimax principle 1.4.1
min-max theorem 1.4.2
multiple basicity 11.4.2
multiple completeness by Keldysh 1.6.5
multiplicity
algebraic 1.1.15
eigen 1.1.15

N

negative subspace 1.3.1
Neumann
problem 1.8.2
series 1.1.17
neutral subspace 1.3.1
nonleaking condition 3.1.1
nonnegative
operator 1.1.10
subspace 1.3.1
nonpositive subspace 1.3.1
norm of
Dirichlet 1.2.4
linear operator 1.1.8
functional 1.1.5
normal
eigenvalue 1.1.18
component of a field on the boundary 2.1.6
nuclear operator 1.1.12
numerical range 1.1.17

O

Oldroit model of visco-elastic fluid 11.1.1
operator
bounded from below 1.1.10
of kinetic energy 3.3.3
of potential energy 4.1.5
pencil 1.6.1
orthocompletable subspace 1.3.1
orthogonal
complement 1.1.3
decomposition 1.1.3
elements 1.1.3
subspaces 1.1.3
sum 1.1.3
orthonormal system 1.1.7
orthoprojector 1.1.10

P

partially dissipative hydrosystem 10.1.1
Parseval identity 1.1.7
pencil linearization 8.2.2, 11.4.1
perturbed operator 1.5.5
Poincare-Ritz principle 1.6.1
Pontryagin space 1.3.6
point spectrum 1.1.18
positive
definite operator 1.1.10
operator 1.1.10
subspace 1.3.1

potential
  energy 4.1.5
  field 2.1.2
  of a field 2.1.2
power of a self-adjoint
  operator 1.1.22
pressure 3.1.2
  dynamical 3.2.1
  statical 3.2.1
problem
  well posed 1.5.2
  of completeness 1.6.4
  uniformly well posed 1.5.2
projection onto a subspace 1.1.3
projector 1.1.10
pure relaxing waves 11.6.3

Q

quadratic form 1.1.10
quite subordinated operator 1.5.5

R

range of operator 1.1.8
Rayleigh functional 1.6.11
region
  of the first type 1.2.4
  of the second type 1.2.6
  of the third type 1.2.7
regular point 1.1.17
relative
  acceleration 3.1.2
  velocity 3.1.1
relaxation 11.6.1
relaxing fluid 11.6.1
resolution
  of the identity 1.1.22
  for a bounded self-adjoint
    operator 1.1.22
  for an unbounded self-adjoint
    operator 1.1.22
resolvent of an operator 1.1.17
  pencil 1.6.1
restriction of an operator 1.1.16
residual spectrum 1.1.18
Riesz projector 1.1.20
root
  elements 1.1.15
  function 1.6.2

S

scalar product 1.1.1
  degenerate 1.1.23
  indefinite 1.3.1
self-adjoint operator 1.1.10
  pencil 1.6.9
separable space 1.1.7

set
  closed 1.1.1
  compact 1.1.1
  dense 1.1.3
  resolvent 1.1.17
symmetrizator 1.6.9
semigroup 1.5.2
  of C_0-class 1.5.2
  holomorphic 1.5.3
  strongly continuous 1.5.2
seminorm 1.1.23
skew-symmetric operator 6.1.2
s-numbers 1.1.12
Sobolev space 1.2.1
solenoidal field 2.1.4
space
  of harmonic functions 1.8.6
  of potentials 2.1.1

spectral
  decomposition 1.1.22
  factorization 1.6.6
  radius 1.1.17
spectrum of
  a bounded operator 1.1.17
  an operator 1.1.17
  an operator-valued function 1.6.1
  an operator pencil 1.6.1
  an unbounded operator 1.1.18
stability 1.5.6
statistically stable in linear
  approximation 4.1.5
stickiness condition 3.1.1
Stokes operator 2.2.5
subordinated operator 1.5.5
subspace 1.1.1
support of a function 1.2.1
surface gravitation waves 8.3.1, 8.3.3
symmetric operator 1.1.21

T

tension tensor 3.1.3
Theorem of
  Banach 1.1.4
  Hilber-Schmidt 1.1.11
  Keldysh 1.6.4
  Riesz 1.1.5, 1.1.6
  Weyl 1.1.19
theorem on spectrum
  localization 10.4.3
thermal convection 7.7.1, 8.9.1
time of relaxation 11.6.3
trace operator 1.8.6
transformer 1.7.3
translation operator 7.2.2

U

unbounded operator 1.1.16
uniformly definite subspace 1.3.2
unitary operator 1.1.8

V

Vandermonde determinant 11.6.3
variable medium characteristic 11.6.5
vector Laplace operator 2.2.3
visco-elastic fluid 11.1.1
Volterra operator 1.5.9

W

weak
  Cauchy problem 1.5.2
  solution 1.5.2, 1.8.4
weakly
  converging sequence 1.1.8
  perturbed operator 1.6.4
wetting line 9.1.1
Weyl
  decomposition 2.1.8
  sequence 1.1.19

Z

Zaremba problem 1.8.2
Zhukovsky potential 3.4.4

MIX
Papier aus verantwortungsvollen Quellen
Paper from responsible sources
FSC® C105338

If you have any concerns about our products,
you can contact us on
ProductSafety@springernature.com

In case Publisher is established outside the EU,
the EU authorized representative is:
Springer Nature Customer Service Center GmbH
Europaplatz 3, 69115 Heidelberg, Germany

Printed by Libri Plureos GmbH
in Hamburg, Germany